Neuroscience of Birdsong

Speech has long been thought of as a uniquely defining characteristic of humans. Yet song birds, like humans, communicate using learned signals (song, speech) that are acquired from their parents by a process of vocal imitation. Both song and speech begin as amorphous vocalizations (subsong, babble) that are gradually transformed into an individualized version of the parents' speech, including dialects. With contributions from both the founding forefathers and younger researchers who represent the future of this field, this book provides a comprehensive summary of birdsong neurobiology, and identifies the common brain mechanisms underlying this achievement in both birds and humans. Written primarily for advanced graduates and researchers, there is an introductory overview covering song learning, the parallels between language and birdsong, and the relationship between the brains of birds and those of mammals. Subsequent sections deal with producing, processing, learning, and recognizing song, as well as with hormonal and genomic mechanisms.

H. PHILIP ZEIGLER is Distinguished Professor of Psychology at Hunter College, City University of New York. He is a behavioral neurobiologist whose research has focused on neural and sensorimotor control of behavior patterns in birds and mammals.

PETER MARLER, Professor Emeritus at the University of California, Davis and Foreign Member of the Royal Society, is recognized as a founding father of bird song research. He has authored numerous books and papers on animal behavior, and his research has set the major directions of the field, providing guidance to several generations of leading researchers.

Neuroscience of Birdsong

Edited by

H. Philip Zeigler
Hunter College, City University of New York

Peter Marler
University of California, Davis

CAMBRIDGE UNIVERSITY PRESS
Cambridge, New York, Melbourne, Madrid, Cape Town, Singapore, São Paulo, Delhi

Cambridge University Press
The Edinburgh Building, Cambridge CB2 8RU, UK

Published in the United States of America by Cambridge University Press, New York

www.cambridge.org
Information on this title: www.cambridge.org/9780521869157

© Cambridge University Press 2008

This publication is in copyright. Subject to statutory exception
and to the provisions of relevant collective licensing agreements,
no reproduction of any part may take place without
the written permission of Cambridge University Press.

First published 2008

Printed in the United Kingdom at the University Press, Cambridge

Portions of this book, as detailed in the Preface, have appeared previously, in somewhat different form, in *Behavioral Neurobiology of Birdsong*, edited by H. Philip Zeigler and Peter Marler, Volume 1016 of the *Annals of the New York Academy of Sciences*, 2004, and are reprinted here courtesy of the New York Academy of Sciences.

A catalog record for this publication is available from the British Library

ISBN 978-0-521-86915-7 hardback

Cambridge University Press has no responsibility for the persistence or
accuracy of URLs for external or third-party internet websites referred to
in this publication, and does not guarantee that any content on such
websites is, or will remain, accurate or appropriate.

To the memory of William H. Thorpe

With gratitude to *Fringilla coelebs*, *Melospiza melodia*, *M. georgiana*, *Taeniopygia guttata*, *Serinus canaria* and other oscine* species for their continuing contributions to the scientific study of birdsong.

---

*Oscine is of Latin origin. Its classical Latin form, *oscin-* or *oscen*, designates birds from whose song augeries/divinations/signs were taken (as opposed to from their flight). The word itself is a combination of the preposition *ob-* (meaning "in the direction of, towards, against, in the way of, in front of, in view of, on account of") and *canere* ("to sing"). (Their combination seems to me to have originally indicated the relationship between the birds' singing/calling and the augurs/auguries.) The use of the word to indicate an order originated when it was adopted as a taxonomic term by Blasius Merrem 1812, in *Abhandl. der Physikalischen Klasse der Königlich-Preuss. Akad. der Wissensch. 1812–13* (1816). (Thomas K. Miller)

Contents

List of contributors	*page* x
Preface	xv
Acknowledgements	xvii

PART I FOUNDATIONS: SINGING AND THE BRAIN — 1

1. Introduction
   H. P. ZEIGLER — 3
2. Birdsong and human speech: common themes and mechanisms
   ALLISON J. DOUPE AND PATRICIA K. KUHL — 5
3. Birdsong and singing behavior
   HEATHER WILLIAMS — 32
4. The Australian magpie (*Gymnorhina tibicen*): an alternative model for the study of songbird neurobiology
   GISELA KAPLAN — 50
5. Songbirds and the revised avian brain nomenclature
   ANTON REINER, DAVID J. PERKEL, CLAUDIO V. MELLO, AND ERICH D. JARVIS — 58
6. The songbird brain in comparative perspective
   MICHAEL A. FARRIES AND DAVID J. PERKEL — 63

PART II SONG PRODUCTION AND ITS NEURAL CONTROL — 73

7. Introduction
   H. P. ZEIGLER — 75
8. From brain to song: the vocal organ and vocal tract
   RODERICK A. SUTHERS AND SUE ANNE ZOLLINGER — 78
9. Peripheral mechanisms of sensorimotor integration during singing
   FRANZ GOLLER AND BRENTON G. COOPER — 99
10. Integrating breathing and singing: forebrain and brainstem mechanisms
    MARC F. SCHMIDT AND ROBIN ASHMORE — 115
11. Birdsong: anatomical foundations and central mechanisms of sensorimotor integration
    J. MARTIN WILD — 136

PART III HEARING AND RECOGNIZING THE SONG — 153

12. Introduction
    PETER MARLER — 155
13. Song selectivity and the songbird brain
    FRÉDÉRIC E. THEUNISSEN, NOOPUR AMIN, SARITA SHAEVITZ, SARAH M. N. WOOLLEY, THANE FREMOUW, AND MARK E. HAUBER — 157

14 Song-selective neurons in the songbird brain: synaptic mechanisms and functional roles
JONATHAN F. PRATHER AND RICHARD MOONEY 174

15 Temporal auditory pattern recognition in songbirds
TIMOTHY Q. GENTNER 187

PART IV LEARNING THE SONG: MECHANISMS OF ACQUISITION AND MAINTENANCE 199

16 Introduction
PETER MARLER 201

17 Comparative aspects of song learning
HENRIKE HULTSCH AND DIETMAR TODT 204

18 Developmental song learning in the zebra finch
SIGAL SAAR, PARTHA P. MITRA, SEBASTIEN DEREGNAUCOURT, AND
OFER TCHERNICHOVSKI 217

19 Auditory feedback and singing in adult birds
SARAH M. N. WOOLLEY 228

20 The anterior forebrain pathway and vocal plasticity
MICHAEL S. BRAINARD 240

21 Circuits and cellular mechanisms of sensory acquisition
ERNEST J. NORDEEN AND KATHY W. NORDEEN 256

22 Chasin' the trace: the neural substrate of birdsong memory
JOHAN J. BOLHUIS 271

23 The template concept: crafting a song replica from memory
PATRICE ADRET 282

PART V MECHANISMS OF MODULATION AND PLASTICITY 301

24 Introduction
PETER MARLER 303

25 Hormonal modulation of singing behavior: methodology and principles of hormone action
CHERYL F. HARDING 305

26 Sex differences in brain and behavior and the neuroendocrine control of the motivation to sing
GREGORY F. BALL, LAUREN V. RITERS, SCOTT A. MACDOUGALL-SHACKLETON,
AND JACQUES BALTHAZART 320

27 Plasticity of the song control system in adult birds
ELIOT A. BRENOWITZ 332

28 Regulation and function of neuronal replacement in the avian song system
CAROLYN PYTTE, LINDA WILBRECHT, AND JOHN R. KIRN 350

PART VI THE GENOMIC REVOLUTION AND BIRDSONG NEUROBIOLOGY 367

29 Introduction
H. P. ZEIGLER 369

30 Studies of songbirds in the age of genetics: what to expect from genomic approaches in the next 20 years
DAVID F. CLAYTON AND ARTHUR P. ARNOLD 371

31	Behavior-dependent expression of inducible genes in vocal learning birds	
	CLAUDIO V. MELLO AND ERICH D. JARVIS	381
32	Genes for tuning up the vocal brain: FoxP2 in human speech and birdsong	
	SEBASTIAN HAESLER AND CONSTANCE SCHARFF	398

PART VII ON A PERSONAL NOTE 407

33	Introduction	
	H. P. ZEIGLER	409
34	William Homan Thorpe	
	R. A. HINDE	411
35	My journey with birdsong	
	MASAKAZU KONISHI	415
36	The discovery of replaceable neurons	
	FERNANDO NOTTEBOHM	425
37	Birdsong and monkey talk: an ethological journey	
	PETER MARLER	449

References	463
Author index	531
Subject index	541

The colour plates are situated between pages 302 and 303.

Contributors

PATRICE ADRET
Department of Organismal Biology and Anatomy, The University of Chicago, Chicago, IL 60637, USA

NOOPUR AMIN
Helen Will Neuroscience Institute, University of California, Berkeley, 3210 Tolman Hall, Berkeley, CA 94720-1650, USA

ARTHUR P. ARNOLD
Department of Physiological Science, University of California, Los Angeles, CA 90095-1606, USA

ROBIN ASHMORE
Neuroscience Graduate Group, Department of Biology, University of Pennsylvania, Philadelphia, PA 19104, USA

GREGORY F. BALL
Department of Psychological and Brain Sciences, Johns Hopkins University, Baltimore, MD 21218-2686, USA

JACQUES BALTHAZART
Center for Cellular and Molecular Neurobiology, Behavioral Neuroendocrinology Research Group, University of Liège, B-4000 Liège, Belgium

JOHAN J. BOLHUIS
Behavioural Biology, Department of Biology and Helmholtz Institute, Utrecht University, The Netherlands

MICHAEL S. BRAINARD
Departments of Physiology and Psychiatry, and Keck Center for Integrative Neuroscience, University of California San Francisco, San Francisco, CA 94143-0444, USA

ELIOT A. BRENOWITZ
Departments of Psychology and Biology and the Virginia Merrill Bloedel Hearing Research Center, University of Washington, Box 351525, Seattle, WA 98195-1525, USA

DAVID F. CLAYTON
Department of Cell and Developmental Biology and the Institute for Genomic Biology, University of Illinois, Urbana, IL 61801, USA

BRENTON G. COOPER
Department of Biology, University of Utah, Salt Lake City, UT 84112, USA

SEBASTIEN DEREGNAUCOURT
Max Planck Institute for Ornithology, Postfach 1564, 82305 Starnberg, Germany

ALLISON J. DOUPE
Departments of Psychiatry and Physiology and Keck Center for Integrative Neuroscience, University of California at San Francisco, San Francisco, CA 94143, USA

MICHAEL A. FARRIES
Division of Life Sciences, University of Texas San Antonio, San Antonio, TX 78249-0662, USA

THANE FREMOUW
Department of Psychology, University of Maine, 301 Little Hall, Orono, ME 04469-5782, USA

TIMOTHY Q. GENTNER
Department of Psychology, University of California San Diego, 9500 Gilman Drive. MC0109, La Jolla, CA 92093-0109, USA

FRANZ GOLLER
Department of Biology, University of Utah, Salt Lake City, UT 84112, USA

SEBASTIAN HAESLER
Institute of Animal Behavior, Free University of Berlin, 14195 Berlin, Germany; Max-Planck-Institute for Molecular Genetics, 14195 Berlin, Germany

CHERYL F. HARDING
Psychology Department, Hunter College Biopsychology Doctoral Program, City University of New York, New York, NY 10021, USA

MARK E. HAUBER
Ecology, Evolution and Behaviour, School of Biological Sciences, University of Auckland, 3 Symonds Street, Auckland, PB 92019, New Zealand

ROBERT A. HINDE
St. John's College, Cambridge, CB2 ITP, UK

HENRIKE HULTSCH
Department of Behavioral Biology, Institute of Biology, Free University of Berlin, 12163 Berlin, Germany

ERICH D. JARVIS
Department of Neurobiology, Duke University Medical Center, Durham, NC 27710, USA

GISELA KAPLAN
Centre for Neuroscience and Animal Behaviour, University of New England, Armidale, NSW-2351, Australia

JOHN R. KIRN
Biology Department, Neuroscience and Behavior Program, Wesleyan University, Middletown, CT 06459, USA

MASAKAZU KONISHI
Division of Biology, California Institute of Technology, Pasadena, CA 91125, USA

PATRICIA K. KUHL
Department of Speech and Hearing Sciences, University of Washington, Seattle, WA 98195, USA

SCOTT A. MacDOUGALL-SHACKLETON
Psychology Department, University of Western Ontario, London, ON N6A 5C2, Canada

PETER MARLER
Division of Neurobiology, Physiology and Behavior, University of California at Davis, Davis, CA 95616, USA

CLAUDIO V. MELLO
Neurological Sciences Institute, Oregon Health and Science University, Portland, OR 97006, USA

PARTHA P. MITRA
Mitra Lab, Freeman Building, 1 Bungtown Road, Cold Spring Harbor, NY 11724, USA

RICHARD MOONEY
Department of Neurobiology, Duke University Medical Center, Durham, NC 27710, USA

ERNEST J. NORDEEN
Departments of Brain and Cognitive Sciences, Neurobiology and Anatomy, University of Rochester, Rochester, NY 14627, USA

KATHY W. NORDEEN
Departments of Brain and Cognitive Sciences, Neurobiology and Anatomy, University of Rochester, Rochester, NY 14627, USA

FERNANDO NOTTEBOHM
The Rockefeller University Field Research Center for Ecology and Ethology, 495 Tyrrel Road, Millbrook, NY 12545, USA

DAVID J. PERKEL
Departments of Biology and Otolaryngology, University of Washington, Seattle, WA 98195-6515, USA

JONATHAN F. PRATHER
Department of Neurobiology, Duke University Medical Center, Durham, NC 27710, USA

CAROLYN PYTTE
Psychology Department, Queens College-City University New York, Flushing, NY 11367, USA

ANTON REINER
Department of Anatomy and Neurobiology, University of Tennessee Health Science Center, Memphis, TN 38163, USA

LAUREN V. RITERS
Department of Zoology, University of Wisconsin, 361 Birge Hall 430 Lincoln Ave, Madison WI 53706, USA

SIGAL SAAR
Department of Biology, The City College of New York, New York, NY 10031

CONSTANCE SCHARFF
Institute of Animal Behavior, Free University of Berlin, 14195 Berlin, Germany; Max-Planck-Institute for Molecular Genetics, 14195 Berlin, Germany

MARC F. SCHMIDT
Neuroscience Graduate Group, Department of Biology, University of Pennsylvania, Philadelphia, PA 19104, USA

SARITA S. SHAEVITZ
Lecturer in the Princeton Writing Program, Princeton University, South Baker Hall, Whitman College, Princeton, NJ 08544, USA

RODERICK A. SUTHERS
Medical Sciences, Program for Neural Science, Department of Biology, Indiana University, Bloomington, IN 47405, USA

OFER TCHERNICHOVSKI
Department of Biology, The City College of New York, New York, NY 10031

FRÉDÉRIC E. THEUNISSEN
Department of Psychology and Neuroscience Institute, University of California Berkeley, Berkeley, CA 94720-1650, USA

DIETMAR TODT
Department of Behavioral Biology, Institute of Biology, Free University of Berlin, 12163 Berlin, Germany

LINDA WILBRECHT
Svoboda Lab, Cold Spring Harbor Laboratory, Cold Spring Harbor, NY 11724, USA

J. MARTIN WILD
Department of Anatomy, Faculty of Medical and Health Sciences, University of Auckland, Auckland, New Zealand

HEATHER WILLIAMS
Biology Department, Williams College, Williamstown, MA 01267, USA

SARAH M. N. WOOLLEY
Department of Psychology, Columbia University, New York, NY 10027, USA

H. PHILIP ZEIGLER
Department of Psychology and Biopsychology Program, Hunter College, City University of New York, NY 10021, USA

SUE ANNE ZOLLINGER
Medical Sciences, Program for Neural Science, Department of Biology, Indiana University, Bloomington, IN 47405, USA

Preface

The study of birdsong has a long past but a relatively short formal history. The fact that many species of songbirds learn their song from their parents was known to Chinese bird fanciers and was the impetus for a thriving trade in tutored birds. That song learning was familiar to Europeans as well is apparent from the report that George Henschel, a late nineteenth-century English conductor, kept a highly trained bullfinch that sang "God Save the Queen." It is reported, perhaps apocryphally, that whenever the bullfinch paused too long in mid-melody, an untrained canary in the next room would pick up the tune and finish it off properly.

In addition to the fact that it is a highly noticeable and aesthetically striking feature of bird behavior, birdsong is of interest to biologists both because it is associated with reproductive behavior (and therefore of interest to students of evolution) and because it is a learned behavior. Song learning became the subject of a long and productive series of studies initiated by William Thorpe of Cambridge University and continued by his student Peter Marler and Marler's many associates and students (see Part VII On a personal note). Research on song learning was quickened by two developments, one technical (see Hinde, this volume; Tchernichovski, 2004) the other neurobiological. The first was the development of the sound spectrogram, which provided a means for the preservation and objective analysis of bird songs. The second was the report, by Nottebohm and his colleagues (1976), that the vocalizations of songbirds reflected the operations of a definable collection of neural structures – a song circuit – originating in the bird's forebrain and capable of influencing the functioning of interneurons and motor neurons involved in the control of the bird's vocal organ (see Figure 5.2 in Reiner et al., this volume)

Subsequent studies have identified features of birdsong that have made it an increasingly fruitful model for research on a number of important problems in behavioral neurobiology. Among these features are (1) the reproducible and quantifiable nature of song behavior, (2) the many similarities between the acquisition of birdsong and of human speech, and (3) the identification and increasingly precise characterization of central neural circuits dedicated to song.

The initial identification of a "song circuit" was followed by studies showing that the structure of male brains differs from that of female brains and that these structural differences are correlated with differences in behavior. Furthermore, the structure of adult brains and the songs they generate exhibit a remarkable degree of plasticity, altering with changes in season and endocrine function. A potential mechanism for this plasticity was suggested by the discovery that that new neurons were born in the adult brain, migrated into position in brain areas involved in learning tasks, and became integrated into functional circuits. Subsequently, many features of the bird song system were also found in vertebrate brains, further increasing the utility of the model. These findings have had a major impact on neuroscience research and fundamentally altered our concepts of brain function.

During the past decade there have been a number of new developments both technical and substantive, which have increased the utility of birdsong as a neurobehavioral model system. These include: (1) novel methods for the computer-assisted analysis of song learning; (2), techniques for the abolition or reversible disruption of auditory feedback from song; (3) methods for the microanalysis of vocal tract activity from the syrinx to the beak; (4) the ability to monitor neural activity in the "song circuit" in awake, singing birds; (5) the molecular analysis of gene expression as a method for functional localization of song-related CNS structures; (6) the identification of song circuit structures putatively homologous to mammalian basal ganglia structures implicated in human motor learning and movement disorders; (7) the continuing contribution of research on the avian song system to the study of neurogenesis in the adult brain.

Given the increasing interest in birdsong as a neurobehavioral model it is somewhat surprising that there exists no single volume that attempts to summarize the current state of research on its neurobiological foundations. This book aims to fill that gap. The book had its origin in a series of papers delivered at an International Conference on Birdsong Neurobiology held at New York's Hunter College in December of 2003 and published as an Annals volume by the New York Academy of Sciences (Zeigler and Marler, 2004). However, the present volume, though it contains chapters by many conference participants, is quite a different book. The differences reflect not only the addition of several new authors but the advances in knowledge and conceptualization that have taken place since the earlier book.

Although, as a comprehensive summary of the current state of birdsong neurobiology, it will be indispensable for birdsong researchers, this book is directed to a much wider audience. Birdsong has become a model system for research on an array of problems of interest to the neuroscience community, to students of development, communication scientists and animal behaviorists. In recognition of that fact, contributors to this volume were exhorted to recognize the diverse nature of their potential readers and to provide a broader perspective on their specialized topics. Several papers from the Annals volume have been omitted as out of date or too specialized; the remaining papers have been extensively revised and several new ones added. These include a lucid introduction to the comparative study of song learning by Hultsch and Todt, a chapter reviewing research on a gene thought to be implicated in song learning in birds and language development in humans (Haesler and Scharff) a new perspective on song memory (Bolhuis) and a chapter by Kaplan which makes a strong case for the Australian magpie as a new model for song system research. Because there are currently several books devoted to ecological and evolutionary aspects of birdsong research we have minimized these aspects in the present volume. However, one feature of the Conference remains evident in the book and that is its breadth of representation. Our contributors comprise a representative cross section of birdsong researchers from the most senior investigators, to a group of young researchers, many of them students of the field's founders. They also reflect the growing internationalization of birdsong research, including not only the American laboratories which tend to dominate the field, but the increasing number of laboratories in Europe and the southern hemisphere.

The book is divided into a number of sections, each with its own introduction and each united by common themes. The first section, which is specifically aimed at the nonspecialist reader, provides an overview of song and singing behavior and introduces some of the issues that have preoccupied birdsong research since its inception. These include the relation of song to human language and the relation of the songbird brain to the mammalian brain. Subsequent sections deal, respectively, with song production mechanisms, from the vocal organ to the brain, with perceptual mechanisms and song recognition, with the behavioral and neural mechanisms mediating song acquisition and with the role of hormones in modulating singing behavior. Several chapters highlight the potential implications, for birdsong neurobiology, of the genomic revolution. A final section ("On a personal note") provides biographical material on W. H. Thorpe, who initiated the scientific study of birdsong, as well as personal reminiscences by his student Peter Marler and by Marler's students Mark Konishi and Fernando Nottebohm who have made critical contributions to "the neuroscience of birdsong." For those readers who are part of the large and growing birdsong research community, these chapters will provide a glimpse into the lives, vicissitudes and passions of the founders of that community.

H. PHILIP ZEIGLER

Acknowledgements

For their original support of the Conference to which this book is, however distantly, related, we thank the following:

Professor Robert Dottin, Center for Gene Structure and Function, Hunter College, New York
Research Centers in Minority Institutions Program, NIH
National Institutes of Health and the National Science Foundation for their original support of the Conference.

We thank the New York Academy of Sciences for publishing the proceedings of the 2003 Conference in their Annals series.

We thank the Royal Society and Professor Robert A. Hinde for permission to reprint a slightly shortened version of his obituary memoir of Professor W. H. Thorpe.

We acknowledge with gratitude the permissions gracefully granted by a number of scientific publishers to reprint materials from publications of contributors to this volume.

Special thanks to Vicky Lawson, who provided invaluable assistance at many stages of this project. I thank my grandson, Thomas Kirkpatrick Miller, for his assistance in clarifying for me the etymological origins of the term *oscine*.

PART I
Foundations: singing and the brain

1 • Introduction

H. P. Zeigler

The chapters in this introductory section focus on two central topics: singing behavior itself and the songbird brain. As Williams points out, what was always a source of aesthetic pleasure for bird watchers has developed into a major area of neurobiological research. Researchers have come to recognize that this apparently simple, "innate" behavior is the product of an ontogenetic learning process; that it is limited to only a very few vertebrate species (including humans) and serves a critical communicative function. Indeed, the initial attraction of birdsong for neuroscientists was as a model system for the study of learned vocalizations, analogous to speech acquisition in humans. For this reason the section opens with a somewhat abridged version of a classic paper by Doupe and Kuhl comparing birdsong and human speech.

The Williams chapter provides an introduction to the essential features of birdsong as a learned vocalization, including those features that qualify it as a model system for studies of the development of human language. But her main focus is on the adult song, the remarkable degree of variability it exhibits and the importance of the social context in which it occurs. She notes that because song is linked to two adaptive behaviors – territory defense and mate attraction – it has evolved specifically to influence listeners' behavior. Thus, while the stereotypy of song is often viewed as one of its defining features, the presence and behavior of conspecific "listeners" can significantly affect both what is sung (song structure) and how (amplitude and tempo). She also calls attention to the role of the visual displays associated with song, an often neglected component of singing behavior. While the links between singing and breathing are already the focus of considerable research effort (chapters by Wild, Suthers and Zollinger, Goller and Cooper, Schmidt and Ashmore), little is known as to the neural circuitry which integrates vocal and visual signals. What is known, however, is that the responsiveness of these circuits is modulated by hormones, and the interplay between neural, hormonal and sensory factors in modulating singing behavior is one of the central themes of this volume (see chapters by Harding, Ball et al., and Brenowitz).

Importantly, Williams emphasizes that much of our knowledge of songbird neurobiology is based upon data from a small number of species selected primarily for their amenability to study in the laboratory (zebra finches, song sparrows, canaries). Singing behavior in these species has a clear reproductive function, is linked specifically to the breeding season and is performed exclusively by males. Moreover, differences in singing behavior between males and females of these species are correlated with differences in the brain circuits mediating singing behavior. As Williams points out, it is all too easy to generalize from male zebra finches (or canaries) to "birds." Thus, our conceptualization of "song" has come to be defined by the song of a small number of avian species, and male brains and behavior have come to be the primary focus of research on brain mechanisms of song learning.

Kaplan's chapter on the Australian magpie provides a thought-provoking challenge to this conceptualization. As described by Kaplan magpie song is complex and its song repertoire enormous, including a capacity for mimicry. However, since both males and females sing, their song circuit structures are of comparable size and song has no obvious reproductive function, magpie vocalizations do not fit our current conceptualization of song. Kaplan argues that the features we think of as defining song, especially sexual dimorphism and breeding-related song, are characteristic of northern hemisphere birds, while the singing behavior of the Australian magpie is, in fact, representative of a very large and diverse range of songbird species, including tropical birds. From this standpoint, she contends, it is certainly as appropriate a model as zebra finch or canary song. Moreover, since in neither humans nor magpies is vocal learning an *exclusively* reproductive behavior, it may even be a more appropriate model for comparative neurobiological studies of language development.

Neuroscience of Birdsong, ed. H. Philip Zeigler and Peter Marler. Published by Cambridge University Press. © Cambridge University Press 2008.

Both Williams and Kaplan would probably agree that the concept of a canonical stereotyped learned song, as in zebra finches, is useful as a tool for studying neural processes and that the ability of a species to thrive in captivity is a useful feature for a model. However, they would probably also agree that the field would certainly benefit from exposure to the widest possible array of vocalization types, including those, like the magpie, that function as exceptions which "prove" (test) the rule.

The function of a model, of course, is to provide access to critical features of a natural phenomenon for experimental analysis. Given the apparent parallels between language development in humans and vocal learning in birds, the extent to which these processes share similar neural substrates in the two groups has been a critical issue for the birdsong community. If the brains of the two groups were obviously different it would be hard to justify the use of birdsong as a model system that might teach us something about the brain mechanisms underlying human language. While much of the brain of birds (from spinal cord to midbrain) reflects an organization common to most vertebrates, the "higher" brain regions, including the forebrain, are clearly different. The avian forebrain is made of large nuclear cell masses and its morphology resembles that of the mammalian basal ganglia (striatum); the mammalian forebrain, while it contains obvious striatal structures, is defined by its unique possession of a laminated isocortex ("grey matter") separated from the underlying basal ganglia by a band of myelinated axons ("white matter"). Moreover, the obvious increase in cortical development from primitive mammals to *homo sapiens* seemed to be correlated with increased possession of that concatenation of abilities grouped colloquially under the name of "intelligence" – something that birds were thought to lack. The absence of a cortex thus became a defining negative taxonomic feature of the avian brain while the notion that the avian forebrain was essentially a "striatal" structure was reflected in the nomenclature assigned to forebrain structures (*hyperstriatum, neostriatum, archistriatum, palaeostriatum*). The widespread use of the term *birdbrain* as a derogatory epithet reflects the view widely held among early twentieth-century comparative anatomists that the avian forebrain represents an evolutionary dead end. Despite the fact that there is no evidence that the organization of forebrain cells into lamina provides any unique advantages for neural processing, this view persisted well into the middle of the twentieth century. It was impossible to directly refute because, in the absence of a fossil record for soft tissue such as brain, the identification of homologies between avian and mammalian brain structures was not feasible. To provide an overview of some of these issues we have reprinted excerpts from a chapter by Reiner and his colleagues which originally provided an introduction and rationale for the recent revision of songbird brain nomenclature. The goal of this revision was to develop a nomenclature both more neutral in its attribution of homologies and more systematic in its application of the latest available anatomical, neurochemical and developmental information.

The advent of techniques for experimental anatomy and genomic analysis has made it possible to carry out detailed comparative studies of the organization of the embryonic and adult avian and mammalian forebrains. These studies have demonstrated many similarities between the connection patterns of structures in the avian and mammalian forebrain and have clarified the relation between basal ganglia structures in the two groups. Farries and Perkel provide an overview of the implications of these findings for both avian and mammalian brain research.

2 • Birdsong and human speech: common themes and mechanisms

Allison J. Doupe and Patricia K. Kuhl

INTRODUCTION

Experts in the fields of human speech and birdsong have often commented on the parallels between the two in terms of communication and its development (Marler, 1970a; Kuhl, 1989). Does the acquisition of song in birds provide insights regarding learning of speech in humans? This review provides a critical assessment of the hypothesis, examining whether the similarities between the two fields go beyond superficial analogy. The often cited commonalities provide the topics of comparison that structure this review.

First, learning is critical to both birdsong and speech. Birds do not learn to sing normally, nor infants to speak, if they are not exposed to the communicative signals of adults of the species. This is an exception among species: most animals do not have to be exposed to the communicative signals of their species to be able to reproduce them. The fact that babies and songbirds share this requirement has intrigued scientists.

Second, vocal learning requires both perception of sound and the capacity to produce sound. At birth, both human infants and songbirds have been hypothesized to have innate perceptual predispositions for the vocal behavior of their own species. We review the nature of the predispositions in the two cases and the issue of whether they are similar. Given that innate predispositions exist, another important question is how subsequent experience alters perception and production in each case. Moreover, vocal perception and production are tightly interwoven in the vocal learning process. We examine what is known about the relationship between perception and production and whether in these different vocal learners it is similar.

In addition, neural substrates of vocal communication in humans and birds have often been compared. Human brains are asymmetric and language tends to be organized in the left hemisphere as opposed to the right. Birds are also often assumed to show similar hemispheric specialization for song. What are the real parallels between the neural substrates in the two cases?

Finally, critical (sensitive) periods are evidenced in both species. Neither birds nor babies appear to learn their communicative signals equally well at all phases of the life cycle. This raises the questions of what causes the change in the ability to learn over time and with experience, and whether the causes are the same in human infants and songbirds. And if the plasticity of the brain is altered over the life cycle, what neural mechanisms control this changing ability to learn?

The research reviewed here relates to ongoing work in developmental biology, ethology, linguistics, cognitive psychology, and computer science, as well as in neuroscience, and should be of interest to individuals in many of these fields. What our review reveals is that although the comparisons between birdsong and speech are not simple, there is a surprisingly large number of areas where it is fruitful to compare the two. Going beyond the superficial analogy, however, requires some caveats about what may be comparable and what clearly is not. In the end, understanding both the similarities and differences will provide a broader spectrum in which to view the acquisition of communication in animals and humans.

Editors note: Adapted and reprinted, with permission from the authors and from the *Annual Review of Neuroscience*, Volume 22, ©1999 by Annual Reviews www.annualreviews.org. The sections on "Specialized neural substrates for song and speech learning (pp. 596–604) on "lateralization" (pp. 606–609) and on "possible neural mechanisms underlying the sensitive period and its closure" (pp. 618–619) have been omitted. The reader is referred to relevant chapters in this volume for reviews of more recent work on neural mechanisms.

Neuroscience of Birdsong, ed. H. Philip Zeigler and Peter Marler. Published by Cambridge University Press. © Cambridge University Press 2008.

SPEECH AND BIRDSONG: DEFINITIONS

Speech and song production

Both birdsong and human speech are complex acoustic signals. Figure 2.1 shows a spectrographic (frequency versus time) display of a spoken human phrase ("Did you hit it to Tom?") and Figure 2.2 a similar display of songs of two different songbird species. In both songbirds and humans, these sounds are produced by the flow of air during expiration through a vocal system. In humans, the process is relatively well understood: air from expiration generates a complex waveform at the vocal folds, and the components of this waveform are subsequently modified by the rest of the vocal tract (including the mouth, tongue, teeth, and lips) (Stevens, 1994).

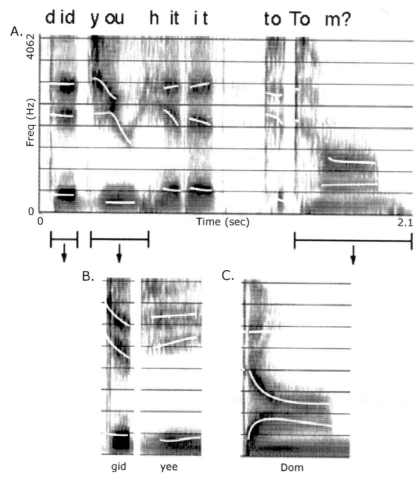

Figure 2.1 Human speech. Three dimensions of speech are shown in a spectrogram: time or duration along the horizontal axis; frequency along the vertical axis; and intensity, which is correlated with loudness, by the relative darkness of each frequency. This spectrogram shows the phrase "Did you hit it to Tom?" spoken by a female (A). White lines are the formants that characterize each individual phoneme. (B–C) Variations on words from the full sentence. (B) A place of articulation contrast using a spectrogram of the nonsense word "gid," which differs from its rhyme "did" (in A) in that it has a decreasing frequency sweep in the second and third formants (between 2000 and 3000 Hz). This decreasing formant pattern defines the sound "g" and a pattern of flat formants defines the sound "d." (C) The words "Tom" and "Dom" contrast in voice onset time (VOT). Notice the long, noisy gap in "Tom" (A), which has a long VOT, compared with the short gap in "Dom."

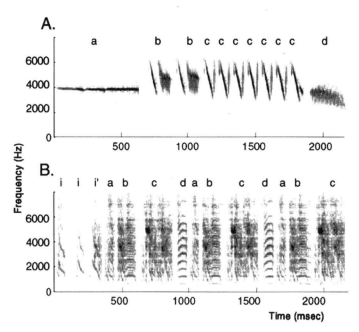

Figure 2.2 Examples of birdsongs from two species. (A) A typical song of a white-crowned sparrow. The smallest elements, the notes, are combined to form syllables (lower case letters), and these are repeated to form phrases. White-crowned sparrow songs typically begin with (a) a long whistle followed by (b, c) trills and (d) buzzes. (B) A typical song of a zebra finch. Note the noisy spectral quality (more like humans) that distinguishes it from more tonal species like the sparrows. Zebra finch songs start with a number of introductory syllables (marked with i), followed by a sequence of syllables (lower case letters), that can be either simple or more complex, with multiple notes (e.g. b, c). Particular sequences of syllables are organized into phrases called motifs, which are repeated.

The vocal tract acts as a filter, creating concentrations of energy at particular frequencies, called formant frequencies (Figure 2.1). Vowels are characterized by relatively constant formant frequencies over time (Figure 2.1A, C), whereas during consonant production the formant frequencies change rapidly (20–100 ms), resulting in formant transitions (Figure 2.1A, B, D).

In songbirds, sounds are produced by the flow of air during expiration through an organ called the syrinx, a bilateral structure surrounded by specialized muscles, which sits at the junction of the bronchi with the trachea. A number of aspects of syringeal function are understood, although the exact mechanism of sound generation is controversial and is under active investigation (Gaunt, 1987; Goller and Larsen, 1997a; Suthers, 1997; Fee et al., 1998). Also, there are indications that the upper vocal tract in bird structures sound in a manner like the upper vocal tract in humans. Recent research suggests that the width of beak opening (known as beak gape) affects sound frequency (Westneat et al., 1993; Suthers, 1997), and there may be some degree of coupling between the syrinx and the vocal tract (Nowicki, 1987). Regardless of differences in component structures, for both birdsong and speech the production of highly structured and rapidly changing vocalizations requires elaborate neural control and coordination of respiration with a variety of vocal motor structures.

The structure of speech and song

It is useful to define the basic terms used in each field, and the various ways in which vocal behavior is described, in order to assess what aspects of each of the signals are comparable. Human speech can be described at many different levels. It can be written, spoken, or signed (using a manual language such as American Sign Language). In all these forms, language consists of a string of words ordered by the rules of grammar to convey meaning. Stucturally, language can be analyzed from the standpoint of semantics (conceptual representation), syntax (word order), prosody (the pitch, rhythm, and tempo of an utterance), the lexicon (words), or phonology (the elementary building blocks, phonemes, that are combined to make up words).

Speech, and especially its development, has been intensively studied at the phonological level. Phonetic units are the smallest elements that can alter the meaning of a word in any language, for example the difference between /r/ and /l/ in the words "rid" and "lid" in American English. Phonemes refer to the phonetic units critical for meaning in a particular language. The phonetic difference between /r/ and /l/ is phonemic in English, for example, but not in Japanese. Each phonetic unit can be described as a bundle of phonetic features that indicate the manner in which the sound was produced and the place in the mouth where the articulators (tongue, lips, teeth) were placed to create the sound (Jakobson *et al.*, 1969). The acoustic cues that signal phonetic units have been well documented and include both spectral and temporal features of sound (Figure 2.1) (Stevens, 1994). For instance, the distinction between /d/ and /g/ depends primarily on the frequency content of the initial burst in energy at the beginning of the sound and the direction of formant transition change (Figure 2.1A, B). An example of a temporal acoustic dimension of speech is voice-onset time (VOT), which refers to the timing of periodic laryngeal vibration (voicing) in relation to the beginning of the syllable (Figure 2.1A, D). This timing difference provides the critical cue used to identify whether a speech sound is voiced or voiceless (e.g. /b/ versus /p/, /do/ versus /to/) and is a classic distinction used in many speech studies.

Which aspects of birdsong can be usefully compared with speech? Birdsongs are distinct from bird calls (which are brief and generally not learned), last from a few seconds to many tens of seconds, and, like speech, consist of ordered strings of sounds separated by brief silent intervals (Figure 2.2). The smallest level of song usually identified is the note or element, defined as a continuous marking on a sound spectrogram; these may be analogous to the smallest units of speech, or phonetic units. Notes can be grouped together to form syllables, which are units of sound separated by silent intervals. When singing birds are interrupted by an abrupt light flash or sound, they complete the syllable before stopping (Cynx, 1990); thus, syllables may represent a basic processing unit in birdsong, as posited for speech.

Another feature that birdsong and language share is the conspicuous timing and ordering of components on a timescale longer than that of the syllable. Song syllables are usually grouped together to form phrases or motifs (Figure 2.2), which can be a series of identical or different syllables. Many songbirds sing several phrases in a fixed order as a unit, which constitutes the song, whereas other species such as mockingbirds and warblers produce groups of syllables in fixed or variable sequences. The timing and sequencing of syllables and phrases are rarely random but instead follow a set of rules particular to a species. In the songbird literature, the ordering of syllables and phrases in song is often called song syntax. The same word applied to human speech, however, implies grammar, i.e. rules of ordering words from various grammatical classes to convey meaning. Therefore, in this review, we avoid using the word syntax for song and simply use "order." Thus, language and song share a dependence on timing on several timescales: a shorter timescale (on the order of tens of milliseconds), as in phonemes and syllables, and a longer one, up to many hundreds of milliseconds (as in syllable, phrase, and word ordering).

Language is also characterized by a boundless and flexible capacity to convey meaning, but this property is not shared with birdsong. The whole set of different songs of a bird is known as its song repertoire and can vary from one (in species such as the zebra finch or white-crowned sparrow) to several hundreds (for review see Konishi, 1985). Numerous behavioral studies, usually using the receiver's response, suggest that songs communicate species and individual identity (including "neighbor" and "stranger"), an advertisement for mating, ownership of territory, and fitness. Some birds with multiple song types use different songs for territorial advertisement and for mate attraction (Catchpole, 1983; Searcy and Nowicki, 1998). Nonetheless, large song repertoires do not seem to convey many different meanings, nor does song have the complex semantics of human speech. The definitions above suggest that the phonology (sound structure), the rules for ordering sounds, and perhaps the prosody (in the sense that it involves control of frequency, timing, and amplitude) are the levels at which birdsong can be most usefully compared with language, and more specifically with spoken speech, and are thus the focus of this review.

VOCAL LEARNING IN HUMANS AND SONGBIRDS

Which animals are vocal learners?

Many animals produce complex communication sounds but few of them can and must learn these vocal signals. Humans are consummate vocal learners. Although there is emerging evidence that social factors can influence acoustic variability among nonhuman primates (Sugiura, 1998), no other primates have yet been shown to learn their vocalizations. Among the mammals, cetaceans are well known to acquire their vocal repertoire and to show vocal mimicry (McCowan and Reiss, 1997); there are also some bats whose vocalizations may be learned (Boughman, 1998). Among avian species, songbirds, the parrot family, and some hummingbirds meet the criteria for vocal learning, but the term birdsong is usually reserved for the vocalizations of passerine (perching) songbirds and that is the focus of this review. The many thousands of songbird species, as well as the parrots and hummingbirds, stand in striking contrast to the paucity of mammalian vocal learners.

Nonhuman primates can, however, make meaningful use of vocalizations: for instance, vervets use different calls to indicate different categories of predators. Production of these calls is relatively normal even in young vervets and does not appear to go through a period of gradual vocal development, but these animals must develop the correct associations of calls to predators during early ontogeny (Seyfarth and Cheney, 1997). What songbirds and humans share is not this development of associations of vocalizations with objects or actions, but the basic experience-dependent memorization of sensory inputs and the shaping of vocal outputs.

Evidence for vocal learning

The basic phenomenology of learning of song or speech is strikingly similar in songbirds and humans. Initial vocalizations are immature and unlike those of adults: babies babble, producing consonant–vowel syllables that are strung together (e.g. bababa or mamama), and young songbirds produce subsong, soft and rambling strings of sound. Early sounds are then gradually molded to resemble adult vocalizations. The result of this vocal development is that adults produce a stereotyped repertoire of acoustic elements: these are relatively fixed for a given individual, but they vary between individuals and groups (as in languages and dialects, and the individually distinct songs and dialects of songbirds within a particular species). This variability is a reflection of the fact that vocal production by individuals is limited to a subset of all sounds that can be produced by that species. Layered on top of the developing capacity to produce particular acoustic elements is the development of sequencing of these elements: for humans this means ordering sounds to create words and, at a higher level, sentences and grammar; in birds this means sequencing of elements and phrases of song in the appropriate order. An important difference to remember when making comparisons is that the numerous languages of humans are not equivalent to the songs of different species, but rather to the individual and geographical variations of songs within a species.

LEARNED DIFFERENCES IN VOCAL BEHAVIOR

That the development of a mature vocal repertoire reflects learning rather than simply the expression of innate programs is apparent from a number of observations. Most important, for both birds and humans, there exist group differences in vocal production that clearly depend on experience. Obviously, people learn the language to which they are exposed. Moreover, even within a specific language, dialects can identify the specific region of the country in which a person was raised. Likewise, songbirds learn the songs sung by adults to which they are exposed during development: this can be clearly demonstrated by showing that birds taken from the wild as eggs or nestlings and exposed to unrelated conspecific adults, or even simply to tape recordings of the song of these adults, ultimately produce normal songs that match those that were heard (Marler, 1970b; Thorpe, 1958, 1961). Even more compelling are cross-fostering experiments, in which birds of one species being raised by another will learn the song, or aspects thereof, of the fostering species (Immelmann, 1969). In addition, many songbirds have song "dialects," particular constellations of acoustic features that are well defined and restricted to local geographic areas. Just as with human dialects, these song dialects are culturally transmitted (Marler and Tamura, 1962).

VOCALIZATIONS IN THE ABSENCE OF EXPOSURE TO OTHERS

Another line of evidence supporting vocal learning is the development of abnormal vocalizations when humans or birds with normal hearing are socially isolated and therefore not exposed to the vocalizations of others. The need for auditory experience of others in humans is evident in the (fortunately rare) studies of children raised either in abnormal social settings, as in the case of the California girl, Genie, who was raised with almost no social contact (Fromkin et al., 1974), or in cases in which abandoned children were raised quite literally in the wild (Lane, 1976). These and other documented instances in which infants with normal hearing were not exposed to human speech provide dramatic evidence that in the absence of hearing speech from others, speech does not develop normally. Similarly, songbirds collected as nestlings and raised in isolation from adult song produce very abnormal songs (called "isolate" songs) (Marler, 1970b; Thorpe, 1958). This need for early auditory tutoring has been demonstrated in a wide variety of songbirds (for reviews see Catchpole and Slater, 1995; Kroodsma and Miller, 1996). Strikingly, although isolate songs are simplified compared with normal, learned song, they still show some features of species-specific song (Marler and Sherman, 1985).

One caveat about studies of isolated songbirds or humans is that many aspects of development are altered or delayed in such abnormal rearing conditions. Nonetheless, the results of isolation in humans and songbirds are in striking contrast to those seen with members of closely related species, such as nonhuman primates and nonsongbirds such as chickens, in whom vocalizations develop relatively normally even when animals are raised in complete acoustic isolation (Konishi, 1963; Kroodsma, 1985; Seyfarth and Cheney, 1997). In combination with the potent effects of particular acoustic inputs on the type of vocal output produced, these results demonstrate how critically both birdsong and speech learning depend on the auditory experience provided by hearing others vocalize.

THE IMPORTANCE OF AUDITION IN SPEECH AND SONG

The importance of hearing one's own vocalizations

Vocal learning, shared with few other animals, is also evident in the fact that both humans and songbirds are acutely dependent on the ability to hear themselves in order to develop normal vocalizations. Human infants born congenitally deaf do not acquire spoken language, although they will, of course, learn a natural sign language if exposed to it (Petitto, 1993). Deaf infants show abnormalities very early in babbling, which is an important milestone of early language acquisition. At about 7 months of age, typically developing infants across all cultures will produce this form of speech. The babbling of deaf infants, however, is maturationally delayed and lacks the temporal structure and the full range of consonant sounds of normal-hearing infants (Oller and Eilers, 1988; Stoel-Gammon and Otomo, 1986). The strong dependence of speech on hearing early in life contrasts with that of humans who become deaf as adults: their speech shows gradual deterioration but is well preserved relative to that of deaf children (Cowie and Douglas-Cowie, 1992, Waldstein, 1990).

Songbirds are also critically dependent on hearing early in life for successful vocal learning. Although birds other than songbirds, e.g. chickens, produce normal vocalizations even when deafened as juveniles, songbirds must be able to hear themselves in order to develop normal song (Konishi, 1963, 1965b; Nottebohm, 1968). Songbirds still sing when deafened young, but they produce a very abnormal, indistinct series of sounds that are much less songlike than are isolate songs; although it varies from species to species, often only a few features of normal songs are maintained, primarily their approximate duration (Marler and Sherman, 1983). As with humans, once adult vocalizations have stabilized, most songbird species show decreased dependence on hearing (Konishi, 1965b; but see below).

The effects of deafness in early life do not differentiate between the need for hearing others and a requirement for hearing oneself while learning to vocalize. In birds, however, there is often a separation between the period of hearing adult song and the onset of vocalizations, and this provided the opportunity to demonstrate that song is abnormal in birds even when they have had adequate tutor experience prior to being deafened (Konishi, 1965b). This revealed that during song learning hearing functions in two ways, in two largely nonoverlapping phases (Figure 2.3B). During an initial sensory phase, the bird listens to and learns the tutor song. After this sensory learning, however, the memorized song, called the template, cannot be simply internally translated into the correct vocal motor pattern. Instead, a second, sensorimotor learning or vocal practice phase is necessary. The bird must actively compare and gradually match its own vocalizations to the memorized template, using auditory feedback. The need for the bird to hear itself is also evident in birds first raised in isolation and then deafened prior to sensorimotor learning. These birds sing abnormal songs indistinguishable from those of deafened tutored birds, demonstrating that the innate information about song that exists in isolate birds also requires auditory feedback from the birds' own vocalizations and motor learning in order to be turned into motor output (Konishi, 1965b). Thus, learning to produce song is crucially dependent on auditory experience of self as well as of others.

Humans likely also have to hear themselves in order to develop normal speech. This issue is more difficult to study in human infants than in songbirds, however, because the need for auditory input from others overlaps substantially in time with when childen are learning to speak (Figure 2.3A). Studies of children becoming deaf later in childhood, however, indicate that speech still deteriorates markedly if deafness occurs prior to puberty (Plant and Hammarberg, 1983). Thus, even though language production is well developed by late preadolescence, it cannot be well maintained without the ability to hear, which suggests that feedback from the sound of the speaker's own voice is also crucial to the development and stabilization of speech production. In addition, special cases in which infants hear normally but cannot vocalize provide relevant data. Studies of speech development in children who prior to language development had tracheostomies for periods lasting from 6 months to several years indicate severe speech and language delays as a result (Locke and Pearson, 1990; Kamen and Watson, 1991). Although these studies cannot rule out motor deficits due to lack of practice or motor damage, the speech of these children, who have normal hearing, is similar in its structure to that produced by deaf children. These studies, and the effects of deafness on older children, provide evidence that, just as in songbirds, both the sounds produced by the individuals themselves and those produced by others are essential for normal speech development.

THE FUNCTION OF AUDITORY FEEDBACK IN ADULTHOOD

In both humans and songbirds, the strong dependence of vocal behavior on hearing early in life lessens in adulthood. Postlingually deaf adults do show speech deterioration (Cowie and Douglas-Cowie, 1992; Waldstein, 1990), but it is less than that of deaf children, and it can be rapidly ameliorated even by the limited hearing provided by cochlear implants (Tyler, 1993). In some songbird species, song deteriorates very little in deafened adults, which suggests song is maintained by nonauditory feedback and/or by a central pattern generator that emerged during learning. In other species, song deteriorates more markedly after deafness in adulthood, both in phonology and in syllable ordering (Nordeen and Nordeen, 1993; Woolley and Rubel, 1997; Okanoya and Yamaguchi, 1997). Even in these cases, in many species song deterioration is often slower in adults than in birds actively learning song and may depend on how long the bird has been singing mature, adult ("crystallized") song. Some birds are "open" learners: that is, their capacity to learn to produce new song remains open in adulthood (e.g. canaries) (Nottebohm et al., 1986). Consistent with how critical hearing is to the learning of song, these species remain acutely dependent on auditory feedback for normal song production as adults.

Moreover, for both human speech and birdsong, incorrect or delayed auditory feedback in adults is more disruptive than the complete absence of auditory feedback. For instance, delayed auditory playback of a person's voice causes slowing, pauses, and syllable repetitions in that subject (Howell and Archer, 1984; Lee, 1950). In addition, presentation to adult humans of altered versions of the vowels in their own speech, at a very short time delay, causes the subjects unconsciously to produce appropriately altered speech (Houde and Jordan, 1998). In songbirds as well, recent results suggest that delayed or altered auditory feedback can cause syllable repetitions or song deterioration (Leonardo and Konishi, 1999; J. Cynx, personal

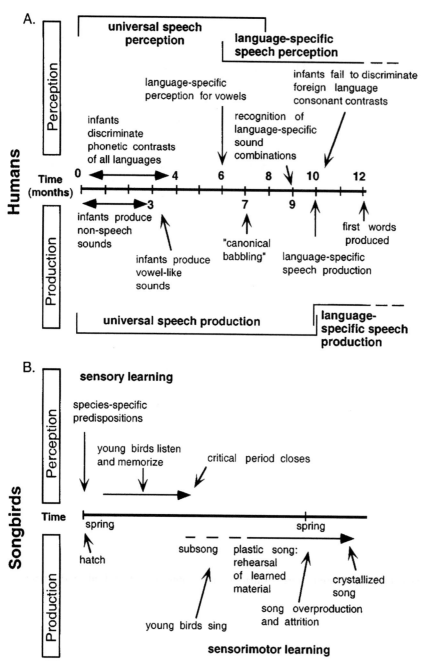

Figure 2.3 Timelines of speech and song learning. (A) During the first year of life, infant perception and production of speech sounds go through marked changes. (A, top) The developmental milestones associated with listening to speech; (A, bottom) the type of sounds produced throughout an infant's first year, leading up to the meaningful production of words. In both aspects of development, infants change from being language-general in the earliest months to language-specific toward the end of the first year. (B) Similar timelines show the early perceptual learning of seasonal songbirds (approximately 2–3 months), followed by sensorimotor learning in the fall and especially the next spring. In zebra finches this entire learning takes place over 3–4 months, with the critical period ending around 60 days of age, and much more overlap between sensory and sensorimotor phases (with singing beginning around 30 days of age).

communication). Thus, although auditory feedback is not as essential for ongoing vocal production in adult birds and humans as in their young, it clearly has access to the adult vocal system, and can have dramatic effects on vocal behavior if it is not well matched with vocal output.

INNATE PREDISPOSITIONS AND PERCEPTUAL LEARNING

Key features of vocal learning are the perception of sounds, the production of sounds, and the (crucial) ability to relate the two. In the next section, two questions, which roughly parallel the course of vocal development and have preoccupied both speech and song scientists, are addressed. What are the perceptual capabilities and innate predispositions of vocal learners at the start of learning? And what does subsequent experience do to perception?

Speech and song perception and production: innate predispositions

Experience clearly affects vocal production in humans and songbirds, but there is compelling evidence that learning in both species does not occur on a *tabula rasa*. Rather, there is evidence of constraints and predispositions that bias the organism in ways that assist vocal learning.

At the most fundamental level, the physical apparatus for vocalization constrains the range of vocalizations that can be produced (Podos, 1996). What is surprising, however, is that motor constraints do not provide the strongest limitations on learning. Both bird and human vocal organs are versatile, and although some sounds are not possible to produce, the repertoire of human and songbird sounds is large.

Looking beyond these peripheral motor constraints, there are centrally controlled perceptual abilities that propel babies and birds toward their eventual goal, the production of species-typical sound. In humans, perceptual studies have been extensively used to examine the initial capacities and biases of infants regarding speech, and they have provided a wealth of data on the innate preparation of infants for language. At the phonetic level, classic experiments show that early in postnatal life, infants respond to the differences between phonetic units used in all of the world's languages, even those of languages they have never heard (Eimas, 1975a, b; Streeter, 1976; for review see Kuhl, 1987). In these studies, infants are tested using procedures that indicate their ability to discriminate one sound from another. These include the high-amplitude sucking paradigm (in which changes in sucking rate indicate novelty), as well as tests in which a conditioned head turn is used to signal infant discrimination. These tests demonstrate the exquisite sensitivity of infants to the acoustic cues that signal a change in the phonetic units of speech, such as the VOT differences that distinguish /b/ from /p/ or the formant differences that separate /b/ from /g/ or /r/ from /l/.

Moreover, as with adults, infants show categorical perception of sounds, a phenomenon initially demonstrated in adults during the 1950s (Liberman *et al.*, 1967). Tests of categorical perception use a computer-generated series of sounds that continuously vary in small steps, ranging from one syllable (e.g. /ba/) to another (/pa/), along a particular acoustic dimension (in the case of /ba/ and /pa/, the VOT). Adult listeners tend not to respond to the acoustic differences between adjacent stimuli in the series but perceive an abrupt change in the category – the change from /ba/ to /pa/ – at a particular VOT (hence the name categorical perception). In adults, categorical perception generally occurs only for sounds in the adult's native language (Miyawaki *et al.*, 1975). Very young infants not only perceive sounds categorically (Eimas *et al.*, 1971, Eimas, 1975a), but also demonstrate the phenomenon for sounds from languages they have never heard as well as for sounds from their native language (Streeter, 1976; Lasky *et al.*, 1975). These studies provided the first evidence that infants, at birth, have the capacity to discriminate any and all of the phonetic contrasts used in the languages of the world, a feature of auditory perception that greatly enhances their readiness for language learning.

Later studies revealed that nonhuman mammals (chinchillas and monkeys) respond to the same discontinuities in speech that human infants do (Kuhl and Miller, 1975; Kuhl and Padden, 1983), which suggested that human speech evolved to take advantage of the existing auditory capacities of nonhuman primates (Kuhl, 1986). Data also showed that human infant sensitivities extended to nonspeech sounds that contained acoustic dimensions critical to speech but that were not identifiable as speech (Jusczyk et al., 1977). These data caused a shift in what was theorized to be innate (Kuhl, 1986, 1994; Jusczyk, 1981). Initial theories had argued that humans were endowed at birth with "phonetic feature detectors" that defined all possible phonetic units across languages (Eimas, 1975b). These detectors were thought to specify the universal set of phonetic units. When data revealed that the categorical perception of speech was not restricted to humans nor to speech, theories were revised to suggest that what was innate in humans was an initial discriminative capacity for speech sounds, rather than a specification of speech sounds themselves. Infants' discriminative capacities are currently viewed as "basic cuts" in auditory perception. Though not precise, they allow infants to discriminate the sounds of all languages (Kuhl, 1994). Evidence supporting this comes from studies showing that, with exposure to language, the accuracy of discrimination increases substantially for native-language sounds (Kuhl et al., 1997b; Burnham et al., 1987). Theorists noted that these innate perceptual abilities, although not unique to humans, provided infants with a capacity to respond to and acquire the phonology of any language.

As with humans, young songbirds begin life endowed with the capacity for responding to the sounds of their own species, before they have done any singing themselves. Studies of changes in heart rate in young birds in response to song playback initially demonstrated that both male and female sparrows innately discriminate conspecific from heterospecific song (Dooling and Searcy, 1980). Measurement of white-crowned sparrow nestling begging calls in response to tape-recorded song also revealed the much greater vocal behavior of young birds in response to their own species' song than to alien song, providing further evidence of inborn sensory recognition of conspecific song (Nelson and Marler, 1993). This assay also used simplified versions of these songs containing single phrases or modified songs with altered order, to begin to define the minimal acoustical cues critical for this recognition (Whaling et al., 1997).

There is a subtle but important difference between most studies of innate predispositions in songbirds and in humans, however. In birds, what has been examined is not discrimination of sounds within a set of possible songs from a particular species, which would be analogous to studies of phonemes from different human languages. Rather, most studies have looked at learning and listening preferences between songs of different songbird species. This is not possible in humans because one cannot isolate humans in order to expose them to the sounds of other species (to macaque monkey calls, for example) to determine whether they would learn such calls. In birds with whom these experiments have been done, both innate conspecific song recognition and preference are evident in the choice of models for learning song. A variety of experiments, using tape playback of tutor songs, showed that songbirds prefer their own species' song over alien songs as tutor models (Marler and Peters, 1977, 1982a). Songbirds are capable of imitating alien songs, or at least producing modified versions of them, especially in situations in which these are the only songs they hear. When given a choice of conspecific and heterospecific song, however, they preferentially copy the song of their own species. They also usually make much more complete and accurate copies of the conspecific model than of the alien song and may take longer to learn heterospecific song (Konishi, 1985; Marler, 1997; Marler and Peters, 1977). The ability to compare different species has provided evidence that there exists some rudimentary model of species-typical song even in the absence of experience. In humans, there is no convincing experimental evidence that infants have an innate description of speech, but only a few preference tests analogous to those in birds have examined the issue (e.g. Hutt et al., 1968), and the results are not conclusive. Moreover, because infants hear their mothers' voices both through the abdominal wall and through bone conduction and have been shown to learn aspects of speech (prosodic cues) while still in the womb (e.g. DeCasper and Spence, 1986; Moon et al., 1993) (see below), it will be difficult to determine whether infants are endowed with an innate description of speech prior to experience.

In birds, where there is an experimentally verified innate song preference, one can then ask what aspect of the song is required for recognition. Marler and Peters (1989) created synthetic tutor songs with syllables from two different species (the closely related swamp sparrows and song sparrows), arranged in temporal patterns characteristic of one or the other species. Using these songs to tutor the two types of sparrows, they demonstrated that predispositions vary across species. For instance, swamp sparrows copied syllables from their own species' song, regardless of the temporal arrangement of syllables in the synthetic tutor song. In contrast, song sparrows could copy swamp sparrow notes, but only when these were ordered in the usual multipart pattern of song sparrow song. Thus, for the swamp sparrow a critical cue (presumably innately specified) appears to be syllable structure, whereas for song sparrows it is syllable ordering as well as syllable structure. Certain acoustic cues may also serve as attentional flags that permit the acquisition of heterospecific notes. For instance, when the calls of ground squirrels were incorporated into tutor songs that began with the long whistle universally found in white-crowned sparrow song, these sparrows could be shown to learn these squirrel sounds, which they would normally never acquire (Soha, 1995).

In addition to the fact that most studies in birds compare species, another difference between the studies of innate predispositions for song and those for language learning is that in many cases the assay in birds is the song that the bird eventually produces. Any deduction of initial perceptual capacities from the final vocal output confounds initial capacities with subsequent sensory learning and motor production. Nonetheless, the studies of sensory capacities in birds with heart rate or begging call measures provide direct support for the idea that birds innately recognize their own species' song. This recognition is presumed to underlie much of the innate predisposition to learn conspecific song evident in the tutoring experiments. Thus, both humans and birds start out perceptually prepared for specific vocal learning. It may be that songbirds also have more complex innate specifications than do humans, or simply that the analogous experiments (pitting speech against nonspeech sounds) have not been or cannot be done with humans.

Another way of examining innate neural biases is to look at vocal production that emerges prior to, or in the absence of, external acoustic influences. For obvious reasons, relatively few data are available from humans. Deaf babies do babble, but their productions rapidly become unlike those of hearing infants. At a higher level of language analysis, there is some evidence that children exposed only to simple "pidgin" languages, and deaf children exposed to no acoustic or sign language, develop some elements (words or gestures, respectively) and order them in a way that is consistent with a rudimentary grammar (Petitto, 1993; Bickerton, 1990; Goldin-Meadow and Mylander, 1998). It remains disputed, however, whether this reflects an innate model specific to language (Chomsky, 1980; Fodor, 1983) or a more general innate human capacity to learn to segment and group complex sensory inputs (Elman *et al.*, 1996; Bates, 1992).

Songbirds again provide an opportunity to study this issue because analysis of the songs of birds reared in a variety of conditions can provide extensive data relevant to the issue of what may be innate in a vocal learner. In normally reared songbirds, the song of every individual bird within a species differs, but there are enough shared characteristics within a species that songs can also be used for species identification. The songs of birds raised in complete isolation vary between individuals but always contain some of the species-specific structure, although these songs are much less complex than those of tutored birds: the songs of white-crowned sparrow isolates tend to contain one or more sustained whistles, swamp sparrow isolates sing a trilled series of downsweeping frequencies, and song sparrow isolates produce a series of notes ordered in several separate sections. Even when white-crowned sparrows have copied alien song phrases, they often add an "innate" whistle ahead of these (Konishi, 1985; Marler, 1997, 1998). Thus, there is innate information that provides rough constraints on the song even in the absence of tutoring experience. Strikingly, almost all these features require auditory feedback to be produced. Because these features must be translated into vocal output via sensorimotor learning, they cannot be completely prespecified motor programs: they must involve some sensory recognition and feedback. Thus, the innate mechanisms that direct isolate song might bear some relationship to the neural mechanisms that allow innate sensory recognition of song. Recent behavioral evidence, however, suggests that there is not complete overlap between isolate song and the features found to be critical for innate conspecific recognition (Whaling *et al.*, 1997).

Innate sensory recognition and learning preferences in both humans and songbirds suggest that there must be underlying genetic mechanisms, perhaps specifying auditory circuitry specialized for processing complex sounds in special ways. An advantage of songbirds is that, unlike humans, there are many different, but closely related, species and even subspecies of vocal learners that show variation in their capacity to learn

(Kroodsma and Canady, 1985; Nelson et al., 1996). An intriguing example is the recent result of Mundinger (1995), who showed that the roller and border strains of canaries, which differ in note types, simply do not learn or retain in their songs the note types most specific of the other strain. However, hydrid offspring of the two breeds readily learn both types, and analysis of the patterns of inheritance of this capacity in these birds and in back-crosses has even begun to point to chromosome linkage (Mundinger, 1998). Comparisons of perceptual and motor learning and their neural substrates in birds like these may facilitate eventual understanding of the neural mechanisms contributing to innate biases for vocal learning.

Perceptual learning and the effects of experience

Although neither human nor songbird brain starts out perceptually naïve, abundant evidence in both fields suggests that innate predispositions are subsequently modified by experience. In addition, both speech and song scientists are grappling with the question of how experience alters the brain. In purely selective models of learning, sensory experience simply selects the sounds to be used to guide vocal learning from an extensive set of pre-encoded possibilities. In purely instructive models, there is no innate information about what is to be learned, and experience simply instructs a wide open brain about what to memorize. In fact, studies of both song and speech are converging on the idea that the mechanisms underlying learning are not described by either of these extreme models but combine aspects of each.

PERCEPTUAL LEARNING IN HUMANS MODIFIES INNATE PREDISPOSITIONS

As described, at the phonetic level of language, infants initially discriminate phonetic units from all languages tested, showing that they perceive and attend to the relevant acoustic features that distinguish speech sounds. By 6 months of age, however, infants have been affected by linguistic experience and show recognition of the specific phonetic units used in their native language. At this age, they respond differently to phonetic prototypes (best instances of phonetic categories) from the native as opposed to a foreign language (Kuhl, 1991; Kuhl et al., 1992). By 9 months, they have learned the stress patterns of native-language words, and the rules for combining phonetic units (Jusczyk et al., 1993), phrasal units (Jusczyk et al., 1992), and the statistical probabilities of potential word candidates (Saffran et al., 1996). Finally, by 12 months of age, native-language learning is evident in the dramatic changes seen in perceptual speech abilities (Werker and Tees, 1992) (Figure 2.3A). Infants no longer respond to speech contrasts that are not used in their native language, even the ones that they did discriminate at earlier ages (Werker and Tees, 1984; Kuhl et al., 1997b). Instead, one-year-old infants show the pattern typical of adult native-language listeners wherein discrimination of foreign-language contrasts has been shown to be difficult: adult English speakers fail to discriminate Hindi consonant-vowel combinations (Werker and Tees, 1984, 1992), American speakers fail on Spanish /b/ and /p/ (Abramson and Lisker, 1970), and speakers of Japanese fail to discriminate American English /r/ and /l/ (Miyawaki et al., 1975). The decline in the language-universal perception of infants has been directly demonstrated for Canadian infants tested sequentially over time with Hindi contrasts (Werker and Tees, 1984) and, most recently, for Japanese infants listening to American English /r/ and /l/ (Kuhl et al., 1997b).

In humans, there is evidence that perceptual learning of the more global, prosodic aspects of language actually commences prior to birth. Studies using the sucking and heart rate paradigms show that exposure to sound in utero has resulted in a preference of newborn infants for native-language over foreign-language utterances (Moon et al., 1993), for the mother's voice over another female's voice (DeCasper and Fifer, 1980), and for simple stories the mother read during the last trimester over unfamilar stories (DeCasper and Spence 1986). This indicates that the prosodic aspects of human speech, including voice pitch and the stress and intonation characteristics of a particular language and speaker, are transmitted to the fetus and are learnable.

All these studies on learning in the first year of life indicate that prior to the time that infants learn the meanings of individual words or phrases, they learn to recognize general perceptual characteristics that describe phonemes, words, and phrases that typify their native language. Thus, as a first step toward vocal learning, infants avidly acquire information about the perceptual regularities that describe their native language and commit them to memory in some form. Understanding the nature of this early phonetic learning and the mechanisms underlying it is one of the key issues in human language development.

PERCEPTUAL LEARNING IN SONGBIRDS

A variety of experiments provide evidence that what occurs in the first, or sensory, phase of song learning is the memorization of the sensory template, which is a subset of all possible vocalizations of the species (Marler, 1970b). This phase is thus in many ways analogous to the early perceptual learning of human infants. The study of perceptual learning in songbirds that is most similar to studies of humans measures vocal behavior of 10- to 40-day-old white-crowned sparrows in response to playback of tutored and novel songs (Nelson et al., 1997). After 10-day periods of tape tutoring with pairs of songs, male white-crowned sparrows not only gave significantly more calls to tutor songs than to novel songs, they also called significantly more to the song of the pair they would subsequently produce than to the nonimitated song of that pair. This suggests that the vocal assay reflected sensory learning that would ultimately be used for vocal production.

Most studies of the sensory learning period in songbirds, however, have assessed what is learned by using adult song production as an assay, after tutoring birds either for short blocks of time beginning at different ages or with changing sets of songs for a long period of time (Marler, 1970b; Nelson, 1997). Measuring learning using song production may underestimate what is perceptually learned. In many of these tutoring experiments, however, the song ultimately produced reflected experiences that had occurred long before the birds had begun to produce vocalizations; these studies, therefore, provide strong evidence that the first phase of learning involves the memorization of song.

In contrast to the emerging data on in utero learning in humans, prehatch or even immediately posthatch experience has not yet been shown to have much influence on song learning. Rather, in the well-studied white-crowned sparrows, the sensory period begins around day 20 and peaks in the next 30 days, with some acquisition possible up to 100 or 150 days (Baptista and Petrinovich, 1986; Marler, 1970b) (Figure 2.3). The timing of sensory learning may be similar for many other seasonal species (Kroodsma and Miller, 1996; Catchpole and Slater, 1995). Studies of zebra finches in which birds were separated from their tutors at different ages suggest that different aspects of the tutor song are memorized in sequence, with the individual component sounds being learned first and the overall order and temporal pattern acquired later (Immelmann, 1969). Careful comparisons of related white-crowned sparrow subspecies under identical learning conditions show that genetics also plays a role in the exact timing of learning, because subspecies of sparrows from harsh climates with short breeding seasons learn earlier and more than their coastal cousins (Nelson et al., 1995). Such differences between birds provide an opportunity to identify the factors governing sensory learning.

HOW DOES EXPERIENCE ALTER PERCEPTUAL ABILITIES IN HUMANS?

The initial studies demonstrating categorical perception of speech sounds in infants and its narrowing with language exposure led many speech theorists to take a strongly nativist or selective view of speech learning. By this hypothesis, infants were thought to be biologically endowed with either phonetic feature detectors that specified all the phonetic units used across languages (e.g. Eimas, 1975b), or with knowledge of all linguistically significant speech gestures (Liberman and Mattingly, 1985). The subsequent decline in speech discrimination was seen as a process of atrophy of the prespecified phonetic representations in the absence of experience. Recent studies of languages and of experience-dependent perceptual maps are changing theories of language learning and the role of innate and learned factors in the acquisition process. Rather than experience only selecting from prespecified categories, experience is thought to establish memory representations for speech that specify the phonetic units used in that language and that alter the perceptual system of the infant (Kuhl, 1994). On this view, experience is instructive as well as selective.

Several lines of evidence support this changing view. For one, cross-linguistic studies show that across languages, even ostensibly similar vowels (such as the vowel /i/) show a great deal of variation (Ladefoged, 1994). This suggests that prestoring all possible phonetic units of the world's languages would not be an efficient process. A second line of evidence against a simple atrophy of phonetic representations from lack of exposure is that, often, listeners are exposed to the categorical variations that they eventually fail to perceive. For instance, approximations of both English /r/ and /l/ are produced interchangeably by Japanese adults, although they do not change the meanings of words (Yamada and Tohkura, 1992). Japanese infants are therefore exposed (albeit randomly) to variants of both /r/ and /l/; similarly, American infants are exposed to variants of Spanish /b/ and /p/. Yet, both groups will eventually fail to respond to those distinctions. Finally, more detailed studies on the changes in infant phonetic perceptions brought about by experience suggest that perceptual learning is not in fact a simple sensory memory of the sound patterns of language. Instead, it seems to be a complex mapping in which perception of the underlying acoustic dimensions of speech is warped to create a recognition network that emphasizes the appropriate phonetic differences and minimizes those that are not used in the language (Kuhl, 1994, 1998, Kuhl and Meltzoff, 1997). This warping of the underlying dimensions is language specific such that no adult speakers of any language perceive speech sounds veridically. Rather, in each language group, perception is distorted to enhance perception of that language: this has been called the perceptual magnet effect (PME).

This last line of evidence results from studying perception of sounds in more detail than simply identifying category boundaries. Kuhl (1998) used large grids of systematically varying consonant–vowel syllables spanning the phonetic boundary between American English /r/ and /l/ to test American and Japanese adults. They asked listeners to rate the perceptual similarity of all possible pairs of stimuli and used multidimensional scaling techniques to create a map of the perceived physical distances between stimuli. The maps for American and Japanese speakers indicated that although the real physical distances between each stimulus in the grid were equal, American and Japanese adults perceived the sounds, and the distances between them, very differently. Americans identified the sounds as belonging to two clearly different categories, /r/ and /l/, whereas Japanese identified all stimuli but one as Japanese /r/ (the only phoneme of this type normally used in Japanese). Moreover, American listeners perceived many sounds as if they were closer to the best, most prototypical examples of /r/ and /l/ (called prototypes) than they really were. This is the origin of the term perceptual magnet effect, meant to describe how prototypes seem to act as magnets for surrounding sounds. Americans also perceived a larger than actual separation between the two categories. Japanese listeners showed no magnet effects, and no separation between the two categories. Thus, neither of the two groups perceive the real physical differences between the sounds. Instead, language experience has warped the underlying physical space so that if certain categories of sounds are used in a language, differences within a category are perceptually shrunk, whereas differences between categories are perceptually stretched. The PME may aid in perception by reducing the effects of the variability that exists in physical speech stimuli.

Critically for theories of speech learning, further studies suggest that these mental maps for speech are being formed or altered early in life as a function of linguistic experience. At 6 months of age, infants being raised in different cultures listening to different languages show the PME only for the sounds of their own native language (Kuhl *et al.*, 1992). Moreover, when American and Japanese infants were tested at 6–8 months of age, both groups showed the ability to discriminate American English /r/ and /l/, as expected from previous studies. By 10–12 months, however, not only did Japanese infants show a dramatic decline in performance, but American infants had also increased their accuracy of discrimination. This suggests that experience is not simply preventing atrophy (Kuhl *et al.*, 1997b). Finally, monkeys do not show the PME, indicating that, unlike categorical perception, it is not an effect that is inherent in the auditory processing of speech stimuli in many animals (Kuhl, 1991). The implication is that magnet effects explain the eventual failure of infants to discriminate foreign-language contrasts. Japanese infants, for example, would form a phonetic prototype for Japanese /r/ that is located between American /r/ and /l/. The magnet effect formed by experience with Japanese would eventually cause a failure to discriminate the American sounds. Although the studies show that magnet effects are altered by experience, it is not yet known whether magnet effects initially exist for all sounds of all languages and are then modified by experience, or whether they do not exist initially and are formed as a function of experience (Kuhl, 1994).

The special kind of speech that adults use when they speak to infants ("parentese") may play a role in the normal infant development of these phonemic maps. It has long been known that adults speak to infants using a unique tone of voice, and that when given a choice, infants prefer this kind of speech (Fernald, 1985, Fernald and Kuhl, 1987; Grieser and Kuhl, 1988). Early work on parentese emphasized the prosodic differences (the increased fundamental frequency or pitch of the voice, its animated intonation contours, and its slower rate). Recent data show, however, that infant-directed speech also provides infants with greatly exaggerated instances (hyperarticulated prototypes) of the phonetic units of language (Kuhl et al., 1997b). When speaking to infants, humans may intuitively produce a signal that emphasizes the relevant distinctions and increases the contrast between phonetic instances.

The studies described above all lend support to the newly emerging view that the initial abilities of infants to discriminate the auditory dimensions employed in speech contrasts are dramatically altered simply by listening to ambient language, resulting in a new and more complex map of the relevant linguistic space. The perception of speech in infants is thus both highly structured at birth, promoting attention to the relevant acoustic distinctions signaling phonetic differences, and highly malleable, allowing the brain to lay down new information, instructed by experience.

HOW DOES EXPERIENCE ACT ON THE SONGBIRD BRAIN?

Studies of perceptual learning in humans suggest that initial basic divisions of sound space are gradually altered by experience with the native language. The same questions about how this occurs that have been raised in humans can be asked about the effects of sensory experience in birds. The two extreme models (instructive and selective) discussed in the case of human speech have also been raised in the case of birdsong (Marler, 1997).

A purely instructive model would suggest that birds have little foreknowledge about the song of their species and are equally ready and able to learn virtually any song to which they are exposed. This is not consistent with innate preferences for learning conspecific song (Marler, 1997; Marler and Peters, 1982a). It also cannot explain isolate songs. These songs vary a great deal between individuals, however, which suggests that the innate template only coarsely defines the species song. The instructive model does account for the fact that prior to the production of the final learned songs, many birds produce copies of syllable types that are not used later in their final songs (Marler and Peters, 1982a). Even the syllables of alien species to which a bird was exposed can be reproduced in this way (Thorpe, 1961; Konishi, 1985). This phemenonon of overproduction of syllables suggests that birds are instructed by experience to memorize and even produce multiple songs, including songs of other species. The instructive model has difficulty, however, explaining the usual attrition later in song learning of syllables not appropriate for the species. A more realistic view of the instructive model would posit that during the impressionable phase, birds memorize a variety of songs, perhaps memorizing more easily or more completely songs that match their prespecified preferences. Later, during sensorimotor learning, birds listen to their vocalizations and use the memorized songs as templates to assess how well their vocal output matches them. They then ultimately elect to produce as adults a subset of those songs; the selection of this subset may be guided by a combination of genetic biases and experience (Nelson and Marler, 1994). Thus, even the simplest instructive model contains some elements of selection, both at the early (sensory) and at the late (sensorimotor) learning stages.

Alternatively, a strictly selective model of song learning can be proposed, in which the songbird brain has extensive innate knowledge about its species song, and this knowledge is then simply activated by experience. Evidence in favor of this includes innate song learning preferences and the surprising lack of variability seen in nature when the song patterns of an entire species are analyzed (Marler and Nelson, 1992; Marler, 1997). In contrast to the drift that might be expected in a culturally transmitted behavior operating by instruction alone, there are a number of features of song that are always shared, so-called species universals. None of these universals develop fully in birds raised in isolation, however. According to the pure selection model, therefore, all possible universals are pre-encoded in the brain, but most of them must be activated by the sensory experience of matched sounds in order to be available for later guidance of motor development, whereas the species universals that are not heard atrophy. Consistent with this idea, although not conclusive, is the surprisingly small number of sensory exposures necessary for learning. For example, white-crowned sparrows can learn from as few as 30 repetitions of a song, and nightingales have been shown to learn songs presented only twice a day for 5 days (Peters et al., 1992; Hultsch and Todt, 1989a).

As with the strict instructive model, however, even highly selective models seem likely to have some elements of instruction, for instance to allow the significant culturally transmitted variation seen within each category of universals (much like the variations in the vowel /i/ in human languages), and the copying of complex sequences, without requiring a multitude of templates. Moreover, because some features of song are produced in isolated birds, there must be two sorts of pre-encoded templates, ones that require no auditory experience of others to be active and a much larger set that do require auditory experience (Marler, 1997). In addition, and perhaps most important, a purely selective and species-based mechanism does not explain why birds can learn songs of heterospecifics, when birds are raised with those songs alone or even sometimes in the presence of conspecific songs as well (Baptista and Morton, 1981; Immelmann, 1969). One must therefore postulate two different learning mechanisms, one for conspecific song and a different one (perhaps a more general sensory learning) when other songs are learned. Although this is possibly consistent with data suggesting that birds take more time to learn alien song, it also necessitates a multiplication of learning substrates and makes it harder to explain why birds may incorporate both conspecific and heterospecific syllables into a single song. Finally, some of the lack of variability in the final crystallized song of many birds could be due not to selection at the early memorization stage, but rather in part to the highly socially controlled selection process active during late plastic song and crystallization, in which birds choose to crystallize the songs most similar to their neighbors (Nelson and Marler, 1994). Clearly, more studies are necessary to resolve the question of how sensory experience acts on the brain. Already it seems likely, however, that some combination of selection and instruction act both in series and in parallel in song learning. In many ways this is strikingly similar to the issues in the speech field, where purely innate and selection-based models are now making way for the idea that initial capacities are revised by instructive effects of experience.

Better understanding of the neural mechanisms underlying learning might also help resolve this issue. For instance, pre-existing circuitry and innate auditory predispositions might be revealed at the neural level, both in humans (using imaging) and in songbirds. The brain of songbirds contains a system of areas devoted to song learning and production and in adult birds these contain numerous neurons that respond selectively to the sound of the bird's own song and poorly to the songs of other conspecifics or to temporal alterations of the bird's own song (Margoliash, 1983, 1986; Margoliash and Fortune, 1992). In young birds just in the process of learning to sing, however, these same neurons are broadly selective for any conspecific songs, and they only gradually develop selectivity for their own song during learning (Doupe, 1997; Solis and Doupe, 1997; Volman, 1993). This suggests that at least this part of the song system contains neurons that are initially nonselective, i.e. without specific foreknowledge of the song the bird will sing, and that are subsequently instructed by experience.

SOCIAL EFFECTS ON SENSORY LEARNING

Both songbirds and humans demonstrate that learning is not solely dependent on innate predispositions and acoustic cues. Social factors can dramatically alter learning. Songbirds have been shown to learn alien songs from live tutors when they would reject the same songs presented by tape playback (Baptista and Petrinovich, 1986), and zebra finches will override their innate preference for conspecific song and learn from the Bengalese finch foster father feeding them, even when adult zebra finch males are heard nearby (Immelmann, 1969). Zebra finches, a more highly social and less territorial species than many songbirds, are particularly dependent on social factors even for selection of a particular conspecific tutor, as demonstrated in a series of experiments from the laboratory of Slater and colleagues. These experiments showed that zebra finches, which do not learn well from tapes, required visual interaction with the tutor in a neighboring cage in order to copy it, even if they could hear it (Eales, 1989; Slater *et al.*, 1988). Zebra finch fledglings prevented by eye patches from seeing, however, would still learn from a tutor if it was in the same cage, allowing the usual local social interactions (pecking, grooming, etc.) seen between zebra finch tutors and young. Finally, Adret (1993) showed that replacing the social interaction with a taped recording that the young zebra finch had to activate with a key press resulted in the zebra finch actively key pressing and then learning from that tape. Thus, the social factors required by zebra finches can come in a variety of modalities, all of which may serve to open some attentional or arousal gate, which then permits sensory learning. Such attentional mechanisms may also explain birds' preferential selection of a conspecific tutor during sensory learning and their choice of a particular song for crystallization.

Social interaction has been suggested to play a critical role in language learning as well (Locke and Snow, 1997; Kuhl and Meltzoff, 1996, 1997), although clearly studies of humans cannot withdraw social interaction to study the effects on vocal learning. Consistent with the importance of social cues are the speech patterns of adults in language addressed to infants. These patterns are greatly modified in ways that may aid language learning. In addition, neglected infants are developmentally delayed in language (Benoit *et al.*, 1996), and much of early word learning is deeply embedded in shared social activities. It is not clear whether a tape recorded or televised speaker would permit language learning in infants, although this could be addressed in studies of second language learning. Infants engaged in social interaction appear to be highly aroused and attentive, which may play a role in their ability to react to and learn socially significant stimuli. As in birds, such arousal mechanisms might help to store and remember stimuli and to change their perceptual mapping (Kilgard and Merzenich, 1998).

VOCAL PRODUCTION AND ITS INTERACTION WITH PERCEPTION

In vocal learning by humans and songbirds, both perception and production of sound are crucial. One must perceive both the vocal models of others and one's own sounds, and one must learn the mapping from one's own motor commands to the appropriate acoustic production. It has been clear for a long time that these two aspects of vocalization interact strongly, and in fact early speech theorists suggested that sound decoding requires creation of a model of the motor commands necessary to generate those sounds (Liberman *et al.*, 1967). In songbirds, however, memorization of sounds clearly precedes their generation. Recently, studies showing that human perception of speech is highly sophisticated at birth and then rapidly sculpted by experience, prior to the emergence of a sophisticated capacity for sound production, have led to a new view in studies of speech that is strikingly similar to that in birdsong. By this hypothesis, acoustic targets that are a subset of all possible species' vocalizations are perceptually learned by the young individual (bird or human) by listening to others. This perceptual learning then powerfully constrains and guides what is (and can be) produced. Subsequent production then aids in creating auditory-articulatory maps; the relationship between production and perception continues to be highly interactive but is derived, at least initially, from perceptual maps.

Production and perception in humans

In humans, the interaction between perception and production has been studied in two ways, by examining the infant's own production of sound and by examining the infant's reactions to the sight of others producing sound. Both assess what infants know about speech production and its relation to perception.

One strategy is to describe the progression of sounds produced by infants across cultures as they mature, examining how exposure to language alters speech production patterns. Characteristic changes in speech production occur as a child learns to talk, regardless of culture (for review see Stoel-Gammon, 1992). All infants progress through a set of universal stages of speech production during their first year: early in life, infants produce nonspeech gurgles and cries; at 3 months, infants coo, producing simple vowel-like sounds; by 7 months infants begin to babble; and by 1 year first words appear (Figure 2.3A). The cross-cultural studies also reveal, however, that by 10–12 months of age, the spontaneous vocalizations of infants from different language environments begin to differ, reflecting the influence of ambient language (de Boysson-Bardies, 1993). Thus, by the end of the first year of life, infants diverge from the culturally universal speech pattern they initially exhibit to one that is specific to their culture, indicating that vocal learning has taken place.

It is not the case, however, that the remarkable ability of infants to imitate the speech patterns they hear others produce begins only toward the end of their first year. Recent laboratory studies indicate that infants have the capacity to imitate speech at a much earlier age. Infants listening to simple vowels in the laboratory alter their vocalizations in an attempt to approximate the sounds they hear, and this ability emerges around 20 weeks of age (Kuhl and Meltzoff, 1996). The capability for vocal motor learning is thus available very early in life. In adults, the information specifying auditory–articulatory relations is exquisitely detailed and allows almost instantaneous reaction to changes in load or position of the articulators in order still to produce the appropriate sound (Perkell et al., 1997). Although speech production skills improve throughout childhood, showing that auditory–articulatory maps continue to evolve over a long period, the early vocal imitation capacities of infants indicate that these maps must also be sufficiently formed by 20 weeks of age to allow infants to approximate sounds produced by others.

A comparison of the developmental timelines relating speech perception and speech production suggests that early perceptual mapping precedes and guides speech production development (Kuhl and Meltzoff, 1996, 1997). Support for this idea comes from a comparison of changing perceptual abilities and production in the infant studies just described: a language-specific pattern emerges in speech perception prior to its emergence in speech production. For instance, although infant vocalizations produced spontaneously in natural settings do not become language-specific until 10–12 months of age, the perceptual system shows specificity much earlier (Mehler et al., 1988; Kuhl et al., 1992). In addition, at an age when they are not yet producing /r/- or /l/-like sounds, infants in America and Japan already show language-specific patterns of perception of these sounds. These data suggest that stored representations of speech in infants alter perception first and then later alter production as well, serving as auditory patterns that guide motor production. This pattern of learning and self-organization, in which perceptual patterns stored in memory serve as guides for production, is strikingly similar to that seen in birdsong, as well as in visual-motor learning, such as gestural imitation (Meltzoff and Moore, 1977, 1997).

A second experimental strategy reveals the link between perception and production for speech. In this case, studies demonstrate that watching another talker's mouth movements influences what subjects think they hear, indicating that representational maps for speech contain not only auditory but visual information as well. Some of the most compelling examples of the polymodal nature of speech are the auditory–visual illusions that result when discrepant information is sent to two separate modalities. One such illusion occurs when auditory information for /b/ is combined with visual information for /g/ (McGurk and MacDonald, 1976; Massaro, 1987; Kuhl et al., 1994). Perceivers report the strong impression of an intermediate articulation (/da/ or /tha/), despite the fact that this information was not delivered to either sense modality. This tendency of human perceptual systems to combine the multimodal information (auditory and visual) to give a unified percept is a robust phenomenon.

Infants 18–20 weeks old also recognize auditory–visual correspondences for speech, akin to what adults do when they lip-read. In these studies, infants looked longer at a face pronouncing a vowel that matched the vowel sound they heard than at a mismatched face (Kuhl and Meltzoff, 1982). Young infants therefore demonstrate knowledge about both the auditory and the visual information contained in speech. This supports the notion that the stored speech representations of infants contain information of both kinds.

Thus, early perceptual learning – primarily auditory but perhaps also visual – may underpin and guide speech production development and account for infants' development of language-specific patterns by the end of the first year. Linguistic exposure is presumably the common cause of changes in both systems: memory representations that form initially in response to perception of the ambient language input then act as guides for motor output (Kuhl and Meltzoff, 1997).

Production and perception in birdsong

The observation that perceptual learning of speech may precede and guide production in humans makes it strikingly similar to birdsong, which clearly does not require immediate motor imitation while the young bird is still in the presence of the tutor. Many seasonal species of birds begin the sensorimotor learning phase, in which they vocally rehearse, only many months after the tutor song has been heard and stored (Figure 2.3B). Thus, birds can remember complex acoustical patterns (that they heard at a young age) for a long time and use them much later to guide their vocal output.

The lack of overlap between the sensory and sensorimotor phases of song learning is not as complete as often supposed, however, and in this sense some songbirds are also more like humans than previously thought. This is most obvious in the zebra finch (Immelmann, 1969; Arnold, 1975a), which is not a seasonal breeder and develops song rapidly over a period of 3–4 months, and in which sensory and sensorimotor learning phases overlap for at least a month. Thus, as in humans, these finches continue to copy new sounds after sensorimotor learning has started. Even in the classical seasonal species, birds often produce the amorphous vocalizations known as subsong as early as 25 days of age, well within the 100-day sensitive phase (Nelson et al., 1995). These early vocalizations of songbirds may allow calibration of the vocal apparatus and an initial mapping between motor commands and sound production, a function similar to that proposed for human babbling (Marler and Peters, 1982a; Kuhl and Meltzoff, 1996, 1997). Moreover, in more complex social settings, the schedule for the onset of singing and sensorimotor learning can be dramatically accelerated (Marler, 1970b; Baptista and Petrinovich, 1986).

Nonetheless, in many species, perceptual learning of the tutor is complete before the so-called sensorimotor stage of learning begins in earnest, usually toward the end of a seasonal bird's first year of life (Figure 2.3B). This stage begins with a great increase in the amount of singing, and soon thereafter, vocalizations show clear evidence of vocal rehearsal of learned material, at which point they are termed plastic song. These are gradually refined until they resemble the tutor song. Along the way, however, birds produce a wide variety of copied syllables and songs, only to drop them before crystallization (Marler and Peters, 1982a; Nelson and Marler, 1994). During the plastic song phase, birds also often incorporate inventions and improvisations that make their song individual. At the end of the sensorimotor phase, birds produce a stable, or "crystallized", adult song, which in most species remains unchanged throughout life (except for open learners; see below). Because tutor learning occurs largely before production, it cannot depend on motor learning. During sensorimotor learning, however, sensory processing of sounds might conceivably change or become more dependent on knowledge of motor gestures. This question could be studied in songbirds raised with normal sensory exposure to others but experimentally prevented from producing sounds.

Just as in humans, not all sensory effects on song learning are mediated solely by auditory feedback. Not only do zebra finches require some sort of visual or social interaction to memorize a tutor, but male cowbirds will choose to crystallize the particular one of their several plastic songs that elicits a positive visual signal, a wingflap, from a female cowbird (West and King, 1988). Thus, visual cues can also affect song learning by acting on the selection of songs during motor learning. Along the same lines, Nelson and Marler (1993) demonstrated that late juvenile sparrows just arriving at the territory where they will settle will choose to crystallize the plastic song in their repertoire that is the most similar to the songs sung in that territory. This result was replicated in the laboratory by playing back to a sparrow just one of four plastic songs that it was singing, which invariably resulted in that song being the one crystallized (Nelson and Marler, 1994). These social effects on crystallization may allow the matched countersinging frequently observed in territorial birds. By allowing visual and auditory cues to influence song selection, birds incorporate the likelihood of successful social interaction into their final choice of vocal repertoire.

PERCEPTION OF SELF AND ITS INTERACTION WITH PRODUCTION

Although auditory processing of the sounds of others is important in speech and song learning, the interaction between perception of one's own sounds and vocal production is also crucial, because vocal learning depends on the ability to modify motor output using auditory feedback as a guide. In both birdsong and speech, the sensory and motor processes are virtually inseparable. One striking demonstration of this is that in frontal and temporoparietal lobes of humans, stimulation at single sites disrupts both the sequential orofacial movements used in speech production and the ability to identify and discriminate between phonemes in perception tasks (Ojemann and Mateer, 1979). This provides more evidence that the traditional description of Broca's and Wernicke's aphasias as expressive and receptive is oversimplified. Likewise, the song premotor nucleus HVC also contains numerous song-responsive neurons (Margoliash, 1983, 1986; McCasland and Konishi, 1981).

The important question of how and where the auditory feedback from self-produced vocalizations acts and how it relates to vocal motor processes remains unclear for both humans and songbirds. In humans, the majority of individual speech-related neurons studied thus far have been active only during either speech production or speech perception (even with identical words presented and then spoken). Thus, the vocal control system seems in some way to inhibit the response of these neurons to the sound of self-vocalized words. More striking, this link between auditory and vocal systems already exists in nonhuman primates: more than half of the auditory cortex neurons responsive to the presentation of calls in squirrel monkeys did not respond to these calls when they were produced by the monkeys (Müller-Preuss and Ploog, 1981). Similarly, in songbirds, despite the strong responses of HVC song-selective neurons to presentation of the bird's own song, these neurons are not obviously activated by the sound of the bird's own song during singing, and in many cases they are clearly inhibited during and just after singing in adult birds (McCasland and Konishi, 1981). Thus, information that there is vocal activity is provided to auditory and even vocal control areas in both primates and songbirds, but it is not clear how the sounds made by this activity are used. This puzzle is evident in some but not all PET studies as well: even though Wernicke's area is strongly activated during auditory presentation of words, a number of such studies have shown surprisingly little activation of the same area from reading or speaking aloud (Ingvar and Schwartz, 1974; Petersen et al., 1989; Hirano et al., 1996; but see also Price et al., 1996). A recent study of vocalizing humans may shed light on this question: this showed much more activation in superior temporal gyri when auditory feedback of the subject's own voice was altered than when it was heard normally (Hirano et al., 1997; McGuire et al., 1996). This raises the possibility that,

at least once speech is acquired, Wernicke's and other high-level speech processing areas may be more active when detecting mismatched as opposed to expected auditory feedback of self. In birds as well, it will be important to test neuronal responses when auditory feedback of the bird's own voice is altered.

As with primates, comparisons of songbirds with closely related species that are not vocal learners have the potential to provide insights into the steps that led to song learning. For instance, the suboscine birds such as flycatchers and phoebes, which are close relatives of the passerine (or oscine) songbirds, sing but show no evidence of dialects or individual variations and produce normal song even when deafened young (Kroodsma, 1985; Kroodsma and Konishi, 1991). These birds also show no evidence of a specialized forebrain song control system, which suggests that another crucial step in the appearance of specialized song control areas may have been the acquisition of auditory input by pre-existing forebrain motor control areas. Likewise, in humans, the capacity to learn speech and the development of specialized cortical systems for its control may have resulted from close interaction of motor control areas for orofacial movements with a variety of areas involved in processing and memorizing complex sounds (Ojemann, 1991). Despite its clear importance, the link between perception and production is surprisingly ill understood in both speech and song systems, and further understanding of how motor control and auditory feedback interact at the neural level will be crucial for progress in both fields.

SENSITIVE PERIODS FOR SPEECH AND SONG LEARNING

A critical period for any behavior is defined as a specific phase of the life cycle of an organism in which there is enhanced sensitivity to experience, or to the absence of a particular experience. One of the most universally known and cited critical periods is that for human language acquisition. Songbirds also do not learn their vocalizations equally well at all phases of life. In this final section we review the evidence suggesting that sensitive periods for vocal learning in these two systems are indeed very similar, and we examine and compare possible underlying mechanisms.

The term critical was initially coined in the context of imprinting on visual objects early in life, in which sensitivity to experience is short-lived and ends relatively abruptly. Many critical periods, however, including those for vocal learning, begin and end less abruptly and can be modulated by a variety of factors, so the term now preferred by many investigators is sensitive or impressionable period. Because critical period is such a commonly recognized term, we use these terms interchangeably, but with the caveat that this does not necessarily imply a rigidly regulated and complete loss of sensitivity to experience.

BASIC EVIDENCE FOR SENSITIVE PERIODS IN BIRDS AND HUMANS

Humans

Lenneberg (1967) formulated the strongest claims for a critical or sensitive period for speech learning, stating that after puberty it is much more difficult to acquire a second language. Lenneberg argued that language learning after puberty was qualitatively different, more conscious and labored, as opposed to the automatic and unconscious acquisition that occurs in young children as a result of mere exposure to language.

Evidence for a sensitive period for language acquisition has been derived from a variety of sources: (1) classic cases of socially isolated children show that early social isolation results in a loss of the ability to acquire normal language later (Fromkin et al., 1974; Lane, 1976); (2) studies of patients who suffer cerebral damage at various ages provide evidence that prognosis for language recovery is much more positive early in life as opposed to after puberty (e.g. Duchowny et al., 1996; Bates, 1992); and (3) studies of second-language learning indicate that there are differences in the speed of learning and ultimate accuracy of acquisition in language learning at different stages of life (Johnson and Newport, 1991; Oyama, 1978; Snow, 1987).

It has been known for a long time that children recover better from focal brain injury than do adults with analogous lesions. Moreover, after major damage to left frontal or parietal lobes, or even hemispherectomies for intractable epilepsy, children can still develop language using the right hemisphere (e.g. Dennis and Whitaker, 1976; Woods, 1983). There is an upper limit to this extreme plasticity, however, with studies suggesting the cut-off occurs sometime between 3 and 6 years of age. In cases of less severe injury, the period after 6–8 years, but before puberty, is still more likely to support learning of speech than the period after puberty (Vargha-Khadem et al., 1997).

Studies on the acquisition of a second language offer the most extensive data in support of the idea that language learning is not equivalent across all ages. For instance, second languages learned past puberty are spoken with a foreign accent, in other words with phonetics, intonation, and stress patterns that are not appropriate for the new language. Comprehension of spoken speech and grammar, as well as grammatical usage, are also poorer for languages learned later in life. Numerous studies show that all these aspects of language are performed poorly by immigrants who learn a second language after the ages of 11–15 years, independent of the length of time the learner has been in the new country (Oyama, 1976, 1978; Johnson and Newport, 1991; Newport, 1991). Even when adults initially appear to acquire certain aspects of language faster than children, they do not end up as competent as children after equivalent amounts of training (Snow, 1987).

Moreover, the capacity to learn may decline in several stages. A number of studies suggest that children who have been exposed to and learned a new language at a very young age, between 3 and 7 years of age, perform equivalently to native speakers on various tests. After 6–8 years of age, performance seems to decline gradually but consistently, especially during puberty, and after puberty (after approximately 15–17 years of age), there is no longer any correlation between age of exposure and performance, which is equally poor in all cases (Tahta et al., 1981; Asher and Garcia, 1969; Flege, 1991). A similar pattern of results is shown in deaf adults who are native-language signers but have learned American Sign Language (ASL) at different ages: a comparison of subjects who had learned either from birth, from 4 to 6 years of age, or after the age of 12 showed a clear progression in both production and comprehension of the grammar of ASL that indicated that earlier learners signed more accurately than later learners (Newport, 1991).

Could the critical period simply be a limitation in learning to produce speech, while perceptual learning is not limited? Studies suggest that the accents adult learners use when attempting to produce a foreign language are not attributable to simple motoric failures in learning to pronounce the sounds of the new language but also involve perceptual difficulties. When students were tested on a foreign language 9–12 months after their first exposure to it, those with the best pronunciation scores also showed the best performance on the discrimination test (Snow and Hoefnagal-Hohle, 1978). Moreover, the numerous studies of perception reviewed earlier (Werker and Polka, 1993; Kuhl, 1994) indicate that adults have difficulty discriminating phonetic contrasts not used systematically in their native language. Interestingly, although the effects of experience on perception are evident early in life (6 months to 1 year of age), these studies of second language learning show that these effects are also reversible, and that plasticity remains enhanced, for a relatively long period. Moreover, even a modest amount of exposure to a language in early childhood has been shown to produce a more native-like perception of its syllable contrasts in adulthood (Miyawaki et al., 1975). Consistent with the idea that perception as well as production is altered, brain mapping studies show that different cortical areas of the brain are activated by the sound of native and second languages when the second language is learned later in life, whereas similar brain regions are activated by both languages if the two are learned early (e.g. Kim et al., 1997). As suggested earlier, perceptual learning may in fact constrain which sounds can be correctly produced. Regardless of whether perception or production is primary, both production and perception of phonology, as well as grammar and prosody, provide strong data in support of sensitive periods for speech.

SONGBIRDS

It has long been realized that songbirds have a restricted period for memorization of the tutor song (Thorpe, 1958; Marler, 1970b). Now that studies of humans show that early perceptual capacities narrow with experience, the parallels between songbird and human critical periods are even more compelling. Despite numerous anecdotal accounts, the number of carefully studied songbird species remains small. The classical study is that of the white-crowned sparrow by Marler (1970b), which shows that as in humans, sparrows have an early phase of extreme plasticity (around 20–50 days of age) with a later gradual decline in openness, with some acquisition possible up to 100 or 150 days (Nelson et al., 1995). After the age of 100–150 days, in most cases, birds did not learn new songs from sensory exposure to new tutors, regardless of whether they had had normal tutor experience or had been isolated. Birds with a classical critical period like this are often called closed learners. Some birds are open-ended learners: that is, their ability to learn to produce new song either remains open or reopens seasonally in adulthood (e.g. canaries [Nottebohm et al., 1986] and starlings [Chaiken et al., 1994; Mountjoy and Lemon, 1995]), although it is still unclear in many cases whether the reopening is sensory or motor in nature. Comparing the brains of these birds with those of closed learners should provide an opportunity to elucidate what normally limits the capacity to learn.

Does the capacity to produce sounds also have a critical period, independent of sensory exposure? That is, if correct motor learning is not accomplished by a certain age, despite timely sensory exposure, or is not closely linked in time with perceptual learning, can it ever be completed or corrected? The studies of tracheostomized children suggest that vocal motor learning may indeed also be developmentally restricted, but this is another question more easily addressed in songbirds than in humans. Songbird experiments provide conflicting evidence, however. One line of evidence comes from hormonal manipulations of birds. Singing of adult male birds is enhanced by androgen, and castration markedly decreases (but does not eliminate) song output in adult birds. Sparrows castrated as juveniles learn from tutors at the normal time, and produce good imitations in plastic song, but fail to crystallize song (Marler et al., 1988a). When given testosterone as much as a year later, however, these birds then rapidly crystallize normal song, which suggests that the transition from plastic to more stereotyped crystallized song does not have to occur within a critical time window. These experiments do not perfectly address the question of a critical period for sensorimotor learning, however, because all young birds vocalized somewhat around the normal time of song onset, giving them some normal experience of sensorimotor matching. Similarly, a castrated chaffinch that had not sung at all during its first year still developed normal song when given hormone later (Nottebohm, 1981). In both these experiments, the absence of androgen, which dramatically decreases singing, might also have delayed motor development and motor sensitive period closure.

Other nonhormonal manipulations suggest that disruptions of motor learning at certain ages are in fact critical. For example, song lateralization primarily involves motor learning and production. Although this lateralization seems to have quite different mechanisms than that of speech, it shares with speech an early sensitive period for recovery from insults to the dominant side. Left hypoglossal dominance in canaries can be reversed if the left tracheosyringeal nerve is cut or the left HVC lesioned prior to the period of vocal motor plasticity when song production is learned, but not thereafter (Nottebohm *et al.*, 1979). This provides evidence that at least in canaries some organization occurs during motor practice that cannot be reversed later. An experiment to address this more directly might be to eliminate or disrupt all vocal normal practice until the usual time of crystallization and then to allow the birds to recover. Recent experiments in songbirds with transient botulinum toxin paralysis of syringeal muscles in zebra finches during late plastic song do suggest that critical and irreversible changes occur during late sensorimotor learning (Pytte and Suthers, 1996).

TIMING AND THE ROLE OF EXPERIENCE: WHAT CLOSES THE SENSITIVE PERIOD?

The question raised by the data on the difficulties of late learning is: what accounts for differential learning of language at different periods in life? By the classical critical period argument, it is time or development that are the important variables. Late experience has missed the window of opportunity for language learning, making it more difficult, if not impossible, to acquire native-language patterns of listening and speaking, or of normal birdsong. This time-limited window presumably reflects underlying brain changes and maturation, which are as yet poorly understood, especially in humans. Lenneberg (1967) thought that at puberty the establishment of cerebral lateralization was complete and that this explained the closing of the sensitive period. The data reviewed in the previous section suggest, however, that the capacity for speech learning declines gradually throughout early life, or at least has several phases prior to adolescence. More striking, both song and speech studies are increasingly converging on a role for learning and experience itself in closing the critical period, as described below.

HORMONES

The approximate coincidence of puberty with closure of the sensitive period points to hormones as some of the maturational factors that limit learning. Surprisingly little has been done to examine this, however, for instance by comparing second language acquisition in boys and girls, or by investigating language development in human patients with neuroendocrine disorders (McCardle and Wilson, 1990). Because dyslexia and stuttering are 10 times more common in boys than in girls, testosterone has been hypothesized to play a role in some forms of dyslexia (Geschwind and Galaburda, 1985), but by their nature, studies of language disabilities may not address normal learning. Recent imaging data suggest that lateralization for speech is less strong in human females than in males (Shaywitz *et al.*, 1995), although whether the origin of this difference is hormonal is unclear, as is its relationship to critical period closure.

Male songbirds provide much more evidence for hormonal effects on learning. The earliest studies of song learning showed that the period of maximum sensitivity was not strictly age dependent but could be extended by manipulations (such as light control or crowding) that also delayed its onset (Thorpe, 1961). Just as in humans, these manipulations suggested a role for hormones, especially sex steroids, in closure of the critical period. This idea was further strengthened by work by Nottebohm (1969). He found that a chaffinch castrated in its first year, before the onset of singing, did not sing and subsequently learned a new tutor song in the second year, when it received a testosterone implant. Although this experiment did not indicate what ended the readiness to learn song, it certainly showed that it could be extended. Because singing often begins in earnest around the time that testosterone rises, and because it can be delayed or slowed by castration, a reasonable possibility is that these developmental increases in male hormones to a high level are also involved in closing the critical period. In an experimental manipulation to test this hypothesis, Whaling and colleagues (1998) castrated white-crowned sparrows at 3 weeks of age and then tutored them long after the normal 100-day close of the critical period. There was a small amount of learning evident in some animals subsequently induced to sing by testosterone replacement, which suggests that the critical period had indeed been extended. The effect was weak, however, perhaps indicating that a single hormone is unlikely to control normal learning.

ACTIVITY-DEPENDENCE: ADEQUATE SENSORY EXPERIENCE OF THE RIGHT TYPE

Although there is much to support a timing or maturational explanation for loss of the capacity for vocal learning, an alternative account is emerging in both humans and songbirds, which suggests that learning itself also plays a role in closing the critical period. In humans, this alternative account has been developed at the phonetic level, where the data suggesting a sensitive period are strongest (Kuhl, 1994). As described earlier, work on the effects of language experience suggest that exposure to a particular language early in infancy results in a complex mapping of the acoustic dimensions underlying speech. This warping of acoustic dimensions makes some physical differences more distinct whereas others, equally different from a physical standpoint, become less distinct; this may facilitate the perception of native-language phonetic contrasts and appears to exert control on how speech is produced as well (Kuhl and Meltzoff, 1997).

By this hypothesis, speech maps of infants are incomplete early in life, and thus the learner is not prevented from acquiring multiple languages, as long as the languages are perceptually separable. As the neural commitment to a single language increases (as it would in infants exposed to only one language), future learning is made more difficult, especially if the category structures of the primary and secondary languages differ greatly (for discussion see Kuhl, 1998). In this scenario, for example, the decline in infant performance is not due to the fact that American English /r/ and /l/ sounds have not been presented within a critical window of time, but rather that the infant's development of a mental map for Japanese phonemes has created a map in which /r/ and /l/ are not separated. This effect of the learning experience could be thought of as operating independently of, and perhaps in parallel with, strict biological timing, as stipulated by a critical period. By analogy to studies in other developing systems, this model might be called experience-dependent.

This view of early speech development incorporates some of the new data demonstrating that children with dyslexia, who have language and reading difficulties and are past the early phases of language development, can nonetheless show significant improvements in language ability after treatment with a strategy that assists them in separating sound categories (Merzenich et al., 1996; Tallal et al., 1996). These children and others with language difficulties (Kraus et al., 1996) often cannot separate simple sounds such as /b/ and /d/. By the activity-dependent model, these children have either not been able to separate the phonemes of language and thus have not developed maps that define the distinct categories of speech, or they have incorrect maps, producing difficulties with both spoken language and reading. The treatment was computer-modified speech that increased the distinctiveness of the sound categories and may have allowed the children to develop for the first time a distinct and correct category representation for each sound, and to map the underlying space. Although children were doing this well after the time at which it would have occurred normally in development, their ability to do so may have depended on the fact that they had not previously developed a competing map that interfered with this new development. This hypothesis suggests that even dyslexic adults might benefit from such treatment, if the lack of normal mapping effectively extended the critical period.

Another test of the experience-dependent hypothesis for critical period closure might be to study congenitally deaf patients, not exposed to sign language, who have been outfitted with cochlear implants at different ages. If the critical period closes simply because of auditory input creating brain maps for sound, the complete absence of input might leave the critical period as open in 8- or 18-year-olds as in newborns. Alternatively, if some maturational process is also occurring, and/or if complete deprivation of inputs has negative effects, the critical period might close as usual, or be extended, but not indefinitely. To date, insufficient data are available to address these issues because cochlear implants of excellent acoustical quality have only recently become available, and relatively few children have been implanted (Owens and Kessler, 1989). However, even though deaf children who learn sign language at different ages are presumably not mapping any other languages prior to acquiring ASL, their decreasing fluency in ASL as a function of the age of learning does suggest that the capacity to learn shows at least some decline with age, even without competing sensory experience (Newport, 1991).

The end of the sensitive period may not be characterized by an absolute decrease in the ability to learn but rather by an increased need for enhanced and arousing inputs. In other systems, such as the developing auditory-visual maps of owls, the timing and even the existence of sensitive periods have been found to depend on the richness of the animal's social and sensory environment (Brainard and Knudsen, 1998). In speech development, the inputs provided by adults who produce exaggerated, clear speech ("parentese" or infant-directed speech) when speaking to infants may be crucial. This speech, which provides a signal that emphasizes the relevant distinctions and increases the contrast between phonetic instances, could be related to the kind of treatment that is effective in treating children with dyslexia. This raises the possibility, for example, that Japanese adults might also be assisted in English learning by training with phonemes that exaggerate the differences between the categories /r/ and /l/. These adults have a competing map, but exaggerated sounds might make it easier to create a new map that did not interfere or could coexist with the original one formulated for Japanese. Studies also show that training Japanese adults by using many instances of American English /r/ and /l/ improves their performance (Lively et al., 1993). Thus, both exaggerated, clear instances and the great variability characteristic of infant-directed speech may promote learning after the normal critical period.

In the songbird field, it has been known for some time that the nature of the sensory experience affects the bird's readiness to learn song. For instance, early exposure to conspecific song gradually eliminated the willingness of chaffinches to learn heterospecific song, or conspecific song with unusual phrase order (Thorpe, 1958). Similarly, birds born late in the breeding season of a year, when adults have largely stopped singing, were able to acquire song later than siblings born earlier in the season and thus exposed to much more song (Kroodsma and Pickert, 1980). More specific demonstrations that the type of auditory experience can affect or delay the closure of the critical period come from studies of other species, especially zebra finches. Immelmann (1969) and Slater et al. (1988) showed that zebra finches tutored with Bengalese finches were able to incorporate new zebra finch tutors into their songs at a time when zebra finches reared by conspecifics would not. This suggested that the lack of the conspecific input most desirable to the brain left it open to the correct input for longer than usual. Even more deprivation, by raising finches only with their nonsinging mothers or by isolating them after 35 days of age, gives rise to finches that will incorporate new song elements or even full songs when exposed to tutors as adults (Eales, 1985; Morrison and Nottebohm, 1993). This is reminiscent of activity dependence in other developing systems, such as the visual system, in which a lack of the appropriate experience can delay closure of the critical period. Although unresolved in birds, it seems likely that the critical period can be extended in this way, although not indefinitely (except perhaps in open learners).

A caveat is that it may not be the sensory experience but the motor activity associated with learning (or, as always, both the sensory and motor activity interwoven) that decreases the capacity to learn. This has been little studied in humans, but in chaffinches, crystallization was associated with the end of the ability of the birds to incorporate new song (Thorpe, 1958). This is also a possibility suggested by Nottebohm's experiment, in which castrated birds that had not yet sung were still able to learn new tutor song. Because testosterone induces singing, perhaps it is not a direct effect of hormones that closes the critical period, but some consequence of the motor act of singing. To dissociate these possibilities will require more experiments, because under normal conditions androgens invariably cause singing and song crystallization (Korsia and Bottjer, 1991; Whaling et al., 1995). Zebra finches raised in isolation do incorporate at least some new syllables as adults even though they have already been singing (isolate) song (Morrison and Nottebohm, 1993; Jones et al., 1996); these studies do not settle the issue, however, because the birds that showed the most new learning were also the least crystallized (Jones et al., 1996).

SOCIAL FACTORS

Closure of the critical period is also affected by social factors. Although young white-crowned sparrows learn most of their conspecific song from either tapes or live tutors heard between days 14 and 50, Baptista and Petrinovich (1986) showed that these birds will even learn from a heterospecific song sparrow after 50 days of age if they are exposed to a live tutor. In zebra finches, social factors interacting with auditory tutoring may explain some of the conflicting results on whether and for how long the critical period can be kept open: birds raised with only their mothers showed extended critical periods, whereas birds raised with both females and (muted) males, or with siblings, did not show late learning (Aamodt et al., 1995; Volman and Khanna, 1995; Wallhausser-Franke et al., 1995). Jones et al. (1996) directly tested the effect of different social settings on learning in finches. They showed that major changes of song in adulthood were rare and were found only in the more socially impoverished groups. It will be crucial to try to tease social and acoustic factors apart. Although the neural mechanisms of social factors (perhaps hormonal in nature) remain unclear, their effects are certainly potent: merely the presence of females caused males to have larger song nuclei than males in otherwise identical photoperiodic conditions (Tramontin et al., 1997).

In both songbirds and humans, it seems likely that a number of factors act in concert to gradually close the critical period, just as a number of factors control the selectivity of learning. Maturation, auditory experience, social factors, and hormones (which could be the basis for the maturational or social effects) can all be shown to affect the onset and offset of learning. When learning occurs in normal settings, these factors all propel learning in the same direction. When some or all of these factors are disrupted, the critical period can be extended, although probably not indefinitely.

CONCLUSIONS

Recurrent themes emerge when the comparable features of birdsong and speech learning are studied: innate predispositions, avid learning both perceptually and vocally, critical periods, social influences, and complex neural substrates. The parallels are striking, although certainly there are differences. Both the commonalities and the differences point to the gaps in our knowledge and suggest future directions for both fields.

The grammar and other aspects of meaning in human speech are the most obvious differences between birdsong and speech. These differences suggest that although human speech is undoubtedly built on pre-existing brain structures in other primates, there must have been an enormous evolutionary step, with convergence of cognitive capacities as well as auditory and motor skills, in order to create the flexible tool that is language. In contrast, it seems a smaller jump from the suboscine birds that produce structured song but do not learn it, to songbirds. Nonetheless, a critical step shared by avian vocal learners with humans must have been to involve the auditory system, both for learning of others and for allowing the flexibility to change the vocal motor map. In fact, the existence of closely related nonlearners as well as of numerous different species that learn is one of the features of birdsong that has allowed more dissection of innate predispositions than is possible with humans. It may seem that more is prespecified in songbirds, with their learning preferences and isolate songs. Many of the analogous experiments, however, cannot be done or simply have not yet been done in humans, for instance examining whether newborn monkeys and humans prefer conspecific sounds over other vocalizations, and if so, what acoustic cues dictate this preference. Neurophysiological analysis of high-level auditory areas in young members of both groups (using microelectrodes in songbirds and perhaps event-related potentials in humans), and comparisons with non-human primates and other birds, should provide insight into what the brain recognizes from the outset, how it changes with experience, and how it differs in nonlearners.

The early perceptual learning in both humans and songbirds seems different from many other forms of learning: it does not require much if any external reinforcement and it occurs rapidly. What mechanisms might underlie this? In songbirds it is known that songs can be memorized with just a few experiences, whereas in humans this area is as yet unexplored. Although human vocal learning seems to be rapid, it is not known if it takes 10 minutes or 4 hours a day to induce the kind of perceptual learning seen in infants, or whether the input has to be of a certain quality or even from humans. Both groups seem to have enormous attentiveness to the signals of their own species and, in most cases, choose to learn the right things. This could be due to a triggering of a prespecified vocal module, as has often been suggested, it could be that auditory or attentional systems have innate predispositions that guide them, or it could simply be that learning in each case is specific to sounds with some regular feature that is as yet undiscovered. The learned "template" of songbirds continues to be much sought after, but it is clear from numerous behavioral studies that the idea of a single sensory template is much too simple: Many birds memorize and produce multiple songs, at least during song development. This presumably means that they have multiple learned templates, or perhaps some more complex combinatorial memory mechanism. And what mediates their ultimate selection of a subset of those songs as adults? It may be guided by a combination of genetics and experience, including potent social effects of conspecifics. Social effects on learning seem crucial in humans as well, but in both cases how social influences may act on the brain is poorly understood. Understanding how to mobilize these, however, could have profound implications for treatment of communication disorders, at any age.

Our understanding of neural substrates of both speech and birdsong should continue to improve as methods for exploring the brain advance, although the study of an animal model such as songbirds will always have a certain advantage. The question of how auditory feedback acts during vocalization is a persisting and important puzzle shared by both fields, and more insight into this would shed light not only on general issues of sensorimotor learning but also on human language disabilities such as stuttering. Finally, brain plasticity and the critical period remain fascinating and important issues. What is different at the neural level about language learning before and after puberty? How does the brain separate the maps of the sounds of languages such as English and Japanese in infants raised in bilingual families? Understanding what governs the ability to learn at all ages may not only advance our basic knowledge of how the brain changes with time and experience, but could also be of practical assistance in the development of programs that enhance learning in children with hearing impairments, dyslexia, and autism, and might aid in the design of programs to teach people of any age a second language. Clearly, studies of songbirds with different types of learning have a remarkable potential to reveal possible neural mechanisms underlying the maintenance and loss of brain plasticity, although this area is as yet largely untapped. These issues all raise more questions than they answer, but research in both fields is progressing rapidly. Continuing to be aware of and to explore the parallels, as well as admitting when they fail, should be helpful to both fields.

3 • Birdsong and singing behavior

Heather Williams

The flowers appear on the earth;
the time of the singing of birds is come,
and the voice of the turtle dove is heard in our land.
(Song of Solomon, 2:12)

Sumer is icumen in, Lhude sing cuccu!
(Anon)

The ousel cock so black of hue,
With orange-tawny bill,
The throstle with his note so true,
The wren with little quill ...
The finch, the sparrow, and the lark ...
(Shakespeare, A Midsummer Night's Dream)

Birdsong has never lacked for admirers drawn by its aesthetic qualities. But to the scientist, birdsong is also of interest because it represents an evolutionary flowering of vocal learning. Among mammals, vocal learning – the imitation of sounds used in communication – is extremely rare, having been demonstrated in one primate species (humans), the cetaceans (whales and dolphins), and two bat species (there is also some evidence for vocal learning in pinnipeds). In contrast, almost 5000 species of birds, distributed among three groups, learn their songs or calls: parrots (350+ species), hummingbirds (300+ species), and oscine songbirds. (The *oscines*, a suborder of the Passeriformes, have a worldwide distribution and demonstrate vocal learning. Taxonomically, the oscines are defined by the complex musculature of the vocal organ, the syrinx.) Since many of these bird species are relatively easy to maintain in the lab, sing readily, and are diurnal, the pre-eminence of birdsong as a model system is understandable.

Beginning in the early 1950s at Cambridge University, William Thorpe's and Peter Marler's work on the development of birdsongs and alarm calls led the way to a systematic analysis of song behavior, yielding a bountiful harvest of insights into the stages of song learning, critical periods, the role of auditory models for song, and song dialects. As understanding of the nuances of birdsong deepened and grew richer, the next step was to begin to use experimental approaches to ask questions about the neural bases and correlates of vocal learning and communication.

Songbirds (and their cousins, the parrots and hummingbirds) offer several advantages in this attempt to understand the brain processes that underlie complex learned behavior. First is the song itself. Because adult song is a complex sound repeated in a stereotyped manner, it can be easily recorded, quantified, and analyzed. Since young birds imitate the songs of adults, it is easy to follow the learning process by recording the tutor's song and successive versions of the young bird's song. As a signal used in communication, song is affected by the conditions and social context the singer experiences – and, in turn, song affects the conspecific listeners' subsequent behaviors; all of these factors can also be observed and quantified. The second major advantage is the ease with which songbirds can be managed. Several species have been domesticated and can readily be raised and studied in the laboratory; some may produce three generations each year. The third advantage is that the neural circuitry which mediates singing behavior is well described (Reiner *et al.*, 2004b), and there is a growing recognition that the avian brain is based on the same organizational scheme as that of mammals (Farries, 2004; see also Farries and Perkel, this volume). Finally, methods and databases that harness the powerful tools of molecular biology for understanding genetic mechanisms are in place for studying the acquisition and expression of singing behavior, and the genome of one songbird will soon be sequenced (Clayton, 2004; see also Clayton and Arnold, this volume). Thus birdsong is currently the best – or perhaps the only – model system that allows the use of neural, genetic, and molecular analyses to be brought to bear on the study of a complex communication system that is learned by imitating a model.

Neuroscience of Birdsong, ed. H. Philip Zeigler and Peter Marler. Published by Cambridge University Press. © Cambridge University Press 2008.

This chapter focuses upon the culmination of vocal learning, the songs and calls of adult birds. Even the simplest and most stereotyped of songs has levels of complexity that can potentially muddy the waters of interpretation. However, this very complexity, evolved by the animals in their natural environment, allows birdsong to serve as a model for a variety of phenomena that cannot be investigated using behavioral protocols designed (by humans) specifically for use in the lab.

THE DIVERSITY OF BIRDSONGS

Much of what we know about how birdsong is related to the bird brain is based upon the study of a very small subset of songbirds (see Kaplan, this volume). One species (the zebra finch, *Taeniopygia guttata*) accounts for approximately half of all studies of the neural basis of vocal learning, and the second most popular subject, the canary (*Serinus canaria*) represents a further 14% of such studies. Six additional species (European starlings *Sturnus vulgaris*, song sparrows *Melospiza melodia*, white-crowned sparrows *Zonotrichia leucophrys*, brown-headed cowbirds *Molothrus ater*, swamp sparrows *Melospiza georgiana*, and Bengalese finches *Lonchura domestica*) account for another 19% of the studies of vocal learning in songbirds. More than two-thirds of all such studies use one of three domesticated species (zebra finches, canaries, and Bengalese finches) as subjects; these species breed readily in captivity and are not stressed by human presence and the laboratory environment. While some wild-caught birds (e.g. European starlings, black-capped chickadees *Parus atricapillus*) can adapt to captivity, other species may not unless they are captured as nestlings and hand-reared (a difficult, time-consuming process). Hence the choice of the primary species used for laboratory-based studies of vocal learning is, understandably, most often based upon criteria other than the properties of the song.

For most songbird species, only the males sing. However, females may sing as much as males in duetting species such as the bay wren (*Thryothorus nigricapillus*; Farabaugh, 1982). Female canaries do not sing during the spring breeding season, but may sing quietly during the fall (Pesch and Güttinger, 1985); female zebra finches do not sing, and differences in the song circuitry of males and females are correlated with the differences in singing behavior (Nottebohm and Arnold, 1976; see also Bolhuis, this volume). Not surprisingly, male birds' songs and brains have been the primary focus of research on the basis of song learning, although comparative studies of females have and will continue to provide insights about hormonal influences on and brain mechanisms for song behavior and learning.

A bias towards studying males of domesticated songbird species may have affected our current view of vocal learning, since it is all too easy to generalize from male zebra finches (or canaries) to "birds." Fortuitously, the songs of the two most commonly used species – the canary and the zebra finch – have different structures as well as different learning trajectories. Even so, these two species' songs cannot begin to span the full range of avian vocal learning. Neither zebra finches nor canaries are mimics; adults of species adept at mimicry (such as the mockingbird *Mimus polyglottus* and mynah *Gracula religiosa*), can produce accurate imitations after hearing a single example of a sound. "Action-based learning", the pruning of excess songs from the repertoire at the end of the song learning period (Marler, 1997), is important for species such as song sparrows and swamp sparrows, but does not appear to occur in zebra finches or canaries. Including mimics and birds that exhibit action-based learning among the species used for studies of the neural basis of vocal learning might yield both new insights and new questions.

THE SONG LEARNING PROCESS

The primary impetus for studying birdsong is its value as a model for vocal learning with parallels to human speech. Song learning has several stages (Marler, 1981): first, a sensory learning period, during which the young male listens to and memorizes the song of a socially salient adult model; second, a subsong stage, akin to the babbling of human infants, during which the young male produces sounds and listens to the results, calibrating his vocal instrument; third, plastic song, during which the young male adjusts his song to approximate the memorized model; finally, crystallization, when the song becomes fixed in its adult form and the components of the song and the order in which they are sung become stereotyped (see Hultsch and Todt, this volume). These stages may be separate and distinct during development, or may overlap (particularly in the case of the first and second stages). For seasonally breeding birds such as canaries, the stages of song learning usually begin during the summer, soon after fledging (when young birds leave the nest) and crystallization occurs the following spring, when young males are ready to breed for the first time (Nottebohm *et al.*, 1986). In contrast, birds that are not seasonal breeders may complete all of the stages of song learning during a more compressed period; zebra finches reach sexual maturity at about 90 days, and song crystallizes at that time (Immelmann, 1969; Price, 1979). Species also differ in whether song is fixed after it crystallizes during the first year (in critical-period learners, e.g. zebra finches and Savannah sparrows *Passerculus sandwichiensis*) or whether new songs may be learned each year (in open-ended learners, e.g. canaries). Notwithstanding these species differences in the timing and repetition of song learning, the four classic stages (sensory learning, subsong, plastic song, and crystallization) are readily apparent in the developmental trajectories of most songbirds.

What material do young birds learn? Most often, it is the sounds and structure of the songs their parent(s) and/or neighbors sing. Some species, such as the canary and white-crowned sparrow, will copy songs played through speakers. Others, such as the zebra finch, must have an interactive relationship with the song model, either in the form of an adult bird (Price, 1979) or a key that is pecked to trigger a song playback (Adret, 1993; Bolhuis *et al.*, 1999; Tchernichovski *et al.*, 2001; see also Saar *et al.*, this volume). In species that learn from tape recordings (such as the chaffinch *Fringilla coelebs* and white-crowned sparrow), young males prefer to imitate the adults with which they interact, even if they first encounter singing males well after the time they would normally have crystallized their songs (Petrinovitch and Baptista, 1987; Nelson, 1998). Song development is also guided by innate predispositions (Marler, 1989). Young male sparrows that are deafened or deprived of the opportunity to listen to any adult tutor (live or recorded) during song development develop songs that deviate substantially from a normal song yet still retain some characteristics of the species-typical song (Marler and Sherman, 1983, 1985). Hand-reared young male chaffinches exposed only to recorded songs from a variety of species tend to copy the songs with conspecific properties (Thorpe, 1958, 1961). Young white-crowned sparrows (Rose *et al.*, 2004) and canaries (Gardner *et al.*, 2005) tutored with songs that have normal notes but defective or incomplete syntax learn the notes from the tutor songs and assemble those notes into a song with species-appropriate syntax. When a young male song or swamp sparrow hears only recorded songs, both the units of sound (syllables, notes, and elements) and the organization of those units within the song, or syntax, may influence the choice of a model to copy (Marler and Peters, 1977, 1988b). So, while exposure to a specific song from a specific tutor strongly influences what a young bird learns, genetic predispositions influence both the choice of a model and how it is subsequently produced.

As well as copied elements, the songs of young males may include altered or novel notes; such notes are considered to be "improvised" or "invented." Learned song notes may also be rearranged to generate a new order. For these reasons, the learned song may not be an exact copy of the tutor's song, but it usually includes many notes that are near or exact matches of the material copied from one or more adult models. Thus the song material that is learned depends both upon genetic predispositions for certain acoustic and structural properties and upon the characteristics of the songs sung by socially salient conspecific birds – with an admixture of "creative" contributions when young birds improvise new song units or arrangements of copied units. Species differ in the relative amounts of imitation and improvisation that contributes to an individual's song – from those that produce near-exact copies of adult models to those that improvise nearly all of their notes (e.g. sedge wrens *Cistothorus platensis*; Kroodsma and Verner, 1978).

Identifying the neural mechanisms that underlie imitative song learning is the "holy grail" of the field (see chapters by Adret and Bolhuis, this volume). Researchers have focused most heavily upon two aspects of birdsong: first, the neural basis for the production of stable, crystallized song, and second, the processes that are responsible for changes in that song during development, during seasonal relearning, or after inducing plasticity in adults. Understanding either of these processes requires a comprehensive analysis of the endpoint of song learning, the adult crystallized song. The remainder of this chapter will examine the production of song by adults – what is sung, how it is sung, and how the social context of singing affects song performance – with reference to the species most commonly used in laboratory studies of birdsong, the zebra finch.

CHARACTERISTICS OF ADULT SONG: STEREOTYPY AND REPERTOIRES

The most salient feature of adult song (for the human and presumably also for the avian listener) is its stereotypy – both in the acoustic structure of the song syllables and, for many species, the sequencing of the syllables. Crystallization, the final stage of learning, is usually accompanied by an increase not only in stereotypy, but also in the volume and amount of singing, particularly for seasonal breeders such as canaries. This is not surprising as song is often used for territory defense and advertisement; a louder song is more effective in establishing a presence over a wide area. Because frequent, loud singing is energetically expensive (Oberweger and Goller, 2001), softer singing during song learning is presumably less costly, and also less likely to attract predators. However, zebra finch song remains relatively quiet, most probably because the object of the song in this highly social species is generally no more than a few meters from the singer. An increase in stereotypy of syllable structure and of the ordering of syllables does occur in both zebra finches and canaries, and appears to be a universal characteristic of adult bird song.

An individual male's crystallized song consists of a distinct repertoire: a set of elements, notes, syllables, phrases, motifs, or songs from which all of his singing is drawn. In some species the repertoire may be limited to a small set of syllables or notes that are each sung in nearly every utterance; for others, recordings of many songs may be needed to define a large repertoire. A zebra finch's repertoire is quite small, ranging from 3 to 15 notes, sometimes called elements or syllables (Zann, 1996). Each song opens with a series of identical repeated introductory notes followed by one or more repetitions of a "motif" that consists of most or all of the repertoire of song notes delivered in a fixed sequence, without any repeated notes (Figure 3.1a). Most of a zebra finch's song output consists of the "canonical motif" (or "canonical song"), a note sequence that is repeated in a nearly invariant fashion and lasts about one second (Figure 3.1a). Other species, such as the canary (Nottebohm and Nottebohm, 1978) or starling (Chaiken et al., 1993), may have a repertoire of dozens of notes grouped into syllables that are repeated to form trills or phrases (3.1b). These trills or phrases are strung together to form the song, which may continue for well over a minute. Each phrase is often associated with a particular successor, but the sequence of phrases sung within a given song is rarely an exact replicate of the previous song. For this reason, it may be necessary to record several minutes of song to accurately describe an individual's repertoire. Still other species (such as song sparrows and red-winged blackbirds) may have a repertoire consisting of several distinct songs, each of which is stereotyped and each of which differs from other songs, or song types, in the repertoire. Such species may sing with "immediate variety", changing their song type with each successive song, or with "eventual variety" (Kroodsma, 1982), repeating a song type several times before changing to another song type in the repertoire.

The patterns of song behavior represented by the two most often studied species, zebra finches and canaries, do not span the entire range of songbird performance, but do represent relatively distant points within that spectrum. There are important aspects of song repertoires that are not well represented in studies of brain mechanisms for singing based on these two species: neither canaries nor zebra finches have repertoires with multiple song types, and domesticated birds do not form song dialects (distinct songs associated with specific geographic regions), nor do they mimic recently heard sounds or provide a clear example of action-based learning (the pruning of learned material based on the observed behaviors of conspecifics). Nevertheless, the two species have two different types of repertoires organized in different ways and learn their songs according to different schedules. Despite these differences, both species (like all songbirds except mimics) have song learning trajectories that culminate in a stereotyped and crystallized adult song. Although the salient characteristic of the crystallized song is its stereotypy, there does remain some potential for variation. The types and sources of this variability are interesting in their own right and also provide a bridge between studies of crystallized song and song learning.

ADULT SONG VARIABILITY

Crystallized zebra finch song is among the least variable songs, consisting as it does of a relatively few notes that are delivered in a fixed sequence. Yet even in these songs a limited form of variability exists: notes can be omitted from the beginning or end of the song

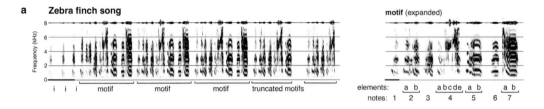
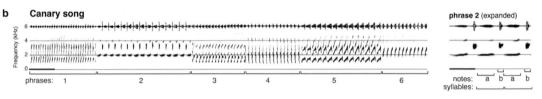

Figure 3.1 Structure of zebra finch and canary song. The sonograms show recordings of a portion of one zebra finch's song (a) and one canary's song (b); expanded segments from each of these two recordings are also shown. The dark bar along the 0 kHz line denotes 0.5 seconds for the longer song segments and 0.1 seconds for the expanded sections of the song. Sound amplitude, in the form of an oscillogram, is shown along the 8 kHz (zebra finch) or 6 kHz (canary) line. The terminology for song structures differs slightly from some existing practices to ensure that similar units in the two species' songs have consistent labels.
(a) In the zebra finch song, introductory notes (i) precede a series of motifs, some of which may be truncated. A motif consists of a repeated sequence of unique notes. A note is defined as an uninterrupted sound; each note may include one or more elements. An element is defined as a unit of song that has a coherent time/frequency structure that distinguishes it from neighboring elements within a note. For example, note 7 has an initial element with strong frequency modulation, followed by a constant-frequency element. In such constant-frequency elements the harmonic nature of zebra finch vocalizations can be clearly seen: element 7b has a fundamental frequency of approximately 650 Hz, and eleven harmonics (integer multiples of the fundamental frequency) can be readily distinguished. Note: this song included five additional motifs that are not shown here.
(b) Canary song consists of a string of phrases, each made up of many repeats of a single syllable. A syllable may consist of a single note (as in phrase 6) or may be made up of two or more notes. The syllable that is repeated to form phrase 2 (see expanded version) is made up of two notes, an extended nearly constant-frequency 2 kHz note and a short higher-frequency note. The short sound at approximately 4 kHz is the second harmonic of the loudest portion of note a and does not represent a third note in this syllable. (The sonogram shown here was drawn from a recording provided by Dr. Fernando Nottebohm's laboratory.)

(Sossinka and Böhner, 1980; Williams and Staples, 1992). This type of variability is akin to flexibility in the order of phrase delivery in canary songs and in the order of song types in species with multiple song type repertoires. The way notes or song types are ordered during crystallized adult song performance may reflect both events that occurred during song development and events that occur during the performance of the song.

Although the father's song is often the only or most important source of material for the son's song, the evolution of vocal learning has made it possible for young birds to acquire song material from males other than their fathers, presumably thus increasing their ability to form dialect groups and to attract mates (Nottebohm, 1972a). In the lab, however, song may be learned exclusively from the father (as when offspring are raised in single-family cages), or from a song chosen by the researcher and played back through a speaker (canaries) or a dummy male that "sings" when a key is pecked (zebra finches). These social and song environments are impoverished compared with those experienced by wild birds; after fledging, a wild young zebra finch male interacts with many adults within the colony. A breeding aviary with multiple pairs can be used to expose a young male zebra finch to multiple song models, and in such an environment young zebra finch males often copy song elements from more than one adult male (Williams, 1990; Figure 3.2).

Copying from more than one male occurs in many species, particularly when males acquire large repertoires and/or are exposed to widely varying song models before breeding. When multiple tutors contribute to a song, elements copied from a specific tutor are associated with each other within the structure of young male's song. Nightingales (*Luscinia luscinia*) have repertoires of up to 100 songs, which they learn as

Birdsong and singing behavior 37

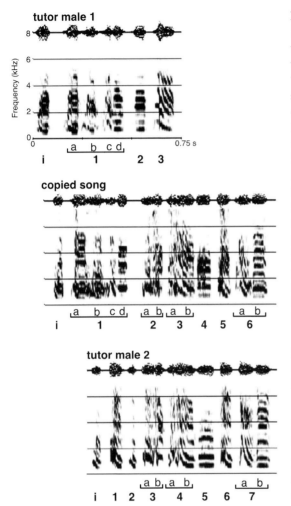

Figure 3.2 Multiple tutors for a single zebra finch song. The song of a young male zebra finch raised in an aviary with eight adult males. The presumptive father of the young male sang the song denoted "tutor male 1," but because of the potential for extra-pair copulations and egg dumping, paternity cannot be conclusively attributed to the adult in whose nest the young male hatched. The young male copied the first note (four elements) from the social father, and copied the remainder of the song (5 notes made up of 8 elements) from a second male (tutor male 2). The young male's introductory note was not copied from either tutor.

likely to be sung in succession (Hultsch and Todt, 1989b; Hultsch, 1991b). The structure of canary song, with its strings of repeated syllables and predictable transitions between syllables, may represent a similarly "packaged" structure that could also reflect units of song learned from individual tutors. Zebra finches, despite their much less complex song, have a similar within-song structure: "chunks" of approximately three notes are learned as units (Williams and Staples, 1992). When multiple tutors provide source material for a young male's song, the notes copied from each tutor are not intermixed, but rather chunks of notes from each song are sung in sequence. Because the order of learned chunks is often rearranged, the learned song may be novel even when a young male copies all of its song notes from a single tutor (Figure 3.3a). The performance of the song also reflects this chunked structure; if the singer truncates a motif, the breaks in the normal sequence most often occur at chunk boundaries (Figure 3.3b).

Thus, like the repertoire of the nightingale (and possibly that of the canary) the adult song of the zebra finch has a hidden level of organization that reflects processes that occurred during song learning.

INDIVIDUAL VARIATION IN THE STEREOTYPY OF ADULT SONG

Individuals' songs are strikingly different from each other. Even birds tutored by the same adult will sing different songs, because of variation in the number of units copied, the accuracy of the copies, and the ordering of the subunits of song. Several factors act in concert to generate these individual differences. Young male zebra finches mutually influence each other's copy accuracy; the fastest learner in a nest sings the most accurate copy of the model song, while his brothers diverge from the model in different ways (Tchernichovski et al., 1999). Individual differences in brain characteristics influence learning; birds with larger forebrain song centers tend to have larger song repertoires in both canaries (Nottebohm et al., 1981) and zebra finches (Ward et al., 1998; Airey and DeVoogd, 2000).

"packages" of approximately four songs copied from a single tutor (Hultsch and Todt, 1989c, and this volume). As adults, the young males sing the songs within a learned package in the same order the tutor used, and packages copied from a given tutor are

Another source of individual variation arises from differences in the way adults perform their songs. Some males sing readily, others do not; some unmated male zebra finches sing only to particular birds, while others sing readily to many potential mates; some males sing

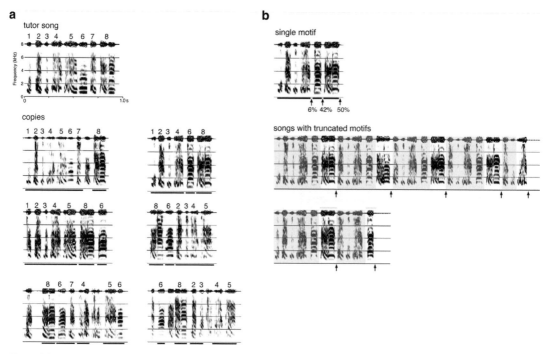

Figure 3.3 Chunks of zebra finch notes are copied and sung as units.
(a) The motif of an adult male zebra finch that was copied by several young males in an aviary environment (the same male's motif appears as tutor male 2 in Figure 3.2). The motifs of six young males that copied the adult are also shown; notes that were copied from the adult male are numbered to correspond with the notes in the tutor's song. Beneath each of the copied motifs, bars denote notes that were copied as continuous sequences, or chunks.
(b) One of the copied song motifs shown in (a), with bars indicating the copied chunks. Arrows and associated percentages denote the proportion of motifs ending after each note during an extended series of recordings. Two song recordings including a total of seven motifs are shown; arrows denote the end of each motif. The final note in each motif is highlighted. Truncated motifs ended between, and not within, the chunks of notes copied as sequences from the adult male's song.

long bouts, while others may sing equal numbers of motifs in shorter bouts; some males' songs may vary somewhat between renditions, while others' will be more stereotyped; some male zebra finches sing several introductory notes before each song bout, while others may sing only one introductory note or forego these notes altogether.

Such variability in the performance of crystallized adult song, which is apparent even in critical-period learners such as zebra finches, has largely been ignored in studies of the neural basis of behavior because the presence of a "canonical" song is valuable for studying the neural patterns associated with multiple examples of the same complex behavior. However, variability does exist in the adult songs of both canaries (where the sequences of phrases is not fixed) and zebra finches (in the form of alternative motifs that omit or add particular notes). One way to analyze individual variability in adult male zebra finch song is to compare the numbers of unusual syllable transitions and canonical syllable transitions within a sample of approximately 100 songs (Scharff and Nottebohm, 1991). Another measure calculates the minimum number of different motif types needed to account for 85% of all song motifs in a sample of at least 100 motifs. Either of these measures demonstrates that there is substantial variability between normally reared males in song stereotypy (Figure 3.4).

The fact that such variation in adult song exists, even in zebra finches – a species that sings one of the simplest and most stereotyped of songs – indicates that variability may play an important role in behavior. One possible reason for adult song variability might be to allow the singer to tailor his stereotyped song to fit the preferences of a specific listener, such as a rival or a potential mate.

Birdsong and singing behavior 39

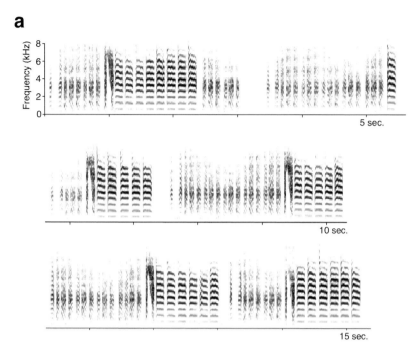

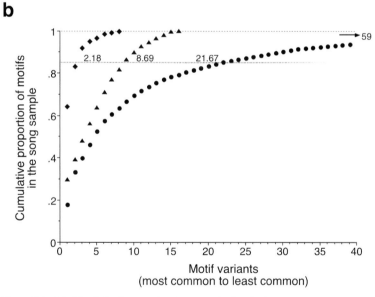

Figure 3.4 Motif variation in tutored and untutored zebra finch song.
(a) A continuous recording of 15 seconds of untutored male zebra finch song. Each of the seven motifs sung in this segment is a different variant. In comparison, Figure 3.1a shows five motifs from a song of a normally reared male; the complete song included ten motifs (seven of them the canonical motif) and three motif variants.
(b) Motif variability differs among normally reared birds, and is lower than in untutored birds. Within a recording of approximately 100 motifs, the frequency with which each motif variant was sung was tabulated and motif variants were ordered from most common to least common. This plot shows the cumulative proportion of the total number motifs sung that are included as each successive motif variant is added to the sample. The number of motif variants needed to generate 85% of all motifs sung (a level represented by the dotted line) can

THE INFLUENCE OF A LISTENER UPON SONG DEVELOPMENT

A male that never hears a song tutor or model develops an abnormal song. This "untutored song" retains many characteristics of normal zebra finch song but is marked by unusual note structure and decreased stereotypy (Price, 1979; Williams *et al.*, 1993). High frequency notes and upsweeps are much more common in untutored song, as are repeated notes and a highly variable note order (although it is usually very difficult to define a canonical pattern in tutored zebra finch song). It has been generally assumed that the unusual properties of untutored song are due to the absence of a song model, but some of the abnormal features of untutored song may arise in part because of the absence of feedback from the behavioral responses of a conspecific listener.

Hearing a model is clearly crucial for normal song development. Nevertheless, in the absence of a model, listeners can guide song development towards a normal outcome – which is the case in brown-headed cowbirds (which are brood parasites and so do not hear appropriate song models until relatively late in development). An adult female listening to a young male responds selectively by flipping her wing or changing her posture in response to particular versions of a young male's attempts at song. The male's observations of the female's actions strongly influences what he sings subsequently, and so the female directs the trajectory of song development (West and King, 1988). Although evidence for powerful listener effects in other species is not as compelling as for cowbirds, zebra finches, because they grow up in social colonies and their early songs are heard by many other birds, are a species in which listeners' responses could affect song development. In the laboratory, untutored males raised with deaf adult females sing more frequently than do untutored males raised with hearing adult females. The songs of the males raised with deaf adult females are similar to those of untutored males, with less stereotypy and more atypical syllables than in the songs of males raised with hearing adult females (Figure 3.5). In contrast, the presence of a hearing female alters the course of untutored song development so that the outcome approximates normal song – possibly because the hearing female provides feedback to the young male about the content of his song. Siblings also affect the outcome of song development, although it is not clear whether the effects are due to his hearer's responses or to what a young bird hears his siblings sing. Young males housed together converge on the same version of an untutored song (Volman and Khanna, 1995), and males within a normally reared clutch influence the accuracy of each others' copies of the model (Tchernichovski *et al.*, 1999).

Conspecifics' influence on song development can thus be exerted in two ways: first, by providing a model song to copy, and second, by listening to the developing song and providing behavioral feedback about what was sung. The relative importance of these two contributions may vary depending upon the species and the circumstances, but the potential role of a listener during song development should not be ignored.

THE INFLUENCE OF A LISTENER UPON ADULT SONG PERFORMANCE

Crystallized adult male birdsong has two well-documented functions – territory defense and mate attraction – that have evolved specifically to influence listeners' behavior. Thus it is not surprising that listeners' behaviors in turn affect adult song performance in several ways. The visual and auditory presence as well as the gender of conspecifics affects the amplitude and tempo of song delivery. Singing male zebra finches adjust the amplitude of their songs as the social context changes (Cynx and Gell, 2004), as do nightingales (Brumm and Todt, 2004). The tempo of undirected zebra finch song is slightly slower than that of directed song (Sossinka and Böhner, 1980), and song tempo also varies systematically throughout the day, increasing during periods when the birds sing frequently (Glaze and Troyer, 2006). Repeated exposure to females over a

Figure 3.4 (continued)
be used to describe the variability of an adult male's song. A typical pattern of motif variation for a normally reared male is shown by the diamonds (the 85% level is reached with 2.18 motif variants), the extreme for normally reared males by the triangles (8.69 motif variants were required to reach the 85% level), and a distribution seen only in untutored males by the circles (21.67 motifs were required to reach the 85% level).

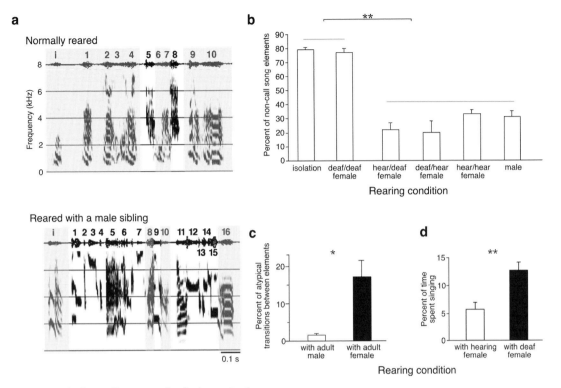

Figure 3.5 Audience effects upon zebra finch song development.
(a) Examples of the songs of males reared with and without tutors (the untutored male was raised with a male sibling of the same age). Elements that are similar to zebra finch calls are shaded. Non-call elements (bold) include high-frequency elements with few harmonics, upwardly modulated elements, and click trains.
(b) The presence of a hearing female during development decreases the proportion of non-call elements incorporated into a young male's song. In recordings made after birds reached 90 days of age, males reared with at least one hearing companion (an adult male tutor, a hearing mother or a hearing adult female companion) used significantly fewer non-call elements in their songs than did males that were housed alone or with deaf conspecifics (deaf mothers and deaf adult female companions; $p < 0.001$). The condition of the mother is designated first, followed by the condition of the adult female companion. Young males were housed with their mothers for the first four weeks after hatching, and then housed with the adult female companion until reaching at least 90 days of age.
(c) Males raised with an adult male song model sang a highly stereotyped sequence: fewer than 5% of transitions between elements are atypical (those not present in the canonical motif). In contrast, males raised with adult females sang a less stereotyped song with a significantly greater number of atypical transitions ($p < 0.05$). Untutored males were recorded at 200 days, well after the age when song normally crystallizes (90 days); normal males were recorded at 150 days.
(d) During song learning (60–75 days), males reared with deaf females sang more frequently than did males housed with hearing females ($p < 0.001$). Singing rates were measured as the average time spent singing during 30-minute recording sessions.

period of weeks, as when directed song is recorded at frequent intervals, is associated with an increase in song tempo (Williams and Mehta, 1999). "Listener effects" are not restricted to these relatively simple song parameters, but extend to the order and identity of the notes or songs that are sung.

Species with multiple songs, such as the nightingale (and potentially those with multiple phrases, such as the canary or starling) may have preferred song orders, but these are strong statistical trends rather than invariant rules. The order in which syllables or songs are sung can be influenced by the responses of other birds. When two rival males "countersing" from adjacent territories or as part of a territorial incursion, the delivery of a male's repertoire often shifts. "Song matching" is a form of song performance in which a male sings the song in his repertoire that best matches the song being sung by the rival of the same species (e.g. the indigo bunting *Passerina cyanea*, Payne, 1983; meadowlark *Sturnella magna*, Falls, 1985; song sparrow, Stoddard *et al.*, 1992). Countersinging males may also shift the timing of their songs so that they overlap the songs of the rival male, as in the nightingale (Naguib and Todt, 1997; Naguib, 1999). The chickadee not only overlaps its rival's song (Mennill and Ratcliffe, 2004) but also shifts the pitch of the song so that it assumes a specific relationship to that of the other male (Weisman and Ratcliffe, 2004). Since wild canaries sing many syllables in a variable order and their songs function in territorial advertisement and defense, they are likely to engage in countersinging – and indeed domesticated males are more likely to sing in the presence of other singing males or similar sounds. Canaries might then perform a type of song matching by shifting the usual syllable sequence in order to best match that of the countersinging male. If shifts in syllable order could be triggered in response to specific playback tapes, canaries could provide a valuable model for studying variation in behavioral sequences, a phenomenon that has not been examined at the level of neural mechanisms.

In contrast to species that broadcast high-volume song from a large, defended territory that provides the male with access to resources needed for reproduction, zebra finches sing a low-amplitude song intended for listeners within a few meters (Zann, 1996). This song has two forms associated with different social contexts: a courtship song that is "directed" at a specific bird, and "undirected" song, which is sung without an obvious object (Sossinka and Böhner, 1980). The two forms of the song consist of the same syllables given in the same order, but undirected song is characterized by greater sequence variability. Undirected songs are more likely than directed songs to include truncated motifs, start in mid-motif, or use atypical transitions between notes (Figure 3.6). Chipping sparrows (*Spizella passerina*) also adjust the variability of their songs according to the social context: the dawn song, directed at close range to other males, is variable in length, and the day song, broadcast to distant birds, is of fixed length (Liu and Nottebohm, 2005). These audience-dependent differences in the variability of song structure are associated with differences in gene expression in the song basal ganglia circuitry. For both zebra finches (Jarvis *et al.*, 1998) and chipping sparrows (Liu and Nottebohm, 2005), singing the more variable form of the song activates expression of a transcription factor (ZENK) in the basal ganglia portion of the song circuit, while the less variable form of the song does not (see Mello and Jarvis, this volume). Although the role of social context in generating the two types of neural activation patterns and associated song variability is not yet clear, it is of considerable interest.

Behavioral feedback from nonsinging females can influence the form of adult male song. Brown-headed cowbirds provide the classic example of this phenomenon (King and West, 1983, 1989). After observing (and comparing) female responses to his own and other males' songs, an adult cowbird male may shift his singing to approximate a successful male's song. As brood parasites that are raised by other species, cowbirds are a special case in which flexibility in adult song may be at a premium, but female responses to adult males also affect the delivery of zebra finch song. The probability that a male will truncate or add a note to his canonical motif is affected by any calls the female gives while he sings (Figure 3.7). Since individual females respond differently to a male's song, repeated exposure to a specific female may result in a shift in a male's canonical motif when particular notes within the repertoire are either omitted or sung more often. In this way, an adult male with a crystallized, stereotyped song may still tailor its contents to fit the fashion favored by the female.

VISUAL DISPLAYS ASSOCIATED WITH SONG

Many songbirds broadcast their songs over a wide area from a hidden location, and such songs can be accurately represented as a purely auditory signal. Other songs are accompanied by stylized movements, or "dances", and displays of plumage, ranging from simple adjustments of feather posture to complex and dramatic deployment of specialized plumes, as in birds of paradise. Such displays may be delivered by solitary singers, or be performed by groups of males at leks (a lek is a traditional display area consisting of a cluster of small territories each occupied by a courting male; lekking species include the black grouse and several manakins). More subtle displays are common as well. Canaries sing with a raised head and loosely fluffed throat feathers, and zebra finches adjust their plumage so that the normally

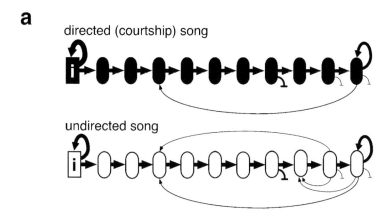

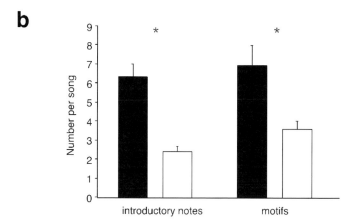

Figure 3.6 Differences between directed (courtship) and undirected song in the zebra finch. During courtship song the male orients himself to and directs a display, or dance, at a conspecific individual. A male singing undirected song neither orients towards an individual nor performs a courtship display. Song stereotypy also differs in these two contexts. (a) Transitions between elements are less stereotyped during undirected song than during courtship song. Thicker lines in this diagram represent transitions that are more frequently observed. The box denoted "i" represents introductory notes. (b) Males singing courtship song (black bars) use more introductory notes ($p < 0.01$) and sing more motifs in each song ($p < 0.01$). (The data for this figure and analysis were drawn from Sossinka and Böhner, 1980).

rounded head becomes a "flat top." Zebra finch males also perform a courtship dance, a series of hopping and bobbing movements oriented to a female, as part of directed song (Morris, 1954).

While singing, many birds open and close their beaks to change the gape, or beak aperture. The primary function of variations in gape is most probably to adjust the resonance of the vocal tract, and so to emphasize sounds of a particular frequency; an open beak yields a shorter vocal tract that emphasizes higher-frequency sounds (Westneat *et al.*, 1993; Hoese *et al.*, 2000; see also Goller and Cooper, this volume). Changes in beak gape are tightly coupled to specific song syllables, and beak movements are also clearly visible to a nearby observer. In a similar fashion, throat movements that reflect changes in the volume of the upper respiratory tract that track changes in syllable frequency (Riede *et al.*, 2006) can be visually emphasized by the "flashing" of contrasting throat feathers. For a nearby listener, changes in the beak gape and throat feathers provide visual correlates of some of the structural features of song, and may serve to accentuate parts of the song.

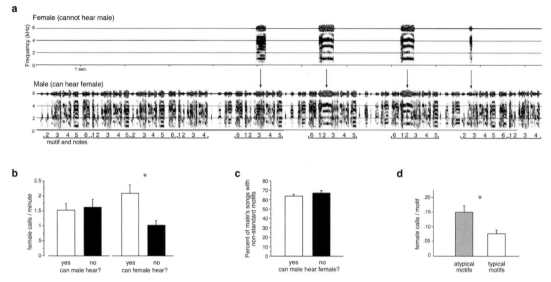

Figure 3.7 Female calls affect male song delivery in zebra finches.
(a) Individual male and female zebra finches were placed so they could see but not hear each other through a lucite partition, and the experimenter controlled which bird could hear the other by relaying a recording to a speaker on the opposite side of the partition. Here the female could not hear the male, but the female's vocalizations were played to the male (note the presence of the female's calls in the recording of the male). Motifs and notes are designated, respectively, by bars and numbers under the sonogram of the male's song (calls are not numbered). The unusual calls and motif structure sung when the female first called were repeated for several motifs before the male returned to a more standard motif structure.
(b) The average female call rate was not affected by males' ability to hear the calls, but female calls increased significantly ($p < 0.05$) when females could hear the males' vocalizations.
(c) Whether or not males could hear a female did not significantly affect the number of atypical motifs the males sang.
(d) Female call rate was significantly higher ($p < 0.05$) during atypical motifs, indicating that when males could hear a female's call, they were more likely to sing atypical motifs.

The dance steps of a courting zebra finch male (hops, head movements, and wing movements) vary somewhat from song to song, and are not as tightly coupled as beak movements to the acoustic structure of specific syllables. Like the patterns of changes in beak gape, however, the overall pattern of dance movements that accompanies an individual's song is very similar to the pattern of the bird that provided the auditory model for the song (Figure 3.8; Williams, 2001). Thus these courtship dance patterns must either be learned from a model or be regulated by a rhythm tied to the acoustic structure of the song.

The reliable coupling of dance and beak movements to the song suggests that the traditionally defined song circuitry also engages other motor patterns, perhaps using respiratory circuits and other pathways outside the song system (Wild, 2004a, and this volume). It further suggests that listeners' responses to a particular song may be affected by the visual signals associated with the song as well as by the acoustic parameters of the song; advances in monitor technology now make it possible to use video stimuli to test such relationships (Galoch and Bischof, 2007). Although the acoustic properties of song are far more tractable for analysis and for research into vocal learning and motor patterns, the biological function of the song may be difficult to separate from the movements that make up an associated visual signal.

HORMONES, MOTIVATION AND SONG PERFORMANCE

As a behavior that is affected by day length, sex steroids and by the presence of specific individuals that are potential rivals or mates, bird song is presumably under the influence of brain areas associated with emotion and

Birdsong and singing behavior   45

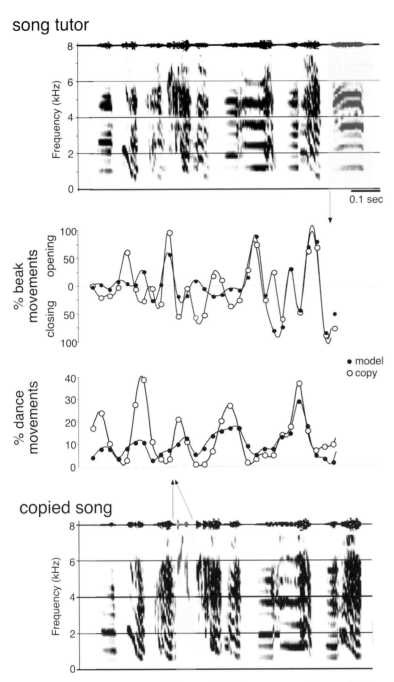

Figure 3.8. Coordination of zebra finch song with beak movements and the courtship dance. The motifs of an adult tutor and a young male that copied the song are shown at the top and bottom, respectively. The final note in the tutor's song (shaded) was not copied, and the copied song includes an improvised note (shaded). The beak and dance movement data from the copied song were shifted to the left to align copied notes to corresponding notes in the tutor's song (as indicated by arrows above the copied song). Videotapes of approximately 100 motifs were examined, and the positions of beak movements (opening or closing) within each motif were determined, as were the positions within each motif of the beginning of "dance movements." Beak movements were highly stereotyped within the motif, and dance movements showed less rigid but distinctive patterns coupled to what was being sung. Movement patterns in the tutor's song and the copied song were significantly correlated (beak movement $p < 0.0001$; dance movement $p < 0.001$).

motivation (Arnold, 1975b; Marler et al., 1988; Nowicki and Ball, 1989). Studies of the role of photoperiod and hormones on song circuitry have tended to focus upon structures related to the mammalian cortex and basal ganglia (Smith et al., 1997b; Tramontin et al., 1999; Ball et al., 2002), but motivational influences on song are likely to be mediated by structures such as the hypothalamus and limbic system (see Ball et al., this volume). Links between song circuitry and the hypothalamus and limbic system have not yet been fully defined, but are likely to be similar to those in nonsongbird species such as the ring dove (Cheng and Durand, 2004).

The effects of sex steroids (testosterone, and its metabolites) on song performance are well known. In mammals, receptors for sex steroids are found in the hypothalamus, limbic system, and midbrain (McEwen et al., 1979), all of which are thought to affect motivational states. Sex steroid receptors are found in the same areas of songbird brains as well as in the cells of song circuit nuclei (Arnold et al., 1976; Gahr, 2001). In the canary and other open-ended learners, seasonal changes in song circuitry are associated with fluctuations in day length and circulating testosterone levels (Nottebohm, 1980b; DeVoogd and Nottebohm 1981b; Nottebohm et al., 1987; Canady et al., 1988; Rasika et al., 1994; Bernard and Ball, 1997; Ball et al., 2004a). Testosterone plays an especially important role in mediating the rapid increase in stereotypy that accompanies song crystallization. Artificially high testosterone levels in development cause early crystallization of song in zebra finches (Korsia and Bottjer, 1991), and reducing testosterone levels or blocking testosterone receptors delays song crystallization in sparrows (Marler et al., 1988). In zebra finches, the song itself is stable after crystallization at 90 days, but manipulating circulating testosterone levels does affect singing (Arnold, 1975b; Adkins-Regan and Ascenzi, 1987). Castrated male zebra finches sing less than do their intact counterparts, and implanting testosterone-filled silastic capsules is widely and effectively used to induce higher singing rates in intact adult males. Testosterone supplementation does, however, affect what is sung as well as how often the bird sings: the fundamental frequency of zebra finch song notes may drop substantially (Cynx et al., 2004) and it is possible that tempo and stereotypy are also increased.

Adult male zebra finch song has the potential for a latent form of plasticity. When auditory feedback is disrupted during song production by deafening (Nordeen and Nordeen, 1992) or introducing auditory interference (Leonardo and Konishi, 1999; Cynx and von Rad, 2001), adult song structure and stereotypy are lost over a period of weeks or months (see chapters by Woolley and Brainard, this volume). Adults may also reconfigure crystallized songs after their vocal output is disrupted by any of a number of interventions, including syringeal nerve section (Williams and McKibben, 1992), muting (Pytte and Suthers, 2000), or mechanical interference with phonation (Hough and Volman, 2002). At least one of these forms of adult plasticity is sensitive to circulating testosterone; adult male zebra finches with high testosterone levels are less likely to change their songs after syringeal nerve section (Figure 3.9; Williams et al., 2003). Thus testosterone's effects on adult song performance and plasticity in zebra finches are consistent with the better-understood role of testosterone during development.

The powerful relationships between testosterone and song rate, stereotypy, and the potential for adult plasticity all point to a role for testosterone in reducing variability in song performance while at the same time increasing song production. However, testosterone's effects cannot account for all of the variation in adult song stereotypy and performance (Boseret et al., 2006). Brain mechanisms unrelated to sex steroids most probably contribute to motivational control of song variability, and factors other than motivation may also account for aspects of song variability. Given the variety of factors with a demonstrated relationship to variability in song performance, it is likely that additional mechanisms for song variability are yet to be described.

CALLS

In addition to their songs, songbirds' vocal repertoires include a number of calls – shorter, simpler vocalizations. The number and type of calls varies from species to species, but most songbirds have contact calls (used to identify and track other individuals), alarm calls (to give warning of a predator), agonistic calls (to denote intention to supplant or attack) and food-associated calls (such as begging calls emitted by young). Some call types are used year-round, while others may be given only within the breeding season; some calls may be given only by one sex, while others are used by all members of a

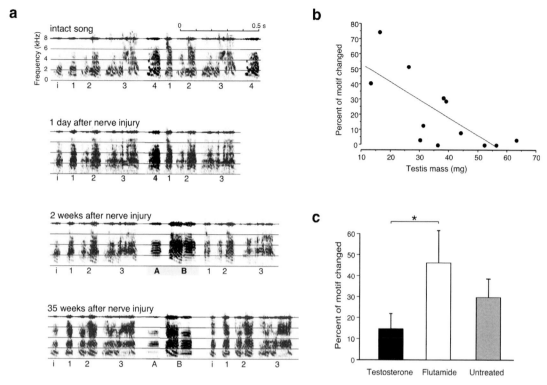

Figure 3.9. Testosterone reduces adult song plasticity in zebra finches.
(a) An example of adult song plasticity. The four distinct notes in the original motif were all retained (in a noisier form) in recordings made one day after one of the nerves serving the vocal organ (syrinx) was severed. Notes that did not change subsequently are shaded throughout. By the time of a recording made two weeks after the injury, note 4 had been deleted and two new notes (A and B) had been inserted at the end of the motif. These changes persisted in all subsequent recordings. The lengths of the deleted and added notes were summed and expressed as a the percentage of the length of the original motif.
(b) The proportion of the song motif that changed after nerve injury varied widely among males. Birds with larger testes were significantly less likely to change their songs ($p < 0.05$).
(c) Circulating hormone levels were manipulated by implanting testosterone or flutamide (a testosterone receptor blocker). Birds with testosterone implants altered a smaller proportion of their song motifs than did birds with flutamide implants ($p < 0.05$).

species. Although the ontogeny of calls has not been studied as systematically as song, it is clear that some calls are learned while others appear to be innate (Güttinger and Nicolai, 1973; Zann, 1985). Both learned and unlearned calls may vary geographically, and the variation can take the form of dialects (Rothstein and Fleischer, 1987; Miyasato and Baker, 1999). Because of the variety of calls and the variety of contexts in which they are used, calls may be more useful than song for exploring how information is communicated; because they are usually simple acoustically (in comparison to song), learned calls may also provide a simpler model for studying the brain mechanisms used for imitating an auditory model (Vicario, 2004).

Many species' call repertoires are relatively easy to characterize: chaffinches have eight different call types, each associated with a specific behavioral context (Marler, 2004). In contrast, some call repertoires are larger and may have complex syntactic structures. Chickadees have a simple, two-noted song, but have two complex calls, the "gargle" and the characteristic "chick-a-dee" call, which may vary from utterance to utterance. The "chick-a-dee" call consists of four different note types, each of which may be repeated several times before proceeding to (or skipping) the next note type (Hailman et al., 1987) (Figure 3.10b). Responses to variants of these combinatorial calls suggest that different configurations convey specific

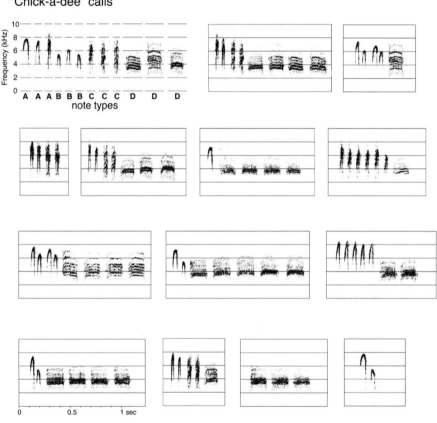

Figure 3.10 Zebra finch and black-capped chickadee calls.
(a) Among the most common zebra finch calls are the long, or "contact" or "distance" call, which differs between males and females, and the softer stack and "tet" calls (which are similar in males and females). All of these calls may be given singly or in clusters.
(b) Combinatorial syntax in black-capped chickadee (*Parus atricapillus*) calls. The "chick-a-dee" call from which this species gets its common name is distinctive and recognizable but is also highly variable. It consists of four notes, labeled A, B, C, and D, with structure that varies from bird to bird (three examples of each note type are shown). "Chick-a-dee" calls tend to follow simple, semi-Markovian rules to generate sequences from these three syllable types: a syllable can be followed by another syllable of the same type or by a type with a letter later in the alphabet. The 13 examples shown here, drawn from a single recording of four wild birds, show some of the variety of utterances that can be assembled from these four syllables and the simple syntactical rules (note that these examples include two "violations" of the rules).

information about the intentions of the caller, such as proceeding towards or away from a food source (Freeberg and Lucas, 2002).

Adult zebra finches have nine call types (Zann, 1996): one alarm call, three calls given while nest building, an agonistic call, a distress call, and three calls given frequently in the course of daily activity (the long, stack, and tet calls) (Figure 3.10a). More than one type of call may be given within a short interval, but the calls are not combined to form complex sequences. The best-studied zebra finch call is the long call, which is the loudest sound a zebra finch makes and is given when a bird is isolated or responding to another bird at a distance (for this reason the long call is also called the "contact call"). The long call is sexually dimorphic: males' calls are shorter and consist of two parts, a harmonic "tonal" portion and a rapidly modulated portion, while females' calls include only a long tonal portion (Zann, 1984). Males learn their long calls, but females do not (Zann, 1985), and the sexually dimorphic circuitry that underlies song learning is also important for call learning (Simpson and Vicario, 1990). Variation in both sexes' calls can convey information about individual identity, and the presence and identity of an audience affects how males respond to their mates and other familiar females (Vignal et al., 2004b).

Because zebra finches' long calls can be elicited by placing an individual out of sight and hearing of other birds and playing back recorded long calls, a "dialogue" of sorts between an experimenter and a bird can be generated. The length of the long call, the sex of the caller, and the sex of the responder all affect the probability that an isolated zebra finch will respond to a long call with a long call of its own (Vicario et al., 2002). This sex difference in responses to long calls disappears after lesioning the RA, a nucleus in the vocal motor pathway that is important for the production of learned songs and calls (Vicario et al., 2001). Studies such as these show the potential of using the system of songbird calls to investigate the behavioral and brain rules for vocal communication. Applying this approach to other call types – perhaps especially to combinatorial calls such as those of the chickadee – may provide an avenue towards understanding the meanings of vocalizations as they are decoded and understood by the birds.

SUMMARY AND CONCLUSION

Because the primary impetus for studying bird song is as a model for vocal learning, we sometimes forget that the endpoint of that process is the *adult* song. Adult song provides the reference that allows for retrospective understanding of the trajectory of song development, and manipulations that skew the outcome of the process, such as rearing young birds without a song model, allow for better understanding of what is required for appropriate song learning. But it is the normal adult song that plays the biologically important roles that have been shaped by natural and sexual selection – primarily territory defense and advertisement for a mate. Adult song, however much it represents a completed process in its crystallized and stereotyped form, is nevertheless variable and subject to a variety of influences – including steroid hormones, visual displays, conspecific listeners, and the identity of tutors for the elements of the song repertoire. Adult song variability may be normally limited to relatively few degrees of freedom, as in the restricted possibilities for delivering alternative motifs in the zebra finch, but this limited situational plasticity may nevertheless have important insights to offer. The original motivation for studying the neural bases of bird song, its unique qualities as a model for a complex behavior that is acquired through imitative learning, is still of central importance to neuroscience. Adult birdsong, because it is a well characterized and easily analyzed behavior, also provides an opportunity for examining mechanisms responsible for variations in song related to the unique experiences of the singer and the social context in which they occur.

ACKNOWLEDGEMENTS

This work was supported by grants from the Essel Foundation and the Howard Hughes Medical Institute to Williams College.

4 • The Australian magpie (*Gymnorhina tibicen*): an alternative model for the study of songbird neurobiology

Gisela Kaplan

INTRODUCTION

During the past half century the song control system has become a model for the study of vocal development in birds and humans. One species in particular, the Australian zebra finch, because it adapts easily to a laboratory environment, has been most widely used as a subject in these studies. Indeed the properties of zebra finch songs, and those of a few other species (see Williams, this volume) have come to define what most researchers mean by birdsong.

Based on such models, birdsong is generally understood as vocalizations "crystallized" into adult form and performed by males during the breeding season (Williams, 2004, and this volume). Rehearsal for such song acquisition by male juveniles is usually described as "subsong", developing into very variable song, known as "plastic song", before stabilizing into its adult form in the next breeding season (Hultsch and Todt, 1989c, and this volume).

In sexually dimorphic birds, sex differences in singing behavior are often paralleled by large sex differences in the morphology of the song control nuclei. The nuclei in males increase in volume before seasonal singing occurs (Arnold *et al.* 1986; Ball *et al.*, 1994; Bernard *et al.*, 1999) and the song nuclei in the forebrain of females are very small in sexually dimorphic songbirds in which only the male sings. There are also differences in sizes and structures of nuclei between songbirds. Song performance varies widely, even amongst vocally dimorphic birds of the higher latitudes.

A link has sometimes been made between size of song nuclei and song complexity. For example, it is said that song nuclei tend to be larger in those species that have more complex songs and that the HVC is larger in individuals with larger repertoires (DeVoogd *et al.*, 1993; but see Bolhuis, this volume). The association between song acquisition, viewed as a process for perfecting a male song for use during the breeding season, and the close functional relation between song and breeding has led to a definition of song as "a male reproductive behavior" (Bottjer and Johnson, 1997).

Although zebra finch song is one of the models upon which our characterization of learned "song" is based, the singing behavior of the zebra finch itself is *not* representative of a substantial number of avian species. Although indigenous to the southern hemisphere its singing behavior is more characteristic of many northern hemisphere birds, being both sexually dimorphic (sung by males only) and functioning exclusively within the context of breeding. Sexually dimorphic vocal behavior is not the norm for tropical avian species or for many species of the southern hemisphere even outside the tropics (Kroodsma *et al.*, 1996; Morton, 1996; Slater and Mann, 2004). Here I review research on the vocalizations of the Australian magpie (*Gymnorhina tibicen*), and suggest that the properties of its singing behavior are representative of a much wider array of songbirds than the zebra finch, including those of the southern hemisphere and the tropics. I argue further that several of the properties of magpie song, including the fact that it is sung by both sexes and is not obviously related to breeding, may make it a particularly appropriate model for studies of human language development.

MAGPIE SINGING BEHAVIOR AND THE DEFINITION OF BIRDSONG

The Australian magpie, the subject of this review, though unrelated to the black-billed magpie (*Pica pica*) of the North American and European continents, is about the same size, also black and white and is broadly akin to corvids (but see Schodde and Mason, 1999). The magpie, like the zebra finch, is endemic to Australia and has a distribution across almost the entire continent (Barrett *et al.*, 2003), thriving in many

Neuroscience of Birdsong, ed. H. Philip Zeigler and Peter Marler. Published by Cambridge University Press. © Cambridge University Press 2008.

different geographic regions, habitats and climate zones, from tropical to temperate, from desert to woodlands and from high to low latitude (15–35° latitude; 122–155° longitude). They are seasonal breeders with a long breeding season (from June to January, across Australia) and within specific ecological contexts (Schodde and Mason, 1999).

In many ways, magpies behave more like tropical birds. They form long-lasting pairs, even duet and cooperate in certain circumstances and, unlike zebra finches, have no specific song for breeding enticements. Magpies are vocally monomorphic and both sexes sing throughout the year. Significantly, "song" is not motivated by reproduction and not used during the breeding season. In fact, magpie vocalizations in general decline during the breeding season (see below). The characteristics of the magpie's vocalizations including lifelong plasticity, a large repertoire, "referential" signaling (Kaplan, 2006a, 2007) and the protracted stages of vocal development, make the magpie an excellent alternative songbird model. Unfortunately, publications on this species' singing behavior are relatively limited (e.g. Brown *et al.* 1988; Brown and Farabaugh, 1991; Kaplan, 2004, 2005a, b, 2006a, b; Sanderson and Crouch, 1993). One aim of the present review is to make that behavior better known to the birdsong community both in order to widen theoretical debates and to argue the case for the magpie as a useful song-learning model.

A BRIEF SURVEY OF THE MAGPIE VOCAL REPERTOIRE

Magpie males do not have a breeding song, hence 'song' is not a male reproductive behavior. Both males and females, like any other accomplished singers, have several classes of vocalizations (Jurisevic and Sanderson, 1994, 1998; Kaplan, 2006a). The class of vocalization that is the most highly variable and is the closest to "song," as typically defined, is the category of warbling. Warbling is the predominant vocalization of magpies; its sound defines the Australian bush and can be heard across the entire Australian continent. Another form of vocalization, carolling, is most closely associated with territorial defense (Carrick, 1972), but has additional functions of pair or group bonding (Kaplan, 2005a).

Both forms of vocalization (see Figure 4.1) have been traced throughout the seasons (Kaplan, 2005a). Carolling shows strong seasonal variations in its frequency of production (Figure 4.2). In August, the female incubates the eggs on her own and this is the only time when carolling events (still at a high rate) derive exclusively from the male. Plasma testosterone rises substantially in male magpies at the onset of the breeding season (Schmidt *et al.*, 1991). Hence, at this time only, high rates of carolling are concurrent with and possibly related to reproductive behavior. Warbling also undergoes marked seasonal change, declining to extremely low levels around breeding time and throughout the entire period of incubation and parent feeding of fledglings (for 3 months post fledging) (Figure 4.3).

Warbling in males and females has often been inappropriately analogized to "subsong." However, it is not a preliminary rehearsal state that leads to crystallized song, but a type of variable and improvised song that is developed by juveniles, is practiced within the nest (Kaplan, 2005a) and then remains the dominant form of vocalization of both sexes for the rest of their lives (see Figure 4.1). The vast number of warbling elements suggests a specific rehearsal process of "overlearning," in which many more elements are initially learned and included in song than will later remain in final song (Nelson and Marler, 1994). However, no such "crystallization" takes place and large numbers of these elements persist into adulthood.

Another characteristic of warbling is the occurrence of mimicry exclusively within warbling sequences (Kaplan, 1999, 2003). Magpies mimic extensively in the wild (other birds, vertebrates, including horses, dogs and even human voices) suggesting a high degree of plasticity within the vocal learning process.

Warbling is not linked to specific activities or to interactions with conspecifics. Individual birds (females slightly more than males) develop individually specific elements without adult tutors, producing sequences that are extremely variable. A single warbling repertoire of an individual female may contain as many as 740 elements. The individual singer is thus recognizable by its specific vocalization (possibly helped by vocal resonance; Suthers, 1994). Twenty-five percent of the magpie's warbling is shared with neighbors (Brown *et al.*, 1988; Brown and Farabaugh, 1991). This sharing of "song" was measured after the magpies in their study group had established permanent territories (Brown *et al.*, 1988), which usually occurs at or about 5 years

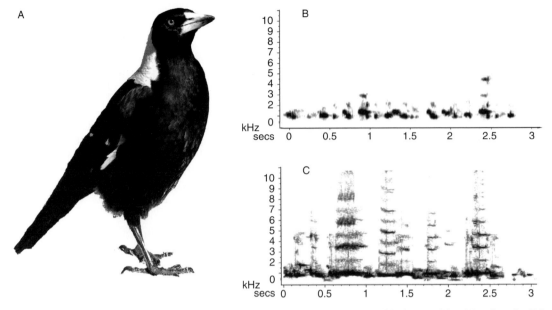

Figure 4.1 (A) Image of an adult magpie (*Gymnorhina tibicen tibicen*), black-backed, one of the largest of the eight subspecies. It is found on the east coast of Australia (from Brisbane in Queensland to the Victorian border in the south) and throughout New South Wales and the Australian Capital Territory. The weight of the adult male is about 400–480 g, size 40–44 cm. (B) A typical warbling sequence. Note the minipauses within the sequence. Duration of warbling sequences is highly variable, usually up to ten minutes per bout but may extend to an hour or more for one bout. (C) Carolling phrase used in territorial defense and in pair and group bonding. In this sample two magpies are carolling: the one closer to the speakers has louder carols, up to 8 kHz, three calls located on the x-axis between 0.5–1 s, 1.1–1.5 s and 2.2–2.7 s; the fainter calls, up to 6.5 kHz, belong to the second bird more distant from the microphone (see x-axis at 0–0.5 s and 1.4–2.2 s).

of age. These observations suggest that magpies, unlike many other songbirds, possess a high level of plasticity for vocal learning based on improvisation (Kaplan, 2006b) and may retain such vocal plasticity throughout life. Such a strategy would be especially functional for a species that is likely to experience one or even several enforced territorial changes during its lifetime (Carrick, 1972), each of which would require a repertoire change or repertoire adaptation to neighbors.

While its vocalizations differ in many respects from those of birds such as zebra finches and canaries, the magpie's brain contains the same forebrain structures (HVC, Area X, RA, LMAN and MMAN; Deng et al., 2001) as have been identified and linked to song control in the canary and the zebra finch (Nottebohm and Arnold, 1976; Nottebohm et al., 1982; Bottjer et al., 1989; Fortune and Margoliash, 1995). These "song system" nuclei are about equal in volume in both males and females (Figure 4.4) and there are no sex differences in cell types in HVC, RA or Area X. All of these nuclei were well developed in the juveniles sampled aged two to three months post fledging. Comparison of adult and juvenile structures reveals some interesting differences including a decrease in size by some 19% in Area X from juvenile to adulthood (Deng et al., 2001). A similar relationship between age and the size of Area X was reported for male Cassin's finches (*Carpodacus cassinii*) by MacDougall-Shackleton et al. (2005).

Although the magpie's singing behavior is not obviously linked to reproduction and differs in multiple respects from that of the "classical" song model, it is in fact representative of a very large and diverse range of songbird species, including tropical birds. In this sense, it is certainly as appropriate a model as zebra finch song. One could even argue that some of its features make it a more appropriate model for studies of the development of human language. First, in neither humans nor magpies is vocal learning an exclusively reproductive behavior. Second, magpies, like humans,

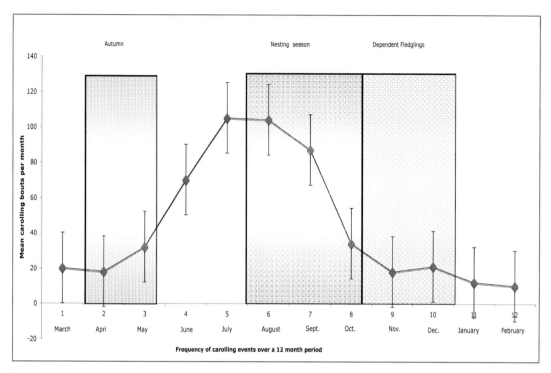

Figure 4.2 Seasonal variations of carolling: mean incidence of carolling over a year, ± standard errors. Carolling remains rather constant from November to April (late spring, summer, early autumn) but increases sharply prior to the onset of the breeding season reaching a peak mid winter (July/August). Means and standard errors are plotted. Carolling declines while the nestlings are raised (September/October) and steadily increases in frequency through late autumn and early winter (May/June) when offspring have fledged and they feed independently (June/July). Note that mid summer records the lowest call rate and mid winter the highest. The gray areas mark autumn (dispersal of young from previous breeding season, and the nesting season in late winter/early spring in the temperate climate of the New England Tableland, Eastern Australia).

have long stages of dependency and do not form permanent relationships for many years post sexual maturity. Third, magpies are likely to have life-long vocal plasticity and, fourth, their vocal behavior appears to follow developmental stages similar to that of humans (Kaplan, 2005b). In the following sections I explore some of the evolutionary, ecological and behavioral considerations that can help provide a framework for the evolution and function of magpie song.

TERRITORIALITY

The biological significance of territories in birds has long been a subject of interest. Its suggested functions include male competition or female selection of males (e.g. Hinde, 1956). However, the processes by which territory is actually established and secured are still not well understood (Stamps and Krishnan, 1999; Dunham et al., 1995). Catchpole (1982) developed a "diagnostic checklist" on the relationship between song, territory and mate selection styles. In northern hemisphere species, *small* syllable repertoires are associated with sedentary avian species and large repertoires with non-territorial, often even migratory species. This interrelationship makes sense in the context of strong seasonal variations and among sexually dimorphic species.

Magpie vocal behavior does not fit easily into this model. In magpies, territoriality is associated with a *large* repertoire and lack of stereotypy. Yet Australian magpies are sedentary and tend not to disperse very far from their home territory (Baker et al. 2000). Males usually remain closer to their natal environment (i.e. are philopatric), but in magpies the reverse is the case. Females are more philopatric while males disperse

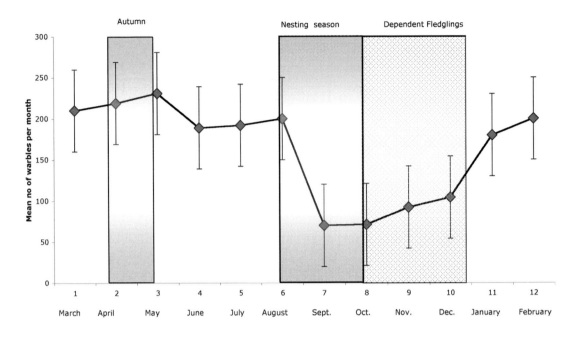

Figure 4.3 Seasonal variations of warbles: mean number of warbles per month recorded from three different magpie groups (5–7 individuals per group) in three consecutive years and one repeat measure of one group in a fourth year (4 day per week sampling of 8 hours per day). Note that territories are large (between 10–100 hectares per group) and it is not possible to record vocalizations from more than one group at any given time. Mean and standard errors are indicated. The seasonal variations of warbles are marked. The decrease of warbles during the breeding season (nestling and fledgling stages of offspring) is significant.

further afield (Veltman and Carrick, 1990). Moreover, both males and females sing and they do so more outside the breeding season, particularly in summer and autumn, well after the juveniles have fledged and have become independent feeders. These characteristics are inconsistent with a role for song in male competition or female selection (West and King, 1988; Langmore, 1998).

Convergence of song in male and female

The existence of female song and of female song repertoires equal to or greater than that of males is characteristic of many tropical species (Riebel 2004). Slater and Mann (2004) proposed two main hypotheses applied to tropical songbirds: (1) the need for mutual stimulation to achieve breeding synchrony in a relatively aseasonal environment and (2) sex role convergence arising from more long-term relationships and greater fidelity than is usual among north temperate species. Given that local environmental cues may be just as effective as triggering stimuli as photoperiod (Moore et al., 2004b) mutual stimulation does not seem to be critical. The hypothesis of sex-role convergence in bonded pairs is supported by studies demonstrating that long-time bonded pairs such as the magpie lark (Hall, 2000; Hall and Magrath, 2000), the Eastern whip bird, *Psophodes olivaceus* (Watson, 1969; Davidson and Langmore, 1991) and others show plumage convergence and often also vocal convergence (same repertoire size and type). On the other hand, the monomorphic barbary dove (*Streptopelia risoria*) exhibits sex differences in vocal production despite the presence of the same song system structures in the brain of this species (Fusani et al., 1997).

Based on the presence of different neuronal phenotypes in the song system of a species which produces songs of similar complexity, Gahr et al. (1998) have argued that it may be misleading to assume a causal relation between sex difference in vocal behavior and in the size of brain areas associated with singing (see

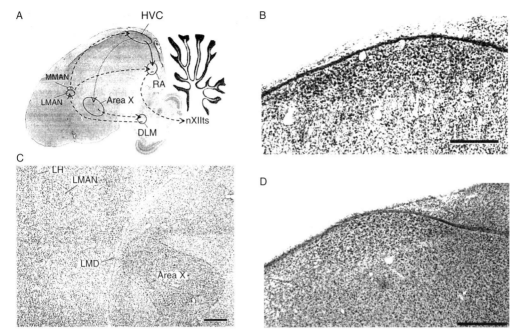

Figure 4.4 Mapping the song system of the Australian magpie. (A) Location of song nuclei identified in the forebrain of the Australian magpie, as seen in a longitudinal section. The dotted line from the high vocal centre (HVC) to Area X of the striatum refers to projections found in studies on zebra finches that could not be seen by DiI and DiA anterograde tracing in the magpie. Dashed line indicates connections seen between nuclei in the magpie, as also found in studies of the canary; MMAN is the medial magnocellular nucleus of the anterior nidopallium; RA, the robust nucleus of the arcopallium; nXIIts, the tracheosyringeal portion of the hypoglossal nucleus. (B) A transverse section of the rostral part of the pallium of a juvenile male magpie showing areas LMAN (lateral magnocellular nucleus of the anterior nidopallium), LH (lamina hyperstriatica), LMD (lamina medullaris dorsalis) and Area X. (C) HVC in cross section in adult female magpie. (D) comparable HVC section in adult male zebra finch. (D was kindly provided by P. Adret.) Bars in A to C indicate 500 μm and in D 100 μm.

Bolhuis this volume). On the other hand, in a species like the magpie, whose vocalizations are monomorphic, there are no obvious differences between males and females in the volume or cell morphology in HVC, RA or Area X (Deng et al., 2001). Long-term bonding, so far as has been studied – and more typical of southern hemisphere species – appears to lead to a convergence of male and female singing not only at the level of behavior but also in underlying brain structure.

CLIMATE AND THEORIES OF BIRDSONG

Almost all theories about birds and birdsong are based upon the behaviors of northern hemisphere birds so that birds whose behavior does not fit this model, e.g. birds of the tropics, are viewed as oddities (Martin, 1996). Yet recent hypotheses on the evolution of birds suggest that songbirds may have originated in Gondwana and, as they radiated, moved from there to the northern hemisphere (Boles, 1995). In fact, the southern hemisphere and tropics are far more bird rich and largely occur at lower latitudes than the bird poorer avian regions of the northern hemisphere, which include land masses at very high latitude (which have snow and ice in winters, largely unknown in the southern hemisphere). For this reason, I would argue that it is the current models that are aberrant; i.e. unrepresentative of the much broader range of breeding behaviors, including vocalization. I would argue further that the magpies' singing behavior reflects a lack of those constraints upon its breeding strategies imposed upon high-latitude birds by climate, but is constrained, instead, by the characteristics of the mini-environments in which breeding takes place.

For example, the importance of photoperiod in birdsong neurobiology (Smith et al., 1997b; Whitfield-Rucker and Cassone, 1996) derives from its role in the control of seasonal song production in sexually dimorphic birds (see, for example, Brenowitz, this volume), but only in species that breed in high latitudes (Moore et al., 2004c). In the tropics, by contrast, temperature remains relatively constant and photoperiod cannot be invoked to explain seasonal change in song production (Kunkel, 1974; Stutchbury and Morton, 2001). However, there is now substantial evidence (1) that the tropics, far from being unchanging, are actually relatively unpredictable environments (Ahumada, 2001; Bendix and Rafiqpoor, 2001), (2) that small environmental cues (rain, humidity levels) can have the same effect as photoperiod (Moore et al., 2004a, b), and, (3) that some changes can be independent of photoperiod in tropical birds (Moore et al., 2006). Indeed, Hau (2001) proposed to use tropical birds as model systems specifically to account for variable environments. Interestingly, Moore et al. (2006) have argued that the ability to respond to a multitude of environmental cues might be part of an ancestral condition, and that the subsequent obligate photoperiodism in high-latitude congeners could reflect a loss of flexibility in response to environmental signals.

It is not clear whether magpies evolved in the tropics and only later colonized the more temperate climate zones of Australia but they show some of the characteristics of tropical birds. Climatic unpredictability (particularly with respect to rainfall) characterizes most of the Australian continent and may have led to the same adaptations in breeding behaviors, including vocalization, as are found in tropical birds. For example, magpie subpopulations in different regions may, like the rufous-collared sparrows (*Zonotrichia capensis*; Moore et al., 2004c), adjust their breeding season to local environmental cues.

Two reproductive strategies associated with tropical species may be particularly relevant to magpies. One is cooperative breeding (Brown, 1987) and the other duetting (Thorpe, 1972). Magpies occasionally breed cooperatively and they occasionally duet, making them similar to many tropical species and to other Australian avian species.

Cooperative breeding

While the predominant model in birdsong neurobiology is a competitive one, a substantial number of species, especially among topical birds (Fry, 1972), do not compete either intra specifically or between the sexes. Indeed, Australia, across several climate zones, has an unusually high share of cooperative breeders among passerines worldwide (Boland, 1998; Cockburn, 1996). In fact, according to Brown (1987), most cooperative breeders occur in the tropics and in Australia (67 of the 222 cooperating species in Australia alone), a remarkable accumulation in a region (Ford, 1989; Stacey and Koenig, 1990).

Magpies may or may not be cooperative breeders (Veltman, 1984, 1989; Farabaugh et al., 1992), depending on region. Breeding arrangements in magpies are a mix of monogamous unassisted pair birds, some polygamy, some groups with helpers and some communally breeding groups including offspring that have not left the natal territory (Veltman and Carrick, 1990). Unlike many other Australian cooperative breeders (Boland et al., 1997; Boland, 1998; Cockburn, 1996), magpies are not always cooperative but their social system is flexible enough for rearrangements evident in variability of group sizes and in dispersal patterns (Baker et al., 2000). It is clear that cooperation in breeding (Dunn and Cockburn, 1999; Dugatkin, 1997; Poiani and Pagel, 1997) is an important topic within birdsong. One would expect to see some differences in vocal behavior between those (male) songbirds that compete (Morton, 1996; Heimovics and Riters, 2005) versus those that cooperate (male and female alike).

Duetting

Farabaugh (1982) has described duetting as consisting of overlapping bouts of sound of two animals. Duetting has two components, simultaneous singing and antiphonal song (one singing the other replying but usually with a short latency; Catchpole and Slater, 1995). Farabaugh (1982) pointed out that when comparing geographic areas with similar numbers of songbirds, as in North America (temperate/high latitude) and Panama (tropical), 13% duet in the tropical region (over 100 species) while only 3% (23) duet in the temperate region. Although magpies are not a true duetting species, magpie pairs may alternate in carolling, though carolling may involve several males and females as a chorus after acts of cooperative territorial defense.

Importantly, some of these duetting species also show convergence of repertoire (Rogers and Kaplan, 2000; Kaplan and Rogers, 2001) and, so far as is known, vocally monomorphic characteristics usually also preclude mate choice by song, as in the magpie and the magpie lark (Hall, 2000). Functions of duetting may include mate guarding either by males or females (Sonnenschein and Reyer, 1983; Slater et al., 2002) and pair bonding (Hall and Magrath, 2000; Hall, 2004; Mann et al., 2003). In the bay wren (*Thryothorus nigricapillus*), according to Brenowitz and Arnold (1986), females sing as much as males and have a similar repertoire size. Like the magpies, these birds show no significant difference between HVC and RA sizes (Brenowitz et al., 1985; Brenowitz and Arnold, 1986; Deng et al., 2001).

CONCLUSION

Although the magpie's singing behavior is not obviously linked to reproduction and differs in multiple respects from that of the "classical" song model, its forebrain contains the nuclei typical of those models. Moreover, it is representative of a very large and diverse range of songbird species, including tropical birds. I suggest that many of the "atypical" characteristics of magpie vocalizations could reflect the conservation of earlier adaptations to geographical and climatic conditions including extremely long isolation and climatic conditions that allowed for extended breeding seasons and few time constraints in mate selection. Moreover, long-term partnerships (and cooperative breeding) reduce the need for ongoing vocal competition, as is the case in those species of songbird that have to find breeding partners every season, but, perhaps, increase the need for mate guarding as seen in pair-bonded duetting species. The Australian magpie may have conserved vocal and auditory features that had evolved prior to the constraints imposed by changing climate, high latitude and dramatic differences in seasons, including the absence of an obvious link to reproductive behavior.

On the other hand, while there may be no functional link between one kind of song, "warbling" and reproductive behavior, there is some evidence for such a link for "carolling." Perhaps the highly defined seasons of the northern hemisphere forced a change of strategy. Over evolutionary time, two song types resembling carolling and warbling could have been functionally combined to create a song that became more closely tied to the breeding season (particularly for migratory birds) as the most efficient and reliable strategy for securing a mate.

Whatever its evolutionary history, the singing behavior of the Australian magpie is certainly an appropriate model for birdsong neurobiology. Indeed, some of its features make it a particularly appropriate model for studies of the development of human language. Magpies, like humans, have long stages of dependency and do not form permanent relationships for many years post sexual maturity. Their vocal behavior appears to follow developmental stages similar to that of humans (Kaplan, 2005b) and its plasticity may be life long. Finally, as noted earlier, in neither humans nor magpies is vocal learning an exclusively reproductive behavior.

And yet, based upon current song models, the rich and varied repertoire of magpie vocalizations might not qualify as "song." If the present chapter leads to reconsideration and broadening of our definition of birdsong, it will have achieved its purpose.

5 • Songbirds and the revised avian brain nomenclature

Anton Reiner, David J. Perkel, Claudio V. Mello, and Erich D. Jarvis

A BRIEF HISTORY OF AVIAN TELENCEPHALIC NOMENCLATURE

The advent of improved techniques for cutting and staining brain tissue resulted in a wealth of new knowledge on brain structure in various vertebrate species at the turn of the nineteenth century and the beginning of the twentieth century (Northcutt, 2001). Based on his interpretation of such material, Ludwig Edinger formulated a theory of cerebral evolution (Edinger, 1885, 1908; Edinger et al., 1903) that, as further developed by his colleague C. U. Ariëns-Kappers (1922, 1928) and subsequently refined and widely promulgated in Ariëns-Kappers et al. (1936), became the dominant view, and led to an avian telencephalic nomenclature that continued to be used into the early years of the twenty-first century (Figure 5.1, upper panel). According to this view, birds and mammals inherited from their fish ancestors, via the fish to amphibian to reptile lineage, an old basal ganglia structure that was called the paleostriatum (old striatum; corresponding largely to the globus pallidus of mammals), and a newer structure from their reptilian ancestors that Ariëns-Kappers called the neostriatum (new striatum; including most of the caudate and putamen in mammals). Reptiles were thought to have elaborated the paleostriatum further into two distinct parts, one Ariëns-Kappers called the paleostriatum primitivum (comparable to a primitive mammalian globus pallidus) and another part he called the paleostriatum augmentatum (i.e. an augmentation of globus pallidus), and both subdivisions were assumed to have been passed on to birds. Similarly, the neostriatum was also thought to have become enlarged in birds, and have given rise to a novel overlying territory that Edinger et al. (1903) and Ariëns-Kappers (1922, 1928) called the hyperstriatum, in the belief that it was entirely "striatal" in nature and a hypertrophy of the neostriatum. Thus by this view the avian telencephalon was thought to consist nearly entirely of an enlarged basal ganglia (i.e. what are now commonly called caudate, putamen and globus pallidus in mammals; Figure 5.1, upper panel). Finally, mammals, birds and reptiles were also thought to have inherited an additional subcortical structure that Edinger and Ariëns-Kappers called the archistriatum (in the belief that it was also part of the basal ganglia) from their amphibian ancestors. This brain region in mammals is now called the amygdala, and it is no longer commonly regarded as entirely part of the basal ganglia.

In contrast to the basal ganglia expansion thought to characterize birds, mammals were thought to have expanded the upper, outer part of the telencephalon (the pallium) into a six-layered cortex from a small dorsal cortical region present in the reptile ancestors of mammals (Ariëns-Kappers, 1909, 1922, 1928; Edinger, 1885, 1908; Edinger et al., 1903). The novel cortical region in mammals was referred to as neocortex, to distinguish it from the presumed older cortices represented by the olfactory cortex (which was called paleocortex) and hippocampus (which they called archicortex). Ariëns-Kappers et al. (1936) slightly modified the position of Ariëns-Kappers' earlier works by concluding that a small upper part of the hyperstriatum (largely corresponding to what we now call the Wulst) provided birds with a meager pallial territory comparable to mammalian neocortex. Nonetheless, the view espoused by Ariëns-Kappers et al. (1936) and by other influential authors (Johnston, 1923; Craigie, 1932; Herrick, 1948, 1956) was that the avian telencephalon consisted mainly of greatly expanded basal ganglia. Except for a dissenting minority (Rose, 1914; Kuhlenbeck, 1938; Källén, 1953), this accretionary theory of vertebrate brain evolution became the prevailing view for the first two thirds of the twentieth century. This led to the predominant use of the terms neostriatum, archistriatum and hyperstriatum to refer to the major sectors of the avian telencephalon above the so-called paleostriatum.

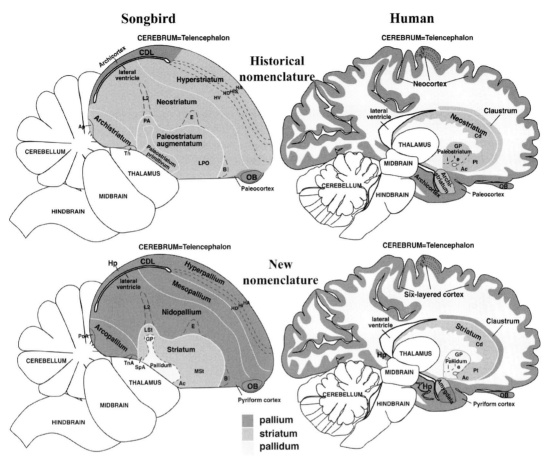

Figure 5.1 (Top) Classical view of avian and mammalian brain relationships according to the historical nomenclature. Although past authors had differing opinions as to which brain regions are part of the pallium versus subpallium, the images are shaded according to the meaning of the actual names given to these brain regions. White lines represent laminae, cell-sparse regions separating brain subdivisions. Large white areas in the human cerebrum are the fiber bundles making up the white matter. Dashed lines divide regions that differ by cytoarchitecture. The abbreviations PA and LPO designate regions as defined by Karten and Hodos (1967), while the spelled-out term paleostriatum augmentatum designates this entire area as defined by Ariëns Kappers *et al.* (1936). (Bottom) Modern view of avian and mammalian brain relationships according to the new nomenclature. In birds, the lateral ventricle is located in the dorsal part of the pallium, whereas in mammals much of the ventricle is located near the border of the pallium with the subpallium. Abbreviations, classical view: Ac, accumbens; Ap, posterior archistriatum; B, nucleus basalis; Cd, caudate nucleus; CDL, dorsal lateral corticoid area; E, ectostriatum; GP, globus pallidus (i, internal segment; e, external segment); HA, hyperstriatum accessorium; HIS, hyperstriatum intercalatum superior; HD, hyperstriatum dorsale; HV, hyperstriatum ventrale; L2, Field L2; LPO, lobus parolfactorius; OB, olfactory bulb; PA, paleostriatum augmentatum; Pt, putamen; Tn, nucleus taeniae. Abbreviations, modern view where different from top panel: E, entopallium; B, basorostralis; HA, hyperpallium apicale; HI, hyperpallium intercalatum; HD, hyperpallium densocellulare; Hp, hippocampus; LSt, lateral striatum; MSt, medial striatum; PoA, posterior pallial amygdala; TnA, nucleus taeniae of the amygdala; SpA, subpallial amygdala.

NOMENCLATURE AND THE PROBLEM OF HOMOLOGY

As commonly used in biology, structures in two or more species are considered to be homologous if they are thought to derive from the same antecedent structure in their common ancestor (Campbell and Hodos, 1970). Major difficulties arise, however, in identifying homologous brain structures because brain, being a soft tissue, does not fossilize in sufficient detail to make it possible to use the fossil record to trace the natural history of given brain structures. The only remaining approach that can be taken is comparing a variety of features of the structures in question in extant species, including embryological origin, location within the adult brain, afferent and efferent connections, and neurochemical phenotype. In the simplest case, if candidate avian and mammalian homologues (to use sample groups of present interest) arise from the same developmental primordium and have similar adult features, and if a similar structure is found in extant reptiles, then a convincing case can be made that the stem amniote common ancestor had an equivalent structure. If, on the other hand, the structures are dissimilar in birds and mammals and/or a comparable structure is not evident in living reptiles, then the compared structures in birds and mammals cannot be said to be demonstrably homologous. It also cannot be automatically said with authority, however, that two morphologically dissimilar structures in birds and mammals are not homologous, since homologous structures can evolve different morphologies (Campbell and Hodos, 1970). Nonetheless, if the dissimilarities are numerous and living reptiles clearly lack a structure resembling either the compared structure in mammals or the compared structure in birds, then the conclusion that the compared structures in birds and mammals are not homologous is the most likely interpretation.

Terms, such as "analogous," "functionally analogous," or "functionally homologous" have also been used in comparing brain structures. The first two terms mean the same and refer to a circumstance in which structures in different species perform the same function (e.g. bird wings and insect wings), even if they are morphologically different and have evolved independently (Carcraft, 1967; Campbell and Hodos, 1970; Lauder, 1986; Schmitt, 1995). "Analogous" would be the appropriate word to use in this context, and some authors consider the term only to refer to structures of the same function that are independently evolved (Carcraft, 1967; Campbell and Hodos, 1970). Note that bat wings and bird wings are analogous as wings but not homologous, since the wings subserve flight in both, but the wingedness of the forelimbs was independently evolved. Nonetheless, the forelimbs of bats and birds are homologous as forelimbs, since both inherited their forelimbs from their stem amniote common ancestor.

The term "functionally analogous" is redundant with the term "analogous," the latter already implying a functional comparison. The term "functionally homologous" can be ambiguous, meant either as a synonym for analogous (which would be an incorrect use of the word homologous) or to suggest a common origin of a function in two or more species from a function in the common ancestor. The latter misapplies a term commonly used to refer to common ancestry of a morphological entity, i.e. "homologous," to a functional context. The complexities of trying to identify homology at the functional level have been discussed elsewhere (Hodos, 1974, 1976; Lauder, 1986; Striedter and Northcutt, 1991).

IMPLICATIONS FOR THE STUDY OF SONGBIRD NEUROBIOLOGY

While research on all avian species was affected by the outdated terminology for the avian telencephalon, the confusion was especially acute for those studying songbirds, for two major reasons. First, researchers on song control mechanisms now constitute the largest single group within the avian brain research community. Secondly, several major cell groups involved in song perception, learning or production are located within the part of the brain that in birds has been called the neostriatum. These findings have been habitually misinterpreted by researchers on the mammalian brain, for whom the term *neostriatum* refers to part of the basal ganglia, as pertaining to the functioning of the basal ganglia.

As neurobiologists have gained deeper insights into the evolution, development and function of avian and mammalian brains, it has become clear that the accretionary theory of vertebrate telencephalic evolution is incorrect (Parent, 1997; Striedter, 1997; Swanson, 2000; Northcutt, 2001). Being flawed, the homologies implied by the classical nomenclature have greatly hindered communication between avian and mammalian brain specialists by perpetuating the view that the telencephalon in birds differs qualitatively in structure and function from that in mammals. In particular, the presumed necessity of neocortex for adaptive behavior and higher order cognition (Herrick, 1956) and the presumed absence of neocortex in birds have continued to make many believe that birds lack such behavioral abilities. Since the basal ganglia were thought to control instinctive motor behavior and the

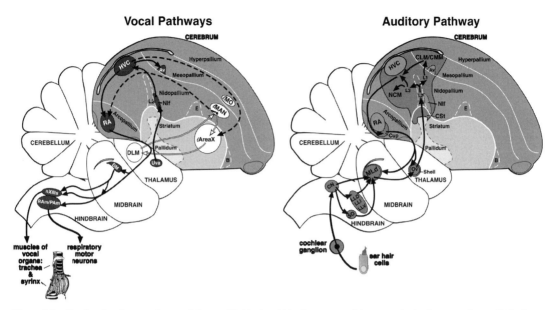

Figure 5.2 Vocal and auditory pathways of the songbird brain within the context of the new avian brain nomenclature. Only the most prominent and/or most studied projections are indicated. For the vocal pathways (left), black arrows show connections of the components (in dark gray) of the posterior vocal pathway, white arrows show connections of the components (in white) of the anterior forebrain pathway, and dashed lines connections between the two pathways. For the auditory pathway (right), most of the hindbrain connectivity is extrapolated from nonsongbird species. For clarity, only the lateral part of the anterior vocal pathway is shown, and the connection from Uva to HVC and reciprocal connections in the pallial auditory areas are not indicated. Note that the NCM and CMM are shown for schematic purposes, as they actually lie in a sagittal plane medial to that depicted, and the pathway from NCM to CMM is not depicted.

Abbreviations: Av, avalanche; CLM, caudal lateral mesopallium; CMM, caudal medial mesopallium; CN, cochlear nucleus; CSt, caudal striatum; DM, dorsal medial nucleus; DLM, dorsal lateral nucleus of the medial thalamus; E, entopallium; B, basorostralis; HVC (no formal name other than HVC); LLD, lateral lemniscus, dorsal nucleus; LLI, lateral lemniscus, intermediate nucleus; LLV, lateral lemniscus, ventral nucleus; MLd, dorsal lateral nucleus of the mesencephalon; LMAN-lateral magnocellular nucleus of the anterior nidopallium; Area X, Area X of the medial striatum; MO, oval nucleus of the mesopallium; NCM, caudal medial nidopallium; NIf, nucleus interface of the nidopallium; nXIIts, nucleus XII, tracheosyringeal part; Ov, ovoidalis; PAm, paraambiguus; RAm, retroambiguus; RA, robust nucleus of the arcopallium; SO, superior olive; Uva, nucleus uvaeformis.

avian telencephalon was thought to be largely a hypertrophied basal ganglia, all complex behavior in birds had widely been thought to be instinctive (Edinger, 1908; Herrick, 1956). As a result of the misconceptions abetted by the Ariëns-Kappers-based terminology, the relevance of the many findings on the avian brain to understanding the functioning of the mammalian brain has been obscured.

It is now evident that birds are not uniformly impoverished in their adaptive learning skills. Songbirds, parrots, and hummingbirds show vocal learning abilities not paralleled by any mammals other than humans and cetaceans (Thorpe, 1951; Marler, 1970a; Baptista and Schuchmann, 1990; Hile et al., 2000; Jarvis et al., 2000; Pepperberg, 2002). Crows, members of the oscine songbird family, show the ability to make and use tools (Weir et al., 2002; Hunt and Gray, 2003), and parrots are capable of learning to communicate with human words and show cognitive skills otherwise evident only in apes and cetaceans among nonhuman species (Pepperberg, 2002). In parallel with the growing awareness of avian behavioral sophistication, it has become clear that the neural substrate for such behavior is not a hypertrophied basal ganglia but the same general brain region used for such tasks as in mammals (i.e. the pallium), albeit without the laminar morphology characteristic of mammalian neocortex, in combination with a smaller basal ganglia region (Karten, 1969, 1991; Nottebohm et al., 1976;

Paton et al., 1981; Striedter, 1994; Durand et al., 1997; Gahr, 2000; Jarvis et al., 2000; Jarvis and Mello, 2000).

To address the problems inherent to the old terminology in the light of our new knowledge, a revised nomenclature for the avian brain was adopted by an international and multidisciplinary group of neuroscientists meeting at Duke University in July 2002.

Figure 5.2 presents schematic drawings illustrating the application of the revised nomenclature to the vocal and auditory pathways of songbirds. A fuller discussion of these issues and a description of the new nomenclature may be found in Reiner et al. (2004a), in Zeigler and Marler (2004), and in Jarvis et al. (2005a).

6 • The songbird brain in comparative perspective

Michael A. Farries and David J. Perkel

INTRODUCTION

The last 50 years have seen the rise of the songbird vocal system from a curiosity studied by only a handful of neurobiologists to a major area of research in neuroscience. Birdsong has been proposed as a model system for the study of such diverse problems as communication, neural development, plasticity and learning, behavioral endocrinology, and motor control. For neuroscientists, birdsong is attractive as a model system because it is a naturally learned behavior that is easily recorded and quantified, and its neural substrate consists of discrete brain structures that appear to be functionally specific to song and are quite distinct from surrounding brain regions. Yet the very distinctiveness of the oscine (i.e. songbird) song system endangers its utility as a model system whose features can be readily generalized to neural systems in other vertebrate taxa, particularly mammals. If we are to make the most effective use of birdsong as a model system, we must understand how it is related to the avian brain as a whole and how avian brain organization, in turn, compares with the organization of the mammalian brain.

This chapter provides an overview of song circuitry in oscine songbirds and its relationship to vertebrate brain organization. Comparison between mammalian and avian forebrains will provide a framework for understanding how the song system relates to the overall structure of the avian brain, and will suggest how data obtained in the song system can be applied to mammals (and vice versa). We also compare song circuitry to structures in surrounding regions of the oscine forebrain and to corresponding regions in nonoscine birds, with a view toward clarifying the relation between the song system and other avian neural circuits. One aim of the chapter is to show how much of the organization of the song system can be accounted for simply by inheritance from the nonoscine structures from which it evolved. Its primary aim is to provide a firm comparative foundation for the continuing contribution of birdsong neurobiology to our understanding of vertebrate brain function.

OVERVIEW OF THE BRAINS OF BIRDS AND MAMMALS

The central nervous systems (CNS) of all vertebrates share a common organizational plan, consisting of four major subdivisions along the rostrocaudal axis (see, for example, Butler and Hodos, 1996). For three of these divisions, the basic vertebrate organization is highly conserved. The caudalmost division is the spinal cord, characterized by ventrally located motor neurons and motor nerve roots, and a dorsal sensory zone containing fibers of sensory neurons that enter the spinal cord through dorsal nerve roots and whose somata are located in the dorsal root ganglia. Immediately rostral to the spinal cord is the hindbrain, which resembles the spinal cord in many ways but has more specialized and elaborate cranial motor and sensory systems associated with organs in the head, as well as major viscerosensory and visceromotor centers. The rostrally adjacent midbrain receives inputs from several sensory systems, including the retina, and can integrate these inputs to generate relatively complicated and well-coordinated behaviors. The midbrain also contains monoaminergic and cholinergic systems that innervate the rostralmost region of the CNS, the forebrain. It is when we reach the fourth division, the forebrain, that substantial differences between mammals and birds begin to appear.

The vertebrate forebrain is divided into the telencephalon and diencephalon, and in both birds and mammals the diencephalon contains regions labeled hypothalamus and thalamus. Even a cursory inspection of the avian diencephalon suggests differences between birds and mammals and such differences only become more pronounced when one examines the telencephalon. The vertebrate telencephalon consists of the olfactory bulb and two universally recognized subdivisions:

Neuroscience of Birdsong, ed. H. Philip Zeigler and Peter Marler. Published by Cambridge University Press. © Cambridge University Press 2008.

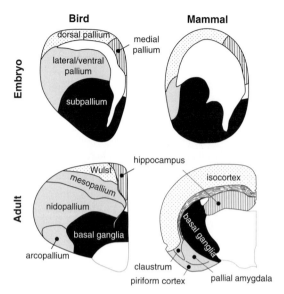

Figure 6.1 Schematic coronal sections through the telencephala of mammals and birds at embryonic and adult stages, illustrating their basic subdivisions. Black areas denote subpallium, while the gray, stipples and striped areas are lateral, dorsal and medial pallial areas, respectively. This figure presents only one view among several on the relationship between the embryonic pallial subdivisions and structures in the adult mammalian brain. In particular, some or all of the claustrum may be effectively a part of isocortex, and thus would lie within the dorsal pallium in this scheme. Furthermore, some hypotheses hold that isocortex includes some lateral pallium, either through cell migration or because the border between lateral and dorsal pallia in adults lies within the isocortex.

a dorsally situated pallium, containing mainly glutamatergic projection neurons, and a ventrally located subpallium, containing mainly GABAergic and cholinergic projection neurons (Figure 6.1). Although these subdivisions are presumed to exist in all vertebrates, identifying homologous regions within these broad divisions can be problematic. Designation of structures in the avian and mammalian brain as "homologous" implies inheritance from a common ancestor, and thus carries a heavy burden of proof (see Reiner et al., this volume). However, since nervous systems leave no fossil record, questions of homology cannot be resolved definitively. Nonetheless, some parallels between avian and mammalian brains may be derived from comparisons of brains of contemporary species using similarities in location, connectivity and embryonic development. This task is complicated by the fact that

telencephala of birds and mammals are quite different in their gross appearance: the mammalian telencephalon is dominated by a laminated isocortex (a major component of the mammalian pallium) that is clearly separated from the underlying basal ganglia (subpallium) by a thick band of myelinated axons ("white matter"), whereas the entire avian telencephalon has a pseudolaminar arrangement of wider cell masses that bear a superficial resemblance to the mammalian basal ganglia. Indeed, for many years the orthodox view held that most of the avian telencephalon actually *is* basal ganglia, with a comparatively tiny pallium (for review, see Reiner et al., 2004a; Jarvis et al., 2005a). This theory became untenable as more concrete information about the anatomy and histochemistry of the avian brain accumulated. But identifying homologous regions in the avian and mammalian telencephalon remains challenging and contentious.

Of the two major subdivisions of the telencephalon, the subpallium presents the fewest difficulties. The subpallium consists of the basal ganglia and basal forebrain, which are now known to comprise only about one third of the avian telencephalon, as they do in mammals. Modern tract-tracing and histochemical studies have shown that the organization of the subpallium is fairly well conserved across amniotes (Figure 6.2). In both birds and mammals, the input structure of the basal ganglia, the striatum, receives topographically organized glutamatergic input from almost the entire pallium and is heavily innervated by midbrain dopaminergic neurons. The striatum makes inhibitory GABAergic projections to the globus pallidus and substantia nigra, output structures of the basal ganglia, which are also reciprocally connected to the subthalamic nucleus. These output structures make GABAergic projections to parts of the thalamus and brainstem, through which the basal ganglia influence the rest of the brain (the substantia nigra is actually a part of the midbrain, and also contains dopaminergic neurons that project far more widely). Although the organization of the amniote basal ganglia is highly conserved overall, there are some significant differences between mammals and birds. First, the mammalian globus pallidus, unlike its avian counterpart, is divided into "internal" and "external" segments with distinct connections and histochemical traits. Furthermore, parts of the striatum in at least some avian taxa bypass the globus pallidus and project directly to the thalamus,

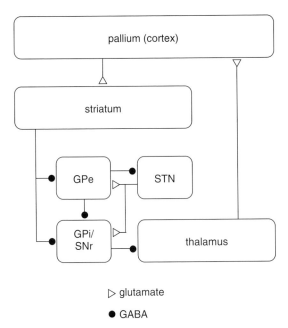

Figure 6.2 The amniote basal ganglia. The canonical basal ganglia circuit consists of a glutamatergic projection from the pallium to the striatum, a GABAergic projection from the striatum to the globus pallidus (GP) and substantia nigra pars reticulata (SNr), and a GABAergic projection to the brainstem (not shown) and thalamus. In mammals, the GP is divided into internal and external segments (GPi and GPe) with distinct connections. This division is not found in birds, but the avian GP may contain distinct cell types with anatomical and histochemical differences corresponding to this division.

a difference that is pertinent to the song system (see below).

In contrast to the subpallium, the identification of homologous structures in the avian and mammalian pallia remains contentious. The difficulties in comparing the pallia of birds and mammals stem largely from the great disparity in their organization, at least at a superficial level. In most mammalian species, the pallium consists largely of a six-layered isocortex. This is supplemented by a trilaminar hippocampus and piriform cortex, the claustrum (a thin band of subcortical gray matter whose functions remain poorly understood), and a group of small ventrolaterally located cell masses known as the amygdala (actually, the mammalian amygdala contains both pallial and subpallial elements). The avian pallium, on the other hand, contains little that could readily be called "cortex." Birds do have a hippocampus, but the rest of their pallium consists of four major subdivisions known in the current nomenclature as "hyperpallium", "mesopallium", "nidopallium", and "arcopallium" (see Reiner *et al.*, this volume). None of these structures has a clear counterpart in the mammalian brain, and this uncertainty has given rise to several competing proposals on their homology to mammalian structures. Such hypotheses are difficult to test based solely on the anatomy and histochemistry of adult forms.

Some difficulties can be mitigated by comparing the developmental precursors of the pallium, because the embryonic telencephala of mammals and birds are much more alike than their adult forms. However, this developmental comparison can resolve questions of adult homology only if the developmental primordia of adult structures are themselves readily identifiable, and this issue remains a matter of some debate. The vertebrate embryonic pallium is classically divided into three parts – dorsal, medial and lateral – and recent studies using several new (and highly conserved) molecular markers of vertebrate telencephalic development suggest that these pallial domains can be reliably identified in the embryos of both mammals and birds (Puelles *et al.*, 1999, 2000). In mammals, a common view holds that the dorsal pallium develops into the isocortex, the medial pallium becomes the hippocampal formation, and the lateral pallium gives rise to the piriform (olfactory) cortex, parts of the amygdala, and perhaps the claustrum.

The avian hyperpallium is derived from embryonic dorsal pallium, and resembles the mammalian isocortex in several respects. For example, caudal hyperpallium receives visual input from a proposed avian homologue of the dorsal lateral geniculate nucleus and is therefore anatomically, functionally, and topologically similar to mammalian visual cortex. The rostral hyperpallium, on the other hand, receives somatosensory input from the thalamus, has descending projections to the brainstem and spinal cord, and is reciprocally connected with the hypothalamus. Based on these connections, the rostral hyperpallium seems to combine features of mammalian somatosensory, motor and prefrontal cortices, further bolstering the impression that avian hyperpallium may be a simplified and relatively undifferentiated version of the mammalian isocortex (Medina and Reiner, 2000). All major proposals of avian–mammalian pallial homology agree that hyperpallium is homologous to at least a portion of mammalian isocortex. In contrast to mammalian isocortex, however, the hyperpallium in the best-studied avian species (pigeon, chicken, songbird, parrot) comprises a relatively small part of the avian pallium. The bulk of the pallium in these species is composed of the mesopallium, nidopallium, and arcopallium – structures for which there are no clear mammalian homologues. Some hypotheses claim that most of these regions, along with the hyperpallium, are collectively homologous to mammalian isocortex. One such view, championed by Karten, argues that the divisions of avian pallium are homologous to different layers of isocortex (e.g. Karten, 1991; Reiner et al., 2005), while an alternative hypothesis proposed by Butler and Molnar suggests that these divisions correspond to different regions of isocortex. In contrast to both of these hypotheses, Striedter, Puelles and collaborators argue that only the hyperpallium is homologous to mammalian isocortex, and that the other regions of avian pallium are greatly expanded homologues of the mammalian amygdala and claustrum (Striedter, 1997; Striedter and Beydler, 1997; Puelles et al., 2000). Regardless, we can say that these avian structures are organized quite differently from mammalian isocortex, yet serve similar functions, whether because of convergent evolution or common inheritance.

OVERVIEW OF THE OSCINE SONG SYSTEM

At the heart of the song system lies a motor pathway in the caudal telencephalon that innervates vocal and respiratory control centers in the brainstem (Figure 6.3). This pathway begins with nucleus HVC in the caudal nidopallium (HVC is the proper name, not an abbreviation). HVC innervates the robust nucleus of the arcopallium (RA) which, in turn, projects to a variety of brainstem motor centers concerned with vocalization and respiration, including a prominent projection to the motor neurons that innervate the syrinx, the principal vocal organ of songbirds (see Part II, this volume). Lesions of the motor pathway disrupt singing, and microstimulation within this pathway can evoke vocalizations, but not full-fledged song, and can disrupt or modulate singing that is already under way (Nottebohm et al., 1976; Vu et al., 1994; Vicario and Simpson, 1995; Ashmore et al., 2005). Thus, the results of anatomical, lesion and microstimulation studies all imply that HVC and RA function as the song system analogue of motor cortex, a claim that is further supported by the observation of motor-related neural activity in both nuclei during singing. Auditory responses, often selective for the bird's own song, have also been recorded throughout the motor pathway in both awake and anesthetized birds (see Prather and Mooney, this volume) and are present in an attenuated form in HVC of awake birds, indicating that the motor pathway may also be an important site for sensory-motor integration.

The forebrain song system contains another major pathway, the anterior forebrain pathway (AFP; Figure 6.3b), which provides a second, indirect connection between HVC and RA. The AFP begins with a projection from HVC to Area X, a specialized region of the anteromedial striatum. Area X projects to the medial part of the dorsolateral anterior thalamic nucleus (DLM), and this projection is GABAergic (Grisham and Arnold, 1994; Luo and Perkel, 1999a). This is an example of a direct striatothalamic projection not found in mammals. DLM then projects to the lateral magnocellular nucleus of the anterior nidopallium (LMAN), and LMAN projects to RA and Area X. DLM also receives a small projection from RA. The AFP is not absolutely required for singing, but is required for song learning (Bottjer et al., 1984; Sohrabji et al., 1990; Scharff and Nottebohm, 1991) and adult song plasticity (Williams and Mehta, 1999; Brainard and Doupe, 2000a). Neurons in the AFP often exhibit song-selective auditory responses, suggesting that the AFP is involved in processing song-related auditory information. Although the AFP is not required for producing song, recent work has revealed that its output can affect song quality in both juveniles and adults (Kao et al., 2005; Ölveczky et al., 2005). Stimulation of LMAN during singing increases song acoustic variability, and disrupting the input to RA from LMAN reduces song variability, especially when the bird is singing alone. These results have led to the proposal that, rather than processing auditory feedback, the AFP could generate song variability, helping the bird explore motor space during song learning (see Brainard, this volume).

In addition to these core components, there are several other forebrain nuclei associated with the song system whose functions are poorly understood. The nucleus interfacialis (NIf) is a nidopallial nucleus that projects to HVC; it may transmit premotor signals or

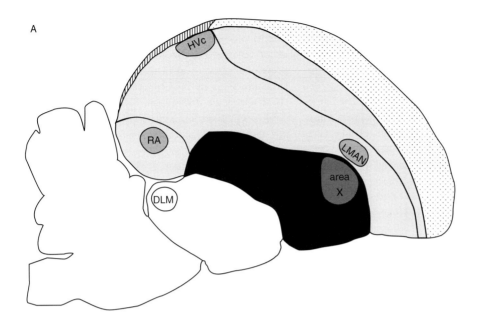

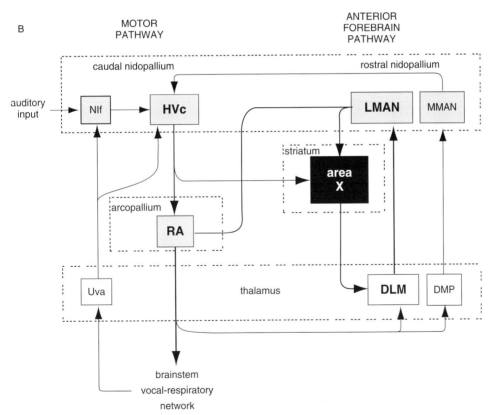

Figure 6.3 The oscine song system. (A) Schematic parasagittal section through the brain of a songbird showing the approximate location of the principal song system nuclei. The shading indicates the different subdivisions of the telencephalon, as in Figure 6.1. (B) Diagram of the connections among song system nuclei. Thick arrows show the connections of the "core" song system for emphasis; thin arrows do not necessarily symbolize weak projections. See text for abbreviations.

auditory information. Indeed, NIf may be the main conduit of auditory information to HVC (see Wild, and Prather and Mooney, this volume). Both NIf and HVC receive input from the thalamic nucleus uvaeformis (Uva) (Nottebohm et al., 1982), which in turn receives input from brainstem vocal and respiratory centers in addition to some visual and somatosensory inputs. Uva may be involved in coordinating song-related motor activity on both sides of the brain (Striedter and Vu, 1998; Coleman and Vu, 2005). In the anterior forebrain, a nucleus adjacent to LMAN, the medial magnocellular nucleus of the anterior nidopallium (MMAN), projects to HVC and receives input from the posterior nucleus of the dorsomedial thalamus (DMP), adjacent to DLM (Foster et al., 1997). DMP in turn receives input from the arcopallium (including part of RA) and the hypothalamus. MMAN appears to play some role in song production and perhaps learning (Foster and Bottjer, 2001).

THE AFP AS A BASAL GANGLIA CIRCUIT

Since the goal of much of the research on the song system is to extract lessons that might be applicable to mammals, it is important to identify commonalities between the song system and mammalian systems in organization, physiology, and function. The clearest point of contact between these systems can be found in the basal ganglia, in part because the characteristics of the amniote basal ganglia in general are so highly conserved. The only major part of the song system that lies within the basal ganglia is Area X, the first nucleus of the AFP. The other components of the AFP, the thalamic nucleus DLM and the pallial (cortex-like) nucleus LMAN, can be thought of as parts of a larger system characteristic of the mammalian basal ganglia: the basal ganglia–thalamocortical loop. Superficially, however, the AFP differs profoundly from the mammalian basal ganglia. The most glaring disparity is the division of the mammalian basal ganglia (and much of the avian basal ganglia as well) into two basic parts: an input structure, the striatum, that receives input from the pallium but generally does not project outside of the basal ganglia, and a set of output structures, the globus pallidus (GP) and substantia nigra (specifically, the pars reticulata, SNr), which receive striatal input and project out to the thalamus and brainstem. Area X is a part of the avian striatum, but it breaks this anatomical rule by projecting directly to the thalamus. A second major difference can be found in the nature of these projection neurons: in mammals, more than 90% of striatal neurons are projection neurons, and they have characteristically densely spiny dendrites, whereas in Area X only a small proportion of the neurons project out of the nucleus, and those neurons are aspiny (Luo and Perkel, 1999a).

A closer examination of the cell types in Area X reveals that the seemingly important differences between it and the mammalian basal ganglia are relatively trivial. Although the projection neurons of Area X are not spiny neurons, the vast majority of neurons in Area X are spiny, and they have characteristic electrophysiological properties virtually identical to those of mammalian striatal spiny neurons (Kawaguchi, 1993; Farries and Perkel, 2002). The projection neurons of Area X, on the other hand, do not resemble any cell type identified in the mammalian striatum, yet they are quite similar to cells found in the mammalian GP, cells that project to the thalamus (among other places; Nakanishi et al., 1990). Furthermore, the spiny neurons of Area X innervate the GP-like neurons that project to the thalamic nucleus DLM. Although Area X is not divided into spatially distinct striatal and pallidal parts, it contains the cellular elements of both, and they appear to have the same basic functional organization as in the mammalian basal ganglia (Farries and Perkel, 2002; Farries et al., 2005b). Viewed in this way, the two major differences between Area X and the mammalian basal ganglia – the projection of "striatal" neurons to the thalamus and the properties of those projection neurons – effectively cancel each other out (Farries et al., 2005b). With those differences disposed of, the list of similarities is quite striking: Area X and the mammalian basal ganglia share the same cell types, including rare interneuron types (Farries and Perkel, 2002), express a similar set of neuropeptides (Carrillo and Doupe, 2004; Reiner et al., 2004a), both receive heavy dopaminergic input from the midbrain (Lewis et al., 1981; Bottjer, 1993; Soha et al., 1996; Gale and Perkel, 2005), and have similar physiological responses to dopamine (Ding and Perkel, 2002, 2004; Ding et al., 2003).

The similarities between Area X and the mammalian basal ganglia are impressive given that the lineages giving rise to birds and mammals have been evolving separately for more than 300 million years, not to mention the highly specialized nature of the song system itself, and they encourage us to believe that these structures share some fundamental operating principles. However, when comparing the basal ganglia component of the song system with mammalian systems it is important to recognize that there *are* some differences and to remain cautious when applying results obtained in one system to the other. One potentially significant difference is the division of the mammalian GP into internal and external segments (GPi and GPe). In mammals, it is the GPi (along with the SNr) that conveys the output of the basal ganglia to the rest of the brain, while the GPe has powerful reciprocal connections with the subthalamic nucleus and projects to the GPi. Area X may in fact have a partial analogue to the GPe, since it appears to have a set of GP-like neurons that do not project to DLM but do heavily innervate the GPi-like DLM-projecting neurons. Nevertheless, these putative GPe-like cells do not project to or receive input from the subthalamic nucleus, an important difference with mammals. Such differences must be kept in mind, but on the whole the close resemblance of the AFP to mammalian basal ganglia pathways suggest that the song system is an excellent tool for studying the contribution of the basal ganglia to motor learning in all amniotes, including mammals.

EVOLUTIONARY CONSIDERATIONS: THE PLACE OF THE SONG SYSTEM WITHIN THE OSCINE BRAIN

The pallial components of the song system (HVC, RA, NIf, LMAN and MMAN) cannot be unequivocally regarded as homologues of isocortex – they lie within the arcopallium and nidopallium, regions whose relationship to isocortex is still debated. The study of these nuclei probably can shed light on the workings of the mammalian isocortex, but their relationship to isocortex may be at the level of functional analogy rather than homology. On the other hand, these song system nuclei do have a close relationship to nonvocal telencephalic circuits found in nonoscine birds and in other parts of the oscine brain. All too often, song-system nuclei have been studied without regard to the regions of the avian brain they inhabit. Paying attention to the relationships between song system nuclei and these other regions is likely to yield insights into why the song system is organized as it is, how nonvocal systems in the avian brain function, and how the song system itself evolved. We will review the organization of forebrain pathways identified in pigeons and chickens – and probably found in most or all birds – that resemble the oscine song system, and suggest how the song system might have evolved from such circuitry (for review, see Farries, 2001).

RA, the motor pathway nucleus that directly innervates motor neurons controlling the syrinx, lies at the medial end of a belt of arcopallial tissue known as the "intermediate arcopallium" (Ai). This structure is found in all birds studied so far, including nonoscines such as pigeons, chicks, and ducks, although only songbirds possess RA. Ai has descending projections to the brainstem in all these species (Zeier and Karten, 1971; Davies *et al.*, 1997; Dubbeldam *et al.*, 1997), including projections to brainstem premotor networks, suggesting that at least parts of Ai facilitate the control of motor functions by the telencephalon. One of the most prominent inputs to Ai arises from a caudolateral belt of nidopallium (NCL) that projects topographically to Ai (Wild *et al.*, 1985; Metzger *et al.*, 1998; Kröner and Güntürkün, 1999). HVC, the motor pathway nucleus innervating RA, lies at the medial end of oscine NCL (Bottjer *et al.*, 2000). This suggests that song system motor pathway might simply be a specialization of a general set of NCL–Ai circuits found in all birds, circuits that already play a role in motor control (but not, of course, learned vocalizations). This idea is further supported by a closer examination of the anatomical connections of the medial region of the NCL–Ai pathway. In nonoscines, the medial NCL receives input from the proposed homologue of the oscine thalamic nucleus Uva and from a nidopallial region that might be comparable to NIf (Metzger *et al.*, 1998; Kröner and Güntürkün, 1999); both Uva and NIf project to HVC. Medial Ai has descending projections to the midbrain vocal center DM (Wild *et al.*, 1993), which in oscines receives input from RA and the adjacent arcopallium. Finally, medial regions of both NCL and Ai receive auditory input from Field L (Metzger *et al.*, 1998; Kröner and Güntürkün, 1999). Although neither HVC nor RA receives strong input from Field L, regions immediately adjacent to them do (Kelley and Nottebohm, 1979; Fortune and Margoliash, 1995; Vates *et al.*, 1996). The song system motor pathway appears to be a specialization of a medial NCL–Ai circuit found in nonoscines, characterized by two major deviations from that ancestral circuit: exclusion of direct input from Field L, and the addition of new direct projections to vocal–respiratory circuits in the brainstem.

While the case for a general nonvocal "motor pathway" in nonoscines is fairly strong, the existence and nature of a nonvocal AFP is less clear. Ai appears to receive a projection from a region of rostral nidopallium that could be comparable to LMAN (Dubbeldam et al., 1997; Bottjer et al., 2000). This area of rostral nidopallium also projects to NCL (Bottjer et al., 2000), unlike LMAN but similar to MMAN, which projects to HVC. Furthermore, this rostral nidopallial domain receives input from the dorsomedial thalamus (Kitt and Brauth, 1982; Wild, 1987; Bottjer et al., 2000), the area in oscines that contains the LMAN-projecting DLM and the MMAN-projecting DMP. Finally, the same nidopallial region projects heavily and topographically to nearby rostral striatum, like the projection from LMAN to Area X (Veenman et al., 1995; Bottjer et al., 2000). Thus, nonoscines do appear to have regions comparable to all components of the oscine AFP. However, no heavy projection from medial NCL to rostral striatum has been reported (corresponding to the projection from HVC to Area X), and while there is evidence for a projection from rostral striatum directly to the dorsomedial thalamus in chickens (Székely et al., 1994), that projection appears to be comparatively weak and has not been confirmed through retrograde tracing. Like the motor pathway, the oscine AFP is probably a specialization of pre-existing circuitry found in nonoscines, but the extent to which its anatomical connections are inherited from this older circuitry (i.e. evolving in parallel rather than arising *de novo* with the evolution of the song system) remains undetermined.

Although the song system nuclei are highly specialized, are easily distinguished from the surrounding brain even in unstained tissue, and serve functions unique to songbirds and a few other avian groups, their anatomical organization is quite similar to structures found in nonoscines, structures that presumably are not devoted to vocal control. This suggests that the song system evolved as a specialization of neural circuitry that was already present in the oscine ancestor and that is shared widely across avian groups. The song system appears to be a relatively modest modification of this pre-existing circuitry, and much of the organization of the song system is simply inherited from an established pattern. Because the song system is associated with a well-defined behavior and has a clear relationship to other neural structures, birdsong may prove to be a good model system for studying the evolution of novel behavioral capacities. The similarities between the song system and other avian systems also suggests that the study of birdsong could have fairly direct implications for the functions of the avian brain in general, and that a comparative approach to studying the song system could bring substantial rewards.

CONCLUSION

The song system is a highly specialized set of structures, several of which appear to be almost exclusively devoted to the various sensory and motor aspects of one specific behavior – the production of learned vocalizations. Yet it evolved from neural circuits which are found in many other taxa and which perform many other functions. It may come as some surprise that the song system retains many of the organizational features of the ancestral circuits, in spite of its extreme specialization. This evolutionary conservatism by no means detracts from the song system's position as a uniquely valuable model system, however. Rather, it provides an opportunity to apply what is discovered in the song system to other systems and to use what is known about other systems to generate testable hypotheses concerning the neural basis of birdsong. This is most true of circuits found in the forebrain of nonoscine birds, which may closely resemble the immediate evolutionary precursors of the song system. But it can also be true of neural systems found in other vertebrate classes, including mammals, so long as a careful consideration of comparative neurobiology is used to define the appropriate scope of such applications. As things stand now, there appear to be enough similarities between the pallial components of the song system and mammalian isocortex (whether from common inheritance or convergent evolution) to support a number of useful parallels between them, although the available data also argue for some circumspection in this area. In the basal ganglia, the parallels between the song system and mammalian neural systems can be stronger – the basal ganglia are well conserved across all amniotes, and comparisons with mammals have proven very fruitful in predicting the histochemical and electrophysiological properties of neurons in Area X, the song-related component of the oscine basal ganglia. Overall, comparative studies suggest that an understanding of the neural basis of singing and song learning would tell us much about the workings of neural systems in mammals, including humans.

PART II
Song production and its neural control

7 • Introduction

H. P. Zeigler

The chapters in this section deal with two critical aspects of singing behavior, the structure and operation of the syrinx – the songbird's sound producing organ – and the role of the bird's respiratory system in the generation of song. As any singer will tell you, accurate note production (phonation) is only half the challenge faced by a singer; the other, and equally difficult, task is breath control. In fact, the real theme of this section is the integration of phonation and respiration. In the first two chapters, the focus is on peripheral mechanisms, including both the syrinx and vocal tract structures (trachea and beak); in the third, the emphasis is on central mechanisms; i.e. on the identification of central neural structures integrating breathing and singing. The last chapter provides an overview combining both approaches and reviewing the organization of circuitry that links central to peripheral structures controlling respiration and phonation. Several chapters raise the issue of the role of sensory feedback in the control of singing behavior – not the auditory feedback, which has been shown to be so critical for the development and maintenance of song – but somatosensory and viscerosensory feedback that may be generated during song production.

In contrast with a pipe organ, in which the fundamental frequency of a note is determined by the resonance frequency of the column of air in the pipe, the fundamental frequency of a song note (i.e. its pitch) is controlled by the syringeal muscles that exert tension on the labia (the vibrating tissues which are the songbird's equivalent of the human vocal folds). Considering the size of the syrinx and its location deep within the thorax near the heart, studies of syringeal function, such as those from the Suthers and Goller labs, often represent technical *tours de force* involving simultaneous monitoring of electrical activity in several pairs of syringeal muscles, rate of respiratory air flow and air pressure during the bird's singing behavior. These studies have clarified the contribution of distinct syringeal muscle groups to the control of the timing and frequency of note production and demonstrated a remarkable degree of independence of the two sides of the syrinx. They have provided the structural foundation for the observation that many birds are capable of generating separate (i.e. independently modulated) frequencies on each side of the syrinx. It is now clear that what we think of as one continuous flow of sound actually reflects the bilateral integration of song components originating from different sides of the syrinx. The other major contribution of such studies is their demonstration of an extraordinary degree of coordination between the respiratory and syringeal systems. Nearly all songs are produced during expiration, which provides the airflow through the syrinx required for phonation.

Since respiration has other functions beside phonation (e.g. gas exchange) and since the volume of air available for phonation is limited by the capacity of the lungs and air sacs, coordination between the two systems becomes crucial. Thus, in order to produce songs of long duration, birds may take short breaths ("minibreaths") which allows them to replenish the air supply used during phonation. Similarly, birds may sing while in a variety of positions and postures or during displays involving movements of body parts. All of these factors generate constraints to which the systems must adjust and the nature of those adjustments is one focus of these two chapters. Another is the nature of the sensorimotor mechanisms by which those adjustments are achieved, including the role of peripheral somatosensory and viscerosensory feedback operating on neural structures at more central levels. Finally, both chapters devote considerable attention not only to syringeal mechanisms but also to the role of the vocal tract – the trachea, oropharyngeal cavity and beak. At its simplest level, the vocal tract may be thought of as a tube of variable length whose resonant frequencies, like those of a brass or woodwind instrument, are related to its length. The authors review a growing body of studies demonstrating a correlation between vocal tract properties (including those of the oropharynx and beak) and frequency that suggest important contributions of vocal tract structures to song properties. However, their review of recent work suggests that the relationship among these structures is likely to be a complex one since the resonance tuning of the tract may be actively modulated by the bird itself.

A focus on the syrinx is quite appropriate because, as Suthers and Zollinger point out, "the peripheral vocal system is the link between the brain and vocal behavior." Thus, the discovery of a sensorimotor circuit linking the forebrain directly to the syrinx opened the way to studies of the role of higher brain areas in the control of vocalization (see Nottebohm, this volume). Given the critical contribution of auditory feedback to the acquisition and maintenance of singing behavior, one direction of subsequent research on brain mechanisms has been the identification of sources of auditory input to the song control systems (see Part III Hearing and recognizing the song). A second has been the delineation of circuitry linking the syringeal (phonatory) and respiratory systems and the identification of forebrain structures that, through their descending connections, might accomplish the task of integrating breathing and singing. The final chapters in this section review our current knowledge of this issue and identify possible anatomical foundations for such integration.

Schmidt and Ashmore begin their chapter by re-emphasizing the constraints on central control that are dictated by the interdependence of syringeal and respiratory mechanisms in generating the basic elements of song (the individual notes). However, their main concern is not with the individual notes but with the next level of song structure, its temporal organization. That organization is determined, in part, by the parameters of inspiration and expiration since the length of expiratory pulses and the duration of inspiratory minibreaths are responsible, respectively, for syllables and the silent intervals between them. Central control of song structure therefore requires some higher-level mechanisms which coordinate the activity of syringeal and respiratory motor commands during the production of song.

One reason for the widespread interest in the "song circuit" described by Nottebohm and his colleagues is that it appeared to provide an anatomical substrate for such mechanisms. Schmidt and Ashmore provide an excellent review of the lesion, stimulation and recording studies that have implicated the telencephalic components of that circuit in the production of learned vocalizations. They point out that these studies have also led to a conceptualization that views the telencephalic structures (HVC, RA) as the apex of a hierarchical, descending system directly controlling song production. However, the identification of a group of brainstem premotor structures with descending projections upon both respiratory and syringeal motoneurons and ascending connections with more rostral parts of the song system (see Wild, this volume) added a new dimension to the story.

In the light of this new information, Schmidt and Ashmore argue that a hierarchical conceptualization of central control of song is incomplete. Building upon anatomical data for ascending brainstem projections to the forebrain (see Wild, this volume) they suggest that, instead of functioning as the top of a hierarchy, telencephalic structures such as RA and HVC should be viewed as components of a recurrent vocal (syringeal)–respiratory network (VRN) that links brainstem and forebrain structures. In a series of lesion, recording and stimulation experiments these investigators have provided evidence for the existence of recurrent circuitry that functions to modulate the activity of syringeal and respiratory premotor neurons during singing behavior and to coordinate that activity on the two sides of the brain. As they note, this view of the relation between brainstem and forebrain structure is a novel one suggesting that "rather than simply following instructions from the forebrain, the brainstem might inform forebrain vocal centers about key temporal features of song."

Many of the anatomical findings upon which their work is based are described in more detail in Wild's chapter. The chapter provides an excellent example of how an initial "bottom-up" research strategy can eventually produce important data for researchers interested in "top-down" control. It begins with a description of the various motoneuron groups innervating three main components of what might be called the final common path for songbird vocalization – the muscles of the syrinx, respiratory system and upper vocal tract. Such information provides a starting point for studies designed to locate premotor structures which might mediate the integration of breathing and singing. These studies led to the identification of a medullary premotor nucleus (n. retroambigualis, RAm) which originates projections to both respiratory and syringeal motoneurons. Moreover its projections are bilateral so that it could also contribute to the coordinated control of vocalization by the two sides of the brain. Finally, RAm itself is among the targets of a descending pathway from the telencephalic nucleus RA. As Wild points out, "songbirds are probably unique in having a single telencephalic nucleus that directly controls both vocal motoneurons and respiratory premotor neurons" (RA). He concludes that because of its location and connections RA is in a position to coordinate the respiratory and phonatory systems in the service of vocalization. Finally, Wild addresses the issue of nonauditory feedback and its possible contribution to the control of the moment-to-moment adjustments in breathing that may be involved in the patterning of singing behavior. Currently, neither the syringeal nor pulmonary receptors putatively involved have been anatomically identified. However, anatomical and physiological studies suggest that afference from these structures may be relayed from the brainstem to a thalamic nucleus (Uva; nucleus uvaeformis). Interestingly, this nucleus also receives both somatosensory and auditory input, making it an excellent potential locus for the integration of auditory and somatosensorimotor feedback in the control of singing behavior, a function consistent with that ascribed to it in the Schmidt and Ashmore chapter.

8 • From brain to song: the vocal organ and vocal tract

Roderick A. Suthers and Sue Anne Zollinger

WHY STUDY PERIPHERAL VOCAL MECHANISMS?

The peripheral vocal system is the link between the brain and vocal behavior. Indeed, the forebrain nucleus generating song motor programs in songbirds is the source of a descending pathway that terminates directly upon neurons controlling the muscles of the bird's vocal organ – the syrinx. A knowledge of how the peripheral vocal system works is therefore an important component in understanding and interpreting the neural mechanisms involved in singing. The vocal periphery also has an important role in defining the acoustic possibilities and limitations of vocal signals. Whereas neural activity can change within a time frame of milliseconds, the mechanical actions of muscles that translate neural signals into song have a longer time constant due to their mass, inertia, and other physical or physiological properties. Knowing how song is produced can thus provide an insight into constraints on vocal performance that have shaped the evolution of vocal communication and song diversity. The location of the syrinx deep in the thorax, close to the heart, poses a challenge for studies of its function, but an understanding of how vocal motor patterns are converted into sound is an important piece in the neuroethological puzzle of birdsong.

SYRINGEAL AND RESPIRATORY CONTROL OF SOUND PRODUCTION

The avian vocal organ

Syrinxes vary in their anatomical design across different avian taxa. Some are located in the trachea and a few are in the primary bronchi. The most common type of syrinx, however, lies at the junction between these structures. This tracheobronchial syrinx is also present in many nonsongbirds, but it is most highly developed in the oscine songbirds, which are the principal focus of this chapter (suborder Passeri or Oscine in the Passeriformes).

The songbird syrinx consists of modified cartilages at the cranial end of each bronchus and the caudal end of the trachea together with typically six bilaterally paired syringeal muscles, including the tracheolateralis and sternotrachealis muscles that extend from the syrinx to other structures (King, 1989) (Figure 8.1a, b). All of these muscles are innervated by the tracheosyringeal branch of the ipsilateral hypoglossal nerve. One of their important actions is to control the movement, and probably the tension, of a pair of small connective tissue pads, the medial (ML) and lateral (LL) labia, located at the cranial end of each bronchus. Each medial labium is intimately associated with a thin medial tympaniform membrane (MTM) that is continuous with its caudal edge (King, 1989).

Endoscopic observations by Goller and Larsen (1997a), of labial motion during spontaneous vocalizations by a crow (*Corvus brachyrhynchus*) and vocalizations elicited by brain stimulation in northern cardinals (*Cardinalis cardinalis*) and brown thrashers (*Toxostoma rufum*) show that during vocalization the medial and lateral labia are adducted into the bronchial lumen to form a slit and vibrate. The destruction of both MTMs in zebra finches (*Taeniopygia guttata*) and cardinals had only a small effect on their song, involving syllable fine structure and harmonic emphasis (Goller and Larsen, 1997a). Subsequently, these researchers (Larsen and Goller, 1999) used an optical vibration sensor to measure the average motion of the labia in an anesthetized hill myna (*Gracula religiosa*) and motion of the lateral tympaniform membrane (LTM) in pigeons (*Columba livia*) and cockatiels (*Nymphicus hollandicus*) during sounds elicited by brain stimulation. They found that the dominant frequency of vibration matched that in the sound being produced. Observations such as these indicate that the medial and lateral labia are the primary

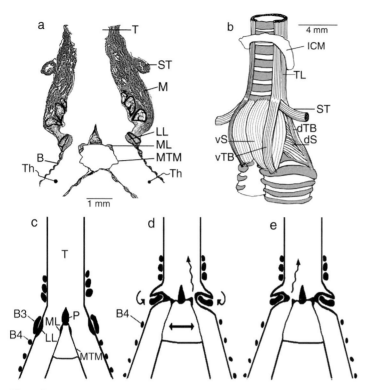

Figure 8.1 The songbird syrinx is a bipartite structure located at the tracheobronchial junction. (a) Frontal section through the syrinx of a northern mockingbird, showing the dual nature of the vocal organ and placement of microbead thermistors (Th) for recording airflow. (b) Ventrolateral external view of syrinx illustrating syringeal musculature. (c–e) Schematic ventral views of songbird syrinx during quiet respiration (c); phonation on the left side with labial valve closed on right side (d); phonation on right side with left side closed (e). In preparation for phonation the syrinx moves rostrad. Contraction of the ipsilateral dorsal syringeal muscles (dS and dTB) rotates the bronchial cartilages (curved arrows) into the syringeal lumen, moving the labia into the air stream, where they are set into vibration, producing sound (wavy arrows). Phonation may be bilateral (not shown) or unilateral (shown).
Abbreviations: T, trachea; M, syringeal muscle; ML, medial labium; LL, lateral labium; MTM, medial typaniform membrane; B, bronchus; ICM, membrane of the interclavicular air sac; TL, m. tracheolateralis; ST, m. sternotrachealis; vS, m. syringealis ventralis; vTB, m. tracheobronchialis ventralis; dTB, m. tracheobronchialis dorsalis; dS, m. syringealis dorsalis; B3 and B4, third and fourth bronchial cartilages; P, pessulus. (c–e, modified from Suthers and Goller, 1997.)

sound source in the songbird syrinx, vindicating the intuition of Setterwall (1901) who had earlier named them the inner and outer vocal cords.

Songbirds use their labia to produce sound in much the same way humans use their vocal folds to generate voiced speech. During vocalization, respiratory airflow through the syrinx causes the labia to oscillate. The oscillation modulates the airflow, resulting in sound waves that propagate through the vocal tract. The medial and lateral labia are generally believed to bump into each other in the manner of a vibrating valve, resulting in a complex waveform with prominent higher harmonics (but see Zaccarelli et al., 2006).

For a history of theoretical accounts of songbird sound generation see, for example, Ballintijn and ten Cate (1998), Casey and Gaunt (1985), Greenewalt (1968). For a detailed description of some of the methods used in the analysis of sound production mechanisms in birds see Goller and Cooper (2004), Suthers and Zollinger (2004). Biomechanical models of labial oscillation can be found in papers by Fee and colleagues (Fee et al., 1998; Fee, 2002), Fletcher (1988), Elemans et al. (2003), Mindlin and Laje (2005) and Zaccarelli et al. (2006).

Motor control of song in the duplex syrinx

Syringeal and respiratory function during spontaneous song with both sides of the syrinx functionally intact has been investigated by placing a heated microbead thermistor into each primary bronchus while also measuring the subsyringeal or tracheal pressure and recording song (Suthers, 1990). The amount of current required to maintain the thermistors at a constant temperature provides a continuous measure of the rate of airflow through each side of the syrinx. In some experiments, electromyograms (EMGs) of syringeal or respiratory muscles were also recorded.

During song bouts, the songbird labia tend to operate in one of three different functional modes, depending on the changing vocal and respiratory demands associated with singing. These three configurations of the labia, which are subject to independent motor control on each side of the syrinx, are associated, respectively, with (1) inspiration between syllables, (2) expiration with sound production and (3) muting one side of the syrinx by preventing ipsilateral sound production during the expiration. During inspiration (Figure 8.1c) the labia are withdrawn from the air stream, reducing the resistance to airflow and facilitating rapid replacement of air exhaled during vocalization. Phonation is initiated by adducting the labia into the syringeal lumen at the beginning of expiration (Figure 8.1d bird's left or, e, right side). The adductive force is opposed by the positive subsyringeal expiratory pressure which tends to push the labia apart. The balance between these forces is important in maintaining the labia in a phonatory position where airflow can sustain oscillation. Rare cases of phonation during inspiration are noted below.

The labia also serve as pneumatic valves that control airflow, and hence phonation, independently on each side of the syrinx, within the constraints of the respiratory rhythm. If the *ad*ductive force on the labia exceeds that *ab*ducting them, or pushing them apart, they will close the syringeal lumen and mute that side of the syrinx by preventing air from flowing through it (Figure 8.1d bird's right or, e, left side). Independent motor control of each side, via motor neurons in the tracheosyringeal branches of the hypoglossal nerve, permits one side of the syrinx to be muted while simultaneously phonating on the other side. Thus in crystallized song, some notes or syllables may be sung on the left side of the syrinx and others on the right side. If both sides of the syrinx are in a phonatory configuration at the same time, each side may produce different, independently controlled sound that is not harmonically related to that coming from the contralateral side. The same syllable is normally produced in the same way during each rendition, but there are interspecific differences in the contribution each side makes to the total repertoire.

Electromyographic recordings from syringeal muscles during song shows that the activity of the dorsal syringeal muscles and dorsal tracheobronchialis muscles increases with increasing syringeal resistance, indicating that they play an important role in adducting the labia to initiate phonation or to silence one side of the syrinx. Direct observation of the biomechanical effects in response to electrical stimulation of individual syringeal muscles in anesthetized brown thrashers and cardinals (Larsen and Goller, 2002) supports the inferences based on electromyography. In addition, they suggest separate roles for the two dorsal muscles in which the medial portion of the dorsal syringeal muscle controls the adduction of the ML and the dorsal tracheobronchialis muscle adducts the LL. The ventral tracheobronchialis muscle opens the syringeal lumen by abducting the LL. The lateral portion of the ventral syringeal muscles may also participate in labial adduction (Goller and Cooper, 2004). These muscles are thus primarily responsible for determining when each side of the syrinx produces sound during a song bout.

Whereas the timing of phonation within the appropriate phase of the respiratory cycle is primarily controlled by dorsal syringeal muscles, the ventral syringeal muscles are primarily responsible for controlling the fundamental frequency of labial oscillation. This is presumably accomplished by varying the tension they exert on the labia. The EMG amplitude of these ventral muscles increases exponentially with the fundamental frequency of the ipsilateral sound ($R^2 = 0.71$ to 0.95) (Figure 8.2) (Goller and Suthers, 1995, 1996b) and appears to increase tension in the ML (Larsen and Goller, 2002). The activity of other syringeal muscles also tends to increase with sound frequency and the ventral tracheobronchial and sternotrachealis muscles may also assist in frequency regulation, but the correlation of EMG activity with sound frequency is much lower for the dorsal than it is for the ventral syringeal muscles (Goller and Suthers, 1995, 1996b).

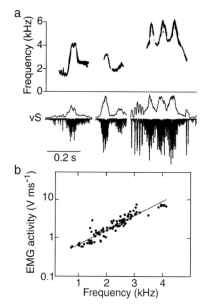

Figure 8.2 The role of ventral syringeal muscles in the control of sound frequency in the brown thrasher. (a) EMG amplitude of ipsilateral ventral syringeal muscle (vS) is positively correlated with frequency modulation of sounds generated on the ipsilateral side. EMG envelope is shown integrated (upward, time constant 5 ms) and rectified (downward). (b) Amplitude of EMG activity is correlated exponentially with fundamental frequency of ipsilaterally produced sounds. EMG activity was averaged over segments of syllables having a relatively constant frequency. (Modified after Goller and Suthers, 1996b.)

The important role of syringeal muscles in gating and modulating vocalizations often requires rapid contractions. Elemans et al. (2004, 2006) showed that the contractile performance of syringeal muscles in ring doves (*Streptopelia risoria*) is comparable to that of superfast muscles involved in sound production by other vertebrates such as the toadfish and rattlesnake. In view of their often complex songs and temporal patterns of EMG activity, it is likely that songbirds also have syringeal muscles adapted for very fast contractions.

Studies on gray catbirds (*Dumetella carolinensis*) and brown thrashers indicate that during song, ventral muscles on both sides of the syrinx are active, even though one side may be fully adducted and mute. It appears that each side of the brain continues to send a song motor program to the ipsilateral side of the syrinx regardless of whether its labia are in a phonating or muted configuration. Lateralization of song production is determined by the motor program or subprogram sent to the dorsal muscles. Strong activation of these muscles on one side silences it by closing the labial valve and prevents the ongoing ipsilateral song motor program from being converted into an acoustic signal (Goller and Suthers, 1995, 1996a). The lateralization of birdsong thus involves peripheral gating of central motor programs that is fundamentally different from the left hemisphere (central) dominance of human speech (Broca, 1861).

Respiratory adjustments for singing

Song requires major adjustments in respiratory ventilation which must continue to meet the needs for pulmonary gas exchange while at the same time providing appropriate rates and patterns of syringeal airflow required for phonation (see Goller and Cooper, this volume). The body cavity of birds is not divided by a muscular diaphragm as it is in mammals (Scheid and Piiper, 1989). A thin post hepatic septum anterior to the abdominal air sacs separates the coelomic cavity into two compartments, but this septum has few muscle fibers and appears to play no significant role in silent respiration. Inspiratory muscles expand the air sacs by moving the sternum ventrally and cranially (Figure 8.3a). This motion is reversed by expiratory muscles, which compress the air sacs (King and Molony, 1971). In the absence of song, peak expiratory air sac pressure is only +0.05 to +0.3 kPa in the canary, but increases about an order of magnitude during song, often reaching 1 to 1.5 kPa in the anterior thoracic air sac and peaking at about 3 kPa ($=$ 30 cm H_2O) for some syllables (Hartley and Suthers, 1989). Nearly all song is produced during expiratory airflow. Two known exceptions are occasional inspiratory syllables in the songs of some zebra finches (Goller and Daley, 2001; Leadbeater et al., 2005) and an inspiratory "wah" sound associated with coos of doves (Gaunt et al., 1982).

During song, the respiratory motor pattern has to be coordinated with that of the syrinx. This can potentially involve the action of up to 11 principal respiratory muscles that participate in normal breathing, plus 8 additional muscles that become active during labored breathing (Fedde, 1987). Electromyograms during singing have been recorded from only a few abdominal expiratory muscles (primarily the external oblique and transverse abdominal muscles) and from two thoracic

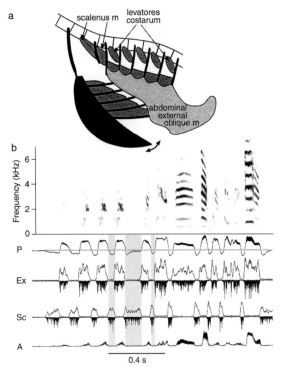

Figure 8.3 Activity of inspiratory and expiratory muscles is correlated with song production. (a) Some of the inspiratory (scalenus m, levatores costarum) and one of the four abdominal expiratory muscles (abdominal external oblique m). (b) EMG activity of inspiratory and expiratory muscles during zebra finch song, showing that inspiratory muscle EMG activity occurs only during inspiration, and does not overlap temporally with expiratory muscle EMG activity. Shaded bars mark inspirations. P, pressure; Sc, EMG activity in m. scalenus; Ex, EMG activity in abdominal expiratory muscles; A, rectified and integrated sound amplitude (time constant 2 ms). (a, modified from Wild, 1997a; b, modified from Wild et al., 1998.)

activity in the same way on both sides when generating syllable-specific patterns of pressure and airflow, even during unilateral phonation (Goller and Suthers, 1999). The similar timing and amplitude of EMGs on both sides, regardless of which side of the syrinx is producing sound, suggest bilateral motor control. The temporal pattern of respiration changes markedly during song and sets the song's basic tempo (Vicario, 1991b). Depending on syllable repetition rate, songbirds use one of two basically different respiratory motor patterns. At moderate syllable repetition rates a brief inspiratory minibreath (Calder, 1970) is taken after each syllable to replace the volume of air exhaled to produce the sound (Hartley and Suthers, 1989) (Figure 8.4). Syllable duration generally has an inverse relationship to repetition rate. Since long syllables use more air, the minibreaths must also be larger. In waterschlager canaries (*Serinus canaria*), for example, a syllable 119 ms long, sung at a repetition rate of 6.5 s^{-1} required 0.25 ml of air to produce, which was also the volume of its associated minibreath. The corresponding volumes for an 11 ms syllable sung 30 times per second were about 0.04 ml (Hartley and Suthers, 1989).

By using minibreaths a bird could in theory maintain an almost constant respiratory volume throughout its song. This might provide important advantages in producing precise, stereotyped respiratory movements needed during vocalization, since the inspiratory and expiratory muscles could always operate near their resting length and with predictable forces of thoracic elastic recoil. The direction of elastic recoil reverses on either side of the neutral position of the sternum at rest. At smaller air sac volumes elastic recoil supplements inspiratory muscle effort returning the sternum toward its neutral point, but at larger volumes the elastic recoil force is in the opposite direction and supplements expiratory effort. A constant respiratory volume during song could allow the sternum to oscillate around its neutral point so that elastic recoil forces supplement the action of both expiratory and inspiratory muscles in reversing the respiratory phase, perhaps permitting faster minibreaths and higher syllable repetition rates.

It is also possible that the syntax of a song may be partially dictated by respiratory needs. At high syllable repetition rates the minibreath volume is smaller than the tracheal dead space (Hartley and Suthers, 1989). It is not known if these minibreaths provide oxygen to the lungs. A bird could conceivably adjust the syntax

inspiratory muscles, the scalenus and levatores costarum (Figure 8.3a). The role of the other respiratory muscles in singing is not known. In zebra finches, brown-headed cowbirds (*Molothrus ater*) and northern cardinals the amplitude of the EMG in the muscles studied increases 5- to 12-fold during singing, compared with quiet respiration, presumably reflecting the recruitment of additional motor units (Hartley, 1990; Wild et al., 1998; Suthers et al., 2002). Electrical activity in expiratory and inspiratory muscles does not overlap in time (Wild et al., 1998) (Figure 8.3b). The activity of expiratory muscles is not lateralized. Expiratory muscles modulate their

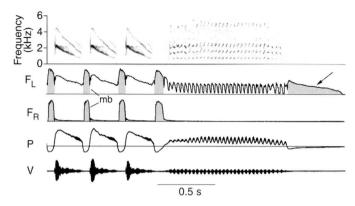

Figure 8.4 Songbirds use different respiratory motor patterns during song, depending on syllable repetition rate. This segment of cardinal song includes the last 3 syllables of a phrase sung at about 3 syllables s^{-1} using a minibreath (mb) after each syllable. Note that both sides open for each minibreath during negative air sac pressure. These low frequency syllables are sung entirely on the left side with the right side closed. This is followed by a trill at 30 syllables s^{-1} produced by pulsatile expiration in which the left side of the syrinx is repetitively opened while the right side remains closed and air sac pressure is positive during the entire phrase. Cardinals switch from a minibreath to a pulsatile pattern at about 16 syllables s^{-1}. Note the longer inspiration (arrow) immediately following the pulsatile portion of the trill. F_L and F_R, rate of airflow through left and right sides of syrinx, airflow associated with positive pressure is expiratory; shaded flow is inspiratory (corresponds with negative pressure). P, pressure in cranial thoracic air sac. V, oscillograph of vocalizations. Horizontal lines indicate zero airflow and ambient pressure. (From Suthers and Zollinger, 2004.)

of its song to meet its respiratory needs by periodically inserting a phrase of low repetition rate, long syllables with correspondingly large minibreaths that ventilate its lungs. Syntactical adjustments of this sort might allow some species to sing longer songs, which otherwise would have to be interrupted by a pause for a large breath.

The motivational state of a bird can also affect the tempo of its song. Cooper and Goller (2006) found that male zebra finches sing faster to females than when they sing alone. Directed song was accompanied by a higher heart rate and had a shorter duration than the same song sung alone. This difference in song duration was due to shorter expirations during vocalization. Since the duration of inspiratory minibreaths between syllables did not change, the expiratory and inspiratory phases of zebra finch song must be controlled by separate neural oscillators, one of which is affected by the social context.

For phrases sung at very high syllable repetition rates a different respiratory motor pattern of pulsatile expiration is used (Figure 8.4). The rate is increased by eliminating minibreaths at the cost of placing an upper limit on phrase duration as either the air reserve or oxygen is exhausted. During a phrase produced by pulsatile expiration, the subsyringeal air sac pressure is maintained at a positive level by expiratory muscle activity. Typically, one side of the syrinx is closed during the entire phrase and the other side opens periodically to release a puff of air that produces the syllable (Hartley, 1990; Hartley and Suthers, 1989). Rarely, both sides of the syrinx contribute to the syllable. The repetition rate at which a bird switches from minibreath to pulsatile respiratory pattern decreases with increasing body size, being about 30 syllables s^{-1} in an 18 g waterslager canary and 10 s^{-1} in a 50 g northern mockingbird (*Mimus polyglottos*). This inverse relationship with body mass is most likely due to the increased mass and inertia of the thoracic and abdominal body wall that must oscillate at the respiratory frequency.

Zollinger and Suthers (2004) investigated respiratory constraints on syllable repetition rate in the northern mockingbird by tutoring this vocal mimic with high repetition rate trills sung by a canary using minibreaths. Although the mockingbird copied the canary's repetition rate of 22 syllables s^{-1} (Figure 8.5a, d), he could not take a minibreath between syllables. Instead he divided the trill into groups of syllables. Each group was sung using pulsatile expiration with a pause between groups for an inspiration (Figure 8.5b, e). At

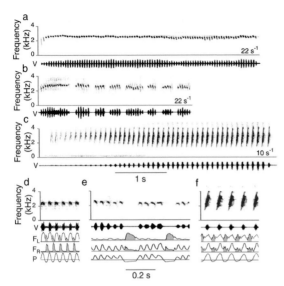

Figure 8.5 Long duration trills produced by canaries and mockingbirds. The canary tutor song (a; expanded view in d) is an uninterrupted trill lasting 4 seconds. At this repetition rate ($22\,s^{-1}$), the canary uses a minibreath respiratory pattern (d). The much larger mockingbird could not achieve the canary syllable repetition rate using minibreaths, but instead broke the trill into short segments each containing several syllables (b; expanded view in e). Each segment was produced using a pulsatile expiratory pattern. Within each segment the mockingbird accurately copied the repetition rate of the tutor song ($22\,s^{-1}$), but without minibreaths to replenish the air supply was forced to periodically interrupt the trill by opening the nonphonating side for a breath (e). The overall duration of the interrupted trill is 1 second shorter than that of the tutor. At lower repetition rates ($<11\,s^{-1}$), mockingbirds were able to use a minibreath respiratory pattern (c; expanded view in f) and, like canaries, were able to sing longer uninterrupted trills (up to 9 s). Inspiratory airflow is shaded. Syllable repetition rates are indicated in the lower right hand corner of each spectrogram (a–c). Abbreviations as in Figure 8.4. (a, b, and d modified from Zollinger and Suthers, 2004.)

lower syllable repetition rates ($<11\,s^{-1}$) there is a longer interval between syllables and the mockingbird can sing minibreath trills rivaling canary trills in length (Figure 8.5c, f). These results support the hypothesis that the maximum syllable repetition rate at which songbirds can produce long duration trills is limited by the inverse relationship between body mass and maximum frequency of respiratory movements.

An understanding of how avian respiratory patterns are generated is also being advanced through the use of mathematical models. Trevisan et al. (2006a, b) have proposed a model that predicts canary respiratory patterns based on air sac pressure and central neural interactions.

Motor stereotypy and sensory feedback

In the songbirds studied, each syllable type in the adult repertoire is always produced in a similar way, as judged by stereotyped temporal patterns of syringeal airflow, air sac pressure and the contribution each side of the syrinx makes to the vocalization. The time-varying pattern of airflow and pressure on each side of the syrinx reflects the combined activity pattern of respiratory muscles driving airflow and syringeal muscles controlling the resistance on each side due to constriction of the syringeal aperture at the labia. In brown-headed cowbirds, different song types in an individual's repertoire have characteristic patterns of pressure and airflow that are repeated with each repetition of the song (Allan and Suthers, 1994). Even in open-ended learners such as brown thrashers, the repetition of a syllable, either immediately to produce a "couplet" or after an intervening period of song composed of different syllables, is accompanied by a similar motor pattern characteristic of that syllable (Suthers et al., 1996b). When juvenile zebra finches copy song syllables from adult male tutors they usually also copy the air sac pressure pattern used by the tutor to produce the syllable. If a strobe flash is used to interrupt zebra finch song, the bird normally stops singing at the end of an expiratory pressure pulse, suggesting that these pulses represent units of motor production in which the syllable is the smallest unit (Franz and Goller, 2002).

This motor stereotypy is consistent with the hypothesis that each syllable type in the crystallized song repertoire is represented by a central motor program (Konishi, 1965b, 1985, 1994; Vu et al., 1994). But, even during adult song, sensory feedback continues to influence the motor output. Various experiments have shown that auditory feedback is necessary for long-term maintenance of song (Nordeen and Nordeen, 1992; Okanoya and Yamaguchi, 1997; Leonardo and Konishi, 1999). Somatosensory feedback also modulates, in real time, the activity of expiratory and syringeal muscles during song (see Wild, this volume). Both deaf and hearing adult cardinals adjust the contraction of their abdominal expiratory muscles to compensate for unpredictable changes in respiratory pressure caused by injection of a small volume of air into an air sac (Suthers et al., 2002). Small increases in air sac pressure are followed, after a latency of about 50 ms, by a reduction in the amplitude of the abdominal expiratory EMG which tends to stabilize expiratory airflow during the remainder of the song syllable (Figure 8.6). Syringeal muscles also respond to the injection of air into an air sac during a syllable by increasing their contraction, presumably to maintain an appropriate sound frequency in the presence of an unpredicted change in pressure and airflow (Suthers and Wild, 2000). Just as juveniles learn

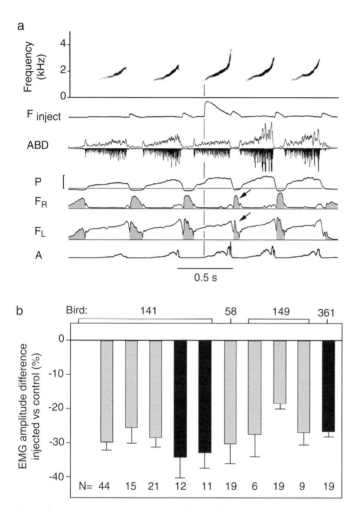

Figure 8.6 Real-time somatosensory feedback of perturbation in air sac pressure to abdominal expiratory muscles aids birds in precise ventilatory control necessary for song. (a) A puff of air injected into the air sac during the third syllable in this northern cardinal song results in a compensatory reduction in EMG amplitude in the abdominal expiratory muscle. The air sac pressure increases as a result of the injection and the following minibreath inspiration (arrows) is smaller than normal, presumably because of the added volume of air. (b) Reduction in mean amplitude of the abdominal expiratory muscle EMG, compared with control, after injection of air puffs. Seven different syllable types in hearing birds (gray bars), and three syllable types from deafened birds (black bars) are shown. Mean ± SE. Syllables sung by hearing birds: $p < 0.013$; syllables sung by deaf birds: $p < 0.001$; paired t test. F_{inject}, rate of airflow through the injection cannula from the picospritzer. The large upward inflection in the flow rate indicates the time course of the injected air puff. ABD, abdominal expiratory muscle EMG activity shown rectified (time constant 0.1 ms) (downward) and integrated (time constant 5 ms) (upward); P, subsyringeal air sac pressure (bracket = 1 kPa or 10 cm H_2O). A, sound amplitude rectified (time constant 0.1 ms) and integrated (time constant 1 ms). Other abbreviations as in Figure 8.4. (Modified from Suthers et al., 2002.)

to sing with the aid of auditory feedback, they may also learn to use the proprioceptive or mechanoreceptive feedback associated with each syllable. In addition to adjusting the central motor pattern for changing conditions at the vocal and respiratory periphery, somatosensory feedback might enable the bird to preconfigure motor parameters for a syllable prior to the onset of phonation when auditory feedback is not available or to adjust these parameters more quickly during phonation.

Bilateral motor skills and vocal complexity

Songbirds have taken advantage of their duplex vocal organ in several ways to increase their vocal versatility. In species studied to date, although there is usually considerable overlap between sides in mid-range frequencies, the left side of the syrinx has a lower frequency range than the right side (Suthers, 1999; Zollinger and Suthers, 2004). The presence of lateral specializations for producing low or high frequencies increases the range of frequencies that can be included when both sides of the syrinx contribute to song. In brown thrashers and catbirds, the right side of the syrinx is responsible for more of the frequency modulation (FM) and amplitude modulation (AM), particularly for rapid AM, than the left side (Suthers *et al.*, 1994) (Figure 8.7). This functional lateralization creates a division of labor that allows lateral specialization of vocal skills and increases the acoustic diversity of sounds that can be produced using both sides of the syrinx.

Borror and Reese (1956) and Greenewalt (1968) were among the first to demonstrate that the songs of many birds contain two simultaneous, independently modulated frequencies. They suggested that the components of these two-voice syllables originated on opposite sides of the syrinx. This view received further support from experiments on chaffinches (*Fringilla coelebs*) (Nottebohm, 1971a, b) canaries and white-crowned sparrows (*Zonotrichia leucophrys*) (Nottebohm and Nottebohm, 1976), which showed that disabling either side of the syrinx by section of the ipsilateral tracheosyringeal nerve caused the loss of certain specific song elements. In these and subsequent similar experiments by other investigators (reviewed by Suthers, 1997), denervating the left side of the syrinx resulted in more syllables being lost or altered than when the right side was denervated. This lateralization of song production was most prominent in the Waterslager strain of

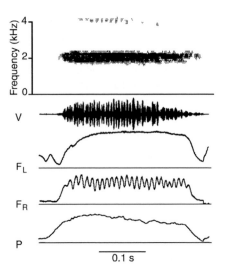

Figure 8.7 Rapid, cyclical amplitude modulation in brown thrasher song produced by modulating airflow through right side of syrinx. Abbreviations in legend of Figure 8.4. (Modified from Suthers *et al.*, 1994.)

canaries, bred selectively for their song. About 90% of the Waterslager song repertoire is lost after disabling the left side of the syrinx, but only about 10% is eliminated if the right side is disabled (Nottebohm and Nottebohm, 1976; Hartley and Suthers, 1990). A similar strong left side syringeal dominance is present in Waterslager canaries singing spontaneously with both sides of their syrinx intact (Suthers, 1992). This extreme left dominance of song production is not a universal characteristic of the species, however. The conspecific outbred domestic canary strain produces about an equal number of notes on each side (Suthers *et al.*, 2004). In similar experiments on zebra finches, in which one side of the syrinx was disabled, one study (Williams *et al.*, 1992) found a modest right side dominance for some song parameters, but another (Floody and Arnold, 1997) reported a left dominance. New direct measures from birds with a bilaterally intact vocal system reveal complex bilateral contributions with little lateralization except for a right side dominance in some calls (see Goller and Cooper, this volume).

In addition to the coordinated patterns of syringeal and respiratory motor systems that actively control most phonatory behavior in songbirds, passive biomechanical properties of the syrinx can contribute to the diversity and complexity of birdsong. The left and right side of the syrinx do not always function independently. The "dee" syllable of the black-capped chickadee (*Parus atricapillus*) contains a complex array of sum and difference frequencies produced by cross-modulation of sound from the two sides of the syrinx (Nowicki and Capranica, 1986a, b). A similar non-linear interaction between the two sides of the syrinx also occasionally occurs in the northern mockingbird (S. A. Zollinger, T. Riede and R. A. Suthers, personal observation) and in the magpie tanager (*Cissopis leveriana*) (Mindlin and Laje, 2005). Bilateral interactions are surprisingly uncommon in birds that have been studied.

Vocal complexity can also arise from a single side of the syrinx when the intrinsic nonlinear properties of the labia can cause abrupt jumps between different oscillatory modes to produce phenomena such as biphonation (in which a single pair of oscillators generate two different frequencies), abrupt frequency jumps, subharmonics or deterministic chaos (Fee *et al.*, 1998; Fletcher, 2000; Fee, 2002; Beckers and ten Cate, 2006; Suthers *et al.*, 2006) (Figure 8.8). Because these complex vocalizations do not require active motor control, two seemingly contradictory hypotheses for their communicative significance have been proposed. One idea is that because they may be "unplanned", involuntary by-products of nonlinear dynamics, they might be indicators of syringeal or respiratory motor constraints or of poor motor control. Alternatively, if a songbird can learn to control the peripheral conditions that trigger them, they may be able to actively exploit these passive biomechanical properties of their syrinx to increase vocal complexity without the need to learn a complex mechanism for motor control.

Songbirds have also exploited their ability to independently control sound production on each side of the syrinx in ways that enhance particular acoustic effects characteristic of their species-specific songs (Suthers, 1997, 1999, 2004). The brown thrasher, for example, includes in its song many two-voice syllables to which independently modulated simultaneous contributions from each side add a dissonant quality (Figure 8.9a). The songs of brown-headed cowbirds begin with two or three note clusters separated by a brief inspiration, followed by a high frequency "whistle." The first note in each note cluster is of very low frequency and is sung on the left side of the syrinx. Subsequent notes are produced in rapid succession on alternate sides with successive notes on each side starting at a higher frequency than the previous one (supplementary video http://www.indiana.edu/~songbird/research/Cowbird%20Movie%207.mov) (Figure 8.9c). Northern cardinal song includes many frequency modulated sweeps spanning a broad range of frequencies. The portion of each sweep that extends above about 3.5 kHz is generated by the right side of the syrinx and the portion below this frequency is generated by the left syrinx. In most syllables the (supplementary video http://www.indiana.edu/~songbird/research/Cardinal%20Movie%207.mov) two sides are coordinated so that their combined output forms a single continuous sweep (Figure 8.9d). Domestic canaries also use both sides of their syrinx to produce about one third of their repertoire. Each side of the syrinx contributes distinct, unconnected notes that are sung sequentially, not simultaneously. Even in syllables where the two sides sweep over complementary frequency bands, the notes are not connected, as they are in cardinal song, but remain temporally distinct (Suthers *et al.*, 2004) (Figure 8.9b). It thus appears that the distinctive acoustic features that characterize each species' song depend on similarly distinctive species-specific vocal motor patterns.

Did the vocal motor patterns in these species diverge as a way to increase the complexity of their songs despite production constraints on how sounds can be produced, or did they arise through developmental or evolutionary pressures unrelated to acoustic production? To answer this question juvenile northern mockingbirds, which are vocal mimics, were tutored with heterospecific song and the motor pattern they used to copy that song when they became adults was compared with the motor pattern of the tutor. If the tutor species' motor pattern has become specialized through selection to produce the salient acoustic features of that song then the mockingbird's success in copying the song should depend on the accuracy with which he reproduces the tutor's vocal motor pattern. If on the other hand there is not an obligatory relationship between a particular motor pattern and a certain acoustic output then, in the process of trial and error motor learning, the mockingbird should sometimes develop a different motor pattern for copying the tutor's song (Zollinger and Suthers, 2004).

Mockingbirds tutored with cardinal or cowbird song generally used the same motor pattern as the tutor when they copied his song or acoustically similar

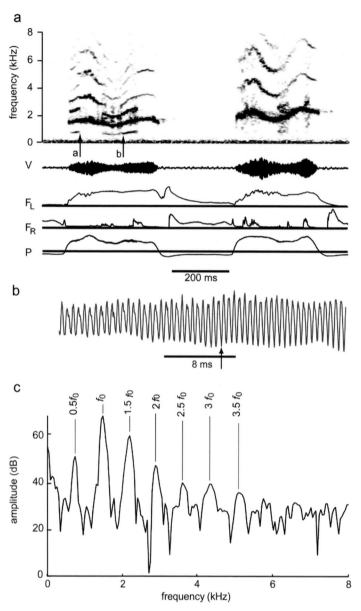

Figure 8.8 Nonlinear phenomena are not uncommon in the song of northern mockingbirds. During most of each syllable there is airflow through only the left side of the syrinx, ruling out the possibility of two voices. (a) Subharmonic frequencies are visible in the spectrogram (top panel, arrow a) as sound energy at half the fundamental frequency ($0.5\,f_0$) and its harmonics (e.g. $1.5\,f_0$, $2.5\,f_0$, etc.). Note the abrupt onset of these period doublings. (b) An expanded view of the sound waveform centered on arrow a, illustrates the sudden transition (c. 10 cycles) in oscillatory modes as subharmonic ends. (c) Power spectra taken at arrow b, showing spectral peaks at f_0 and its higher harmonics ($2\,f_0$, $3\,f_0$ etc.) as well as subharmonics at intervals of $\tfrac{1}{2}\,f_0$ ($0.5\,f_0$, $1.5\,f_0$, etc.). See Figure 8.4 for abbreviations.

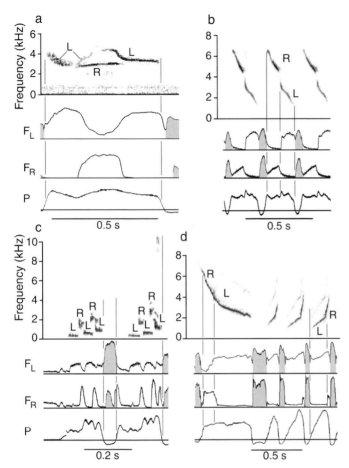

Figure 8.9 Species differences in patterns of syringeal lateralization. (a) Independently modulated, simultaneous contributions from the right and left sides of the syrinx in brown thrasher song (two-voice). (b) Domestic canary syllables which, like cardinal FM sweeps, are produced by sequential contributions from the right and left sides of the syrinx, but unlike the cardinal, the notes are not connected to form a continuous sweep. (c) Brown-headed cowbird song note clusters are produced by rapid alternation between phonation on the right and left sides, giving rise to abrupt frequency steps between notes. Final whistle at end of song (not shown) is sung on the right side. (d) Broadband, frequency-modulated (FM) sweeps by northern cardinals are produced by sequential coordinated contributions of the right and left sides. Frequencies higher than ~3.5 kHz are generated on the right side, lower frequencies are produced by the left side. Most cardinal syllables are formed by coordinated switches mid-syllable between sides. (a, modified from Suthers *et al.*, 1996b; b, modified from Suthers *et al.*, 2004; c, modified from Allan and Suthers, 1994.) Abbreviations as in Figure 8.4.

computer synthesized sounds. When the mockingbird's motor pattern differed from that used by the tutor, the vocalization was also a less accurate copy (Figure 8.10). It is particularly interesting that the mockingbird seemed to have the most difficulty reproducing the "special", species-typical acoustic features of other species' songs. Examples of these features include the cardinal's smoothly coordinated switch from one side of the syrinx to the other in the middle of its frequency sweep and the cowbird's ability to sing successive notes, having little or no silent interval between them, at abruptly different frequencies without connecting them by a slurred FM transition (Figure 8.10). The developmental convergence of model and mimic onto the same motor pattern when producing similar sounds, suggests that natural selection has pushed the envelope of motor performance in different directions for different species, according to

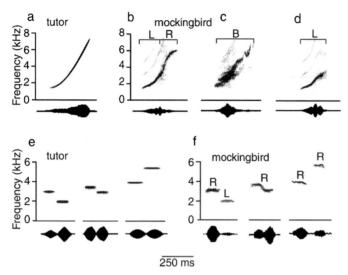

Figure 8.10 Mockingbird copies of synthesized tutor sounds. When copying cardinal-like synthesized FM sweeps (a), mockingbirds were able to reproduce both the tonal quality and wide bandwidth of the sweep only when they used a cardinal-like motor pattern, switching smoothly mid-syllable from the left to right side of the syrinx (b). If mockingbirds attempted to sing the sweep with both sides of the syrinx open, the resulting copies were not tonal (c). If sweeps were produced using only one side of the syrinx (d), the bandwidth of the sweep was greatly reduced (mean BW unilateral = 2.96 ± 0.715 kHz, mean bilateral = 4.7 ± 0.52 kHz. $t = 8.761$, df = 38, $p < 0.001$, $N = 198$ sweeps). Abrupt, step-like frequency jumps (e) were only accurately copied by mockingbirds that used a cowbird-like alternating phonation pattern, switching between sides to produce successive tones (first tone pair in f). If mockingbirds did not switch between sides, they could either retain the immediate onset of the second frequency, but introduce a slur between tones as tension on labia changed (center tone pair in f), or insert a short silent interval between tones (last tone pair f), preserving the spectral contrast between tones, but altering the temporal pattern. (Modified from Zollinger and Suthers, 2004.)

the biologically significant acoustic properties of their respective songs. Each species has become an acoustic specialist for producing its style of song with a skill that the mockingbird, as a vocal generalist, has difficulty equaling.

Modeling song complexity

With all the potential sources for complexity, it may seem like an insurmountable task to understand how birds produce their songs. Each syllable type requires a unique, precisely timed, complex motor pattern that, especially at high syllable repetition rates, must often push the performance limits of both the bird's vocal system and of the neuroscientists striving to understand it! In this context it is noteworthy that some recent theoretical models for synthesizing birdsong (Gardner *et al.*, 2001; Laje *et al.*, 2002; Mindlin *et al.*, 2003) suggest that a significant portion of song diversity can be generated by controlling the temporal relationship between the smoothed oscillations of as few as three control parameters consisting of the respiratory pressure, the position of the labia (as indicated by the activity of the dorsal syringeal muscles) and the tension on the labia (represented by activity of the ventral syringeal muscles). Investigations of the central neural mechanisms that determine the temporal relationship between these putative control parameters may yield new insights into the motor control of song diversity that are not apparent from the neural correlates of specific acoustic features (Suthers and Margoliash, 2002).

FROM SYRINX TO SONG: THE VOCAL TRACT

The vocal tract consists of the air-filled passages through which sound must pass as it travels from the vocal organ to the outside environment. In terrestrial mammals, which generate vocalizations by vibration of vocal folds in the larynx at the cranial end of the trachea, the vocal tract includes the pharyngeal, nasal and oral cavities. Since the vocal organ in birds is located at the caudal end of the trachea, the trachea is included as part of the avian vocal tract. Here we consider the role of the avian vocal tract in determining the acoustic properties of vocal signals.

Behavioral importance of the vocal tract acoustics

The vocal tract has an important role in determining the amplitude spectrum (i.e. the distribution of sound energy as a function of frequency) of vocalizations. The syringeal muscles in birds control the sound's fundamental frequency by varying the tension they exert on the labia, which in turn determines the frequency at which they vibrate. The resulting changes in fundamental frequency are perceived as changes in pitch. The vocal tract, on the other hand, controls the relative intensity of different harmonics or overtones, which determine the timbre of the vocalization. The vocal tract thus modifies the broadband syringeal sound by amplifying some frequencies and attenuating others. Birds can control which frequencies are emphasized in their songs or calls and which are suppressed by varying the shape of their vocal tract.

The way in which a bird uses its vocal tract to modulate the spectral properties of its vocalizations can determine the perceptual salience of different acoustic parameters that may be used to convey information from the sender to the receiver (Williams *et al.*, 1989; Fitch and Kelley, 2000). In many birds the vocal tract amplifies the fundamental frequency and attenuates higher harmonics to produce a more or less pure-tonal, whistle-like vocalization having most of its energy in the fundamental. Supporting the transmission of this dominant frequency component of the sound produced by the syrinx is an efficient way to increase the loudness and tonal purity of the emitted vocalization (Nowicki, 1987; Fitch and Kelley, 2000; Beckers *et al.*, 2003; Riede *et al.*, 2006). When given a choice, many birds seem to prefer pure-tonal syllables over those with strong harmonics (Nowicki *et al.*, 1992, 1989; Peters and Nowicki, 1996; Strote and Nowicki, 1996). Songs of this type are good for communicating over long distances since they are relatively resistant to degradation by environmental noise (Dabelsteen *et al.*, 1993) and repertoire size may be increased by different modulation patterns of the dominant frequency (Williams *et al.*, 1989; Fitch and Kelley, 2000).

Songs of other species, however, are not whistle-like, but contain prominent upper harmonics, or even nonperiodic components having a broad bandwidth, as in the "screech" of some parrots (Fletcher, 2000). Instead of emphasizing the fundamental in these vocalizations, the vocal tract often selectively transmits energy in certain higher frequency bands called *formants*. Formants are the primary means for conveying meaning in human speech (Fant, 1960; Titze, 1994; Fitch and Kelley, 2000). English vowels, for example, are perceived on the basis of their unique formant frequencies, which are determined by articulatory movements of the tongue and lips (Peterson and Barney, 1952), not by the larynx. Formants are present in many avian vocalizations (Williams *et al.*, 1989; Cynx *et al.*, 1990), but surprisingly little is known about their possible importance in vocal communication (reviewed by Fitch and Kelley, 2000). The flexible, dynamic control of formants is something that birdsong shares with speech, and that differentiates it from most other mammalian vocalizations, in which formants are static. Parrots and mynahs imitate human formants when they copy speech sounds (Klatt and Stefanski, 1974; Nottebohm, 1976; Warren *et al.*, 1996) and some birds can be conditioned to respond to differences in the formants of human speech (Heinz *et al.*, 1981; Dooling and Brown, 1990; Dooling, 1992).

The ability of birds to perceive and respond to formants in their own calls has so far been demonstrated only in whooping cranes (*Grus americana*) (Fitch and Kelley, 2000). These and some other cranes are among several species of nonsongbirds that have evolved elongated tracheas which form loops or coils inside their body. Fitch (1999) showed that the amount of tracheal elongation is inversely correlated with the spacing between the formants in their calls and hypothesized that tracheal elongation evolved as a means of exaggerating the sender's body size.

Vocal tract resonance and damping

A detailed discussion of how the vocal tract modulates sound to produce a pure-tonal or formant structure is beyond the scope of this chapter. An introduction to basic vocal tract acoustics can be found in Fletcher

(1992), Mindlin and Laje (2005) and Titze (1994). Briefly, when the sound pressure wave traveling up the vocal tract from the syrinx reaches the bird's open mouth, or encounters changes along the way in vocal tract shape, some of its energy is reflected back down the vocal tract toward the syrinx. If the phase relationship between the pressure wave emitted by the syrinx and the reflected wave is such that the positive and negative pressure peaks of one coincide with the corresponding pressure peaks of the other, these waves undergo constructive interference resulting in a standing wave of relatively high amplitude in the vocal tract. The frequency at which the maximum constructive interference occurs is called the natural or resonant frequency of the vocal tract. At frequencies that are not close to a vocal tract resonance, positive pressures in the syringeal waveform may coincide with negative pressures in the reflected wave, resulting in destructive interference, i.e. damping, and reduce the amplitude of that frequency in the vocalization. Vocal tract resonance thus behaves like an acoustic filter that selectively amplifies sounds near the resonant frequency and attenuates other frequencies.

The importance of vocal tract resonance in birdsong was once a controversial subject (Hersch, 1966; Greenewalt, 1968; Gaunt and Gaunt, 1985; reviewed by Nowicki and Marler, 1988). This view changed, however, after it was shown (Nowicki, 1987) that when songbirds sang in a light gas mixture consisting of 80% He and 20% O_2 (heliox) the highest amplitude frequency component of their song shifted from the fundamental, which was most prominent in air, to the second harmonic (Figure 8.11a, b, e). This shift occurs because heliox increases the speed of sound about 74%, compared with air, making the wavelength of the second harmonic almost as long as that of the fundamental in air.

Relationship between the syringeal source and the vocal tract resonance

In most mammalian and avian vocalizations the resonant oscillation of the air column in the vocal tract is not believed to directly affect the motion of the vocal folds, labia or other vibrating structures that generate sound in the vocal organ. Studies of vowel production in human speech show that the acoustic or biomechanical coupling between the vocal tract resonance and the vocal folds is normally very weak so that the frequency and amplitude of vocal fold oscillation is not affected by the resonance filter in the vocal tract. This acoustic independence between the vocal organ and vocal tract is the basis of the source-filter theory of speech production (Fant, 1960). In human speech, separation between the motion of the vocal source and the acoustics of the vocal tract allows the vocal organ to control the fundamental frequency and amplitude of vocal fold oscillation independently from articulatory changes in the shape of the vocal tract (Titze, 1994) that modulate the timbre of the vocalization. This independence, though important in speech production, is overridden in many songbirds which adjust their vocal tract resonance to match the fundamental frequency of their relatively stereotyped song. Interestingly, human professional sopranos, like some songbirds, also adjust their vocal tract resonance to match high fundamental frequencies in order to be heard over the orchestra (Sundberg, 1975; Joliveau et al., 2004).

An alternative to source–filter separation is source–filter interaction in which the resonant oscillation of the air column in the vocal tract drives the labia to vibrate at the same frequency. A model developed by Laje and Mindlin (2005) suggests that acoustic interaction between the source and the vocal tract involving acoustic feedback might occur in some avian vocalizations, and could provide a mechanism for increasing vocal complexity, such as the appearance of subharmonics. Paulsen (1967) found evidence for acoustic interaction between the syrinx and vocal tract in an excised preparation of goose syrinx and trachea. Goller et al. (2004) (and Goller and Cooper, this volume), suggest that an abrupt beak opening associated with non periodic oscillation during sudden acoustic transitions in the songs of zebra finches might produce an impedance change that could allow acoustic feedback to couple the vocal tract to the labia.

Neither the source–filter separation nor the source–filter interaction hypothesis has been rigorously tested in songbirds or any other passerine species. An experiment designed to discriminate between them has as yet only been performed on doves. Beckers et al. (2003) found that when ring doves (*Streptopelia risoria*) or collared doves (*S. decaocto*) produce the almost pure tone portions of their coos, the sound in the interclavicular air sac surrounding the syrinx contained prominent harmonic overtones originating in the syrinx, but which are not present in the vocalization. This indicates

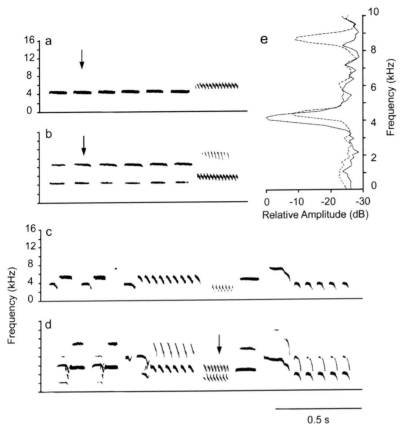

Figure 8.11 Spectrographic illustration of the effect of a light gas mixture (heliox) on the harmonic composition of song sparrow song. In normal air, both song types are pure tone, with only the fundamental present (a, c), but in heliox the second harmonic also becomes prominent (b, d). The arrow in (a) and (b) marks the note for which amplitude spectra (e) are shown in normal air (solid line) and heliox (dashed line). Arrow in (d) shows second harmonic in heliox at same frequency as the fundamental of the next syllable. In air (c) this second harmonic is suppressed, indicating that the bird changes the resonant tuning of its vocal tract between these two syllables. (Redrawn after Nowicki and Marler, 1988.)

that doves have an independent vocal tract filter (Gaunt et al., 1982; Fletcher et al., 2004; Riede et al., 2004), which attenuates all but one of the harmonics present in the sound coming from the syrinx. If the vocal filter was coupled to the syringeal sound source, the spectrum of syringeal sound should be similar to that of the vocalization. It remains to be determined if this finding can be generalized to other avian groups since, unlike songbirds, doves vocalize with their mouth and nares closed and there are important anatomical differences in the syringeal oscillators (King, 1989; Ballintijn et al., 1995; Goller and Larsen, 1997b), respiratory dynamics and vocal tract acoustics (Gaunt et al., 1982; Fletcher et al., 2004; Riede et al., 2004).

Tuning the vocal tract filter

A vocal tract with fixed resonant frequencies could be maladaptive. The fundamental frequency of birdsong often changes between notes and individual notes may have frequencies changing over an octave or more. In these cases a fixed-frequency vocal tract filter might attenuate more song than it amplifies! Nowicki's (1987) experiments with birds singing in light gas show that songbirds have avoided this problem by evolving an ability to vary, or tune, the frequency of their vocal tract resonance during the course of their song so that it matches the song's changing fundamental frequency. Nowicki noted that when birds sing in air, the second harmonic is suppressed regardless of its absolute frequency (Figure 8.11c, d). Even when the fundamental frequency of one note was the same as the second harmonic of an adjacent note, the second harmonic, but not the fundamental, was suppressed in both notes when sung in air. A major factor in determining the resonant frequencies of the vocal tract is the relationship between dimensions of the tract and the wavelength of the sound. The ability of at least some birds to adjust their vocal resonance so that it tracks their song's fundamental frequency implies that song is accompanied by a motor pattern that appropriately adjusts the dimensions of the vocal tract based on the frequency of the fundamental being generated in the syrinx.

Despite some recent advances, much remains to be learned regarding the anatomical and physiological basis of this adjustable vocal tract filter and the means by which it is coordinated with syringeal motor patterns controlling the fundamental frequency. It nevertheless has important implications for the integration of diverse motor programs required for singing, including the role of sensory feedback during song production. The speed with which a bird can change the shape of its vocal tract in order to adjust the resonant frequency of its filter may impose a performance constraint on song in the form of a trade-off between the bandwidth and repetition rate of notes in trilled phrases, which in some species affect female mate choice (Podos, 1997; Ballentine *et al.*, 2004; Podos and Nowicki, 2004b; Ballentine, 2006).

Resonance filters in the songbird vocal tract

The filter characteristics of the intact avian vocal tract depend on the acoustic properties of its various components (Figure 8.12a). Although there is some acoustic interaction between these different compartments, it is useful to consider them individually.

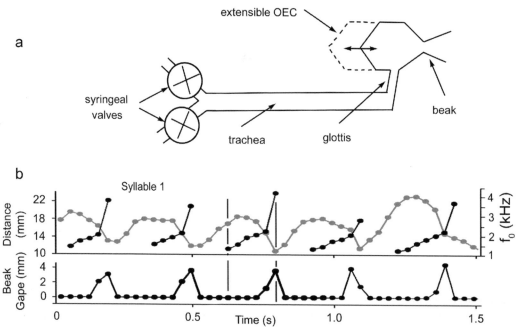

Figure 8.12 (a) Schematic diagram of the cardinal's vocal tract showing its major components. (b) Each song syllable is accompanied by coordinated movements of the larynx and hyoid that maintain an inverse relationship between the size of the OEC and the song's fundamental frequency. Upper panel shows five upward sweeping FM syllables (black; right axis) superimposed on cyclical movements of the larynx (gray, left axis). Upward deflection of gray curve indicates increasing dorsoventral displacement of larynx and increased volume of the oropharyngeal-esophageal cavity (OEC). Lower panel indicates beak opening, which coincides with high fundamental frequency. When f_0 was less than about 2 kHz, beak gape was usually too small to measure on the fluoroscopic images and, though slightly open, was recorded as zero. Each data point represents measurement from one frame of x-ray movie at 30 frames s^{-1}.

Our attention here focuses on the suprasyringeal vocal tract, but it should be noted that the syrinx is suspended in the interclavicular air sac, which is pressurized during song and separated from the syringeal labia only by the thin-walled primary bronchi and the medial tympaniform membranes. The latter are continuous with the posterior edge of the medial labium and form the lateral walls of the interbronchial lumen. The interclavicular air sac is thus separated from the vibrating labia by structures that must be almost transparent acoustically yet there are no data on if or how the resonant properties of the air sac affect vocalization.

The trachea

The resonant properties of the trachea, which is the first component of the suprasyringeal avian vocal tract, can be modeled acoustically as a simple tube that is closed at the syrinx and open at the glottis. The primary resonant frequency of such a tube corresponds to the sound frequency that has a wavelength about four times the length of the tube. This primary resonance is accompanied by a series of higher resonances close to the odd numbered harmonics (Titze, 1994).

The predicted quarter wavelength primary tracheal resonance of a northern cardinal or eastern towhee, for example, is about 1.8 to 1.9 kHz (Nelson et al., 2005; Fletcher et al., 2006; Riede et al., 2006) and that of a zebra finch is about 2.6 kHz (Daley and Goller, 2004). The resonant frequency of the trachea could in theory be raised or lowered by shortening or lengthening it, respectively, but tracheal length appears to change very little during song. A singing zebra finch, for example, changes its tracheal length a maximum of only 3%, altering its resonant frequency (~3–6 kHz) by less than 200 Hz (Daley and Goller, 2004).

The glottis

A larger change in the tracheal resonance might be achieved by constricting the glottis at the cranial end of the trachea. Glottal constriction should lower the trachea's resonant frequency. This might be useful in tuning the vocal tract resonance to a low fundamental frequency that is below the tracheal resonance when the glottis is open. The ring dove's coo, for example, has a fundamental frequency around 600 Hz whereas the calculated primary resonance of its open-ended trachea is about 1180 Hz (Riede et al., 2004). Unfortunately, there are no data on glottal aperture in birds during song.

The oropharynx and esophagus

The oral cavity and pharynx in birds form a single chamber called the oropharynx. Sound from the trachea enters the oropharynx through the glottis. Despite the prominent throat movements that often accompany birdsong, until recently nothing was known about the shape of the oropharyngeal cavity during vocalization or its acoustic significance in songbirds. Riede et al. (2006) used x-ray cinematography to measure changes in the dimensions of this portion of the vocal tract of singing northern cardinals. These movies revealed large song-related cyclical changes in the shape of the vocal tract between the glottis and the beak, such that its volume is inversely correlated with the fundamental frequency of the song (Riede et al., 2006) (Figure 8.12b); (supplementary video http://www.indiana.edu/~songbird/multi/supplementary-video1.mov).

At low fundamental frequencies the oropharyngeal cavity is enlarged by a downward and forward movement of the larynx and the hyoid apparatus to which it is attached. The cervical portion of the vertebral column also moves slightly backward. At low frequencies the volume of the vocal tract between the glottis and the beak is further increased by an expansion of the cranial portion of the esophagus. These movements produce a single large oropharyngeal-esophageal cavity (OEC) which reaches its maximum volume of about 2 ml in the cardinal when the fundamental frequency is lowest (between 1 and 2 kHz) (Figure 8.13).

The expansion of the cranial esophagus contributes about half of the total volume. When, during an upward sweeping FM note, the fundamental frequency rises, the cranial end of the esophagus collapses and the oropharynx becomes smaller as the larynx returns to its pre-phonatory position. At 5 kHz the volume of the OEC is about 0.6 ml. This sequence of events is reversed during syllables that start at a high frequency and sweep downward.

Changes in the OEC are presumably controlled primarily by muscles of the hyoid apparatus, which is highly mobile in birds (Homberger, 1999). The specific contribution of particular muscles has not been determined. The means by which this song-related oropharyngeal motor pattern is coordinated with motor patterns to respiratory and syringeal muscles awaits further study (see Wild, this volume).

A computational acoustic model of the cardinal vocal tract (Fletcher et al., 2006), based on the dimensions

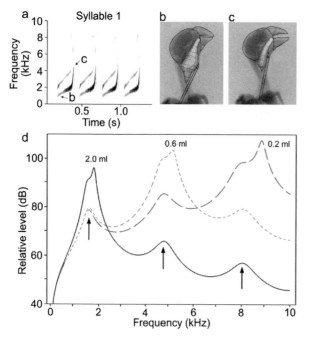

Figure 8.13 (a) Sound spectrogram of syllable 1 that is also shown in Figure 8.12. (b and c) Three dimensional reconstructions of oropharyngeal-esophageal cavity (OEC) for syllable 1, showing it (b) at its maximum volume of 2 ml at beginning of syllable (about 1.5 kHz) and (c) at a volume of 0.6 ml at the end of syllable (about 5 kHz). (d) Predicted resonance curves when volume of OEC is 2 ml and beak gape is small at start of syllable 1 and at 0.6 ml with a large beak gape at end of syllable 1. In both cases, the primary resonance peak closely matches the fundamental frequency. By further reducing the volume of the OEC to 0.2 ml the bird could tune the most prominent resonance of its vocal tract to the highest fundamental frequency observed in some other syllable types. Arrows indicate tracheal resonances.

obtained from cineradiographic images during song, indicates that cardinals actively adjust the volume of the OEC as they sing so that its primary resonance tracks (i.e. matches) the fundamental frequency of the song (Figure 8.13d). Riede et al. (2006) estimate that tuning the most prominent resonance of the OEC to the fundamental frequency may increase its level about 11 to 15 dB, significantly increasing the loudness of the fundamental in the vocalization and the extent to which it dominates the frequency spectrum of the radiated song.

The bird's ability to actively vary the volume of the OEC as the song's fundamental frequency changes appears to provide an important part of the physiological mechanism responsible for the tracking resonance filter. Expansion of the esophagus, which is not normally part of the vocal tract, extends the primary resonance to lower frequencies than might otherwise be possible for a relatively small bird.

Inflatable vocal sacs associated with the esophagus have evolved independently in several nonoscine groups where they probably function as resonant chambers and in some species may even generate percussive sounds (Dantzker and Bradbury, 2006), but none is known to track the fundamental frequency generated in the syrinx. Except for doves (Fletcher et al., 2004; Riede et al., 2004) and sage grouse, their acoustic function, though undoubtedly important, is poorly understood. In some species inflation of a bare sac provides a concomitant visual display. Male greater sage grouse (*Centrocercus urophasianus*) repeatedly inflate a pair of bare vocal sacs that direct sound laterally during their strut display on the lek. Since displaying males do not face females, this laterally directed beam may maximize sound intensity in the direction of the female (Dantzker et al., 1999).

The tongue

Tongue movement related to vocalization has not been measured in songbirds. However, vocalizing parrots move their relatively large tongues in ways suggestive of an acoustic role in vocalization (Patterson and Pepperberg, 1994, 1998; Warren *et al.*, 1996; Homberger, 1999). Formants are prominent features of many parrot vocalizations. Beckers *et al.* (2004) showed that small changes in the tongue position of euthanized monk parakeets (*Myiopsitta monachus*), in which the syrinx was replaced by a miniature speaker, altered the formant frequencies of the vocal tract.

Beak opening

Birdsong is often accompanied by prominent beak movements. Nowicki and Marler (Nowicki, 1987; Nowicki and Marler, 1988) suggested birds might use beak opening to adjust the resonance of their vocal tract. Westneat *et al.* (1993) were the first to show that beak gape in songbirds (i.e. the distance between the tip of the mandible and the tip of the maxilla, or upper beak) is correlated with the song's fundamental frequency over at least part of its frequency range, being large at high frequencies and small at low frequencies. Changes in beak gape alter the length of the vocal tract, modeled as a simple tube, and by altering the acoustic impedance between the mouth and environment, change the tract's end-correction. Increasing the gape reduces the effective acoustic length of beak (Fletcher and Tarnopolsky, 1999; Fletcher *et al.*, 2006), which should raise the resonant frequency of the vocal tract.

Since the initial studies by Westneat *et al.* (1993) on white-throated sparrows (*Zonotrichia albicollis*) and swamp sparrows (*Melospiza georgiana*), a similar positive relationship between beak gape and fundamental frequency has also been reported in other species studied, including song sparrows (*Melospiza melodia*) (Podos *et al.*, 1995) and Bengalese finches (*Lonchura striata*) (Moriyama and Okanoya, 1996). In northern cardinals beak gape is correlated with fundamental frequency below about 3.5 kHz, but not at higher frequencies (Suthers, Goller, Bermejo, Wild and Zeigler reported in Suthers and Goller, 1997). A positive correlation between gape and frequency was also present in two thirds of the individuals representing seven species of Darwin's finches, although there was substantial individual variation in the slope of the regression within some species (Podos *et al.*, 2004b). In her videographic study of courtship dances by male zebra finches, Williams (2001) observed a correlation between beak aperture and peak frequency (the harmonic containing the most energy in a syllable composed of a harmonic stack). A more detailed investigation of zebra finch beak dynamics by Goller *et al.* (2004) further revealed that the beak is nearly closed during the introductory notes and open less than 1 mm during most of the song. This small beak opening is modulated during song such that gape is positively correlated with the fundamental frequency of all different note types except harmonic stacks, during which normalized beak gape was correlated with the peak frequency, as reported by Williams, instead of with the fundamental (Goller *et al.*, 2004).

When the beak is held at a fixed opening by a bite block or open wire frame, the amplitude spectrum of the vocalization is altered in a manner consistent with the hypothesis that a small gape attenuates high frequencies. If a sparrow's beak is immobilized at a relatively small gape, high frequencies are attenuated more than low frequencies (Hoese *et al.*, 2000). Fixing a cardinal's beak at a large opening (e.g. 7 mm) is accompanied by a 10 to 20 dB increase in the magnitude of the second harmonic, relative to the fundamental for fundamental frequencies below about 3 kHz (Suthers *et al.*, 1996a; Suthers and Goller, 1997). In zebra finches, a gape fixed at either a small or large opening shifted the emphasis in harmonic syllables to a lower or higher harmonic, respectively, but had little effect on other syllable types (Goller *et al.*, 2004). Beak movements can be distorted by attaching weights to the mandible. In the canary, this caused the mandible to overshoot its normal maximum gape and increased the relative amplitude of the second harmonic (Hoese *et al.*, 2000).

As the interface between the vocal tract and the environment, beak gape may influence acoustic variables other than vocal tract resonance, such as the amplitude (see Goller and Cooper, this volume) and directionality of song. In the European blackbird (*Turdus merula*) (Larsen and Dabelsteen, 1990) and eastern towhee (Nelson *et al.*, 2005) increasing gape was accompanied by a frequency-dependent increase in the directionality of the sound radiating from the beak and an increase in the sound pressure level that is most pronounced for frequencies above about 4 kHz.

A quantitative assessment of the effect of gape on vocal tract resonance requires knowledge of the source spectrum. Nelson *et al.* (2005) removed the syrinx from euthanized eastern towhees and replaced it with a miniature speaker that fit into the base of the trachea and generated a known source spectrum. They found that changes in beak gape alone had little effect on vocal tract resonance below about 4 or 5 kHz, but higher frequencies up to 7.5 kHz were attenuated as gape was reduced. The data suggest that instead of causing the vocal tract resonance to track the fundamental frequency, reducing beak gape functions as a low pass filter with a fixed cut-off that attenuates higher harmonics of fundamental frequencies below about 3.5 kHz. The subsequent discovery of frequency-related changes in the volume of the OEC requires that these data be interpreted with caution since the volume of the OEC in the towhee preparation remained small and essentially constant across all beak gapes. Nevertheless, this result is compatible with evidence from live songbirds that changes in gape primarily affect frequencies above about 3.5 kHz (Nowicki, 1987; Suthers and Goller, 1997; Hoese et al., 2000; Goller et al., 2004) and with the prediction, based on a theoretical acoustic model of the beak (Fletcher et al., 2006), that its effective acoustic length is a small fraction of its actual length unless it is nearly closed.

Evidence for a tracking filter in songbirds comes from Nowicki's light gas experiments (Figure 8.11) showing that the production of song that is nearly pure-tonal is achieved by the suppression of higher harmonics, regardless of the absolute frequency of the fundamental. The sequence of song development in young birds indicates this is accomplished by some mechanism that does not involve changes in beak gape. Juvenile song sparrows learning to sing acquire adult-like tonality during plastic song when their maximum gape is <1 mm and long before their beak movements become coordinated with the fundamental frequency, which does not occur until about the time of song crystallization (Podos et al., 1995). It would be interesting to know if the appearance of tonality in juvenile song coincides with that of coordinated changes in the dimensions of the OEC during song.

Further experiments are needed to clarify the relationship between beak gape and the OEC motor patterns. The acoustic effect of the beak on vocal tract resonance does not depend only on its geometrical dimensions and gape, but also to an important degree on its relationship to the OEC (Fletcher et al., 2006). The presence of these two, often coordinated but independent, song-related motor patterns underlines the important part motor control of the suprasyringeal vocal tract has in providing flexible, sophisticated modulation in avian vocalizations.

CONCLUSIONS

Research on how birds sing continues to make important contributions at multiple levels of the peripheral vocal system to the neuroscience of birdsong. A number of recent investigations have benefited from an interdisciplinary approach that reaches beyond the traditional life sciences to include, for example, physicists and acousticians. The availability of quantitative physiological data from the vocal periphery during song has facilitated the development of theoretical models of sound production and motor control that in turn pose new questions in need of experimental answers. It is apparent that an understanding of song production must include the suprasyringeal vocal tract where the challenge of integrating the diverse motor programs required for singing has been made even more interesting by the discovery of a new motor program that coordinates changes in the volume of the oropharyngeal-esophageal cavity with the fundamental frequency generated in the syrinx. Further research is needed to determine the motor pathways, sensory mechanisms, and muscles that mediate this active resonant tuning of the upper vocal tract; however, it is clear that the avian upper vocal tract cannot adequately be modeled as simply a tracheal tube terminated by the beak.

ACKNOWLEDGEMENTS

We thank W. Tecumseh Fitch and Neville H. Fletcher for their constructive comments on portions of this article. The authors' research was supported by the NIH and NSF in the USA.

9 • Peripheral mechanisms of sensorimotor integration during singing

Franz Goller and Brenton G. Cooper

INTRODUCTION

Young songbirds, like young humans, need to experience their species-typical vocalizations early during ontogeny in order to develop their characteristic songs. The developmental path mediating song acquisition in birds is similar to that involved in human speech learning. This commonality has led investigators to ask how complex vocal behavior in birds is controlled by the brain and how these motor programs are formed during ontogeny (e.g. Doupe and Kuhl, 1999). However, while most research has focused on the brain, a thorough understanding of the *peripheral* motor mechanisms mediating song production is also necessary. Identifying the specific movements that lead to the generation of song is the first step in characterizing the central motor instructions needed for generating the behavior. Another step is to examine the contribution of acoustic feedback to the maintenance of song behavior (see Woolley, this volume). However, we know little about the role of nonauditory feedback mechanisms, (e.g. somatosensory and viscerosensory) that may be generated during the song production process (see Wild, this volume). Identifying the potential sources of such feedback and the pathways through which the information flows will help us to understand how the brain monitors and assesses feedback during singing behavior. This bottom-up approach to understanding motor control complements the top-down approach, which focuses on central control patterns (e.g. Yu and Margoliash, 1996; Hahnloser *et al.*, 2002; Schmidt *et al.*, 2004).

As is the case with human vocal behavior, song production in birds results from the complex interaction of a number of motor systems. The most obviously important of these systems is respiration, despite the fact that it receives relatively little attention. Phonation, the production of vocal sounds, is initiated by generation of a stream of air past the vocal organ, where vibrations are induced through a combination of active and passive processes. The controlled generation of this air stream requires the coordinated activity of different groups of inspiratory and expiratory muscles, the muscles of the avian vocal organ, the syrinx, and muscle systems involved in adjusting the filter characteristics of the upper vocal tract and beak (see also Suthers and Zollinger, this volume). Figure 9.1 presents a schematic diagram of the components of the song production system, their central control areas and potential feedback mechanisms.

The reliance of the phonatory mechanism on an air stream dictates that vocal behavior is both intimately linked to and constrained by the properties of the respiratory system. For example, the volume of air available in the respiratory system will limit the duration of continuous vocal elements and constrain temporal features of song. In addition, vocal behavior has to be integrated into the functional requirements of the respiratory system, such as the need for gas exchange. Similarly, because upper vocal structures, for example the beak and larynx, contribute to a variety of behaviors in addition to singing, their possible functions in song modification may be constrained by the functional requirements of these other nonvocal behaviors (Podos, 1997; Podos and Nowicki, 2004a, b). In this chapter, we review our current knowledge of how the activity of the respiratory and upper vocal tract systems is integrated into the process of song production and identify gaps in our current knowledge of that integration.

RESPIRATORY–SYRINGEAL COORDINATION

Song is typically produced during the expiratory phase of respiration by increasing pressure several fold over that driving quiet breathing (Figure 9.2). Contraction of abdominal muscles builds up air sac pressure at the onset of each expiratory pulse, while, simultaneously,

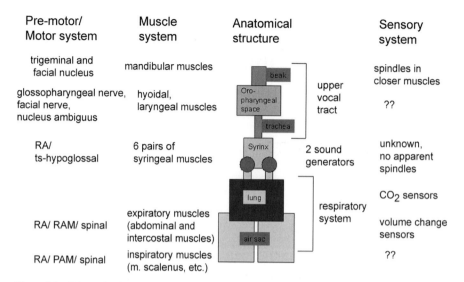

Figure 9.1 Schematic of main motor systems involved in birdsong production. Motor and premotor control areas of each system are listed together with possible sensory feedback mechanisms (based on Wild, 2004). RA, robustus arcopallialis; ts, tracheosyringeal; RAM, retroambigualis; PAM, parambigualis.

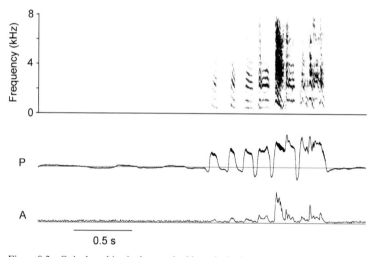

Figure 9.2 Quiet breathing is characterized by a rhythmic pressure pattern, which changes both in amplitude and temporal pattern when the bird starts to sing. One song of a male zebra finch is displayed spectrographically (top) with the corresponding air sac pressure (P; horizontal line marks ambient pressure, air sac pressure below this line is inspiratory and above expiratory, respectively) and the rectified and integrated sound amplitude (A). (Modified from Wild et al., 1998, and reprinted with permission.)

the syringeal valves are partially closed by action of the syringeal adductor muscles (Figure 9.1). Therefore, the measured subsyringeal air sac pressure represents the combined activity of the pressure generating respiratory system and the regulation of airflow by the syringeal valves (resistance to airflow). The increased pressure in combination with the constriction of the airways induces vibrations of the syringeal labia, the main sound-generating structures in the songbird syrinx (Goller and Larsen, 1997a; Goller and Suthers, 1996a, b). This syringeal–respiratory interaction initiates sound production, typically within a few milliseconds after the onset of expiratory pressurization. In contrast to the regular and rhythmic pattern of normal quiet breathing, the duration of expiratory pulses during singing varies greatly between different syllables

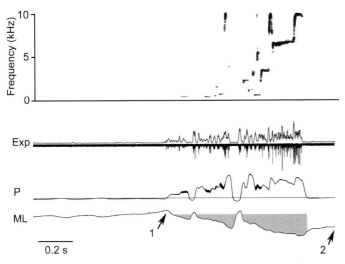

Figure 9.3 Song of the brown-headed cowbird (top, spectrogram) and motor correlates of respiratory activity: Air sac pressure, P, (horizontal line marks ambient pressure) electromyogram of the expiratory muscle sheet (Exp, rectified original trace downward and integrated upward), and sonomicrometric measurements of the length of the muscle sheet (ML, relative voltage change associated with change of distance between crystals, upward lengthening, downward shortening). Song is accompanied by a visual display, which starts with "puffing" seconds before the song. During this puffing display the expiratory muscle sheet is stretched. Muscle length increases during puffing before the song and is at its greatest length before the onset of the song (marked 1). At the end of the song muscle length does not increase to the level present during the display (marked 2). (Modified from Cooper and Goller, 2004, and reprinted with permission.)

within a species and between species (e.g. Figures 9.2, 9.3), in many cases reflecting the complex temporal patterns of songs (e.g. Hartley and Suthers, 1990; Allan and Suthers, 1994; Suthers et al., 1994; Wild et al., 1998). In most species expiratory pressure pulses are not simple square pulses, but pressure is often dynamically modulated during the course of the pressure pulse (Figures 9.2, 9.3) in a manner which suggests that expiratory effort is continuously adjusted for specific acoustic features. Such "dynamic" modulation raises the more general question of how respiratory control is integrated into the song motor program and suggests that feedback mechanisms may be operating.

To address the specific contributions of the expiratory system to the fine control of airflow past the syrinx, we examined correlations between changes in air sac pressure and two variables of muscle activity which might control such changes: electrical activation of expiratory muscles (as measured electromyographically) and resulting length changes of the abdominal muscle sheet. Electromyograms (EMG) provide information on both the timing of abdominal muscle contractions and the degree of muscular effort. Muscle length data indicate how this effort is translated into contraction of the muscle sheet. Taken together with air sac pressure, these measurements clarify the manner in which respiratory motor commands during song contribute to generation and modulation of air sac pressure, and how the relationship between air sac pressure and muscle activation varies during different parts of the song. Measurements of electrical activity, muscle length change and the resulting relative pressure at various points during long syllables in the zebra finch indicate that respiratory effort is different depending on timing within the pulse (Goller and Cooper, 2004).

At the onset of expiratory pulses, as the bird switches from inspiration to expiration, air sac pressure is still low, but EMG activity is very high (Figures 9.4, 9.5). This suggests that the pressurization at the beginning of a phonatory expiration requires additional muscular effort. Such additional recruitment of motor units may be required to quickly overcome inertial forces and elastic components of the abdominal wall, and thus effect rapid reconfiguration of the respiratory system from inspiration to expiration. Furthermore, the

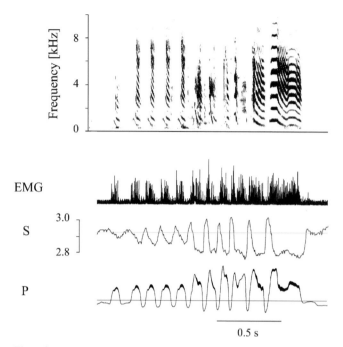

Figure 9.4 Simultaneous measurement of song (top, spectrogram), air sac pressure (P), electromyographic activity (EMG, rectified with time constant 0.1 ms) and muscle length (S, distance between crystals in mm) of expiratory muscles in a zebra finch. Muscle length changes correspond to the EMG activity and air sac pressure. Inspiratory activity causes a lengthening of the abdominal muscle sheet beyond resting length (dotted line). (Reprinted with permission from Goller and Cooper, 2004.)

generation of long pressure pulses may require a continuous adjustment of expiratory effort because air sac volume decreases as air is exhaled during a song syllable. Consistent with this expectation, EMG activity during the last third of zebra finch expiratory pulses is higher, but results in lower air sac pressure, than during the first third of a pulse. This decrease in pressurization efficiency could be produced by a combination of the decreased volume of the air sacs and the length change in the muscle sheet (Figure 9.5). To achieve such regulation, the same amplitude of air sac pressure may have to be generated with varying degrees of muscle effort depending on the structure of the song. In birds that sing a stereotyped song, the motor program for such adjustments in expiratory activity might be easily acquired during song development. This seems less likely for birds with variable song syntax where a particular note may appear at variable positions within a song bout. An alternative possibility is that online monitoring of air sac pressure during song provides a mechanism for reaching a target pressure during "improvised" (i.e. not fixed during song ontogeny) syllable combinations.

It is unknown whether a direct feedback mechanism is used to determine the correct expiratory effort or whether, even in songs with highly variable syntax, the motor commands for each combination are learned and produced stereotypically. Recent experiments in the northern cardinal involving experimental perturbation of air sac pressure during song showed appropriate adjustments in expiratory muscle EMG, providing evidence for the existence of an online feedback system (Suthers *et al.*, 2002). However, it is unknown whether this system plays a role in pressure generation during normal (i.e. unperturbed) song production.

How much do the expiratory muscles contribute to modulations of air sac pressure within pulses? The presence of modulated pressure itself does not conclusively show that expiratory effort is adjusted, because pressure fluctuations could also be caused by changes in syringeal resistance to airflow that are controlled by syringeal muscles. However, indirect evidence from EMG recordings of abdominal expiratory muscle activity indicates that electrical activity is indeed modulated in a manner consistent with pressure fluctuations of

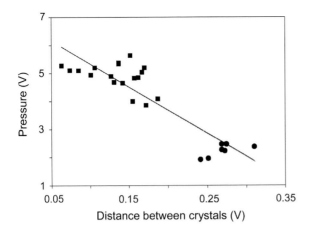
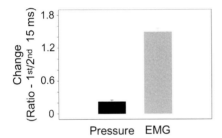
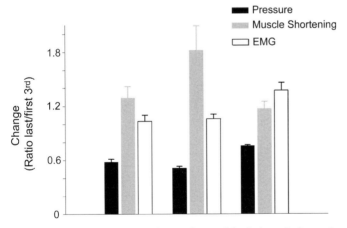

Figure 9.5 (Top) Length change of the expiratory abdominal muscle sheet and resulting air sac pressure are correlated for expiratory pulses of song in a zebra finch. The distance between the sonomicrometry crystals (expressed as relative voltage) decreases as air sac pressure increases. The variance not explained by the relationship is probably a result of variable syringeal resistance during different syllables of the song. Data points represent peak pressure and the corresponding muscle shortening for individual syllables (circles, introductory notes; squares, syllables of the motif).
(Center) Mean ratio ± 1 s.e.m. between measurements of air sac pressure and expiratory EMG activity for the first 15 ms of a syllable and the second 15 ms bin ($n = 12$; data are from 3 individuals). These data illustrate that at the onset of expiratory pressure pulses EMG activity is greater than for the second 15 ms bin, but results in lower amplitude of the air sac pressure pulse. The EMG activity precedes pressure generation by 8–12 ms; we used a 10 ms shift of EMG activity to align it with the air sac pressure.
(Bottom) At the end of syllables increased EMG activity and increased muscle shortening result in lower air sac pressure, presumably because the volume of air is reduced in the air sac. The change in all three parameters is expressed as the ratio of the respective mean for the last third divided by the mean for the first third of a given syllable ($n = 5$; data from the longest syllable of each individual; duration varied between 120 and 250 ms. (Reprinted with permission from Goller and Cooper 2004.)

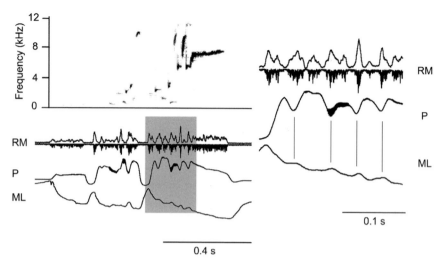

Figure 9.6. Many modulations of air sac pressure (P) are paralleled by appropriate length changes in the abdominal expiratory muscle sheet (recorded with sonomicrometry, ML, increasing length is upward) as displayed by the motif of one zebra finch. The most pronounced modulations in pressure that show concurrent muscle length change are marked by lines. Only one pressure peak was not clearly accompanied by a decrease in muscle length.

expiratory pulses (e.g. last syllable in Figure 9.4). Increased air sac pressure is preceded by increased EMG activity in those muscles. This interpretation is confirmed by *direct* measurements of muscle length in three species, zebra finch, brown-headed cowbird and European starling (Goller and Cooper, 2004, and unpublished data). As illustrated in this example from cowbird song, pressure modulations typically coincide with changes in the acoustic structure of the song (Figure 9.6). Some modulations are oscillatory changes that accompany amplitude and frequency modulations of sound. Frequently, these modulations of respiratory effort are also accompanied by rhythmic activation of various syringeal muscles (Goller and Suthers, 1996a; Goller and Cooper, unpublished results), again indicating an intricate coordination between vocal and respiratory motor systems.

Song production and respiration: constraints on song structure and duration

The volume of air that is available for phonation is limited by the capacity of the lung air sac system. Birds can sing long songs by taking short minibreaths in between song syllables to replenish the air supply (e.g. Calder, 1970; Hartley and Suthers, 1989; Suthers *et al.*, 1994; Allan and Suthers, 1994; Wild *et al.*, 1998; Zollinger and Suthers, 2004). However, the time required for switching from expiration to inspiration limits the maximal sound pulse rate that is possible with this mechanism (see Suthers and Zollinger, this volume). Such respiratory constraints may be important for the large number of species whose song exhibits an elaborate temporal structure with rapid trill rates and long durations. These temporal features play a role in species recognition and may also be important in mate choice, which suggests that they are under strong sexual selection (e.g. Podos, 1997; Searcy and Nowicki, 2005). One major selective force might be a trade-off between the temporal complexity of the song and respiratory constraints, such as availability of air and the need for gas exchange. In order to understand the limitations of air supply it is important to know how requirements for air supply vary with the acoustic characteristics of song syllables.

Airflow and acoustic structure are intimately linked. To begin with, syringeal resistance itself most likely varies with changing sound frequency, as a result of increased muscle tension for frequency control and lower vibration amplitude of the labia (Goller and Suthers, 1996b; Goller and Larsen, 1997). This change in syringeal configuration generates a connection between sound features and air sac pressure. Although little is known about this interaction, it is obvious from just a few observations that no single, simple relationship exists. In cardinals, for example,

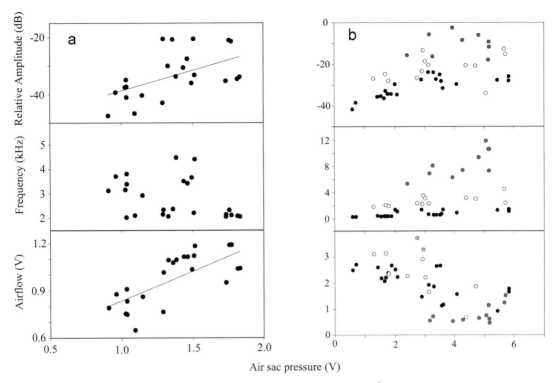

Figure 9.7 (a) In cardinal song air sac pressure and airflow show a positive relationship ($r^2 = 0.49$), but sound frequency is not tightly correlated with air sac pressure in this individual (but is in one other individual, data not displayed). The amplitude generally increases with increasing air sac pressure ($r^2 = 0.28$). Data points represent measurements from 7 syllable types of 1 individual. (b) The relationship between air sac pressure and airflow and acoustic parameters of cowbird song is complex. Mean values are plotted for individual notes in the introductory note clusters according to the side of the syrinx on which they were produced (left, black circles; right, open circles). All notes in the final whistle (gray circles) are produced on the right side. Side of production was inferred from Allan and Suthers (1994). The only two-voice segments are characterized by the highest airflow (gray circles in the bottom panel). Data represent measurements taken from 3 songs of each of 3 song types sung by 1 individual. (Reprinted with permission from Goller et al., 2006.)

air sac pressure and total airflow are positively correlated, but air sac pressure and fundamental frequency of sound are not consistently correlated (Figure 9.7a; Goller et al., 2006). In brown-headed cowbirds air sac pressure is negatively correlated with airflow during song. High-frequency syllables are generated with relatively low flow rates but with high pressure peaks and are much louder than the low-frequency elements (Figure 9.7a). These interspecific differences in the relationship between airflow and air sac pressure may indicate potential differences in the syringeal and labial biomechanics between species. Syringeal resistance to airflow will be determined, for example, by how uniformly the labia close during oscillations along their dorsoventral contact zone. Endoscopic images of the syrinx during just a limited range of vocalizations suggest that labial closing does vary between different sounds (Goller and Larsen, 1997a, and unpublished observations). Clearly, we need more information on the detailed biomechanics and vibratory behavior of the labia for a wider range of sound characteristics in songbirds. A full understanding of the relationship between airflow and driving air pressure will require detailed insight into the specific behavior of the sound-generating structures across the remarkable spectrum of acoustic diversity.

A second potential limitation on temporal song characteristics is the need for gas exchange, i.e. uptake of oxygen and release of carbon dioxide. Maintenance of gas exchange will be particularly important in birds, which sing long songs. Alternatively, birds could put gas exchange "on hold" during the period of song and power metabolism with anaerobically gained energy. This strategy would lead to the accumulation of an oxygen debt, which would have to be repaid by elevated oxidative metabolism after the song.

It is also possible that the respiratory pattern of song enhances ventilation of the lungs. In this case hyperventilation (i.e. enhanced release of carbon dioxide) can result, which will typically, via feedback mechanisms, slow or stop respiration until the proper carbon dioxide levels in the blood are restored. The respiratory system of birds is highly efficient at low partial pressures of oxygen, making flight possible at very high altitudes (e.g. Tucker, 1968; Powell and Scheid, 1989; Maina, 2000; Scott and Milsom, 2006). In addition and unlike in mammals, increased ventilation of the lung and the resulting low carbon dioxide levels in the blood do not cause decreased cerebral blood flow (Grubb et al., 1977; Faraci and Fedde, 1986; Butler, 1991; Taylor et al., 1999). This general tolerance of variability in gas concentrations poses the interesting question of how the respiratory patterns of song are integrated into the metabolic functions of breathing.

Long songs are made possible by minibreaths, which replenish the air expelled during phonation. It is unclear, however, whether this rapidly exchanged air volume is also maintaining gas exchange in the lung. This question specifically arises in cases such as the canary where the exchanged volume of air during some syllable types is less than the air volume contained in the trachea (tracheal dead space, approx. 0.09 ml) (Hartley and Suthers, 1989). This raises the question whether syllable types with minibreaths of less than 0.09 ml allow oxygenated air to reach the lung during this period.

Direct measurements of gas exchange during a dynamic behavior such as song are not easily possible, but indirect evidence suggests that oxygen supply is not limited by the respiratory patterns of song. Measurements of oxygen consumption during and after song do not indicate that an oxygen debt builds during song even in species with very long song sequences and temporally rich acoustic structure, such as the canary and European starling (Oberweger and Goller, 2001; Franz and Goller, 2003). However, it is possible that the need to maintain gas exchange may constrain song syntax. For example, if ventilation is low during a particular phrase, the lung may not be ventilated with oxygen-rich air, thus reducing oxygen uptake during this part of the song. Could this variation in oxygen uptake during song cause the bird to switch to another syllable type with enhanced perfusion of the lung? Current studies of oxygen consumption during song do not provide sufficient temporal resolution to directly measure such variability in gas exchange. However, detailed analysis of ventilatory patterns and song syntax may reveal whether or not needs for gas exchange influence song syntax in this predicted way.

Surprisingly, rather than resulting in reduced oxygen uptake, respiratory exchange during song is enhanced. In some species, the major problem arising may be hyperventilation and larger than normal release of carbon dioxide. In canaries, some individuals did not resume normal respiratory activity immediately after songs, suggesting that they hyperventilated during song (Hartley and Suthers, 1989). Similarly, some male zebra finches had pronounced apneic periods (i.e. ceased to breathe) after song bouts, and there was a positive correlation between bout length and duration of apnea (Franz and Goller, 2003; Figure 9.8). It is possible therefore that some respiratory patterns of song increase release of carbon dioxide above that of normal respiration and therefore limit the duration of sustained song. Because the degree of apnea varies between individuals, it would be interesting to know whether respiratory patterns that lead to apnea result in temporal and acoustic song properties that are preferred by females. For example the amount of air exchanged during song potentially affects sound amplitude, syllable duration or frequency characteristics and therefore might be perceived as acoustic quality by a receiver. If this hypothesis is correct, males, which can sustain long song bouts of such high quality songs, must be able to counter the effects of hyperventilation and therefore may signal fitness.

Integrating respiration and movement during song

Although we often picture birds singing while perched on a branch, they may, in fact, assume a variety of different positions, producing a range of postural variations. Some species sing while turning on or hanging from a branch, and they may also deliver their song during flight. In addition, many species combine song with a visual display, which is often optional and can range from simple raising of feathers to elaborate wing

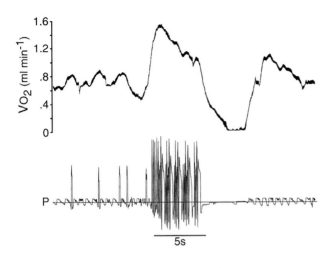
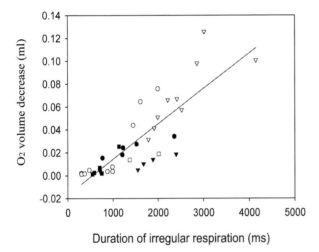

Figure 9.8 (Top) Some zebra finches hyperventilate during song. As is illustrated by this example, oxygen consumption (VO$_2$) drops to zero after the song, which coincides with a reduction of respiratory pressure (pressure in the anterior thoracic air sac, P). (Bottom) The degree of hyperventilation is positively correlated with the duration of the period during which air sac pressure indicates a reduced respiratory activity (irregular respiration). (Reprinted with permission from Franz and Goller, 2003.)

displays and "acrobatic" movements (see Williams, this volume). Such postural changes and movements are likely to affect respiration, and respiratory adjustments must therefore be integrated into the song motor program to achieve the normal vocal characteristics of song. In brown-headed cowbirds, for example, song is always accompanied by a puffing display which starts a few seconds before vocalizations are generated. During the song a wing display may also be performed, during which the wings are spread and moved up and down several times, with a final bowing display. These movements are likely to produce postural changes such as movements of the sternum, which, in turn, will influence the biomechanics of respiratory movements. The situation may be even more complex for birds which produce elaborate songs while in flight.

To determine whether respiratory patterns are different between songs with and without a wing display we measured air sac pressure, expiratory muscle activity and changes in expiratory muscle length during songs with and without the wing display (Cooper and Goller, 2004). The results indicated a very close

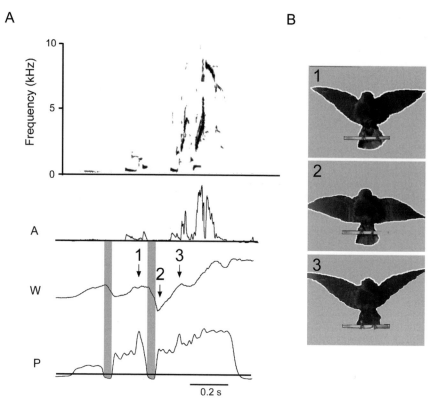

Figure 9.9 (A) Cowbird song is synchronized with movements of the visual display. The largest amplitude downward movements of the wings (W, measured with impedance subcutaneously implanted impedance electrodes) occur during the transition from inspiration to expiration (P, indicates subsyringeal air sac pressure, horizontal line indicates atmospheric pressure) corresponding to silent periods in the song (sound is displayed as spectrogram and as integrated and rectified time waveform, A). (B) Frames from high-speed video recordings of singing and displaying birds illustrating wing position at various points during the song (as marked in panel A). (Reprinted with permission from Cooper and Goller, (2004.)

integration of the visual and vocal displays with the respiratory pattern. First, the wing movements are synchronized to the respiratory pattern (Figure 9.9). The lowest wing position occurs near the point in the respiratory cycle where inspiration shifts to expiration and therefore at a time when the bird is not phonating. In addition, during the first part (up to one third) of each expiratory pressure pulse in which wing movements occur, the bird closes its syrinx and does not generate sound. This extended silent period has not been observed for adult song in species that do not have wing movement displays (e.g. Suthers and Goller, 1997; Suthers et al., 1999; Cooper and Goller, unpublished; exceptions during song development: Suthers, 2004). In cowbirds, this silent period may allow additional time to complete the most rapid wing movements and thus avoid their potential interference with sound production. Second, birds that did not spread their wings generated higher EMG activity to induce the same muscle length change and generate the same air sac pressure as birds that showed a full wing display. This suggests that performance of the wing display may facilitate the expiratory effort for singing (Cooper and Goller, 2004).

The variable visual display therefore leads to variation in the expiratory effort required to keep air sac pressure and airflow of song constant. Whereas this experimental approach investigated the presence or absence of wing movements, during natural behavior the wing display is highly variable, from gradual differences in wing movements to asymmetric use of the left and right wing. This high variability in wing display imposes a similarly diverse biomechanical variation onto the respiratory system. If air sac pressure is maintained at the same level during natural song, as was the case in our experiments, we must postulate a sensory mechanism that allows rapid adjustments in expiratory effort. It is unlikely that adjustments to a gradually changing biomechanical influence on song respiration could be anticipated without feedback mechanism.

Sensorimotor integration during song displays

How is this integration between respiration and song achieved? For birds with highly stereotyped songs it is possible that this coordination is perfected during song ontogeny and then the stereotyped motor programs assure coordinated activity without further need for sensory feedback. For birds with more complex and variable songs, this solution may be inadequate. Moreover the experimental evidence from the perturbation study in cardinals (Suthers *et al.*, 2002) and the integration of singing and display movements in cowbirds (Cooper and Goller, 2004) suggest that some respiratory variables are under direct feedback control even *during* song.

The first step in the analysis of any feedback system is the identification of receptors that could provide sensory input during movement. The avian respiratory system contains both pulmonary carbon dioxide receptors and mechanoreceptors that are most likely situated outside the lung in the air sac system or, possibly, even outside the respiratory system (e.g. other organs in the thoracoabdominal cavity; Gleeson and Molony, 1989; see Wild, this volume). These mechanoreceptors are sensitive to inflation and fire at a continuous rate during the inspiratory phase in normally breathing chickens (Molony, 1974). They also respond to increased intra-pulmonary pressure in unidirectionally ventilated ducks (Fedde *et al.*, 1974), but artificial inflation of the respiratory system in chickens does not show such a clear relationship (Molony, 1974). These observations suggest that some mechanical event related to volume changes provides the sensory stimulus for these mechanoreceptors (Gleeson and Molony, 1989). Although no comparable information on sensory feedback mechanisms in songbirds is available, these mechanoreceptors are the most likely sensory system responsible for the observed compensatory effect of respiratory effort following air injections into the anterior thoracic air sac in cardinals (Suthers *et al.*, 2002).

MOTOR INTEGRATION OF THE UPPER VOCAL TRACT

Watching birds as they are singing, an observer is struck by the amount of activity in the head and throat, including "puffing" of the "throat" and beak movements (Figure 9.10). How does this activity relate to the production of the song?

Some answers to these questions may emerge as we consider the different structures of what is collectively called the upper vocal tract. In human speech the upper vocal tract makes a substantial contribution to the acoustic structure of individual sounds. This is achieved by filtering the sounds generated by the vocal folds. The filter properties arise from combined effects of the length and volumes of supralaryngeal airways, which can be dynamically adjusted by tongue, lip and other cranial movements. This dynamic modulation of filter properties gives rise to the characteristic harmonic structures (formants) of speech sounds. Because most vowels are generated by similar vibration frequencies at the laryngeal source, the filter characteristics of the upper vocal tract structures account in large part for the perceptual meaning.

Unlike in human speech, it was originally assumed that in birds modification by the upper vocal tract of the sounds generated by the syrinx does not play a significant role (Greenewalt, 1968). However, experiments with birds singing in helium indicated that modification of sound by filter properties of the upper vocal tract is important in the generation of pure tonal sounds (e.g. Nowicki, 1987).

The sound that is generated in the syrinx radiates through the trachea and mouth of the bird to the outside. As it travels along this path, the spectral composition of the sound (i.e. energy distribution across the harmonic frequencies), can be changed by the filter characteristics of the upper vocal tract. The resonance properties of the trachea itself are determined by its length, such that higher frequencies are enhanced by shorter tracheal length. It has been suggested that the drastically elongated tracheae of some bird species serve the purpose of providing filter characteristics that allow more closely spaced formant frequencies and, thus, give an acoustic impression of larger body size (Fitch, 1999). Aside from this static length parameter the question arises as to whether or not birds are able to adjust tracheal length dynamically in order to match

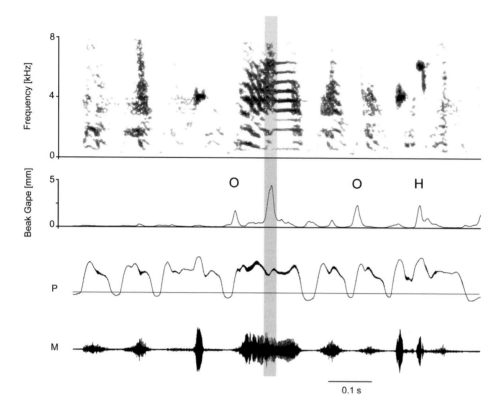

Figure 9.10 Distinct beak movements accompany the song of the zebra finch. The vocalizations are displayed oscillographically (M) and spectrographically (top) together with air sac pressure (P) and calibrated beak opening (values refer to the distance between the tips of upper and lower beak). Whereas beak gape is small throughout most of the song, there are pronounced peaks at the onset of syllables (O), at one high-frequency syllable (H) and at a rapid acoustic transition (gray bar). (Reprinted with permission from Goller et al., 2004.)

filter properties to the variable frequency of sound during singing.

The most obvious mechanisms for such adjustments are the trachea and the beak. Measurements of tracheal length changes during song in the zebra finch do not show adjustments in tube length that may be significant for altering resonance (Daley and Goller, 2004). Initial research on beak movements, however, found a consistent correlation between gape size and frequency of song syllables in a number of species (e.g. Westneat et al., 1993; Suthers and Goller, 1997; Williams, 2001; Podos et al., 2004b; Goller et al., 2004). These findings were interpreted as indicating that the beak perhaps acts as a continuation of the trachea, such that beak opening effectively shortens tube length and thus favors high frequencies (Nowicki, 1987; Nowicki and Marler, 1988; Podos and Nowicki, 2004a, b). However, in most of the investigated species, the relationship between sound frequency and beak aperture is not a simple linear correlation. Although beak aperture is typically greater for high frequency sounds, the amount of beak opening may reach a plateau at a given frequency and no longer follow increasing sound frequency (Suthers and Goller, 1997). In the zebra finch, whose song is composed of many syllables with rich upper harmonic structure, there is a still more complex relationship between beak aperture and frequency characteristics. Energy distribution across the harmonic spectrum is frequently concentrated in higher harmonics, suggesting that upper vocal tract resonance plays an important role. Beak aperture is correlated with the frequency component of highest energy, such that larger beak gapes occur for syllables in which peak energy is shifted towards higher harmonics (Williams, 2001; Goller et al., 2004). However, as found in other birds, the relationship

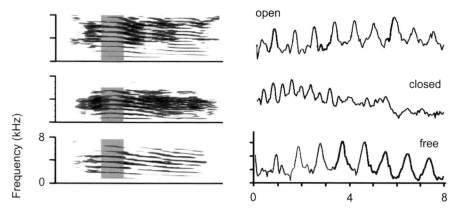

Figure 9.11. Experimentally fixed beak aperture results in shifted harmonic emphasis of zebra finch syllables. Power spectra (right panel) show an upward shift in harmonic content for the open beak condition (6 mm aperture, top trace and top spectrogram) and a downward shift for the closed beak condition (<0.5mm aperture, middle trace and spectrogram) as compared with the unmanipulated control (bottom trace and spectrogram). In the closed beak condition, the harmonic structure was also affected relative to the control segment. (Reprinted with permission from Goller et al., 2004.)

between beak aperture and sound frequency is not tight enough to conclude that beak aperture tracks fundamental frequency. Often, high-frequency syllables are not accompanied by the largest beak gape (Figure 9.10). Instead, peaks in beak aperture occur at transitions and the onset of syllables. In syllables with frequency modulation, changes in beak aperture do not track the modulation pattern. Resonance changes associated with beak movements are therefore unlikely to track the full range of fundamental frequencies (Nowicki, 1987; Nowicki and Marler, 1988; Suthers and Goller, 1997; Fletcher and Tarnopolsky, 1999). Because tonality of sounds occurs across the full frequency spectrum of song, birds must have additional mechanisms for controlling the filter properties of the upper vocal tract.

Confirmation of this interpretation can also be found in experiments in which gape size was manipulated during song (Figure 9.11). In these studies, beak gape was either fixed at a constant high or low aperture (zebra finch, cardinal, white-throated sparrow, swamp sparrow), or in canaries was influenced by increasing gape through added weight on the lower mandible (Suthers and Goller, 1997; Hoese et al., 2000; Goller et al., 2004). Under these conditions, the most prominent effect of the manipulation was seen at low frequency syllables. These are normally sung with low beak aperture, but when gape was experimentally kept high, upper harmonics were emphasized thus reducing the tonality of the sounds. These findings suggest that the role of beak movements in adjusting filter properties of the upper vocal tract appears to be more of a facilitating one (threshold effect). For example, at high sound frequencies, an open beak will support tonal sounds, but, beyond a threshold value, the absolute aperture no longer affects high frequency sounds. Beak aperture therefore does not control the tuning of the upper vocal tract filter, but acts as a filter by attenuating frequencies. This interpretation is supported both by theoretical models (Fletcher and Tarnopolsky, 1999) and by experiments in which the syrinx was replaced with a speaker and beak gape was manipulated during playback of frequency sweeps (Nelson et al., 2005).

Given the acoustic evidence that songbirds adjust the filter properties of the upper vocal tract for the entire range of frequencies, what other possible mechanisms may be involved? In a recent experiment in cardinals modification of the oropharyngeal-esophageal space was identified as an important mechanism for adjusting filter properties. Movements of the hyoid apparatus, which were visualized during song with x-ray cinematography, were used to estimate the volume of the oropharyngeal-esophageal space. Volume is adjusted dynamically and shows a strong inverse relationship with sound frequency. This indicates that the upper vocal tract may be tuned to the specific acoustic frequency of the sound (Riede et al., 2006; Fletcher et al., 2006; see Suthers and Zollinger, this volume).

Interesting questions for future research remain. For example, the active regulation of volume most likely involves the hyoidal and laryngeal muscles, but the identity of the specific muscles involved and their activity patterns are not known. Similarly, while it can be assumed that beak movements influence the control of oropharyngeal space, the biomechanical systems involved remain to be clarified.

Compared with humans, the tongue of most birds is rigid, and its role in modification of sound is incompletely understood. Parrots have a more flexible tongue, and tongue positioning may be important in speech imitation and natural calls of parrots by contributing to vocal tract filtering (e.g. Patterson and Pepperberg, 1994; Beckers et al., 2004). Placement of the tongue in monk parakeets was varied while sound was played from a speaker, which was attached to the base of the trachea. Different positions of the tongue, especially along the forward-backward axis, affected the distribution of sound energy across the frequency spectrum suggesting a role in upper vocal tract filtering (Beckers et al., 2004). Because tongue movements affect the position of the larynx, it is not clear whether tongue position itself changes filter properties or whether its effect is through movements of the larynx. In either case, movements of the tongue may also affect upper vocal tract filtering in other groups of birds, but no information is available on tongue position during vocal behavior.

The mechanisms discussed above involve adjustments of filter properties which modify the spectral distribution of energy. It is unclear how these motor actions are integrated into the song motor program. Some insight into the coordination of beak movements and respiration during song comes from another, unexpected relationship. In one species, the zebra finch, a clear correlation between beak aperture and sound amplitude has been reported. In distance calls sound amplitude and beak aperture are correlated with subsyringeal air sac pressure. Expiratory effort and beak aperture may be linked in the motor program for song production such that increased effort is accompanied by relatively larger beak aperture and increased sound amplitude (Figure 9.12). This possible interaction suggests a complicated relationship between acoustic characteristics of the song and beak aperture during singing. Because of the strong relationship between beak aperture and frequency, a link between sound amplitude and beak aperture may be masked if data from different syllable types within a song are pooled for analysis. However, when the same song is produced softly or loudly by an individual, the absolute beak aperture should be greater for the loud version of the same song although in both cases the same temporal dynamics of movement will be present (Goller and Cooper, unpublished results). This same relationship may explain beak aperture patterns during song development, where the beak is mostly held closed during soft subsong and, as sounds become louder, beak opening is observed (e.g. Podos et al., 1995; Goller and Cooper, unpublished observations).

In addition to the strong evidence that the upper vocal tract may filter sound generated at the source, it is also possible that feedback from the upper vocal tract might itself modulate the action of the source. One indirect suggestion of such an interaction came from a study that looked at the relationship between beak movements and song in the zebra finch. The largest beak aperture was often observed when very short duration beak openings (approx. 25 ms) accompanied rapid transitions in the sound (Figure 9.10). These sudden transitions were affected by experimental manipulation of the beak aperture. One possible explanation for these observations is that feedback from the upper vocal tract changes the vibratory behavior of the sound-generating structures in the syrinx. Although a nonspecific effect of the fixing of the beak might explain the observed change at the rapid sound transition, the short duration and high aperture of the beak movement remains a striking observation. Theoretical evidence supports this notion that a rapid change of impedance at the beak may provide feedback to the sryinx. Computational models show that a feedback mechanism can generate sudden transitions to subharmonic frequencies (Laje and Mindlin, 2005; Mindlin and Laje, 2005). It has been suggested that nonlinear behavior occurs at these rapid sound transitions in zebra finch song syllables (Fee et al., 1998). The combination of experimental and theoretical studies suggests the exciting possibility that the vibratory behavior of the syringeal sound generators may not be completely uncoupled from the upper vocal tract filter mechanisms (see also Fletcher et al., 2006). Such coupling is typically not observed during human speech, and its possible occurrence in birds presents a good opportunity for studying how beak aperture control is integrated with the song motor program.

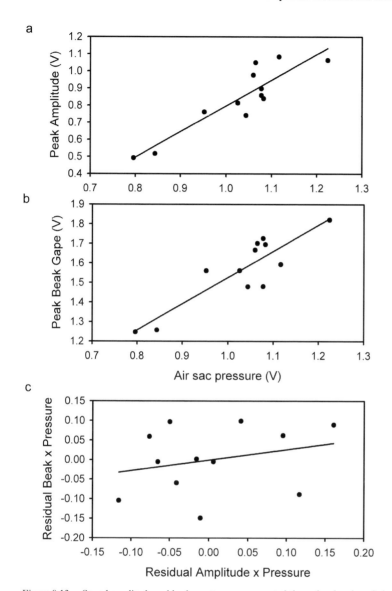

Figure 9.12. Sound amplitude and beak aperture are connected through subsyringeal air sac pressure for zebra finch distance calls. Subsyringeal air sac pressure shows a positive correlation with peak amplitude (Panel a, $r=0.9$, $F=43.36$, $p<0.0001$) and peak beak gape (Panel b, $r=0.88$, $F=34.13$, $p=0.0002$), but the residuals of these two relationships are no longer strongly correlated (Panel c, $r=0.27$, $F=0.77$, $p=0.4$). (Reprinted with permission from Goller et al., 2004.)

In conclusion, we know very little about the way in which mechanisms for the control of the upper vocal tract are integrated into the general motor programs and neural pathways for producing and varying song (see Wild, this volume). Nor have we yet devoted sufficient attention to the way in which the motor behaviors involved in song (e.g. beak movements) are developed during ontogeny, the possible role of feedback in their maintenance or the potential sources of such feedback. In zebra finches, for example, beak movement patterns do not appear to be affected by alteration of acoustic feedback (e.g. Goller and Cooper, 2005). This suggests that motor programs for beak movements, like respiratory and syringeal motor programs, may be fixed at the end of vocal development. The correlation between beak aperture and respiratory effort suggests that these two are either linked at the central motor planning stage or through an instantaneous peripheral feedback mechanism. Further experiments will be required to decide between these alternatives.

FUTURE STUDIES OF PERIPHERAL SENSORIMOTOR INTEGRATION OF BIRDSONG PRODUCTION

Despite significant progress over the last decade, our understanding of sensorimotor integration of birdsong motor control is still in its infancy. This becomes immediately clear, if we compare our knowledge of the avian system with that of sensory receptors and feedback mechanisms in the human speech system (e.g. Lass, 1996; Kent 1997). At the level of each motor system involved in speech production, we have detailed knowledge of sensory receptors, their activity during speech and motor adjustments to perturbation of the normal systems. In contrast, sensory receptors engaged in avian song motor control have often not been identified, or their activity has not yet been studied during song production. However, simple comparisons, such as the presence (mammals) or absence (birds) of muscle spindles in vocal muscles (Wild, 2004a), suggest that there must be differences in sensorimotor control of the vocal organs between birds and mammals. Similarly, respiratory control during sound production, as dictated by the need for maintaining gas exchange, must be influenced by the anatomical and physiological differences in the respiratory systems of birds and mammals. Unraveling these mechanisms in birds and studying similarities and potential differences to the human speech system are important goals for future research into the peripheral motor control of birdsong.

ACKNOWLEDGEMENTS

The authors' research is funded by NIH grants DC 04390, DC 06876 and NIH Ruth R. Kirschstein Postdoctoral Fellowship 05722. We thank Dr. Phil Zeigler for critical reading and editing of the manuscript.

10 · Integrating breathing and singing: forebrain and brainstem mechanisms

Marc F. Schmidt and Robin Ashmore

INTRODUCTION

Song production in birds offers a tremendous opportunity for studying the underlying neural bases of complex, experience-dependent motor sequence generation. In many species of birds, song is made up of many different stereotyped motor gestures, known as notes or syllables, which are produced in precise sequential order (see Williams, this volume). The production of song motor sequences therefore requires not only the generation of motor commands that specify the features of each element (e.g. spectral content, duration, amplitude, etc.) but also the order, or sequence, with which these elements are produced. In the zebra finch, the subject of choice for many song researchers, both the features of individual syllables as well as the sequence with which they are produced are highly stereotyped across song renditions (Sossinka and Böhner, 1980; Cardin *et al.*, 2005) (Figure 10.1). This species is therefore ideally suited for deciphering the neural mechanisms underlying the specification of the features that make up syllables. It also offers the opportunity for understanding the neural basis of motor sequence generation and how these different control mechanisms (feature specification and sequence generation) might be coordinated.

The avian song control system is made up of a discrete interconnected network of brain structures that together act to control the syrinx (the avian vocal organ), the muscles of respiration, and a number of secondary structures that might also affect acoustic quality (Nottebohm *et al.*, 1976, 1982; Wild *et al.*, 2000). At the output end, this network is made up of distinct brainstem nuclei that directly control muscles of the syrinx (hypoglossal nucleus; nXIIts) and indirectly control muscles of expiration (nucleus retroambigualis; RAm) and inspiration (nucleus parambigualis; PAm) (Wild, 1993a, 2004a, and this volume). These brainstem nuclei in turn receive motor commands from the forebrain nucleus RA (nucleus robustus arcopallialis), which is itself innervated by nucleus HVC (used as proper name) (Figure 10.2).

It has generally been assumed that HVC and RA specify the combination of respiratory and syringeal motor commands necessary to produce the acoustic features of individual song elements. However, the existence of several anatomical pathways that link the brainstem vocal respiratory network (VRN, which is made up of nuclei PAm, RAm and DM, see Figure 10.2) back to song control nuclei, such as HVC, in the forebrain (Vates *et al.*, 1997; Reinke and Wild, 1998; Striedter and Vu, 1998) suggests that the song control system is not organized as a simple hierarchical descending pathway (Ashmore *et al.*, 2005). It appears, instead to be organized as a recurrent pathway with no identifiable single structure at the top of a motor hierarchy.

In this chapter, we first review the functional organization of the descending motor pathway, then present anatomical and physiological evidence linking the brainstem back to forebrain vocal control nuclei. We then describe an integrative model for song production that views song motor control as distributed along a recursive pathway where the brainstem vocal respiratory network plays a central role in determining key features of the song's temporal structure. Finally, given the paucity of available evidence of bottom-up influences on motor control (Wurtz *et al.*, 2005), we discuss how the avian song motor control system might serve as a powerful model system for understanding general principles of brainstem to forebrain interactions in the context of motor control of complex learned behaviors.

FUNCTIONAL ORGANIZATION OF THE DESCENDING VOCAL MOTOR PATHWAY

Peripheral control of vocal production

Birds produce a wide range of vocalizations that include song as well as shorter vocalizations known as calls. In

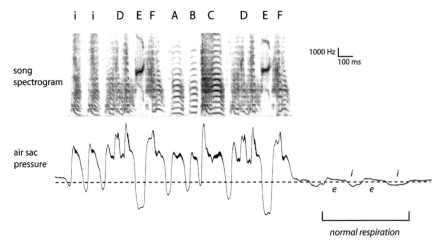

Figure 10.1 Respiratory profile of a male zebra finch song. This figure illustrates the acoustic features of a typical adult zebra finch song and its associated changes in air sac pressure. The top trace represents the acoustic sonogram of the song and the bottom trace the air sac trace. This particular song is only two motifs long and therefore allows a comparison of the fluctuations in air sac pressure during the song and those produced during normal respiration. In this example, normal respiration is shown following song offset. The majority of syllables in this song are produced during expiration (syllables A, B, C, D and F) except for syllable "E" which is produced during the inspiratory phase, a notable exception to the general rule that syllables are always produced during expiration (Goller and Daley, 2001). Note that individual expiratory (e) and inspiratory (i) pulses during normal respiration are significantly smaller than those produced during singing. The dotted line in the air sac pressure trace denotes ambient atmospheric pressure.

laboratory settings, male zebra finches usually produce two primary call types, known as contact (or distance) calls and shorter "tet'" calls (Zann, 1996; Vicario et al., 2001). During normal respiration birds typically alternate between inspiratory and expiratory phases approximately 2–3 times per second. Once the bird initiates a vocalization, the pressure amplitude of individual respiratory pulses increases approximately 6–20 fold (Goller and Cooper, 2004, and this volume). During vocal production, expiratory and inspiratory pulses decrease significantly in length compared with normal respiration with expiratory pulses typically lasting longer (70–300 ms) than inspiratory pulses (30–100 ms) (Figure 10.1). Syllables are usually only produced during expiration, although there are a few exceptions in the zebra finch song where some syllable types are produced during inspiration (Goller and Daley, 2001). Inspiratory pulses, or minibreaths, serve to replenish the air supply between vocal expiratory pulses produced during a song (Goller and Cooper, 2004).

Singing is associated with a dramatic shift in respiratory control which is reflected in the marked changes in the amplitude and temporal pattern of respiratory activity (Suthers et al., 1999). In many cases, expiratory pulses contain temporal fluctuations in their amplitude envelope which are sufficiently stereotyped to allow individual syllable identification based simply on each syllable's pressure pattern (Franz and Goller, 2002). The exact source of these rapid modulations in air sac pressure remains unclear; they could be mediated either directly by respiratory muscles or caused by syringeal-mediated gating of airflow, or both (Goller and Cooper, 2004). Aside from the rapid modulation of air sac pressure, the overall temporal pattern of air sac pressure is a direct consequence of respiratory muscle activity which receives motor commands via the intermediary of nuclei in the brainstem VRN (Wild et al., 1998; Suthers and Zollinger, 2004). Activity in these areas is therefore assumed to directly influence song acoustic features, by regulating air sac pressure via activation of expiratory and inspiratory muscles. The brainstem VRN, however, also directly contributes to the temporal features of song since it determines the length of expiratory pulses (i.e. syllables) and the duration of the intervening inspiratory minibreaths (i.e. silent intervals). In addition to the changes in respiratory pattern, song is also associated with changes in syringeal motor activity. During the production of sound, the syringeal lumen moves rostrally and syringeal muscles rotate the third bronchial

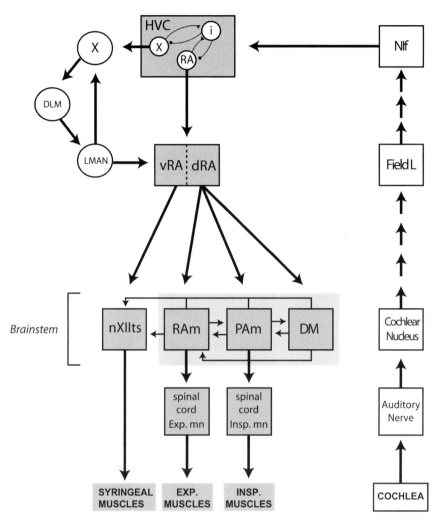

Figure 10.2 Simplified top-down representation of the avian song control system. Vocal control, and song control in particular, is driven by a specialized circuit referred to in the text as the "descending vocal motor pathway" (shown here in gray). As is typically depicted in most representations of this circuit, the forebrain nucleus HVC (used as proper name) lies at the top of this pathway and projects to the forebrain vocal control nucleus RA (robust nucleus of the arcopallium). This nucleus is functionally divided into two parts. The ventral portion (vRA) projects to the tracheosyringeal portion of the hypoglossal nucleus (nXII) in the brainstem. This nucleus contains the motoneurons that innervate the syrinx. The dorsal portion of RA (dRA) projects to nuclei in the brainstem that form part of what is referred in the text as the vocal-respiratory network (VRN). This network is highlighted by the light gray box and includes nuclei RAm, PAm and DM. RAm projects to expiratory motoneurons in the lower thoracic and upper lumbar regions of the spinal cord whereas PAm projects to inspiratory motoneurons in the lower brachial and upper thoracic regions of the spinal cord. In addition to forming part of the descending motor pathway, HVC, which is made up of at least three distinct cell types (X, I and RA), also projects to the anterior forebrain pathway and receives auditory inputs from nucleus NIf. The anterior pathway, which is made up of Area X, DLM, and LMAN is not necessary for song production but plays an important role in song learning and maintenance. Note that while HVC is only 3 to 4 synapses away from output musculature, it is at least 8 or 9 synapses (depicted by each arrow in the ascending auditory stream) away from the cochlea. Anatomical names: DLM, medial part of the dorsolateral thalamic nucleus; LMAN, lateral magnocellular nucleus of the anterior nidopallium; Field L is the primary auditory forebrain structure in birds; Area X, Area X of the medial striatum; NIf, nucleus interfacialis of the nidopallium; RAm, nucleus retroambigualis; PAm, nucleus parambigualis; DM, dorsomedial nucleus of the intercollicular complex.

cartilages, causing the medial and lateral labia to move into the air stream, where the positive pressure in the air sac causes expiratory airflow to induce labial vibrations and produce sound (Goller and Larsen, 1997a; Larsen and Goller,1999, 2002; Suthers and Zollinger, 2004 and this volume).

The avian syrinx is a bipartite structure and muscles in both halves of the syrinx often show differential activation patterns, especially during sound production (Suthers, 1990, 1997; Wild et al., 1998; Suthers and Zollinger, 2004). Specific acoustic features, for example, are therefore often determined by the differential activation of both halves of the syrinx in a highly coordinated manner (Goller and Suthers, 1995; Floody and Arnold, 1997; Suthers and Zollinger, this volume). This allows for a range of phonic configurations: from both sides producing the same sound to both sides producing different sounds. Sound can even be produced exclusively unilaterally by allowing air to flow through only one half of the syrinx and not the other (Goller and Suthers, 1995). During the production of song bouts, sound production can therefore result from rapidly alternating airflow in both halves of the syrinx (Goller and Suthers, 1996a, 1996b; Suthers, 1997). Each syringeal half is controlled by six different muscle groups (see Suthers and Zollinger, this volume, for details) and recordings from these different muscles in a number of songbird species are beginning to reveal predictive relationships between specific muscle groups and the types of acoustic elements that result from their activation (Goller and Suthers, 1996b; Suthers and Zollinger, this volume).

Given the large number of different ways sound can be produced, an important requirement for the production of acoustically precise elements is that there is exquisite coordination between the syringeal and respiratory control systems (Sturdy et al., 2003; Kubke et al., 2005; Wild, this volume) Such coordination is not necessarily trivial given that syringeal and respiratory motor commands originating from nucleus RA in the forebrain are primarily ipsilateral in nature (Wild, 1993a, 1997a; Wild et al., 2000). Because mismatches in syringeal motor activation would likely result in profound distortions of the acoustic signal, the rapid switching between sides without acoustic distortions suggests the existence of highly sophisticated neural mechanisms for coordinating the motor commands that reach each half of the syrinx.

Telencephalic control of vocal production

Syringeal and respiratory motor areas in the brainstem are strongly innervated by nucleus RA, a telencephalic structure that forms part of the descending vocal motor pathway (Wild, 1993a, 1997a; Vates et al., 1997). RA is itself innervated by nucleus HVC, a higher order vocal structure, often assumed to lie at the top of a hierarchically organized descending vocal motor pathway (Yu and Margoliash, 1996; Margoliash, 1997; Fiete et al., 2004a; Solis and Perkel, 2005). It is also the first nucleus in the motor pathway to receive auditory inputs (Cardin and Schmidt, 2004a; Coleman and Mooney, 2004; Cardin et al., 2005). HVC is itself made up of three general classes of neurons which are typically defined by their projection pattern (see Figure 10.2) (Nixdorf et al., 1989; Dutar et al., 1998; Mooney, 2000). These include (1) HVC interneurons (HVC_i), whose axon terminals are primarily restricted to the nucleus, (2) RA-projecting neurons (HVC_{RA}), which can be thought of as motor command neurons since they are the only neurons projecting directly to the descending motor pathway and (3) Area-X projecting neurons (HVC_X), which project to a part of the basal ganglia (Area X) that plays an essential role in song maintenance and learning (Bottjer et al., 1984; Brainard and Doupe, 2000b; Brainard, this volume).

Although the concept of a hierarchically organized descending pathway is, in its most simplistic form, likely to be incorrect, there is nevertheless a wealth of data to suggest that HVC and RA play a major role in the production of learned vocalizations (Margoliash, 1997; Hahnloser et al., 2002; Suthers and Margoliash, 2002; Ashmore et al., 2005a). Here we critically review evidence linking HVC and RA to song production.

Lesions of the descending motor pathway prevent the production of learned vocalizations

Like peripheral control mechanisms, the telencephalic neural pathway for song control is organized in a bilaterally symmetrical fashion with identical vocal control nuclei present in both hemispheres. Bilateral lesions of HVC or RA will completely abolish the production of song in male songbirds (Nottebohm and Arnold, 1976; Simpson and Vicario, 1990; Williams et al., 1992). These lesions will also remove the learned component of the normally stereotyped male contact call, transforming it into a female-like call, which is acoustically simpler and highly variable in duration (Simpson and Vicario, 1990). Interestingly, these lesions do not appear to affect the production of nonlearned short "tet" calls even though robust premotor activity is observed in HVC and RA during the production of these calls (see next section). Studies in which HVC$_{RA}$ projection neurons are selectively ablated using laser targeting, have shown similar song deficits to those elicited by HVC lesions (Scharff et al., 2000). Deficits in the production of contact calls are also observed and are generally consistent with a feminization of the normally stereotyped male contact calls (C. Scharff, pers. comm., September 2006). Interestingly, lesions of telencephalic vocal control nuclei that abolish song production do not appear to eliminate secondary motor behavior associated with singing. Several studies have noted, for example, that birds still assume normal singing posture and beak movement despite the lack of phonation (Nottebohm et al., 1976).

While the effects of bilateral lesions on vocal production are unambiguous, the observed effects of a unilateral lesion on vocal behavior are much more variable, ranging from subtle distortion of song output (Nottebohm et al., 1976; Williams et al., 1992) to complete elimination of song (Ashmore et al., 2008). Results obtained from unilateral HVC or RA lesions are often difficult to assess because the intact contralateral nucleus is potentially able to compensate for the lesion. Our own studies in the zebra finch suggest that unilateral removal of RA in adult birds completely eliminates the ability to produce normal song (Ashmore et al., 2008). Lesioned birds attempt to sing but only produce long strings of stuttered introductory notes or short contact calls. Other reports show that unilateral lesions of HVC only cause partial deficits to song. These results have been interpreted as evidence that different hemispheres contribute differentially to song output (Williams et al., 1992). A primary difficulty with such interpretations, however, is that song is often assessed many days after the lesion. Given the great capacity for functional recovery following perturbation (Cardin et al., 2005), this extended recovery period might mask direct effects caused by the lesion. This is particularly true for HVC, which has the capacity for regeneration of its HVC$_{RA}$ projection neurons (Kirn and DeVoogd, 1989; Alvarez-Buylla et al., 1990a; Scharff et al., 2000) and may therefore be able to functionally reorganize following lesions that do not completely eliminate all of HVC. Because neuronal replacement has not been shown to occur in RA, this structure is less likely to be subject to functional reorganization.

While lesion studies suggest that both HVC and RA are necessary for song production, it should be noted that the presence of song nuclei in the forebrain is not a necessary requirement for the production of complex song. A number of passerine birds of the suborder suboscine, such as the eastern phoebe (*Sayornis phoebe*) for example, are able to produce complex songs but lack the discrete telencephalic nuclei that make up the song system (Kroodsma and Konishi, 1991). Because the primary distinction between oscine and suboscine song is that oscines learn their song while suboscine birds do not (Kroodsma, 1984; Kroodsma and Konishi, 1991), a possible interpretation of these findings is that forebrain song nuclei are only necessary for the production of learned vocalizations. The existence of HVC and RA might therefore serve the purpose of integrating auditory feedback, which is critical for song learning and maintenance in oscine songbirds, with the motor commands that control song output.

HVC population activity during learned and nonlearned vocalizations

Because zebra finches chronically implanted with electrodes in song control nuclei will produce a normal vocal repertoire (both spontaneously and elicited by the presence of a conspecific bird), this technique can be used to record neural activity patterns during the production of their various vocalizations (McCasland, 1987; Yu and Margoliash, 1996; Hessler and Doupe, 1999a; Hahnloser et al., 2002; Schmidt, 2003). In addition, these electrodes can also be used to apply brief electrical stimuli to temporarily perturb neural activity in these structures and test the effect such perturbations have on vocal output (Vu et al., 1998; Ashmore et al., 2005a).

If premotor activity in HVC is necessary for the production of learned vocalizations, then perturbing activity in this structure should cause distortions to the vocal output. Several studies have tested this hypothesis by applying brief electrical stimuli to HVC (or RA) during song (Vu et al., 1994, 1998; Ashmore et al., 2008). Such perturbations cause changes in the song temporal pattern as well as short latency perturbations in both the acoustic and the respiratory pattern of the ongoing song. These results suggest that activity in HVC and RA can directly influence the vocal output by perturbing both the syringeal and the respiratory system. This conclusion is supported by experiments showing that stimulation in HVC or RA during periods of normal respiration cause short latency activation of syringeal muscles (Goller and Cooper, 2004) as well as changes in air sac pressure (Ashmore et al., 2005).

Using these same electrodes, neural activity can be sampled from small populations of neurons (20–100 neurons) and activity patterns can be assessed in relationship to the production of different vocalizations. At typical HVC recording sites, neural activity is low during nonvocalizing periods but increases dramatically prior to song onset (Yu and Margoliash, 1996; Schmidt, 2003) and remains generally elevated during the entire duration of song (Figure 10.3a). Activity levels do fluctuate, however, and careful analysis reveals that activity, for the most part, decreases at the end of each syllable and increases again 40–50 ms prior to the onset of each syllable in the song (Figure 10.3b) (Yu and Margoliash, 1996; Schmidt, 2003). These fluctuations in spike rate are nearly identical at all recording sites in HVC, whether in the same hemisphere or in different hemispheres (Schmidt, 2003), suggesting that the overall firing rate pattern in left and right HVC is driven by a common input. These premotor activity patterns occur in both normal and deafened birds. Interestingly, the sustained increase in premotor neural activity and the tight temporal correlation with acoustic onset observed during song is also present during the production of contact calls and short "tet" calls (M. Schmidt, unpublished observations), even though neither of these vocalizations are abolished following complete lesions of HVC and RA (Simpson and Vicario, 1990). Because the contact call contains learned components that disappear following ablations of HVC and RA, the presence of premotor activity during the contact call suggests that HVC activity might be specifically associated with the production of the learned components of that call. The presence of activity during the "tet" call is more difficult to account for, however. Because "tet" calls are not learned and lesions of HVC and RA do not appear to affect their acoustic structure, production of these calls appears therefore to occur independently of any motor command that might be initiated in either of these nuclei.

Because multiunit activity in these studies reflects activation patterns for a large heterogeneous population of HVC neurons, it is not possible to distinguish between the different types of neurons that might be activated during vocal production. Recorded neural activity patterns, for example, could potentially be caused mostly by the activation of non-RA projecting neurons and therefore not reflect activity related to the motor commands that are sent out by the HVC_{RA} projection neurons. In this context, one possible explanation for the finding described above is that multiunit "premotor-like" activity recorded in HVC during the production of "tet" calls might represent a copy of the motor commands (i.e. efference copy) generated in the vocal-respiratory brainstem (see next section) rather than the motor commands generated in HVC.

HVC motor output is represented by a sparse code that specifies each moment in time

The evidence presented above suggests that HVC sends song motor commands to RA which, in turn, activates brainstem areas that control respiratory and syringeal muscle activity. By recording exclusively from HVC_{RA} projection neurons, which represent the motor output of HVC, Fee and colleagues (Hahnloser et al., 2002) have been able to gain direct insight into the nature of the motor command generated in HVC. While the number of neurons sampled in that study was small, the general trend suggests that individual HVC_{RA} projection neurons exhibit only a single short burst of three to six action potentials lasting approximately 4–8 ms during each song motif (Hahnloser et al., 2002). Any given HVC_{RA} projection neuron is therefore only active for a short window of time associated with a specific syllable in the song. Multiple repetitions of that syllable, as would occur during the production of multiple motifs, result in that neuron firing a single burst for each repetition at exactly the same time in each motif. This firing pattern implies that any given HVC_{RA} projection neuron only has a "sparse" representation during any given song motif.

Based on these findings, Fee and colleagues have proposed that each consecutive 10–15 ms time window in the song's motif is encoded by a distinct population ($\sim$200) of HVC_{RA} projection neurons (Fiete et al., 2004) and the entire duration of a song motif can therefore be broken down into discrete time windows each represented by a distinct population of HVC_{RA} projection neurons. If each of these populations projects onto a distinct set of neurons in RA, each population of

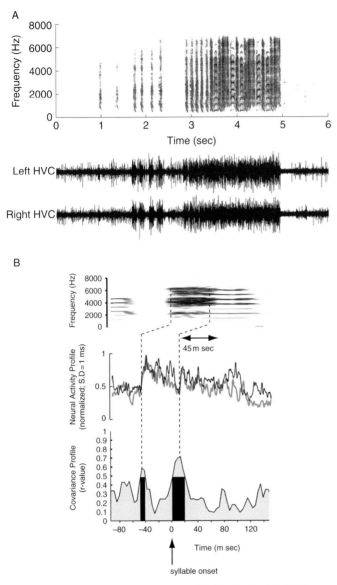

Figure 10.3 Multiunit neural activity in HVC during singing. (A) Simultaneous recording of premotor activity in left and right HVC during singing of an adult zebra finch. This example of raw neural traces reveals that premotor activity obtained from multiunit electrodes is generally elevated in both hemispheres during the entire song bout. Premotor bursts prior to the song can be observed during production of short calls and introductory notes. (B) Evidence for synchronization of premotor activity across hemispheres during short segments of the song trace. Simultaneous recording in left and right HVC during production of a two-note syllable (top panel). In this representative example, premotor activity (middle panel) in both left (gray) and right (black) HVC is elevated but highly modulated during the entire syllable. The onset of premotor activity occurs approximately 45 ms prior to the acoustic onset of the syllable. Using sliding window cross-covariance analysis (lower panel), premotor activity is shown to be highly correlated ($r > 0.5$) across hemispheres during two short periods of the motor trace. The black bars highlight these periods. When shifted by a 45 ms premotor delay time, the first period of correlated activity corresponds to the acoustic onset of the syllable and the second period to the transition between the first and second note of the syllable. (Reproduced with permission from Schmidt, 2003.)

HVC_{RA} projection neurons could be in a position to specify the acoustic output associated with each time window (Figure 10.4).

Ensemble coding in RA driven by the sparse temporal code in HVC

Based on the projection of its output neurons, RA can be functionally divided into a dorsal (dRA) and a ventral (vRA) part (Vicario and Nottebohm, 1988; Vicario, 1991a; Wild, 1993a; Vates et al., 1997) (Figure 10.5). Neurons in dRA project to brainstem areas involved in vocal-respiratory control such as RAm, PAm and DM (Wild, 1993a, 1997a; Wild et al., 2000 and this volume) Neurons in ventral RA project almost exclusively to the tracheosyringeal portion of the hypoglossal nucleus (nXIIts), and this part of RA is organized as a myotopic map of the different muscles that innervate the syrinx (Vicario and Nottebohm, 1988; Vicario, 1991a). Nucleus RA can therefore be viewed as containing two functional compartments, one more directly associated with respiratory control and one that is directly linked to syringeal muscle control (Ashmore et al., 2005).

There is no evidence of functional segregation in the projections of HVC to RA and it is therefore currently assumed that neurons in both dRA and vRA are activated by sparse inputs from HVC (Hahnloser et al., 2002). In contrast to HVC, single neurons in RA produce multiple bursts during each motif (Leonardo and Fee, 2005). The precision with which these bursts are produced is quite remarkable, with individual bursts occurring at nearly exactly (submillisecond accuracy) the same time in the motif from one rendition to the next (Chi and Margoliash, 2001). Thus in contrast to the sparse representation in HVC, the signal in RA becomes transformed into a temporally precise population representation where every neuron is active multiple times during the motif.

Unresolved issues for a simple HVC to RA sparse code model

Although a "sparse" code for song representation in HVC suggests a novel and exciting way in which higher-level motor commands might be encoded, in its current form it raises a number of as yet unresolved issues.

Temporal scaling of song

While zebra finch song is often used as an example to illustrate the precision and reliability with which songs can be produced, careful analysis reveals that the speed of delivery, or tempo, of even so stereotyped a song can vary significantly. It is well known, for example, that the male song is produced at a higher tempo when singing toward a female (directed-song) than when singing alone (undirected song) (Sossinka and Böhner, 1980; Cooper and Goller, 2006; Kao and Brainard, 2006). Additionally, changes in tempo can also be observed within a given song when one compares the tempo of the first motif with the tempo of the third or fourth motif.

It is generally agreed that increasing song tempo is achieved by shortening or compressing the length of individual syllables (Cooper and Goller, 2006; Glaze and Troyer, 2006). Changing tempo therefore involves changing the length of the individual song elements within a motif. If each population of HVC_{RA} projection neurons were to encode a fixed window of time, as implied by the simplified "sparse coding" model (Fiete et al., 2004), changes in song tempo would have to scale equally across the entire motif. Several recent studies (Cooper and Goller, 2006; Glaze and Troyer, 2006) suggest, however, that scaling is not linear throughout the song. While the requirement for such nonlinear scaling is not inconsistent with a sparse code model per se, it is inconsistent with a model where the entire song sequence, syllables and gaps included, are scaled identically.

Functional divisions within RA

Given the functional division of RA into respiratory-related and syringeal-related components, it is possible that each portion of RA might be responsible for different aspects of the song motor command (Vicario, 1991a; Ashmore et al., 2005). It would follow that neurons forming part of the respiratory-directed motor outputs (dRA) would have different properties from those intended for syringeal muscles (vRA). Ignoring these functional subdivisions during recording in RA might directly impact the interpretation of the coding studies of HVC. If the projection of HVC neurons has some topographic relationship to the two divisions of RA – something currently unknown – then placement of the stimulating electrode in RA used to identify RA-projecting neurons could bias the sample of HVC_{RA} projections neurons recorded in HVC. Unfortunately, in both the HVC and RA studies described above (Hahnloser et al., 2002; Leonardo and Fee, 2005), the authors do not distinguish between the functional subdivisions of RA and so it is not clear

Breathing and singing: forebrain and brainstem 123

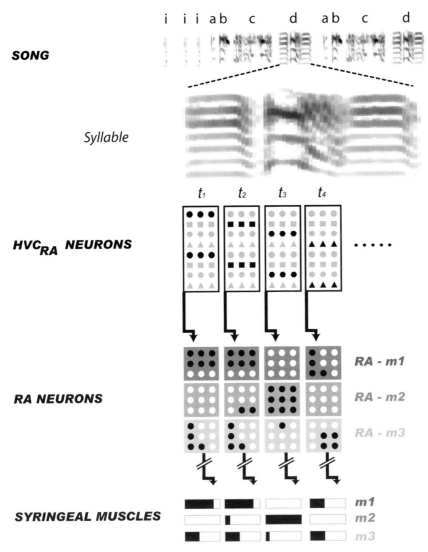

Figure 10.4 Conceptual representation of the "sparse code" model for song production. During the production of song in the zebra finch (top panel: "Song"), single unit recordings from RA-projecting neurons (HVC$_{RA}$ neurons) have shown that each neuron only fires a single burst during the entire duration of the motif and that this burst occurs at exactly the same acoustic transition from one motif to the next. Given the typical duration of a typical motif, the overall number of neurons in HVC and the duration of each burst (~10 ms), it has been hypothesized that about 200 neurons will simultaneously be active (i.e. produce a single burst) during each of these time windows (t) (Fiete et al., 2004). The second panel ("HVC$_{RA}$ neurons") shows a schematic representation of this concept by dividing syllable "d" into discrete time windows ($t_1, t_2, t_3, t_4 \ldots$) where different subpopulation of HVC$_{RA}$ neurons (round cells during t_1, square cells during t_2, etc.) are active during each time window. Each of these subpopulations of HVC$_{RA}$ neurons (e.g. square cells in window t_2) is thought to activate a discrete population of neurons in RA. This population of neurons is shown in the third panel ("RA neurons") by the black filled circles. To represent the known myotopic map in the ventral part of RA, which is known to project to nXIIts, RA in this figure is divided into three sections (RA-m1, RA-m2 and RA-m3) to schematically represent these functional subdivisions. For simplicity, the nXIIts layer has been omitted and only three out of the six muscle groups are represented (lowest panel "Syringeal muscles"). Activation of a specific subset of neurons in RA would therefore recruit various muscles groups and result in a distinct acoustic output for that specific window in time. In the case of window t_1, for example, activation of the round cell population in HVC leads to activation of a subset of neurons in RA which, in this example, activate muscle 1, a little bit of muscle 3 and not muscle 2.

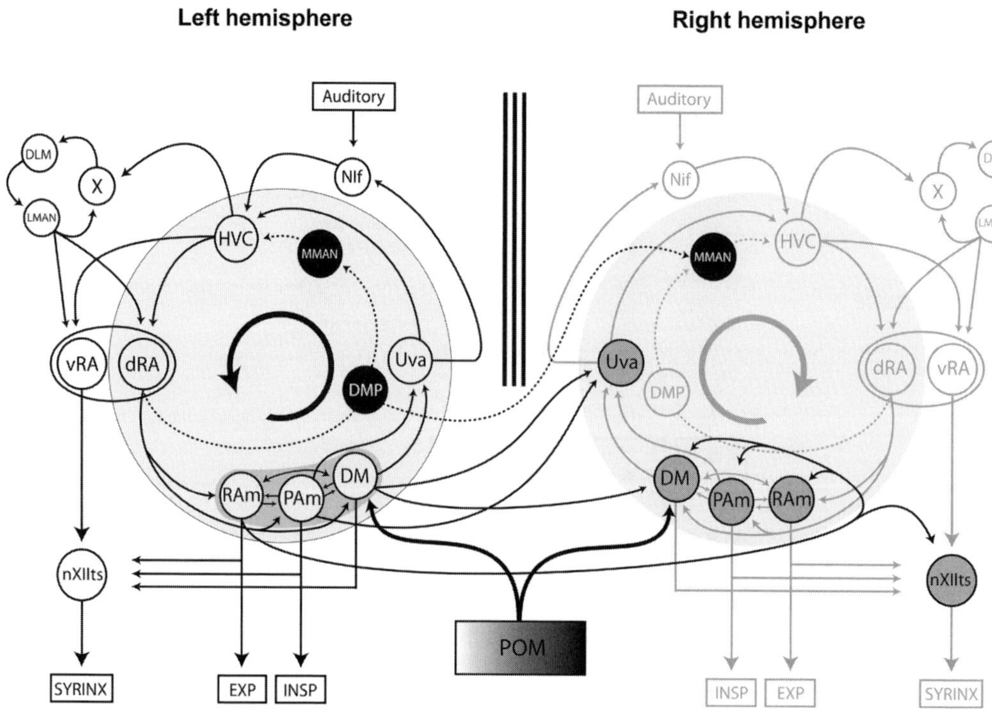

Figure 10.5 Diagram of the avian song system emphasizing its bilateral organization and the bilateral projections from the brainstem to the forebrain. Highlighted in gray is the portion of the song system that is thought to be involved in song pattern generation. This loop consists of the forebrain nucleus HVC, the dorsal portion of RA (dRA), the vocal-respiratory network (highlighted in darker gray), and the thalamic nucleus uvaformis (Uva). Nucleus DM of the vocal-respiratory network receives bilateral projections from the medial preoptic nucleus (POM) which might provide the "drive" into the system to initiate song. An additional loop consisting of the thalamic nucleus DMP and MMAN (filled black circles connected by a dotted line) has also been shown to play a role in song motor production (Foster and Bottjer, 2001; Coleman and Vu, 2005). The connection between vRA and nXIIts serves as an output pathway to the syrinx rather than as part of the song pattern generator. The three parallel lines between hemispheres illustrate the lack of commissural connections between forebrain song control nuclei. Many of the nuclei in the vocal motor system (DMP, RAm, PAm, and DM) project directly or indirectly to vocal motor nuclei in the contralateral half. These projections are bilateral but we only illustrate projections from the left to the right for simplicity. Nuclei receiving contralateral inputs are highlighted in dark gray. The anatomical connections shown here represent the major projections in the song system and have been compiled from different sources (Stokes *et al.*, 1974; Nottebohm *et al.*, 1982; Vates *et al.*, 1997; Reinke and Wild, 1998; Striedter and Vu, 1998; Sturdy *et al.*, 2003; Riters and Alger, 2004; Wild, 2004a). Weak projections have been left out. Anatomical names: DMP, dorsomedial posterior nucleus of the thalamus; MMAN, medial magnocellular nucleus of the anterior nidopallium; NIf, nucleus interfacialis of the nidopallium. INSP and EXP represent respectively the inspiratory and expiratory motor neurons. For other abbreviations, see Figure 10.2.

whether the sparse motor output observed in HVC can be generalized to all HVC_{RA} projection neurons.

Linking inspiration with HVC motor output
The sparse code in HVC is proposed to represent each time window in a song sequence. It is unclear, however, whether this output encodes only acoustic elements in the song or whether it encodes the entire song sequence, including the silent transitions between syllables as well as between consecutive motifs. Given that song occurs during the expiratory portion of the respiration cycle, it is possible that periods of inspiration could provide time markers for the occurrence of silent periods in the song. If so, one might expect some indication of this in the neural activity recorded from HVC during song. However, from the recordings of single unit activity shown in Hahnloser *et al.* (2002), there appears to be a lack of HVC_{RA} projection neurons that burst during time windows associated with intersyllable silent intervals (Hahnloser *et al.*, 2002). While this might be a

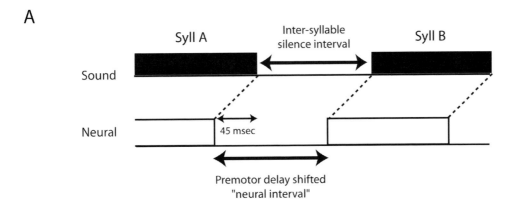

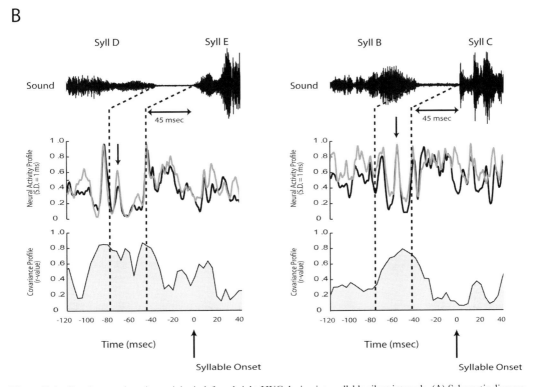

Figure 10.6 Synchronous bursting activity in left and right HVC during intersyllable silent intervals. (A) Schematic diagram illustrating the HVC neural equivalent ("neural interval") of the intersyllable silent interval. (B) Evidence that the HVC neural equivalent of the intersyllable silent interval contains stereotyped neural patterns that are highly correlated between hemispheres. Shown below are examples from two different birds showing the presence of one, or two, clear bursts of activity occurring coincidently in both hemispheres during the silent interval period. The dotted lines are aligned respectively to the syllable onset and offset and shifted by a 45 ms premotor delay. (Reproduced with permission from Schmidt, 2003.)

consequence of undersampling of these neurons, it contrasts markedly with the strong neural responses we observe in HVC during these silent periods using multi-unit electrode recording techniques (Schmidt, 2003).

We show that approximately 80% of intersyllable silent intervals are associated with a robust, short-duration burst of activity in HVC (Figure 10.6) that is highly synchronized across recording sites, both in the same

HVC as well as across hemispheres (Schmidt, 2003). One interpretation is that rather than representing motor output, these synchronized bursts reflect the common synchronous input that drives activity in HVC in both hemispheres (see next section). This activity might therefore represent a timing command signal from the brainstem which functions to specify the inspiratory phase and, possibly, the onset of the next syllable.

BRAINSTEM STRUCTURES AND THE INTEGRATION OF BREATHING AND SINGING: EVIDENCE FOR A RECURRENT VOCAL MOTOR CIRCUIT

Bottom-up projections from brainstem respiratory centers to forebrain vocal motor command centers

The segregation of forebrain vocal control nuclei in each hemisphere and the simultaneous need for tight interhemispheric coordination requires mechanisms that ensure the tight coordination between the respiratory system and syringeal motor output in both halves of the syrinx (Williams, 1985; Schmidt et al., 2004). Consistent with this notion, all three vocal respiratory nuclei in the brainstem (RAm, PAm and the dorsomedial nucleus of the intercollicular complex (DM)) are tightly interconnected ipsilaterally and two of these nuclei (RAm and DM) also have strong projections to their contralateral counterparts. All of the nuclei in this interconnected vocal respiratory network (VRN) project to the ipsilateral nXIIts and one of these nuclei (RAm) also projects to the contralateral nXIIts (Sturdy et al., 2003; Kubke et al., 2005) (Figure 10.5). Thus despite the segregation of vocal control nuclei in the forebrain, there is a significant amount of bilateral interconnectivity between the respiratory and syringeal motor system at the level of the brainstem. This connectivity could serve to coordinate respiratory and syringeal muscle activity on both sides of the syrinx.

In addition to this putative brainstem-level mechanism of coordination, there exists at least one additional way by which the respiratory system could coordinate vocal output. Specifically, two of the vocal respiratory brainstem nuclei (PAm and DM) project back to the forebrain nucleus HVC via the intermediary of the thalamic "relay" nucleus uvaeformis (Uva) (Vates et al., 1997; Reinke and Wild, 1998; Striedter and Vu, 1998; Ashmore et al., 2005). Because these connections are bilateral in nature they may serve to coordinate premotor activity in both hemispheres according to the respiratory demands of the system. Moreover, the connections between the vocal respiratory brainstem and the forebrain vocal control nuclei also close an anatomical loop. Thus the vocal motor control system, rather than being represented as a simple top-down motor pathway, can be represented instead as a recurrent circuit where no single structure stands at the top of a motor hierarchy (Ashmore et al., 2005) (Figure 10.5). This view differs from most current models of song motor control where HVC is given a prominent and autonomous role in the production of the song motor sequence and where HVC is proposed to act as the main clock driving song behavior (Vu et al., 1994; Yu and Margoliash, 1996; Fee, et al, 2004). The idea of a recurrent song circuit is further strengthened by a second pathway linking motor outputs from dRA back up to HVC via the intermediary of the dorsomedial posterior nucleus of the thalamus (DMP) and the forebrain nucleus MMAN (medial magnocellular nucleus of the anterior nidopallium) (Vates et al., 1997) (Figure 10.5). Each hemisphere therefore potentially receives a rich source of information concerning both the respiratory state of the system (from bilateral PAm and DM projections) and the state of motor commands produced in RA (from bilateral DMP projections).

Bottom-up synchronization of forebrain premotor activity in both hemispheres

In addition to the anatomical evidence, recording and ablation studies have provided additional support for the idea of a recurrent song circuit. First, simultaneous recordings from the HVC of both hemispheres during singing reveal a high degree of synchronization of premotor activity during both song (Schmidt, 2003) and call production (M. Schmidt, unpublished observations). Not only is premotor activity highly correlated with the onset of individual syllables (Figure 10.3b), and in some cases with individual notes, but most silent intervals between syllables contain pronounced bursts of premotor activity that are highly synchronized across recording sites both within HVC as well as across hemispheres (Schmidt, 2003) (Figure 10.6). These results suggest that synchronization of premotor activity in HVC is provided by a common input. In the absence of any known midline-crossing projections between left and right HVC and any other forebrain

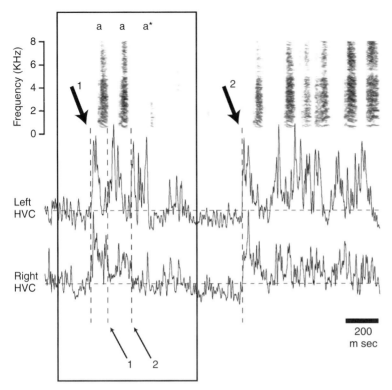

Figure 10.7 Partial loss of bilateral HVC synchronization following unilateral lesions of Uva. Premotor activity was recorded simultaneously in both left and right HVC while the bird attempted to sing (top trace). As is typical for birds recorded shortly after unilateral Uva lesion, singing attempts typically result in stuttered iterations of a few introductory notes or syllables. In the sequence of syllables highlighted by the box, the bird produces two identical song elements (a) followed by a truncated version of a third element (a*). In this example, where premotor activity was recorded 19 days after Uva was lesioned in the right hemisphere, HVC premotor activity in the intact left hemisphere shows distinct bursts of premotor activity that precede individual syllable elements by approximately 45 ms. HVC premotor activity on the side of the Uva lesion (right HVC) shows an initial premotor burst during production of the first syllable that is tightly correlated with the onset of the premotor burst in the contralateral hemisphere (thick arrow #1). A similar correlation is observed during a second attempt to sing (thick arrow #2). In contrast, however, the second burst (small arrow #1) is temporally delayed relative to the premotor pattern recorded on the intact contralateral side. The expected third burst (small arrow #2) on the lesioned side never occurs. This uncorrelated pattern of premotor activity across hemispheres is never observed in intact birds. In this example, neural activity has been rectified and smoothed with a wide Gaussian filter (S.D. = 10 ms) and normalized to the maximum firing rate. (Figure adapted from Coleman and Vu, 2000, and is identical to figure 6 in Schmidt et al., 2004. Reproduced with permission from Schmidt et al., 2004.)

song control nuclei (Wild, 2004a), such inputs are likely to originate from either of the pathways linking the vocal respiratory network to the forebrain (VRN/Uva pathway) or the DMP to MMAN pathway.

Support for a synchronizing role for the VRN/Uva pathway is suggested by the observation that unilateral lesions of Uva are followed by a temporary failure to produce normal song (Coleman and Vu, 2005). Following lesions, song attempts consist primarily of introductory notes followed by the first syllable of the first motif. These are then sometimes followed by several unformed syllables but mostly results in aborted song attempts. Interestingly, while normal song function recovers after several weeks, adding a unilateral lesion of MMAN prevents such recovery even though MMAN lesions alone have only minor effects on song (Foster and Bottjer, 2001; Vu and Coleman, 2001). Simultaneous neural recordings from left and right HVC suggest that the inability to produce syllables beyond the first syllable in Uva-lesioned birds (Figure 10.7) is caused by the lack of synchronization of HVC activity patterns between hemispheres (Coleman and Vu, 2000). The idea that Uva plays an important role in song production and that it might serve to synchronize premotor activity in HVC is further supported by recordings showing bursts in Uva during singing that appear time-locked to the production of individual song syllables (Williams and Vicario, 1993).

One of the strongest arguments against a recurrent circuit for song control has been the evidence that brief electrical stimulation in HVC, but not its downstream target RA, could disrupt the song temporal pattern (Vu et al., 1994). This suggested that HVC, possibly in combination with afferent inputs, was responsible for determining the temporal structure, including syllable sequencing, of song. In this scheme, RA simply followed commands initiated in HVC. More recent results from our laboratory, however, show clearly that stimulation in RA, as well as its downstream target PAm, causes a resetting of the song temporal pattern (Ashmore et al., 2008). Importantly, stimulation in nXIIts, which is also a target of RA but does not form part of the loop that projects back to the forebrain, produces only temporary acoustic distortions without having any effect on the temporal pattern (Ashmore et al., 2005). This suggests that disruption of neural activity within the recurrent loop of song control nuclei can influence premotor activity throughout the whole song system. This in turn implies that activity along the ascending pathway from the brainstem to the forebrain has the ability to influence premotor activity in both left and right HVC.

Bilateral bottom-up information flow between the respiratory system and forebrain song control nuclei

The results from anatomical, lesion, and stimulation studies suggest that the VRN/Uva pathway transmits ascending information bilaterally into the forebrain song system during song production. This hypothesized model of song system network structure makes an important prediction, namely that neural transmission along the VRN–thalamus–forebrain pathway should be functionally robust. To evaluate whether the respiratory brainstem is capable of reliably relaying signals back to HVC in both hemispheres, we investigated whether brief electrical stimuli delivered to forebrain and brainstem vocal control nuclei in one hemisphere can transsynaptically evoke activity in forebrain vocal control nuclei of the contralateral hemisphere. Brief electrical stimuli (as few as one pulse) applied to either HVC or RA in one hemisphere cause short latency (20–30 ms) neural responses in the contralateral HVC or RA. The reliability of these responses is remarkable given that the shortest pathway between the HVC in one hemisphere and the RA in the other involves at least five synapses. The short latency of about 5 ms per synapse suggests that this motor pathway is dedicated for rapidly relaying signals across hemispheres. We have also shown that stimulation in either DM or PAm, two nuclei forming part of the VRN, can drive short latency transsynaptic responses (∼15 ms) in the contralateral RA. These responses are completely abolished following contralateral lesions of Uva, confirming a role for this structure as a relay nucleus between the VRN and HVC (Figure 10.8).

The ability of brief electrical stimuli to elicit neural responses in contralateral song control nuclei suggests that spontaneous activity in the respiratory brainstem might also be able to influence baseline neural patterns in forebrain vocal control nuclei. While the majority of neurons in PAm are phase locked to the inspiratory phase of respiration, we have identified a class of neurons in this structure whose output is not tied to the respiratory pattern but rather to the bursting activity in the contralateral RA (Ashmore et al., 2008). In fact, despite the existence of many PAm neurons that are precisely active during the inspiratory phase of respiration, and the ability of single stimulation pulses in PAm to elicit activity in the contralateral RA, RA never shows any sign of rhythmic activity time-locked to the bird's inspiratory pattern (Figure 10.9). If, as this observation suggests, activity linked directly to inspiration is prevented from flowing toward the forebrain, then PAm activity may influence forebrain vocal control structures via this distinct newly identified class of nonrespiratory neuron. Consistent with this view, Wild has recently identified a distinct class of neurons in PAm that projects directly upstream to Uva rather than downstream via the bulbospinal tract to inspiratory motorneurons in the spinal (Wild, this volume).

The data presented above are consistent with a major role for the VRN/Uva pathway in synchronizing HVC activity in both hemispheres. As mentioned previously, however, there exists an additional recurrent

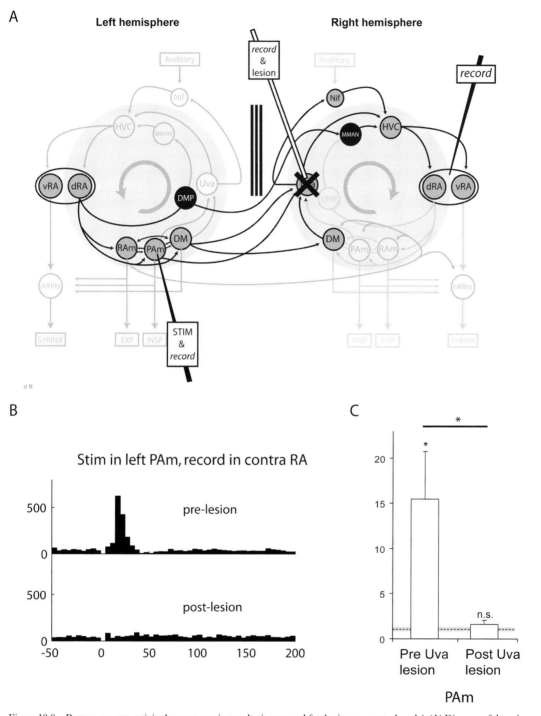

Figure 10.8 Bottom-up connectivity between respiratory brainstem and forebrain song control nuclei. (A) Diagram of the avian song system emphasizing the two known pathways crossing from one hemisphere to the other. The first is the connection from RA through the vocal respiratory network (DM and PAm) and Uva. The second, filled in black, is the pathway from RA to DMP to the contralateral MMAN. Both pathways converge on the contralateral HVC. Sites where either stimulation (RA, DM, and PAm), recording (Uva and

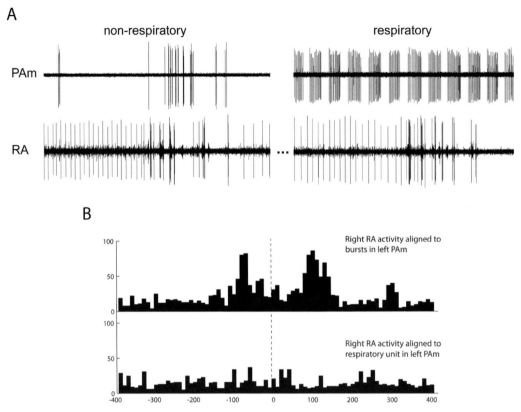

Figure 10.9 Intrinsic activity in a subset of paraambiguus (PAm) neurons is correlated with bursting in the contralateral robust nucleus of the arcopallium (RA). (A) Simultaneous recordings from PAm and the contralateral RA under anesthesia. Two separate units recorded in PAm are shown in the top panel. The neural trace to the left is from a nonrespiratory type neuron whereas the trace on the right is recorded from a respiratory-type neuron whose bursts are time locked to the inspiratory phase. Both neurons were recorded within close proximity to each other during the same penetration. The bottom trace represents neural activity recorded simultaneously from an RA neuron recorded in the contralateral hemisphere. Note that RA activity in both traces transition from a steady tonic rate to spontaneous bursting. Transitions typically observed in anesthetized RA recordings. Burst activity in the nonrespiratory PAm neuron is tightly coupled to the burst pattern in RA whereas no temporal relationship is observed between RA and the respiratory PAm neuron. (B) Two peri-event time histograms (PETHs) are shown, representing activity recorded in the right RA of one bird. In the top PETH, RA activity was aligned to bursts in nonrespiratory PAm units. In the bottom PETH, RA activity was aligned to the first spike of each rhythmically active burst. Peaks in the top histogram indicate that RA activity is high around the time of nonrespiratory bursts in PAm, but shows no correlation to respiratory activity. (Reproduced with permission from Ashmore et al., 2008.)

Caption for Figure 10.8   (cont.)

RA), or lesions (Uva) were performed are illustrated with an electrode. (B) Effects of lesions of the intermediate Uva between PAm and the contralateral RA. The PSTH shows activity following stimulation in the left PAm, recorded in the right RA, before and after electrolytic lesions of the contralateral Uva. In this example, lesioning Uva completely eliminates the evoked response in RA caused by brief electrical stimuli in PAm. (C) Quantification of evoked responses in the right RA from left PAm stimulation, before and after Uva lesions. The y-axis values indicate the mean peak-to-baseline ratio for each group (the value of the peak bin within 100 ms following stimulation divided by the mean of all bins in the 100 ms preceding stimulation, $n = 4$ birds). The dotted line represents a mean control peak-to-baseline value calculated from all birds prior to lesions (the peak bin prior to stimulation divided by the mean of all bins in a 100 ms window prior to stimulation), and the gray area represents one standard deviation from this mean. Error bars represent standard error. * indicates significance for the group by one-tailed paired t-tests, wherein each bird was paired with either its own control peak-to-baseline value, or paired between values obtained before and after Uva lesions. (Reproduced with permission from Ashmore et al., 2008.)

circuit that might also play an important role in hemispheric coordination. This pathway consists of projections from ipsilateral RA to the thalamic nucleus DMP. DMP then projects to both the ipsilateral and contralateral telencephalic nucleus MMAN which then projects back to HVC. Several recent findings support an important coordinating role for this circuit. First, in the experiments where RA stimulation caused a short latency response in the contralateral RA, Uva lesions did not completely abolish this functional connection. This suggests that the alternate pathway through DMP and MMAN may contribute to the functional connectivity between left and right RA (R. A. Ashmore et al. unpublished data). Second, in birds with unilateral lesions of Uva, onset of premotor activity recorded in left and right HVC, while desynchronized for the second and third syllables in the motif, was nevertheless synchronized for the first introductory notes and the first syllable of the song (Figure 10.7; Coleman and Vu, 2000). This observation implies that synchronization of HVC activity across hemispheres is still intact for introductory notes and the first syllable even in cases where the VRN/Uva pathway is eliminated. If confirmed and extended, these findings suggest that the DLM/MMAN circuit forms an additional important bottom-up pathway in the control of song production.

INTEGRATIVE MODEL OF SONG PRODUCTION

Functional relationship between brainstem and forebrain: a tale of two clocks

At its most fundamental level, song is composed of acoustically distinct elements, known as notes, which are either produced in isolation or attached to one or more additional notes to produce a syllable (Sossinka and Böhner, 1980; Ho et al., 1998; Tchernichovski et al., 2000). Given the precision and general temporal and acoustic invariance of note production in species such as the zebra finch, it is easy to imagine that note elements are controlled by the execution of temporally precise motor commands delivered to both the respiratory and syringeal system (Chi and Margoliash, 2001; Leonardo and Fee, 2005). However, songs do not simply consist of isolated notes or syllables. Songs are made up of sequences of syllables that are separated by silent intervals. While song patterns vary greatly among species (Catchpole and Slater, 1995), even in birds with highly stereotyped songs like the zebra finch, the tempo and sometimes the order with which syllables are produced within these sequences can vary (Cardin et al., 2005; Glaze and Troyer, 2006). This variability may depend on the social context in which songs are produced (Jarvis et al., 1998; Castelino and Ball, 2005; Cooper and Goller, 2006) and the existence of such variability suggests that the temporal pattern with which syllable sequences are executed can be differentially modulated (Cooper and Goller, 2006; Glaze and Troyer, 2006).

The finding that HVC output to RA consists of discrete bursts of action potentials that are precisely time-locked to single acoustic events in the song provides a compelling model for HVC to act as a clock that regulates the temporal output of song (Hahnloser et al., 2002; Fiete et al., 2004). While recent evidence points to the possibility that inputs from LMAN to RA might be able to modulate the temporal output of RA (Kao and Brainard, 2006) and possibly even influence syllable sequencing (Olveczky et al., 2005), it is unclear whether this sparse temporal output of the HVC to RA pathway is sufficient to explain all of the more global changes in song temporal patterning. The existence of bottom-up pathways that serve to synchronize song premotor activity in both hemispheres suggests the existence of a second clock system that might be responsible for setting overall song tempo (Figure 10.10) (Schmidt, 2003; Ashmore et al., 2008). The observed pattern of synchronized premotor activity in HVC is certainly consistent with the hypothesis that HVC in both hemispheres receives precisely timed synchronizing inputs that act to specify the onset for each song syllable (Schmidt, 2003; Schmidt et al., 2004). When the syllable is complex, such inputs might even specify the onset of the note components that make up these syllables. These timing signals would therefore serve the purpose of a second clock that would initiate the cascade of timed bursts in HVC which ultimately determine the specific acoustic features of syllables (Figure 10.4). By controlling syllable onset, such a bottom-up system could ultimately control the duration of inspiratory gaps between syllables and therefore control song tempo. Control of timing by two separate clocks, one for syllable/note duration (HVC clock) and one for overall song tempo (clock associated with bottom-up control system) might

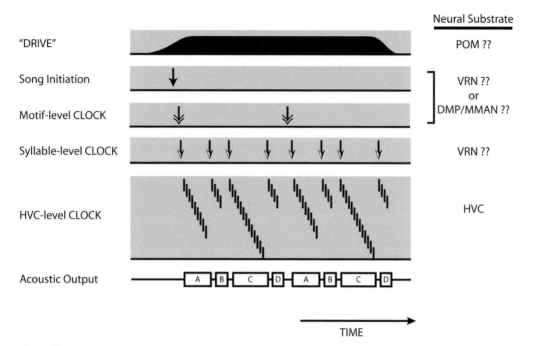

Figure 10.10 Conceptual model describing the various timescales involved in song production and their associated neural control systems. Zebra finch song (bottom row) can be described as a sequence of acoustically distinct syllables, known as motifs (schematized as boxes A–D), which are repeated multiple times to produce a song bout. Song can therefore be broken down into multiple timescales and it is therefore plausible that different aspects of song are under the control of different neural substrates within the song control system. In this scheme, song might be initiated by activation or "drive" of the song pattern generating circuit (top row) by structures such as the medial preoptic nucleus (POM). Such activation might then modify the brainstem vocal-respiratory network (VRN) to transition from a respiratory network into a song producing network. Timing of song onset (black arrow) and motif onset (double arrows) might therefore be under the control of either the VRN network or alternatively by the DMP/MMAN network, since DMP appears to be necessary for the synchronization of introductory notes and the first syllable in the song (see Figure 10.7). The onset of individual syllables (black/white arrows) in the motif is proposed to be determined by timing inputs generated in the VRN. These inputs then activate a cascade of "sparse" events (black tick marks) in both HVCs, which encode the acoustic features of each syllable on a 10 ms window scale.

more easily allow for the observed nonlinear scaling of song duration (Cooper and Goller, 2006; Glaze and Troyer, 2006).

PAm as a possible integrator of descending motor commands and respiration

How the brainstem vocal-respiratory network, and the respiratory nucleus PAm in particular, might serve such clocking functions is unknown. Certainly respiration is key to all vocalization, and respiratory tempo is arguably key to vocalizations with complex temporal structure. However, the rate of respiration is often under the influence of multiple factors, both central and environmental. Low blood oxygen can increase respiratory rate (Lahiri et al., 2006), as can behavioral states associated with excitement, agitation, or startle (Feldman and Del Negro, 2006). In addition, due to the physical structure of the respiratory apparatus in birds, converging signals governing expiratory and inspiratory cycles must be coordinated in order for efficient breathing to occur. Thus, nuclei governing respiration are ideally situated to integrate descending premotor signals with feedback from the periphery (Figure 10.11). Such integration might be crucial in song production, where commands imposed by the song system may make taxing demands of the periphery. If the peripheral musculature cannot meet those demands, or if they jeopardize oxygen homeostasis generally, then the demands of the song system may need to be tempered, or aborted. Preliminary evidence has suggested that

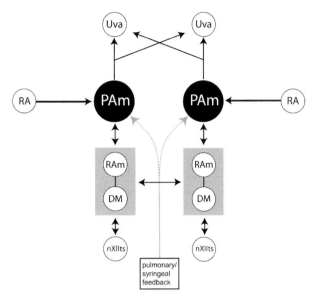

Figure 10.11 PAm as an integrator of vocal motor commands and respiratory feedback. Schematic showing a hypothesized model of signal integration in PAm. Four signal pathways are shown, based on existing anatomical and electrophysiological evidence. The first is the descending motor input from RA carrying premotor commands from the forebrain. The second group of pathways, projecting away from and also toward PAm, forms dense interconnections with other brainstem nuclei (RAm and DM; gray box), and signals along these may mediate peripheral coordination. The third pathway impinges on PAm from the periphery, carrying signals representing pulmonary or other somatosensory feedback (dotted gray). The fourth pathway is the ascending projection from each PAm into the thalamus (Uva), which carries timing input and signals necessary for interhemispheric coordination at the forebrain level. Not shown are PAm's bulbospinal projections onto respiratory neurons in the spinal cord. This model illustrates the potential of PAm to integrate premotor signals with peripheral feedback in order to generate coordinated signals for the vocal–respiratory network and forebrain vocal control centers in both hemispheres. Abbreviations as in Figure 10.2.

PAm receives peripheral feedback from the air sacs via the nucleus of the tractus solitarius (Wild, 2004b, and this volume), information that presumably reflects aspects of the current respiratory state. PAm thus lies at a vital intersection of descending motor commands and ascending respiratory feedback, two sources of information whose integration could provide useful input to both hemispheres of the song system. Since the rate of song production is an important variable in respiratory demand, it is perhaps not surprising that this hypothesized integration function of PAm might make this nucleus a critical component in determining the clocking rate of song production.

The vocal-respiratory brainstem as a possible integrator of motivational drive and song control
Song length in zebra finches, and many other species, is related to the arousal level of the bird and is longer in instances where the bird is highly motivated to produce song (Jarvis et al., 1998; Cooper and Goller, 2006). These longer song bouts, which in zebra finches translate into the production of more motifs, presumably serve the purpose of impressing a potential mate since female zebra finches (Neubauer, 1999) and starlings (Eens et al., 1991b; Gentner and Hulse, 2000b; Gentner et al., 2001; Sockman et al., 2002) have demonstrated a preference for long song bouts over short song bouts. The neural correlate of the "drive" to sing in the presence of a sexually desirable individual has putatively been identified as the medial preoptic area (POM) (Riters and Ball, 1999; Riters and Alger, 2004; Riters et al., 2004). The POM is involved in sexually motivated behaviors in a number of species and lesions to the POM in male starlings affect sexually related behaviors. Such lesions also result in little to no production of female directed song in these birds, while not affecting the level of undirected

song. Interestingly, recent anatomical evidence suggests that nucleus DM within the vocal respiratory network receives a strong projection from the POM (Figure 10.5) (Riters and Alger, 2004; see also Ball et al., this volume). This nucleus might therefore provide the drive into the VRN (Figure 10.10) and, therefore, by controlling the number of motifs a bird sings, might also play a role in determining overall song bout duration.

THE AVIAN SONG SYSTEM AS A MODEL FOR BRAINSTEM INSTRUCTION OF FOREBRAIN MOTOR ACTIVITY

With the exception of relatively simple rhythmic motor patterns such as breathing, chewing and locomotion (Marder and Calabrese, 1996; Marder and Bucher, 2001; Ramirez et al., 2004; Feldman and Del Negro, 2006), models of motor control have often placed most of their emphasis on the forebrain (Tanji, 2001; Krauzlis, 2004). This has certainly been true for the production of learned vocal behaviors (Geschwind, 1970; Yu and Margoliash, 1996; Margoliash, 1997) where the brainstem has been primarily viewed simply as an output pathway for motor commands generated in the forebrain. In this article, we have presented evidence suggesting that the brainstem plays an active role in the generation of vocal motor commands. Rather than simply following instructions from the forebrain, the brainstem informs forebrain vocal centers about key temporal features of song.

Study of brainstem influences on motor nuclei of the forebrain is not new. A broad literature exists focusing on modulatory influences from dopaminergic, serotonergic, and noradrenergic brainstem centers such as the ventral tegmental area, raphe nucleus, and locus coeruleus (Schultz et al., 1997; Berridge and Waterhouse, 2003; Aston-Jones and Cohen, 2005). These systems are increasingly viewed as shaping the details of motor production by regulating behavioral state, attentional focus, or selection from multiple possible actions. Likewise, many studies have examined the role of recurrent loops of connections in motor production, frequently involving projections from the forebrain (the mammalian cortex or the avian pallial regions) through the basal ganglia or cerebellum, through the thalamus, back to the forebrain (for mammalian/avian comparisons see Farries, 2004; Perkel, 2004). However, only a handful of studies to date have examined the role of brainstem nuclei in providing instructive input to motor nuclei of the forebrain.

The only well characterized example of brainstem to forebrain contribution to motor control is the system controlling eye saccade generation. In monkeys, there exist robust projections from the superior colliculus (SC) to the frontal eye fields (FEF), through the mediodorsal nucleus of the thalamus (MD) (Sommer and Wurtz, 2004a, b; Wurtz et al., 2005). Recording and stimulation along this pathway has suggested that it conveys corollary discharge signals back to the forebrain, which potentially contribute to coordinating sequential saccades. There are obvious parallels with the account presented here in which brainstem nuclei close to the periphery transmit signals of impending action back to the forebrain, through the thalamus, in order to guide the production of sequential behavior in a temporally precise manner.

Vocal control systems in mammals are also tightly linked to the respiratory system (Jurgens, 2002; Smotherman et al., 2006). One structure in particular, the parabrachial area, plays a key role in vocal control as it receives descending vocal motor commands and is known to significantly influence respiratory rhythm generation. Interestingly, recent work in the macaque has described ascending projections from the medial parabrachial area to cortical areas responsible for laryngeal control (Simonyan and Jürgens, 2005). Like the avian brainstem nuclei discussed in this chapter, the parabrachial area might therefore also form part of a possible recurrent network. Although functional characterization of these pathways has not been performed, these findings suggest the exciting possibility that motor circuits designed to integrate vocal production with respiratory control might share similar constraints in circuit architecture.

Our current work provides only indirect evidence for brainstem instructive mechanisms. It clearly demonstrates, however, that the birdsong system could serve as an important model for deciphering how the brainstem integrates multiple signals, and in turn influences forebrain motor output in a bottom-up fashion. Because a crucial component of vocal production is the coordination between vocal output and respiration, the vocal respiratory brainstem might play an important role in instructing the forebrain in all vocalization-related behaviors including human speech.

ACKNOWLEDGEMENTS

We would like to thank Dr. Brent Cooper and members of the Schmidt lab for insightful comments on this manuscript. This research was supported by a grant from National Institutes of Health (RO1 DC006102).

11 • Birdsong: anatomical foundations and central mechanisms of sensorimotor integration

J. Martin Wild

INTRODUCTION

Vocalization in general and singing in particular requires the control and coordination of at least three sets of muscles, those acting on the vocal organ itself (the syrinx in birds), those of the respiratory system, and those acting on the structures of the upper vocal tract (trachea, larynx, pharynx, tongue, and beaks). Although birds have a larynx, it does not contain vocal membranes and does not appear to be involved in sound production. The syrinx in songbirds contains six or seven pairs of small muscles grouped around the confluence of the lower end of the trachea and the two primary bronchi (See Figure 8.1 in Suthers and Zollinger, this volume). The nuclei and pathways mediating neural control of the syrinx, from the telencephalon to the vocal (syringeal) motoneurons, were described in a seminal paper by Nottebohm *et al.* in 1976 (see Fig. 5.2 in Reiner *et al.*, this volume). More recently the nuclei and pathways that mediate control of breathing and, to a lesser extent, those that mediate control of upper vocal tract structures, have been described (Wild, 2004a). What is not yet understood however, is how the diverse sets of muscles involved in singing are coordinated, i.e. how the different neural control systems interact for the purposes of vocal production.

In this chapter I describe the various groups of motoneurons belonging to the three systems – vocal, respiratory and upper vocal tract – and some of their premotor neurons. The descending pathways that act on these motor and premotor neurons will then be outlined, and this will be followed by a description of a nucleus in the caudal medulla called retroambigualis (RAm), which is hypothesized to play a significant role in the coordination of the different groups of motoneurons innervating syringeal and respiratory muscles – and, indirectly, upper vocal tract muscles as well. Finally, I briefly consider some of the feedback pathways that transmit sensory information back into the song system. The respiratory–vocal nuclei and pathways of the brainstem, their interconnections, and their afferent and efferent projections are schematically illustrated in Figure 11.1, which will serve to guide the reader through the complex anatomical connectivity to be discussed.

MOTOR CONTROL

The vocal organ (syrinx) and its motor nucleus (XIIts)

Birds are apparently unique in the animal kingdom in having a vocal organ situated at the distal end of the trachea, rather than at the proximal end where the larynx is situated in mammals. Recently, some distinct similarities have been discovered between the phonatory mechanisms of the avian syrinx and the mammalian larynx (Goller and Larsen, 1997a, b; Larsen and Goller, 1999, 2002), but the two structures are not homologous as vocal organs. The mammalian vocal organ, the larynx, is innervated by the tenth cranial nerve (vagus); the avian larynx is innervated by the nineth cranial nerve (glossopharyngeal); but the syringeal muscles – like the bird's intrinsic tongue muscles – are innervated by the twelfth cranial nerve (hypoglossal). This suggests that the syringeal muscles, like the intrinsic tongue muscles, are developmental derivatives of occipital myotomes and thus share the same cranial nerve and nucleus.

The organization of the various syringeal muscles and their basic functions are described in Suthers and Zollinger, this volume. Here we can simply note that different pairs of syringeal muscles in songbirds have different functions in the respiratory–vocal mechanism, especially those located dorsally versus those located ventrally (Goller and Suthers, 1996a, b; Larsen and Goller, 2002; Suthers and Zollinger, 2004). A simplified view of these differences is that the dorsal muscles are separately concerned with adduction of the syringeal labia, and hence phonation on each side of the syrinx

Neuroscience of Birdsong, ed. H. Philip Zeigler and Peter Marler. Published by Cambridge University Press. © Cambridge University Press 2008.

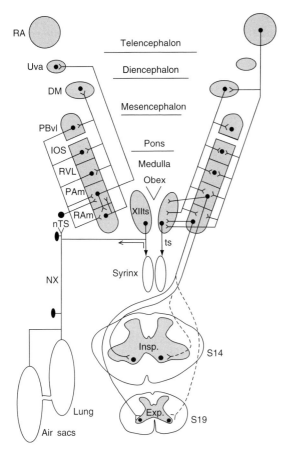

Figure 11.1 Nuclei and some of the interconnections of the respiratory-vocal system in a songbird. Abbreviations from top down: RA, robust nucleus of the arcopallium; Uva, nucleus uvaeformis; DM, dorsomedial nucleus of the intercollicular complex; PBvl, ventrolateral part of the parabrachial nucleus; IOS, nucleus infra-olivarus superior; RVL, ventrolateral nucleus of the rostral medulla; PAm, nucleus parambigualis; nTS, nucleus of the solitary tract; RAm, nucleus retroambigualis; XIIts, tracheosyringeal part of the hypoglossal nucleus; ts, tracheosyringeal nerve; NX, cranial nerve X (vagus); Insp, inspiratory motoneurons at the level of spinal cord segment 14 (lower brachial); Exp, expiratory motoneurons at the level of spinal cord segment 19 (lower thoracic). Dots represent cell bodies and inverted arrow heads terminations. The two sides are symmetrical, of course, but for clarity the projections of RA and DM to all the nuclei of the ventrolateral pons and medulla, as well as to XIIts, are shown on the right, and the cascade of descending projections is shown on the left, as are the ascending recurrent pathways from RAm and PAm. Note the sensory input from the lungs, air sacs and syrinx that reaches nTS, and hence PAm and Uva – see also Figure 11.5.

(i.e. they are functionally lateralized), while the tracheobronchialis ventralis muscle functions as an abductor. The largest ventral muscle (M. syringealis ventralis) is probably concerned with control of fundamental frequency and is not functionally lateralized. These basic differences in the functions of the syringeal muscles are reflected in the organization of the vocal motoneurons, such that those innervating ventral muscles are located rostrally in XIIts, whereas those supplying dorsal muscles are located caudally in XIIts (Vicario and Nottebohm, 1988; Ruan and Suthers, 1996; Roberts et al., 2007). Consistent with this myotopic organization of the vocal motor nucleus is the fact that the premotor nucleus, robustus arcopallialis (RA), is also organized functionally such that distinct parts of it separately innervate these two functional groups of motoneurons (Vicario, 1991a) (see further below).

The role of the tongue in songbird vocalizations is largely unknown. Songbirds do not articulate their vocal sounds as we do, for they have no lips or teeth, and their beaks are nondeformable. Thus, despite the fact that the motoneurons supplying the intrinsic tongue muscles (lingual motoneurons) and the tracheosyringeal muscles (tracheosyringeal motoneurons) form part of the same nucleus, they are organized into distinct subnuclei, reflecting their functional separation. In songbirds lingual motoneurons occupy approximately the rostral third of the hypoglossal nucleus in the medulla, and tracheosyringeal motoneurons occupy approximately the caudal two thirds of the nucleus. The lingual motoneurons overlap slightly the tracheosyringeal motoneurons in the rostrocaudal direction and are displaced slightly ventral to them (see Fig. 7 in Wild, 1993a). Their functional and physical separation from the tracheosyringeal (ts) motoneurons is reflected in the fact that, unlike the latter, lingual motoneurons do not receive descending projections directly from either telencephalic or mesencephalic respiratory–vocal control nuclei – at least in songbirds (Striedter, 1994).

In the transverse plane, XIIts motoneurons form a small, dorsoventrally flattened oval cluster at caudal levels of the nucleus, but a larger, more circular cluster at rostral levels (see Figure 11.4, 11.5). The cell bodies of the hypoglossal motoneurons are typical, polygonal alpha motoneurons. Those innervating the syrinx (XIIts) tend to be larger and less densely packed in males than in females (DeVoogd et al., 1991; Roberts et al., 2007), reflecting a similar situation in the premotor nucleus RA (Nottebohm and Arnold, 1976). The XIIts nucleus is surrounded by relatively cell-free neuropil, especially dorsally and dorsomedially, a region that has been called suprahypoglossal (SH) and noted as a site of particularly concentrated terminations of descending axons of song control nuclei on the dendrites of XIIts motoneurons (Wild, 1993a, 1997a; Roberts et al., 2007). Interestingly, although SH, like XIIts, is larger in male then in female zebra finches, the dendrites of XIIts motoneurons are more extensive and more branched in females than in males (Roberts et al., 2007). Since male zebra finches sing, but females don't, this difference in dendritic architecture may reflect a gender difference in the number or proportion of projections originating in different vocal control nuclei (e.g. RA versus DM) and/or the type of terminations made by these projections on the dendrites of XIIts motoneurons. In addition to male XIIts motoneurons being larger than their female counterparts, rostral XIIts motoneurons are larger than caudal motoneurons, in both males and females (DeVoogd et al., 1991; Roberts et al., 2007). Since the ventral syringeal muscles are larger in males than in females (Luine et al., 1980; Wade and Buhlman, 2000; Wade et al., 2002), this could suggest that the larger rostral motoneurons innervate more muscle fibers of the larger ventral syringeal muscle in the male, i.e. they comprise larger motor units than the smaller XIIts motoneurons located more caudally in the nucleus.

The axons of the hypoglossal motoneurons exit the medulla ventrally in three roots. The most rostral root carries lingual efferents, while the caudal two roots carry most, if not all, the tracheosyringeal efferents. The three hypoglossal roots merge in the upper neck, form an anastomosis with the vagus nerve there, and then proceed as a single main nerve to the trachea. The nerve then divides, with the lingual branch turning rostrally to the tongue and the tracheosyringeal branch caudally to proceed down the neck on each side of the trachea to innervate the tracheolateralis muscle and the intrinsic syringeal muscles on the same side (Figure 11.5). That is, in songbirds the tracheosyringeal innervation is strictly ipsilateral. The trachaeosyringeal (ts) nerve carries a small afferent component, the cell bodies of which are located in a vagal ganglion associated with the anastomosis mentioned above (Figures 11.1, 11.5). An important implication of the ipsilateral syringeal innervation is that bilateral coordination of the syrinx must involve crossed connections above the level of XIIts (Schmidt et al., 2004).

Extracellular single unit and multiunit recordings from XIIts in anaesthetized birds reveal a respiratory rhythm that is in phase with expiration (Figure 11.2). The source of this rhythm is thought to be external to the nucleus, probably the nucleus retroambigualis (RAm) lying in the ventrolateral medulla. Neural recordings from this nucleus also reveal a pronounced respiratory rhythm in phase with expiration (Figure 11.2) and RAm projects, via separate neurons, to both spinal motoneurons that innervate expiratory muscles and XIIts (see Figures 11.1, 11.4, and below).

Respiratory motor and premotor nuclei

Unlike mammals, birds possess neither a diaphragm nor expandable lungs, and it is the air sacs, not the lungs, that inflate and deflate during breathing and singing. Inspiration is effected by the intercostal, scalene and other muscles, which act on the ribs to assist thoracic recoil, expand the thorax, and fill the air sacs under negative pressure (Fedde, 1987). Also in birds, unlike mammals, expiration is an active process which is effected by abdominal and intercostal muscles compressing the thoracic and abdominal air sacs. Although these air sacs are independent, paired structures on either side of the body, they communicate across the midline with the rest of the air sac system. In contrast to the distinctly lateralized function of the syrinx (Suthers and Zollinger, 2004), for which they supply the pressure head, they are compressed by the abdominal expiratory muscles during singing apparently in a bilaterally symmetrical way (Goller and Suthers, 1999).

Motoneurons innervating inspiratory muscles (e.g. Mm. scalenus and levatores costarum) are located in medial lamina IX at lower brachial levels and at the tip of the ventral horn at upper thoracic levels. Motoneurons innervating abdominal expiratory muscles occupy a similar position at the tip of the ventral horn at lower thoracic and upper lumbar levels (Wild, 1993b; Reinke and Wild, 1997, 1998). All these motoneurons have two sets of very extensive dendrites, one extending dorsolaterally into the white matter, and another extending dorsomedially toward the central canal and column of Terni, which houses the

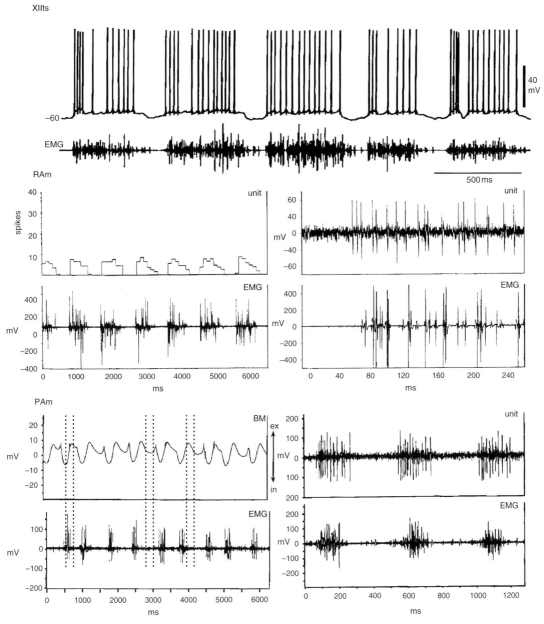

Figure 11.2 Rhythmical activity in XIIs (upper), RAm (middle) and PAm (lower). XIIts is an intracellular in vivo record from an anesthetised zebra finch showing bursts of spikes that are correlated with EMG activity recorded simultaneously from the abdominal expiratory muscles. Note in the single unit record the apparent absence of inhibition at the termination of each burst of spikes. RAm (left) is an extracellular single unit recording in which the spikes (unit) have been binned during each burst that is correlated with EMG activity obtained simultaneously from the abdominal expiratory muscles. At right is shown a single typical burst of spikes (unit) to show that it actually commences before the EMG activity, thereby implying that the former drives the latter. The unit spike traces have been clipped owing to sampling limitations. In PAm (left) EMG activity obtained from a scalene muscle is correlated with inspiration as indicated by the downward direction (in) of a breath monitor (BM) trace. At right is an extracellular single unit recording (spikes, again clipped) of bursts of spikes that are correlated with EMG activity in a scalene muscle.

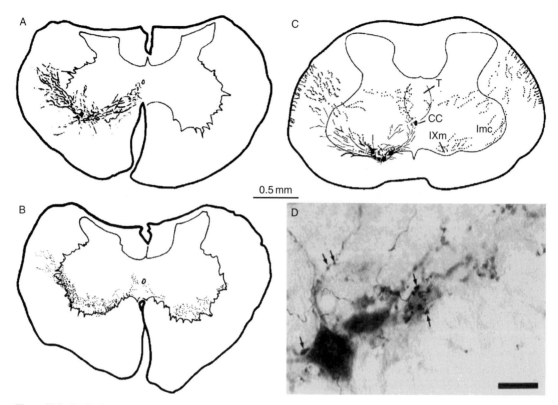

Figure 11.3 Projections of RAm and PAm to the spinal cord and their terminations in relation to expiratory and inspiratory motoneurons, respectively. (A) Camera lucida drawing of a transverse section of lower thoracic spinal cord of a greenfinch showing a couple of cell bodies at the edge of the ventral horn and their extensive dendrites that were retrogradely labeled from an injection of CTB-HRP into the ipsilateral abdominal expiratory muscles. Note the two groups of dendrites, one extending dorsolaterally and the other medially. (B) A similar drawing of a spinal cord section from a similar level showing the terminations (dots) of axons that were anterogradely labeled by an injection of CTB-HRP into RAm. Terminations in three adjacent sections have been superimposed. (C) Camera lucida drawing a single spinal cord section of a zebra finch at the lower brachial level showing cell bodies and dendrites retrogradely labeled from an injection of CTB into a scalene muscle and axons and their terminations labeled by an injection of BDA into PAm. CC, central canal; T, column of Terni; IXm, medial group of lamina IX motoneurons; lmc, lateral motor column. (D) Photomicrograph showing what was described in (C); small arrows point to the BDA labeled profiles. Scale bar = 20 μm.

avian sympathetic preganglionic neurons (Figure 11.3). Although the separation of the dendrites might suggest differential access to descending pathways, both sets appear to receive projections from the same premotor nucleus in the medulla (Wild, 1993b; Reinke and Wild, 1997, 1998).

Apart from the lack of a phrenic nucleus (because there is no diaphragm), the organization of the premotor nuclei for respiratory control appears basically similar to that in mammals (Wild, 1993b, 2004a; Reinke and Wild, 1997, 1998). A ventral respiratory group (VRG) occupies the ventrolateral medulla and is made up of nucleus retroambigualis (RAm) situated most caudally, with neurons that fire in phase with expiration, and a more rostral nucleus parambigualis (PAm), whose neurons fire in phase with inspiration (Figure 11.2). Neurons of both RAm and PAm occupy a narrow arc that extends laterally and ventrally from XIIts and widens as it approaches the lateral edge of the medulla. RAm and PAm both project to the cord bilaterally, with a contralateral predominance, which is in contrast to the predominantly ipsilateral projections of RA to all the respiratory–vocal nuclei of the brainstem (Wild, 1993a; Wild et al., 2000). The axons of RAm and PAm travel in the dorsolateral funiculus, apparently immediately under the pia, until they reach their level of termination. They then turn ventromedially, make contact with the dorsolaterally and dorsomedially directed dendrites of the respiratory motor neurons, and terminate on their cell bodies, apparently monosynaptically (Figure 11.3).

Laryngeal, tongue, hyoid and jaw motoneurons

Although the larynx is not directly involved in phonation in birds, its possible involvement in some form of vocal modulation is supported by the suggestion that glossopharyngeal motoneurons innervating the larynx – which are located in nucleus ambiguus caudal to the obex (Wild, 1981; Grabatin and Abs, 1986; Wild, 1993a) – appear to receive a descending projection from the vocal premotor nucleus, RA (Wild, 1993a). Lingual motoneurons, however, do not receive such a projection, which is consistent with the apparent absence of a direct lingual contribution to singing. Nevertheless, the tongue could be involved in a complex mechanism related to vocal modulation because it is an integral component of the hyoid apparatus, which appears to be involved in changing the shape and size of the oropharynx during singing (Riede et al., 2006). Tongue motoneurons comprise several groups; as in mammals, the intrinsic muscles are innervated by lingual (hypoglossal) motoneurons, as noted above, but unlike in mammals, the muscles acting to produce tongue protrusion and retraction in birds are innervated by glossopharyngeal (IX) and facial (VII) motoneurons, respectively. Tongue protrusion is brought about by the actions of M. geniohyoideus, whose motor neurons are located in the ventrolateral caudal pons and upper medulla. Tongue retraction and elevation are brought about by the actions of Mm. serpihyoideus and stylohyoideus, whose motoneurons are located close to the superior olive, rostral and more ventrolateral to those innervating M. geniohyoideus (Wild and Zeigler, 1980; Zweers, 1982; Dubbeldam and Bout, 1990).

Although jaw movements during singing are probably only part of a more complex oropharyngeal modulation of sound production, beak gape (distance between upper and lower beak) is to some extent correlated with the acoustic frequency of the sound emitted, such that higher or peak frequencies – up to about 3–4 kHz – are produced with wider gapes (Hausberger et al., 1991; Westneat et al., 1993; Podos et al., 1995; Suthers et al., 1996a; Hoese et al., 2000; Williams, 2001; Goller and Cooper, 2004; Goller et al., 2004). It is possible, therefore, that jaw motoneurons may be influenced by the telencephalic song control circuitry. The myotopic organization of jaw muscles within the trigeminal and facial nuclei in the pons has been worked out only for pigeons and ducks (Wild and Zeigler, 1980; den Boer et al., 1986), but not for songbirds. In pigeons and ducks the beak opener muscles, M. depressor mandibulae and M. protractor quadratus are innervated by a composite subnucleus consisting of facial and trigeminal motoneurons, respectively, while the several beak closer muscles are innervated by other subnuclei of the trigeminal complex. The jaw closers, but not the openers, have abundant spindles, which receive their sensory innervation from neurons in the mesencephalic trigeminal nucleus (Bout and Dubbeldam, 1991; Bout et al., 1997). Neither tongue, hyoid nor jaw motoneurons appear to receive a direct RA projection. Pathways that could mediate their involvement in singing are discussed later in this chapter.

Telencephalic components of the premotor song control pathway: HVC and nucleus robustus arcopallialis (RA)

For the purposes of this chapter, we will consider that HVC is the origin of the premotor control pathway for song, while recognizing that localizing the origin of song, like localizing the origin of any behavioral act, is more problematic. HVC projects to RA, and RA projects directly to the vocal motor nucleus (XIIts) and the respiratory premotor nuclei (RAm and PAm) in the medulla (Nottebohm et al., 1976; Wild, 1993a) (Figure 11.1, 11.4). HVC, RA and their descending projections therefore comprise the premotor control pathway for song. HVC also projects to Area X, and is thus the origin of the anterior forebrain pathway (AFP) as well. Since much will be said about HVC in other chapters of this book; only a brief description of the nucleus will be given here. It is a generally lens-shaped nucleus (as seen in the frontal plane) located in the dorsocaudal part of the nidopallium, just under the lateral ventricle, and often produces a small bump on the surface of the brain. It has a narrow, medial extension called para-HVC that extends down the medial margin of the hemisphere (Foster and Bottjer, 1998).

HVC is made up of two kinds of projection neurons and several kinds of intrinsic neurons. Projection neurons are those that project to RA and those that project to Area X of the basal ganglia. No neurons have been found to project to both targets. X-projecting neurons are born before hatching, whereas RA-projecting neurons are largely born post hatching and in some species, such as the canary, continue to be recruited from the subependymal layer to become new neurons each year (see Pytte et al., this volume). Intrinsic HVC neurons express GABA, and may also express one or more of the calcium binding proteins parvalbumin, calretinin and calbindin (Wild et al., 2005). In conjunction with auditory afferents (Janata and Margoliash, 1999; Coleman and Mooney, 2004), the projection neurons and intrinsic neurons make up the microcircuitry of HVC that is thought to mediate auditory–vocal integration (see Mooney, 2004; Prather and Mooney, this volume). The output of this integration, as far as song production is concerned, is directed via axonal fascicles that pass ventrally and slightly caudally to target nucleus robustus arcopallialis (RA).

Songbirds are probably unique amongst all animal species, including other birds, in having a single telencephalic nucleus that directly controls both vocal motoneurons and respiratory premotor neurons for the production of learned vocalizations, as in singing. In many species of adult songbirds RA is a sexually dimorphic, spherical-to-oval, semi-encapsulated nucleus in the medial part of the arcopallium. Like HVC, RA is made up of projection neurons and intrinsic neurons, or interneurons (Spiro et al., 1999; Wild et al., 2001). RA is not obvious in either nonpasserines, or suboscine passerines (Kroodsma and Konishi, 1991), although there is always a possibility that in the latter, at least, this nucleus, and perhaps other nuclei of the song system of oscines, remain cryptic, i.e. that versions of them might be visualized by means yet to be discovered.

RA's role in the control of song was discovered by Nottebohm et al. (1976) in canaries. The great majority of RA projection neurons project via the occipitomesencephalic tract (OM) to subtelencephalic targets in the thalamus, midbrain and rhombencephalon as far as upper C1, and to the vocal motoneurons making up the tracheosyringeal portion of the hypoglossal nucleus in the medulla (XIIts), but it is not known whether any single neuron projects to more than one target. Vicario (1991a) then showed in zebra finches that RA is organized functionally, such that projections to DM arise from a dorsal "cap" of neurons, whereas projections to XIIts arise from the remainder of the nucleus (see also Vates et al., 1997). Vicario (1991a) also found that projections to syringeal motoneurons innervating either dorsal or ventral syringeal muscles arise from largely separate horizontal slabs of RA: ventral RA projects to caudal XIIts where motoneurons innervating dorsal syringeal muscles are located, while middle RA projects to rostral XIIts where motoneurons innervating ventral syringeal muscles are located (Vicario and Nottebohm, 1988). Further analyses of RA projections found them to be considerably more extensive than previously thought, to be similarly distributed in both males and females, and, in some species, such as the canary, to have an appreciable contralateral component (Vicario, 1993; Wild, 1993a; Wild et al., 2000, 2001). Thus, in addition to projections to DM and the vocal motoneurons in XIIts, RA was found to project to certain dorsal thalamic nuclei (DMP and dorsolateral regions of DLM) that project back upon the anterior forebrain pathway (Wild, 1993a; Foster et al., 1997; Vates et al., 1997) and to a series of nuclei in the ventrolateral rhombencephalon, including nucleus ambiguus (which innervates laryngeal muscles) and the respiratory premotor nuclei parambigualis (PAm) and retroambigualis (RAm; Wild, 1997a; 2004a; Reinke and Wild, 1998). The projections of RA to PAm and RAm, as well as to XIIts, show how it might be possible for RA to coordinate syringeal and respiratory muscle activity in the service of vocal production, although the cellular and pharmacological analysis of this coordination has begun only recently (Sturdy et al., 2003; Kubke et al., 2005). A specific aspect of this coordination, perhaps related to timing signals for the expiratory phase of vocalization, may be accommodated by a small proportion of RA neurons situated at the dorsocaudal edge of the nucleus. These interesting neurons have one axon that projects downstream, probably upon RAm, and a collateral axon that projects back upon HVC (Roberts et al., 2006) (see Figure 11.4).

In male, but not in female, zebra finches, all the descending projection neurons, their axons and terminations can be labeled with an antibody directed against parvalbumin, a marker that can therefore be used to define the respiratory–vocal control pathway (Wild et al., 2001). RA interneurons are GABAergic and they probably also express one or both of the calcium-binding proteins parvalbumin and calbindin (Braun et al., 1985; Grisham and Arnold, 1994; Spiro et al., 1999; Wild et al., 2001). They provide for long-range inhibition in RA and can synchronize the firing of RA projections neurons in vitro (Spiro et al., 1999). Therefore, various RA projection neurons with distinct respiratory and vocal targets may be coordinated by RA interneurons.

The dorsomedial nucleus of the intercollicular complex (DM), the ventrolateral parabrachial nucleus (PBvl) and the descending cascade

Although it has long been recognized that in all birds there is a component of the vocal control system that lies close to the inferior colliculus at mesencephalic levels (Phillips and Peek, 1975; Nottebohm et al., 1976), the specific role of the nucleus called DM in singing is still unclear. Stimulation of DM with trains of electrical currents as low as 2 μA can drive vocalizations accompanied by what appears to be a fully integrated respiratory pattern (Vicario and Simpson, 1995; Wild et al., 1997). The nucleus has descending projections and brainstem targets that mimic those of RA, i.e. DM projects upon RAm, PAm and XIIts in both songbirds and nonsongbirds (Wild et al., 1997), Whether DM and RA have differential access to the dendrites and somata of vocal motoneurons is not known. In songbirds – which have both an RA and a DM, and in which RA projects upon DM – it is not known whether DM neurons have firing patterns during singing that are correlated with the temporal or acoustic aspects of syllabic production, or, if they do, whether they contribute to the control of vocal output in a way different from that of RA. DM projects back upon nucleus uvaeformis (Uva) of the posterior thalamus, which is a key source of input to the song control circuitry (Reinke and Wild, 1998; Striedter and Vu, 1998; see also Schmidt and Ashmore, this volume). From the comparative and hodological points of view, DM seems very much akin to lateral regions of the periaqueductal grey (PAG) in mammals that are involved in vocal control by way of their projections upon nucleus retroambiguus (Holstege, 1989; Gerrits and Holstege, 1996; Vanderhorst et al., 2000; Jurgens, 2002).

The ventrolateral parabrachial nucleus (PBvl) in the lateral pons is thought to be the avian equivalent of the mammalian Kollicker-Fuse (K-F) nucleus (Wild et al., 1990) that is involved in the regulation of respiratory phase. Like K-F, PBvl receives ascending projections from a part of nucleus tractus solitarius (nTS Arends et al., 1988; Wild et al., 1990) that, in turn, receives a primary afferent input from the lung (Katz and Karten, 1983). Like RA and DM, PBvl projects to all the respiratory–vocal nuclei located more caudally in the brainstem (Wild and Arends, 1987; Wild et al., 1990) and is thus ideally suited to play a role in the temporally regulated aspects of vocalization by mediating signals related to the inspiratory phase of respiration.

Two other nuclei in the ventrolateral rhombencephalon are also thought to be part of the respiratory-vocal control complex. One (infra-olivarus superior, IOS) is a nucleus located between the superior olive and the spinal lemniscus, and the other (ventrolateral nucleus of the rostral medulla, RVL) is located rostral to PAm. Almost nothing is known about these nuclei, although on the basis of their relative positions and some of their connections, it is possible that IOS is similar to the retrotrapezoid nucleus and RVL to the pre-Bötzinger complex of mammals, two other key nuclei in the ventrolateral medulla that are involved in the complex network of respiratory control. As shown in Figure 11.1, both IOS and RVL receive projections from all the more rostral components of the respiratory–vocal control complex and project to each of the more caudal components of the complex, including XIIts (Wild, 1994b, 1997a). Taken together these structures constitute a cascade of descending projections upon the medullary respiratory–vocal nuclei (Wild, 1994b), which is defined by both parallel and serial pathways (Figure 11.1). The discovery of the way in which these pathways interact for the purposes of vocal control remains an ambitious goal for the future.

Nucleus retroambigualis (RAm), a nucleus that coordinates respiration and vocalization?

In mammals nucleus retroambiguus in the caudal medulla has been regarded as a nexus for the integration and distribution of vocal control signals emanating from the midbrain periaqueductal grey (PAG) to all the nuclei involved in vocalization, such as the laryngeal, peri-oral and respiratory motor nuclei (Holstege, 1989). A very similar situation holds for the avian nucleus retroambigualis, which may be homologous with mammalian nucleus retroambiguus on the basis of its relative position, connections and function. However, RAm in birds not only receives direct descending projections from the telencephalon (i.e. RA), in addition to the midbrain, but it also has a much greater mediolateral extent than retroambiguus in mammals, stretching in an arc from XIIts medially almost to the lateral edge of the medulla (Figure 11.4). The borders of the nucleus are not well defined in Nissl-stained material, but the nucleus as a whole is readily delineated in experimental tracing studies which label the terminal fields of RA (Wild, 1993a; Wild et al., 2001). The delineation of RAm and XIIts by RA axon terminals points to the prime importance of telencephalic influence on the patterning and coordination of respiratory and vocal activity during singing in songbirds.

RAm is composed of several different types of projection neurons and possibly intrinsic neurons, although it is not yet clear whether all the different types of projection neurons project to separate targets. Projection neurons include: (1) bulbospinal neurons that project to motoneurons innervating abdominal expiratory muscles (Wild, 1993b) and probably to those innervating expiratory intercostal muscles; (2) a separate group of XIIts-projection neurons, some of which project bilaterally, but most ipsilaterally, and provide a dense, largely glycinergic (inhibitory) innervation of the vocal motor nucleus (Wild et al., 2000; Sturdy et al., 2003; Kubke et al., 2005); (3) a second group of XIIts-projecting cells that are excitatory (Sturdy et al., 2003); (4) a group of large, multipolar, vagal neurons in the more ventrolateral parts of RAm, the specific peripheral target(s) of which are presently unknown, but may include respiratory structures; (5) neurons that project in an ascending fashion to more rostral components of the respiratory-vocal network, viz. PAm, IOS, RVL, PBvl and DM (Wild, 1993b; J. M. Wild, M. F. Kubke and R. Mooney, unpublished observations). These ascending projections are bilateral with an ipsilateral predominance (Figures 11.1, 11.4).

Extracellular single unit and multiunit recordings from RAm in anesthetized birds establish the presence of a respiratory rhythm that is in phase with expiration (Figure 11.2). Each burst of action potentials commences approximately 30–50 ms before the start of EMGs recorded simultaneously from abdominal expiratory muscles and ceases with the commencement of inspiration (Wild, 1993b, 1994b), consistent with the idea that these neurons are expiratory premotor in nature. Such rhythmic activity may also be expressed by XIIts-projecting RAm cells; in fact it is this type of cell that may account for the rhythmical activity of XIIts neurons that fire in phase with expiration (Figure 11.2) (Manogue and Paton, 1982; Williams and Nottebohm, 1985; Wild, 1993b; Sturdy et al., 2003). Ultimately, this model suggests that the nonglycinergic, excitatory type of XIIts-projecting RAm cells are responsible for the XIIts expiratory-related rhythm recorded during quiet breathing. The role of the expiratory-related rhythmical activity recorded in XIIts is not entirely clear, but it may be related to the maintenance of a patent airway during nonvocal breathing. In fact, if the ts nerves are severed bilaterally in zebra finches, they wheeze with every breath.

Despite the unequivocal evidence of robust excitatory expiratory drive, XIIts neurons do not appear to be inhibited during inspiration (Figure 11.2) (Sturdy et al., 2003). Therefore, the glycinergic, inhibitory input to XIIts from RAm is thought to serve other purposes, probably having to do with shaping XIIts motoneuronal discharge during singing. One idea is that the RAm inhibitory input to XIIts motoneurons delays the latter's discharge long enough to coincide with expiration, the result being a precise coordination of expiratory and syringeal muscle activity that is required to produce the rise in subsyringeal pressure necessary for phonation and syllabic production. When the bird sings, the intrinsic respiratory rhythm of quiet breathing is rapidly changed to one which supports syllabic production, frequently on the basis of one expiration to one syllable or even part-syllable, unless the syllable repetition rate is too high, and then pulsatile respiration takes over (Hartley, 1990). Interspersed during the phrase are minibreaths, making for a highly dynamic respiratory pattern that is intricately coordinated with syringeal activity and hence vocal output (Hartley and Suthers, 1989; Wild et al., 1998).

The telencephalic song premotor nucleus, RA, is likely to be the dominant forebrain structure that engages the RAm–XIIts network to achieve such exquisite coordination. Probably, different RA neurons project to XIIts and to RAm, enabling RA to exert influence on both XIIts motoneurons and RAm neurons (as well as all the other components of the respiratory-vocal network), the end result of which is a coordination of syringeal and respiratory muscle activity that produces the amazing phenomenon of song. These and other considerations of multiple descending inputs from, and outputs to, other parts of the respiratory–vocal network (e.g. DM, PBvl, etc.) suggest that RAm is a true integrator of respiratory–vocal activity. Moreover, its bilateral projections to XIIts, contralateral RAm, and more rostral components of the network, also suggest that it plays an important part in coordinating activity on the two sides of the brain during singing (see also Schmidt et al., 2004). This would seem to be an essential role in a brain that is otherwise dominated by ipsilateral projections from the telencephalic vocal control nucleus and ipsilateral projections to each side of the bipartite but bilaterally and functionally interacting syrinx (Suthers, 1997; Suthers and Zollinger, 2004).

Control of the jaw and other upper vocal tract structures

A number of structures, though not directly involved in respiration or phonation, may contribute to the modulation of vocalization by the upper vocal tract. Beak gape, or aperture, it will be remembered, is correlated to some extent with the acoustic frequency of the notes sung, which, together with other evidence, suggests that beak movements have an important role in modulating the filter properties of this region (Hoese et al., 2000; Williams, 2001; Goller and Cooper, 2004; Goller et al., 2004). These properties are also likely to be importantly influenced by changes in the shape and size of the oropharynx that are brought about by the action of the hypoid apparatus and associated muscles (Riede et al., 2006).

Jaw and other upper vocal tract muscles are, of course, not used exclusively for singing. That, is, they are multifunctional, like many other muscles in the body and are components of behaviors involving jaw movements (e.g. feeding, preening) that may have a higher adaptive priority for the animal than vocalization. It is likely, therefore, that the control of vocalization involves co-optation of neural circuitry that already existed for the control of the jaw, tongue and hyoid apparatus, just as the song system co-opted the respiratory control circuitry that probably predated it evolutionarily (Figure 11.4).

In cranial as in spinal systems, there are in the brainstem monosynaptic arcs for reflex control of jaw muscle activity involving sensory input from jaw muscles and jaw muscle motoneurons. The best known of these is the projection from muscle spindle afferents originating in jaw closer muscles, via trigeminal mesencephalic primary sensory neurons, directly to jaw closer motoneurons (Bout et al., 1997) (Figure 11.4). But there also appears to be a less well known monosynaptic reflex arc for the jaw that involves jaw premotor neurons (Wild, unpublished observations; Figure 11.4). A major group of these premotor neurons – which project specifically to the contralateral trigeminal and facial motor nuclei, as well as to geniohyoid and lingual motoneurons – is located medially adjacent to the arc of RAm, with many of the neurons having dendrites that extend into the RA terminal field that encompasses RAm (Figure 11.4). In the zebra finch, these jaw premotor neurons receive a direct projection from sensory neurons in the trigeminal ganglion (Figure 11.4). The peripheral structures innervated by these ganglionic neurons are not known, but in the context of beak gape and vocalization, it is interesting that the primary afferent projection directly to jaw premotor neurons is present in a songbird but is absent or much diminished in a nonsongbird, such as pigeon, which vocalizes with a closed beak (Riede et al., 2004; Wild, unpublished data).

Superimposed on this basic circuitry are long loops that allow the telencephalon to modulate reflex activity. Figure 11.4 shows, in a highly schematic way, the long sensorimotor loops that originate in the principal sensory trigeminal nucleus (PrV), and other pontine nuclei. Unlike the case in mammals, these projections bypass the thalamus and terminate directly in nucleus basorostralis (Bas) in the anterior telencephalon, a nucleus that has been proposed as the avian equivalent of the mammalian thalamic VPM nucleus (Cohen and Karten, 1974). From Bas projections pass, via several synapses, to the lateral arcopallium, which then originates, in a similar way to RA in the medial arcopallium, long descending projections to the brainstem. The lateral arcopallial projections do not overlap those of RA, however, but terminate in other parts of the rhombencephalic lateral reticular formation and spinal trigeminal tract nuclei that house a host of premotor neurons involved in feeding and associated behaviors. In particular, they target the group of jaw premotor neurons located medially adjacent to RAm (Figure 11.4; Wild and Farabaugh, 1996), which implies that lateral

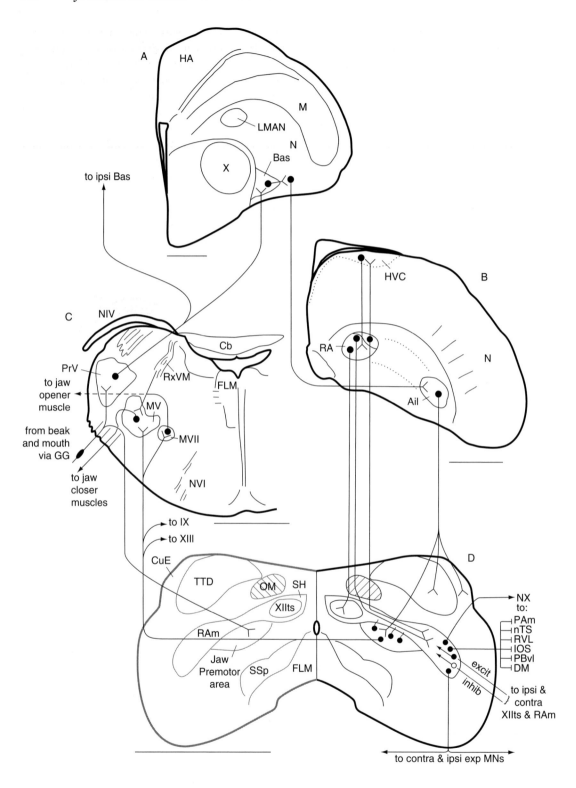

arcopallial projections are able to modulate the activity of the trigeminal reflex circuitry involving jaw premotor neurons, an anatomical scenario that is common to many descending systems that influence reflex circuitry at lower levels of the neuraxis.

Given the apparent absence of direct projections from RA to jaw, lingual and hyoid motoneurons (Wild, 1993a) (and see Figure 11.1) and of the apparent absence of direct projections from RAm to jaw, lingual and hyoid premotor neurons (Wild, unpublished observations), movements of upper vocal tract structures during singing are likely to be effected indirectly, possibly via synaptic interactions between RA terminations in RAm and the dendrites of jaw, lingual and hyoid premotor neurons that extend into RAm. Whether RA is actually capable of driving upper vocal tract motoneurons via this route requires physiological verification.

SOMATOSENSORIMOTOR AND AUDITORY FEEDBACK FOR THE CONTROL OF SONG

Most complex motor tasks, such as singing, are generally thought to entail some form of sensory feedback that has as its functional consequence the refinement and optimization of the motor output, or at least the continuance or completion of the very act itself. In the case of singing, because several otherwise separately functioning sensorimotor systems are involved, this feedback could conceivably take a variety of forms, including, most obviously, the auditory input that the bird's own song inevitably produces, and, least obviously, the generalized and apparently unlocalized feedback that derives from the respiratory apparatus, including the lungs, air sacs and associated muscles during breathing, and from various parts of the upper vocal tract such as the beak, tongue, larynx and pharynx. The role of auditory feedback has long been a focus of research interest (Konishi, 2004; Woolley, this volume). We review here the sensory structures and the neural pathways mediating nonauditory (e.g. somatosensory and visceral) feedback to the song system, and then provide an update on a possible auditory feedback pathway that may serve a rather special function in vocal control.

Proprioceptive feedback from the respiratory and syringeal systems

Konishi (1965b) first considered the possibility that proprioceptive feedback might be used to control vocal output, either during development or after auditory feedback has contributed to and perhaps enabled the crystallization of the species typical song (Konishi, 2004). There is now some evidence that certain forms of somatosensory feedback do play a part in adjusting output variables during singing (Suthers et al., 2002). Some of this evidence comes from studies of the effects upon song of experimental manipulation of respiratory and syringeal structures.

Caption for Figure 11.4

Connections of motor and sensory components of respiratory-vocal and jaw control circuits in the brain of a songbird. (A) is through a level of the rostral telencephalon, (B) and (C) are at the levels of the pons and caudal telencephalon, respectively (sectioned in the same transverse plane), and (D) is at the level of the caudal medulla. Cell bodies are denoted by black dots (except for the inhibitory neuron in RAm) and the terminations of their axons are indicated by inverted arrowheads. Ail, lateral part of intermediate arcopallium; Bas, nucleus basorostralis; Cb, cerebellum; CuE, external cuneate nucleus; DM, dorsomedial nucleus of the intercollicular complex; FLM, medial longitudinal fasciculus; GG, Gasserian (trigeminal) ganglion; HA, hyperpallium apicale; HVC, HVC (its border is dashed because the nucleus actually belongs at slightly more rostral levels than the one depicted; the medial extension denotes para-HVC); IOS, nucleus infra-olivarus superior; LMAN, lateral part of the magnocellular anterior nidopallial nucleus; M, mesopallium; MNs, motoneurons; MV, trigeminal motor nucleus; MVII, that part of the facial motor nucleus that innervates M. depressor mandibulae; N, nidopallium; nTS, nucleus of the solitary tract; NIV, Cranial Nerve IV (trochlear); NVI, Cranial Nerve VI (abducens); NX, Cranial Nerve X (vagus); OM, occipitomesencephalic tract; PAm, nucleus parambigualis; PBvl, ventrolateral part of parabrachial nucleus; PrV, principal sensory trigeminal nucleus; RA, robust nucleus of the arcopallium; RAm, nucleus retroambigualis; RVL, ventrolateral nucleus of the rostral medulla; RxVM, root of the mesencephalic trigeminal nucleus; SH, suprahypoglossal area; SSp, supraspinal nucleus; TTD, nucleus of the descending trigeminal tract; XIIts, tracheosyringeal part of the hypoglossal nucleus; IX, motor nucleus of Cranial Nerve IX (glossopharyngeal); XII l, lingual part of hypoglossal nucleus. X, Area X. Note that neurons in the jaw premotor area have processes that extend into RAm and have axons that project to MV and MVII primarily contralaterally, while the collaterals of trigeminal ganglion cells project to and terminate in the jaw premotor area ipsilaterally. Scale bars = 1 mm.

Input from the air sacs

One possible source of such feedback could arise from respiratory structures. Certainly the bird needs to know how much air is available in order to regulate the timing, duration and frequency of song phrases or motifs, but it may also require information about the availability of air for the temporal control of phrase structure, in so far as minibreaths are involved (Franz and Goller, 2003; Goller and Cooper, 2004). Because ventilation in birds involves the air sacs and their associated muscles, it is these rather than the lungs that could possess sensory receptors that inform the nervous system about subsyringeal air pressure and/or volume. Thus, injections of brief pulses of air into the air sacs of northern cardinals as they sang, resulted in a significant decrease in the amplitude of EMGs recorded from the expiratory muscles during the initial part of the syllable being sung at the time of air injection, with a minimum latency of about 30 ms (Suthers et al., 2002). The suggestion is that air injection elicits a decrease in abdominal expiratory muscle EMG to compensate for an increase in air sac pressure. In turn, this implicates some kind of sensory receptor either in the air sac or associated muscles and a reflex arc that mediates adjustments of motor output to those muscles. Some fibers in the cervical vagus have been shown to fire during air sac inflation (Ballam et al., 1982) and vagally innervated complex sensory receptors have been identified in the air sac walls of zebra finches (Kubke et al., 2005). However, the specific function of these receptors remains unknown, and whether there are functionally associated receptors in the muscles and other tissues that are stretched as a result of air sac inflation is also unknown.

Syringeal inputs

Air injected into the air sac during singing also brings about a compensatory reflex change in the tension of the syringeal adductor muscles, except that, in this case, the tension is increased rather than decreased – with the result that a constant subglottal pressure is maintained by narrowing the syringeal aperture and preventing air loss (Suthers and Wild, 2000). The syrinx does not appear to receive a spinal innervation, but it does receive a sensory innervation via the tracheosyringeal nerve (NXIIts; Bottjer and Arnold, 1982; and see Figure 11.1). The syringeal receptor(s) have not been identified but muscle spindles appear to be absent.

Syringeal afferents terminate centrally in nucleus tractus solitarius (nTS) in close proximity to air sac and pulmonary afferents (see below, and Figures 11.1, 11.5), and possibly in the spinal trigeminal nucleus (Wild, 2004b). The projections of the recipient parts of nTS have not been worked out in detail, but they appear to be similar to those receiving signals from the lung – see below.

Input from the lung

Since the avian lung does not expand during inspiration, there are no pulmonary stretch receptor afferents in the vagus nerve that act to inhibit inspiration, as there are in mammals. Instead, birds have intrapulmonary CO_2 receptors that are also innervated by vagal afferents. These produce a similar inhibitory effect to mammalian lung stretch receptors by increasing their firing rate during inspiration and terminating on GABAergic neurons in nTS (Gleeson, 1987; Gleeson and Molony, 1989; Fortin et al., 1994). As shown for the pigeon and zebra finch, pulmonary afferents terminate in nTS and specifically in nucleus parasolitarius lateralis (lPs) located lateral to the tract at levels just caudal to the obex (Katz and Karten, 1983) (Figures 11.1, 11.5). In zebra finches and canaries, pulmonary afferents also terminate around the tract at these same levels (Wild, 2004b). The specific targets of lPs neurons appear to be similar in songbirds and nonsongbirds, and include nucleus parambigualis (PAm), some of whose neurons discharge in phase with inspiration and project, apparently monosynaptically, upon spinal motoneurons innervating inspiratory muscles, such as the scalenes (Fedde, 1987; Reinke and Wild, 1997, 1998). Other PAm neurons project rostrally to terminate densely, specifically and bilaterally in nucleus uvaeformis (Uva; Reinke and Wild, 1998; Striedter and Vu, 1998), a major source of input to both HVC and NIf of the telencephalic vocal control circuitry. Another target of lPs is the ventrolateral parabrachial nucleus (PBvl), which then projects back upon the respiratory–vocal nuclei in the medulla (Wild et al., 1990). Thus, pulmonary afferents appear to have intimate connections with nuclei involved in the duration and timing of respiratory phase and are therefore likely to be of vital importance during normal breathing and, perhaps,

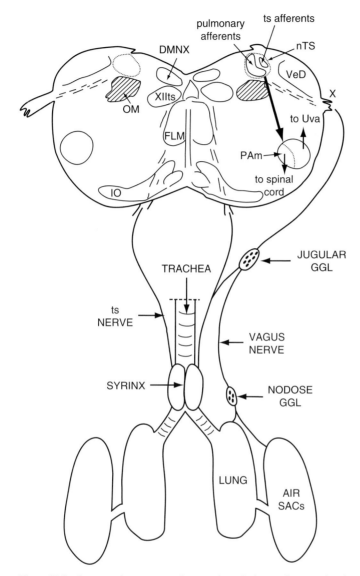

Figure 11.5 Access to the song control system from the lungs, air sacs and syrinx. DMNX, dorsal motor nucleus of the vagus; FLM, medial longitudinal fasciculus; GGL, ganglion; IO, inferior olivary nucleus; OM, occipitomesencephalic tract; PAm, nucleus parambigualis; nTS, nucleus of the solitary tract; ts, tracheosyringeal; XIIts, tracheosyringeal part of the hypoglossal nucleus; Uva, nucleus uvaeformis; VeD, descending vestibular nucleus; X, vagus nerve.

during singing as well (Franz and Goller, 2003). Electrical microstimulation in PAm during singing produces similar disruptive effects on song as does similar stimulation in HVC, implying that feedback via PAm plays an important role in song control (Ashmore et al., 2005; Schmidt and Ashmore, this volume). In cats vagal afferent feedback is essential for phonation, signaling the CNS about the volume of air in the lungs (Nakazawa et al., 1997). Perhaps phonation in birds is similarly dependent on the activity of intrapulmonary CO_2 receptor afferents in the vagus nerve.

Input from upper vocal tract structures

Although structures such as trachea, larynx, and tongue might be expected to generate feedback that could contribute to the control of song, there is little convincing evidence that they do. In contrast, beak gape is known to influence oropharyngeal shape and size, and is known to play a significant part in vocal production, such that higher pitched notes are accompanied by a wider gape (Westneat et al., 1993; Podos et al., 1995; Hoese et al., 2000; Goller et al., 2004). Moreover, weighting the beak during singing changes the harmonic content in canary song (Hoese et al., 2000), and fixing the beak open during singing in sparrows and zebra finches alters filter properties of the upper vocal tract (Hoese et al., 2000; Goller et al., 2004). Both these procedures could suggest that proprioceptive feedback from the jaw during singing is important for maintaining the acoustic structure of song, There are several possible sources of jaw proprioception, including the beak itself, the craniomandibular joint, the protractor hinge at the base of the upper beak, as well as muscle spindles of the jaw muscles themselves. Pathways linking some of these structures to the song system have been reviewed above and one might speculate that it is such structures that are innervated by the ganglionic neurons whose central processes terminate on the jaw premotor neurons adjacent to RAm (Figure 11.4; see above).

Integration of auditory and somatosensorimotor feedback for the control of song

When the bird sings, somatosensory feedback from a variety of structures of the upper vocal tract – including skin and feathers of the head and neck that move extensively during singing – is likely to be associated with auditory feedback from the song (Brainard and Doupe, 2000b; Konishi, 2004). Whether and where the auditory and somatosensory feedback that is produced during singing is integrated in the brain has not been determined conclusively, although there are candidate nuclei.

Nucleus basorostralis (Bas) in the telencephalon is a possible candidate because it receives sensory input from the beaks and oral cavity, via the principal sensory trigeminal nucleus (Figure 11.4), and a distinct auditory input from the intermediate and ventral nuclei of the lateral lemniscus (Wild and Farabaugh, 1996; Wild, 1997b; Wild, unpublished observations). However, it is not clear that cross-modal integration takes place in this nucleus, or whether the distinct inputs remain separate until combined at a later stage of processing. All parts of Bas reach the surrounding nidopallium, which then projects to the caudolateral nidopallium and to the lateral arcopallium (Figure 11.4). The trajectory and targets of projections arising in the lateral arcopallium have been charted for the zebra finch (Wild and Farabaugh, 1996), and they specifically include the jaw premotor nuclei, as described above (Figure 11.4).

A second candidate for auditory-somatosensory integration is nucleus interfacialis (NIf), especially since this nucleus, together with one of its sources of thalamic input, namely Uva, projects directly to HVC (Nottebohm et al., 1982). NIf receives a somatosensory input from Uva and the rostral Wulst (Wild, 1994a; Wild and Williams, 1999, 2000), and auditory inputs via intratelencephalic circuitry (Vates et al., 1996). Whether auditory and somatosensory inputs actually converge on the same NIf neurons, or whether these two kinds of inputs remain separate within the nucleus, is unclear. Also unclear is the extent to which all parts of the body supply somatosensory input to NIf; so far only body, and not head (trigeminal), inputs have been recorded there (Wild, 1994a).

A third candidate is Uva itself, since both somatosensory and auditory responses can be recorded there. The former arise in the dorsal column nuclei (Wild, 1994a), while the latter appear to arise from the ventral nucleus of the lateral lemniscus, a pontine component of the auditory brainstem in receipt of ascending projections from the cochlear nucleus angularis and nucleus laminaris (Krützfeldt et al., 2007; Coleman et al., 2007). The role of the auditory input conveyed to Uva from LLV is not entirely clear, despite the known direct projections from Uva to both HVC and NIf. Inactivating Uva with a local anesthetic does not affect the efficacy of BOS to elicit responses in HVC, which are presumably effected via the classical auditory pathway through the inferior colliculus, auditory thalamus and telencephalon and, penultimately, NIf. Thus, Uva could play an important role in state-dependent gating of auditory activity in telencephalic sensorimotor structures important to learned vocal control (Coleman et al., 2007; see also Prather and Mooney, this volume). Since Uva also receives a very specific input from a brainstem respiratory-related nucleus (PAm; Reinke and Wild, 1998; Striedter and Vu, 1998) (Figures 11.1, 11.5), part of this "state-dependency" may have to do with the state of the respiratory system on a breath-by-breath basis (Wild, 2004b; Ashmore *et al.*, 2005; see also Schmidt and Ashmore, this volume).

Conclusion: a systems approach to the sensorimotor control of singing

In recent years there have been concerted efforts to model the function of the songbird syrinx and its neural control (Gardner *et al.*, 2001; Laje *et al.*, 2002; Mindlin *et al.*, 2003). The ultimate success of such models, however, will inevitably depend on the incorporation of a host of variables, some of which have been considered in the present chapter, namely those related to respiration, respiratory–vocal interactions, and upper vocal tract activities, about which we have as yet very limited knowledge. Between the vocal organ and the forebrain song control nuclei is a collection of brainstem nuclei, including the vocal motor nucleus itself, about which there is little information regarding their detailed contributions to singing in particular or vocalization in general. Initial studies of the relationship between XIIts and RAm (Sturdy *et al.*, 2003; Kubke *et al.*, 2005) underscore our limited understanding of the complex interactions between brainstem respiratory–vocal nuclei and the way these are modulated by the forebrain during singing, and by peripheral inputs that function in a feedback manner. Although further investigations of these interactions and the resolution of their underlying neural mechanisms will not be easy, they have the potential to yield important information about how, during the learning and performance of a skill such as singing, the forebrain engages and temporarily controls the vegetative, vital respiratory centers in the brainstem. Perhaps in no other model system than the songbird can this be done with respect to a learned communicative skill of such obvious relevance to humans (Doupe and Kuhl, 1999; Kuhl, 2003; Doupe and Kuhl, this volume).

PART III
Hearing and recognizing the song

12 • Introduction

Peter Marler

The chapter by Theunissen and his colleagues begins with a systematic review of the avian auditory system. The general design of the neural pathway by which information about sounds heard is carried into the songbird brain is much like that in mammals, passing through stations in the medulla and the midbrain, to the pallium, and into the song system. Thus the pallial region identified by the early anatomists as Field L has been thought of as the bird's equivalent to the primary auditory cortex. As the authors indicate, aspects of the comparison are contentious and there are some clear differences, such as the lack in birds of the interhemispheric connectivity between primary auditory cortex present in mammals, but the many parallels invite comparisons between song perception and the processing of speech sounds in the human brain.

Given the critical contribution of auditory feedback to the acquisition and maintenance of singing behavior, one direction of subsequent research on brain mechanisms has been the identification of sources of auditory input to the song control systems. However, the problem of auditory processing was brought into sharper focus by the observation that neurons in the songbird brain could be selectively responsive, not merely to the songs of other birds, but to the bird's own song (BOS). This is a central topic of the first two chapters of this section, the first of which describes in detail the discovery of this phenomenon and the way in which it emerges as adult song matures. In some respects BOS seems to function as a sensorimotor template for the current song, although there are still many questions about its precise functional significance (discussed by Prather and Mooney). Theunissen and his colleagues differentiate between three levels of sound processing, and arrange them in a kind of hierarchy. At the most general level, neurons in the auditory forebrain are tuned in such a way as to favor conspecific song, as compared with sounds that lack this particular spectrotemporal structure. At the opposite extreme is the exquisitely precise tuning of neurons in the song system to details of the adult bird's own song. Somewhere between in terms of specificity are neurons in the caudal medial nidopallium (NCM) and caudal medial mesopallium (CMM) tuned to complex sounds that are familiar and have particular behavioral relevance, including those of potential tutors. The authors describe a valuable new method for estimating and classifying what the authors call spectrotemporal receptive fields (STRFs) of individual auditory neurons, throwing new light on how the auditory system conducts a kind of feature analysis of song structure.

Taking more of a top-down approach, Gentner utilizes the methods of experimental psychology to study the problem of song recognition. In the laboratory he trained European starlings by operant conditioning to discriminate between songs of different individuals. Each bird has a repertoire of fifty or more distinct motifs, mostly unique to the individual, that are delivered in long strings. Birds were able to identify individuals on the basis of single motifs, doing so even more efficiently when given a series of them. Trained birds were then tested for the responsiveness of neurons in CMM to familiar and unfamiliar songs. The great majority of those that responded selectively to song stimuli preferred one of the training songs, suggesting that CMM may be directly involved in the identification process.

Gentner was also interested in the kinds of motif sequences that starlings respond to. As would be expected if they are sensitive to normal motif sequencing, birds had more trouble with random sequences. Experiments with sequences patterned according to different organizational rules led to the finding that starlings can classify strings arranged according to what linguists call a context-free grammar, an interesting step in the direction of recursive syntactical patterning that is so fundamental to our own linguistic abilities.

It is more than 20 years since the discovery by Katz and Gurney that neurons in HVC, presumed to be primarily involved in motor activity, were also responsive to sound, particularly to the bird's own song in all of its detail, an observation that continues to fascinate researchers to this day. Prather and Mooney are specifically concerned with the mechanisms by which such selectivity arises. They focus on three questions: where does the BOS effect originate, what are its underlying mechanisms, and what kinds of experience give rise to this highly specific pattern of acoustic responsiveness within the song system?

Prather and Mooney note that the fact that BOS-sensitive neurons are responsive to both spectral and temporal features makes them especially well suited to the analysis of ways in which complex sound sequences are processed, including of course our own speech. They take as a starting point the current evidence pointing away from the tutor song and towards auditory feedback and auditory–vocal interactions as the major determinant of the selectiveness of the BOS phenomenon. It thus becomes important to disentangle the patterns of activity within the neural network that give rise to BOS responsiveness. This has proved to be extraordinarily challenging, both conceptually and methodologically. One aspect of their research strategy is the identification of sources of auditory inputs to HVC, a critical song system nucleus, afferent to the "motor" components of the circuit which appears to be a focal point for "sensorimotor" integration. A second is the use of intracellular neurophysiology to study the synaptic interactions within the circuit that generate BOS-evoked firing in song system structures. Finally, they extend their analysis to awake behaving birds by studying the patterns of single-unit activity evoked by song playback in awake birds and to identify the specific classes of responsive neurons within the nucleus. Using the swamp sparrow, a bird with a mature repertoire, not of just one song type like the zebra finch but of several song types, they found robust and highly selective auditory responsiveness for each of these types in the HVC of awake birds. This observation is of particular interest since previous work in awake zebra finches found little or no responsiveness. The implications of such striking differences in two relatively closely related species have yet to be explored but they testify, once again, to the importance of studying a variety of species. Who knows what other surprises more comparative studies may hold in store.

13 • Song selectivity and the songbird brain

Frédéric E. Theunissen, Noopur Amin, Sarita Shaevitz, Sarah M. N. Woolley, Thane Fremouw, and Mark E. Hauber

INTRODUCTION

Auditory perception is crucial for reproductive success in both male and female birds. Listening to others allows bird to classify them as conspecific or heterospecific, neighbor or stranger, mate or nonmate, kin or nonkin (Marler, 1970b). Birds are known to be particularly adept at recognizing conspecifics based solely on their vocalizations, and often in very unfavorable acoustical environments (Aubin et al., 2000; Appeltants et al., 2005; Leonard and Horn, 2005). How the auditory system uses acoustical features of song to recognize individuals and how this discrimination task is affected by environmental noise and competing acoustical signals from other animals is an unresolved mystery in auditory science (Bergman, 1990). In songbirds, listening serves an additional important function: feedback for song learning (see Hultsch and Todt, this volume). Juvenile songbirds listen to adult conspecifics to form a template of normal song. Depending upon the songbird species, songbirds begin vocalizing either while a memory is forming or once a memory is formed. During this rehearsal period, they use auditory feedback to evaluate their progress towards producing their own stable song (see Williams, this volume). For this reason, song learning in songbirds has been used as a model system for understanding the neural basis of vocal learning and production in other animals, including speech learning in humans. More recently it has been shown that song learning is not only restricted to the production of stereotypical acoustical patterns but also extends, in certain birds, to learning what can be considered grammatical rules (Gentner et al., 2006). Perceptual and vocal behavior in songbirds has also been shown to be affected by complex social cues (Vignal et al., 2004b). The science of birdsong has already provided valuable insight about the process of speech learning in humans and now has the potential of providing insight on the broader faculty of language (Hauser et al., 2002a).

This chapter reviews the progress that has been made in the last 25 years in understanding the neural basis of song perception both in the context of song learning and in the larger cortex of purely perceptual auditory tasks by examining the nature of auditory responses in the songbird brain. Our review can be thought of being top down in both functional and anatomical terms. First, we will review briefly some properties of the set of specialized interconnected forebrain regions called the "song system" – a system of specialized brain structures found only in birds that learn to sing (Kroodsma and Konishi, 1991). Although this system plays a crucial role in both song production (see Schmidt and Ashmore, this volume) and song learning and maintenance (see chapters by Brainard, and Woolley) the song system also exhibits auditory responses and is clearly involved in the processing of the bird's own song (see also Prather and Mooney, this volume). Second, we will examine the neural responses in primary and secondary avian auditory areas and their role in coding not only the bird's own song but also other behaviorally relevant sounds. These general brain areas are found in all birds and have a more direct correspondence with mammalian auditory regions. In keeping with the theme of vocal learning, one of the goals of this second line of research is to understand the role of auditory feedback in vocal learning (see Woolley, this volume). More specifically, we ask how an individual bird's auditory system processes not only the model or tutor song that he will imitate (see also Adret, this volume) but also his own vocalizations, as well as how the auditory and vocal systems are connected (see Prather and Mooney, this volume).

The top-down organization of our chapter also reflects the historical progression of research in this field. In the period following the anatomical and functional characterization of song system nuclei (Nottebohm et al., 1976; Bottjer et al., 1984), much of the auditory research on the neurobiology of birdsong focused on the auditory responses found in the song

Neuroscience of Birdsong, ed. H. Philip Zeigler and Peter Marler. Published by Cambridge University Press. © Cambridge University Press 2008.

system itself. The discovery of neurons that were selective to the sound of the bird's own song (BOS) (McCasland and Konishi, 1981; Margoliash and Konishi, 1985) shifted attention to the question of how these selective responses could participate in both vocal learning and in song perception. In contrast to the obvious relevance of research on auditory processing by the song system, studies of sound processing in the classical auditory areas of songbirds yielded few clues as to how the tutor song was recognized and stored or, more generally, about how overall song perception was mediated (Sachs et al., 1980).

That original picture has now changed. As our knowledge of the properties of neurons within the song system has increased, areas outside the song system have been implicated in both the perception of familiar conspecific songs and the storage of the tutor song. It is now clear that to understand how the selectivity for the BOS found in the song system is generated, we need to understand how song and other important sounds are processed in auditory areas that are presynaptic to the song nuclei and how those areas are connected to song system structures.

THE AUDITORY SYSTEM OF SONGBIRDS AND ITS CONNECTIVITY WITH THE SONG SYSTEM

Figure 13.1 shows the ascending auditory system and its known connections with the song system. Whereas a specialized brain circuit for song production and learning has evolved fairly recently in evolutionary time and in only a few avian orders (Jarvis et al., 2000) (see also Reiner et al. this volume), the avian auditory system is much older and shares features with all avian groups and with other vertebrates, including mammals (Butler and Hodos, 1996). The most noticeable similarity among vertebrate forms is seen in the number of auditory nuclei (or neural processing stages) and the pattern of feedforward connections from the cochlear nucleus to the auditory forebrain. Afferents from the cochlea project to the cochlear nucleus in the medulla. As in mammals, input from medulla to midbrain has both a direct route and an indirect route through, in succession, the superior olive and the lateral lemniscus. In the midbrain, these pathways converge in the dorsal lateral nucleus of the mesencephalon (MLd), which is functionally homologous to the inferior colliculus (IC) in mammals. The auditory midbrain projects to a relay nucleus in the thalamus, ovoidalis (Ov), just as the IC projects to the medial geniculate body (MGB) in mammals. Ov in turn sends projections to the primary auditory area in the pallium, Field L (Zaretsky and Konishi, 1976; Vates et al., 1996). For this reason, principally, Field L can be thought of as the primary auditory cortex of the bird. Field L has been further divided into subregions (L1, L2a, L2b, L3, L) based on differences in cyto-architecture and connectivity (Fortune and Margoliash, 1992; Vates et al., 1996)

Input from the auditory thalamus goes to subregions L2a and L2b, which in turn project to L1 and L3. Subregions L1 and L3 make bidirectional connections with two secondary auditory areas in the pallium: the nidopallium caudal medial (NCM) and the caudal lateral mesopallium (CLM). The NCM and CLM are in turn interconnected via the caudal medial mesopallium (CMM) as shown in Figure 13.1. While it has been suggested that these forebrain auditory regions may be functionally analogous to some areas of auditory cortex, the issue remains contentious (see chapters by Bolhuis, and Mello and Jarvis, this volume). Moreover, there are some important differences between the mammalian and the avian auditory system with respect to their feedback and interhemisphere connectivity patterns. In mammals the primary auditory cortex shows strong feedback projections to the thalamus and more limited ones to the midbrain (Winer, 2005). In birds this feedback circuitry exists but it involves two additional processing stages in the forebrain, found, in songbirds, in the shell regions of song system structures HVC and the robust nucleus of the arcopallium (RA), as illustrated in Figure 13.1 and in similar anatomical locations in non-songbirds. The direct interhemispheric connectivity between primary auditory cortex present in mammals (Lee et al., 2004b) is also missing in birds.

The song nuclei are shown in gray in Figure 13.1. The connectivity pattern in the song nuclei is mostly unidirectional: nucleus NIf (in the nidopallium) and Uva (in the thalamus) project to HVC. HVC projects to RA via two distinct pathways, and RA projects to vocal and respiratory motor nuclei. Note that for the purpose of classification in this chapter, NIf has been classified as part of the sensorimotor song system and not the general auditory forebrain system (for more details see chapters by Wild, Prather and Mooney, and Reiner et al.).

Song selectivity and the songbird brain 159

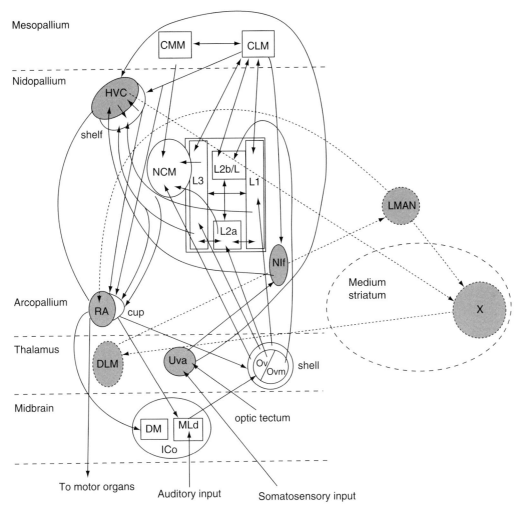

Figure 13.1 The ascending auditory system and its connectivity with the song system. Song nuclei are shown in gray. The dashed song nuclei and connections make the anterior forebrain pathway. See text for definitions of abbreviations.

The exact flow of information from the auditory system to the song system is not completely understood, although great progress has been made recently. NIf is a key player as it shows robust and selective auditory responses and projects strongly to HVC (Fortune and Margoliash, 1995; *et al*., 1996; Janata and Margoliash, 1999; Coleman and Mooney, 2004). More significantly, inactivating NIf eliminates the auditory selectivity to BOS responses observed in HVC (Cardin and Schmidt, 2004a; Coleman and Mooney, 2004; Cardin *et al*., 2005). NIf, in turn, receives input from the secondary auditory area CLM (Vates *et al*., 1996). For all these reasons, NIf currently appears the most likely source of auditory input to the song system (see also Prather and Mooney, this volume). On the other hand, the shelf region of both HVC and RA receive auditory projections from auditory forebrain areas (L1, L3 and CLM) and input from the shelf region of HVC might enter the nucleus proper. Moreover, direct sparse projections from L1 and L3 to HVC have been described anatomically (Fortune and Margoliash, 1995). The functional importance of this second sparse route for sensory input into the song system is not known. In addition, nucleus Uva (in the thalamus) is known to project to both HVC and NIf (Wild, 1994a) and appears to play a functional role in initiating and terminating the neural activity in HVC and correspondingly in song output (Williams and Vicario, 1993). For a discussion of Uva's potential role in auditory processing see the chapter by Prather and Mooney, this volume.

AUDITORY RESPONSES IN THE SONG SYSTEM

Selectivity for the bird's own song

The auditory properties of neurons in the song system, particularly in HVC, have been studied extensively in playback experiments. A striking property of neurons in the song system of adult male songbirds is their "preference" i.e. their selective responsiveness to the sound of the bird's own song (BOS) over the BOS played in reverse or to other conspecific songs (CON). The differences in responsiveness to these three stimuli can be drastic, as illustrated in the recording from HVC neurons in an adult male zebra finch shown in the top panel of Figure 13.2. In this example, there is a strong response to the bird's own song and a lack of response to the song played in reverse or to a conspecific song. This response property has been labeled "song selectivity." The use of that phrase is somewhat misleading since sensory neurons in the song system are in fact selective for only one particular song, the BOS,

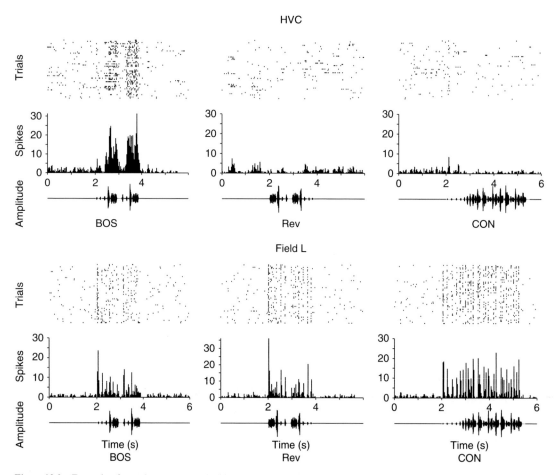

Figure 13.2 Example of neural responses to the bird's own song (BOS), the BOS played in reverse (Rev) and conspecific song (CON) in the song system (HVC) and in the primary auditory forebrain (Field L). These recordings were obtained in urethane anesthetized adult male zebra finches. The recordings in HVC and Field L were obtained simultaneously in a double electrode experiment. Fifty trials to each stimulus were acquired. The figure shows the spike raster on the top line (Trials), the peristimulus time histogram (PSTH) in the middle line (Spikes) and the oscillogram of the stimulus (Amplitude) on the bottom line.

as compared with other conspecific songs. Moreover, a more general form of song selectivity is found in the forebrain auditory areas where selective responses are found to familiar conspecific song relative to nonfamiliar conspecific songs or to conspecific song relative to matched synthetic sounds. We will therefore refer to the auditory property of neurons in the song system as BOS-selectivity.

BOS-selectivity was first observed in awake restrained songbirds (canary, white-crowned sparrow, and zebra finch) using a chronic recording technique that allowed the comparison of sensory and motor responses in HVC (McCasland and Konishi, 1981). The specificity of the BOS-selective response in HVC was then explored in detail in awake restrained or urethane-anesthetized white-crowned sparrows and zebra finches by Margoliash and his collaborators (Margoliash, 1983, 1986; Margoliash and Konishi, 1985; Margoliash and Fortune, 1992; Sutter and Margoliash, 1994). These studies clearly demonstrated the specificity of the selectivity to the BOS and highlighted some of the acoustical features responsible for the selective responses. The auditory responses in HVC were shown to be sensitive to the particular temporal and spectral combinations of sounds found in the BOS. This sensitivity is exquisite: the spectrotemporal amplitude envelope of the bird's own song had to be preserved at a 98% level to obtain neural responses that were indistinguishable from those of the original sound (Theunissen and Doupe, 1998). The presence of sensory responses in the song nuclei confirmed both their place as a center for sensorimotor integration and their role in song learning and maintenance. However, the specific role of BOS-selective responses in song learning or song perception remains unclear. Some of the additional properties of BOS-selectivity and their implications for sensory-motor learning and auditory perceptual behavior are discussed below.

The development of BOS-selectivity and the tutor template

The remarkable BOS-selectivity observed in adult songbirds leads quite naturally to the hypothesis that these BOS-selective neurons might play a role in the storage of the tutor template during learning: more precisely, BOS-selective neurons could in fact be tutor-selective neurons (i.e. selective for the version of the tutor song that was memorized) since these two songs are so similar acoustically. The response of such putative tutor-selective neurons could then be used to assess the quality of the young bird's immature songs relative to this stored template. Experiments that followed the development of BOS-selectivity in song nuclei during the two phases of song learning refuted that hypothesis, at least in its strongest form (see chapters by Williams, Adret, and Prather and Mooney). BOS-selectivity in neurons of the song nuclei emerges during the vocal practice phase of song learning (Doupe, 1997; Solis and Doupe, 1997) and is absent at an earlier time when the young bird has not begun to sing but has been exposed sufficiently to the tutor song for model matching (Volman, 1993). Moreover, the development of the response properties of the BOS-selective neurons in the song system appear to follow the vocal output of the young birds such that the neurons become selective for the plastic song that the bird is producing (see Table 13.1; Solis and Doupe, 1997), even when this song is highly distorted and quite distinct from the tutor song (Solis and Doupe, 1999). The nature and development of the majority of BOS-selective neurons are therefore inconsistent with early memory storage of the tutor song. If the tutor song is in fact stored in the song system, this storage would have to be in a minority of neurons (Solis and Doupe, 1999) or in a form that is not revealed by single unit recordings in the anesthetized songbirds (see Bolhuis, this volume).

The selectivity of auditory responses in the song nuclei appears to follow the development of the vocal motor program: the largest response is obtained for the song that the bird is currently producing. This selectivity can therefore be interpreted as a memory for the produced vocalization, a sensorimotor template for the current song output. If the auditory feedback deviates from the currently produced song, a lack of response could be used to modify the motor program in order to re-establish the previously learned output. Such a mechanism, which could be used for the continuous evaluation of the produced BOS in the adult bird, has been demonstrated in behavioral experiments. Alteration of auditory feedback leads to song changes which can be reversed once normal feedback is restored (Leonardo and Konishi, 1999; Brainard and Doupe, 2000b). The same mechanism could be also used in young birds during song learning in order to stabilize the motor program that has already been learned. Additional factors would then be involved in guiding the plastic output of the young birds towards a tutor-like song.

Table 13.1.

Area	Stimuli	Neuron type	d'	N	Comment	Reference
NIf	BOS-CON	Unknown	1.5	15	14/15 > 0	Janata and Margoliash, 1999
NIf	BOS-CON	HVC projecting and not identified (mixed)	1.3	14	Intracellular recording	Coleman and Mooney, 2004
	BOS-rev		1.3	12		
HVC	BOS-CON	Unknown	2.3	54	98% > 0	Theunissen and Doupe, 1998
	BOS-rev	Unknown	1.7	63	99% > 0	
HVC	BOS-rev	RA-projecting	1.3	10	Intracellular recording	Mooney, 2000
		X-projecting	1.3	16		
		Interneuron	3.0	11		
		X-projecting	1.0	15	Inhibitory potential	
HVC	BOS-rev	Interneuron (putative)	3.5	9	Sleeping bird	Rauske et al., 2003
			2.2	21	Awake bird	
			2.3	13	Urethane	
HVC	BOS-rev	Unknown	−0.5		Awake/aroused	Cardin and Schmidt, 2003
			2.7		Awake/sedated	
HVC	BOS-CON	RA-projecting	1.7	8	Intracellular recording	Coleman and Mooney, 2004
		X-projecting	0.7	18	(simultaneous recordings with NIf study on second row)	
		Interneuron	0.3	?		
	BOS-rev	RA-projecting	1.5	14		
		X-projecting	0.6	19		
		Interneuron	2.5	9		
HVC	BOS-CON	All	1.05	98	Long days	Del Negro et al., 2005
			−0.06	98	Short days	
	BOS-rev	All	0.54	98	Long days	
			−0.02	98	Short days	
X	BOS-CON	Unknown	1.0	28		Doupe, 1997
	BOS-rev		1.3	21		
X	BOS-CON	Unknown	0.6	47	60-day-old birds. BOS is plastic song; CON is adult song	Solis and Doupe, 1997

Table 13.1. (cont.)

Area	Stimuli	Neuron type	d'	N	Comment	Reference
LMAN	BOS-rev	Unknown	0.9	40		Doupe, 1997
	BOS-CON		1.3	48		
	BOS-rev		1.3	41		
LMAN	BOS-CON	Unknown	1.0	47	60-day-old birds. BOS is plastic song; CON is adult song	Solis and Doupe, 1997
	BOS-rev		1.1	47		

d', normalized distance measure of the difference in responses to two stimuli (see text).

For example, the degree of BOS-selectivity has been shown to be correlated with the quality of the produced BOS relative to the tutor song both in young birds (see Table 13.1 and Solis and Doupe, 1997) and in adult birds that sing a distorted song (Solis and Doupe, 2000). In a very different experiment but with similar conclusions, the selectivity of HVC neurons was seen to decrease during the nonbreeding season in the canary, a seasonal breeder with open song learning (Del Negro et al., 2005). Thus, in a situation where the current motor output deviates substantially from the song template the reduced BOS-selectivity would allow for greater motor plasticity. In other words, the degree of BOS-selectivity, in adult birds with poor song or in adults who are learning a new song (canaries) or in young birds who are learning to sing for the first time, could be a marker of the goodness of fit of the vocal output with a desired model. In young birds deviations of vocal output in direction of the tutor template would then be favorably reinforced by an additional signal (see Saar et al., this volume). The neural basis of such a reinforcement signal, of the comparison between the vocal output with tutor template, and of storage of the tutor song (which must precede the sensory-motor template in the song system) remains to be determined.

The similarity between motor and sensory responses in the song system and sensorimotor integration

A second striking property of the auditory responses to BOS in the song system is their similarity to the motor response. In the adult bird, the temporal pattern of the sensory responses in RA, to the playback of the BOS is similar to the pattern of motor activity produced by the same neuron during singing (Dave and Margoliash, 2000). This striking property provides further evidence that BOS-selective neurons in the song system may play a crucial role in integrating auditory feedback information with the motor circuitry for vocal learning and maintenance (Troyer and Doupe, 2000a; Margoliash, 2002; Mooney et al., 2002).

The modulation of BOS-selective responses

Most of the original recordings of auditory responses in the song system were performed in anesthetized or awake but restrained songbirds. More recent work has shown convincingly that the auditory responses in the song system are sensitive to the arousal state of the animal. In particular, in the zebra finch, strong sensory responses are only observed in the urethane-anesthetized animal or in the sleeping animal (Dave et al., 1998; Schmidt and Konishi, 1998; Nick and Konishi, 2001). Weaker and less selective responses are found in the awake bird for a subclass of interneurons in HVC (Rauske et al., 2003) and these awake responses are modulated by the arousal state of the animal where a higher arousal state leads to a suppression of the sensory responses (see Table 13.1 and Cardin and Schmidt, 2003, 2004a). BOS-selective responses in awake birds have also been observed in the song nuclei X and LMAN of the anterior forebrain pathway (Hessler and Doupe, 1999a; Margoliash, 2002). The modulation of neural activity in HVC and in the song system in general appears to be mediated by neuromodulators implicated in arousal and attention such as acetylcholine and norepinephrine. Both NIf and HVC receive cholinergic and noradrenergic innervation (Ryan and Arnold, 1981; Bottjer, 1993; Soha *et al.*, 1996; Mello *et al.*, 1998b). Injections of norepinephrine into HVC (Dave *et al.*, 1998) or NIf (Cardin and Schmidt, 2004b) interfere with the modulation of the activity. Similarly, injections of cholinergic agonists into HVC or RA eliminate the selectivity and strength of auditory responses in those song nuclei (Shea and Margoliash, 2003).

Because BOS-selective responses are observed to some extent both in the awake (non-aroused) and in the sleeping bird, it is possible that these auditory responses play a role both in online and offline feedback for song maintenance and learning. In the context of conspecific song recognition, the decrease in selectivity of sensory responses observed during a calm awake bird might also be an adaptive feature. The high degree of selectivity of auditory responses makes them poor candidates for mediating conspecific song recognition other than their own song. It has, however, been argued that the decrease in selectivity observed in awake recordings would allow the song system circuitry to participate in conspecific song recognition (Cardin and Schmidt, 2003; Rauske *et al.*, 2003).

In addition to short-term modulation by arousal, a seasonal modulation of neural activity and selectivity has been measured in open song learners (Del Negro *et al.*, 2005): the selectivity of HVC neurons for the BOS is greatly reduced when birds are housed under short day rather than long day housing conditions (see Table 13.1 for numbers). This functional modulation goes hand in hand with the seasonal structural modulations observed in HVC in these birds (see also Pytte *et al.*, this volume) and is almost certainly mediated by gonadal hormones (DeVoogd and Nottebohm, 1981b; Gahr *et al.*, 1987). For detailed discussions of hormonal modulation of song behavior see chapters by Harding, Ball *et al.*, and Brenowitz, this volume.

BOS-selectivity and hierarchical processing in the song system

The selective response to the BOS has been observed in all of the nuclei of the song system (Williams and Nottebohm, 1985; Doupe and Konishi, 1991), which leads to the question of whether any significant auditory processing is occurring as one moves up in the processing chain of the song system nuclei. This question is particularly relevant since song selectivity is present in the two song nuclei, HVC and NIf, that receive input from the primary and secondary auditory areas (Janata and Margoliash, 1999; Coleman and Mooney, 2004) (see Figure 13.1). Since, as we shall see, these auditory areas are not selective for the BOS relative to CON, BOS-selectivity seems to be a characteristic of the song system and must initially appear in one processing stage at the interface between the auditory system and the song system (see Prather and Mooney, this volume). Is there any further refinement of song selectivity beyond NIf and HVC? The degree of selectivity of single neurons can be quantified by the signal detection measure of d': a normalized distance measure of the difference in responses to two stimuli (Theunissen *et al.*, 2004b). A d' distance greater than one corresponds to differences in responses that are greater than the standard deviations of the responses of single neurons and therefore highly discriminable. Table 13.1 shows a summary of mean d' values obtained for the BOS-CON or BOS-rev comparison in a series of studies from different laboratories. Unless specified, the data are from urethane-anesthetized adult male zebra finches.

Two somewhat contradictory conclusions can be drawn from these studies. First at a gross level, there seems to be little or no hierarchical processing of song across the different nuclei in the song system. As shown in Figure 13.1 and Table 13.1, the feedforward flow of processing is from NIf to HVC to X and LMAN. The d' values for NIf are smaller than those for interneurons in HVC for the BOS-rev comparison but are similar to the projection neurons in HVC and also of the same order as the neurons recorded in the song nuclei of the anterior forebrain pathway: Area X and LMAN. Based on that analysis, it would be difficult to argue that one role of the song system circuitry is to process auditory information with the purpose of increasing the selectivity for BOS relative to other sounds. On the other hand BOS-selectivity differs across different neurons within HVC and auditory response properties are clearly different across different song nuclei. Within HVC, the projection neurons to RA and Area X appear to have different signs: the RA-projecting neurons are depolarized and fire action potentials during the presentation of song. The Area X-projecting neurons are mostly hyperpolarized during song and fire action potentials at the offset of sound (Mooney, 2000). The spike output of both RA-projecting and X-projecting neurons is selective for song but at different temporal phases. The inhibitory interneurons in HVC have the highest firing rates and are the most selective in the zebra finch (see Table 13.1 and Mooney, 2000). Blocking the inhibition results in both an increase in the selectivity of X-projecting neurons and a drastic change in the temporal properties of the response (Rosen and Mooney, 2003). In addition, in a study involving a songbird that sings multiple song types, the interneurons responded to many exemplars of the bird's own songs while the projection neurons responded selectively to one or two of the songs in the repertoire. In that case, there was a refinement of selectivity within nucleus HVC (Mooney et al., 2001). Therefore, the circuitry within HVC plays a crucial role in shaping the neural representation of BOS and these different representations are potentially mediating different roles in error correction (Mooney, 2000; Rosen and Mooney, 2000, 2003, 2006, see also Prather and Mooney, this volume).

Although the intrinsic circuitry and corresponding physiology of other song-system nuclei have not been so thoroughly examined, it is clear their neurons have different auditory response properties. For example, auditory neurons in NIf respond in a sustained fashion to many conspecific songs or other complex sounds while showing higher firing rates to the BOS. In HVC one can observe a similar differential response between BOS and conspecific song but with different response properties to each than those observed in NIf: responses to conspecific often become below threshold and responses to BOS can be phasic. Thus a transformation of the neural representation of the BOS is obtained between NIf and HVC (Coleman and Mooney, 2004). Along similar lines, auditory neurons in Area X have high firing rates and will respond to nonsong sounds such as white noise or pure tones, albeit with lower rates than those emitted in response to BOS. Auditory neurons in LMAN have low spontaneous rates and respond much less frequently to sounds other than the BOS (Doupe, 1997). Just like the responses in the different neuron types in HVC, the different representation of sounds in Area X and LMAN might play a significant role in vocal learning (see Brainard, this volume).

Do BOS-selective sensory responses in song system nuclei play a role in the perception of conspecific song?

Besides their potential role in vocal learning, a second postulated role of the BOS-selective responses has been to participate in conspecific song perception. It has been proposed that the BOS-selective responses of song nuclei neurons could mediate such purely perceptual tasks by using the BOS as a reference point (Margoliash, 1986; Nottebohm et al., 1990). In this scheme, conspecific songs would be categorized by how similar or dissimilar they are from the BOS. Support for this hypothesis comes principally from a series of lesion studies that have pointed to song system nuclei HVC and LMAN as being required for certain types of acoustical discrimination tasks both in male and female birds (Brenowitz, 1991; Scharff et al., 1998; Burt et al., 2000; Gentner et al., 2000).

However, although robust gene expression of *ZENK* and *c-fos* is seen in the song system nuclei following *singing* behavior, no significant activity has been measured after passive *listening* to the song (Mello and Clayton, 1994; Kimpo and Doupe, 1997). This lack of gene expression correlates with the greatly reduced auditory responses observed in song nuclei in the awake and aroused bird.

Moreover, the properties of BOS-selectivity measured in electrophysiological recording, immediate early genes (IEG) studies, and the anatomy of the sexually dimorphic song system raise some questions. First, the extreme selectivity for the BOS and the typical lack of spiking response to many conspecific songs (Margoliash, 1986; Theunissen and Doupe, 1998; Mooney et al., 2001) is inconsistent with a role for these neurons in discriminating among different conspecific songs. Second, auditory responses in song nuclei to the BOS are greatly reduced in the awake bird and further reduced in the awake aroused bird, which is presumably a more attentive state (but see Prather and Mooney, this volume). Third, the output of the song system drives almost exclusively vocal control areas and not more general motor areas (Wild, 1993a). It is therefore unclear how nonsinging behavioral responses could be triggered by a neural recognition of song in song system nuclei. Finally, although most female songbirds in temperate species do not sing and their song system atrophies during development, these females can discriminate song as well or better than males.

A stronger correlation between the auditory responses of neurons in the song nuclei and auditory discrimination could be achieved in experiments combining awake neurophysiological recordings with discrimination tasks. The selectivity of individual neurons or ensemble of neurons as measured by their d' could then be compared with behavioral performance. Such experiments as well as further research in the flow of auditory information through the song system nuclei and beyond will be necessary to demonstrate the role of the song system in discrimination or recognition of conspecific songs.

AUDITORY RESPONSES TO SONG AND OTHER COMPLEX SOUNDS IN THE AUDITORY MIDBRAIN AND FOREBRAIN

This review focuses on responses in MLd, Field L, NCM and CM (both medial and lateral) because there are few neurophysiological studies of Ov (Bigalke-Kunz et al., 1987). Most of the recent data from the auditory forebrain in songbirds have been obtained from urethane-anesthetized adult male zebra finches. Although anesthesia is known to affect the temporal profile of auditory responses in Field L (Capsius and Leppelsack, 1996), the relatively large modulation of responses across states of deep anesthesia, sleep, sedation and wakefulness that is observed in the song system is absent in the auditory forebrain (Cardin and Schmidt, 2003). IEG studies to song playback also show strong activation in all auditory forebrain areas mentioned above, except for Ov and L2 (Mello, 2002a).

Absence of BOS-selectivity in the auditory forebrain

In contrast to song system nuclei, auditory neurons in Field L and CLM are not selective for the BOS relative to conspecific song. The lower panel in Figure 13.2 shows the responses of a single unit in Field L in response to BOS, conspecific song and the BOS played in reverse. This response can be compared with the simultaneous recording that was obtained from HVC shown in the upper panels. The neuron in Field L responded strongly to all three stimuli, whereas the neuron in HVC responded strongly only to the BOS. The auditory responses in Field L compared with HVC were also more consistent across trials and neural activity lines up (or phase-locks) to particular acoustic features of the sound at a much finer timescale in L than in HVC.

The lack of selectivity for BOS in Field L was first observed in the white-crowned sparrow (Margoliash, 1986) and later verified in a second study in zebra finches (Lewicki and Arthur, 1996; see also Janata and Margoliash, 1999). To determine the extent to which the degree of song selectivity in auditory nuclei might be related to its position in the sound processing hierarchy we sampled neuronal response to BOS or tutor song in all the subareas of Field L as well as of the secondary area CLM (Amin et al., 2004). A second goal of that study was to examine, for the first time, the selectivity for BOS or tutor song in CLM. Neural tracing studies have characterized CM (lateral and medial) as an intermediate processing stage between the auditory system and the song system (see Figure 13.1) while IEG and lesion studies (MacDougall-Shackleton et al., 1998a) and recordings from awake behaving birds (Gentner and Margoliash, 2003) have implicated CMM in the perception of conspecific song. Based on these observations, we expected to find selective neural representation of the tutor song or the BOS in this region.

We found little evidence that the subregions of Field L or CLM were selective for BOS. The average d' values for the BOS-CON and BOS-rev comparisons for neurons in Field L and in CLM are shown in Figure 13.3, where it is contrasted with the selectivity that we had measured in HVC as part of a previous study (Theunissen and Doupe, 1998). In contrast to the d' values for the song system (Table 13.1), we found that Field L and CLM neurons show on average no positive selectivity for the BOS: the mean or median d' values are very close to 0. Our results also show a small preference for BOS over reverse BOS both in Field L and in CLM. We also found that the preference for BOS over reverse is of a similar magnitude in CLM. For reasons to be detailed below, we interpret the intermediate selectivity for BOS over reverse BOS as a preference for the natural spectral-temporal structure found in natural sounds and particularly in conspecific song, rather than as a sign of an intermediate stage of processing for BOS-selectivity. Finally, neither Field L nor CLM exhibited selectivity for the tutor song.

If we assume that auditory input enters the song system via Field L and/or CLM yet we find no intermediate stages of selectivity for BOS over conspecific

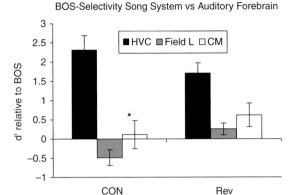

Figure 13.3 Average d' values for the BOS-CON comparison and the BOS-rev comparison for neurons in HVC, Field L and CM in adult male zebra finches. (The data for HVC is from Theunissen and Doupe, 1998. The data for Field L and CM is from Amin et al., 2004.)

song prior to NIf, how then do we account for the initial development of selectivity in the song system? One possibility is that some auditory neurons in CLM or Field L exhibit stronger responses to certain syllables present in the BOS and selectively project to NIf or HVC. To test this hypothesis, simultaneous recordings of CLM or Field L neurons and NIf or HVC neurons could be performed to assess functional connectivity and sensory selectivity in the same experiment (Shaevitz and Theunissen, 2007; E. Bauer and R. Mooney, in press). Further selectivity could then be generated within NIf or HVC by temporally and nonlinearly integrating this selective input (Lewicki and Konishi, 1995). This hypothesis does not ignore the fact that additional processing has been shown to occur in HVC leading to different song representations, all selective for BOS, which might be crucial for different aspects of vocal production and learning (as explained above); however, a quantal leap between selective and nonselective BOS responses would occur in one step at the interface of the auditory system and the song system. Alternatively, NIf or HVC could also receive auditory input from an auditory region other than CLM or Field L, which might show intermediate selectivity for BOS. Auditory information also affects the neural responses of the song nucleus Uva (M. J. Coleman and R. Mooney, unpublished data), which is known to project to both NIf and HVC (Wild, 1994a). For a discussion and review of the neural mechanisms involved in song selectivity see Prather and Mooney, this volume.

Selectivity for familiar songs in secondary auditory areas NCM and CM

Although the neurons in the auditory forebrain areas are not selective for the BOS, neurons in secondary areas NCM and CM are selectively responsive to certain types of songs. For example, IEG expression in NCM has been shown to be the largest for songs that have greater behavioral significance such as conspecific song in males and "sexy" features of conspecific song in females (Mello et al., 1992; Gentner et al., 2001). Recent fMRI data are in agreement with this finding (Van Meir et al., 2005). In addition, both IEG expression and neural recordings in NCM habituate to repeated presentation of the same conspecific song (Mello et al., 1995) and the degree of adaptation is correlated with song familiarity (Chew et al., 1995, 1996a; Stripling et al., 1997). The IEG and electrophysiological experiments are consistent with the idea that NCM is involved in the discrimination of familiar songs relative to novel songs. It has also been suggested that this discrimination of familiar songs extends to the tutor song and that therefore NCM could be the site where the neural trace of the tutor template is found. Bolhuis and coworkers found that IEG expression was stronger for the tutor song than for an unfamiliar conspecific song and that the strength of the response was correlated with the degree of learning of the tutor song (Bolhuis et al., 2001; Terpstra et al., 2004; Bolhuis, this volume). Using neural activity as a marker Vicario and coworkers (Phan et al., 2006) compared the habituation of the responses to tutor song with those observed in response to unfamiliar conspecific song or the bird's own song. The different adaptation curves show that the tutor song was familiar although it had not been heard for a prolonged period of time. The degree of familiarity was also correlated with how well the bird copied the song. How the adaptation characterization in this neurophysiological study relates to the IEG expression remains to be elucidated. More importantly investigations on how a differential adaptation for the tutor song can be used to guide vocal learning and how that guiding information will reach the song system will have to be elucidated to fully interpret these very interesting findings.

Lesion, IEG and neurophysiological studies have also implicated the secondary auditory area CMM in perception of familiar conspecific song (Gentner et al., 2001). A lesion study in female zebra finches showed that CMM but not HVC was important for song discrimination for mate choice (MacDougall-Shackleton et al., 1998a). Similarly, an IEG study in female zebra finches showed that the ZENK response in CMM to the female birds' father song (the tutor song) correlated with the degree of learning measured in behavioral preferences tests (Terpstra et al., 2006). Gentner and colleagues (Gentner and Margoliash, 2003) have combined behavior and chronic recordings to demonstrate that single neurons and the ensemble of neurons in CMM become more responsive to conspecific song that is being learned in a perceptual discrimination task. After training birds to recognize individual conspecific songs in either a two-alternative forced choice (AFC) or a go/no-go paradigm, neurons in CMM became selective for the songs used in the experiment. This selectivity is observed at the population level: whereas selectivity for the BOS in the song system can be observed in most neurons and selectivity for other sounds is for all purposes nonexistent, the neurons in CMM retained a range of preferences for both novel and familiar sound elements. But when the responses are summed together in the population one observes a preferential response for the familiar songs. Interestingly, both of the two songs in the AFC paradigm elicited enhanced responses whereas in the go/no-go paradigm, the response to the "go" song was enhanced relative to the response to the no-go song. Therefore this learned selectivity cannot simply be explained by bottom-up effects (enhanced response for a repeated stimulus) but must also involve top-down processing (enhanced response for stimulus with greater valence versus lower valence but equal familiarity). For a detailed discussion of these experiments and the role of CM in conspecific song perception see chapters by Gentner and by Mello and Jarvis, this volume.

Selectivity for conspecific songs over synthetic sounds

Although the primary auditory forebrain of songbirds does not appear to be tuned to particular conspecific songs, many studies have now shown that it is sensitive to the *acoustic* structure commonly found in natural sounds, particularly to the spectral and temporal structure found in conspecific song or calls. This selectivity for natural sounds is also already present in the auditory midbrain.

Several studies have shown that neurons in Field L and CLM of zebra finches show a small preference for the BOS played forward versus reverse but not for BOS over CON (Lewicki and Arthur, 1996; Janata and Margoliash, 1999; Amin et al., 2004). Given the absence of BOS selectivity, we would predict that the same preference would be observed for any unfamiliar conspecific song relative to the same sound played in reverse. On the other hand, there is little or no difference between the response to the song played forward and the response to the song with the order of the syllables reversed (reverse order song) (Lewicki and Arthur, 1996; Janata and Margoliash, 1999; Amin et al., 2004). One can therefore conclude that it is the order of the temporal and spectral modulations within a single syllable of song that determines the selectivity for forward song over reverse. Consistent with this interpretation, a statistical analysis of the spectral and temporal modulations of sound in a 300 ms timescale shows that zebra finch song was asymmetrical with more down-sweeps than up-sweeps (Singh and Theunissen, 2003). In other words, neurons tuned for down-sweeps will be on average more excited by the syllables of the zebra finch song played in the natural order than the reverse order.

It has also been shown in many neurophysiological studies that subsets of neurons in Field L do not respond well to simple synthetic sounds (e.g. tones) but will respond selectively to particular conspecific vocalizations (Leppelsack and Vogt, 1976; Leppelsack, 1978; 1983; Muller and Leppelsack, 1985). These observations are reminiscent of the preference for vocalizations (i.e. natural sounds) over pure tones, tone complexes or white noise that is also found in subsets of neurons in the mammalian auditory cortex (Newman and Wollberg, 1978; Rauschecker et al., 1995; Wang et al., 1995). Some researchers have argued that because the response to "simple" tones cannot be used to predict the response to natural sounds they have identified a population of neurons selectively tuned to natural conspecific vocalizations. An alternative interpretation of these data is that the observed "selectivity" for the animal's vocalizations is simply a consequence of more selective tuning as one moves up in the auditory processing stream. At higher levels of auditory processing, auditory neurons become responsive to more and more complex spectrotemporal combinations of sounds. If these sound structures are often found in complex vocalizations but not in simple tones or white noise, one will observe a "preference" for natural sounds. A careful comparison of acoustical properties of natural (i.e. conspecific) and matched synthetic sounds and of the response properties of the neurons to these sounds is required to confirm the specificity of the tuning of the neurons to conspecific vocalizations (Schafer *et al.*, 1992). Two such studies are discussed below.

The first study used IEG expression in NCM to show that the neural representation of the whistle syllables found in canary song was drastically different from the representation of synthetic whistles or guitar notes that were matched in intensity and pitch (Ribeiro *et al.*, 1998). The representation of the natural syllables in NCM was significantly more clustered, suggesting that these natural sounds were better discriminated than the synthetic sounds or the guitar notes. The second study came from our laboratory (Grace *et al.*, 2003). We designed synthetic stimuli that had identical power spectra and similar amplitude modulation spectra as the zebra finch song. The synthetic sounds consisted of train of tone pips ("pips"), combination tones ("tones") and harmonic stacks ("stacks"). The tone pips and the combination tones had the same frequency spectrum and amplitude modulation in each frequency band as song. The harmonic stacks consisted of fundamental frequencies taken from the distribution of frequencies found in song. In addition, the duration of each stack and inter-stack interval was matched to the duration of zebra finch syllable and intersyllable silence. If the neuron's response could be explained by their frequency tuning and their amplitude modulation tuning, then similar responses would be found for these matched synthetic sounds and for conspecific song. The results showed that, despite the fact that the synthetic and natural sounds were acoustically matched, neural responses were greater for the natural song (Figure 13.4). Studies such as this, while confirming the existence of neurons which are selectively responsive to natural song, also suggest that understanding the mechanisms underlying such selectivity will require careful studies of the tuning properties of single neurons for specific temporal and spectral features of sound.

Spectrotemporal receptive fields of auditory neurons in MLd, Field L and CLM

Progress in the analysis of sensory processing in all modalities has been facilitated by the development of

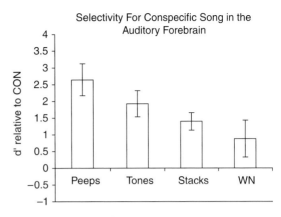

Figure 13.4 Average d' values for the consynthetic comparisons for neurons in Field L and CM. The synthetic sounds consisted of train of random tone pips constructed to have the same power spectrum and pip duration as song (pips), random combinations of sums of tone pip trains (tones), synthetic harmonic stacks (stacks) and white noise (WN). (Data from Grace *et al.*, 2003.)

methods for estimating a neuron's receptive fields: i.e. the range of stimulus parameters over which a neural response is elicited. Since receptive fields have a temporal dependence, a complete analysis requires the estimation of spatiotemporal receptive field (Ghazanfar and Nicolelis, 2001). In audition the spatial dimension is given by the frequency decomposition initially performed in the cochlea. Thus, we have approached the analysis of auditory neurons by estimating their spectro temporal receptive fields (STRFs). The STRF is a method for visualizing and quantifying the linear relationship between the presentation of sounds of particular frequencies at particular times and the increase or decrease in the probability of spiking in the neuron under study. The STRF therefore combines, in a time-frequency plot similar to a spectrogram, information about the frequency sensitivity (spectral tuning) and the temporal sensitivity (temporal tuning) of the neuron. It enables us to grasp rather intuitively the nature of the stimulus features (temporal, spectral and joint spectrotemporal) which would most effectively drive the neuron. The temporal sensitivity will tell us whether the neuron is tuned for fast sounds (e.g. the onset of syllables), slow sounds (the amplitude envelope of songs) or specific successions of sounds (e.g. the combination of two or more syllables). To make the distinction between fast or slow tuning, it is convenient to look at the Fourier Transform of the STRF and characterize the tuning in terms of modulations: the temporal modulation tuning is measured in Hz and describes the neural sensitivity to the rate of temporal changes; the spectral modulation is measured in cycles/kHz and describes both the bandwidth of the spectral tuning and the tuning for particular harmonic stacks. We call the STRF, expressed in terms of its spectral and temporal modulation tuning, the modulation transfer function (MTF). Similarly, the average acoustical properties of a sound, such as song, can be characterized by its spectral and temporal modulations in what we call the modulation spectrum: the modulation spectrum is the Fourier Transform of the spectrogram of the sound (Singh and Theunissen, 2003). The STRF in terms of its MTF can then be directly compared with the modulation spectrum of a sound to test whether the neural tuning is indeed matched to the spectral and temporal features of sounds found in song.

A second advantage of the STRF description is that it represents a quantifiable model of the stimulus-response function of neuron: the STRF can be used to obtain a prediction of the neural response to novel stimuli. We have developed methods that allow one to estimate and validate the STRFs of neurons in higher level auditory areas from their responses to complex sounds (Theunissen et al., 2000, 2001; Hsu et al., 2004b; Gill et al., 2006; Woolley et al., 2006).

We and others have begun to apply that methodology to estimate the STRFs of neurons in MLd, Field L and CLM from their responses to conspecific song and other complex sounds (Sen et al., 2001; Cousillas et al., 2005; Woolley et al., 2005, 2006). Figure 13.5 shows examples of the different types of STRFs that we have found in Field L (left column), the corresponding modulation transfer function (right column) and the typical responses to song (middle column). Most neurons could be classified according to their STRF as belonging to one of these types and it is clear these different subsets can subserve different auditory computations. The response of the neuron label FBOn (Fast Broad Band Onset) marks, with high temporal precision, the onset of each syllable. These onset coders generate a precise neural representation of the song's rhythm. The SB (Slow Broad Band) neuron integrates the energy over longer periods of time than FBOn neurons yielding a more sustained firing rate that can be used to obtain a good representation of the amplitude envelope of the sound. The detail of the shape of the amplitude envelope of a complex sound affects the perceived textural quality of the sound, called the timbre. For example, the difference in the sound of a piano and organ are in part due to the difference in the slope of the attack and decay of individual notes. The SI (Side Inhibition) neuron also integrates the energy over relatively long periods but shows more selectivity than the SB neurons. For the illustrative song shown in Figure 13.5, the SB neuron is particularly responsive to the repeated longer syllable that has more energy in the higher frequency range, and is therefore able to detect the longer complex syllable in each motif. The NB neuron shows a high degree of temporal and spectral precision and can be used to detect specific spectral patterns in song. For this example, one can imagine setting a threshold line to detect the isolated middle note present two times during the song. The SSI (Slow Sideband Inhibition) and C/FS (Constant/Frequency Sweep) neurons provide further specialization. The response to the SSI is mostly sustained during song while being maximally enhanced for harmonic stacks and thus could be used as a detector of harmonic sounds. The C/FS cluster responds maximally to all syllables that show frequency down-sweeps, which are quite common in zebra finch song. What this illustrative analysis demonstrates is that – just as in mammalian visual cortex specific classes of cells are selectively responsive to such features as form, movement and color – there is an analogous division of labor occurring in the auditory forebrain, with different neuron types extracting specific acoustical features of sound. Each of these features can be understood as representing specific information-bearing aspects of song and all of which can also be related to auditory perceptual classes in human listening.

Analyzed as an ensemble, we have also shown that the average tuning in the auditory system is for the spectrotemporal modulations that are particularly informative in natural sounds (Singh and Theunissen, 2003; Woolley et al., 2005). This tuning can explain in part the increase in information rates and the selectivity found through the avian auditory system for songs over synthetic sounds (Hsu et al., 2004a). Along similar lines, Narayan et al. (2005) showed that the delayed inhibitory regions in the STRF (seen for example in our FBOn, SB, NB cell types) yield responses that

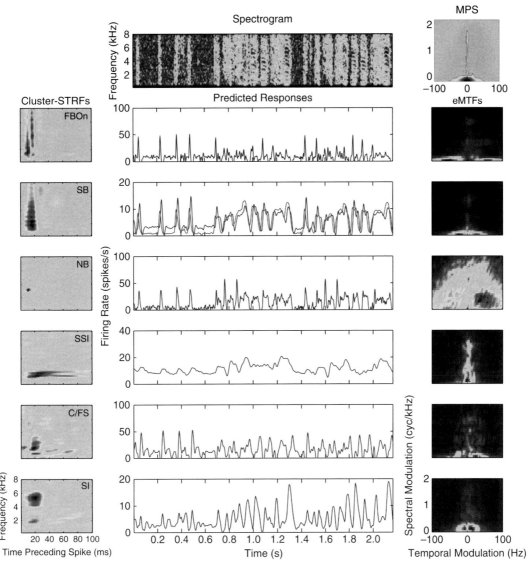

Figure 13.5 Examples of the six functional classes of neurons found in the zebra finch auditory forebrain region, Field L. Classes are defined by both spectral tuning (e.g. broadband versus narrowband) and temporal tuning (e.g. onset versus sustained responses) as determined from their spectrotemporal receptive field (STRF). FBOn, fast broadband onset; SI, sideband inhibition; SB, slow broadband; NB, narrowband; SSI, slow sideband inhibition; C/FS, constant/frequency sweep. The majority of cells (71 out of 105; 68%) fit into one of these classes. The left column shows the STRF in time-frequency space. The corresponding plots on the right column show the modulation transfer function for all neurons in that functional class in temporal-spectral modulation space. The middle column shows the response of each type of neuron to the song shown in the top row, in blue. The amplitude envelope of the song is shown in red along with the SB prediction. The plot on the right next to the spectrogram of song is the modulation spectrum of an ensemble of zebra finch song. The modulation power spectrum and the modulation transfer functions can be compared with each other to study which features of sound are represented in the different classes. (Data from Woolley et al., unpublished.) (See also color plate section.)

improve the discriminability of song relative to model neurons that would lack such delayed inhibition.

We have also observed changes in tuning in neurons in MLd while processing song versus processing a special filtered noise that covered the same temporal modulations and spectral modulations found in song. These changes can be quantitatively analyzed by measuring the two STRFs obtained for the same neuron for the two different types of sounds. On average, the STRFs when processing song were tighter in time and more spread out in frequency showing that the neurons became less sensitive to spectral features and more sensitive to temporal features. Moreover the STRFs of different neurons are more similar to each other while processing song, resulting in a synchronization of the neural response across cells. This synchronized response allowed for a very accurate population code for the temporal features of the song (Woolley et al., 2006).

Tonotopy and other classical characterizations of response properties in the auditory forebrain

The response properties of auditory forebrain neurons have also been characterized using more classical techniques. One of the principal results of those studies is the preservation of tonotopy in the auditory caudal nidopallium, as characterized by the measurement of the frequency of best response (BF) to pure tone stimuli. In subarea L2 and in the neighboring auditory nidopallium and mesopallium, the frequency axis runs along the dorsal-caudal (low frequencies) rostral-ventral (high frequency) directions (Zaretsky and Konishi, 1976; Muller and Leppelsack, 1985; Rübsamen and Dörrscheidt, 1986). The tonotopy in the auditory forebrain of songbirds follows the same organization as in other avian species (Bonke et al., 1979b; Heil and Scheich, 1985). In a more recent study, this simple tonotopic picture of the caudal auditory neostriatum was questioned: Gehr et al. (1999) found multiple functional areas in the neostriatum of the male zebra finch each with their own tonotopic gradient. These functional areas overlapped only loosely with anatomically defined subareas of Field L. The tonotopic organization in Field L and neighboring areas will have to be revisited.

The temporal and spectral response properties have also been characterized using sinusoidal amplitude modulation and sinusoidal frequency modulation of pure tones, band-passed Gaussian white noise and other pseudo-random stimuli (Muller and Leppelsack, 1985; Knipschild et al., 1992). As is the case in the mammalian auditory cortex, there is a significant fraction of neurons in the auditory forebrain of songbirds that do not respond to pure tones but do respond to sounds with frequency or amplitude modulation. The best modulation frequency (BMF) appears to be higher in birds than in mammals and, in the starling, frequency modulation is a more effective stimulus than amplitude modulation.

As noted above, STRF methodology can be used to relate the tuning properties of auditory neurons in songbirds to the acoustic properties of adaptively significant signals like conspecific vocalizations. The comparison between the response properties that can be extracted from STRFs and those obtained using classical methods is a work in progress (Schafer et al., 1992). Ultimately, it will be desirable to understand the distribution of neuron types defined by their functional properties obtained both with classical methods and with the STRF methodology. To further understand the underlying circuitry mediating this auditory processing, it will then be important to assess whether these different neuron types are topographically organized in the auditory nidopallium and/or whether they can be correlated with different anatomically defined neuron types.

CONCLUSIONS

There appear to be three gross levels of auditory selectivity in the songbird forebrain. First, at the highest level, sensorimotor neurons in the song system are selectively responsive to song that the bird is currently singing. The development of this selectivity during song learning and the striking similarity between the motor and sensory responses for the BOS-selective neurons strongly suggest that these responses can be used by the bird to evaluate the auditory feedback from its own current song for comparison with what the bird has been singing in the recent past. The degree of selectivity for the BOS is also correlated with the degree of plasticity in vocal learning in both juvenile and adult birds, suggesting that such selectivity plays an important role in song production maintenance. Whether such selectivity contributes to perceptual learning is, in our opinion, currently unclear and remains to be determined.

At the second level of selectivity, one finds neurons in NCM and CMM that are tuned to behaviorally specific, complex natural sounds such as songs to which the bird has been exposed recently or which have particular behavioral relevance. These neurons are potentially involved in perceptual tasks requiring the recognition of familiar or recently heard song. Recent data also points to NCM as a potential place for the storage of the tutor song but that conclusion needs further analysis and verification.

At the third level, in primary and secondary auditory forebrain areas, neurons are on average tuned to the spectrotemporal sounds that are found frequently in conspecific song. The result of this tuning is a population preference for conspecfic song over synthetic stimuli that lack this characteristic spectrotemporal structure but that are otherwise matched acoustically. These response properties suggest that complex natural sounds such as conspecific song, and perhaps heterospecific songs with similar acoustical structure, are preferentially represented in the neural activity of the auditory forebrain relative to other background sounds that are commonly present in the bird's environment. Moreover, the neural representation of these characteristic acoustical features of song appears to be extracted by a small number of functional types of neurons.

These three levels of selectivity are suggestive of a coarse hierarchical processing of auditory information. Conspecific song and other similar sounds are efficiently encoded in the auditory midbrain and the primary auditory forebrain, Field L. In the secondary forebrain areas of NCM and CMM, the effect of recent experience becomes crucial and neurons can be selective for particular familiar songs or sounds. Finally among those specific songs, the bird's own song plays a very special role and is selectively represented in the song system nuclei. However, there are significant gaps in this picture. A major gap is our lack of understanding of the mechanisms that generate these different levels of selectivity, and of the links between these different auditory areas and the song system. We badly need further studies of both the anatomical and functional connectivity between all these areas. For example, it would be very interesting to investigate whether the functional types of neural responses observed in Field L correspond to different types of neurons and, if so, to delineate the local microcircuitry as it has been done so successfully in HVC (see Prather and Mooney, this volume). We also need to understand the nature of the reward or reinforcement signal that is involved in both the perceptual learning of arbitrary conspecific song and in the memorization of the tutor song. We do not know, for example, how the auditory information in Field L is combined with a reward or recognition signal that leads to song specific neural recognition in NCM or CMM. Equally unknown is the process by which BOS-selective neurons in the song system obtain their selectivity and whether or not this processing involves the secondary areas NCM and CM (both medial and lateral) or is achieved via a separate pathway. Finally, the "holy grail" of birdsong research, the precise neural substrate for the tutor template and how this tutor template interacts with feedback from the young bird's vocalization to guide vocal learning, remains undiscovered.

In terms of more general questions in auditory perception, there remains the problem of characterizing the neural mechanisms that generate the songbird's unique perceptual world: that is, of identifying the correlation between the functional properties of auditory neurons and the perception of songs. To solve this problem requires a much better understanding of the bird's auditory perception than is currently available. What are the acoustically informative features of songs and other sounds for birds? How are these features organized in perceptual classes? Do birds perceive the rhythm, pitch and timbre of the song as we do? If so could FBOn neurons actually mediate the sense of rhythm and SB neurons generate the sense of timbre, as we suggested above? How is sound source separation, so important for recognition under realistic conditions, achieved? Gentner (this volume) has combined behavioral testing and neuronal recording to begin to characterize the neural mechanisms by which songbirds process complex acoustic communication signals. It is clear from his work and from the studies reviewed in this chapter that answers to these questions will require the integrated application of behavioral, physiological, and computational methods. Aspiring birdsong researchers can look forward to solving these exciting puzzles.

14 • Song-selective neurons in the songbird brain: synaptic mechanisms and functional roles

Jonathan F. Prather and Richard Mooney

INTRODUCTION

Birdsong is learned and maintained via auditory experience, a process requiring interactions between auditory and song motor areas. For this reason the report that simple acoustic stimuli could evoke auditory activity in the song "motor" nucleus HVC of anesthetized zebra finches (Katz and Gurney, 1981) caused considerable excitement in the birdsong community and initiated a flurry of research activity which continues to this day. Subsequently, McCasland and Konishi (1981) reported that auditory activity could be evoked in the HVC of awake canaries by playing back a recorded version of the bird's own song (BOS). A more extensive subsequent characterization of auditory selectivity of HVC neurons, conducted by Margoliash (1983), showed that HVC neurons in the anesthetized white-crowned sparrow were song selective, firing vigorously to forward but not reverse BOS playback, and in some cases only responded to specific note combinations in the BOS.

These remarkable findings and the relative ease of analyzing synaptic connectivity in the songbird's brain make song-selective HVC neurons extremely attractive candidates for addressing mechanisms for the generation of stimulus-specific sensory responses (see Theunissen, this volume). They are of additional interest to the neuroethologist because of their established role in song recognition (Brenowitz, 1991; Del Negro *et al.*, 1998; MacDougall-Shackleton *et al.*, 1998a; Gentner *et al.*, 2000; Halle *et al.*, 2002, 2003) and because these neurons could mediate auditory–vocal interactions important to vocal mimicry.

In this chapter, we address three important questions about song selectivity. Where in the brain does song-selectivity originate? What are the synaptic mechanisms underlying the remarkably selective auditory responses recorded in HVC? What forms of experience – auditory, vocal motor, tutor song or auditory feedback – shape BOS-selective responses in HVC and other parts of the song system? In describing answers to these questions, we focus on the analysis of auditory selectivity in the anesthetized bird, an approach that provides superior recording stability and minimizes potential confounds due to changes in the animal's state of attention or arousal. These studies have provided an increasingly detailed picture of where BOS-selective responses arise in the brain, the synaptic mechanisms that contribute to BOS-selective responses in HVC, and the role that auditory experience plays in shaping song-selective responses in HVC and other parts of the song system.

Despite the advantages to recording in an anesthetized preparation, answers to some questions of great functional relevance can only be sought in awake, freely behaving songbirds. The issue of the potential functional significance of auditory activity in HVC hinges on the degree to which HVC auditory activity is manifested in the awake bird. Here we discuss results from studies using chronic recording methods in awake songbirds, including those we have undertaken in the swamp sparrow. These studies reveal that some of the HVC neurons that project to a basal ganglia pathway important to song learning and perception display robust and highly selective auditory activity in the waking animal.

WHAT IS BOS-SELECTIVITY?

BOS-selectivity is defined as a stronger neuronal response to forward playback of the BOS than either to temporally altered versions of the BOS or to conspecific songs. Such selectivity can be relative, in which both the BOS and the non-BOS stimuli evoke firing rate increases, or absolute, such that only the BOS evokes a response. Figure 14.1 depicts auditory responses recorded from different song system neurons exhibiting relative and absolute BOS-selectivity. Neurons

Neuroscience of Birdsong, ed. H. Philip Zeigler and Peter Marler. Published by Cambridge University Press. © Cambridge University Press 2008.

Song-selective: synaptic mechanisms and functions 175

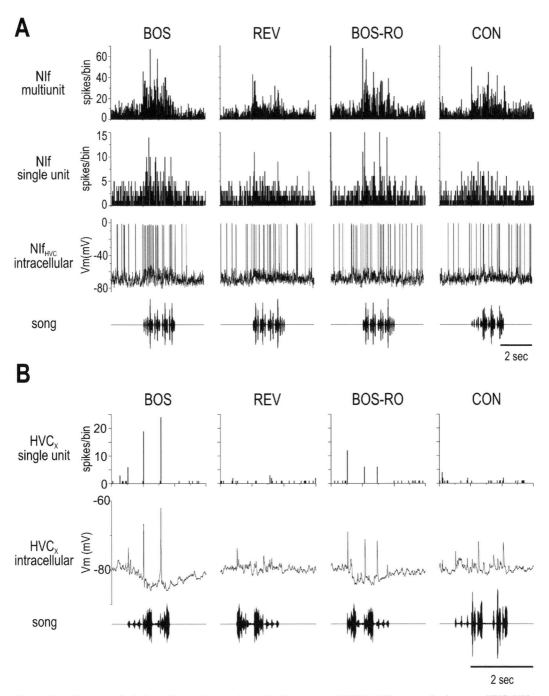

Figure 14.1 Examples of relative auditory selectivity for the bird's own song (BOS) in NIf neurons that innervate HVC (NIf$_{HVC}$; panel A) and absolute auditory selectivity for the BOS in HVC$_X$ neurons (panel B).
(A) Comparison of song-evoked multiunit NIf activity with single-unit activity in NIf$_{HVC}$ neurons recorded from urethane-anesthetized zebra finches. Stimuli presented are the bird's own song (BOS), the reversed BOS (REV), individual syllables of the BOS each played forward but assembled in the reversed order of the natural BOS (BOS-RO), and conspecific song (CON).

that display absolute song selectivity afford especially compelling examples of an auditory "grandmother neuron," a hypothetical cell so selective that it would respond only in the presence of one's grandmother (Marr, 1982; Gross, 2002). Many neurons in the song system of the anesthetized songbird exhibit relative BOS-selectivity (Volman, 1996; Doupe, 1997; Livingston and Mooney, 1997; Theunissen and Doupe, 1998; Mooney, 2000; Rosen and Mooney, 2000; Grace et al., 2003), and a sizable minority appears to respond exclusively to the BOS. When neurons respond to only a single song, they are referred to as song-specific neurons (Margoliash, 1983).

The fact that BOS-selective neurons respond much more strongly to forward over reverse BOS indicates that they must be sensitive to temporal features of song, because these two stimuli have equivalent spectral content but contrasting local and global temporal features. That many BOS-selective neurons actually discriminate longer timescale features of the song is indicated by their tendency to respond more strongly to the BOS than to artificial versions of the BOS in which the note or syllable order has been reversed (Figure 14.1B) (Lewicki and Arthur, 1996; Volman, 1996; Doupe, 1997). Indeed, some song-specific neurons have been shown to respond in an all-or-none fashion to specific note or syllable combinations naturally present in the BOS, a feature requiring temporal integration over many tens to hundreds of milliseconds. From a functional standpoint, such long timescale integration is well suited to detect the variations in syntax that distinguish the BOS from other similar conspecific songs and to reinforce global aspects of the learned song, namely the syllable sequence. Additionally, neurons sensitive to specific harmonic combinations have been detected in songbirds with spectrally complex songs, such as the zebra finch (Margoliash and Fortune, 1992). These various findings underscore that BOS-selective neurons are sensitive to complex temporal and spectral features of the song and thus are well suited for characterizing the neuronal mechanisms that detect complex learned vocal sequences, including human speech.

As with other studies addressing sensory coding of natural stimuli in other systems, studies designed to assess auditory selectivity in HVC and other areas in the songbird brain have employed both natural stimuli, including the BOS and the songs of other conspecifics, and synthetic acoustical stimuli (Theunissen and Doupe, 1998; Mooney, 2000). However, regardless of which of these two approaches is used to identify them, a far greater proportion of BOS-selective neurons is found within the song system than in primary and secondary regions of the avian auditory telencephalon (Lewicki and Arthur, 1996; Theunissen and Doupe, 1998; Janata and Margoliash, 1999; Grace et al., 2003; Amin et al., 2004; Theunissen et al., 2004a). Indeed, single unit recordings made from Field L and HVC in individual zebra finches show that the proportion of neurons displaying sensitivity to forward over reverse BOS playback, as well as note and syllable order, increases from Field L to HVC (Lewicki and Arthur, 1996; Janata and Margoliash, 1999).

WHERE IN THE BRAIN DOES SONG SELECTIVITY ARISE?

As noted above, the highly selective auditory responses of HVC neurons stand in stark contrast to the relatively nonselective auditory responses exhibited by neurons in Field L, the primary auditory telencephalon of the bird and a likely source of either direct or indirect

---

Caption for Figure 14.1 (cont.)
Relative selectivity is evident in the strong response to the BOS and the weaker responses to BOS-RO, CON and REV. Bottom panel, oscillogram of song stimuli; third panel, response of a NIf_{HVC} neuron to a single playback of each song stimulus. Second panel, peristimulus time histogram (PSTH) of the action potential response of this NIf_{HVC} neuron to 20 iterations of each song presentation. Top panel, PSTH of the multiunit responses to 20 iterations of each song presentation; data taken from the same region of NIf and in the same bird as the corresponding single-unit data. (Adapted from Figure 11 in Coleman and Mooney, 2004.) (B) Comparison of single-unit activity in HVC_X neurons evoked by different song stimuli. Absolute selectivity is evident in the strong response to the BOS, the weaker response to BOS-RO, and a lack of suprathreshold response to other stimuli (REV, CON). Bottom panel, stimulus oscillogram; middle panel, median-filtered averaged intracellular membrane potential record taken from 20 repetitions of each stimulus in an HVC_X neuron; top panel, histogram of action potentials evoked by 20 repetitions of each stimulus in the same HVC_X neuron as the corresponding middle panel.

auditory drive to HVC (Lewicki and Arthur, 1996; Janata and Margoliash, 1999). These differences reinforce the impression that auditory neurons in HVC are predominantly concerned with processing song-related information and that BOS-selectivity is largely the result of neuronal computations performed above Field L. Important issues to resolve are the exact path via which auditory information flows from Field L to HVC and the extent to which BOS-selectivity arises in structures interposed between these two areas.

Our understanding of the specific pathway that links Field L to HVC remains imprecise. An early idea was that Field L formed a direct (i.e. monosynaptic) connection with HVC. However, studies where injections of anterograde tracers were made into Field L resulted in little or no terminal label in HVC itself, but did heavily label the "shelf", a region just ventral to HVC into which a small number of HVC neurons extend dendrites (Kelley and Nottebohm, 1979; Vates et al., 1996; Benton et al., 1998). Indeed, the apparent route via which much or all auditory information reaches HVC is more complex and may include additional processing steps between Field L and HVC. A compelling clue in support of this indirect auditory route was the detection of BOS-selective auditory responses in NIf, a sensorimotor nucleus that provides dense axonal terminations in HVC (Vates et al., 1996; Janata and Margoliash, 1999). This finding lends strong support to the idea that HVC is not the site where BOS-selective responses originate, although full confirmation of this idea rested on an analysis of functional connectivity between NIf and HVC (Cardin and Schmidt, 2004a; Coleman and Mooney, 2004).

HVC FILTERING OF AUDITORY INPUT: METHODOLOGICAL CHALLENGES

Experimental analyses of the functional interactions between HVC and its putative auditory afferents face significant challenges. The first is the high degree of anatomical convergence of axonal inputs onto HVC, including those originating from auditory as well as nonauditory areas (Nottebohm et al., 1982; Fortune and Margoliash, 1995; Foster and Bottjer, 1998; Shea and Margoliash, 2003). A second is the necessary distinction between anatomical and functional connectivity. Thus although HVC receives input from many areas, some may play a more or less important role in driving HVC's auditory activity or may differ in the types of auditory information they provide to HVC. A third is that HVC contains, in addition to interneurons, projection neurons (PNs) innervating respectively either the song premotor nucleus RA (HVC_{RA} neurons) or the basal ganglia homologue Area X (HVC_X neurons) (Mooney, 2000). Because extracellular recordings made in HVC are biased towards sampling from interneurons, characterizing the activity of identified HVC projection neurons requires intracellular methods or extracellular methods applied in conjunction with antidromic stimulation (Mooney, 2000; Hahnloser et al., 2002; Rauske et al., 2003). Ultimately, pinpointing the origins of BOS-selectivity and establishing the exact nature of any auditory transformations in HVC requires techniques that can assess functional connectivity between neurons in different areas, probe the synaptic mechanisms that underlie highly selective suprathreshold patterns of activity, and track cellular identity. These requirements are best met with in vivo intracellular recording methods. Using these methods, we and others have made several observations that together establish that NIf provides BOS-selective input to HVC. First, reversible pharmacological silencing of NIf activity abolishes much if not all of the spontaneous and auditory activity in HVC, supporting the idea that HVC derives much or all of its auditory drive from NIf (Cardin and Schmidt, 2004a; Coleman and Mooney, 2004). Second, dual electrode recordings coupled with spike-triggered averaging methods show that action potentials in NIf neurons slightly precede membrane depolarizations in HVC neurons, consistent with a monosynaptic excitatory linkage between NIf and HVC neurons (Coleman and Mooney, 2004). Third, we have used intracellular recordings from NIf neurons projecting to HVC (NIf_{HVC}) and HVC projection neurons (i.e. HVC inputs and outputs) to compare song selectivity before and after HVC processing. These studies reveal that the relative bias to the BOS versus other acoustic stimuli is established in NIf, but that an absolute bias to the BOS (i.e. song-specific responses) arises in HVC (Figure 14.1) (Coleman and Mooney, 2004). More specifically, a direct quantitative comparison of BOS-selectivity shows that NIf_{HVC} neurons and HVC PNs are equally BOS-selective (Coleman and Mooney, 2004). However, NIf_{HVC} neurons fire to most non-BOS stimuli, including conspecific songs and white noise bursts, and fire in a sustained fashion to the BOS (Coleman and Mooney, 2004), whereas HVC PNs fire little or not at all to non-BOS stimuli, and fire in a highly phasic manner to the BOS (Figure 14.1) (Mooney, 2000; Coleman and Mooney, 2004). In fact, some HVC PNs fire only during a very narrow time window during the song motif, and thus can be said to be "temporally sparse" in their firing patterns. These findings indicate that BOS-selectivity is generated prior to HVC and also characterize more precisely how auditory activity in HVC differs from that in its major auditory afferent, NIf. These differences indicate that circuit interactions between NIf and HVC, or within HVC itself, alter the temporal pattern of BOS-evoked activity.

WHAT ARE THE SYNAPTIC MECHANISMS UNDERLYING BOS-SELECTIVE RESPONSES IN HVC?

A variety of observations point to circuit interactions within HVC as the mechanism underlying temporally sparse, song-specific firing patterns of HVC PNs. First, HVC_{RA} and HVC_X neurons show similarly sparse patterns of BOS-evoked action potential activity, but differ in their underlying subthreshold response patterns: HVC_{RA} neurons show sustained depolarizing subthreshold responses, whereas HVC_X neurons show a complex mixture of hyperpolarizing and depolarizing responses (Mooney, 2000), similar to those reported by Lewicki (1996) for certain note-combination sensitive HVC neurons (Lewicki, 1996). Second, dual recordings and spike-triggered averaging methods reveal that NIf neurons make functionally excitatory connections with both PN types and interneurons in HVC (Coleman and Mooney, 2004). This pattern of excitatory connectivity from NIf to HVC suggests that the contrasting subthreshold responses of different HVC PN types to BOS playback are not simply due to differences in extrinsic input; instead, BOS-evoked hyperpolarizing responses in HVC_X cells most likely arise due to interactions with inhibitory neurons local to HVC (Mooney, 2000; Mooney and Prather, 2005; Rosen and Mooney, 2006). Third, in direct support of this idea, the firing patterns of interneurons in HVC are sustained throughout BOS playback and closely correlate with the BOS-evoked membrane hyperpolarizations of HVC_X cells (Mooney, 2000), and these interneurons are immunopositive for parvalbumin (PV) (Kawaguchi et al., 1987; Kawaguchi, 1993; Mooney and Prather, 2005; Wild et al., 2005), a calcium-binding protein expressed at high levels in inhibitory interneurons in other systems. Fourth, intracellular recordings made in brain slices from synaptically coupled cell pairs show that these PV+ interneurons make inhibitory synapses on HVC_X cells and that both HVC_{RA} and HVC_X cells make excitatory synapses onto these PV+ interneurons (Mooney and Prather, 2005). These results suggest a model where BOS-selective inputs from NIf provide monosynaptic excitation to all three HVC cell types, and ultimately drive feedforward and feedback inhibition mediated by the HVC network to generate BOS-evoked hyperpolarizing response in HVC_X cells.

This model predicts that inactivating the local HVC circuit should cause the BOS-evoked hyperpolarizations in HVC_X neurons to disappear, whereas their depolarizing responses should persist. To measure more directly whether and how HVC transforms its extrinsic auditory input, we compared the auditory-evoked subthreshold activity of HVC neurons with the local circuit either "on" or "off." We made this comparison by intracellularly recording the auditory-evoked synaptic activity of individual HVC neurons before and after we pharmacologically silenced the local circuit (Rosen and Mooney, 2006). Synaptic responses of both HVC PN types remained BOS-selective even when the local HVC circuit was inactivated, confirming that HVC receives an extrinsic source of BOS-selective input. However, with the local circuit inactive, the subthreshold response patterns of HVC_X neurons were rendered purely depolarizing and closely resembled the response patterns of HVC_{RA} neurons recorded in the same bird. In contrast, the shape of the subthreshold response patterns of HVC_{RA} neurons changed very little with local circuit inactivation. This result supports the idea that the differences in subthreshold response patterns of the two HVC PN types is due to selective targeting of HVC_X cells by inhibitory interneurons.

Though useful, these local circuit inactivation experiments silenced action potential activity in all HVC neurons, and hence could not address whether or how inhibition shapes BOS-evoked firing patterns of HVC_X cells. For example, it would be useful to know whether inhibition renders the suprathreshold responses of HVC_X cells highly phasic, or whether the excitatory inputs by themselves can evoke highly phasic firing. Furthermore, it remains unclear whether inhibitory and excitatory inputs onto HVC_X cells are effectively balanced, with both being recruited most strongly by the BOS, or whether non-BOS stimuli preferentially recruit inhibition, masking excitatory responses to non-BOS stimuli and ultimately resulting in greater stimulus specificity in HVC_X cells. Another important question is whether inhibition helps to regulate the temporal precision of BOS-evoked firing in HVC_X cells? Finally, one idea is that syllable combination-sensitive responses in song-specific HVC neurons might arise when hyperpolarizing responses evoked by the first note deinactivates a low-threshold calcium current, "priming" the cell to fire a burst of action potentials to a second note that when played in isolation evokes only a subthreshold depolarization (Lewicki, 1996). If blocking inhibition onto HVC cells diminished combination-sensitive action potential bursts, this would lend support to this priming model.

To address these issues, a method was needed that could remove some of the inhibitory input onto the cell while leaving its action potential machinery intact. Furthermore, it would be best if this method deprived only the impaled cell of its inhibitory synaptic input, to avoid runaway excitation triggered by silencing the entire inhibitory network. Fortunately, substantial in vitro work from David Perkel's group had established the inhibitory repertoire of HVC_X cells, which are targeted by a wide variety of potent inhibitory inputs, including those that activate ionotropic chloride currents and G-protein coupled inward rectifying potassium currents (i.e. GIRKs) (Dutar et al., 1998, 1999, 2000; Schmidt and Perkel, 1998). These different forms of inhibition can be disrupted at the intracellular, single cell level either by chloride loading the cell or by dialyzing it with compounds that block G-protein signaling.

When we disrupted either ionotropic chloride currents or GIRK signaling in individual HVC_X cells, BOS playback evoked more sustained firing patterns, qualitatively resembling those seen in NIf_{HVC} neurons and HVC interneurons (Rosen and Mooney, 2003). Therefore, interactions between excitatory and inhibitory inputs are necessary to generate highly phasic firing patterns in HVC_X cells. Blocking inhibition also unmasked excitatory responses to non-BOS stimuli, but these responses were strongly graded, suggesting that inhibition and excitation onto HVC_X cells are balanced. This behavior may reflect feedforward inhibitory architecture from NIf to HVC, which would be expected to activate HVC interneurons most strongly in response to the BOS. This synaptic behavior, wherein the preferred stimulus as defined by excitatory responsiveness also activates the strongest inhibitory response, resembles the balanced excitatory and inhibitory synaptic interactions described in the primary auditory cortex of the rat (Wehr and Zador, 2003), suggestive of common mechanisms underlying auditory selectivity in higher level sensory and sensorimotor areas of the vertebrate brain. The variability of BOS-evoked spiking in HVC_X cells also increased when inhibition was disrupted, indicating that inhibition is important to regulate precise spike timing in these neurons. Finally, some of the phasic excitatory responses that were initially preceded by membrane hyperpolarizations increased when hyperpolarizing inhibition was blocked, whereas others were diminished, lending partial support to the priming model of combination sensitivity.

We know less about the mechanisms that contribute to the temporally sparse patterns of BOS-evoked firing in HVC_{RA} neurons, although they appear to differ from those operating in HVC_X cells. As mentioned earlier, HVC_{RA} neurons are unlike HVC_X cells in that they display sustained depolarizing membrane potential responses in response to BOS playback, even though these sustained depolarizations drive only temporally sparse patterns of firing (Mooney, 2000). The depolarizing responses are not simply a mix of excitatory postsynaptic potentials (EPSPs) and depolarizing inhibitory postsynaptic potentials (IPSPs), because tonically depolarizing an HVC_{RA} neuron causes it to fire in a sustained fashion to BOS playback (Mooney, 2000). Furthermore, inactivating the local HVC circuit has little effect on the shape of the BOS-evoked depolarizing response, suggesting that HVC_{RA} neurons are targeted less heavily than HVC_X neurons by local inhibition (Rosen and Mooney, 2006). However, BOS-evoked hyperpolarizations can be detected when the cell is depolarized substantially above action potential threshold (Mooney, unpublished observations), suggesting that these cells do receive some, albeit weak, inhibition, consistent with the known synaptic connections between interneurons and HVC_{RA} cells (Mooney and Prather, 2005). The relatively weak BOS-evoked inhibition in HVC_{RA} cells suggests that the sparse action potential output of these neurons evoked by song playback may involve mechanisms that do not depend heavily on inhibition. These mechanisms could include postsynaptic thresholding, perhaps mediated via the precise regulation of the cell's resting membrane potential or the weight of its excitatory synaptic inputs, or cooperative interactions between excitatory inputs.

In summary, in vivo intracellular recordings have revealed many interesting features of the synaptic mechanisms underlying BOS-selective firing patterns in HVC, especially in HVC_X cells. These include the finding that BOS-evoked activity in interneurons drives inhibition in HVC_X cells, transforming tonic patterns of synaptic excitation from NIf into a highly precise and phasic output. In addition to such suppressive inhibitory interactions, inhibition onto HVC_X cells can also augment BOS-evoked excitatory peaks through a priming mechanism, providing a mechanism for combination sensitivity. These various inhibitory interactions help generate auditory activity in HVC_X cells that is highly precise in its timing and more exclusively responsive to a single stimulus, namely the BOS. A fascinating feature of this process is that remarkably selective sensory-evoked responses are generated in HVC, an area traditionally associated with motor aspects of singing. Indeed, this sharpening of stimulus specificity in HVC may reflect motor-driven effects on sensory processing, as has been described for the refinement of whisker representations in the rodent motor and somatosensory cortices (Kleinfeld et al., 2002; Polley et al., 2004). The generation of such temporally precise and stimulus-specific responses in HVC may be especially important for matching auditory to motor representations of song, a process that could facilitate vocal mimicry and communication.

OTHER SOURCES OF AUDITORY INPUT TO HVC

In addition to NIf, HVC receives input from at least three other areas that display auditory activity: Uva, a thalamic nucleus afferent to both NIf and HVC, (Nottebohm et al., 1982; Wild, 1994a); CM, the secondary auditory telencephalic region CM (Vates et al., 1996; Vates et al., 1997) and the anterior telencephalic nucleus MMAN (Nottebohm et al., 1982; Foster and Bottjer, 2001). The first two of these receive input from identified auditory structures, while MMAN is thought to receive its auditory drive indirectly from HVC, and may function as a recurrent auditory pathway (Vates et al., 1997).

Anatomical pathway tracing studies we have undertaken with Martin Wild show that Uva receives ascending input from the ventral part of the lateral lemniscus, which in turn receives monosynaptic input from the cochlear nucleus (Coleman et al., 2007). Thus, the auditory pathway through Uva to NIf and HVC comprises relatively few synapses, providing a potentially short-latency source of auditory input to the song system. Extracellular recordings we made in Uva found that many neurons respond in a robust fashion to BOS playback (Coleman et al., 2007). However, although most Uva neurons are nonselective, a small fraction shows highly selective responses to BOS, suggesting that Uva could transmit selective as well as nonselective auditory information to NIf and HVC. Whether Uva actually contributes to auditory activity in HVC remains less certain: in the urethane-anesthetized zebra finch, reversibly inactivating Uva has no discernible effect on HVC auditory activity, even though low frequency electrical stimulation in Uva elicits EPSPs in HVC (Coleman et al., 2007). One possibility is that Uva's functional interactions with HVC are suppressed by anesthetic, and/or high levels of Uva activity are necessary to influence HVC responsiveness, a conclusion consistent with the finding that high frequency electrical stimulation in Uva can transiently suppress spontaneous and auditory activity in HVC (Coleman et al., 2007). Alternately, Uva's functional interactions with HVC may be more purely modulatory in nature, a view supported by the observation that Uva neurons can be excited by visual and tactile stimuli as well as auditory stimuli (Williams and Vicario, 1993; Wild, 1994a). Perhaps such integrative properties may enable Uva to respond to environmental stimuli that either favor or discourage singing.

Another source of auditory input to HVC is the secondary auditory telencephalic region CM (E. Bauer et al., in press), an area implicated in experience-dependent auditory plasticity (Gentner and Margoliash, 2003). CM axons make sparse terminations in HVC as well as NIf which are functionally important to auditory activity in the song system: reversibly inactivating CM strongly suppresses auditory activity in both NIf and HVC, although it does not suppress spontaneous levels of activity in either area (E. Bauer et al., in press). Both BOS-selective and non-selective cell types can be found in CM, and both types appear to make functional connections with NIf. Because the auditory selectivity of many CM neurons can be strongly modified by operant conditioning (Gentner and Margoliash, 2003), CM may provide NIf and HVC with input that has been shaped by auditory experience, including song-related feedback or even experience of the tutor song. Thus, an important goal of future studies will be to understand whether HVC receives input from the same neurons in CM that have been shown to exhibit learning-dependent changes in auditory selectivity.

The projections from Uva and CM may convey different types of auditory information to HVC. The projection from CM likely provides highly processed auditory information to the song system: CM is densely interconnected with primary and secondary regions of the auditory telencephalon, including Field L and NCM (Vates et al., 1996), and all three areas contain neurons with complex response properties (Theunissen et al., 2000, 2004a; Sen et al., 2001; Grace et al., 2003; Amin et al., 2004). NCM is an important site of experience-dependent auditory plasticity, and some NCM neurons are selective for the tutor song (Chew et al., 1995, 1996a; Bolhuis et al., 2000, 2001; Terpstra et al., 2004). Thus the connections that NCM makes with CM may provide a potential route for tutor-selective information to enter the song system. In contrast, Uva relays mostly nonselective auditory information directly from the lower levels of the auditory brainstem to the song system, and this information may serve more of a modulatory role. Although the functional significance of this convergence of auditory information onto HVC is unknown, one possibility is that salient auditory or visual cues in the animal's environment activate Uva, which in turn modulates auditory information relayed to HVC through experience-dependent perceptual filters in CM and NIf. Another possibility is that Uva provides a short latency pathway for conveying auditory as well as proprioceptive feedback to HVC, which is then compared with experience-dependent auditory representations (i.e. auditory memories) transmitted through NCM, CM and NIf to HVC.

FUTURE DIRECTIONS AND CHALLENGES TO ASSESSING FUNCTIONAL NETWORKS PRIOR TO NIF

Continued work is needed to fully divine the origins of BOS-selectivity. Although it has often been assumed that BOS-selectivity is generated in the song system, BOS-selective neurons can be found in different proportions throughout the auditory telencephalon. This distributed organization raises the possibility that interconnected subsets of selective cells in the auditory telencephalon form a segregated channel that routes BOS-selective information to the song system. Therefore, a major goal of future analyses is to assess functional connectivity in auditory areas presynaptic to NIf, such as CM, NCM and Field L, a goal which is likely to be more challenging to accomplish than the analysis of functional connectivity in the song system. Whereas NIf and HVC are each spatially compact, synaptically interact with each other in a largely feed-forward manner, and contain neuronal populations relatively homogeneous in their auditory selectivity, the primary and secondary regions of avian auditory telencephalon are vast regions, characterized by highly reciprocal interconnections (Vates et al., 1996) and populated by a wide variety of selective and nonselective neurons (Theunissen et al., 2000, 2004a; Sen et al., 2001; Grace et al., 2003; Amin et al., 2004). As a result, some circuit-analysis techniques that worked effectively in the song system, such as reversible inactivation, may be less useful in assessing functional connectivity of different regions in the auditory telencephalon. Similarly, the distributed and possibly sparse pattern of synaptic connectivity between these different auditory regions will likely make the assessment of functional connections between their constituent neurons more challenging. Despite these challenges, an important goal will be to carefully analyze how secondary auditory regions implicated in tutor song imprinting and adult forms of experience-dependent auditory plasticity communicate with sensorimotor areas necessary to the learning and maintenance of song.

THE ROLE OF EXPERIENCE IN SHAPING BOS-SELECTIVITY IN THE SONG SYSTEM

Because the bird's song is learned, BOS-selectivity must at a fundamental level reflect the effects of experience. But what forms of experience actually contribute to the development of BOS-selectivity? Do BOS-selective neurons constitute an auditory memory of self-produced songs, or do they also encode memories of the tutor song? Do they encode persistent auditory memories of any kind, or merely track the current auditory feedback? Furthermore, because HVC and other song nuclei display song motor as well as auditory activity, is BOS-selectivity purely a reflection of the auditory feedback experienced by the bird when it sings, or is it a product of interactions between the motor and auditory systems?

Much of what we know about the role of experience in shaping song selectivity comes from studies of neurons in the song nucleus LMAN, the output of an anterior forebrain pathway (AFP) necessary to juvenile and adult forms of audition-dependent vocal plasticity (Bottjer et al., 1984; Doupe, 1997; Livingston and Mooney, 1997; Kittelberger and Mooney, 1999; White et al., 1999; Rosen and Mooney, 2000; Boettiger and Doupe, 2001; Livingston and Mooney, 2001). Because HVC_X neurons are the putative source of auditory and singing-related activity in the AFP, experience-dependent effects on LMAN auditory selectivity are likely to reflect changes in selectivity initiated in HVC. For the purposes of this chapter, we will focus on some of the common themes that have emerged from studies of experiential factors that contribute to song selectivity in both HVC and LMAN. (For a review of the anterior forebrain pathway see Brainard, this volume.)

To a great degree, developmental studies point away from the tutor song and towards auditory feedback or auditory–vocal interactions as the major determinant of song selectivity. First, in contrast to what would be expected for a persistent tutor song memory, the vast majority of BOS-selective neurons in the HVC and LMAN of adult birds respond more strongly to the BOS than to the tutor song (Margoliash and Konishi, 1985; Volman, 1993; Solis and Doupe, 1997, 1999; Nick and Konishi, 2005b). Second, recordings made in the HVC and LMAN of anesthetized juvenile songbirds reveal that song selectivity is only manifested after the bird begins to sing; prior to this time, auditory responses in HVC and LMAN are typically weaker and nonselective, despite substantial auditory experience of the tutor song (Volman, 1993; Solis and Doupe, 1997). These studies also reveal that even in juvenile birds singing plastic song, the majority of HVC and LMAN neurons are selective for the BOS rather than the tutor song. Although there is one report that some HVC neurons in the awake juvenile zebra finch are tutor song-selective (Nick and Konishi, 2005b), this same study found that by early adulthood, most HVC neurons are BOS-selective. These findings suggest that BOS-selective neurons in the adult do not represent a persistent auditory memory of the tutor, and raise the possibility that the tutor song excites HVC and LMAN neurons in the juvenile because it resembles the BOS in its acoustical structure.

A potential confound to this conclusion is that it is difficult to assess the fidelity of tutor song memory independently of the juvenile's vocal imitation of the tutor. That is, incomplete copying of the tutor song may reflect a deficit in the auditory memory of the tutor song or inaccurate song motor learning. In the former case, BOS-selectivity could reflect a flawed auditory memory of the tutor, rather than a neuronal correlate of the bird's song performance (Solis and Doupe, 1999). One way to test this idea is to artificially maximize the acoustical distance between the bird's own song and the possibly imperfect tutor song memory. In juvenile birds sustaining severe spectral degradation of the song produced by syringeal denervation, Solis and Doupe (1999) found that most LMAN neurons develop a strong selectivity for the distorted BOS over the tutor song by the midpoint of sensorimotor learning ($\sim$ PHD 65). Nonetheless, some neurons responded equally well to the distorted BOS and to the tutor song, raising the possibility that they encoded different BOS and tutor song features. However, the features in the distorted BOS and the tutor song that evoked responses were not characterized and were not necessarily those judged to be dissimilar in the two songs. Thus, it remains plausible that selectivity in LMAN is shaped by the bird's experience of its own song and dual-selective neurons in "dysphonic" birds respond to features common to the BOS and tutor song.

Another idea is that song-selective neurons are initially influenced by experience of the tutor, but that this early experience is overwritten by feedback as the bird sings its own song. According to this model, tutor selectivity is replaced by BOS-selectivity, which then serves as a permanent referent that in the adult serves to maintain a stable vocal output (see Woolley, this volume). Is there any evidence that songs learned early in development leave a lasting imprint on the auditory responses of song system neurons? One way to answer this question is to allow juvenile birds to sequentially learn from two different tutors, then assess whether neuronal selectivity for the early renditions of the BOS, or the tutor model from which it was copied, is maintained following copying from the second tutor. Experiments using this approach show that, at least in LMAN, selectivity does develop in juvenile zebra finches for the initially learned song and the model from which it was copied (Yasaki-Sugiyama and Mooney, 2004). However, intracellular recordings made in adult birds that were tutored sequentially show that all synaptic as well as suprathreshold responsiveness to early versions of the BOS and the first tutor song is lost after the bird copies from the second tutor (Yazaki-Sugiyama and Mooney, 2004). This indicates that song-selective neurons in LMAN track the current learned repertoire, rather than providing a persistent memory of early auditory experience either of the tutor or discarded versions of the BOS.

A remaining possibility is that, following crystallization, BOS-selectivity does become an indelible feature of the song system, perhaps serving to maintain a stable song. If this model is correct, then selective responses for the crystallized song should be maintained even in the face of long-term exposure to distorted feedback. To test this idea, we unilaterally severed the vocal nerve in adult zebra finches singing crystallized songs, then measured auditory selectivity at different postoperative times (Roy and Mooney, 2007). Over the course of one to two weeks, neurons in LMAN, as well as NIf and HVC, could be detected that were selective for the distorted BOS over the pre-nerve-cut, crystallized song. Thus, there is little evidence that BOS-selective neurons furnish a permanent auditory memory against which the bird's song is maintained. Instead, even following song crystallization, song selectivity can shift in the AFP to track the most current renditions of the bird's current vocal performance, regardless of how well they may match either the crystallized BOS or the tutor song.

Although these various observations reinforce the idea that experience of one's own vocalizations is the major factor influencing song selectivity, they do not fully distinguish whether the effects of singing experience are purely feedback driven, or instead arise through auditory–motor interactions. Both HVC and LMAN exhibit song motor-related activity (McCasland, 1987; Hessler and Doupe, 1999a; Solis et al., 2000; Hahnloser et al., 2002) and although auditory responses in both areas can be evoked by song playback (Margoliash and Konishi, 1985; Volman, 1993; Solis and Doupe, 1997, 1999; Nick and Konishi, 2005b), there is little evidence that these responses can be recruited by auditory feedback. Indeed, at least in the zebra finch, the singing-related activity of HVC_X or LMAN neurons is not acutely altered by distorted auditory feedback (Leonardo, 2004; Kozhevnikov and Fee, 2007), suggesting that activity recorded in these areas during singing is predominantly motor-related. An important related observation is that auditory responses of LMAN neurons are depressed in adult birds following chronic experience of distorted feedback (Solis and Doupe, 2000). Such depression of auditory-evoked activity would not be expected if auditory responses were influenced solely by auditory feedback and instead may indicate that responsiveness is determined by the quality of the match between actual and expected feedback. The source of such an expected feedback signal is unknown in the song system, but in other systems, estimates of expected sensory feedback are computed using corollary discharge from the relevant motor system (Sawtell et al., 2005; Poulet and Hedwig, 2006). A better understanding of this issue in the songbird AFP will require resolving the auditory- and motor-related properties of HVC_X neurons, because these neurons are the major source of auditory and motor-related drive to the AFP.

AUDITORY REPRESENTATIONS OF SONG IN HVC NEURONS OF AWAKE BIRDS

Although one might assume that the auditory selectivity of neurons in HVC is shaped at least in part by auditory experience, investigation of the activity of HVC neurons in waking zebra finches has revealed only little or no auditory activity (Nick and Konishi, 2001; Rauske et al., 2003; Cardin and Schmidt, 2004a;

Nick and Konishi, 2005a), with the small amount of detectable activity apparently restricted to only a subset of HVC interneurons (Rauske et al., 2003). In contrast, robust auditory responses have been detected in HVC of awake birds of other species, although these studies relied on multiunit recording methods, and thus the identity of the responsive cells remained uncertain (McCasland and Konishi, 1981; Nealen and Schmidt, 2002, 2006). If only interneurons are responsive in the HVC of the awake songbird, then it is unclear how auditory activity could propagate from HVC to other brain areas and ultimately affect behavior. Furthermore, the observed differences in the state dependence of HVC auditory responsiveness observed across species raise questions as to whether zebra finches are more the rule or the exception in this regard.

To further explore to what extent HVC neurons exhibit auditory activity in the waking state and to investigate the auditory representations of song in the different classes of neurons in HVC, we chose to record from swamp sparrows (J. F. Prather et al., unpublished data). Two primary considerations influenced our choice of swamp sparrows as the subjects for these experiments. First, the repertoire of an individual male swamp sparrow consists of a few song types, with each song type comprising 10 to 20 repetitions of a single, multinote syllable (Marler and Pickert, 1984). The presence of multiple song types allows the investigator to ask whether there are subpopulations of neurons dedicated to single song types. Second, the highly stereotyped trilled structure of swamp sparrow song is advantageous because a single song bout yields many samples of the repeated vocal unit that defines the song type. Together, these features make swamp sparrows ideal subjects in which to address questions such as whether auditory responses are present and how distinct song types are represented in HVC during wakefulness.

We used a lightweight microdrive (Fee and Leonardo, 2001) to record auditory activity in the HVC of the freely behaving male swamp sparrow. A previous study showed that HVC neurons are responsive to auditory stimuli in the urethane-anesthetized swamp sparrow, as in the zebra finch, and are highly selective for individual song types in the bird's repertoire (Mooney et al., 2001). Thus we first asked whether HVC neurons in the awake sparrow respond to auditory stimuli. Following methods developed by Michale Fee's group, we used antidromic stimulation methods to identify different cell types in HVC (Hahnloser et al., 2006). Song playback experiments revealed that interneurons and HVC_X neurons display robust auditory activity in the awake sparrow, while HVC_{RA} neurons lacked auditory activity altogether (Figure 14.2). Further, we found that HVC_X cells, but not interneurons, display remarkably selective auditory activity, responding to only a single song type in the bird's repertoire (Prather et al., 2008). Furthermore, these song type specific responses of HVC_X cells involve temporally precise firing patterns with typically one action potential per syllable (Figure 14.2), similar to that seen in anesthetized birds (Mooney et al., 2001). These results indicate that highly selective auditory representations are manifested in one class of projection neuron in HVC of the awake swamp sparrow, and thus have the potential to influence behavior. Indeed, the AFP has been implicated in song perception (Scharff et al., 1998) and thus it is likely that these auditory HVC_X neurons are important to this role.

We also found that different HVC_X cells respond to playback of different song types, suggesting that the bird's full repertoire is represented across the entire population of HVC_X cells. The contrast in the amount and selectivity of auditory responses that can be observed in the HVC of the awake swamp sparrow and zebra finch are remarkable, and may reflect species differences in the function of song. Swamp sparrows are largely solitary, highly territorial birds that broadcast their songs over long distances to attract females and in response to hearing the songs of conspecific males (Ehrlich et al., 1988). In contrast, zebra finches are colonial nomads that sing at close range to attract females, rather than to defend territory from neighboring males (Zann, 1996). Perhaps the robust auditory activity that can be detected in the HVC of the awake swamp sparrow reflects behavioral demands on territorial birds to be ever-ready to detect and respond to the songs of neighboring conspecifics. Alternatively, auditory activity in HVC neurons of awake swamp sparrows but not zebra finches may reflect some difference in the functionality of HVC associated with learning and maintenance of a more extensive song repertoire, as waking auditory responses are also observed in HVC of other species that express song repertoires, rather than a single song, like the zebra finch (McCasland and Konishi, 1981; Nealen and Schmidt, 2002, 2006).

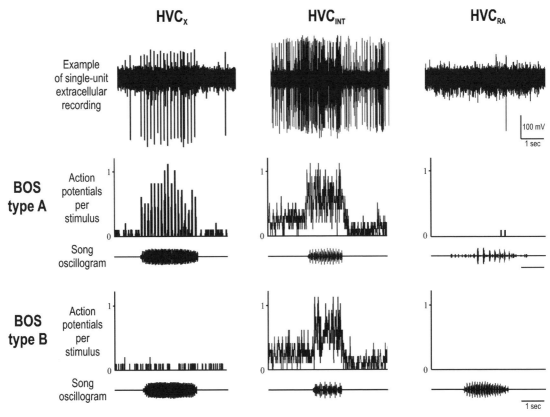

Figure 14.2 Auditory responses in identified HVC neurons of awake, unrestrained swamp sparrows. The strength and selectivity of auditory responses vary across the three classes of HVC neurons. Responses in HVC_X neurons (left column) are reliable and highly phasic, and are evoked by only a single song type in the bird's repertoire. In contrast, HVC_{INT} cells (middle column) show tonic responses to both song types in the bird's repertoire. HVC_{RA} neurons (right column) had no auditory response to any stimulus and were almost entirely inactive in the absence of stimulus presentation (< 0.05 Hz). Each cell type was recorded from a different adult male swamp sparrow. Top panel, example of extracellular recordings from individual neurons of each class, as identified using antidromic stimulation methods (not shown). Middle histogram and associated oscillogram, single-unit response of the neuron to 10 repetitions of one song type in the bird's repertoire (10 ms bin size in all histograms). Bottom histogram and associated oscillogram, single-unit response of the same neuron to 10 repetitions of a second song type in the bird's repertoire.

CONCLUSIONS AND FUTURE DIRECTIONS

The analysis of functional connectivity in the song system reveals that HVC receives at least three sources of auditory input, including NIf, CM and Uva. Of these, at least some of the input from NIf and CM provides BOS-selective information, whereas Uva provides a nonselective auditory input and may also convey other sensory information. Uva receives its auditory input from the brainstem, whereas NIf and CM receive their major auditory input from primary and secondary regions of the auditory telencephalon.

Thus, HVC receives both relatively direct and non-selective auditory input via Uva and more indirect but more selective input from CM and NIf. Although NIf is a dominant source of auditory input to HVC in the anesthetized bird, we still know very little about how NIf, CM and Uva contribute to HVC auditory activity in the awake bird, and whether these inputs are differentially regulated with changing behavioral state or over development. We also know very little about the connectivity in primary and secondary regions of the auditory telencephalon, and the precise nature of the connections these areas make with NIf and HVC. What are the earliest stages in the auditory system where BOS-selectivity can be detected? Is there a dedicated pathway for conveying song-selective information to the song system, possibly involving the minority of highly selective cells that can be detected in primary auditory regions, or does song selectivity develop only in penultimate stages of processing, within or immediately presynaptic to NIf?

Because song is a learned behavior, BOS-selectivity must in some sense reflect the bird's auditory and/or vocal experience. Forms of auditory experience that may be important to BOS-selectivity include the tutor song and auditory feedback. The current view is that BOS-selectivity reflects the bird's experience of its own song, because BOS-selectivity in HVC and the AFP can continue to shift following crystallization, at least when birds are made to sing chronically distorted songs. However, whether this shift in selectivity is driven solely by auditory feedback or by changing patterns of song-motor-related corollary discharge transmitted from HVC to the AFP remains uncertain. The finding in zebra finches that the singing-related activity of HVC_X cells, as well as those in LMAN, is unaffected by distorted auditory feedback suggests that motor-related activity may play a prominent role in shaping BOS-selectivity in the AFP. More generally, the apparent insensitivity of HVC_X and LMAN neurons in the zebra finch to acute distortions in auditory feedback raises the fundamental question of where singing-related auditory feedback is registered in the brain. Thus, an important goal of future research will be to better characterize the singing-related activity of HVC neurons, as well as their inputs from NIf, CM and Uva, in order to identify potential conduits for auditory feedback. Furthermore, our present observation that HVC_X neurons in the awake swamp sparrow respond selectively to playback of certain song types in the bird's repertoire raises the possibility that the auditory properties of these neurons are also activated by singing-related feedback. If so, this would suggest that HVC_X neurons function in different capacities in different songbird species.

Although NIf neurons projecting to HVC and the projection neurons of HVC are equally BOS-selective, NIf projection neurons fire to most non-BOS stimuli, including conspecific songs and white noise bursts, and fire in a sustained fashion to the BOS, whereas HVC PNs fire little or not at all to non-BOS stimuli, and fire in a highly phasic manner to the BOS. Thus, HVC is the site where song-specific and temporally sparse and precise auditory representations of the BOS become dominant. The emergence of this representation involves local circuit interactions in HVC, including inhibitory shaping and priming in HVC_X cells. What could be the functional significance of generating a temporally sparse, highly selective and precise pattern of BOS-evoked activity in HVC_X cells? One idea is that this transformation places the auditory representation of the BOS in the same temporally sparse and precise framework as the motor code used to produce it, which could establish an exact sensorimotor correspondence that facilitates mimicry and perception of communication gestures (Prather et al., 2008).

15 • Temporal auditory pattern recognition in songbirds

Timothy Q. Gentner

Language is uniquely human. Yet, the source of this uniqueness is poorly understood. Some theoretical positions advocate the existence of a small number of uniquely human capacities that form the core of linguistic ability (e.g. Hauser et al., 2002a). Others contend that, like any complex behavior, language is best understood as a confluence of subprocesses and emergent features, some of which may be shared with other nonlinguistic animals (Bates et al., 1998). Empirical validation of either position requires comparative studies that examine the capacities of nonhuman animals in several language-related domains.

Human speech and birdsong share several important features. Both communication systems entail large, acoustically rich repertoires of temporally patterned vocal signals, and both must be learned early in ontogeny (Marler, 1970a, 1975; Kuhl, 2003, 2004). Over the last 20 years, these similarities have helped to establish birdsong as an important model system for understanding the neurobiological bases of vocal learning and production (Doupe and Kuhl, 1999; Margoliash, 2002; Zeigler and Marler, 2004). But communication is not a solitary endeavor. The target of the songbird's elaborate vocal-motor production system is not the song itself, but rather the behavior of nearby conspecifics (Kroodsma and Miller, 1996). In short, songbirds sing to be heard; and the interplay between song perception and behavior provides an excellent model to study the mechanisms that underlie the processing of complex acoustic communication signals (Gentner and Margoliash, 2002; Zeigler and Marler, 2004; Theunissen and Shaevitz, 2006; also see Theunissen et al. and Mello and Jarvis, this volume).

The songs of the prototypical temperate zone songbird modulate aggressive territorial encounters among neighboring males and attract prospective female mates (see Kaplan, this volume). For those on the receiving end of song, these broad (though not exhaustive) evolutionary functions imply a host of auditory perceptual and cognitive processes that must be engaged, from signal detection to decisions based on pattern recognition (Gentner and Margoliash, 2002). While these processes are interesting in their own right, recent advances (Gentner et al., 2006) suggest that some of the acoustic pattern recognition abilities of songbirds may inform current debates on the evolution of language-relevant skills in nonhumans. This chapter describes those recent advances in the context of a larger endeavor to understand the biological basis of temporal pattern recognition in one species of songbird, European starlings.

The chapter begins by discussing the behavioral and neural basis for song recognition in starlings. These studies establish that as adult starlings learn to recognize different songs, a rich comprehension of each song's component structure emerges that is reflected in the birds' behavior and the responses of single neurons in the auditory forebrain. Building upon this understanding of the functional units of song (called motifs), the chapter then describes the abilities of starlings to attend to temporal sequences of motifs, and to arbitrary rules that describe patterns among motifs. Along the way we note what we consider important similarities and differences between birdsong and human speech perception.

STUDYING SONG RECOGNITION IN THE LAB

As any birder can attest, vocalizations are a vital diagnostic cue to species identity. But songbirds can discriminate not only between the songs of their own and other species, but also between those of individual conspecifics. Various forms of individual vocal recognition have been observed in nearly every species of songbird studied to date (Stoddard, 1996) and have been examined more extensively here than in any other group of animals. Knowledge of which individuals sing which

Neuroscience of Birdsong, ed. H. Philip Zeigler and Peter Marler. Published by Cambridge University Press. © Cambridge University Press 2008.

songs serves as the basis for decisions in more elaborate social behaviors such as female preference and choice (e.g. Wiley *et al.*, 1991; Lind *et al.*, 1997; O'Loghlen and Beecher, 1997), kin recognition among communally breeding birds (reviewed by Beecher, 1991), and male territoriality (Peek, 1972; Falls and Brooks, 1975; Falls, 1982; Godard, 1991).

To study individual vocal recognition we devised an operational definition that permitted direct laboratory study, whereby the "meaning" of a song could be controlled experimentally. To do this, we use operant conditioning techniques that require subjects to make one response to the songs of a specific bird and a different response to the songs of one or more other birds (Hulse, 1995; Gentner and Hulse, 1998). Typically, the birds are trained to obtain food by pecking buttons on a panel mounted on the side of their cage (Figure 15.1). They are then reinforced with food for pecking one button, say the left, every time they hear a song from male "A", and for pecking another button, in this case the right, every time they hear a song from male "B". Tasks such as this, in which two sets of stimuli (songs) are associated with similarly reinforced behaviors (peck left/peck right), are called two-alternative choice tasks (2AC). In a close variant, the "go/no-go" (GNG) procedure, behavioral responses to only one set of stimuli are reinforced, leading the subject to cease responding to the nonreinforced stimuli. With both training procedures subjects become proficient at recognizing the songs in each class. Once the basic song recognition is learned we can vary the stimuli in myriad ways to ask questions about the precise acoustic features and associative processes that guide individual song recognition.

Subjects in these experiments were European starlings (*Sturnus vulgaris*). Male starlings tend to sing in long continuous episodes called *bouts*. Song bouts are composed of temporally shorter acoustic units referred to as *motifs* (Adret-Hausberger and Jenkins, 1988; Eens *et al.*, 1991a) (Figure 15.2) that, in turn, are composed of still shorter units called *notes*. Notes can be broadly classified by the presence of continuous energy in their spectrotemporal representations. Although a motif may consist of several notes, the note pattern within a motif is usually stereotyped between successive renditions of that motif. Commonly, each motif is repeated two or more times before the next one is sung. Thus, starling song appears (acoustically) as a sequence of changing

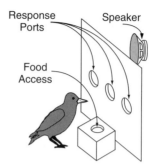

Figure 15.1 Operant apparatus used to demonstrate classification of song stimuli. Subjects start a trial (presentation of a song stimulus selected at random) by pecking the center response port. After the stimulus completes, the subject can either peck at the left or right response ports depending on the class from which the stimulus was drawn (2AC, see text), or peck the center port (GNG; see text). Correct responses yield food reward. Incorrect responses lead to a short "time out" during which the house-light is extinguished and food is inaccessible.

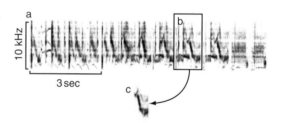

Figure 15.2 Sonogram of starling song segment. Power across the frequency spectrum is shown as a function of time. Darker regions show higher power. Starling song bouts are organized hierarchically. Normal bouts of song can last over a minute, and comprise a series of repeated motifs. A short sequence of motifs (a) as they might appear in a much longer song bout. A single motif is outlined in (b). Complete song bouts contain many different motifs. Motifs comprise stereotyped note patterns. An example of one note is shown in (c).

motifs, where each motif is an acoustically complex event (Figure 15.2). Different motifs can vary in duration from roughly 200 ms to 1000 ms and the number of unique motifs that a mature male starling can sing (i.e. his repertoire size) can exceed 50 or more. Consequently, different song bouts from the same male are not necessarily composed of the same set of motifs. Over time, however, the songs of a specific male can be characterized by a typical set of motifs. Although some sharing of motifs does occur among captive males (Hausberger and Cousillas, 1995; Hausberger, 1997), the motif repertoires of different males living in the wild are generally unique (Adret-Hausberger and Jenkins, 1988; Eens et al., 1989, 1991a; Chaiken et al., 1993; Gentner and Hulse, 1998).

Given that motif repertoires are largely unique between males, learning which males sing which motifs can provide a diagnostic cue for individual recognition. Indeed our data indicate that this strategy is very close to what starlings actually do. Using our operant trainings techniques, we have shown that starlings can easily learn to recognize many songs sung by different individuals, and can maintain this accurate recognition even when tested with novel songs from the same singers (Gentner and Hulse, 1998; Gentner et al., 2000). The question, then, was how do they recognize these novel songs? One strategy is to use idiosyncratic source and/or filter properties of each individual's vocal apparatus, or "voice" characteristics. The use of voice characteristics (e.g. vocal timbre, the frequency of glottal pulsation, and spectral contours imparted by laryngeal morphology) is well documented for individual talker recognition in humans (Bricker and Pruzansky, 1976). To examine the role of "voice characteristics" in song recognition, we trained birds to recognize isolated motifs shared by two different males (e.g. bird A singing motif 1, and bird B singing motif 2), and then watched recognition fall to chance for recordings of the same birds singing the opposite motifs (i.e. bird A singing motif 2, and bird B singing motif 1). These results rule out a critical role for voice characteristics in individual vocal recognition in starlings.

If starlings don't rely on voice characteristics to recognize the songs of individual singers, then the most reasonable (though not the only) hypothesis is that they learn to recognize the songs of other starlings by memorizing sets of motifs sung by each individual. Consistent with this hypothesis, when we tested the recognition of novel song bouts that have *no* motifs in common with the training songs, performance fell to chance (Figure 15.3). If the motif-memorization hypothesis is true, then it should be possible to control song recognition systematically by varying the proportions of motifs in a "target" song that come from two vocally familiar males. That is, if subjects memorize a large set of motifs from each singer, recognition behavior should be correlated with the relative proportions of familiar motifs from different males independent of the specific motifs comprising a given song. If they attend to the presence (or absence) of a single motif or a small set of motifs, recognition should not follow relative motif proportions, and should not generalize between songs in which different motifs make up similar proportions. To test these ideas, we again trained starlings to recognize sets of songs from different individual males, and then watched as subjects classified novel song bouts in which motifs from the training songs were combined in different ways (Gentner and Hulse, 2000a). Consistent with the motif memorization hypothesis, we observed an approximately linear relationship between song classification and the relative proportions of familiar motifs from different singers composing each bout (Figure 15.3).

Taken together, the results of these behavioral studies suggest that when starlings learn to recognize conspecific songs from different singers, they memorize large numbers of unique motifs corresponding to each individual singer. From a human perspective, this might seem a suboptimal strategy to solve the problem of individual vocal recognition. Indeed the rich semantic content of human language necessitates that most words are shared among speakers, and so precludes a similar strategy. Instead, the voice characteristics used in speaker recognition appear to be coded in acoustic parameters of the signal that are predominantly nonlinguistic (Remez et al., 1997). It is not clear whether similarly independent communication channels exist in any nonhuman. If such independence is rare, a vocal communication signal, such as that of starlings, that codes individual identity and other semantic information using variation along the same acoustic dimensions may represent the ancestral condition from which language evolved. In any case, for starlings, their solution to vocal recognition is functionally parsimonious. Under natural conditions individual starlings possess unique motif repertoires. Thus, sets of motifs sorted into disjoint perceptual classes or categories, although memory intensive, will correspond to individual identity.

NEURAL CORRELATES TO SONG RECOGNITION

The large-scale architecture and pattern of connectivity within the starling auditory forebrain is shared with other songbird species (Vates et al., 1996) and with

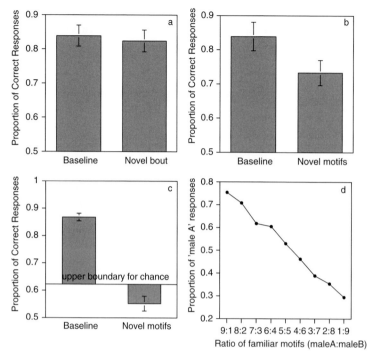

Figure 15.3 Vocal recognition behavior in European starlings. (a) Mean (± SEM) proportion of correct responses given during asymptotic performance on an operant recognition task ("Baseline"), and during initial transfer to novel songs containing familiar motifs ("Novel bout"). (b and c) Mean (± SEM) proportion of correct responses during transfer from the baseline training to novel songs from the same singers composed of "Novel motifs"; when (b) subjects were exposed to the training and test songs outside of the operant apparatus, or (c) after controlling for all previous song experience where recognition falls below chance. (d) Data showing the close (and approximately linear) relationship between the proportions of familiar motifs from two different singers in a song and individual vocal recognition.

vertebrates in general (Carr, 1992). The Field L complex is the primary telencephalic target for auditory information arriving via parallel pathways from the thalamus. Further within the ascending sensory hierarchy, NCM and the caudal mesopallial structures (CLM, CMM) are analogous to secondary auditory cortices in mammals. Neurons throughout the starling auditory telencephalon show complex patterns of tonotopic organization (Leppelsack and Schwartzkopff, 1972; Rübsamen and Dörrscheidt, 1986; Capsius and Leppelsack, 1996; Haüsler, 1996), including selectivity to species-specific vocalizations (Leppelsack and Vogt, 1976; Bonke et al., 1979b; Muller and Leppelsack, 1985; see also Theunissen and Doupe, 1998; Theunissen et al., 2004 and this volume). In other songbirds, the general pattern of increasing response selectivity along the sensory hierarchy (Hsu et al., 2004a; Woolley et al., 2005) continues into NCM and CM (Sen et al., 2001) suggesting that these regions are involved in the extraction of complex features (Leppelsack, 1983; Chew et al., 1995, 1996a; Stripling et al., 1997; Sen et al., 2001; Grace et al., 2003). Additional support for the role of NCM and CM in the processing of conspecific song comes from studies of stimulus-driven expression of the immediate early gene (IEG) *zenk*, a putative marker for song-induced experience-dependent plasticity (Mello et al., 1992; Jarvis et al., 1995; Mello and Clayton, 1995; Ribeiro et al., 1998; Jones et al., 2001; Mello et al., 2004). In starlings, the IEG response in NCM appears tied to stimulus novelty, whereas IEG activity in CMM correlates with novelty and the ongoing recognition of familiar songs (Gentner et al., 2004). Finally, lesions to CMM in both zebra finches (MacDougall-Shackleton et al., 1998a) and starling (Gentner, unpublished) elicit specific deficits in song-based behaviors.

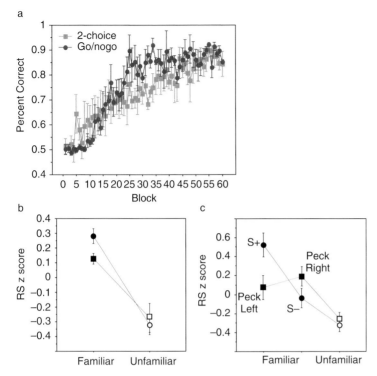

Figure 15.4 CMM behavior and physiology. (a) Acquisition curves showing mean performance (as the proportion of correct responses) over the first 60 blocks of recognition training (100 trials/block) with GNG and 2AC procedures. (b) Mean normalized (z-scores) response strengths of CCM single units to familiar and unfamiliar songs, split by training regime (2AC, squares; go/no-go, circles). (b) RS z-scores as in (a) but with the responses associated with each of the two sets of training stimuli (2AC, S+ left, S+ right; GNG, S+, S−) shown separately.

To examine the role of CMM in the representation of learned conspecific song in adult birds, we trained starlings to recognize two sets of conspecific songs using both two-alternative choice (2AC) and go/no-go (GNG) operant procedures (Figure 15.4). Following acquisition, we anesthetized each subject with urethane and recorded the extracellular responses of single neurons in the CMM to an ensemble of acoustic stimuli, including songs used during operant recognition training ("familiar" songs) and novel conspecific songs ("unfamiliar" songs). To control for biases to any specific song, we varied the stimuli across subjects so that while the stimulus ensemble was similar for each subject, the familiarity of each song differed (Gentner and Margoliash, 2003).

As a population, CMM neurons responded selectively with a significantly larger increase in firing rate to the class of familiar compared with unfamiliar songs. Moreover, the strong response bias for familiar songs was consistent in animals trained under both the two-alternative choice and go/no-go operant regimes (Figure 15.4). Since all the animals were tested with the same sets of song stimuli, the only difference between birds was their relationship to each song. A song familiar to one bird was unfamiliar to another. Therefore, the differential responding to the familiar and unfamiliar songs observed across birds cannot reflect trivial acoustic differences between the songs. Instead, the different responses to the familiar and unfamiliar songs appears to be the result of a broad representational plasticity that coincides with learning.

This apparent plasticity in CMM could be driven solely by exposure to patterns of acoustic variation in the stimulus such that, in the extreme, all the information represented by the cell's response is present in the acoustic variation of the signal. Alternatively, attention, motivation and/or reward mechanisms might also help to shape a CMM cell's selectivity in a manner independent of any particular signal acoustics. Consistent with the role of associative learning in individual song recognition, a significant portion of the spiking activity in CMM neurons appears to be influenced by different reinforcement signals. Whereas the subjects trained using the two-alternative choice procedure showed no reliable difference between response strengths associated with the two sets of training songs, those trained with the S+ and S− classes in the go/no-go procedure did. That is, songs associated with positive reinforcement (S+ stimuli) elicited significantly stronger responses than those associated with no reinforcement (S− stimuli; Figure 15.4). This suggests that both "bottom-up" activation driven by stimulus acoustics and "top-down" modulation mechanisms, perhaps through reinforcement, act to shape the responses of CMM neurons during song recognition learning.

Response selectivity

The strong population response bias for familiar songs indicates that CMM neurons do not respond equally to all songs. In fact, roughly 64% of the cells in our sample of CMM gave a selective response to one of the test stimuli (Gentner and Margoliash, 2003); and of these cells, almost all (93%) preferred one of the training songs (Figure 15.4). For nonselective cells, the song that elicited the strongest response was not significantly more likely to be either familiar or unfamiliar.

Motif selectivity

For many of the song-selective cells, responses were restricted to one or a small number of repeated motifs within one or a few songs, typically with suppression of background activity for all other motifs. These responses appear to be driven by acoustic variation at the level of the motif (Gentner and Margoliash, 2003), responding on average to about 8 motifs, whereas nonselective cells respond to significantly more motifs on average (roughly 20). The fact that the "song" selectivity described in this population of CMM neurons is derived from selective tuning for spectrotemporally complex features centered at the level of the motif closely matches the results of behavioral work described above showing the importance of variation at the level of the motif for individual song recognition.

On a broader scale, we now recognize that experience-dependent plasticity guides many forms of learning (e.g. Bakin and Weinberger, 1990; Kay and Laurent, 1999; Gilbert *et al.*, 2001; Kilgard, 2003) across a variety of vertebrate sensory systems and brain regions, including the primary sensory cortices once thought to provide static stimulus representations (Schoups *et al.*, 2001; Nagel and Doupe, 2006). The data from starlings suggest that the plasticity mechanisms engaged during song recognition learning give rise, perhaps through hierarchical processing (Sen *et al.*, 2001), to selective neural representations for acoustic features diagnostic of individual (or small sets of) motifs. More simply, songbirds build specific and idiosyncratic representations for the functionally relevant units of their communication signals. These observations suggest a system that is simultaneously constrained in its immediate representational capacities by each animal's history, yet is tremendously adaptive in its ability to acquire a broad (currently undetermined) range of complex representations. Based on these observations, it is tempting to speculate that similar experience-dependent, hierarchical, processes may give rise to the phonetic, phonemic, syllabic and word-level representations that must be correlated with language experience in humans. Unfortunately, a cellular-level understanding of speech representation in the human auditory system is not yet possible. It is our hope that by the time such experiments do become empirically tractable, the birdsong system will be sufficiently advanced so as to provide a set of testable hypotheses that can be used to examine the similarities and differences (both qualitative and quantitative) across these representational systems.

SENSITIVITY TO TEMPORAL MOTIF PATTERNS

The prior sections on individual song recognition in starlings establish the close correspondence between neural representation and behaviorally relevant variation in song. This correspondence highlights the functional importance of the motif in song organization. Under normal conditions, however, motifs are almost never produced in isolation but typically occur as part of long and elaborate song bouts in which 25–30 (or even more) different motifs may be strung together in close succession. In the following sections, we examine what starlings know about the temporal patterning of motifs within song bouts. The results suggest a surprising sophistication in their abilities.

Motif sequences

Perhaps the simplest notion regarding sensitivity to temporal patterning is that it doesn't matter for individual vocal recognition. We have described above how starlings learn to recognize songs by attending to the familiar motifs within a bout, and so perhaps the order in which these motifs appear (are sung) is not important to the receiver. In contrast, a bird might learn to recognize the motifs "a", "b", "c" and "d" (where letters denote different motifs) *and* the sequence where "d" follows "c" which follows "b" which follows "a". If so, then presenting the same motifs in a different sequence may affect song recognition.

To test this idea, we trained starlings to recognize normally patterned songs from different singers using a 2AC operant procedure. We then had the birds classify synthetic motif sequences in which the sequencing of now familiar motifs was varied systematically (Gentner *et al*., 1998). If starlings are sensitive to the temporal sequencing of motifs in a song, then randomizing the learned sequence of motifs (or altering it in other ways) should affect recognition. That is exactly what we found. The randomly ordered song bouts were significantly more difficult to recognize than those that followed the natural transition probabilities in each male's songs (Figure 15.5).

It is not difficult to imagine how such an ability to attend to motif sequences might be advantageous. Even if what starlings are trying to do is keep track of single recognizable motifs, as suggested by the data in the first part of the chapter, then knowing something about which motifs are likely to occur will aid the recognition process by limiting the number of potential matches the listener must find for any given motif. That is, if the to-be-recognized male has just sung motif "b", it is useful to know that the next motif is likely to be "c", "d" or "e". Thus, information about explicit motif sequencing, while not the dominant cue, might make each motif a kind of "prime" or cue for recognition of the motif that follows and thereby contribute to individual vocal recognition of normal starling songs. It is interesting to note that human infants attend to the transition probabilities between speech sounds from a very

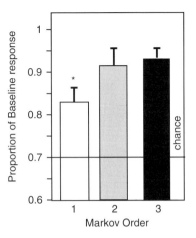

Figure 15.5 Sensitivity to motif sequence. Performance during partial transfer sessions with synthetic songs that varied sequence reliability, normalized for each individual according to their baseline performance. Markov order 1 corresponds to randomly sequenced motifs, and is significantly lower than the other types of sequences that follow the natural ordering of song motifs (Markov order 2 and 3).

young age (e.g. Saffran *et al*., 1996), and this "statistical learning" is hypothesized to play an important role in language learning. Our data suggest that the general mechanisms for these auditory learning abilities may have a long evolutionary history.

Motif patterns

In humans, particularly for speech, temporal pattern perception goes well beyond the sensitivity to transition probabilities between sounds. In an effort to probe the upper bound of starlings' auditory pattern recognition, we have begun a second line of research that moves beyond transition probabilities between adjacent motifs to ask if starlings can acquire abstract rules that describe the patterning of familiar motifs (Gentner *et al*., 2006). To understand the distinction between sequence and patterning rule it is helpful to consider an example. Imagine that a starling learns to give a response every time it hears the motif sequence *ab* (i.e. where motif *b* follows motif *a*) and the same response when it hears the sequence *cd*. This same pair of sequences could also be described by a rule that says "the first element of the pair is always *a* or *c*, and the second element is always *b* or *d*." A bird that learns the rule for our simple example should have no

194 TIMOTHY Q. GENTNER

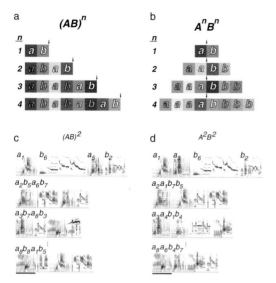

Figure 15.6 Grammatically generated motif patterns. Both the (a) FSG and (b) CFG generate relatively simple temporally patterned sequences of elements (lowercase letters) of the sets denoted by "A" and "B". Increasingly longer strings of the form $(AB)^n$, where n gives the number of AB iterations, are produced by rules that append elements to the end of a shorter, $n-1$ order, string. An equivalent set of rules could append elements to the start of the $n-1$ string. In contrast, increasingly longer strings with the form $A^n B^n$ are produced by rules that embed elements into the center of an $n-1$ sequence. Sonograms (frequency range: 0.2–10.0 kHz; scale bars = 1 s) showing four of the eight sequences constructed from (c) the finite-state grammar, $(AB)^n$, and (d) the context-free grammar, $A^n B^n$, used in the initial FSG versus CFG pattern classification training with $n = 2$.

problem giving the same learned response to the sequences *ad* and *cb*, whereas one that has memorized the explicit sequences *ab* and *cd* should stumble on *ad* and *cb*. Why might anyone (or any bird) even bother to extract a rule? Well in our simple example, where the number of possible sequences is only four, the advantage may not be clear. Add the possibility of a few more elements in each position, however, and the number of possible sequences gets very large very fast, while the rule remains comparatively simple. In this way, the use of a rule allows for the identification of patterns over a large corpus of sounds (or any other set of discrete objects). For humans, these patterning rules provide a basis for the comprehension of novel utterances in one's language and hence the communication of new ideas – one of our most powerful tools. The comparative study of pattern rule learning, therefore, has important ramifications for understanding the evolution, uniqueness, and biological mechanisms of language.

To examine pattern rule learning in starlings, we need to be more precise about the sorts of rules we test, about their relationship to one another, and ultimately to the kinds of rules that underlie human language. These types of patterning rules have a rich history in the mathematical theory of formal grammars, about which it is helpful to know a little as we proceed. In short, a formal grammar is a collection of rules that produce patterned strings from some set of elements – just like the simple rule above that described strings of *a*s, *b*s, *c*s, and *d*s. Not all grammars are equal, however. Some can only generate relatively simple patterns whereas others can generate sequences with much richer complexity and structure. To capture these differences in generative capacity, grammars can be organized into a strict hierarchy (Chomsky, 1957; Hopcroft and Ullman, 1979). At the lowest level in the hierarchy are the so-called finite-state grammars (FSG; Figure 15.6). These are the most limited types of patterning rules, and have traditionally been thought to describe all animal communication systems (but see Suzuki *et al.*, 2006).

While finite-state grammars are capable of generating many patterns, they cannot capture one important feature that occurs in many human languages. A finite-state grammar cannot support "recursive embedding," or "recursion." To understand what is meant by recursion, consider the following. In the course of a conversation I might say, "The students are smart", and then to avoid any confusion say, "The students the university admits are smart." Both statements are perfectly understandable and all I've done is to insert the phrase "university admits" into the initial sentence "The students are smart." This insertion into the middle of an otherwise grammatical sentence is the essence of "recursion." The capacity to generate a sequence, or in this case a sentence, that has this recursive structure requires something more powerful than a finite-state grammar, it requires a context-free grammar (CFG; Chomsky, 1957; Hopcroft and Ullman, 1979).

By itself, and in its most abstract sense, recursion captures just one of many possible ways that the elements in a patterned sequence might be organized. To recognize recursion within a pattern, however, requires specific computational devices (Hopcroft and Ullman,

1979), and thus perhaps specialized cognitive mechanisms. Moreover, because recursive structures appear in many human languages, the capacity for recursion is often seen as a strong requirement of any formal theory of syntax (Chomsky, 1957). Indeed, some prominent linguistic theories go so far as to hypothesize that the capacity for recursion is one of, if not the only, unique component to human syntactic abilities (Hauser *et al.*, 2002a; Fitch *et al.*, 2005).

There is of course much more to the theory of formal grammars than briefly outlined here. For our purposes, however, it is enough to concentrate on the contrast between finite-state grammars and context-free grammars in their ability to generate recursive sequences. Importantly, such sequences are not confined to human speech. Although it is helpful to think about these sorts of patterning rules in the context of human language, there is no reason why they cannot be used to produce patterned sequences from any set of discrete elements (e.g. Ramus *et al.*, 2000; Fitch and Hauser, 2004; Newport *et al.*, 2004; Fitch *et al.*, 2005) including birdsongs. Our goals in examining the abilities of songbirds to extract these sorts of patterns from motif sequences are twofold. First, we want to determine if *any* patterning rule, including relatively simple finite-state rules, can be extracted from a set of stimuli. Second, we want to know if the ability to extract such rules extends to *any* context-free patterning rule, as this must be a prerequisite to the recursive syntactic processing capacities proposed by some to be uniquely human (Hauser *et al.*, 2002a).

Recently, we have shown that European starlings (*Sturnus vulgaris*) can indeed learn to classify song motif patterns generated by both a finite-state grammar (FSG) and a context-free grammar (CFG). Both grammars generated patterned songs sequences from the same set of eight "rattle" and eight "warble" motifs. These two types of starling song motifs are acoustically distinct, and starlings can easily learn to tell the difference between individual rattles and warbles (unpublished data). For the ease of notation we label the eight rattle motifs as $a_1 \ldots a_8$ and the eight warble motifs as $b_1 \ldots b_8$. Our CFG generated strings of the form A^2B^2, while the FSG generated strings of the form $(AB)^2$ (Figure 15.6), where A and B refer to the rattles and warbles, and the superscript (^{2}) denotes the number of times motifs from each class are repeated. Thus, an example string generated by the CFG might go $rattle_3 - rattle_1 - warble_6 - warble_8$, while one generated by the FSG (using the same subset of motifs) might go $rattle_1 - warble_8 - rattle_3 - warble_6$.

Using a go/no-go operant conditioning procedure we trained European starlings to classify subsets of sequences from each grammar (Gentner *et al.*, 2006). Although many more have since been trained (Gentner, unpublished), 9 of the original 11 starlings eventually learned to classify the FSG and CFG sequences accurately (Figure 15.7). The most trivial strategy that the birds might use to solve this difficult classification task would be to simply memorize the explicit training stimuli. To rule out rote memorization of the CFG and FSG patterns, we transferred some of the subjects from the baseline training stimuli to a novel subset of sequences from the same two grammars (A^2B^2 and $(AB)^2$). If subjects had memorized the training motif patterns rather than learning a rule, then recognition of the novel sequences should fail. Subjects correctly classified the novel CFG and FSG sequences (Figure 15.7), however, suggesting that they had acquired some general knowledge about the two patterns. Given that the same elements (motifs) composed the sequences in each class, this knowledge must be related to the patterning of elements by each grammar. This result does not tell us, however, how the birds have solved the discrimination. To find out about the patterns the starlings learned, we need to conduct several additional tests.

Classification of novel grammatical and agrammatical sequences

One strategy to classify the two patterns correctly is to learn only the finite-state patterning rule and treat the patterns generated by the CFG as the complement set. Thus, the sequence in question on any given trial is either generated by the FSG or not. Indeed, there is ample precedent that animals can learn finite-state patterns (Hauser *et al.*, 2001, 2002b; Fitch and Hauser, 2004). To examine this possible solution strategy, we constructed a set of motif sequences based on four different agrammatical patterns (*AAAA*, *BBBB*, *ABBA*, *BAAB*; using the same rattle and warble motifs as for the FSG and CFG training patterns) and presented them as probe stimuli along with novel grammatical sequences. If the subjects had learned only the finite-state pattern, then they should treat the agrammatical stimuli the same as the novel CFG sequences.

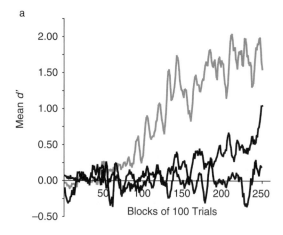

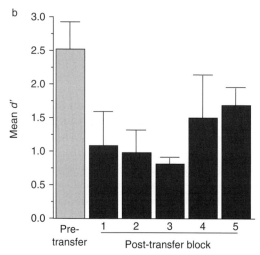

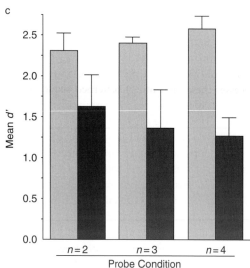

Instead, the subjects treated the agrammatical sequences differently from both the novel CFG sequences and the novel FSG sequences, indicating that they recognized some difference between the agrammatical stimuli and both patterns. That is, subjects learned both the CFG and the FSG patterns rather than only one or the other. Consistent with their general grasp of both patterning rules acquired during training, correct recognition of grammatical patterns extended to instances in which the different rules were used to generate "grammatical" patterns longer than the training stimuli (i.e. A^3B^3, $(AB)^3$, and A^4B^4, $(AB)^4$; Figure 15.7). Although time and memory capacity must constrain the functional length of any grammatical string, part of the power of a formal grammar is its capacity to describe strings of arbitrary length. This last set of results suggests that such generalizations, at least across some pattern recognition domains, may emerge even from relatively impoverished learning environments.

Alternate solution strategies

There are, of course, a host of additional strategies that might permit starlings to solve the pattern recognition task we have posed to them and account for the data as described so far. For instance, they might attend to either the initial or terminal pair of elements in a sequence, count the number of transitions between As and Bs, or use a host of other strategies. These possibilities are very important to consider. Rather than review here the long list of explicit control tests and stimulus comparisons that rule out these alternate solution strategies the curious reader is directed to a full account in the published literature (Gentner et al., 2006). These controls include a strong test ruling out the possibility that starlings learned a finite-state

Figure 15.7 Pattern recognition. (a) Acquisition curves for the baseline FSG/CFG classification, showing mean d' over the first 250 blocks (100 trials/block) for birds that learned quickly and were subjected to further testing (gray), and birds that learned slowly or not at all (black). (b) Mean d' just prior to and over the first 100 trials after transfer to novel motif sequences in the FSG and CFG training patterns, showing significant discrimination. (c) Subjects correctly classified novel A^nB^n and $(AB)^n$ sequences when $n = 2, 3,$ and 4 during probe sessions (see text).

approximation to our context-free grammar of the form $A^n B^m$, where $m \neq n$, and that indicate subjects had indeed learned $A^n B^n$. Although it is theoretically impossible to rule out all possible finite-state solution strategies that might describe the observed classification of the $A^n B^n$ patterns, those that remain following our extensive controls would need to be heavily contrived and are largely implausible (Gentner et al., 2006). On the whole, the data demonstrate that at least one species of animal besides humans is capable of acquiring a context-free (or functionally equivalent) patterning rule.

PATTERN RECOGNITION SUMMARY

The fact that starlings appear able to recognize patterns in conspecific song that are described by formal grammars, including at least one that entails recursive embedding, bears directly on our understanding of what makes human language unique. It has been hypothesized that recursion forms the computational core of a uniquely human narrow faculty for language (Hauser et al., 2002a; Fitch et al., 2005). The abilities of starlings to recognize $A^n B^n$ patterns indicate that at least a basic form of recursive syntactic pattern processing is shared between humans and other animals.

Of course, the $A^n B^n$ pattern that starlings can learn is only one of several possible context-free patterns that entail recursion (Chomsky, 1957), and there may well be more complex recursive context-free languages that starlings cannot learn to recognize (Gentner et al., 2006). Indeed, compared with natural (human) languages, the FSG and CFG patterns learned by the starlings are quite simple. In our earlier example (*The students the university admits are smart*) the recursive embedding also carries with it the requirement that each specific noun is tied to a specific verb. The $A^n B^n$ patterns used here lack these nested dependencies, and it is not known whether starlings could learn them. Therefore, to be clear, the abilities of starlings to recognize $A^n B^n$ do not, as some popular press accounts suggest, show that starlings can represent or learn grammars that are anything like human languages, nor do they suggest that starlings can learn *any* context-free grammar. What we have done in finding that starlings can recognize at least a simple form of recursion is to move the bar of human uniqueness a notch higher. If humans do possess a unique ability to process recursive syntactic structures (and that remains an open question) it must necessarily apply to only a subset of all the context-free grammars.

The development of algorithms with the capacities to learn CFGs that can approximate natural-language syntax is still very much an open research area in computational linguistics, and the empirical evidence that might begin to show how the actual computational mechanisms are instantiated at the neural level is sorely lacking. The research on temporal pattern recognition in starlings, while interesting in its own right, holds the potential to contribute to our understanding at the neurobiological level of pattern recognition abilities shared among species.

CONCLUSIONS: PATTERN RECOGNITION MECHANISMS AND THE EVOLUTION OF LANGUAGE?

The ultimate goal of this work is to move beyond the initial phenomenological descriptions of behavior, and to uncover the underlying neural mechanisms that control the representation and processing of complex temporally patterned vocal communication signals. To that end we have laid a strong foundation that builds up from a basic description of how the functional units of starling song, motifs, are represented in the responses of single cells. Although our understanding of the representation of song motifs is only in its earliest stages, one can easily imagine how such representations might converge at higher-level neurons or change across a population of neurons to give rise to even longer temporal patterns of the sort that we know starlings can recognize. Understanding how the system processes these (and other) patterned sequences will require a broader understanding of the projection patterns and systems-level organization of high-level auditory forebrain regions, and the coordinated responses among large populations of neurons throughout these regions.

Together with the recent observation of neurons selective for learned sequences of motifs in starlings (Gentner and Margoliash, 2003), these results open the way to a comparative exploration of the physiological and cellular-level brain mechanisms for at least some forms of syntactic processing. What might that mechanism for temporal pattern recognition and simple recursion look like? From a computational standpoint, the recursion entailed by grammars such as $A^n B^n$ are actually fairly modest. The system needs to be able to count and recognize different items (in our case motifs) as they occur in time, and it needs a memory so that it can compare the number of As in the first part of the sequence with the current number of Bs (Hopcroft and Ullman, 1979). There is evidence that animals already possess such capacities. They have memories for acoustic objects (e.g. Gentner and Margoliash, 2003), can achieve simple counting (Brannon and Terrace, 1998), and have single neurons with acoustic response properties that are sensitive to the number of distinct elements in a sequence (Edwards et al., 2002). Moreover, relatively simple neural networks that can learn $A^n B^n$ patterns have been described (Rodriguez, 2001), and these might provide theoretical neural architectures that are now open to empirical study.

Given the selective advantages with which language is thought to have imbued humans, it is tempting to believe that all of the underlying abilities and processes carry some portion of that advantage with them. This need not be, however. For if the computational and anatomical constraints on the precursors to language are as modest as those for learning $A^n B^n$ then the presence of such abilities may not require an adaptive explanation. Although a role for syntactic rules has been suggested in structuring chickadee and some wren songs (Hailman and Ficken, 1986; Holland et al., 2000; Clucas et al., 2004), the existence of complex patterning rules in songbird vocalizations has not been broadly observed across songbirds. Yet starlings, and perhaps other songbirds, can recognize surprisingly complex syntactic patterns. This apparent lack of function suggests these basic abilities may be evolutionary "spandrels", or phenotypic characteristics that evolved as a side effect of a true adaptation (Gould and Lewontin, 1979). As such, these basic forms of pattern recognition may come "for free" (i.e. at no functional cost) with the elaboration of more general cognitive skills. Basic recognition, classification, and memory capacities are likely under strong selection pressures (Gentner and Margoliash, 2002), for reasons unrelated to syntactic pattern recognition, in species such as vocal learners that produce long temporally arrayed strings of component signals. In this scenario, linguistically primitive pattern recognition abilities need to be paired with more sophisticated cognitive or vocal production systems as a permissive step to the subsequent evolution of language in its modern forms; and the emergence of language marks the unique confluence of several more generalized (i.e. shared) cognitive abilities. Thus, in the pursuit of understanding the neurobiology of acoustic communication systems, including human speech and language, it may prove more useful to consider many species differences as quantitative rather than qualitative distinctions in cognitive mechanisms.

PART IV
Learning the song: mechanisms of acquisition and maintenance

16 • Introduction

Peter Marler

Any account of the natural history of birdsong makes clear the extraordinary range of variation, from the basic trill of the grasshopper warbler to the orchestrated splendor of thrush song. Yet this great diversity is underlain by a set of unique features that all songbirds share. They all have the same elaborate syringeal apparatus, the primary taxonomic criterion for classification as an oscine songbird. During ontogeny we find that when song develops in a young oscine, it does not begin as a simplified version of adult song as in most birds. Instead it starts as the soft, highly variable vocalization, subsong, quite different from the mature song that will eventually emerge. There are species differences in subsong but, as Hultsch and Todt have graphically displayed in the next chapter, subsong always has a kind of amorphous experimental quality, as the bird first conquers the difficult task of using the two-voiced oscine syrinx as a vocal instrument, becoming more orderly in plastic song and eventually stabilizing in crystallized song. Aspects of the actual song learning process are shared by all songbirds. Song acquisition from a tutor generally precedes production, often with a phase of maximal sensitivity, when the rate of acquisition can be extraordinarily rapid. For example, a nightingale learns something from only five exposures to a sequence of tutor songs, and by 20 exposures can remember both the structure of individual songs in a long string, and also the order in which 50 or more song types occur, surely one of the most remarkable memory-based achievements of any animal. Interestingly they employ strategies not unlike some of those we employ in learning long strings of words (Hultsch and Todt, this volume).

The process by which a bird builds on the first vocal experiments in subsong to achieve mature singing has always been somewhat mysterious, in part because its amorphous structure has made analysis difficult. The chapter by Saar, Mitra, Deregnaucourt and Tchernichovski describes new and powerful computer-based techniques that make it possible for the first time to map moment-to-moment changes, even in the most variable vocal sequences. Their work has already provided a more quantitative picture of how a syllable type gradually emerges and converges on the learned target than was possible with even the most sophisticated qualitative methods. Especially interesting are the hints that birds may pass through intermediate stages that are a kind of ontogenetic shortcut to the production of species-specific song, as though even learned songs have innate underpinnings.

More than 40 years ago Konishi showed that auditory feedback plays a crucial role in converting a learned song "template" into a song imitation. But for a long time it was thought that auditory feedback becomes less important, and perhaps even unnecessary, once adult song has crystallized. It was tempting to assume that something like an "endogenous motor tape" had taken over. As Woolley (this volume) shows, this view is mistaken. We were misled in part by species differences in the stability of so-called crystallized song, and also because, as Nordeen and Nordeen (this volume) have shown, we did not wait long enough for effects of feedback perturbation to emerge. Exploiting the predictable regrowth of hair cells in the avian cochlea after damage by antibiotics, Woolley has derived new insights by disrupting auditory feedback at controlled sound frequencies for varying periods. After perturbation by this method, or by delayed auditory feedback, song syllables and syntax become more variable, and birds stutter. Although some developmental plasticity is reintroduced, the most striking consequence of restoring normal feedback is the almost complete spontaneous recovery of the original song. This process of song reinstatement from memory provides Woolley with a valuable preparation for exploring the neurobiology of one of the information-storage systems involved in song learning.

Brainard's chapter reviews the current status of research on the anterior frontal pathway of songbirds.

Neuroscience of Birdsong, ed. H. Philip Zeigler and Peter Marler. Published by Cambridge University Press. © Cambridge University Press 2008.

We have known for 20 years that this circuit plays an important role during the early phases of vocal development. By complex processes that are still not fully understood a delicate balance is maintained, apparently by interplay between Area X and LMAN, between achievement of a stable learned song, and the exploratory benefits of maintaining a certain degree of variability in the motor patterns of subsong and plastic song.

Efforts to understand how the bird monitors its own song production by auditory feedback have encountered many complications. For example, Brainard discusses two competing models of the role of the AFP that have emerged, one instructive, the other permissive. The former assumes that feedback signals guide adaptive changes in the motor pathway, aiding progress towards a match with the template: the changes might be directed by the feedback signals, or more exploratory in nature, freed by deviation from the desired target to become more variable. The alternative, permissive model assumes that the AFP has a kind of trophic effect on the motor pathway, either augmenting or reducing its potential plasticity, without necessarily influencing the direction of change. Brainard shows that we still cannot fully distinguish between these two models. Indeed, both may be valid, each more or less relevant to different phases of the developmental process. Brainard argues that permissive factors may even have an effect after song crystallization, perhaps declining as mature singing becomes more stable. This is one of many complications with which birdsong researchers must now deal.

Song memorization is the first stage in the process of learning from a tutor. As Bolhuis (this volume) has made clear, to understand the neural basis of this first step it is necessary to look to brain regions afferent to the classical "song system," especially NCM and CMM (see also chapters by Mello and Jarvis, and by Theunissen *et al.*, this volume) Intensive research has revealed much about their contributions, although potential confusion between memory traces from song stimulation on the one hand and auditory representations of the bird's own song on the other has complicated the search, as has the state dependence of some of the data recorded, differing strikingly in birds that are anesthetized or awake, and singing or silent (see Prather and Mooney, this volume). Nevertheless recent electrophysiological experiments by Vicario and colleagues using devocalized birds have identified in NCM long-term consequences of early tutoring that, although weak, suggest that this is one candidate location for the auditory template that guides the song learning process (Phan *et al.*, 2006).

The remarkable rapidity with which a young bird commits a tutor song to memory invites exploration of the neural basis of this process of sensory acquisition and the possibility that pharmacologically distinctive mechanisms of synaptic plasticity are involved. Focusing especially on processes of long-term potentiation, Kathy and Ernest Nordeen review developmental changes in glutamate receptor physiology in certain parts of the song system, especially LMAN and Area X, that are clearly implicated in the process of song acquisition. Interpretation of their data is challenging and complex, especially since, as the Nordeens make clear, developmental changes occurring simultaneously both within the anterior forebrain pathway and in its afferent auditory circuits must be taken into account. Yet the fact that pharmacological interventions have the predicted impact on the timing of song acquisition provides reassurance that this approach to understanding song learning mechanisms has great promise.

The presumption that birds are able to memorize songs they hear is basic to virtually all research on song learning. But this ability to memorize sounds heard has other functions as well. Adret (this volume) makes a valuable distinction between recognition memory and production memory. The first is the basis by which birds are able to identify companions by acoustic means: although it is little studied we can presume that it operates throughout life. Production memory, on the other hand, establishes the auditory templates in the brain that guide vocal development and maintain the stability of mature song. It operates in a more limited timescale, in particular phases of the life cycle, with its own set of physiological parameters.

Adret provides a thoughtful overview of the history of the template concept. This emerged originally from Konishi's deafening experiments, showing that a bird must hear its own voice if the acquired memory trace, or "template", of a learned song is to yield a vocal imitation. The notion of a memory trace might itself have been sufficient, but for the additional complication that the song of a bird raised in social isolation also seems to be template-dependent. This "innate" song contains normal features that fail to develop in naïve birds that have been deafened; the "innate auditory template" that they seem to possess has been invoked as the basis for the species-selective property of memorization of songs for production that Thorpe described in the chaffinch almost 50 years ago. Subsequent experiments on the selectivity of song learning in different sparrow species leave us in no doubt that the establishment of production memories is not a fully open process, but is subject to innate constraints that vary from species to species.

Thus the interplay between nature and nurture is manifest even in learned birdsongs, a class of behaviors in which nurture would at first sight appear to play a completely dominant role. As Adret makes clear, the establishment of an internalized representation of an acquired song and its transformation into a motor output is no less subject to genetic constraints than behaviors regarded as innate. It follows that we can expect the new genetic methodologies now becoming available to yield new insights into how templates are put to use in the process of learning to sing (see Clayton and Arnold, this volume).

17 • Comparative aspects of song learning

Henrike Hultsch and Dietmar Todt

INTRODUCTION

The development of singing in birds ranks as a prime biological model for understanding how nature and nurture interact, and how they jointly affect the growth of behavioral competence. Learning to sing is a somewhat different accomplishment from that involved in either associative or operant conditioning. To begin with listening and responding to the singing of conspecifics is not a year-round activity, but has a clear seasonal distribution. There is also an age-related pattern in a bird's readiness to memorize and then to develop its song vocally. With some exceptions, the first year is the most important time for song learning in a bird's life. Within this year, there is usually an early and limited period of time, the "sensitive phase", when birds are especially receptive to song stimuli and when they memorize them most readily.

The discovery that perceptual aspects of song acquisition are linked to a sensitive phase had great impact on the progress of learning research and drew attention to similarities between sexual imprinting and song learning. Although these two kinds of learning serve rather different functions – the formation of a social preference in one case, and the acquisition of behavior for communication in the other – they share a number of common mechanisms (Immelmann and Suomi, 1981; Slater *et al.*, 1988, 1993; ten Cate, 1989; ten Cate *et al.*, 1993). For example, both sexual imprinting and song learning are biased towards species-typical signals, both need relatively few exposures to stimuli (Bischof, 1998; Hultsch *et al.*, 1999a), and both use highly specialized neural circuitry to acquire and store stimulus representations.

Learning to sing takes some time. Usually it begins when a young bird leaves the nest and then continues for several months. It comes to an end only shortly before the bird establishes his own territory and flags its turf by singing. Although some species also learn new songs later in life, the first year of song development is crucial in all songbirds. Their many accomplishments can be subdivided into two major stages: the auditory phase of song memorization and the motor phase of song development. In this chapter we treat the various aspects of learning to sing in some detail, using as an organizational framework the time course of an individual's vocal development. Our approach will be selective, without attempting to be comprehensive, focusing especially on recent birdsong research. Of necessity, many of our conclusions are drawn from laboratory studies. However, because the relevance of laboratory-based findings to song development under completely natural conditions is not always clear (Beecher, 1996) we propose, when possible, to re-examine conclusions drawn from laboratory research in the light of subsequent studies in the field.

THE MEMORIZATION PHASE

Young songbirds are often exposed to a great variety of different voices. How do they avoid memorizing "wrong" stimuli while managing to pick up and later develop their species-typical song pattern? Several behavioral adaptations and intrinsic mechanisms help to cope with such problems. For example the sensitive phase of song acquisition is a kind of time window that facilitates stimulus memorization during a species-typical age period and restricts the ability to learn new songs later on. The window typically opens around fledging, when appropriate song stimuli are perceived, and closes after a limited amount of time. The timing of sensitive phases provides a further mechanism designed to help a young bird to avoid learning the wrong song. There is coordination between the timing of sensitive phases and the time when adult songsters of the bird's own species are vocally active (Figure 17.1a).

The risk that a fledgling's song learning is misled by non species-typical auditory stimuli is further minimized by a mechanism called pattern selectivity – an

Neuroscience of Birdsong, ed. H. Philip Zeigler and Peter Marler. Published by Cambridge University Press. © Cambridge University Press 2008.

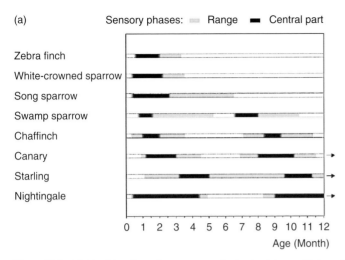

Figure 17.1 (a): Sensitive phases for song memorization in eight songbird species. Species with a single phase are given first. Other species where learning extends into the second year are indicated by arrows. References (top-down): (1) Böhner, 1990; Zann, 1997; (2) Marler, 1970b; Petrinovich and Baptista, 1987; (3) Nelson et al., 1995; Marler and Peters, 1987; (4) Dooling and Searcy, 1980; Marler and Peters, 1982b; (5) Thorpe, 1958; Ince and Slater, 1985; (6) Marler and Waser, 1977; Nottebohm and Nottebohm, 1978; (7) Chaiken et al., 1994; Böhner and Todt, 1996; (8) Hultsch and Kopp, 1989; Todt and Böhner, 1994.

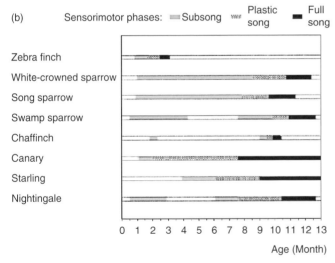

Figure 17.1 (b): Sensorimotor phases of eight songbird species. References (top–down): (1) Zann, 1997; (2) Nelson et al., 1995; (3) Marler and Peters, 1982b; (4) Marler and Peters, 1982c; (5) Nottebohm, 1971; (6) Nottebohm et al., 1986; (7) Chaiken et al., 1993, 1994; (8) Hultsch 1991a, b, 1993, b; Geberzahn and Hultsch, 2004.

inherent predisposition to listen and memorize species-typical sound patterns selectively. This concept implies that even birds without prior experience are able to recognize such patterns as relevant and to distinguish them from the song patterns of other species. (Konishi, 1965b; Konishi and Nottebohm, 1969; Marler, 1976a). In an early study swamp sparrows were allowed to hear songs of their own species, of the closely related song sparrow and, as a control, the domestic canary (Dooling and Searcy, 1980). Only species-typical songs were processed in such a way as to elicit clear physiological responses, in this case a change in heart rate. Such predispositions play a crucial role as guidelines escorting a bird through the first steps of song learning (Marler and Peters, 1980; Soha and Marler, 2000).

The closing of sensitive periods is not merely due to a clock-like mechanism that constrains learning after reaching a certain age. Rather, the evidence suggests that the accumulation of a certain amount of experience with the song patterns to be learned plays a crucial role. In several species experiments wherein males were prevented from hearing song stimuli early in life, subjects extended the period for acquiring songs (Nottebohm, 1989; Eales, 1987a; Slater et al., 1993; Nelson, 1997). Flexibility in the timing, especially with regard to the closure of sensitive phases, is a biologically significant characteristic of the song learning process, allowing birds to cope with some of the variation and uncertainty that exists in their social and ecological environments (Hultsch, 1993a, b; Hultsch et al., 1998, 1999a, b; Todt and Geberzahn, 2003).

Effects of social experience on song memorization

Social factors are substantial in some species and minimal in others (Baptista, 1996). For instance, indigo buntings and zebra finches can acquire their songs from the father, from flock members, or from territorial neighbors (Payne, 1981; Böhner, 1983; Clayton, 1987a; Williams, 1990). Such social selectivity is affected by prior experience with individual conspecifics or individuals of other species that are accepted as tutors by mimics (Chaiken et al., 1993, 1994; Todt and Böhner, 1994). Factors contributing to these effects include parental care by a potential tutor, aggressiveness, or just visual and acoustical contact (Todt et al., 1979; Payne, 1981; Böhner, 1983; Kroodsma and Pickert, 1980; Clayton, 1987a; Eales, 1987; Williams, 1990). Prominent among the prerequisites of tutor acceptance is the age of the pupil, while some biological properties of the tutor seem to be less important. Young bullfinches (Nicolai, 1959), nightingales (Hultsch and Kopp, 1989) and starlings (Böhner and Todt, 1996) readily learn songs even from a human provided that hand-rearing has begun before they open their eyes. In the field, for example, male bullfinches (*Pyrrhula pyrrhula*) produce a very simple song. But when trained by a social tutor who presents them with more complex learning stimuli than those from their biological father, the birds will readily memorize and later perform song patterns that, instead of being three to five notes long, contain series of more than 30 elements (Nicolai, 1959). They are so skillful at this that in the past, they were kept as pets and trained to whistle simple melodies (see the Preface, this volume). Findings like these demonstrate the extent to which natural behavior can be radically transformed by changing the circumstances in which a young songbird grows up.

The significance of social factors is illustrated by the differential effectiveness of tape versus live tutoring. In song sparrows and swamp sparrows these are equally effective. (Marler and Peters, 1987, 1988a; Baptista and Petrinovich, 1984). In canaries and starlings the amount of song material acquired is much smaller when birds are kept isolated and hear only tape recordings (Marler and Waser, 1977; Chaiken et al., 1993). Finally, there are species, such as tree-creepers of the genus *Certhia*, which do not seem to learn at all from tape recordings (Thielcke, 1970). Moreover, the impact of social factors may shift over the course of development. White-crowned sparrows, for example, will learn readily from tape until 50 days of age (Marler, 1970b), but thereafter will only accept live tutors as song models (Baptista and Petrinovich, 1986). In contrast, four-month-old starlings learn better from live tutors than from tape recordings alone, and tape tutoring is more effective at 12 months than at 4 months of age (Chaiken et al., 1993). A similar shift is seen in nightingales. During the first part of their sensitive phase, from day 13 to day 40 post hatching, nightingales learn only if their familiar, live tutor is present during a playback presentation. But later on and especially after having experienced typical nightingale song, they also memorize species-typical songs heard when they are alone (Todt et al., 1979; Todt and Böhner, 1994). One reason for such differences between white-crowned sparrows and nightingales may be the clear differences in repertoire size, imposing weaker or stronger constraints on the selectivity of learning. Individual white-crowned sparrows rarely sing more than one song as adults (Baptista, 1975), while the normal repertoire of a nightingale consists of up to 200 different song types (Hultsch, 1980). In this species, it might be too restrictive to limit learning to one or two familiar tutors if a bird is still to develop a large repertoire. Nevertheless, social selectivity early in life may serve to refine global innate preferences for the type of songs which a male will acquire later on and subsequently attend to as an adult.

A review of studies in which the quality, amount or timing of social stimulation was monitored or manipulated suggests that the concept of social selectivity is better conceptualized as a graded phenomenon, with social factors facilitating rather than necessarily initiating learning (Böhner, 1983; Baptista and Petrinovich, 1984; Eales, 1985; Clayton, 1987a; Williams, 1990; Chaiken et al., 1993; Slater et al., 1993; Todt and Böhner, 1994). Moreover, the wide range of contextual variables found to be inducive to selective learning, like parental care, aggression or proximity of the tutor, leads us to question whether it is based on a truly social mechanism designed, for example, to tag song stimuli as significant or non-significant depending on the tutor's presence or behavior. Rather than being a social mechanism designed to enhance the "salience" of some stimuli over others, the effects of social factors could be mediated by perceptual mechanisms such as arousal or attention. This conclusion is consistent with the finding by Hultsch et al. (1999b) that young nightingales will preferentially acquire songs that were paired, during human tutoring, with synchronous stroboscopic stimulation.

THE SENSORIMOTOR PHASE

The ontogeny of singing shows a number of characteristic traits that are widespread across oscine birds (Marler, 1991a). For example, in the typical songbird the early phase of auditory learning usually precedes the phase of vocal learning by an interval of several weeks. Such observations have led to the notion of song acquisition as a two-stage learning process in which the bird first develops an auditory representation of the tutor's songs and then acquires a motor program translating the memorized sound patterns into behavior. This phase of vocal development ushers in a new stage in a bird's life, often covering a span of several weeks or months and divisible into three prominent stages: subsong, plastic song and crystallized full song (Marler and Peters, 1981, 1982b; Figure 17.1b).

Subsong: first steps in song development

During the song memorization phase, while listening to song, birds may themselves vocalize, performing coherent sequences of special vocalizations from which mature song only emerges gradually, with many changes of timing and structure. First to emerge

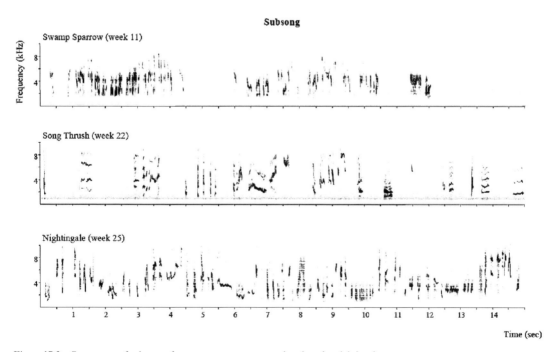

Figure 17.2 Sonograms of subsong of a swamp sparrow, a song thrush and a nightingale.

are soft, rambling sounds which are classified as "subsong" (Figure 17.2). In structural terms, subsong is never the same from moment to moment, and is thus rather difficult to analyze. Although subsong may be stimulated by hearing song, its acoustic structure seems not to be specifically related to that of the stimuli. Its great variability invites comparison with the vocalizations of human babies before they start babbling. The functional significance of subsong is unclear. It may serve to train the vocal apparatus and thus to improve sensorimotor control, perfecting the art of two-voiced sound production (see Suthers and Zollinger, this volume).

Plastic song: imitations begin to emerge

Subsong gradually merges into plastic song, the next phase of song ontogeny (Figures 17.3, 17.4, 17.5). We know that plastic song has started when rehearsal of previously memorized song patterns begins, and birds shift gradually to more stereotyped vocalizations. As the phonetic morphology of plastic song is elaborated, the first precursors of acquired imitations emerge. Increasing numbers of syllables and phrases can be discerned, reoccurring in a similar form, making it possible to arrange them in categories and to make measurements. The progress in song quality can be documented with measures like pattern stereotypy, vocal amplitude and repertoire size (Marler and Peters, 1981; Podos et al., 1999; Brumm and Hultsch, 2001; Hughes et al., 2002).

The study of developmental trajectories is a particularly promising approach (Clark et al., 1987; Tchernichovski et al., 2001; also see Saar et al., this volume). For example, it has been shown that in some birds note and syllable structure take on adult form ahead of song syntax. Thus, for some time after achieving the stereotyped production of song constituents, birds may still exhibit plasticity in arranging these

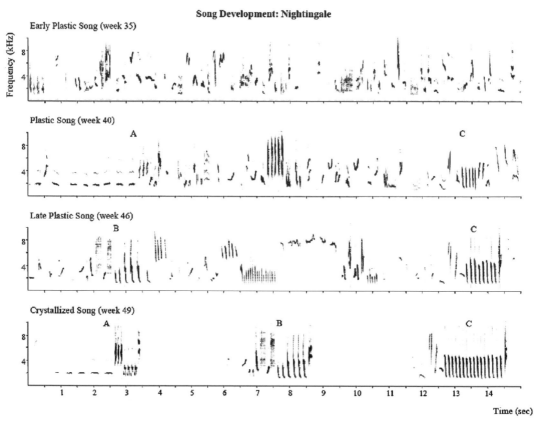

Figure 17.3 Song development of a nightingale (week 35 to week 49). At the bottom are three crystallized imitations (A2, B2, and C3) with precursors shown in the second and third row.

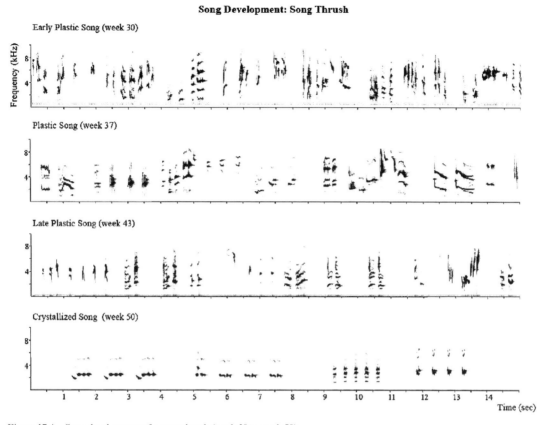

Figure 17.4 Song development of a song thrush (week 30 to week 50).

constituents into sequences. They may sing incomplete songs, with some song constituents missing, or a typical succession of song phrases may be inverted. Another finding is the coupling of different trajectories during ontogeny, as with changes in vocal amplitude and pattern quality. As they proceed through plastic song, birds usually sing not only better but also louder. Brumm and Hultsch (2001) found that in young nightingales not only is there a long-term developmental coupling between loudness and song stereotypy, there are also clearly correlated short-term changes in song quality and volume. Even within a given performance that contains song material of different pattern quality the correlation with loudness remains significant.

Song crystallization: adult song

By comparison with the long period of vocal activity that precedes it, the progression from plastic song to mature singing is fairly rapid (Todt and Geberzahn, 2003). During this phase, called "song crystallization", the various, highly variable types of song, so typical of preceding stages of development, come to an end and the previously plastic structure of vocalizations gets "frozen." The time structure of singing often changes, with most birds shifting from more or less continuous song sequences to a temporally segmented performance. This is done by the insertion of progressively longer silent intervals at the boundaries between two successive song patterns (Kopp, 1996). Finally, birds cull from their originally vast range of diverse sound patterns a limited number, assembled into species-typical songs. The extent of this attrition varies. It is less drastic in species like nightingales, with large song repertoires, than in birds with smaller repertoires like song sparrows, swamp sparrows, or white-crowned sparrows. In nightingales it averages about 8%, whereas it can go as high as 80% in sparrows.

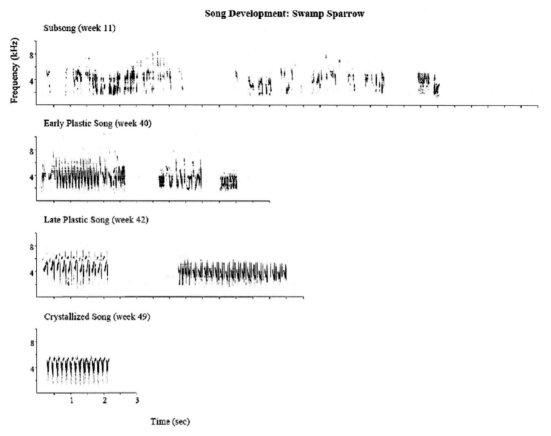

Figure 17.5 Song development of a swamp sparrow (week 11 to week 49).

Studies of this process by Marler and his associates have led to the concept of "action-based learning" (Marler and Nelson, 1993; Nelson and Marler, 1994). The core idea is that social experience at the time of crystallization, as the male establishes his first breeding territory, determines which songs a bird will select and produce in his adult repertoire. In the field, such social experience will be provided by interactions like countersinging with neighboring adults. As a result of action-based learning a juvenile bird will retain those songs in his own repertoire that match the species-typical structure and even the particular song types of his neighborhood. However, some vocal material not incorporated in his actual repertoire, may nevertheless be retained in memory, as a kind of "dormant" pattern reservoir that can get reactivated later in life, as during social interaction with new neighbors (Marler and Peters, 1982b, 1988a; Nelson, 1992; Geberzahn and Hultsch, 2003, 2004). Evidence in support of this proposition was obtained by exposing nightingales, at an age of 11 months, to playbacks of songs with which the birds had previously been tutored, but which they had not imitated, either entirely or in part, during their vocal development (Geberzahn et al., 2002). Exposure to these songs instantly elicited song matching, triggering the selective production of imitations that had not previously been uttered. This outcome suggests that the nightingales had indeed memorized these previously unsung song patterns.

Mechanisms of development

We are still grappling with many questions about the mechanisms that underlie song development (Székely et al., 1996; Tchernichovski et al., 2001; Brainard and Doupe, 2002; Saar et al., this volume). A seemingly basic issue concerns how birds proceed when they are uttering their first and often fragmented precursors of song imitations. Do they "hit" on them by chance, or do they have more efficient strategies of memory retrieval than this? Studies of the temporal progression of pattern development in nightingales suggest that new precursors of imitations do not emerge randomly distributed over the singing, but in close sequential proximity to those developed earlier (Todt and Hultsch, 1996). Moreover, from early on during the process of repertoire development, precursors of imitations occur as nested clusters within the performance of other song material (Hultsch et al., 1999a, b; Brumm and Hultsch, 2001).

We may also ask whether and how far the sensorimotor phase can be modified by environmental or physiological variables, and what the consequences may be for singing later in life. Does the relatively long duration of the ontogenetic process have any functional implication for the resulting singing behaviors (Korsia and Bottjer, 1991; Whaling et al., 1995)? Males were given testosterone implants at different ages, thus inducing an artificially accelerated song ontogeny and reducing both the length of the pre-imitation period, and the duration of the sensorimotor phase by several months. After such accelerated development the structure of crystallized songs was clearly abnormal and similar to the songs of individuals raised in acoustic isolation. This finding suggests that the storage phase may not be a time of passive retention, but may involve intrinsic processes of active consolidation and/or maturation of song material.

THE PHENOMENOLOGY OF VOCAL LEARNING: A COMPARATIVE PERSPECTIVE

As has often been pointed out, there are some fascinating parallels between song learning in birds and the development of speech in humans (Doupe and Kuhl, 1999). First, both behaviors have to be learned to achieve the normal signal repertoire of the species. Second, such learning relies on the auditory perception, memorization and imitation of sound patterns. In both, perception precedes the production of vocal material. Third, acquisition is best accomplished early in life, during sensitive periods, and is apparently guided by specific predispositions. Finally, vocal expertise is successfully reached only after progression through particular stages of development, in which vocal practice plays an essential role. As a general framework, these three parallels help us in dealing with some more specific issues.

Song learning and the acquisition process

During the early phase of development in which patterned sounds are memorized, young birds and human infants are confronted with a rather similar problem: instead of hearing a single auditory stimulus, they are exposed to a medley of many vocalizations. Children need to look out for cues that help them to parse the speech flow produced by their caretakers into segments. Once this is accomplished, it obviously helps them to identify and store information about particularly frequent segments more efficiently, especially words and combinations of words (Jusczyk et al., 1992). Do young birds apply a similar strategy?

We have examined the process in nightingales, which, like humans, memorize and develop a large repertoire of different vocal patterns. To identify their song learning strategies we composed learning programs comprising particular songs from our catalogue of master song types and recorded successively on tape to form song strings. In a standard master string, each successive song in the sequence is of a different song type. Likewise, each of the different sequences to which a young bird is exposed during the tutoring periods consists of a unique set of song types. We thus label a particular tutoring regime by the particular song sequence which birds hear in that situation (Hultsch et al., 1984). The acquisition success of the tutored males, assessed by how many songs they imitate, then allows us to see whether a particular tutoring regime influences their singing. Audiovisual recordings allow us to measure other aspects of the bird's behavior during a given tutoring experiment (Müller-Bröse and Todt, 1991). With this design we have identified a number of conditions in which nightingales not only imitate single song patterns accurately (Hultsch, 1993a, b), but also readily memorize and produce long song sequences (Figure 17.6).

There is a complex relationship between success at song acquisition and the nature of the training regime. For example, while for songs experienced only 5 times the acquisition success of nightingales is low (approx.

212   HENRIKE HULTSCH AND DIETMAR TODT

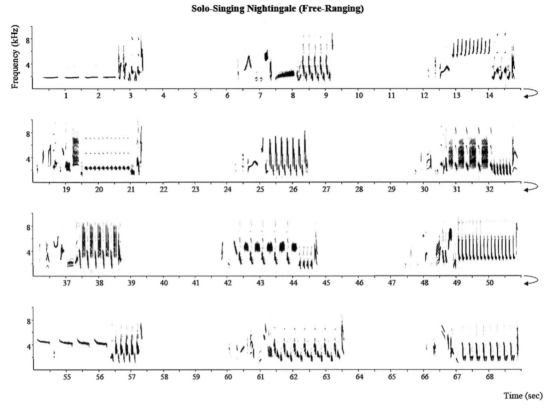

Figure 17.6   A sequence of twelve songs of a nightingale, taken from a longer bout of solo singing.

30%), birds imitate around 75% of song types heard 15 times. Also, the number of song types in a sequence can be increased considerably, say from 20 to 60 song types, without any need to raise the frequency of exposure. The ease with which birds cope with such an increase in the number of songs to be acquired (Hultsch and Todt, 1989c) is not well predicted by classical learning theory which posits a relationship between exposure frequency and the number of stimuli to be acquired (e.g. Crowder, 1976). These findings are relevant to the concept of song acquisition as a kind of special process or template learning (Marler, 1976a; see Adret, this volume).

Moreover, the process of song learning may be expected to involve specific evolutionary adaptations. One such adaptation was revealed by experiments testing how nightingales deal with exposure to song sequences after removal of the silent intersong intervals and other cues that might help birds to recognize each song as a separate sequential unit. The imitations they developed were each uttered separately, showing that even in the absence of such cues, birds were still able to determine boundaries between successive tutor songs. Thus, in contrast with human infants whose parsing of the speech stream is based on experience with their mother tongue, young nightingales may have some innate mental concept of how their song patterns are segmented (Hultsch et al., 1999a).

Memorizing song sequences

Humans, when given a string of different items to memorize, such as words or sentences, split the string into "chunks" approximately four units long. This "chunking" process turns out to be optimal for processing in short-term memory (Bower, 1970; Simon, 1974; Cowan, 2001). Birds, such as nightingales, also perform well in extracting and memorizing information about a string of song types and use a similar process (Hultsch and Todt, 1989a, b, c, 1996). When exposed to a long sequence of say 60 songs, nightingales memorize smaller subsets composed of three to five successive songs and store information about each group of songs separately. In other words, they split the sequence into segments that are more easy to process than a long series. This effect has been termed "package formation." As with humans, the evidence suggests that these processes reflect properties of short-term memory with the songs composing a segment treated as a chunk or single unit.

Young nightingales were exposed to song strings where silent inter song intervals were either reduced or extended in duration (1 s or 10 s, respectively) as compared with the standard 4 s (control) duration. Such "dense" or "spaced out" playback of tutored songs had a clear influence on the size of developed song packages. For example, smaller packages resulted from the "spaced out" song strings than from the "normal" or "dense" song presentation, suggesting that package formation is not only a capacity-constrained process, but is also controlled by a time window playing a role in the segmentation of perceived song sequences (Hultsch, 1992). Such dual constraints on sequence learning are already evident during the bird's very first experience with a tutored song string. This was demonstrated by an experiment allowing young nightingales to hear a set of different songs 15 times, enough for them to memorize each song readily. However, instead of playing the string with stereotyped song order on each trial, the serial succession of song types was altered from one tutoring exposure to another. Remarkably, as adults, not only did all birds form normal packages but their packages were made up of imitations of songs which were sequential neighbors only during the first tutoring session. Evidently package formation can occur during a single exposure and the very first experience seems to play a key role (Hultsch and Todt, 1996).

There are further parallels between humans and birds. For instance, if we are exposed to long strings of different items often enough, we will eventually be able to remember and reproduce the entire sequence, by rearranging items in the right order. Marsh wrens (Kroodsma, 1979) and nightingales (Hultsch and Todt, 1991) behave very similarly. If nightingales are allowed to hear a song string 50 or 100 times, they imitate the entire sequence of tutored songs in proper serial succession. The imitation of serial order first emerges within packages, and only later, with more frequent exposure, comes to reflect the complete succession of packages. It appears that with more opportunities to listen to a song string, order information gradually gets consolidated, suggesting a truly hierarchical process, progressively incorporating higher and higher levels of song organization.

A hierarchical model of song retrieval

Rapid and faithful memorization of significant items can be a real advantage, but it is only one side of a coin. For efficient use of the learned material it has to be supplemented by a system for organizing memories that is also optimally structured for retrieval. In humans, we know that the more information one has to keep in mind the more important this issue becomes. One optimizing principle is to organize mental representations of learned items in a hierarchical manner, a concept that we had in mind in asking how nightingales organize the memory system for their large song repertoire (review in Todt, 2004).

Indeed, at the level of notes and individual songs hierarchical organization is already known. That is, notes combine to syllables, syllables combine to phrases and finally phrases combine to the signal unit "song" (strophe). The freedom of combining units differ for these building blocks of behavior and increases from the bottom up. For the basic compounds, rules are rather rigid and prescribed by species-typic structure, but they are rather flexible for the song level (i.e. open for individual decisions or choices; Todt, 1970; Todt and Hultsch, 1980, 1998a, b). However, hierarchical structure above the individual song level remains a little-explored territory. Functionally, it would make sense to expect that the management of a large repertoire is facilitated by precisely that higher level structure. For instance, if songs are organized from the point of view of retrieval in clusters, retrieval of one of them may facilitate the availability of other members of the same cluster. This might reduce the number of operations the brain has to go through to produce longer series of songs in nonrandom sequences. For instance, song type matching among the members of a population of neighbors may be facilitated if songs learnt in a given context tend to be retrieved together. Empirically, this informal functional hypothesis is supported by our findings of song acquisition in hand-reared nightingales, described in the previous section, indicating that groups of songs

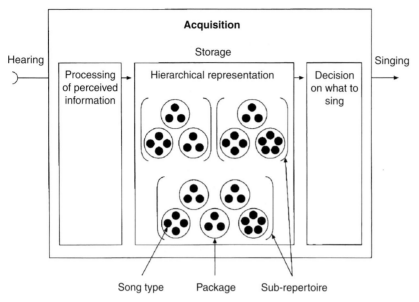

Figure 17.7 Diagram of the mechanisms we postulate to explain the relationships between hearing, memorization, memory retrieval and singing (see Hultsch and Todt, 2004).

may be stored and retrieved as suprasong units that we call packages (Figure 17.7).

Like song packages, song associations occurring in free ranging birds are typically limited in size and encompass about three to four song types only (Todt, 1968, 1971; Verner, 1976; Ince and Slater, 1985). In addition, however, several of these small groups may associate to form larger clusters of sequentially connected song types, termed subrepertoires in nightingales (Hultsch, 1985; Kipper *et al.*, 2004). The size of such subrepertoires is remarkably variable and thus makes another contrast to the small sized groups. For those large and variable song associations the crucial question was whether they were read-outs from the birds' memory of stored song material.

We examined this question experimentally by exposing our birds to different sets of tutor songs and by segregating the presentation of these sets in time. In other words, birds were run on a tutoring schedule which simulated situations in the field where young nightingales, during early dispersal, may experience novel sets of conspecific songs. The trained birds' singing revealed two effects: first, they performed imitations acquired in the same learning context jointly as one large association of song packages. Second, birds sang these associations sequentially separated from imitations developed from another learning context. This achievement ("context effect"; Hultsch and Todt, 1989b) may help to account for subrepertoire formation. Concurrently, it suggests that the internal representation of learned song material in nightingales is organized hierarchically (Figure 17.7). Some of the rules accounting for such memory organization can be discerned already during the ontogeny of singing. Developmental trajectories of the three hierarchy levels – songs, packages and subrepertoires – reflect those levels of order by unfolding at different times in the singing that develops.

When singing becomes functional as a territorial display in adults, a repertoire that is hierarchically prestructured would be a candidate mechanism for facilitating song retrieval in situations demanding rapid vocal responses (Todt and Hultsch, 1998a, b; Todt and Naguib, 2000). During "rapid matching," for example, a male has to identify a neighbor's song and immediately select and retrieve a song of the same type from his own repertoire, doing so with a latency of approximately one second (Todt, 1981; Hultsch and Todt, 1982; Wolffgramm and Todt, 1982). In making decisions about "what to sing next" it is expedient if the bird does not have to do a "serial" search from the whole pool of developed song types. The number of decision steps and the decision time are both reduced by using a search routine that can focus on particular subsets of patterns (Todt *et al.*, 2001). Thus a hierarchically organized format for storing representations of song has clear adaptive value in birds that, like the nightingale, have to manage large repertoires.

The suggestion of a hierarchical organization of song memories led us to a further, more general parallel between the learning of humans and birds (Figure 17.7). If the first step in this "memory hierarchy" is placed at the level of songs, the next involves the packages, composed of a few songs, and at a still higher level are the larger association groups or subrepertoires, composed of packages. Both song order and subrepertoires reflect the serial organization of input during auditory learning and it follows that they are exposure-induced. As such they are to be distinguished from the package groups, which can be characterized as self-induced associations. A crucial question about the formation of exposure-induced song associations is whether they simply reflect the memorization of stimulus chains or whether they are linked to more abstract cognitive abilities, such as the formation of perceptual or conceptual categories. For the subrepertoires, for instance, such categorical concepts might be "songs heard from a particular individual," "at a particular location or time," or "songs specified by a particular quality" (Hultsch *et al.*, 1999a).

Vocal communication: songs and sentences

In an ideal communication, when two individuals are contributing equally to the signal exchange, one can expect that the signaling routines will not be arbitrary, but will follow certain rules or conventions. The significance of such rules has been documented for both the verbal and nonverbal dialogues of humans (Burgoon and Saine, 1978) and for the vocal duels of songbirds (Todt and Naguib, 2000). The suggestion was also made that, in ideal interactions, signal patterns would be used which strike a compromise between two opposing requirements: the signals should be long enough to convey a given message, but at the same time not so long as to delay a potential reply. Most human sentences and most birdsongs that are used in an interactional context are only a few seconds long, and are thus candidates for optimal units of vocal interaction (Hultsch *et al.*, 1998b). With this shared property of songs and sentences in mind, we can examine some of their other features and contrast some of them from formalistic and developmental perspectives.

From a formal perspective, most sentences are composed of several constituents, words or phrases, that follow each other according to syntactical rules. For example, a sentence often begins with a word that is particularly frequent in a given language. Similarly, most birds develop songs with certain syntactical rules. The species-typical composition of most songs requires particular types of notes, syllables, motifs and phrases, occurring at specific positions in a given type of song. Although such structural features are more rigidly fixed than those of sentences and thus are not acceptable as genuine lexical syntax, changes of note positions are meaningful in both cases (Todt, 1974). Finally, many species, however versatile their singing style, may nevertheless start successive songs with the same type of introductory note (Todt and Hultsch, 1998b). In other words, from a strict formal perspective there are certain similarities between sentences and songs.

There are also developmental similarities. At about three months human infants produce only vowel-like sounds. Then, at about seven months, they begin their "canonical babbling" by incorporating consonants into their vocal repertoire. During this period when they produce word-like syllables children rehearse repeatedly and have already begun to imitate the prosodic features of their particular mother tongue. By about 12 months infants usually start to use so-called "one-word-sentences", and at some later time these develop into sentences composed by two or more words (Kuhl, 1989; Todt, 2004). This progression can be compared to the succession of stages in the development of singing in birds, including their subsong, plastic song, and eventually their use of crystallized full song.

Finally, the formation of song type packages shows striking similarities to the chunking of information in human serial learning (Bower, 1970; Simon, 1974). In humans, chunking is related to cognitive processes, so that a chunk can be defined as a functionally meaningful unit, whether it be a simple phrase (a morpheme) or a sentence, or more. In birds, on the other hand, the equivalent to a chunk is a song, and a song can also be regarded as a meaningful unit. This can be inferred from the way songs are used as units of interaction, as during vocal communication between neighbors. Also, there are good reasons for thinking that such interactions reflect underlying cognitive accomplishments, as for instance when a bird's singing performance is organized according to categorical cues acquired from the context in which song memorization occurred. Such contexts may be defined temporally, spatially, or socially. In the exposure-induced song type associations developed by nightingales, the subrepertoires represent such an accomplishment at a higher level of behavioral organization. Finally, some birds, such as warblers, learn the situations in which to use their songs (Kroodsma, 1988; Spector et al., 1989). But the particular cues on which such categorization is based and the mechanisms by which a categorical representation is achieved remain to be explored.

CONCLUSIONS

In the course of studying song learning, mechanisms have been uncovered that are basic to learning in general, including predispositions influencing decisions about what to learn, sensitive phases, when the potential for developmental plasticity peaks, and social factors that motivate learning selectively, all helping a young bird to cope with the challenges of learning to sing. These mechanisms have two main effects: they allow for selection among potential learning stimuli and at the same time they facilitate pattern memorization. Birdsong studies have also revealed many significant properties of sensorimotor learning, demonstrating the value of a stepwise progression through successive stages, as when birds proceed through the different stages of vocal development, subsong, plastic song, and song crystallization, before they achieve the final stage of adult singing. And in many cases, song development still continues, after birds have finished their first year of life and become sexually mature.

Learning to sing is fascinatingly diverse when it is considered in a functional and comparative framework. It provides a male bird with a repertoire that serves both mate attraction and effective territory defense. From a territorial viewpoint, it paves the way for specific vocal interactions among males who have had the chance to learn similar songs beforehand and thus share parts of their vocal repertoires. Finally, from a comparative perspective, the potential of studies of song learning is unique. In the animal kingdom, vocal learning is a rare achievement that merits comparisons with genuinely human accomplishments such as language and speech development (Marler, 1970a; Kuhl, 1989; Pepperberg, 1993; Doupe and Kuhl, 1999; Hultsch et al., 1999a; Todt, 2004; Haesler and Scharff, this volume). Indeed, there are several parallels in the learning mechanisms involved, as evidenced for example in the ways in which a human and a particularly accomplished avian singer like a nightingale copes with the management of huge arrays of songs, and their myriads of constituents, as we deal with our prodigious word repertoires. By using these shared properties as a frame of reference, learning to sing can be exploited as an excellent biological model for research on memory and its underlying neural mechanisms.

ACKNOWLEDGEMENTS

We thank the editors for their valuable help in improving this chapter, and Nicole Geberzahn for preparing most of the figures. The studies on nightingales reported here were supported by a fund of the German Science Foundation DFG (Az: To 13/30–1).

18 · Developmental song learning in the zebra finch

Sigal Saar, Partha P. Mitra, Sebastien Deregnaucourt, and Ofer Tchernichovski

INTRODUCTION

Developmental learning (for example, speech acquisition in human infants) takes place early in life but its effects may last the entire lifetime of the individual. Developmental learning is difficult to study because the behavioral changes involved span many timescales: the challenge is to relate the developmental outcome to behavioral changes, which can occur within hours as well as across daily cycles of wakefulness and sleep and over developmental stages. The study of developmental song learning in birds provides a unique model system for examining this process in detail.

Until recently, most song development research focused on neural activity in brain structures involved in song learning (Bottjer *et al.*, 1984; Scharff and Nottebohm, 1991) and production (Nottebohm *et al.*, 1982), identification of the role of auditory feedback in song development (Konishi, 1965b) and in the study of higher level (perceptual and social) effects on the development process (Price, 1977; Slater, 1983; Marler, 1990). Little attention was paid to the fine (moment to moment) details of song development, in part because the juvenile song appears unstructured and unstable, but also because of the sheer volume of data involved in such studies and the labor intensive nature of the required data analysis. However, without studying the temporal structure of the song development process it is not possible to answer the central question of birdsong developmental learning: how does the bird transform its sensory memory of the song it has heard into a set of complex motor gestures that generate an imitation of that song?

We know, in a very general way, how this sensory-motor conversion might occur (see Adret, this volume). At the sensory level, the zebra finch can perceive and memorize many songs, but only one (or a few) of those are selected to become the song template for imitation. The bird uses auditory feedback to guide vocal changes. This is thought to be achieved by comparing feedback from the bird's own song (BOS) to the song template (Brainard and Doupe, 2000a), i.e. evaluating the acoustic error. We use the term "error," even though we do not know how, and based on what features, the brain evaluates the difference between the memory template and auditory input from the BOS. It is assumed that the error gives rise to an "instructive" signal: an action resulting from detection of the difference (Doupe *et al.*, 2004). Examination of the details of song development gives us an opportunity to assess the structure of this instructive signal: when we observe vocal changes during development, we might assume that some of the changes we observe are driven by the postulated instructive signal. This hypothesis raises two distinct questions. First, what types of vocal change occur? Second, what is the relation of these changes to the instructive signal?

At the outset we should note that not all vocal changes during development are driven by the acoustic error from the acquired template: some features of song development evolve regardless of sensory experience in socially isolated, and even in deaf birds (Price, 1979). In other words, song learning is modulating a more primitive (and perhaps a more fundamental) developmental process. Still, it is sometimes possible to identify vocal changes that are driven by specific features of the song template (Tchernichovski *et al.*, 2001). Tracking those changes could potentially tell us something about the postulated error signal. It might be useful to distinguish between two different types of error signals, a simple and a complex error signal. A simple error signal (e.g. classifying vocal performance as similar or dissimilar to the template) cannot drive any particular type of change, but it can guide "learning by experimentation" (Fiete *et al.*, 2004). Its role would be primarily selective, resulting in the retention of one rather than another syllable or subsyllabic vocal gesture. In contrast, a complex error signal could reflect certain deviations from the template and then attempt to correct them. It could generate a very specific instructive signal to attempt specific corrections (e.g. elevate the pitch of a particular sound). Therefore, vocal changes induced by a complex error signal should appear more structured, or even "planned", compared with vocal changes driven by a simpler error signal.

Vocal changes can be detected in song (or subsong) even in very early stages of development. For example, the juvenile chaffinch gradually manipulates the acoustic structure of its innate begging calls (Nottebohm, 1968) until they become similar to its tutor's syllables. Detailed analysis of vocal changes in the zebra finch revealed that syllable types observed in the fully developed song can be traced back in time, often until early development. For each syllable type we observed structured trajectories of vocal change. Interestingly, it was often difficult to interpret each specific change without looking at the overall cascade of vocal changes over development. For example, in some cases we could see that the pitch of a syllable increases from one day to the next, becoming less similar to the pitch of the song model syllable (which was of lower pitch). However, the increase in pitch error was then followed by an abrupt event of period doubling: a nonlinear effect of halving the pitch that corrected almost the entire pitch error in a single step. The finding that some individual changes do not reduce the global acoustic error, but their cumulative influence does reduce the error, supports the notion of a complex error signal. However, because we know so little about song perception in birds, and because vocal changes are not predestined (Liu et al., 2004), it is difficult to make conclusive inferences about the features of the instructive signal based on particular examples of vocal change.

Other evidence suggests that simple reinforcement learning could play an important role in song development. For example, early observations led to the discovery that the juvenile white-crowned sparrow (Nelson and Marler, 1994) imitates songs from several tutors, but only one of its songs is retained in adulthood, termed selective attrition (Nelson and Marler, 1994). Selective attrition could play a role also during early song development, when syllabic structure is highly variable. Recent studies (Scharff and Nottebohm, 1991; Doupe et al., 2004, Ölveczky et al., 2005) have shown that inactivating LMAN, the final component of the anterior forebrain pathway, promptly reduces the degree of variability in song structure of a juvenile bird compared with that of an adult bird. This finding suggests that the variability of the juvenile song is actively "injected" into the premotor song nuclei via the forebrain song pathway. Introducing variability (Doupe et al., 2004; Deregnaucourt et al., 2005) into the song could allow vocal learning by experimentation, namely by exploring different versions of song syllables while attending to the changes in the vocal output, and then selecting the most appropriate path (in the "production space") towards the model (Tchernichovski et al., 2001).

One measure of our understanding of the song development process is our ability to survey the entire trajectory of development, formulate some hypotheses about the specific process in a given bird and, after describing the "current state" of its song structure, attempt to make predictions – what is the bird going to do next? For example, comparing different versions of a syllable in a current production, can we predict, perhaps based on similarity to the template, the direction in which the trajectory of vocal changes will proceed?

Nothing quite of this sort has been attempted before, perhaps due to the assumption that juvenile song structure is highly unstable, so much so as to make it impractical to compare moment-to-moment vocal changes based on an entire account of the production at any given time. This chapter summarizes recent findings from our laboratory which challenge this view. Our data show that if we look only at spectrograms, which present the time-frequency (spectral) structure of individual songs, individual syllables of the juvenile bird are indeed quite unstable in structure. However, analysis of the distribution of syllable features across several minutes of singing reveals stable organized structures (e.g. clusters) even in the early song. In other words, while juvenile song structure appears disorganized at the millisecond timescale (looking at sonograms), if we expand our timescale to examine feature distributions spanning several minutes we can demonstrate significant structure and stability in juvenile songs. Using an appropriate temporal resolution (e.g. hours) the vocal changes leading to the final production of an individualized conspecific type song can be studied in real time, providing an overview of the entire process of developmental learning.

The distribution of syllable features evolves smoothly over time, allowing us to identify distinct vocal changes, and to track their progression in real time. Across birds, we tracked two categories of vocal change: the emergence of syllable types and the trajectories of changes within each syllable type. While some of these vocal changes reduce the difference (error) between the juvenile song and the target song model, many of them are not related in any simple manner to error reduction. For example, vocal changes that occur during the day, while the bird is "practicing", tend to add structure to the developing song syllables, but vocal changes are also induced by sleep, and those changes tend to deteriorate song structure. Together, they cause oscillations in the developmental trajectories, and the magnitude of those oscillations is a good indication of the eventual accuracy of the adult bird's imitation (Deregnaucourt et al., 2005).

In the present chapter we describe the experimental methods used for the acquisition of large volumes of acoustic data needed to provide an overview of song development in individual birds. We also summarize analytic methods that enable us to construct multidimensional temporal "maps" of song development. We will then examine cases where investigating "what is the bird doing right now", can tell us something about "what will happen next" and even about the final outcome of song development. Then we return to the question of inferring features of the error signal from observing vocal changes. Finally, we will discuss briefly the implications of such an approach for integrative studies which monitor behavioral, physiological and neural activity during song development.

SONG DEVELOPMENT DATABASES

Until recently, data on song development was obtained from birds exposed to either live male tutors or recorded songs under testing conditions which could vary from lab to lab. Singing behavior was sampled sparsely, generating a relatively modest number of samples, and inferences about developmental changes were based upon inspection of sound spectrograms. However, sonograms of short duration present only a static representation of the dynamic developmental process. They can only capture short-term changes in sound (over timescales of milliseconds), whereas song development is a process that spans multiple timescales ranging from minutes to weeks. The failure to achieve a satisfactory description of vocal changes using traditional methods led us to develop methods that would facilitate analysis of the *distribution* of an array of song features over extended periods. These methods permit us to record singing behavior *continuously* under standardized conditions over the entire course of song development and to analyze the resulting comprehensive, longitudinal data sets. To carry out our research program, we need (a) to record and store large amounts of vocal acoustic data, (b) to automate analysis of that data, and (c) to display the results of that analysis as visual representations that highlight key features of the development process as they occur over time.

Data for our studies are obtained from socially isolated zebra finches trained with song playbacks. Subjects are the offspring of zebra finch pairs (male–female), maintained and bred in single cages, and separated from the father one week after hatching. The mother (who does not sing) is left with the chicks, and raises them in song-isolated conditions. At about day 30 – the age at which the bird becomes independent, and song development, i.e. the first recognizable subsinging, is about to start – the young male is placed in a soundproof recording chamber and we begin the continuous recording of songs. At a specified age (e.g. day 43) we turn on a training system. Song playback is initiated by the bird when the bird pecks on a key.

To facilitate the acquisition and analysis of the continuous recording of song learning of individual birds we have developed an open source software program that automates much of the data acquisition, feature calculation and database handling. (Details of the system, instructions for downloading and appropriate documentation can be found at http://ofer.sci.ccny.cuny.edu/html/sound_analysis.html.) Here we provide only a brief description of how the system operates and how feature maps of song learning are generated.

Each bird's vocalizations are monitored continuously, and each song bout is automatically identified and recorded. The software then performs multitaper spectral analysis (Percival and Walden, 1993) of the bout and calculates a set of song features such as pitch, frequency modulation, etc. (Tchernichovski et al., 2000). The features summarize the acoustic structure of the sound. We chose those features to describe the song because they may be related to mechanisms of song production (articulation): song is generated by vibration of sound producing membranes (the medial tympaniform membrane and the medial labium of the syrinx) (see Suthers and Zollinger, this volume). Those vibrations have basic characteristics such as amplitude, period (pitch), and regularity (entropy). We define spectral features of the sound that quantify these characteristics, namely sound amplitude, pitch, and Wiener entropy (an estimate of how broad is the power spectrum, ranging from pure tone to white noise). The change in pitch over time is called frequency modulation (Tchernichovski et al., 2000).

The next step in song analysis is segmentation of the sound into syllable units. In general, sound segmentation is a difficult task, but quite often segmentation by a simple amplitude threshold gives us robust syllable units during development (Tchernichovski et al., 2004). Once we have detected syllable segments, we can summarize the structure of each syllable by an array of acoustic features. For each of the continuous feature measurements mentioned above, we computed the mean and variance across the syllable. Together, the means and variances of the feature values summarize the structure of each syllable in a simple and biologically intuitive manner (e.g. duration, mean pitch, variance of pitch, etc.).

DYNAMIC VOCAL DEVELOPMENT (DVD) MAPS

In the next sections we present images that describe song development, constructed from the entire song development database of each bird. With the entire vocal ontogeny of a bird on file, tracing vocal changes over time becomes straightforward because we can visualize those changes in a single image that captures changes in features distribution over an entire vocal ontogeny. In this chapter, we focus on one-dimensional developmental histograms based on changes in a single feature, but we often plot two- and three-dimensional developmental maps. We call those collectively "dynamic vocal development (DVD) maps" (Deregnaucourt, 2004) because like other maps they represent the distribution of features. However, unlike standard maps which represent the distribution of geographical features in space,

DVD maps represent the distribution of acoustic features over time – i.e. they trace the way in which vocal features change over the course of development. DVD maps are composed of a series of single frames, each of which represents a cross-section of the status of a feature (such as song duration, or song frequency range) at a specific time during the course of song development (e.g. day 43). For this reason, such DVD maps are most informative when viewed as movies since the static single frames are transformed into dynamic representations of vocal changes over time (see example in http://ofer.sci.ccny.cuny.edu/109_new2.avi). For a more detailed discussion of the construction of DVD maps see Tchernichovski et al. (2004). The present chapter illustrates how the construction of DVD maps can be used to trace the progression of the two categories of vocal change we present here: the emergence of syllable types and the trajectory of changes within each syllable type.

The emergence of syllable types

Let's start by looking at one of the simplest features of a syllable – its duration. We will begin at the end of the song learning process, i.e. with the learned adult song. Figure 18.1a presents a smooth histogram of the durations of all the song syllables produced in one day (20 216 syllables) by a 90-day-old zebra finch. The peaks in the histograms indicate that three particular durations are much more frequent than the rest. Those peaks indicate that this bird, as an adult, produces three distinct syllable types; each syllable type can be identified in the sound spectrogram (Figure 18.1b). The DVD map was created by plotting duration histograms for each day during development and joining the histograms together (Figure 18.1c). The three 1D peaks have turned into three 3D ridges. We can easily see when each ridge emerged during development. The first and shortest ridge (corresponding to syllable 1) emerged on day 44–49. Note that training with song playbacks started on day 43. The ridge that corresponds to syllable two (the longest one) shows an interesting trajectory; it emerged at about day 50, and then took a turn to the right. This turn indicates a smooth increase in duration. The less structured histograms of the days prior to training show no apparent continuity with the three ridges that appeared after the onset of training.

Developmental song learning in the zebra finch 221

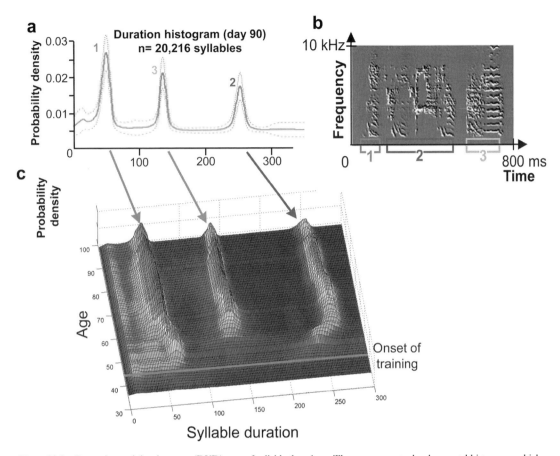

Figure 18.1 Dynamic vocal development (DVD) map of syllable durations. The map presents developmental histograms which represent the distribution of syllable durations over time – i.e. it traces the way in which syllable durations change over the course of development. DVD maps are composed of a series of frames each of which represents a cross section of the status of a feature (such as syllable duration) at a specific time (e.g. over one day) during the course of song development.
(a) A smooth histogram (locfit density plot) of the durations of all the song syllables produced in one day (20 216 syllables) by a 90-day-old zebra finch. Dotted lines represent a 95% confidence interval. The peaks in the histograms indicate that three particular durations are much more frequent than the rest.
(b) Those peaks correspond to the durations of three distinct syllable types. Each syllable type can be identified in the sound spectrogram.
(c) A DVD map was composed by plotting duration histograms for each day during development and joining the histograms together. The three 1D peaks have turned into three 3D ridges. We can easily see when each ridge emerged during development. The first ridge (corresponding to syllable 1) emerged between days 44–49. Note that training with song playbacks started on day 43. The ridge that corresponds to syllable three shows an interesting trajectory; it emerged at about day 50, and then took a turn to the right. This turn indicates a smooth increase in duration, demonstrating that this vocal change occurred mainly to this particular syllable. The less structured histograms of the days prior to training show no apparent continuity with the three ridges that appeared after the onset of training. (See also color plate section.)

We noted earlier that the three peaks in the duration histogram of the adult bird correspond to three syllable types. When we see a ridge emerging in the developmental histogram, we say that this ridge indicates a distinct syllable type. This effect of emergence of syllable types during early song development can be seen even more clearly in higher dimensional DVD maps that show how sparse and

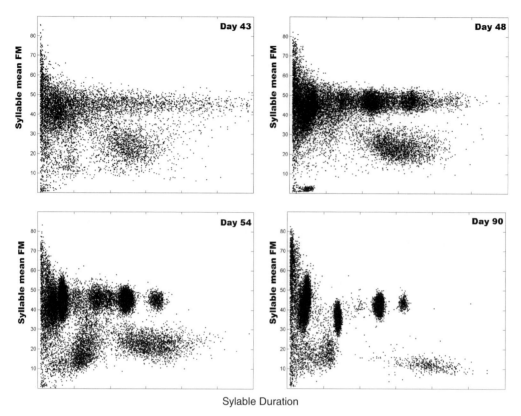

Figure 18.2 Four 2D DVD maps. For each syllable produced during day 43, 48 and 51 (accordingly) we plotted duration versus mean FM. On day 43 features are broadly distributed showing higher densities at the low FM area. Those loose clusters correspond to short and long calls. The distinct clusters at the high FM area of the map, present on day 51, are not present on day 43, but start to form on day 48. The dense cluster at the low FM area shown on day 50 is a long call that is about to evolve into a song syllable of higher FM (see day 90). We see a dynamic transition starting from a graded signal, where features of syllables show a continuous range of values, and evolving into a categorical signal with distinct syllable types. The emergence of distinct syllable types (i.e. clusters in the distribution of syllable features) is an important turning point in song development. Once syllable types can be identified, we can look at them as distinct entities (types) in which the bird can alter either their internal spectral structure or their global syntax and rhythm structure.

relatively unstructured distribution of features change over a few days into a highly organized construction of clusters. The technical details about clustering syllables are provided elsewhere (Tchernichovski et al., 2004).

The emergence of distinct syllable types (i.e. clusters in the distribution of syllable features; Figure 18.2) is an important turning point in song development: in most cases, we describe a biological signal as either a graded signal (e.g. the loudness of an infant cry is a graded signal) or as categorical (the infant is either crying or laughing). Here we see a unique dynamic transition starting from a graded signal, where features of syllables show a continuous range of values, and evolving into a categorical signal with distinct syllable types. Once syllable types can be identified, we can look at them as distinct entities (types) that the birds can alter through their internal spectral structure or by their global syntax and rhythm structure. The emergence of clusters (syllable types) could therefore reflect a transition from a pseudorandom exploration of vocal production space into a more advanced exploration state where each type of syllable has a trajectory of its own.

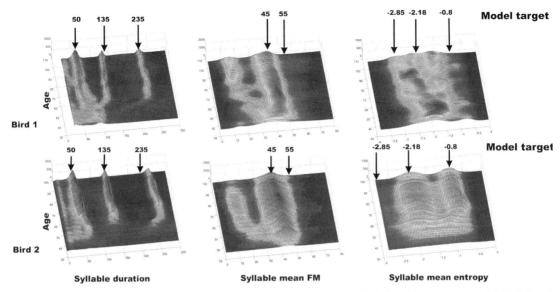

Figure 18.3 Developmental histograms of two birds (top and bottom) that successfully imitated the same song model. We followed the development of this syllable type in 10 birds that were trained by the same song model. We found that the duration of this syllable type increased by at least 20% in 8/10 birds between days 50–70 post hatching. The DVD maps present three features: syllable duration (left panel), mean frequency modulation (middle panel) and mean Wiener entropy (right panel). Note that across features (left to right) the developmental maps are not similar in their details, namely different features that describe the same developmental data show different dynamics. For example, just as we see how the distribution of duration becomes more structured over development, we also see how the distribution of mean frequency modulation changes from a broad range of values into two distinct ridges, indicating that we now have two categories of syllables: one of high frequency modulation and one of low frequency modulation. Across the DVD maps, we see ridges appearing, bifurcating from other ridges, and smoothly shifting right or left. During development, ridges become more distinct, and they tend to move away from each other, i.e. the different syllables tend to differentiate over time. Overall, these images indicate both a differentiation of structure and a closer approximation to a model target. The model target is indicated by the arrows and values above each figure. (See also color plate section.)

Trajectories of vocal change: learning and differentiation

Figure 18.3 presents developmental histograms of two birds that successfully imitated the same song model. The DVD maps plot changes across the course of song learning for three features: syllable duration (left panel), mean frequency modulation (middle panel) and mean Wiener entropy (right panel). Note that across features (left to right) the developmental maps are not similar in details, namely different features that describe the same developmental data show different dynamics. For example, just as we see how the distribution of duration becomes more structured over development, we also see how the distribution of mean frequency modulation changes from a broad range of values into two distinct ridges, indicating that we now have two categories of syllables: one of high frequency modulation and one of low frequency modulation. Across the DVD maps, we see ridges appearing, bifurcating from other ridges, and smoothly shifting right or left. During development, ridges become more distinct, and they tend to move away from each other, i.e. the different syllables tend to differentiate over time (Figure 18.3). Overall, these images indicate both a differentiation of structure and a closer approximation to a model target (see values above arrows in Figure 18.3). The details of the trajectories may vary across birds, but all birds examined so far have shown a similar developmental trend (namely appearance of ridges and then bifurcations and movement of ridges away from each other). In this respect we are claiming that trajectories of song development are not arbitrary, rather they are a modulation of a developmental process (Nottebohm, 1968).

HOW SYLLABLE STRUCTURE APPROACHES ITS TARGET

We noted that syllable types (ridges in Figures 18.1 and 18.3) tend to differentiate (i.e. their features often move away from each other) during development. Syllable types emerge and become more and more distinct. This process can be captured simply by looking at the diversity of feature values, which tend to increase over song development (Tchernichovski et al., 2001). This trend is noticeable also at the subsyllabic level: during early development the relatively unstructured syllables appear rather homogeneous, but as development proceeds we see the emergence of stable subsyllabic structure (sometimes referred to as "notes"). As distinct note types appear within a syllable, the diversity (i.e. variance) of syllable features (viewed at a 1 ms time resolution) also tend to increase over development (Table 18.1). Indeed, this measure captures what we can easily hear when we compare the sounds of the juvenile to those of the adult bird: song syllables of juvenile zebra finches sound "flat" and "scratchy" whereas mature song syllables often sound like rapid transitions between tonal and broadband sounds. One diversity feature that captures this transition very well is the variance of Wiener entropy (EV). The stronger and more abrupt the transition from tonal (low entropy) to broadband (high entropy) sound within a syllable the larger is the EV. In about 91% of vocal changes that we have tracked EV increases over development (Table 18.1), in most cases this increase is pronounced during early stages. Overall, the variance of all the features tends to increase over development, whereas the mean values of song features show weaker trends: mean pitch tends to increase, the harmonic structure of the pitch tends to stabilize (an increase in the mean goodness of pitch) and mean Wiener entropy tends to decrease. These trends merely reflect the increase in song structure, namely they capture the appearance of more and more structured syllable types over the course of development, a process that occurs even in social isolates that are not trained with song playbacks. Therefore, the vocal changes we measure capture an increase in song structure (and perhaps a better performance), which may, or may not, reflect the progression of song imitation. We need to separate this increase in structure (which as we noted may be seen in isolates) from progress in increasing similarity to the song model.

Table 18.1. *Diversity of syllable features*

Syllable feature	Trajectories with positive developmental trend (%)
Mean pitch	66
Mean goodness of pitch	70
Mean Wiener entropy	33
Variance of pitch	77
Variance goodness of pitch	86
Variance Wiener entropy	91

The progression in similarity to the song template can be assessed by measuring the similarity between the developing syllables and the song model used to train the bird (Tchernichovski et al., 2000). We can now ask to what extent the increase in syllabic structure is correlated with the increase in similarity to the song model. The answer to this question turned out to be much more exciting than we had suspected. It turns out that the increase in syllabic structure does not progress monotonically during development. During days of rapid learning, as the juvenile bird is "practicing" his song, we see steep increase in syllabic structure, but immediately after a night's sleep we see strong deterioration in song structure (Figure 18.4; (Deregnaucourt et al., 2005)). This post-sleep deterioration often offsets most of the increase in structure that occurred during the previous day! Whatever the cause of these back and forth oscillations in structure, one might suspect that their presence could affect the learning process. This correlation, however, is not negative, but positive. We observed a positive correlation between the amplitude of these oscillations (measured in the juvenile) and imitation accuracy (measured in the adult). Counterintuitively, birds that lost more song structure after night sleep during critical stages of learning ended up achieving better learning. Interestingly, within each day, birds tend to achieve new records of similarity to the song template during the late morning, when the bird is in late stages of recovery from the post-sleep deterioration effect. We therefore speculate that deterioration of song structure after sleep imposes a learning pattern of repeated refinement of structure, allowing the bird to revisit each morning the song structure that was consolidated

Developmental song learning in the zebra finch 225

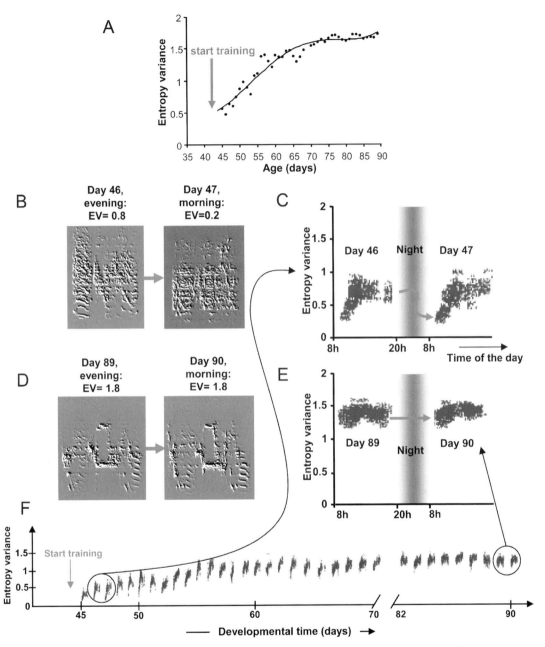

Figure 18.4 Postsleep deterioration. During days of strong vocal changes we often see deterioration of syllable structure after night sleep, which offsets most of the increase in structure that occurred during the previous day. (A) Developmental change in the structure of one syllable type as captured by variance of Wiener entropy (EV). The increase in EV during development relates to the increase of syllable structure shown in the spectral images. (B-E) Changes in the structure of the same syllable during early (B-C) and late (D-E) song development. Tracking EV values continuously shows a decrease in EV values after the night sleep of day 46 (C) but not after the night sleep of day 89 (E). (F) Tracing EV values continuously during development shows strong post night sleep oscillations between days 45 and 60. EV values have been smoothed with a running median (period = 40 data points).

previously. More generally, we suggest that, particularly when learning a complex multidimensional task, there is a tension between retaining plasticity and consolidation of structure, and that the daily oscillations in song structure might alleviate some of this tension (Deregnaucourt et al., 2005).

CONCLUSIONS

Continuous analysis of song development allows us to track changes in the distribution of syllable features and to detect distinctive vocal changes. Developmental maps of song features show the emergence of syllable types. Subsequently, each syllable type goes through a developmental trajectory until a model match is achieved. The better we can describe the "current song structure" during any developmental time, the easier it becomes to test specific predictions and to assess how vocal changes might relate to the postulated error signal.

For example, the duration histograms presented in Figure 18.3 show that one syllable type increased in duration over development until it matched the duration of its template. This increase can be explained by correction of the bird's own song "duration error" in reference to the template. However, this is not the entire story. Across several birds we observed a tendency to develop this complex syllable type from a prototype of shorter duration. This tendency cannot be explained by error correction, but it could mirror a developmental constraint. For example, it could stem from gradual maturation (and expansion) of the subsyllable structure of this complex syllable. We call such postulated constrains "developmentally diachronic" to indicate that those might relate to the developmental stage of the bird.

In addition to imposing constrains, the structure of song development might mirror solutions that evolved to facilitate vocal learning. This might be the case in the post-sleep deterioration effect (Deregnaucourt et al., 2005). The juvenile bird has to master multiple song features at the subsyllabic, syllabic, syntax, and rhythm levels. Because of these embedded layers of structure, any vocal change that reduces error in one feature could potentially cause problems later on, when attempting to match a different feature. Further complications could stem from the nonlinear physics of song production. For example, as in vitro studies of the syrinx have shown, even a single manipulation, such as ramping the air pressure, produces a cascade of complex vocal transitions (Fee et al., 1998). Therefore, while trying to correct one vocal error, the bird might generate additional errors because the timescales of peripheral dynamics might exceed that of the error (Fee et al., 1998). We call such postulated effects "developmentally synchronic" constraints. Post-sleep deterioration of song structure during a period of rapid learning might allow for the removal of such falsely consolidated structures, promoting plasticity over developmental stages and over layers of song structure. Specifically, it might allow the bird to restructure current vocalizations based on consequences of earlier vocal changes (as opposed to "one step at a time" error corrections).

The examples presented above are a mixture of thin observations and strong speculations. We presented them only to suggest that vocal changes cannot be fully explained by any frame-by-frame template matching, while ignoring the existence of constraints in song time (e.g. imitating neighboring sounds, matching rhythms) and in developmental time (e.g. the maturation and acquisition of motor skills). Assuming that vocal changes could be fully explained by template matching is similar to attempting to explain the trajectory of an automobile moving toward a target by postulating the desired operations according to distance and direction error signals – while ignoring the existence of roads, rivers and bridges. We suspect that some aspects of the developmental song learning process may allow the bird many degrees of freedom whereas other aspects may strongly limit the amount of exploration. Cross-level analysis of song development, recording articulatory variables and song-system neuronal activity continuously while vocal changes occur, might provide a better description of vocal changes. Such data could allow the detection of error corrections at articulatory and neuronal levels. Further investigation might tell us more about the structure of the error signal: do trials and errors (within the developmental constrains) fully explain vocal changes, or perhaps the bird has some generalization and anticipation skills that must be taken into account?

We started this chapter by saying that developmental learning is difficult to study because it is difficult to relate the developmental outcome to behavioral changes, which can occur within hours as well as across daily cycles of wakefulness and sleep and over developmental stages. We provided only a few, incomplete answers, but we hope that we demonstrated that this is a tractable problem. We believe that tracking vocal development from moment to moment has an enormous potential and that quantitative behavioral approaches combined with techniques at the molecular, cellular and peripheral levels, are now making it possible to aim at deep, cross-level understanding of developmental sensory-motor learning.

ACKNOWLEDGEMENTS

We would like to thank Olga Feher, Peter Andrews, Kristen Maul, Zvi Devir and Dan Kenigsberg for their help with this work. This research was supported by US Public Health Services (PHS) grants DC04722-07, NS050436 and by a NIH RCMI grant G12RR-03060 to CCNY.

19 • Auditory feedback and singing in adult birds

Sarah M. N. Woolley

INTRODUCTION

In many birds, song learning in juveniles involves memorization of the song of an adult tutor. The natural tutor (or at least one of the tutors) is the father in nonterritorial species such as zebra and Bengalese finches and neighbors in territorial species such as song sparrows. The memorization of an adult's song is followed by a period of vocal practice during which the juvenile uses auditory feedback of his own singing to gradually match his vocal output with the stored memory of his tutor's song. While some songbird species sing normal adult song without the assistance of auditory feedback (Konishi, 1964a, 1965b), others such as adult zebra finches and Bengalese finches require auditory feedback to sing normally. In these birds, singing behavior degrades over time when auditory feedback is removed or severely disrupted (Nordeen and Nordeen, 1992, 1993; Woolley and Rubel, 1997; Yamaguchi and Okanoya, 1997; Leonardo and Konishi, 1999). Disruptions in auditory feedback can also disrupt singing in real time (Cynx and von Rad, 2001; Sakata and Brainard, 2006). Thus, in these species, the auditory system and the vocal motor system are continually exchanging information; auditory cues guide vocal output. Because much of the circuitry involved in song production has been described and because changes in song behavior can be carefully quantified, the songbird makes an ideal model system for investigating the control of a complex motor behavior by sensory feedback. Konishi (2004) has reviewed some of the conceptual issues related to the role of auditory feedback. This chapter focuses on two questions of general interest. First we ask what components of auditory feedback are necessary and sufficient to maintain normal adult song patterns. Second, we ask how auditory feedback is used to control adult song behavior. The first topic concerns the specific acoustic cues from a bird's own voice that are crucial for maintaining normal singing. The second topic addresses how sensory feedback and motor commands meet and interact in the brain.

METHODOLOGICAL CONSIDERATIONS

Studies addressing the relationship between auditory feedback and song behavior require tight control of the bird's experience and careful quantification of songs. They also require subjects that breed readily, and learn and tutor songs under laboratory conditions. Convenient subjects in which to study changes in song behavior are those species in which: (1) each male sings one song type; (2) adult songs are highly stereotyped; and (3) adult songs are normally stable over time/not influenced by exposure to the songs of other birds. These properties are characteristic of both zebra finches and Bengalese finches (Immelmann, 1969; Price, 1979; Dietrich, 1980; Bottjer and Arnold, 1984; Clayton, 1987b, 1988, 1989; Nordeen and Nordeen, 1992; Woolley and Rubel, 1997, 1999, 2002; Zevin et al., 2004). The studies described here focus largely on the Bengalese finch, a domesticated species with Southeast Asian ancestry (Immelmann, 1969; Okanoya, 2004). In this species, there are two lineages. Males from one of these lineages do not show stereotypy in their syllable sequences (Okanoya, 2004). The studies described here used males from the second lineage. These birds sing a stereotyped sequence of syllables called a "motif" (Figure 19.1, upper panel) that is repeated several times in a singing "bout." With respect to number of song elements, the general organization of song, body posturing and the conditions under which song is elicited, these Bengalese finches are similar to zebra finches.

The stereotyped adult songs of both species show acoustic deterioration following surgical deafening (Nordeen and Nordeen, 1992, 1993; Okanoya and Yamaguchi, 1997; Woolley and Rubel, 1997;

Neuroscience of Birdsong, ed. H. Philip Zeigler and Peter Marler. Published by Cambridge University Press. © Cambridge University Press 2008.

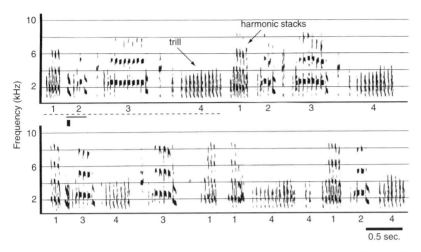

Figure 19.1 Normal Bengalese finch song is composed of stereotyped sequences of syllables (motifs) which are repeated in a singing bout. (Top) A spectrogram of two normal song motifs. (Bottom) The same bird as in top panel but recorded one week after deafening. The syllable order does not match that of the original song and is no longer stereotyped. Syllable types are labeled with numbers below the time axis. The dashed line indicates a motif. The solid line indicates a syllable and the black box indicates a note.

Lombardino and Nottebohm, 2000; Scott et al., 2000). Bengalese and zebra finch song share the same acoustic frequency range (~250–8000 kHz) and some structural characteristics such as the frequent occurrence of harmonic stacks. However, the acoustic structure of Bengalese finch song is more complex than that of zebra finches, containing trills composed of rapid, repeated sequences of short song elements or notes (Figure 19.1, upper panel) grouped into what we call syllables. Zebra finch song lacks this temporal grouping of song elements. Thus, the singing behavior of the Bengalese finch allows us to address the role of auditory feedback in both the structure of song elements/syllables and the temporal organization of syllables.

MANIPULATING AUDITORY FEEDBACK

Song-related auditory feedback in adults may be manipulated *indirectly*, by disrupting the motor control of song. Indirect manipulations that produce a motor deficit, such as severing or damaging the nerves that control the vocal organ, the syrinx, or the vocal motor regions of the brain, produce a situation in which auditory feedback is changed because birds produce and hear abnormal song (Williams and McKibben, 1992; Williams and Mehta, 1999; Kittelberger and Mooney, 2005; Coleman and Vu, 2005). These manipulations affect auditory feedback but have the complication of also affecting the functioning of the vocal motor system.

Auditory function may also be perturbed *directly*, using surgical, pharmacological and/or behavioral manipulations, to disrupt or remove auditory input while leaving the vocal motor system intact. One approach involves surgical deafening, which is complete and permanent. Another is to change the acoustic environment reversibly while keeping the auditory system intact. By presenting sound while birds are singing it is possible to obscure or distort auditory feedback of the song (Marler et al., 1973). Leonardo and Konishi (1999) devised a computer-controlled delayed auditory feedback (DAF) environment in which slightly delayed overlays of a bird's song are externally presented (by a speaker system) while the bird sings. Under these conditions, the bird's auditory feedback becomes a jumbled sequence of temporally overlapping syllables. In modifications of this technique, feedback of specific durations or specific syllables can be altered (Cynx and von Rad, 2001; Sakata and Brainard, 2006). Feedback can also be masked by presenting white noise to birds, either during singing or continuously. This approach has been used to study song development in canaries and zebra finches (Marler et al., 1973; Marler and Waser, 1977; Funabiki and Konishi, 2003) and to examine the role of auditory feedback in the production of adult song in zebra finches (Zevin et al., 2004). All of

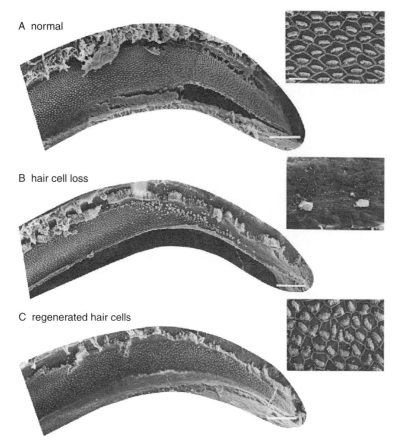

Figure 19.2 Auditory hair cells in the Bengalese finch regenerate after original cells are lost. (A) A scanning electron photomicrograph montage of a normal basilar papilla (avian cochlea). The basal (high frequency) half of the papilla is shown. A high magnification photomicrograph shows the sensory epithelium which is composed of hair cells and support cells. Note the highly organized array of stereocilia bundles on neighboring cells. (B) A basilar from a bird that has been treated with the ototoxic drug Amikacin. Hair cells in the basal end of the epithelium (right) have been killed. A high magnification view shows that the epithelial surface has been denuded of hair cells except for two dead cells lying on the surface. (C) A basilar papilla from a bird that recovered from hair cell loss for 12 weeks. The epithelium is populated with regenerated hair cells. A high magnification view shows the surface structure of regenerated hair cells. The array of regenerated hair cells is less well organized than an array of original hair cells (see A). Scale bars indicate 100 μm on montages and 10 μm on high magnification images.

these methods are effective in disrupting auditory feedback and all cause the degradation of singing behavior.

Auditory feedback may also be manipulated by the selective elimination of the sensory receptor cells for hearing in the basilar papilla (avian cochlea) (Figure 19.2). Like its mammalian counterpart, the basilar papilla of birds is tonotopically organized. High frequency sounds are encoded by hair cells in the base of the sensory epithelium and low frequency sounds by those in the apex or distal end of the epithelium. Accordingly, mid frequencies are encoded by hair cells in the mid region of the epithelium. Thus, the selective elimination of hair cells in specific locations on the basilar papilla results in the selective removal of specific frequency ranges of auditory feedback.

Hair cells are vulnerable to mechanical stimuli, such as prolonged exposure to loud sounds, and to ototoxic drugs such as aminoglycoside antibiotics. The hair cell death and hearing loss caused by auditory overstimulation varies with the intensity, frequency and duration of the sound, with high frequency cells in the base of the inner ear most at risk. High frequency sound will kill

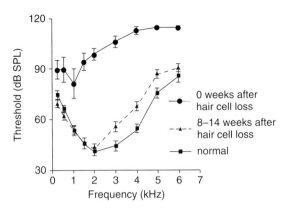

Figure 19.3 Hearing thresholds show profound shifts after original hair cells are killed, and recover to normal (lower frequencies) and near normal (higher frequencies) levels after hair cell regeneration. Auditory thresholds are shown for birds that are normal, have profound hearing loss, and have recovered for 8 or 14 weeks. Error bars represent ± SEM.

hair cells in the base and spare the others. Mid-frequency sound will kill both basal and mid-region hair cells. Low frequency sound alone creates patches of hair cell loss which are highly variable in location and extent of damage (Ryals and Rubel, 1982; Cotanche et al., 1994; Woolley, unpublished observations). Following exposure to ototoxic levels of aminoglycosides, hair cells in the base of the sensory epithelium, the high frequency cells, die and are extruded from the epithelium (Figure 19.2B). With increased dose and duration of drug exposure, cells in progressively more apical (lower frequency) regions of the epithelium also die. The exact pattern and extent of hair cell loss and consequent hearing loss depends on what drug is used, in what dose and for how long. While either chemical or mechanical treatments produce damage to hair cells, combined treatment with ototoxic drugs and sound exposure has a synergistic effect, causing more hair cell loss than either treatment alone (Collins, 1988; Brummett et al., 1992; Cotanche et al., 1987).

Unlike mammals, birds regenerate new hair cells once the original cells have died (Cruz et al., 1987; Corwin and Cotanche, 1988; Ryals and Rubel, 1988; Tucci and Rubel, 1990). In Bengalese finches, even when nearly all original hair cells have been lost, a new population of cells grows and covers the epithelial surface thoroughly by eight weeks after the loss of original hair cells (Woolley and Rubel, 2002). An array of regenerated hair cells can be distinguished from the original hair cell array because it is less precisely organized (compare panels A and C in Figure 19.2). Most importantly, regenerated hair cells become innervated by the brain and restore hearing (Tucci and Rubel, 1990; Marean et al., 1993, 1998; Woolley and Rubel, 1999, 2002; Woolley et al., 2001). Thus, in birds, the hearing loss caused by hair cell loss can be reversed. In Bengalese finches, new hair cells grow mostly over the first four weeks after loss of the original cells, and this regeneration is accompanied by the recovery of auditory response thresholds (measured electrophysiologically) (Figure 19.3; Woolley et al., 2001). The functional recovery, though remarkably good, is not perfect. Hearing thresholds in mid and low frequencies recover to match normal hearing thresholds. But when hair cell loss is extensive, the recovery of high frequency hearing sensitivity suffers; threshold shifts in the higher hearing frequencies persist after hair cell regeneration is complete (Figure 19.3; Tucci and Rubel, 1990; Marean et al., 1993; Woolley and Rubel, 2002). Partial hair cell loss eliminates feedback of some frequencies, leaving feedback of other frequencies intact. This hearing loss is also reversible, with good recovery of hearing thresholds when new hair cells grow in (Woolley and Rubel, 1999, 2002; Woolley et al., 2001). In summary, the tonotopic specificity of hair cell loss and the ability of birds to regenerate hair cells make it possible to eliminate auditory feedback, selectively and reversibly. Using this method, we have examined the effects of selective perturbation of auditory feedback in songbirds whose song control systems are intact.

CONTRIBUTIONS OF AUDITORY FEEDBACK TO SYLLABLE STRUCTURE AND ORDER

In normal Bengalese finches of this lineage, song syllables are produced in a stereotyped temporal order. For example, a bird with four syllable types will sing 1-2-3-4-1-2-3-4-1-2-3-4 (Figure 19.1, upper panel). Within one week of deafening, however, the syllable sequencing becomes nearly random so that the bird may sing 1-3-4-4-4-3-2-1-3-2-2 (Figure 19.1, lower panel). The syllable order changes from song bout to song bout such that no stereotypy over time remains. Interestingly, most singing bouts still begin with introductory notes followed by the first syllable from the normal song, suggesting that initiating a song bout can occur normally without feedback but the production of the next "correct" syllable (#2) requires auditory feedback of the first syllable. Thus, a significant change in syllable *order* occurs early after deafening. In zebra and Bengalese finches, DAF can also disrupt syllable order, after long-term perturbation of feedback and in real time (Leonardo and Konishi, 1999; Cynx and von Rad, 2001; Sakata and Brainard, 2006).

The deterioration of syllable structure begins after the changes in syllable order. One of the ways in which the similarity of sounds is quantified is by crosscorrelating the spectrograms of those sounds. This analysis is equivalent to sliding the time-frequency representations of sounds past each other in time and calculating how well they match. In Bengalese finches, spectrogram cross-correlation analyses of predeafening and postdeafening syllables have shown that the degradation of syllable structure is significant within two weeks of surgical deafening and gradually worsens over time. This pattern of degradation is quite different from that of syllable ordering and suggests that the precision of the motor commands that code syllable structure can persist without feedback for a couple of weeks but then are slowly lost in the continued absence of auditory feedback. In juvenile zebra finches raised for several months in the presence of loud noise to obscure auditory feedback, songs were abnormal and unstable into adulthood (Funabiki and Konishi, 2003). Once normal feedback was reinstated, however, syllable structure during the post-sensitive period developed to match that of the tutor's song. However, when birds were kept from developing their own singing beyond 80 days of age, syllable order did not develop to match the tutor's song. These demonstrations that it is possible to dissociate feedback effects on the changes in syllable order from those affecting syllable structure suggest that these two aspects of song organization may depend on auditory feedback in different ways.

WHAT AUDITORY FEEDBACK REGULATES SONG PRODUCTION?

The ability to produce frequency-specific hearing loss by the combined use of drugs and sound exposure has allowed us to address the question of what kind of auditory feedback (i.e. what frequency components) is necessary for the maintenance of normal adult song behavior (Woolley and Rubel, 1999, 2002). We conducted a study in which we manipulated the frequency ranges of auditory feedback available to birds. In one group of birds, high frequency feedback was abolished by killing the hair cells in the basal one half of the basilar papilla. In another group, hair cells encoding high and mid frequency sounds were killed; only low frequency feedback was preserved. In a third group, hearing was severely impaired across all frequencies by killing hair cells across the entire length of the basilar papilla. Birds with only high frequency hearing were not included because it is not possible to eliminate only low frequency hair cells (see above). The hearing losses were measured electrophysiologically and songs of these three groups of birds were analyzed for changes after the frequency-specific hearing loss.

We found that birds with normal hearing below 1500 Hz maintained normal song patterns. The birds missing low frequency hair cells in addition to the mid and high frequency cells, i.e. those with the largest hearing impairment, showed song degradation similar to that produced by surgical deafening. Therefore, the preservation of apical hair cells encoding only low frequency auditory feedback (1500 Hz and below) is sufficient to maintain normal song, even though the frequency range of song goes up to at least 8000 Hz. This finding suggests that the temporal pattern of the song (which can be obtained from low frequency feedback) and/or the information carried by low frequencies in song contain the crucial auditory feedback information used to maintain normal singing. Thus, even though birds normally hear feedback of the entire song, only a specific subset of the information in the song may be necessary and sufficient to control song production.

The generality of this finding for other song species has not yet been examined. Nor do we know whether there is any correlation between the feedback information used to maintain normal vocal behavior and the feedback that is required for accurate song perception. Note that accurate speech perception in humans appears to be maintained even after removal of a substantial amount of spectral information (Shannon, 1995). Auditory neurons in the zebra finch midbrain and forebrain also appear to be particularly well tuned to the temporal aspects of song (Woolley *et al.*, 2006) and temporal information in general (Woolley and Casseday, 2004, 2005; Woolley *et al.*, 2005). Understanding the components of auditory feedback that are crucial for both vocal production and perception may be helpful for characterizing the neural mechanisms of auditory–vocal integration. For example, are the same spectral and/or temporal cues used to maintain vocal output and to understand the vocal output of another?

SONG MEMORIES AND ADULT SONG

Studies of deafened birds have shown that both zebra finches and Bengalese finches depend on auditory feedback for the maintenance of normal adult song. In both species, song degrades after removal of auditory feedback (Okanoya and Yamaguchi, 1997; Woolley and Rubel, 1997, 1999, 2002). After deafening, adult song degrades more slowly in zebra finches than in Bengalese finches (Nordeen and Nordeen, 1992, 1993; Scott et al., 2000). The time course and extent of postdeafening changes in song vary with the bird's age at deafening, as is true for the chaffinch (Nottebohm, 1968). After deafening older birds maintain normal song patterns longer than younger birds, suggesting that the motor circuitry for song production becomes more fixed and less dependent on sensory feedback as adults age (Lombardino and Nottebohm, 2000). This observation confirms the early studies correlating age with the resistance of song to effects of deafening; younger birds show larger singing deficits after auditory feedback removal (Konishi, 1965b; Nottebohm, 1968).

In the studies discussed above, the removal of auditory feedback has been permanent. But the ability to reversibly deafen songbirds using hair cell manipulations or to temporarily distort auditory feedback using sound interference allows us to ask several questions about how birds use auditory feedback to maintain normal song. Do adult songbirds store memories of their own songs? If so, how long do song memories persist in the absence of feedback? Are these memories stable in the brain regardless of what sounds the bird produces? Can such memories be used to shape song production? And, finally, if disrupting sensory-motor integration destabilizes the circuitry used for song production, do plasticity mechanisms permit the learning of new songs?

Bengalese finches with profound hearing loss due to hair cell loss show song degradation within one week, and the time course of degradation after hair cell loss is similar to that after surgical deafening. Syllable sequences become nearly random immediately after the completion of a one-week treatment to cause hair cell loss (Figure 19.4). Syllable structure significantly degrades by one week after the end of treatment (Figure 19.5). But most important is what happens to the songs during and after the recovery of hearing, as hair cells regenerate. While hair cells regenerate over the four weeks following extensive hair cell loss, syllable order gradually returns toward the ordering in the original, pretreatment songs (Figure 19.4; Woolley and Rubel, 2002). By the end of four weeks following hair cell loss, syllable sequences match those of the original songs. Syllable structure also recovers significantly; but more slowly and less completely (Figure 19.5) with some syllables recovering better than others.

The recovery of original song following hearing restoration indicates that adult birds store neural representations of their own normal songs even when they are unable to produce those songs for some time. Additionally, that "song memory" can be used to shape vocal output when access to it is once again available. Therefore, adult birds may use auditory feedback to compare their own vocal output with a stored model of the "normal" song in a process that is similar to the sensory-motor integration between the tutor song model and vocal output during development.

Additional evidence for the idea that adult birds store memories of their own songs comes from a study of the effects of feedback disruption in adult zebra finches using DAF (Leonardo and Konishi, 1999). After several weeks in the DAF environment, the structure of adult zebra finch song deteriorates significantly; syllables are "stuttered" and deleted, and syllable structure becomes more variable than in normal song. Interestingly, presentation of DAF to humans during speech production also results in disfluency, including stuttering (see Timmons, 1982, for review). Under DAF conditions, the songs of adult zebra finches degrade significantly by roughly six weeks after the onset of the feedback delay. When normal feedback is restored, the songs gradually recover their original forms, suggesting that zebra finches, like Bengalese finches, retain stored models of their normal songs that can be used to shape ongoing adult vocal behavior via auditory feedback.

Much of the evidence on the role of feedback in adult song maintenance comes from studies in which

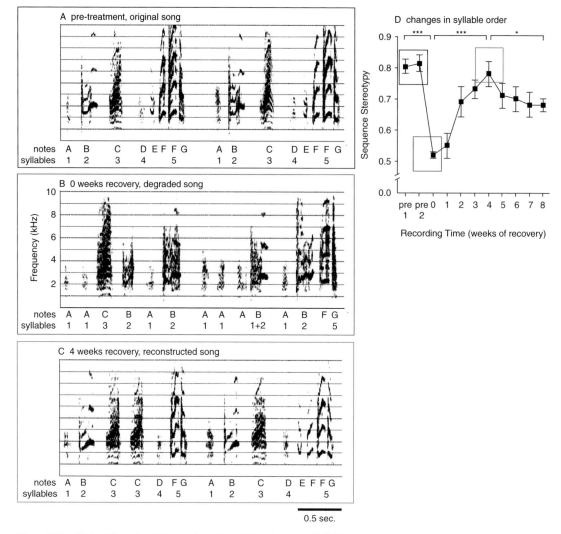

Figure 19.4 Birds with profound hearing loss showed degraded song. Syllable sequences returned to the original, pretreatment order, and some aspects of syllable structure recovered toward that of the original song by four weeks after treatment.
(A) Spectrogram of two song motifs recorded before treatment. (B) Song from the same bird as in (A) but one day after the end of treatment. Comparison of (A) and (B) shows that syllable order has changed and the acoustic structure of syllables has degraded. (C) Two song motifs from the same bird as in (A) and (B) recorded four weeks after treatment. The song recorded after four weeks of recovery appears similar to the pretreatment song and dissimilar to the song recorded immediately after treatment. Individual syllables are labeled with numbers, and notes are labeled with letters below the x-axis. (D) Sequence stereotypy scores for birds singing degraded song after hair cell loss decreased significantly by one day after treatment. By four weeks after treatment, scores had increased significantly and were no longer different from pretreatment scores. Between four and eight weeks after treatment, scores decreased again (see Figure 19.6). $*p < 0.05$, $***p < 0.001$. Sequence stereotypy scores are expressed as: (# syllables types per bout / # transition types per bout) + (sum typical transitions / sum total transitions per bout). Transitions are defined as the progression from one syllable to the next. Typical transitions are defined as those that occur in the normal song.

Auditory feedback and singing in adult birds 235

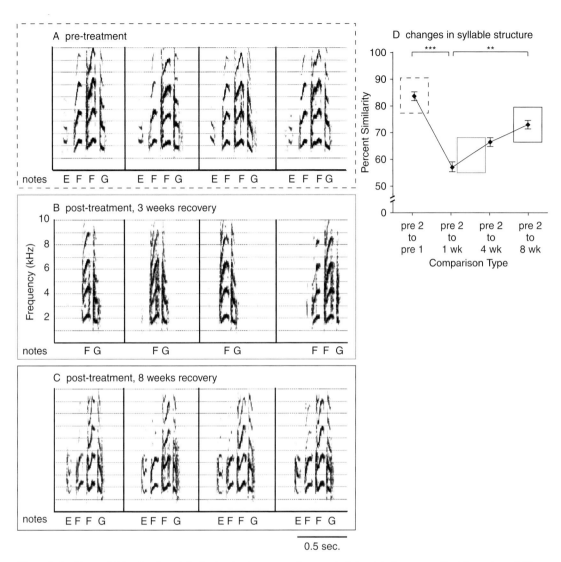

Figure 19.5 Most syllables that degraded after hearing loss returned to match their original structure. By eight weeks, syllable structure was more similar to pretreatment song than to song recorded earlier in recovery.
(A) Spectrograms of the first four iterations of the same syllable type taken from one randomly selected song bout in pretreatment records. The variability of several iterations of the same syllable type in normal song can be seen.
(B) Spectrograms of the first four iterations of the same syllable type from the same bird as in (A) taken from one randomly selected song bout recorded after 3 weeks of recovery.
(C) The first four iterations of the same syllable type from the same bird as in (A) and (B) taken from one randomly selected bout recorded after eight weeks of recovery. (A) and (C) are more similar than (B) is to either (A) or (C). Notes are labeled with letters below each syllable.
(D) Average acoustic similarity between pretreatment syllables and between pretreatment syllables and syllables recorded after treatment show that syllables significantly deteriorated by one week after treatment. Acoustic similarity between syllables from pretreatment recordings and recordings made eight weeks after recovery was significantly improved.
Pre 1 indicates the first pretreatment recording, and pre 2 indicates the second pretreatment recording. 1 wk indicates one week after treatment. Error bars represent ± SEM; $**p < 0.01$, $***p < 0.001$. Acoustic similarity was measured using Sound Analysis (Tchernichovski et al., 2000).

the perturbation of feedback was relatively short term – involving either recovery from hair cell loss or the use of a delayed auditory feedback. However, permanent changes were seen after long-term (5 to 12 months) disruption of auditory feedback in zebra finches by continuous white noise (Zevin et al. 2004) Some birds showed minor changes in song, while others produced severely disrupted song with the loss of whole syllables, degraded syllable structure and decreases in temporal stereotypy that are characteristic of deafened birds (Nordeen and Nordeen, 1992; Brainard and Doupe, 2000a). Song recovery after such long-term perturbation is not complete. Recovery of syllable structure does occur in some cases, indicating that some aspects of a memory of the original song persist through several months of missing feedback. However, unlike the full recovery of song after short-term feedback perturbation (Leonardo and Konishi, 1999), long-term disruption of auditory feedback produces long-term song degradation. In summary, the evidence suggests that song memories exist in the adult brains of both Bengalese and zebra finches, and that they remain stable during short-term removal or disruption of auditory feedback. But the long-term integrity of song memories appears to depend on reinforcement with auditory feedback. Whether the memory of a bird's original song degrades or becomes corrupted over time, or whether access to the memory is lost over time is currently unclear.

SONG PLASTICITY: ADULT LEARNING

Juvenile Bengalese and zebra finches that are deprived of an appropriate song model will continue to be able to learn for several weeks beyond the normal close of the sensitive period, until an acceptable model can be copied (Eales, 1985; Clayton, 1987b). This suggests that the timing of the sensitive period for song learning can be manipulated. But what about the potential for song learning in adult birds? Reversible deafening experiments have allowed us to explore the potential for behavioral plasticity by examining the possibility of song learning during the period of recovery from hair cell loss. Can adult birds be induced to learn a new song by a short-term disruption in normal singing? If so, this would imply that the circuitry required for new song acquisition is present in the adult brain.

We addressed this issue by analyzing song learning in two groups of adult Bengalese finches recovering from a period of auditory feedback perturbation following treatment with a combination of ototoxic drugs and sound exposure (Woolley and Rubel, 2002). One group of birds was housed together during hair cell regeneration and the gradual recovery of hearing. In the other group, each recovering male was isolated with one untreated cagemate singing normal song – to serve as a potential tutor. All birds recovered their original songs and there were no initial differences between the songs of the two groups. However, after song recovery was completed, some (3 of 7) birds from both groups changed their songs away from a match with their original songs. Changes included deleting notes from within syllables, structurally modifying notes into new notes and rearranging those notes to form new syllables (Figure 19.6). Often, such modified or new notes and syllables were sung at the end of a motif. The important aspect of the song modifications was that they gradually emerged as acoustic matches with the notes and syllables of the cagemates' songs. As the match between recovering birds' song notes and their original versions gradually and significantly decreased, between 4 and 8 weeks after hair cell loss (see Tchernichovski et al., 2000; Figure 19.6D), the match between those same notes and the notes of their cagemates' songs increased significantly. After eight weeks of recovery, in these three birds, the new song notes were significantly more similar to the cagemates' notes than to their original versions (Figure 19.6D).

We concluded that the birds with regenerated hair cells had copied the notes and syllables of their cagemates' songs, and concluded that these adult birds had been induced to learn. However, this adult copying process differs in several respects from the learning that occurs during normal song development. During song acquisition, juveniles develop highly structured copies of their tutors' songs starting with only a vocal repertoire of begging calls. In contrast, the new notes that adults acquire are the result of incremental modification of existing notes over time and the new syllables result from the assembly of those notes into patterns that match the cagemate's song but do not match the original song. Thus, new song learning in the adult involves revising and rearranging previously existing song material rather than acquiring new material. Therefore, the song learning exhibited by adults does appear to be

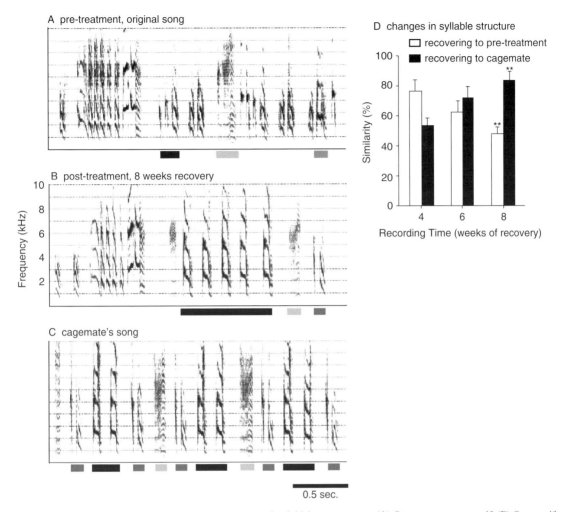

Figure 19.6 An example of song modification that occurs after initial song recovery. (A) One pretreatment motif. (B) One motif from the same bird as in (A) after eight weeks of recovery; the introductory note and the first syllable recovered to their original structures, but the following notes were modified such that they do not match their original structures. These modified notes immediately follow the first syllable, and the rest of the original motif is skipped. (C) Song from the cagemate of the bird shown in (A) and (B). The similarities between (B) and (C) are stronger than between (A) and (B). Bars below the x-axis indicate the similar song portions among panels (A), (B), and (C). (D) Acoustic similarity between modified notes and their original versions (open bars) decreased significantly between four and eight weeks after treatment. Similarity between modified notes and the cagemate's notes (filled bars) increased significantly between four and eight weeks of recovery. Similarities are higher at four weeks than would be expected on the basis of data shown in Figure 19.5 because notes deleted from syllables during song modification could not be included in this analysis. **$p < 0.01$.

limited by comparison with that of juvenile birds. Moreover, Zevin et al. (2004) found no evidence of song learning in zebra finches during recovery from song degradation produced by long-term masking of auditory feedback. It is not clear whether this finding reflects a difference between Bengalese and zebra finches or is related to the use of a long-term rather than a short-term feedback perturbation. In any case, the finding that song learning, albeit in a limited way, can be induced in Bengalese finches after the end of the sensitive period suggests that at least some song learning circuits in some birds are retained past the age at which the sensitive period closes. Because the learning process is quite different from the extension of juvenile song learning, it is possible that it involves different neural circuitry.

The demonstration that adult birds can be induced to learn suggests some new perspectives on song learning. First, there may be forms of song learning that are not limited by a developmental sensitive period (Marler, 1997). Second, initial song learning which occurs during the sensitive period and later (more subtle) song learning may depend on different neural circuits, providing a model system for investigating learning differences in the juvenile and adult brains. Third, the circuitry required for adult song learning may function only (or optimally) under conditions of vocal plasticity, such as are produced by the destabilization (e.g. degradation) of normal song behavior. Marler and Nelson have suggested that song learning in juvenile sparrows involves a "selective" or "action-based" process in which the adult song repertoires of birds are not simply the outcome of an "instructional" process. But, instead, young birds use vocal experimentation and social reinforcement to crystallize their own versions of song types that they hear other birds producing (Nelson and Marler, 1994). Like Marler and Nelson's "action-based learning" adult learning in Bengalese finches occurs during a period of vocal plasticity in which experimentation is possible and in which adult learning may be subject to social reinforcement through interactions with another bird from which the new song elements are learned. The comparative analysis of neural activity during singing in birds with normal, stable song and in birds with abnormal, destabilized song may offer clues as to how sensory and motor circuits interact to accomplish song learning.

CONCLUSIONS: AUDITORY FEEDBACK, ADULT SONG MEMORIES AND NEURAL MECHANISMS OF SONG MAINTENANCE

Studies of a songbird that depends on auditory feedback for the maintenance of normal adult song have provided us with several insights as to how song may be represented in the brain. First, only a subset of the auditory feedback cues available to a bird appear to be necessary for maintaining normal song behavior. Second, adult finches appear to store memories of their own songs that may be used to shape vocal behavior. Third, the observation that limited song learning can be induced in some adults of a species that normally learns only during a sensitive developmental period indicates a greater than expected degree of plasticity in the adult brain.

These studies suggest that there is a stored set of instructions specifying the adult song that *should* be and is normally produced and another set of instructions for the motor commands that actually produce song. These sets of instructions can be decoupled by the removal of auditory feedback. If we consider the motor commands that produce degraded song to contain "errors", then the stored song model in the adult brain could be considered a set of stable circuits that are upstream from those commands. This leads to the question of the physical representation of a song memory. Is this a stored sensory model, similar to the stored representation of the tutor song that is (presumably) formed and used as a template in the juvenile brain (see Adret, this volume)? When auditory feedback is removed in juvenile birds that are in the process of shaping their song output, the normal development of song stops (Konishi, 2004). The tutor song memory is still present in the brain. But the feedback loop between that memory and a young bird's own vocal output is severed by the removal of auditory input. A similar process may take place in adults who store memories of their own songs rather than (or in addition to) the tutor song memory. In this case, removing auditory feedback in adult birds could sever the connection between the stored instructions representing the normal adult song and the vocal motor commands producing the song. Under these circumstances, vocal motor commands would no longer be instructed by the stored song representation and vocal output would degrade.

Understanding how adult song memory is represented in the avian brain will benefit from several lines of research that are currently ongoing. One such line is focused on how song is encoded by neurons in the ascending auditory regions that are presynaptic to the song system (see Theunissen et al., this volume). We are just beginning to describe what song feedback information reaches the sensory/motor structures of the song control system by examining how the auditory midbrain and forebrain process songs (Woolley et al., 2005, 2006). Second, the interactions between sensory activity in the ascending auditory system and motor activity in the vocal motor system are being examined. For example, if the activity of the vocal motor system and the processing of feedback in the auditory system are physiologically monitored at the same time during singing, a functional relationship between auditory feedback and vocal control may be established (Fee et al., 2004). Third, the relationship between song-specific neurons and adult song production can be examined using reversible paradigms for induced plasticity in adult songs. For example, does response selectivity in song-specific neurons change when the song changes? If so, then the song selectivity exhibited by these neurons may be susceptible to training through auditory experience that occurs after the sensitive period for song development, and potentially throughout a bird's life. This is best approached using a delayed feedback (Leonardo and Konishi, 1999) or noise masking (Funabiki and Konishi, 2003) paradigm because normal hearing can be restored soon after song has degraded. Finally, the location(s) of a song memory in the brain must be investigated. Lesion studies combined with manipulations to auditory feedback similar to those designed by Brainard and Doupe (2000a) could be successful in identifying candidate brain regions for memory storage and retrieval. Once some idea of where to look has been gained, the organization of neuron ensembles that represent the memory coding of a complex and learned behavior can be approached. This will be an exciting opportunity and one that demonstrates how understanding the role of auditory feedback in song production in birds can contribute to the basic question of how a memory is represented in the brain.

20 • The anterior forebrain pathway and vocal plasticity

Michael S. Brainard

INTRODUCTION

The anterior forebrain pathway (AFP) is a basal ganglia–dorsal forebrain circuit comprising a number of interconnected nuclei (Area X > DLM > LMAN) whose projections form a loop linking it to the premotor pathway for song production (see chapters by Reiner *et al*. (Figure 5.2) and Farries and Perkel, this volume). Nucleus HVC of the motor pathway sends a projection to Area X, which is homologous to basal ganglia (Bottjer and Johnson, 1997; Reiner *et al*., 2004a). Area X then projects to the medial nucleus of the dorsolateral thalamus (DLM), which in turn projects to the lateral magnocellular nucleus of the anterior nidopallium (LMAN). LMAN completes a loop back to the motor pathway with a projection to the motor nucleus RA. The nuclei of the AFP are prominent in birds that are song learners, and evidence from many sources suggests an important role for this pathway in vocal learning and plasticity.

The initial evidence for a role of the AFP in vocal learning derived from lesion studies in zebra finches. In adult zebra finches with stable, crystallized songs, lesions of the AFP have little overt disruptive effect on song, indicating that the AFP is not critical to the premotor pattern generation that underlies the production of the bird's learned song (e.g. Figure 20.1A). However, in juvenile birds, lesions of the AFP prevent the normal progression of song learning, and result in adult birds with songs that have highly abnormal features (Bottjer *et al*., 1984; Sohrabji *et al*., 1990; Scharff and Nottebohm, 1991). For example, lesions of LMAN, at the output of the AFP, cause an abrupt stabilization of normally variable juvenile song (Figure 20.1B). This stabilization is observed in the first songs produced following lesions (1–3 days post lesion) and results in songs in which both the structure of individual syllables as well as the sequences in which they are produced are often simplified and more stereotyped than in songs prior to lesions (Bottjer *et al*., 1984; Sohrabji *et al*., 1990; Scharff and Nottebohm, 1991). A strikingly similar stabilization of song can be achieved by reversible inactivation of LMAN (Ölveczky *et al*., 2005) (Figure 20.1C). Moreover, infusion of APV (an antagonist of the NMDA subtype of glutamate receptor) into the motor nucleus RA also elicits a rapid and reversible song stabilization (Ölveczky *et al*., 2005) probably because it blocks transmission from LMAN, which is known to be glutamergic (Mooney, 1992; Stark and Perkel, 1999). These data indicate that much of the effect reported for LMAN lesions can be attributed to the immediate consequences of silencing synaptic inputs from the AFP to the motor pathway.

In addition to an immediate stabilizing effect of LMAN lesions on juvenile song, such lesions also prevent the normal progression of song towards a good match with the tutor. Following LMAN lesions in juveniles, song may remain unchanging or, in some cases, regress over time to an even more simplified form (Bottjer *et al*., 1984; Scharff and Nottebohm, 1991). Figure 20.1D illustrates the gradual simplification and deterioration of song following bilateral lesions of LMAN in a 35-day-old zebra finch. Here, the ultimate adult song produced by the bird (at 90 days, bottom panel) is highly aberrant and essentially unrecognizable as zebra finch song. The occurrence of such a gradual deterioration of song in some (though not all) birds suggests that, in addition to the immediate effects of removing patterned neural input from the AFP, there may be slowly developing, possibly trophic, consequences of lesioning LMAN (Bottjer *et al*., 1984; Scharff and Nottebohm, 1991).

Lesions of Area X also prevent a progression of song towards a good match with the tutor song. However, there are qualitative differences in the disruptive effects of the two types of lesions. In contrast with the premature stereotypy and simplification seen after LMAN lesions, the songs of early Area X lesioned birds remain highly variable and fail to develop normal

Neuroscience of Birdsong, ed. H. Philip Zeigler and Peter Marler. Published by Cambridge University Press. © Cambridge University Press 2008.

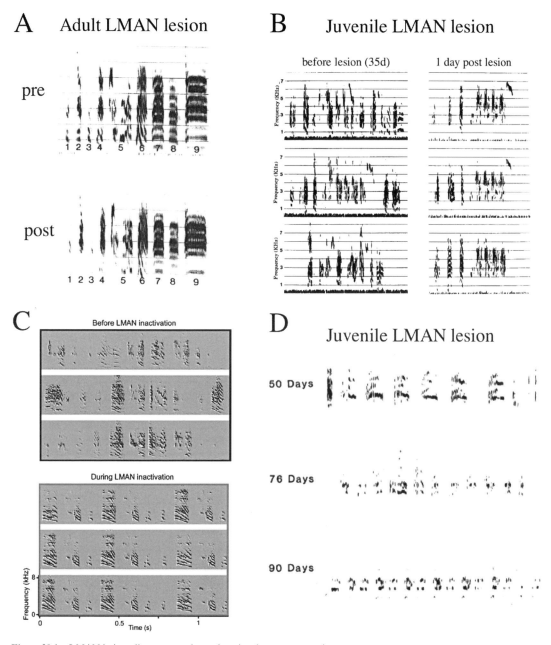

Figure 20.1 LMAN lesions disrupt normal song learning, but not song maintenance.
(A) In adult zebra finches, learned song normally remains stable over time, and lesions of the AFP have little overt effect on song. The top panel shows the learned song motif of a normal adult zebra finch. This bird's motif was composed of 9 distinct syllables (numbered 1–9) that were repeatedly produced in the stereotyped sequence illustrated here. The bottom panel shows the motif produced by the same bird 16 weeks following large bilateral lesions of LMAN. Lesions had no apparent effect on the bird's song. (Data are from Nordeen and Nordeen, 1993.)
(B) In juvenile zebra finches, lesions of LMAN cause an abrupt stabilization of song. The three panels at the left show examples of songs from a 43-day-old zebra finch. At this age there is significant variation in the structure and sequencing of syllables between successive songs produced by the bird. The three panels at the right illustrate the remarkable increase in stereotypy of the same bird's song 1 day

stereotypy with respect to either syllable structure or syllable sequencing (Sohrabji *et al.*, 1990; Scharff and Nottebohm, 1991) (Figure 20.2). These data indicate that the AFP plays a key role during song learning in establishing normal, learned patterns of connectivity within the song motor pathway. Studies in adult birds provide further support for dissociation between AFP lesion effects on song production and song plasticity. Normally, adult zebra finch song remains stable over time. However, manipulations of auditory and/or proprioceptive feedback in adults, such as deafening or tracheosyringeal nerve section, can produce plasticity in otherwise stable song (Nordeen and Nordeen, 1992; Williams and McKibben, 1992; Leonardo and Konishi, 1999). Lesions of the AFP output nucleus LMAN prevent these experience-dependent changes to adult song (e.g. Figure 20.3) (Williams and Mehta, 1999; Brainard and Doupe, 2000a). Finally, in *adult* birds, raised in isolation, a condition that facilitates adult song learning long after the normal learning period is over, such learning is prevented by LMAN lesions (Morrison and Nottebohm, 1993). Collectively, the lesion data implicate the AFP in enabling vocal plasticity in both juvenile and adult birds, but the specific contribution of the AFP to such plasticity remains to be determined.

Instructive versus permissive contributions to vocal plasticity

Figure 20.4 schematizes the various processes that are thought to contribute to sensorimotor learning of song. They include (1) the premotor control of song production ("motor structures"), (2) the encoding and transmission of information about the quality of the bird's own song ("auditory feedback"), (3) the evaluation of auditory feedback relative to a previously memorized song "template" and (4) the generation of instructive signals that impinge on motor structures and drive adaptive changes in song. Given these processes, and the lesion data presented above, the AFP might play either an "instructive" or a "permissive" role in vocal plasticity.

Instructive model

According to an "instructive" model, the AFP is part of the system schematized in Figure 20.4 that participates in evaluation of feedback from the bird's own vocalizations and the generation of instructive signals that guide adaptive changes in the motor pathway so that song progresses towards a better match to the template. Such signals could simply reflect the degree to which the song deviates from the template (i.e. scalar "reinforcement" or "punishment" signals). Alternatively, the information conveyed could be "parameterized", reflecting the direction of deviation of the current song from the template (see Troyer and Bottjer, 2001, and Saar *et al.*, this volume, for further discussion).

In this account, the deficits that follow AFP lesions in juvenile birds arise from aberrant or absent instructive input to the motor pathway. In lesioned juvenile birds, there is no guidance to drive song towards a good match with the tutor song. In contrast, for normal

---

Caption for Figure 20.1 (cont.)
later, after bilateral lesions of LMAN. The postlesion song incorporated a sequence of notes from the prelesion song that is visible in the middle section of the middle spectrogram on the left. (Data are from Scharff and Nottebohm, 1991.) This kind of rapid reduction in syllable and sequence stereotypy is typical for the first songs produced following LMAN lesions (1–3 days postlesion) (Bottjer *et al.*, 1984; Scharff and Nottebohm, 1991).
(C) A strikingly similar stabilization of song follows transient inactivation of LMAN in juvenile zebra finches. The top panel shows three songs recorded from a juvenile zebra finch at 57 days of age. The bottom panel shows songs of the same bird recorded one hour later, following inactivation of LMAN by infusion of TTX. These results indicate that the immediate consequences of LMAN lesions are likely due to the removal of synaptic input to RA, rather than trophic consequences of lesions. (Data are from Olveczky *et al.*, 2005.)
(D) LMAN lesions in juvenile zebra finches lead to abnormal adult songs. The three panels show songs recorded from a zebra finch that received lesions of LMAN at 35 days of age. The songs at ensuing time points all exhibit simpler and more stereotyped structure than is typical of zebra finch song. The adult song produced by this bird at 90 days of age (bottom) consists of syllables that are atypical for adult zebra finches. Moreover, the number of syllables is reduced relative to normal, and these syllables are repeated in an abnormally long string. These features are characteristic of the adult songs that result from juvenile lesions of LMAN (Bottjer *et al.*, 1984; Scharff and Nottebohm, 1991). In this instance it is also noteworthy that a gradual evolution of song towards a simpler form follows the initial effects of lesions. (Data are from Bottjer *et al.*, 1984.)

The anterior forebrain pathway and vocal plasticity 243

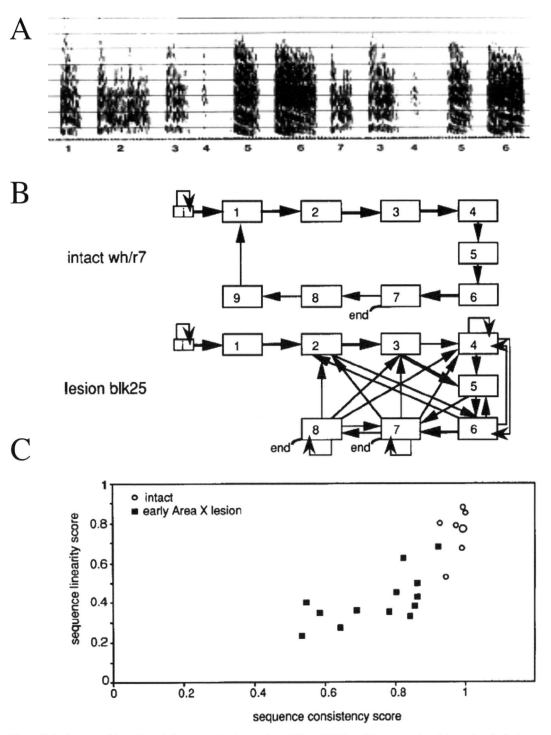

Figure 20.2 Lesions of Area X result in songs with abnormal variability. (A) The adult song produced by a zebra finch that received bilateral lesions of Area X at 31 days of age. Although the number of distinct syllables produced by this bird is typical of zebra finch song, the syllables are abnormally "noisy." In addition, both the structure of individual syllables and their sequencing are

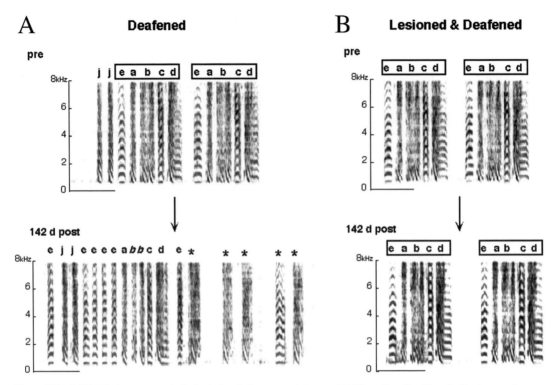

Figure 20.3 LMAN lesions prevent deafening induced changes to adult song. (A) Disruption of auditory feedback normally causes a gradual deterioration of adult zebra finch song (Nordeen and Nordeen, 1992). Here, the song of a normal adult zebra finch is shown before (top), and 142 days after (bottom) removal of auditory feedback by deafening. The bird's stereotyped motif (boxed: eabcd) has degraded both with respect to individual syllables and with respect to the temporal pattern of the song. Asterisks (*) indicate strings of syllables not observed prior to deafening. (B) The song of a second adult zebra finch before (top) and 142 days after (bottom) a combination of deafening and bilateral lesions of LMAN. This bird was a brother to the bird pictured in (A), and initially had a similar song motif (boxed: eabcd), due to shared tutor experience. In this lesioned bird, song remained unchanged following deafening. Horizontal scale bar = 500 ms. (Data are modified from Brainard and Doupe, 2000a.)

adults, after song learning has progressed to completion, the motor pathway is autonomously capable of producing the bird's learned song, and there is no longer a requirement for instructive input to guide changes to song. Hence, lesions in adults have no immediate effect on song. However, in adult birds subjected to manipulations such as deafening, the mechanisms that evaluate auditory feedback interpret the absence of feedback as an error of motor production, leading to the generation of aberrant instructive signals that actively drive deterioration of song. If the changes to adult song that occur following experimental manipulations are actively driven by mechanisms of feedback evaluation, then lesions of the AFP, if they

---

Caption for Figure 20.2 (cont.)
more variable from one song to the next than is normal for adult zebra finch song. (B) Transition diagrams for songs recorded from an intact adult zebra finch (upper) and an adult zebra finch that received bilateral lesions of Area X as a juvenile (lower). Syllables of the birds' repertoires are represented by the numbered boxes and the probabilities of transitions observed between syllables are represented by the thickness of lines connecting boxes. The intact bird produced syllables in a very stereotyped sequence, whereas the lesioned bird exhibited highly variable patterns of transitions between syllables. These features are typical of the adult songs that result from juvenile lesions of Area X (Sohrabji et al., 1990; Scharff and Nottebohm, 1991). (Data are from Scharff and Nottebohm, 1991.)

Feedback-dependent learning

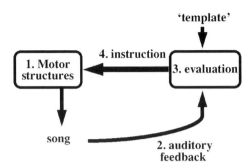

Figure 20.4 Schematic of processes underlying song learning and plasticity (after Konishi, 1965b). (1) Vocal motor structures produce song. (2) Auditory feedback from the bird's own song is conveyed to the CNS and (3) evaluated relative to a previously stored song "template." (4) A mismatch between experienced feedback and the template is presumed to give rise to instructive signals that actively guide changes in the song motor pathway. In juvenile birds, such a mismatch is naturally present during the period of sensorimotor learning, and instructive signals guide improvement in developing song. In adult birds, such a mismatch can be experimentally introduced either by altering or removing feedback. Here, evaluation mechanisms may interpret the altered feedback as an error of production, leading to the generation of aberrant instructive signals that actively drive song deterioration.

do indeed interrupt instructive input to the motor pathway, should prevent these attendant experience-dependent changes to adult song (Figure 20.5A).

Permissive model
According to a "permissive" model, the AFP provides "plasticity factors" that are required for change in the song motor pathway without directly providing signals that guide such change. In this account, the effects of AFP lesions in both juveniles and adults can be understood simply as arising from the removal of factors necessary for the operation of plasticity mechanisms within the motor pathway (Figure 20.5B); song remains stable despite the presence of signals that would otherwise drive song learning (juveniles) or song deterioration (e.g. deafening in adults).

One version of the permissive model suggests that abnormalities of song learning in juvenile birds following lesions of LMAN derive directly from removal of necessary trophic inputs to the motor pathway, and consequent damage to the normal function of this pathway. This hypothesis is consistent with the

A. Instructive model

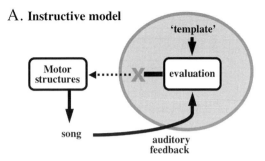

B. Permissive model

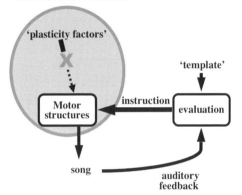

Figure 20.5 Instructive and permissive models of AFP function in vocal plasticity. (A) According to an instructive model, AFP lesions may prevent changes to song by interrupting or eliminating feedback-based instructive signals that normally drive song changes. (B) According to a permissive model, AFP lesions may remove neural or trophic factors that are required in order for the motor pathway to change in response to instructive input that potentially arises from elsewhere.

observation that the AFP is necessary for normal development and physiological function of the song motor pathway, even prior to the initiation of song learning. Lesions of LMAN in juvenile birds cause atrophy of the motor nucleus RA, including extensive cell death (Akutagawa and Konishi, 1994; Johnson and Bottjer, 1994). However, since lesions of LMAN in adult birds have much less effect on RA neurons, and almost no effect on song, the consequences of LMAN lesions for adult song plasticity do not appear to reflect gross disruption of the normal functioning of the motor pathway in song production (Bottjer *et al.*, 1984; Nordeen and Nordeen, 1993; Kittelberger and Mooney, 1999; Williams and Mehta, 1999).

Nevertheless, despite grossly normal functioning of the motor pathway with respect to song production in lesioned adult birds, lesions may produce more subtle changes to physiological properties of RA that could impact plasticity. LMAN projections to RA are known to be glutamatergic and to rely principally on NMDA receptors (Mooney, 1992; Stark and Perkel, 1999). Furthermore, these projections synapse on dendrites of RA neurons that also receive excitatory synapses intrinsic to the motor pathway (Herrmann and Arnold, 1991) and that likely form a key element of the connectivity responsible for producing song. Lesions of LMAN, by physically removing one set of glutamatergic synapses onto RA neurons, might interfere with putative plasticity mechanisms within RA, such as LTP (see Nordeen and Nordeen, this volume).

Similarly, lesions of LMAN may remove from the motor pathway trophic factors that normally support plasticity. A variety of evidence suggests that the AFP may provide BDNF or other neurotrophins to the motor pathway. The cell death within RA that is normally caused by lesions of LMAN can be rescued by infusion of BDNF (Johnson et al., 1997). Moreover, exogenous BDNF supplied to adult RA results in synaptic sprouting within the motor pathway, and overt changes to song (Kittelberger and Mooney, 2005). These results are consistent with the possibility that AFP derived neurotrophins normally regulate the functioning of motor pathway neurons, and that lesions of the AFP may interfere with vocal plasticity by eliminating factors required for structural and functional reorganization. In this case, the normal role of the AFP would be considered permissive to the extent that such factors enable plasticity without directing the nature of song changes.

A related possibility for mediating permissive effects is that the AFP provides inputs that, while not themselves critical to plasticity mechanisms (i.e. LTP or subsequent structural changes) do provide a necessary substrate for those mechanisms. For example, in addition to synaptic sprouting, BDNF has been linked to neurogenesis in the song system (Rasika et al., 1999; Li et al., 2000). It has been suggested that new neurons might form a substrate for the formation of new "memories" and attendant plasticity of song, without providing signals that specify the nature of that plasticity (Goldman and Nottebohm, 1983; Nordeen and Nordeen, 1988; Rasika et al., 1999; Li et al., 2000; Nottebohm, 2002). If, in this case, the birth of new neurons requires trophic inputs from the AFP, then removal of those inputs could prevent the normal incorporation of new neurons and thereby block plasticity. Consistent with this possibility, lesions of LMAN indeed appear to alter the incorporation of new neurons in the motor pathway under conditions of induced vocal plasticity (Wilbrecht et al., 2002a; Pytte et al., this volume).

Current data do not clearly discriminate between instructive and permissive roles of the AFP. Indeed, these two models are not mutually exclusive, and it is possible that the AFP contributes to both kinds of processes. An instructive model would be supported by evidence that the AFP has access to information about the quality of the tutor song and/or bird's own song, and can generate signals that are able to direct change in the motor pathway and song. A permissive model would be supported by evidence that the AFP provides neural or trophic factors that alter the capacity of the motor pathway to undergo change. The following sections present observations about AFP function that provide evidence consistent with one or both models, and that ultimately must be incorporated in any comprehensive account of AFP function.

Possible role of the AFP in developmental regulation of plasticity

Since lesions of the AFP have a conspicuous effect on song plasticity, regulation of plasticity may be one of the normal functions of this pathway. In support of this idea, there are a number of developmental events both within the AFP and in its connections to the motor pathway that parallel at a gross level a developmental decline in the capacity of birds to alter their songs. In zebra finches, the AFP and its connection to RA undergo numerous changes by 60 days, a time when the critical period for zebra finch sensory learning closes. Many of these changes are regressive in nature. During the period in which HVC innervates RA (Herrmann and Arnold, 1991), the synapses from LMAN to the motor pathway, which are mediated largely by glutamate receptors of the NMDA type (Kubota and Saito, 1991; Mooney, 1992; Stark and Perkel, 1999), decrease in number, and the initially coarse topographic projection from LMAN to RA undergoes refinement (Iyengar et al., 1999). During this period, pruning of connections is also prominent within the AFP itself: LMAN neuron spine density decreases between 25 and 55–60 days of age in zebra finches (Wallhausser-Franke et al., 1995). This is accompanied by decreased NMDA receptors in LMAN

(Aamodt et al., 1992; Basham et al., 1999), faster NMDA currents at the synapses from thalamus to LMAN (Livingston et al., 2000), pruning of thalamic arbors in LMAN (Iyengar and Bottjer, 2002a), and loss of activity-dependent synaptic potentiation and depression at synapses within LMAN (Boettiger and Doupe, 2001).

Lesions of LMAN, by removing the projection from the AFP to the motor pathway, directly mimic a process similar to the normal developmental decline in the strength of connections between LMAN and RA. In juvenile birds, LMAN lesions also appear to cause the premature appearance of adult-like properties in the motor pathway. This includes both a loss of dendritic spines within RA and an increase in the speed of synaptic currents at HVC to RA synapses (Kittelberger and Mooney, 1999). These observations suggest that neural or trophic inputs from LMAN may normally serve to maintain the motor pathway in a juvenile, plastic state (Kittelberger and Mooney, 1999). Lesions of the AFP may thus mimic the normal developmental decline in these inputs, resulting in a prematurely crystallized song.

In this context, it is worth noting that the nature of song abnormalities that result from AFP lesions depend on where the pathway is disrupted. While lesions of the AFP output nucleus, LMAN, arrest song development in a manner that is suggestive of a premature crystallization of song (Bottjer et al., 1984; Scharff and Nottebohm, 1991) (Figure 20.1), lesions of the input nucleus, Area X, seem to have the opposite effect (Figure 20.2). Area X lesions prevent the normal progression of the juvenile song towards a good match with the tutor song. However, they appear to leave song in a perpetually variable state; song syllables and overall song structure do not improve, but continue to change from one rendition to the next (Sohrabji et al., 1990; Scharff and Nottebohm, 1991).

Given the known connectivity and physiology of the AFP, different effects of LMAN and Area X lesions are not entirely surprising. Lesions of LMAN necessarily remove the connections from the AFP to RA and any associated neural or trophic factors provided by these inputs. In contrast, lesions of Area X leave synapses from LMAN to RA intact. Moreover, because Area X sends an inhibitory projection to DLM (which in turn most likely excites LMAN), the net effect of Area X lesions may be to increase activity within LMAN (Luo and Perkel, 1999a). In this case, Area X lesions may lead to an excess of those factors that enable or drive plasticity within the motor pathway.

In regulating vocal plasticity, a key function of the AFP may be to introduce variability into the patterns of activity within the motor pathway (Bottjer et al., 1984; Scharff and Nottebohm, 1991; Doya and Sejnowski, 2000; Kao et al., 2005; Ölveczky et al., 2005). Variability of motor output is a requisite component of reinforcement-based learning; in order for song to change in response to a feedback-based reinforcement signal, song must vary at least somewhat from rendition to rendition (Sutton and Barto, 1998). Such variation is required so that evaluation mechanisms can differentially reinforce patterns of motor activity that give rise to better songs (i.e. renditions closer to the tutor song) and/or extinguish patterns that give rise to worse songs. In this sense, variability itself may be a permissive factor for vocal plasticity – such variability does not direct the nature of change to song, but reflects motor exploration that enables better and worse songs to be discovered.

The AFP could potentially influence such variability within the motor pathway on more than one timescale. On a slow timescale, neurotrophins from the AFP might introduce song variability by promoting synaptogenesis or the incorporation of new neurons into the motor pathway (Scharff and Nottebohm, 1991; Johnson et al., 1997; Kittelberger and Mooney, Kittelberger and Mooney, 1999; Rasika et al., 1999; Li et al., 2000). On a faster timescale, neural activity from the AFP has the potential to modulate ongoing patterns of activity within the song motor pathway. The rapid effects of LMAN lesions on song stability (present within the first songs produced 1–3 days post lesion) (Bottjer et al., 1984; Scharff and Nottebohm, 1991) (Figure 20.1B), suggest that synaptic activity from LMAN might indeed be a direct source of modulation of song variability. In juvenile birds, reversible inactivation of LMAN elicits a rapid stabilization of song structure, similar to that observed following lesions (Ölveczky et al., 2005) (Figure 20.1C). Moreover, in adult birds, lesions of LMAN prevent rapid modulation in the variability of song structure that is normally driven by changes in social context (Kao et al., 2005; Kao and Brainard, 2006). Finally, microstimulation of LMAN during singing can alter song structure on a moment-by-moment basis, indicating directly the capacity of neural activity from the AFP to contribute

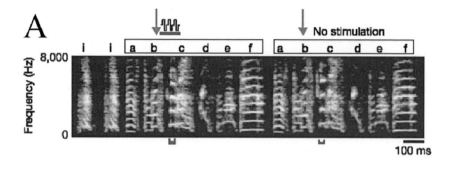

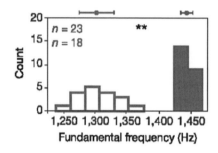

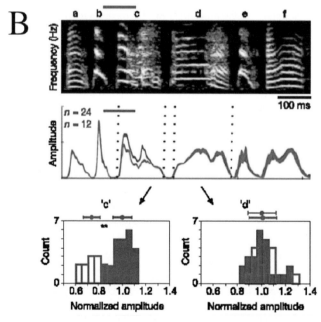

Figure 20.6 Microstimulation of LMAN during singing in adult zebra finches can drive acute changes to learned aspects of song syllables. (A) Example of a stimulation-induced change in fundamental frequency. Top: Spectrogram of adult zebra finch song. On randomly interleaved trials LMAN was stimulated during the production of syllable "c" (30 uA; bar). Bottom: fundamental frequency for syllable "c" for control (shaded bars) and stimulation (open bars) trials. Stimulation in this case caused a 10% decrease in

to modulation of ongoing song (Kao et al., 2005) (Figure 20.6). Hence, these data indicate that the AFP is both necessary and sufficient to generate some forms of variability within the motor pathway and subsequent vocal output. Together, these data support the possibility that one of the critical functions of the AFP in enabling song learning and plasticity may be to generate motor variability that is a substrate for reinforcement learning (Doya and Sejnowski, 2000).

For efficient reinforcement learning, variability should be regulated over time in relation to the current status of motor performance. When song is far from the desired target, high variability would facilitate emission of a large number of variants that, through reinforcement, could come to successively approximate the "target" song. In contrast, as the desired target is approached, low variability ensures that each rendition is near that target; once song is learned, only a low degree of residual variability should be required for its maintenance. Given the potential importance of motor variability or "exploration" to motor learning, it may be that a normal developmental decline in those factors from the AFP that promote variability is a central component of song crystallization.

Synaptic and morphological changes in the connections from LMAN to RA and within the motor pathway are most dramatic in zebra finches up to 60–90 days of age, with little overt change observed subsequently (Herrmann and Arnold, 1991; Kittelberger and Mooney, 1999; Stark and Perkel, 1999). Likewise, the plasticity of song in response to manipulations of sensory experience (such as deafening) progressively declines over this period. Although song plasticity in response to deafening is comparatively attenuated in birds beyond 90 days of age, there is nevertheless a significant residual degree of song plasticity, which further declines over a period of months following the nominal crystallization of song (Lombardino and Nottebohm, 2000; Brainard and Doupe, 2001). It is unknown whether this decline in adult plasticity is paralleled by further, subtle regressive events in the connections of LMAN to RA. However, if this is the case, then LMAN lesions in adult birds may cause a stabilization of song (in response to manipulations such as deafening) by accentuating these events. Such lesions might precipitate a maturation of the song system and attenuation of factors required for plasticity, effectively creating a "hyper-crystallized" song.

SONG SELECTIVITY IN THE AFP AND ITS POSSIBLE CONTRIBUTION TO SENSORY LEARNING

Neural responses to song stimuli have been characterized in several parts of the song system and provide some insights into what mechanisms may contribute to representing the tutor song and feedback of the bird's own song. One intriguing property is "song selectivity," first described in HVC (McCasland and Konishi, 1981; Margoliash, 1983; Margoliash and Fortune, 1992; Mooney, 2000), but now shown to exist throughout the song system, including the AFP (Doupe, 1997; Solis and Doupe, 1997, 1999; Janata and Margoliash, 1999). Song selective neurons respond more strongly to playback of the bird's own song and, in some cases, the tutor song, than to conspecific songs (for reviews see Theunissen et al. and Prather and Mooney, this volume). Such neurons potentially are well suited to process auditory feedback of the bird's own song and to participate in evaluating the similarity of that song to the tutor song. For example, neurons that respond selectively to the tutor song would presumably fire more strongly to the extent that auditory feedback of the bird's own song resembles the tutor song, and hence could contribute to a signal indicating degree of match.

Although song selective neurons are found within the AFP, consistent with the possibility that this pathway is involved in evaluation of song, the presence of similar

---

Caption for Figure 20.6 (cont.)
fundamental frequency. (B) Example of a stimulation-induced change in syllable amplitude. Top: Spectrogram of another bird's song. Middle: mean amplitude waveforms (± SEM) for control (black) and stimulation (shaded) trials. Bottom: Distribution of amplitudes for individual version of syllables "c" and "d" under control (shaded bars) and stimulation (open bars) conditions. Stimulation caused a reduction in the amplitude of syllable "c" but had no effect on the ensuing syllable "d". As illustrated here, stimulation of LMAN during song often caused modulation of the structure of individual syllables. The nature of the modulation often depended on the site of stimulation in LMAN and varied across distinct syllables of the bird's repertoire. These results demonstrate the capacity of signals from the AFP to direct specific, real-time changes to the structure of song. (Data are from Kao et al., 2005.)

selectivity in many song nuclei suggests that mechanisms involved in song memorization and feedback evaluation could be highly distributed or reside elsewhere in the song system. Indeed, some of the specialized auditory processing that subserves song learning may occur in auditory areas that are afferent to the song system. Consistent with this possibility, neurons in the high level auditory areas NCM and cHV exhibit some aspects of song selectivity (Mello et al., 1992; Chew et al., 1996a; Jin and Clayton, 1997a; Stripling et al., 1997; Mello and Ribeiro, 1998; Bolhuis et al., 2000; Gentner and Margoliash, 2003; Bolhuis and Gahr, 2006). For each area exhibiting song selectivity, it will ultimately be important to investigate the extent to which this property reflects sensory learning of the tutor song versus acoustic experience of feedback from the bird's own song during sensorimotor learning (Marler and Doupe, 2000; see also Adret, this volume).

Lesion studies are problematic for testing the specific role of brain regions in the sensory phase of song learning since the main assay for what a bird has memorized is the song that the bird ultimately produces. It is therefore difficult to dissociate song abnormalities due to a disruption of sensory learning from those that may reflect disruption of subsequent sensorimotor learning or song production. To circumvent this problem, reversible inactivation of LMAN has been used to disrupt AFP activity selectively, i.e. during tutoring sessions but not during subsequent sensorimotor rehearsal (Basham et al., 1996). Under these conditions, song learning in experimental birds was significantly reduced. This reduction was not due to nonspecific or lasting damage to LMAN, because learning was comparatively normal in control birds that received similar inactivation of LMAN outside the context of tutor song exposure. Because these control birds were able to memorize tutor song and proceed normally through sensorimotor learning, the deficits in experimental birds seem unlikely to derive from a gross disruption of putative permissive factors such as neurotrophins. This intriguing experiment provides the most direct evidence to date of involvement of a brain area in memorization of tutor song. Experiments along these lines, in which AFP function is manipulated in a temporally controlled fashion, seem likely to provide important insights (see Adret, this volume).

Another way in which sensory versus sensorimotor functions of the AFP can be dissociated is to study the effects of lesions in purely perceptual tasks, outside the context of vocal production. Although such studies have not addressed the issue of tutor song memorization, they have found that lesions of song nuclei, including HVC and LMAN, interfere with the performance of birds in tasks that require song memorization and discrimination (Brenowitz, 1991; Scharff et al., 1998; Gentner et al., 2000).

Regressive developmental changes within the AFP could potentially underlie an experience-dependent narrowing of song responsiveness as birds encode a particular tutor song memory. In zebra finches, however, the period of sensory learning also overlaps with the onset of vigorous singing, sensorimotor rehearsal, and refinement of auditory selectivity for BOS, making it difficult to specifically attribute any changes to sensory learning. Several manipulations, such as isolate rearing and hormonal treatments, have been used to alter the time-course of song learning in an attempt to test the correlation between physiological events and learning (as opposed to learning-independent developmental changes). Thus far, studies that have carefully tested the relationship between behavioral plasticity and physiological changes in the AFP (such as changes in NMDA receptor kinetics or pruning of dendritic spines) have generally concluded that there is not a tight correlation between any of the examined physiological changes and the closure of sensitive periods for learning (Wallhausser-Franke et al., 1995; Livingston et al., 2000; Livingston and Mooney, 2001; Heinrich et al., 2002, 2005). For example, although isolation rearing enables late learning, it delays but does not prevent shortening of NMDA receptor kinetics within the AFP (Livingston et al., 2000; Livingston and Mooney, 2001; Heinrich et al., 2002). Hence, the ability of isolates to learn new songs indicates that changes in NMDA receptor kinetics do not prevent song learning in the way that closure of the sensitive period does. (For a discussion of the relationship between these sorts of developmental changes in LMAN and the timing of learning see the chapter by Nordeen and Nordeen, this volume.)

SENSORIMOTOR ACTIVITY IN THE AFP AND A POSSIBLE ROLE IN FEEDBACK EVALUATION

Song selective neurons have primarily been characterized in anesthetized birds presented with recorded

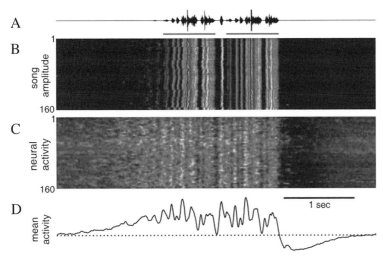

Figure 20.7 Neural activity in LMAN during singing. (A) Amplitude waveform of a typical song produced by an adult zebra finch. The song usually contained two motifs (indicated by black lines) sung in close succession. (B) Representation of the amplitude waveform for 160 song renditions that matched this pattern. Each horizontal line corresponds to a single rendition. Rendition number is indicated on the vertical axis. Waveforms are vertically aligned by the onset of the second motif. (C) Neural activity recorded in LMAN during song production. For each of the 160 renditions of song, corresponding multiunit neural activity, was recorded with a chronically implanted electrode. Neural waveforms were rectified and smoothed to yield a representation of the level of neural activity. Each horizontal line indicates the level of neural activity recorded in LMAN during the corresponding song indicated in (B). (D) Mean level of neural activity recorded in LMAN during a song rendition. The mean trace was calculated by averaging the 160 vertically aligned traces in (C). The horizontal line indicates the average baseline level of activity recorded when the bird was not singing. There is a conspicuous, patterned increase in activity during song. Moreover, activity begins to increase prior to the onset of vocalizations. This indicates that at least a component of the activity present in LMAN reflects premotor activity relayed to the AFP from HVC. (Data are from Hessler and Doupe, 1999a.)

stimuli. In awake birds, some neurons in the song system continue to respond to the presentation of song stimuli (McCasland and Konishi, 1981; Dave et al., 1998; Hessler and Doupe, 1999a). However, responses in awake birds are generally weaker than in anesthetized or sleeping birds, and in some cases entirely absent (Dave et al., 1998; Schmidt and Konishi, 1998; but see Prather and Mooney, this volume). Thus, the strength, and perhaps the nature, of responses to sounds are "gated" by the behavioral state of the bird. In other systems, sensory responses related to a behavior are "gated" by the motor activity that generates the behavior (Pearson, 1993). For songbirds, auditory feedback of the bird's own song is only available when the bird is actually singing. This raises the possibility that the effects of anesthesia or sleep may be to artificially open a gate that normally is operated by the act of singing. To the extent that such processes are operative, the role of auditory feedback may best be clarified by recording neural activity when that feedback is relevant, namely during the production of song.

Recordings from the AFP of adult birds that are singing and listening to their own songs indeed reveal that neural activity during singing is strikingly different from that elicited by playback of the bird's own song (Hessler and Doupe, 1999a) (Figure 20.7). In particular, the level of activity is much greater during singing than during playback, and a significant portion of this activity appears to correlate with motor production rather than sensory feedback. Activity such as this, which may reflect premotor commands without actually being required for motor production, is sometimes referred to as "corollary discharge" or "efference copy" (Jordan, 1995). Such activity may inform sensorimotor structures about the timing and nature of impending movements. Hence, one possible role for corollary discharge in the song system could be to "gate" or otherwise inform the processing of auditory feedback.

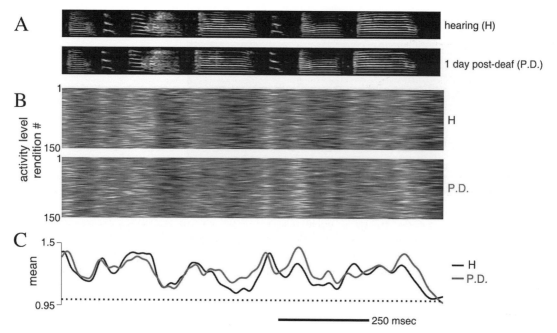

Figure 20.8 Effect of deafening on singing-related neural activity in LMAN. (A) Spectrograms of the song motif produced by an adult zebra finch before (top) and 1 day after (bottom) removal of auditory feedback by deafening. There was no apparent change in song structure, consistent with the slow time-course for song deterioration following removal of auditory feedback. (B) Multiunit neural activity level recorded in LMAN during 150 song renditions when the bird could hear (top) and postdeafening (bottom). The level of activity is plotted with a gray scale where lighter shades indicate greater activity. (C) The mean activity level for 300 renditions of the motif prior to deafening (H, black) and 161 renditions following deafening (P.D., gray). Activity level was normalized by the average background level. Deafening did not elicit a gross change in the overall level of activity in LMAN. Moreover, the pattern of activity changed only subtly, and this change was of comparable magnitude to that observed between control recording sessions without a change in auditory feedback (Hessler and Doupe, 1999a). These data indicate that a significant portion of the singing-related activity in LMAN is independent of auditory feedback. (Data are from Hessler and Doupe, 1999a).

The presence of corollary discharge during singing complicates the investigation of how auditory feedback is processed in the AFP, since sensory and motor-related activity may be intermingled. The most direct way to identify and characterize sensory components of AFP activity would be to alter auditory feedback during singing and look for correlated changes in neural activity. In humans, this technique has revealed that Wernicke's and other high-level speech processing areas are more active when auditory feedback of the subject's own voice is altered than when it is heard normally, indicating that such areas may participate in the evaluation of feedback (McGuire et al., 1996; Hirano et al., 1997). For zebra finches, an approach to detecting signals related to processing of auditory feedback has been to assess the effects of deafening on singing related activity in the AFP. At closely related sites in LMAN, the activity during singing was very similar before and 1–3 days after deafening (Hessler and Doupe, 1999a) (Figure 20.8). Similarly, chronic recordings from singing birds under conditions of reversible feedback perturbation suggest little change in the firing of LMAN neurons in response to acute alterations of auditory feedback (Leonardo, 2004). Thus, the hypothesis that alterations in auditory feedback elicit large and immediate changes in AFP activity seems doubtful. It remains possible, however, that acute changes in activity elicited by such manipulations are small or present only in a minority of neurons, especially in adult zebra finches, in which deterioration in response to feedback disruption is slow.

An alternative hypothesis about the effects of deafening on the AFP is suggested by the presence of corollary discharge activity in this pathway. In other systems, such activity may serve to provide information about the expected sensory consequences of motor commands (Jordan, 1995). Thus, rather than directly evaluating auditory feedback, the AFP may receive a prediction of expected feedback, perhaps created by the association of premotor signals and auditory feedback in HVC (Troyer and Doupe, 2000a, b). One advantage of such a model is that it potentially shortens the delay between premotor activity and evaluation of its sensory consequences. If the AFP receives a predictive signal from HVC, this signal may change only slowly, after consistently altered feedback changes the pattern of association between motor commands and auditory feedback. In this case, the gradual deterioration of song following deafening might not reflect the time required to drive plastic changes in the motor pathway, but rather (or in addition) the time necessary to update an efference copy prediction.

The presence of patterned neural activity in the AFP during singing indicates the availability of signals to potentially instruct changes in the motor pathway for song. Modeling work has suggested at least two possibilities for how such activity might direct changes to song (Troyer and Bottjer, 2001; Margoliash, 2002). First, the AFP could, in principle, provide instructive signals that selectively reinforce patterns of premotor activity that give rise to better songs and weaken others. Second, the AFP could provide signals that bias song towards desired targets, such as the tutor song. Both of these possibilities presuppose that in some sense the AFP is informed about the quality of produced songs. According to the former possibility, this information is used to differentially reinforce patterns of premotor activity present within the song motor pathway. According to the latter possibility, signals from LMAN contribute to premotor drive and direct moment-by-moment change in the motor pathway that "push" vocalizations towards the desired target (i.e. a good match to the tutor song). Persistent biasing in this fashion is presumed to result in the eventual transfer of the appropriate motor commands to the motor pathway itself. Consistent with the possibility that LMAN provides such a biasing signal, recent microstimulation experiments in singing birds indicate that alteration of activity at localized sites within LMAN can drive real-time modulation of rather specific features of song, such as increasing or decreasing the amplitude or pitch of individual song syllables (Figure 20.6; Kao et al., 2005). It remains to be determined whether persistent biasing in this fashion (for example by chronic microstimulation) can drive lasting changes to song.

POSSIBLE ATTENTIONAL AND MOTIVATIONAL FUNCTIONS OF THE AFP

In addition to hearing, it is clear that social interactions contribute to directing and modulating song learning. Social factors can influence which song models are memorized during sensory learning and which of a developing bird's song variants are retained during late stages of sensorimotor learning (Marler and Peters, 1982; West and King, 1988; Nelson and Marler, 1994). Moreover, in adult birds that have completed song learning, there continues to be an influence of social factors on song production; in many species, birds produce more song when singing to others, in courtship or territorial contexts ("directed" song) than when singing alone ("undirected" song), and subtle aspects of song structure vary between these two conditions (Sossinka and Böhner, 1980; Kao and Brainard, 2006).

The AFP may contribute to mediating the effects of these nonauditory factors on vocal learning and production. In adult zebra finches, neural activity in the AFP varies between directed and undirected song (Jarvis et al., 1998; Hessler and Doupe, 1999b) (Figure 20.9). Moreover, lesions of nucleus LMAN at the output of the AFP, prevent social modulation of specific aspects of song structure without eliminating social influences on the motivation to sing (Kao and Brainard, 2006). These data indicate that signals related to social interactions reach and modulate the AFP during adult vocal production. If such signals have similar access to the AFP during development, then the AFP might mediate their influence on sensory and sensorimotor learning. Possible sources of such signals include midbrain dopaminergic neurons that are thought to participate in reward and reinforcement learning in all vertebrates. In songbirds, dopaminergic neurons project heavily to the song system, especially Area X, and are thus well situated to provide signals to the AFP that could modulate or guide song learning (Bottjer and Johnson, 1997; Ding and Perkel, 2002; Ding et al., 2003). Indeed, in vivo microdialysis experiments indicate that levels of

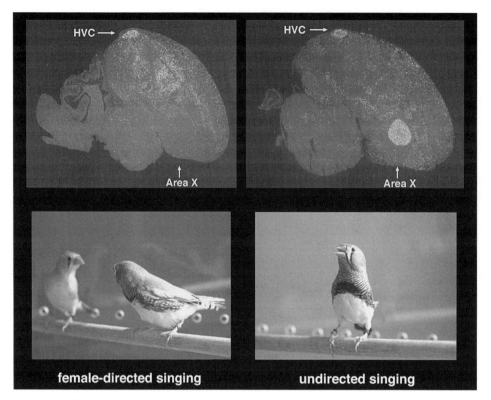

Figure 20.9 Singing-induced expression of the immediate early gene *zenk* differs in a context dependent manner. The top panels show the pattern of ZENK expression from the brain of two male zebra finches following 30 minutes of female-directed song (left) or 30 minutes of undirected song (right). Both birds sang approximately 90 bouts of song during the 30 minute period. The levels of ZENK expression in the motor nucleus HVC were elevated and comparable between the two conditions. However, in Area X and LMAN ZENK expression was elevated only in the undirected condition. These data are consistent with a strong modulation of singing-related neural activity in the AFP between directed and undirected conditions, and indicate that social context modulates activity in the AFP. (Data are from Jarvis et al., 1998.)

dopamine within Area X are modulated by social context (Sasaki et al., 2006).

PARALLELS BETWEEN AFP AND MAMMALIAN CORTICO-BASAL GANGLIA CIRCUITS

Based on numerous anatomical and physiological considerations, the AFP appears to be homologous with mammalian cortical-basal ganglia circuits (Bottjer and Johnson, 1997; Reiner et al., 1998; Luo and Perkel, 1999b; see Farries and Perkel, this volume). Here we briefly consider evidence for *functional* parallels between the avian AFP and cortico-basal ganglia circuits.

Cortical-basal ganglia circuits in mammals are critical for motor control and for motor and reinforcement learning (Graybiel et al., 1994; Houk et al., 1994). Moreover, as in adult zebra finches, the contribution of such circuitry to motor function in humans and monkeys can be more pronounced during conditions of learning and error correction than during previously learned performance (Graybiel et al., 1994; Nakamura et al., 1999; Smith et al., 2000). However, one conspicuous difference between effects associated with basal ganglia damage in mammalian systems and AFP lesions in songbirds is that in mammals, damage to the system is typically associated with overt disruptions of motor performance, as in Huntington's and Parkinson's diseases. This contrasts with the minimal effects of AFP lesions on learned song in zebra finches. While this may reflect real phylogenetic and functional differences, it is also possible that the specificity of deficits in zebra finches reflects in part the extremely stereotyped and "overlearned" nature of adult song in this species. In the Bengalese finch, a related species with similar song ontogeny but more variable adult song, lesions of the AFP nucleus, Area X, have subtle but significant effects on production of learned song (Kobayashi *et al.*, 2001). This suggests that there may be a relation between the effects of AFP lesions and the complexity or variability of song. It therefore may be instructive to examine the functioning of the AFP in a greater diversity of species in order to more fully understand the contribution of this pathway to vocal motor control as well as vocal plasticity. Because the AFP is a specialized basal ganglia pathway involved in the learning of a stereotyped motor output, it may prove a particularly tractable system for revealing basic principles of basal ganglia function in motor control and learning.

21 • Circuits and cellular mechanisms of sensory acquisition

Ernest J. Nordeen and Kathy W. Nordeen

INTRODUCTION

Like human language acquisition, avian vocal learning comprises several distinct processes that occur both sequentially and in parallel. During *sensory acquisition* birds memorize auditory stimuli that will serve as targets for song imitation. *Sensorimotor learning* establishes a map between vocal motor activity and auditory feedback, and serves to adjust vocal output to imitate the song template. *Song maintenance* actively preserves stable adult song patterns by continually evaluating vocal behavior against previously established templates. While song development depends upon the progressive integration of these learning processes, they represent distinct aspects of vocal learning and provide separate opportunities to understand neural mechanisms underlying early perceptual learning, sensorimotor integration, and ongoing motor plasticity. The behavioral and neural features associated with sensorimotor learning and song maintenance are the subject of several other chapters in this volume. Our intent here is to review our understanding of the neural substrates for sensory acquisition and to highlight some of the promising directions, as well as the difficulties, associated with studying this aspect of vocal learning.

The focus of this chapter is "the acquired auditory template," a construct first suggested by the pioneering work of Peter Marler and Mark Konishi (Konishi, 2004). As used in this chapter, the term refers to the representation of song that both develops as a result of auditory experience during sensory acquisition and is used to guide vocal imitative behavior during sensorimotor learning (see Adret, this volume). An important issue is how this template might relate to other representations of song utilized for vocal learning and/or perceptual discrimination. For instance, the existence of an "innate template" is inferred from the observation that birds denied access to an external song model still develop songs with substantial species specificity (Marler and Sherman, 1985). It often is assumed that the acquired template is an embellishment of this innate version, but in fact, the neurobiological relationship between these two representations is obscure. It also is unclear how the acquired auditory template relates to other auditory representations that are used for song discrimination. The later presumably involve circuitry shared by imitative vocal learners and nonlearners (e.g. doves, pigeons), as well as nonsinging members of imitative species (e.g. female zebra finches). Their function in perceptual discrimination (e.g. mate choice, individual/group recognition) does not require access to vocal motor pathways, and thus they may involve circuits and mechanisms that are not identical to those that establish/store the auditory template used for vocal imitation. Alternatively, if such perceptual representations are available to vocal motor circuits at least during sensorimotor learning, they could serve a dual function by supporting both vocal imitation and more general forms of auditory perceptual learning. Finally, there are templates of the bird's own song (BOS) that are firmly established as a result of sensorimotor learning. As described below, evidence is mounting that these templates can be distinguished from those established during acquisition, but this begs the question of whether the underlying neural circuits and cellular mechanisms for each are distinct.

CIRCUITS IMPLICATED IN SENSORY ACQUISITION

Avian species that exhibit vocal learning possess a specialized neural system not evident in closely related species that produce only nonlearned vocalizations (Figure 21.1). This "song system" is typically described as comprising two interrelated circuits: (1) the vocal motor pathway (VMP) that is necessary for song production, and (2) the anterior forebrain pathway (AFP) that is more specifically involved in vocal learning and plasticity. Auditory information is conveyed to this

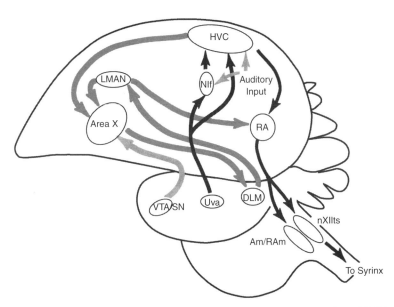

Figure 21.1 Schematic of the avian song system. Nuclei in the vocal motor pathway (VMP) are connected with black arrows. Nuclei in the anterior forebrain pathway (AFP) are connected with dark gray arrows. Auditory inputs and dopaminergic input from VTA are drawn with light gray arrows. Uva, nucleus uvaeformis; NIf, nucleus interface of nidopallium; HVC (proper name); RA, robust nucleus of archopallium; nXIIts, tracheosyringeal portion of hypoglossal nucleus; Am, nucleus ambiguous; RAm, nucleus retroambigualis; X, Area X; DLM, medial portion of the dorsolateral nucleus of the thalamus; VTA/SN, ventral tegmental area and substantia nigra; LMAN, lateral magnocellular nucleus of anterior nidopallium.

neural system from several telencephalic auditory regions that directly or indirectly innervate the VMP (see Theunissen, et al., this volume). In turn, the VMP provides a major input to, and receives the major output from, the AFP. Despite some clear functional distinctions between these two pathways, both exhibit motor activity related to song production (McCasland and Konishi, 1981; Jarvis and Nottebohm, 1997; Hessler and Doupe, 1999a), and both are likely involved in vocal learning since each contains auditory-sensitive neurons whose response selectivity reflects learned features of song (Margoliash, 1997; Brainard and Doupe, 2002). Regarding sensory acquisition, it is presumed that the cellular and synaptic mechanisms that enable this process reside somewhere within this neural system and/or the auditory regions afferent to it. It also is likely that the different stages of song learning engage overlapping pathways within the system.

Methodological issues

Several different methodological approaches have been used to identify neural circuits and cellular/molecular mechanisms contributing to sensory acquisition. Alone, each has its limitations, but over the years they have converged in support of several important notions regarding the neural substrates for song development. The function of specific brain regions has been assessed most directly via lesion and stimulation experiments. These studies help to identify circuits involved in song development versus song production, but they have generally failed to elucidate clearly which participate directly in sensory acquisition as opposed to sensorimotor learning or more general aspects of auditory processing. Furthermore, for those regions essential to song production, traditional lesion approaches obviously cannot demonstrate a specific role in vocal learning. Approaches that temporarily disrupt synaptic function and/or plasticity through targeted manipulations hold more promise in this regard.

Studies employing both electrophysiological and genomic approaches have sought the acquired template by identifying neurons that are selectively activated by the tutor's song. A significant challenge to this approach is created by the existence of neurons selectively responsive to the bird's own song (BOS). Such selectivity emerges over the course of sensorimotor learning, and neurons tuned to BOS are evident in both the VMP and the AFP (Margoliash, 1983, 1997; Margoliash and Konishi, 1985; Vicario and Yohay, 1993; Volman, 1993; Doupe and Solis, 1997; Brainard and Doupe, 2002). Obviously, to the extent that a bird precisely mimics the acquired model in their own vocalizations, it becomes impossible to differentiate between neurons tuned to BOS and those tuned to the model. Several strategies have addressed this dilemma. One is to search for tutor selective neurons before sensorimotor learning has produced an accurate imitation of tutor's song. Another is to compromise the bird's ability to accurately mimic a song tutor (e.g. by disrupting syringeal muscle control) thus allowing one to distinguish neurons tuned to BOS from those encoding the tutor's song. Both of these approaches have yielded important insights, but as we will discuss below, many results require re-evaluation in light of recent evidence that behavioral state can dramatically alter the response of "song-selective" neurons.

Finally, a variety of studies have been designed to test specific hypotheses about the cellular and synaptic mechanisms underlying sensory acquisition. Most of these are based upon the premise that vocal learning depends on specific forms of activity-dependent synaptic strengthening (long-term potentiation (LTP)) and weakening (long-term depression (LTD)) associated with developmental plasticity and learning in other behavioral systems. Accordingly, neurophysiological studies have sought evidence for these forms of synaptic plasticity within relevant neural circuits during periods of song learning. Other studies have used biochemical or molecular approaches to localize receptors or signal transduction cascades necessary for LTP/LTD in other systems. Then, attempts have been made to explicitly link these processes to sensory acquisition. While these studies are largely correlative in nature, the stimulus and temporal constraints associated with song learning allow for some strong inferences regarding the role of specific neuronal populations and cellular mechanisms in this process.

Auditory processing and sensory acquisition

An obvious starting point for studies of sensory acquisition is the auditory system itself (see Theunissen *et al.*, this volume). Two telencephalic auditory regions have been identified as potential storage sites for auditory templates. These regions, the caudomedial nidopallium (NCM), and caudomedial mesopallium (CMM), project directly and indirectly to the VMP, have bidirectional connections with the primary auditory pallium (Field L), and communicate with one another.

Studies of immediate early gene (IEG) activation suggest that NCM and/or CMM may be involved in song-related memory formation (see Mello and Jarvis, and Bolhuis, this volume). Initial studies in adult songbirds revealed that playback of songs (but not tones or white noise) increases, within both NCM and CMM, expression of ZENK and cJun, two IEGs that have been implicated in plasticity (Mello *et al.*, 1992; Nastiuk *et al.*, 1994). Moreover, when a particular song is played repeatedly, genomic and electrophysiological responses to that particular song decline within NCM and CMM, while a novel song evokes a full response (Mello *et al.* 1992, 1995; Chew *et al.*, 1995; Stripling *et al.*, 1997). This response habituation lasts for several days, indicating that a memory of specific song patterns either is stored in, or influences, these high level auditory regions (Mello *et al.*, 1995; Chew *et al.*, 1995, 1996a).

While these characteristics are consistent with NCM/CMM being involved in the creation and/or storage of auditory memories used for perceptual learning, other characteristics of this genomic response do not mirror the selectivity of sensory acquisition and thus suggest that encoding mechanisms for perceptual memories and the auditory template might differ in important ways. First, the genomic response does not differentiate between taped song presentation and live tutoring even in juvenile zebra finches (Jin and Clayton, 1997a), yet these birds do not readily imitate songs heard passively via taped playback (Eales, 1989; ten Cate, 1991; personal observations; but see Houx and ten Cate, 1999). Secondly, song-induced ZENK expression within NCM does not discriminate between conspecific and heterospecific song until sometime after day 30 post hatch in zebra finches (Jin and Clayton, 1997a; Stripling *et al.*, 2001), yet conspecific song is preferentially selected over heterospecific song as a target for vocal imitation, and significant acquisition can occur by day 30 in this species (Immelmann, 1969; Eales, 1987b). An interesting uncoupling between electrophysiological and IEG responses is that spiking activity in NCM does discriminate between conspecific and heterospecific song in juveniles, but not in adults

(Jin and Clayton, 1997a; Stripling *et al.*, 2001). Thirdly, the song-induced expression and habituation of ZENK in NCM occurs also in female zebra finches (who do not produce learned song), and is as robust in adults (no longer capable of imitating new songs) as it is in juveniles (Jin and Clayton, 1997a; Stripling *et al.*, 2001; Mello *et al.*, 2004; Bailey and Wade, 2005; Terpstra *et al.*, 2005). Finally, playback of conspecific vocalizations elevates ZENK expression in ring dove CMM (Terpstra *et al.*, 2005) and pigeon (E. D. Jarvis, personal communication), species that do not produce learned vocal behavior. Of course, age, sex, stimulus, and species-related constraints on vocal learning may be imposed downstream of sensory acquisition, thus limiting template access rather than formation. Until we have a better idea where the template is formed and stored, it will be difficult to decipher where the constraints on vocal imitation are imposed.

Despite such caveats several recent reports have reinvigorated the notion that NCM may be a storage site for the acquired song template (Bolhuis and Gahr, 2006; see also Bolhuis, this volume). As shown in Figure 21.2, while the amplitude of NCM's genomic response in adult zebra finches does not discriminate between tutor song, BOS, or novel conspecific song, the response to playback of tutor song correlates positively with the number of song syllables copied from the tutor (Bolhuis *et al.*, 2000; 2001; Terpstra *et al.*, 2004). The implications of this result are not entirely clear. For example, if a representation of tutor's song is stored in NCM, one might expect BOS to also activate ZENK in proportion to the degree of accurate imitation, since the songs of birds that learned well should be more effective in activating the stored representation of tutor song. However, ZENK expression in response to BOS does not relate consistently to the fidelity of song imitation (Terpstra *et al.*, 2004). Nevertheless, electrophysiological studies also suggest that an enduring representation of tutor song may exist in NCM; the habituation rates of neurons in this region differentiate between tutor song and other conspecific songs, including BOS (Phan *et al.*, 2006). This difference in habituation rate is evident at least 30 days after birds have last heard the tutor's song, so the duration of the memory trace is compatible with the long-term storage of an acquired auditory template. In birds devocalized to prevent accurate imitation, habituation rates are intermediate to normal tutored and untutored birds (for

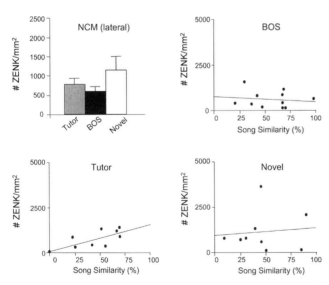

Figure 21.2 Song-induced ZENK induction within the lateral portion of NCM. Overall levels of ZENK protein within the lateral NCM of adult male zebra finches are similar following playback of Tutor song, BOS, and Novel conspecific song (upper left, mean ± SEM). However, only after playback of tutor song is there a significant correlation between ZENK protein expression within the lateral NCM and the percentage of song copied from the tutor (lower left scatterplot). No significant correlation between learning and ZENK induction is evident in lateral NCM after playback of BOS (upper right) or Novel song (lower right). (Data from Terpstra *et al.*, 2004.)

whom the tutor song is novel), so sensorimotor learning clearly plays some role in establishing or maintaining whatever representation underlies the response to tutor song.

Together, the genomic and electrophysiological data strongly support the conclusion that a sensory representation of tutor's song ultimately resides within NCM. Whether or not this representation is used for vocal imitation is still an open question. However, a separate line of investigation argues strongly that neural changes within NCM are not sufficient for encoding the acquired auditory template essential for vocal mimicry. Intermittent disruption of AFP function interferes with vocal imitation when the disruption coincides with song tutoring, but not when it occurs on intervening nontutoring days (Basham et al., 1996). If the acquired auditory template is encoded and stored solely in NCM, such manipulations of downstream circuitry should not compromise vocal imitation as long as sensorimotor learning is spared (as appeared to be the case). Thus, results of pharmacological manipulation of the AFP (discussed in detail below), together with other characteristics of NCM/CMM electrophysiological and genomic responses, suggest that formation and/or storage of the acquired auditory template must at least involve regions that lie beyond NCM/CMM.

The vocal motor pathway

The VMP consists of several hierarchically organized neural regions including Uva→NIf→HVC→RA→nXIIts and respiratory premotor groups (see Figure 21.1). All of these regions exhibit strong premotor activity, and also are responsive to auditory stimuli (see Schmidt and Ashmore, this volume). Since all regions, except for Uva, contain neurons that respond selectively to learned features of song and learning necessarily involves modification of VMP circuitry, it is presumed that the VMP plays some role in vocal learning. However, it has not yet been determined if the VMP contributes to sensory acquisition beyond its conveyance of auditory input to other parts of the song system.

One approach used to explore the role of the VMP in acquisition has been to characterize electrophysiological responses to song stimuli (especially tutor's song). However, the use of this approach to reveal evidence of the acquired auditory template is problematic, due to the inherent similarity between tutor's song and BOS. Nevertheless, initial studies on anesthetized whitecrowned sparrows who had been tutored but were not yet singing suggested that HVC neurons do not acquire a preference for tutor song prior to the onset of vocal practice (Volman, 1993). Instead, selectivity for BOS becomes evident as birds produce plastic song, and this stimulus is preferred even over tutor song. This failure to find a predominance of neurons selective for tutor's song, together with the fact that neurons tuned to BOS emerge as sensorimotor learning progresses, have been used to argue that HVC neurons responsive to BOS are a product of sensorimotor learning rather than a reflection of template acquisition.

Unfortunately, interpretation of electrophysiological data has been complicated by the discovery that auditory responses throughout the song system are strongly modulated by both anesthesia and behavioral state (Dave et al., 1998; Cardin and Schmidt, 2003, 2004a; Rauske et al., 2003). In zebra finches 30–70 days of age, HVC neurons respond preferentially to tutor song during wakefulness, but to BOS during sleep (Nick and Konishi, 2005a). As song development proceeds, responsiveness during sleep changes gradually so that preferences to earlier vocalizations are replaced by a preference for the current BOS. Towards the end of sensorimotor learning, BOS becomes the most effective stimulus during both wakefulness and sleep. Note that the waking preference for tutor song early in development cannot be accounted for by similarity to BOS, since tutor song and BOS are most dissimilar at this stage of song development. Thus, these most current data suggest that a neural trace of tutor song does exist within HVC, at least prior to late stages of sensorimotor learning.

The anterior forebrain pathway

The AFP is a specialized pathway through the avian basal ganglia that polysynaptically connects HVC to RA (Figure 21.1). It includes Area X→DLM→LMAN and constitutes a basal ganglia-thalamo-pallial circuit that is critical for song development but not essential for the production of stable adult song. The HVC conveys auditory and motor information to Area X, and the LMAN projects back to the VMP via a projection to RA and also back to Area X to form a closed loop (Nixdorf-Bergweiler et al., 1995a; Vates and

Nottebohm, 1995; Luo et al., 2001). Like mammalian basal ganglia circuitry, Area X receives a substantial dopaminergic input from the ventral tegmental area and substantia nigra pars compacta (Lewis et al., 1981; Bottjer, 1993; Soha et al., 1996), and also contains circuitry consistent with both a "direct" and "indirect" striatopallidothalamic pathway (Luo et al., 2001).

Given the established role of cortico-basal ganglia loops in motor sequence learning (Hikosaka et al., 2002; Lehericy et al., 2005) it is not surprising that lesion studies clearly implicate the AFP in sensorimotor learning (see Brainard, this volume). Lesioning Area X in juvenile zebra finches prevents them from achieving a stable song pattern, the normal culmination of sensorimotor learning (Bottjer et al., 1984; Sohrabji et al., 1990; Scharff and Nottebohm, 1991). In contrast, LMAN lesions promote immediate stabilization of songs that contain abnormally few syllables, and such lesions become less effective as sensorimotor learning progresses (Bottjer et al., 1984; Sohrabji et al., 1990; Scharff and Nottebohm, 1991; Nordeen and Nordeen, 1993). Because of these effects on sensorimotor learning, other methodologies have been necessary to assess whether the AFP also is essential for sensory acquisition.

Electrophysiological recordings in anesthetized birds have revealed that, within the AFP, selectivity for BOS predominates over that for tutor song, even during early stages of sensorimotor learning. However, both stimuli drive LMAN and Area X neurons more effectively than do other auditory stimuli, and neurons equally responsive to BOS and tutor song were found even when BOS was highly dissimilar to tutor song (Solis and Doupe, 1997; Yazaki-Sugiyama and Mooney, 2004). When tracheosyringeal nerve section is used to disrupt syringeal function in young birds and thereby divert BOS from the acquired template, some LMAN and Area X neurons remain selectively responsive to both the tutor's song and the BOS in adulthood, even when the acoustic structure of these two stimuli differ substantially (Solis and Doupe, 2000). Moreover, BOS selectivity in these birds is less than that found in normal birds, suggesting that previously established representations of tutor's song affect the representation of BOS that subsequently develops during sensorimotor learning. Importantly, the tutor-song representation in the AFP was identified in adults, while a representation of tutor song within HVC has only been detected during early stages of vocal learning (Nick and Konishi, 2005a), To test whether these contrasting results reflect real differences in the longevity of representations in the VMP and AFP, it will be important for future neurophysiological experiments to test for representations of tutor song in HVC of birds prevented from matching BOS to the acquired template. Also, in light of the dramatic impact of anesthesia, age, and state on auditory responsiveness in the zebra finch song system, some of the work done within the AFP should be replicated using awake, behaving juveniles and adults.

It is important to note that our understanding of song-related circuitry is still evolving. For example, AFP organization may be more complex than is usually described. Indeed, it has been proposed that there are two functionally and anatomically separate pallial-basal ganglia-thalamo-pallial pathways that support song behavior; the AFP as considered above, and another that involves HVC-medial Area X-dorsomedial nucleus of the posterior thalamus-medial MAN-HVC (Jarvis et al., 1998). The medial magnocellular nucleus of the anterior nidopallium (MMAN) projects to HVC, receives a disynaptic input from RA (via afferents from the dorsomedial nucleus of the posterior thalamus), contains neurons selective for BOS, and is necessary for normal song development and song stability in adulthood (Foster et al., 1997; Vates et al., 1997; Foster and Bottjer, 2001). The role of this more medial anterior forebrain pathway in sensory acquisition deserves more attention.

CELLULAR MECHANISMS IMPLICATED IN SENSORY ACQUISITION

Several other studies have linked the AFP to sensory acquisition by studying the expression of specific cellular mechanisms involved in synaptic and behavioral plasticity. The encoding of a sensory stimulus is believed to involve enduring changes in synaptic strength that establish a "Hebbian" network of neurons whose spatiotemporal pattern of activity can effectively recreate salient aspects of the stimulus. Most often implicated in the creation of such networks are long-term potentiation (LTP) and long-term depression (LTD), two forms of synaptic plasticity that adjust synaptic weights based on correlations between presynaptic and postsynaptic activation. In the context of sensory acquisition, such synaptic adjustments could establish a representation of song based on the association of temporally structured auditory inputs (i.e. tutor song), as well as neuromodulatory signals conveying stimulus salience (e.g. conspecificity) and/or reward value.

In many neural systems, LTP, LTD and behavioral learning require activation of the N-methyl-d-aspartate glutamate receptor (NMDAR) (Cotman et al., 1988; Burchuladze and Rose, 1992; Davis et al., 1992; Kim and McGaugh, 1992; Brennan, 1994; Basham et al., 1996; Roberts et al., 1998; Lisman, 2003). These receptors are uniquely suited to detect correlations in pre- and postsynaptic activity patterns because their full activation requires both presynaptic glutamate release and sufficient postsynaptic depolarization to promote dissociation of a magnesium ion that blocks the channel in a voltage-dependent manner. Importantly, the fully activated NMDAR channel not only conducts sodium and potassium currents but also permits a significant calcium influx. The resulting rise in intracellular calcium initiates a variety of biochemical cascades that, in turn, promote synaptic modification (Morris, 1989; Fields et al., 1991; Bliss and Collingridge, 1993; Castellani et al., 2001).

NMDAR expression and associated forms of synaptic plasticity are evident within several regions of the song system. These receptors first were identified functionally at the LMAN-RA synapse, where they mediate the majority of the synaptic current in both young and adult male zebra finches (Mooney and Konishi, 1991). Shortly thereafter, receptor binding studies revealed the presence of NMDARs in other song-related brain regions, including HVC, RA, Area X, and LMAN (Aamodt et al., 1992). In these binding studies, LMAN was unique in exhibiting a developmental decline in overall NMDAR binding. While further studies revealed that development regulates the subunit composition and physiology of NMDARs throughout several nuclei of the VMP and AFP (Livingston and Mooney, 1997; Basham et al., 1999; White et al., 1999; Heinrich et al., 2002), LMAN again stood out as being the only region in which early isolation from conspecific song alters these aspects of NMDAR expression (Livingston et al., 2000; Singh et al., 2000). Thus, further studies investigating the role of NMDARs in song learning focused on the AFP.

Within the song system, NMDAR-mediated synaptic plasticity has thus far been studied only within LMAN and Area X (Boettiger and Doupe, 2001; Ding and Perkel, 2004). In LMAN, LTP can be evoked at the recurrent collaterals of principle neurons, and this synaptic strengthening is accompanied by synaptic weakening at the synapses derived from DLM afferents (Boettiger and Doupe, 2001). Both forms of synaptic plasticity require activation of NMDARs, and in zebra finches can be induced only during the sensitive period for song memorization. Interestingly, at this same time there is an increase in synaptic contact length and a conversion of silent (NMDAR only) to active (AMPA-receptor containing) synapses in LMAN (Nixdorf-Bergweiler, 2001; Bottjer, 2005). Additionally, the overall number of synapses and dendritic spines in LMAN declines during song learning, as does the density of DLM → LMAN afferents (Johnson and Bottjer, 1992; Nixdorf-Bergweiler et al., 1995b; Nixdorf-Bergweiler, 2001; Iyengar and Bottjer, 2002a; Bottjer, 2005).

Within Area X, LTP can be evoked at the glutamatergic synapses made by HVC and/or LMAN onto medium spiny neurons, and LTP at these synapses depends on coactivation of NMDARs and the D1-like class of dopamine receptor (Ding and Perkel, 2004). Recall that Area X receives a strong dopaminergic input from the ventral tegmental area (VTA), and the Hebbian nature of LTP could allow for synaptic changes that depend upon the coincident activation of any or all of the inputs to Area X (e.g. both auditory and neuromodulatory).

If NMDAR-mediated forms of synaptic plasticity within the AFP are indeed necessary for sensory acquisition, then interfering with these synaptic processes should impair normal song development. Generally, pharmacological or genetic manipulations are employed to test the functional significance of candidate mechanisms, but their interpretation can be difficult because such manipulations frequently lack adequate specificity and may activate compensatory processes that can obscure the normal physiological role of a particular process. Also, studies aimed at pinpointing mechanisms specifically involved in template encoding and storage must deal effectively with the potential confounds that stem from natural overlap between sensory acquisition and sensorimotor learning in certain species. Despite these caveats, such studies add critical information because correlative studies alone are not sufficient to determine what processes are indeed necessary for learning.

Behavioral pharmacology experiments have demonstrated that activation of NMDARs is critical for sensory acquisition. Birds injected systemically with an NMDAR antagonist just prior to restricted tutoring sessions develop songs that resemble those of birds never exposed to a song model (Aamodt et al., 1996). In contrast, antagonist delivered on nontutoring days that alternate with the tutoring sessions do not disrupt vocal imitation. Subsequent experiments extended these findings to show that similarly timed disruptions of NMDAR function specifically within the AFP also are sufficient to impair song imitation. In these studies, NMDAR antagonists were targeted to LMAN. As shown in Figure 21.3, birds that receive intracranial infusions of an NMDAR antagonist immediately prior to tutoring sessions ultimately reproduce significantly less of the tutor's song than do various different control groups, including birds that receive identical infusions of the antagonist on nontutoring days (Basham et al., 1996).

Although these data suggest that NMDAR activation in the AFP is involved directly in encoding and/or storing the song template, other interpretations are possible. For instance, the behavioral deficits resulting from NMDAR blockade could reflect impairments of sensorimotor learning as opposed to acquisition, since both of these phases of song learning overlapped with drug treatments. However, because behavioral disruptions were only evident when the antagonist was given on tutoring days, this interpretation requires the added presumption that vocal practice during (or within hours after) tutoring is critical for normal vocal imitation. There currently is no evidence for such a privileged role of vocal practice that occurs in close proximity to tutoring. Moreover, successful tutor imitation can occur in zebra finches when tutoring ends before (or within days of) the onset of vocal practice (Immelmann, 1969; Böhner, 1990), or if sensorimotor learning is disrupted for several months after song tutoring (Funabiki and Konishi, 2003). Also, in some songbird species, sensory acquisition normally is separated from the onset of sensorimotor learning by many weeks or months (Marler, 1987). Thus, the most parsimonious interpretation of the data described above is that NMDAR activation within the AFP is critical to template encoding.

Implicating the regional activation of NMDARs in sensory acquisition does not necessarily imply that NMDAR-mediated synaptic plasticity is occurring within that region during template encoding. This is because NMDAR antagonists can significantly disrupt fast excitatory transmission in some neural circuits (Daw et al., 1993). While this issue is difficult to resolve, in vitro recordings from LMAN neurons suggest that significant synaptic transmission persists even in the presence of NMDAR antagonists (Livingston and Mooney, 1997; Boettiger and Doupe, 1998; Bottjer et al., 1998). Moreover, other experiments (see below) indicate that opportunities for sensory acquisition also provoke within the AFP intracellular signaling cascades associated with LTP.

As noted previously, NMDAR-mediated increases in intracellular calcium activate a variety of biochemical cascades important for long-term changes in synaptic function. One such cascade involves the phosphorylation (activation) of calcium/calmodulin-dependent protein kinase II (CaMKII). This kinase is abundant in the postsynaptic compartment where it complexes with NMDARs, and its phosphorylation is critical for the induction of NMDAR-dependent LTP and various

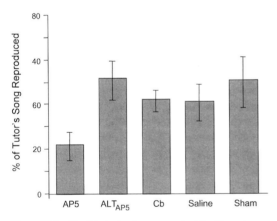

Figure 21.3 Vocal imitation is compromised by disrupting NMDAR function in LMAN during song tutoring. Tutoring sessions (90 min) occurred every other day from posthatch day 32–52 and adult songs were evaluated at posthatch day 90. Two groups received bilateral injections of the NMDA receptor antagonist AP5 targeted towards the LMAN; either 10 minutes prior to tutoring (AP5) or on nontutoring days (ALT$_{AP5}$). Other birds received injections of either AP5 into the cerebellum (Cb) or saline into the LMAN (Saline) 10 minutes prior to tutoring, or were left uninjected (Sham). The NMDAR antagonist significantly reduced the amount of song material imitated only when delivered into the LMAN just prior to tutoring. (Reprinted with permission from Basham et al., 1996.)

forms of learning and developmentally regulated synaptic plasticity (Lisman et al., 2002; Otmakhov et al., 2004). CaMKII activation elevates synaptic strength through a variety of mechanisms, including increasing conductance of AMPA receptors, facilitating the recruitment of synaptic AMPA receptors, and promoting production of retrograde messengers that increase presynaptic transmitter release (Lisman, 2003; Colbran and Brown, 2004). Importantly, more modest NMDAR-mediated calcium currents can promote activation of phosphatases facilitating synaptic weakening.

CaMKII is expressed at high levels in the avian striatum, and the precise stimulus conditions optimal for vocal imitation promote CaMKII phosphorylation within Area X (Singh et al., 2005). In young male zebra finches isolated before the onset of sensory acquisition, just 2 hours of interaction with a live, singing tutor on posthatch day 35 results in a three fold increase in levels of phosphorylated CaMKII (pCaMKII) within Area X (Figure 21.4, top). In contrast, this song tutoring does not alter the total amount of CaMKII (phosphorylated + unphosphorylated forms). The tutoring-induced elevation in pCaMKII occurs in the absence of vocal practice by the pupil, and is not evident in young female zebra finches (who do not exhibit vocal learning), or in young males exposed to a nonsinging male tutor (Figure 21.4, bottom left). Together, the data are consistent with the hypothesis that the pCaMKII signal relates to the encoding of a template associated with vocal mimicry.

Another intriguing aspect of the tutoring-induced pCaMKII signal within Area X is that prior experience with the song tutor "tunes" this biochemical response. That is, in young male zebra finches allowed access to their father's song until day 30, re-exposure to that same tutor at day 35 provokes a 10-fold increase (relative to untutored controls) in pCaMKII with Area X (Figure 21.4, bottom right). In contrast, unlike early isolates, these "pre-exposed" birds exhibit no elevation in pCaMKII levels after brief exposure to an unfamiliar tutor at day 35. Again, the song-induced pCaMKII response to familiar song occurs in the absence of vocal practice and is absent in females and in males who are visually isolated from their father during tutoring. This latter observation regarding males is particularly significant because zebra finches do not readily imitate a tutor under such conditions of visual isolation (Eales, 1989; personal observations).

The data summarized above suggests that song exposure prior to 30 days post hatch is sufficient to bias circuitry afferent or intrinsic to Area X in male zebra finches. Our working hypothesis is that familiar song activates CaMKII (and downstream biochemical events) powerfully in Area X because it reactivates connections that have already undergone some synaptic strengthening. If such strengthening is accompanied by weakening of other (nonactivated) synapses, then presenting unfamiliar song to previously tutored birds may activate pathways that have undergone LTD and therefore produce insufficient postsynaptic activation to elicit a detectable pCaMKII signal. Consistent with this view, song exposure until day 30 tunes the auditory response properties of LMAN neurons such that they favor songs previously heard, and yet subsequent tutoring with unfamiliar song (for 30 days) can overwrite this initial tuning and promote vocal imitation of that second tutor's song (Yazaki-Sugiyama and Mooney, 2004). Within Area X, similar effects of early experience have not been explored; however we predict that electrophysiological outcomes would be similar to those observed in LMAN, and that extended tutoring with unfamiliar song also would "retune" the pCaMKII response within this region.

If sensory acquisition does indeed engage LTP-like processes within the AFP, identifying the inputs and cell types contributing to this process could shed insight into how this pallial-striatal-thalamic-pallial loop processes information important for template encoding and storage. As noted earlier, LTP within Area X can be elicited at the glutamatergic synapses made by both HVC and LMAN neurons onto medium spiny neurons. These same medium spiny neurons receive substantial DA input from the VTA, and the LTP described above depends upon both NMDAR and DA receptor activation (Ding and Perkel, 2004). Recent results suggest that this same circuitry also is involved in the tutoring-induced phosphorylation of CaMKII within Area X. Virtually all of the Area X cells that express CaMKII (and thus are capable of mediating the pCaMKII response) also are immunoreactive for DARPP-32 (Hein et al., 2005), a protein involved in dopamine receptor signaling and whose expression identifies at least one prominent class of striatal medium spiny neuron (Svenningsson et al., 2004). Furthermore, infusion of a dopamine receptor antagonist into Area X just prior to tutoring blocks

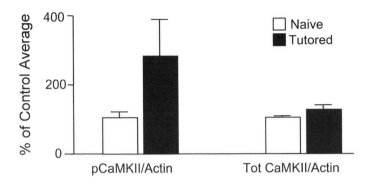
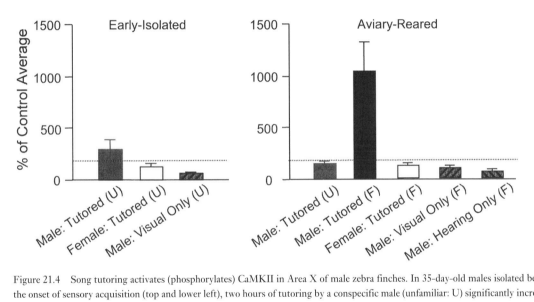

Figure 21.4 Song tutoring activates (phosphorylates) CaMKII in Area X of male zebra finches. In 35-day-old males isolated before the onset of sensory acquisition (top and lower left), two hours of tutoring by a conspecific male (unfamiliar: U) significantly increased levels of phosphorylated CaMKII (pCaMKII) relative to untutored controls. Total CaMKII (phosphorylated + unphosphorylated) was unaffected by treatment. In contrast to the early isolates, males exposed to song until day 30 (aviary-reared: lower right) did not mount a pCaMKII response when tutored by an unfamiliar male on day 35. However, tutoring by a familiar (F) conspecific resulted in a 10-fold increase in pCaMKII expression within Area X. Regardless of rearing condition, pCaMKII levels were unaffected in females, and in males exposed to either a nonsinging tutor (visual only), or a tutor that could be heard but not seen (hearing only). Data expressed are group means ($\pm$ SEM) expressed relative to the average of values measured in untutored controls run on the same gel. Dotted line is 100% of control average. (Data from Singh et al., 2005.)

the tutoring-induced increase in pCaMKII expression (Hein et al., 2005). In this context, it is especially interesting that preliminary results from our lab show that social tutoring also promotes c-fos expression (a marker of neuronal activation) in VTA and substantia nigra neurons, the primary source of dopaminergic innervation to Area X (Lewis et al., 1981).

Together, the data support a working model of sensory acquisition first formulated by Doya and Sejnowski (2000). Synaptic strengthening in the AFP would encode song patterns when specific auditory input (carried by HVC and/or LMAN afferents to Area X) coincides with appropriate neuromodulatory signals (e.g. DA release from the VTA/substantia nigra) triggered by a combination of species-specific auditory, visual and social cues. An observation challenging this model is the failure of Ding and Perkel (2004) to evoke LTP in Area X slices from birds younger than 37 days. Instead, synaptic depression was elicited by the same stimulus parameters that successfully induced LTP in slices from birds > 37 days. While this leaves open the possibility that the pCaMKII response to tutoring reflects changes in neuronal function unrelated to LTP, it may also be that the LTP/LTD threshold in Area X is developmentally regulated, and that the initiation of molecular cascades provoking striatal LTP requires the specific patterns of activity achieved by social tutoring. To resolve this issue, it will be important to search more directly for evidence that tutoring promotes synaptic strengthening within Area X. Also needed are direct tests of whether NMDAR activation is necessary for the pCaMKII signaling in Area X, whether the pCaMKII signal is essential for normal song learning, and whether song tutoring stimulates dopamine release within Area X.

TEMPORAL CONSTRAINTS ON ACQUISITION

Many songbirds can only imitate songs heard during a distinct "sensitive period." Such temporal regulation of learning could occur through developmental changes in any aspect of the biological cascades translating sensory experience into lasting neural and behavioral change. Among these are (1) upstream circuitry that provides information to the learning mechanism (e.g. balance of inhibitory and excitatory inputs or neuromodulatory pathways), (2) the molecular machinery that provokes synaptic plasticity (e.g. receptors, intracellular signaling molecules, trophic and neuromodulatory molecules), or (3) downstream events involved in expression of plasticity cascades (e.g. specific patterns of gene expression, synapse formation, modification or elimination). Thus, another approach to understanding the neurobiology of sensory acquisition is to identify, from among the mix of cellular and molecular changes that take place during acquisition, candidate biological substrates whose developmental regulation coincides with the sensitive period for learning. One could then determine which are necessary, and if any are sufficient, to account for the sensitive period. To address these issues, experiments can exploit tremendous species diversity in the timing, and even the very existence, of sensitive periods for avian vocal learning. Also, one can extend sensitive periods for sensory acquisition by limiting early exposure to conspecific song, or providing only suboptimal learning stimuli (Kroodsma and Pickert, 1980; Eales, 1985, 1987b). Such approaches enable one to dissociate chronological age from the system's ability to support vocal learning.

Developmental changes in NMDAR function

Although there are many cellular and molecular phenotypes whose expression is developmentally regulated within the song system in concert with the capacity for sensory acquisition, only a few have been tested to determine whether their timing indeed limits the capacity for song imitation. One such developmental change involves the expression of NMDARs within the LMAN. Studies in a variety of other neural systems have revealed robust developmental changes in NMDAR structure and function that generally coincide with sensitive periods for neural and behavioral plasticity (Carmignoto and Vicini, 1992; Hestrin, 1992; Fox and Daw, 1993; Hofer et al., 1994; Crair and Malenka, 1995). Native NMDARs consist of the NR1 subunit, essential for channel activity, and one or more modulatory subunits (NR2A-E) that determine the biophysical properties of the receptor channel (Monyer et al., 1992; Buller et al., 1994; Cull-Candy et al., 2001). For example, increased NR2A expression reduces the duration of NMDAR currents (Monyer et al., 1994; Flint et al., 1997; Tovar and Westbrook, 1999; Kuehl-Kovarik et al., 2000), thereby reducing synaptic integration times and postsynaptic calcium influx (Hestrin, 1992; Hoffmann et al., 2000). Also, changes in NR2B expression can impact synaptic plasticity because this subunit associates preferentially with signaling and scaffolding proteins in the postsynaptic density that are involved in synaptic strengthening (Strack and Colbran, 1998; Nicoll and Malenka, 1999; Lisman and McIntyre, 2001; Barria and Malinow, 2005; Bayer et al., 2006).

Within the LMAN, maturational changes in NMDAR subunit expression occur during the period of sensory acquisition. Specifically, increased NR2A expression in this region coincides with early phases of acquisition (Basham et al., 1999; Heinrich et al., 2002), and this increase is accompanied by a shortening of

NMDAR-mediated currents (Livingston and Mooney, 1997; White et al., 1999). Also in LMAN, there is a reciprocal, albeit more protracted, decline in NR2B subunit expression (Basham et al., 1999; Heinrich et al., 2002) that could alter the efficiency of biochemical cascades that mediate changes in synaptic plasticity. Therefore, the changes in NMDAR expression within the song system are consistent with the observation that the ability to induce LTP and LTD at synapses within the LMAN normally declines in juvenile zebra finches as development proceeds (Boettiger and Doupe, 2001). In fact, the same stimulus parameters that induce LTP in young birds elicit LTD in adults, suggesting that the LTD/LTP modification threshold has shifted in favor of LTD (Abraham and Bear, 1996).

Several of the maturational changes in NMDAR expression and physiology within LMAN are affected by manipulations that alter the timing of the sensitive period for vocal imitation. That is, early isolation from song (before the onset of sensory acquisition) delays developmental decreases in both NR2B mRNA expression (Singh et al., 2000) and NMDAR current durations within LMAN (Livingston et al., 2000), and extends the sensitive period for song acquisition (Eales, 1987b; Morrison and Nottebohm, 1993). Importantly, these effects of early experience are region and subunit specific: early isolation does not delay the maturation of NMDAR currents in RA, does not alter NR2A mRNA expression in any region examined (LMAN, Area X, HVC, or RA), and does not delay the developmental decrease in NR2B mRNA expression in Area X, HVC, or RA.

These data, together with the observation that NR2B expression in LMAN (and RA) in male canaries covaries with photoperiod-induced changes in vocal plasticity (Singh et al., 2003), fueled the idea that developmental/seasonal regulation of NMDAR expression within LMAN may contribute significantly to the closure of sensitive periods for vocal learning. However, recent studies do not support this hypothesis. As shown in Figure 21.5, late isolation (beginning at 25–30 days post hatch) extends the sensitive period for acquisition in zebra finches, even though both NMDAR current durations and subunit expression in LMAN have achieved an adult phenotype by the time birds engage in late learning (Livingston et al., 2000; Heinrich et al., 2003).

On a cautionary note, it would be premature to conclude from these studies that developmental regulation of NMDARs has no impact on vocal learning. The measures of song imitation taken in the experiments described above were relatively crude, and it is possible that small, but nevertheless important, changes in learning propensity result from changes in NMDAR physiology and/or subunit expression. Nonetheless, it seems clear that the developmental changes in LMAN NMDAR function/structure do not close the sensitive period for sensory acquisition.

Synapse rearrangement in the AFP

"Synapse selection" theories of neural plasticity postulate that plasticity is decreased as an initially exuberant set of connections is reduced through activity-dependent synaptic rearrangement (Edelman, 1993; Tononi et al., 1996; Lichtman and Colman, 2000). Consistent with this idea, filial and sexual imprinting in birds are associated with large-scale, experience-dependent pruning of dendritic spines in forebrain regions associated with these behaviors (Bischof et al., 2002; Bock and Braun, 1999a, b; Rollenhagen and Bischof, 1998). Likewise, in the song system, glutamatergic inputs to LMAN arising from DLM (Livingston and Mooney, 1997) expand greatly in young zebra finches between 20 and 35 days post hatch and then steeply reduce their terminal fields (Johnson and Bottjer, 1992; Iyengar and Bottjer, 2002a). As the DLM terminals are pruned within LMAN, there is a corresponding reduction in the density of dendritic spines (postsynaptic specializations) on LMAN neurons (Nixdorf-Bergweiler et al., 1995b) as well as a reduction in the proportion of "silent" synapses within this nucleus (Bottjer, 2005). These anatomical and functional modifications in zebra finches overlap with both sensory acquisition and sensorimotor learning. But the observation that pruning of LMAN dendritic spines is delayed when young birds are isolated from conspecific song, a manipulation that also extends the sensitive period for acquisition, suggested that overlap between sensory acquisition and synaptic pruning in LMAN may be critical for effective learning (Wallhausser-Franke et al., 1995).

However, as in the case of LMAN NMDAR function, the normal time course of LMAN spine elimination has been dissociated from the capacity for sensory acquisition (Figure 21.6). Although zebra finches isolated by day 30 can successfully imitate a novel tutor heard between days 65–90 (beyond the normal close of

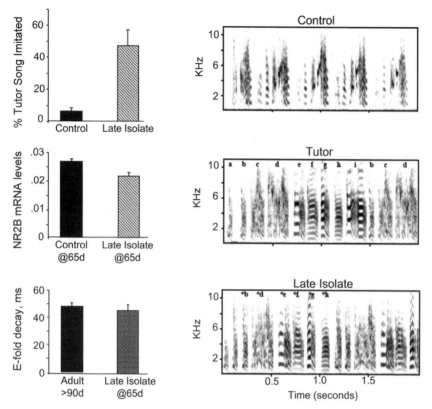

Figure 21.5 Maturation of NMDAR structure and function do not close the sensitive period for sensory acquisition. NR2B subunit expression and NMDAR current durations within LMAN decline during normal song acquisition in zebra finches, and isolation from conspecific song can delay these developmental changes and also delay closure of the sensitive period for acquisition (see text for further details). Zebra finches isolated from conspecific song beginning at 25–30 days (late isolates) imitated significantly more of a tutor's song heard between 65–90 days than did aviary-reared controls (top left and right panel; data from Heinrich et al., 2003). Despite this group difference in the capacity for extended learning, NR2B transcript levels in LMAN at 65 days were not elevated in late isolates relative to controls (middle left; modified from Heinrich et al., 2003). Similarly, at 65 days the duration of NMDAR currents in late isolates had declined to adult levels (lower left; modified from Livingston et al., 2000).

the sensitive period), such birds do not differ significantly from normally reared birds in the density of dendritic spines on LMAN neurons measured at either day 65 or day 120 (Heinrich et al., 2003). Evidently, the developmental loss of LMAN spines that normally occurs between days 25–65 is not sufficient to close the sensitive period for sensory acquisition. Also, extended learning is not accompanied by enhanced spine loss, thus the capacity for spine pruning does not appear to limit sensory acquisition.

While there remains no good account of the biology underlying sensitive periods for sensory acquisition, the established role of NMDAR activation in song learning coupled with the insight provided by work on a variety of other instances of developmental plasticity suggest some avenues for future research. Since the timing of sensitive periods can be manipulated by early experience, focusing on features of song system development that likewise depend on early song exposure will remain a powerful initial strategy for identifying events that represent or constrain learning. For example, expression of protein kinase C (PKC), an enzyme associated with synaptic plasticity and early aspects of memory formation in mammals (Micheau & Riedel, 1999), transiently increases in the RA (LMAN's target nucleus in the vocal motor pathway) during the sensitive period for song learning. Moreover, when young birds are deprived of song experience by either early deafening or isolation,

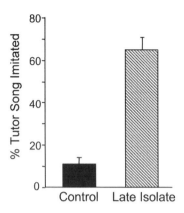 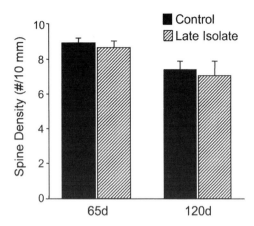

Figure 21.6 Developmental pruning of LMAN spines does not close the sensitive period for sensory acquisition in zebra finches. The density of dendritic spines on LMAN neurons decline during normal song acquisition in zebra finches, and early isolation from conspecific song can delay this anatomical pruning and also delay closure of the sensitive period for acquisition (see text for further details). Birds isolated at 30 days post hatch imitated significantly more of a tutor's song heard between 65–90 days than did aviary-reared controls (left). Despite this group difference in the capacity for extended learning, spine density was similar in the two treatment groups at the age when late tutoring began (65 days) and at the age when song imitation was evaluated (120 days) (right; modified from Heinrich et al., 2005).

PKC expression is depressed (Sakaguchi and Yamaguchi, 1997). Also, the projection from LMAN to RA spatially segregates early during song development (Iyengar et al., 1999) and the development of this topographical organization is delayed in young deafened males (Iyengar & Bottjer, 2002b). However, our developmental investigations of NMDAR function and spine loss in LMAN emphasize the need for caution in relating these experience-dependent changes to the capacity for acquisition. The early auditory, visual, social and endocrine experiences so important to vocal learning undoubtedly impact a variety of developmental processes not specifically involved in regulating the capacity for vocal imitation. When these stimuli are withheld completely (early isolation), a host of maturational consequences can be expected, only some of which may contribute to sensitive period regulation. Gentler manipulations of early input are needed (e.g. later isolation or exposure to suboptimal song models), so as to alter the timing of learning without necessarily altering all developmental changes that depend on that input.

CONCLUSIONS AND FUTURE DIRECTIONS

This review has raised a number of issues with implications for future studies of sensory acquisition. For example, it is clear that multiple representations of song-related auditory information are formed during song development, both in regions within and outside the traditionally defined song system. Apparently, these different representations reflect sequential, hierarchical processing of auditory input. What remains at issue is which of these representations function to structure vocal motor behavior. We have seen, for example, that some properties of NCM/CMM auditory responsiveness seem well suited to perceptual discrimination learning and species-specific song recognition, but lack some of the stimulus and temporal constraints expected of the acquired auditory template. Yet, these same regions provide auditory input to the song system through their projections to NIf and the HVC shelf. While the processing that occurs within NCM/CMM is presumably critical for the tutor selective responses apparent within HVC and, in turn, the AFP, it may be that only the representations at these latter stages of processing are directly related to vocal mimicry. Alternatively, it may be that multiple representations are utilized for vocal mimicry. Resolving these issues will require reversible neuropharmacological manipulations that specifically target regions during opportunities for sensory acquisition. Such approaches will be particularly powerful if also combined with subsequent electrophysiological/genomic studies that can assess effects on song-selective responses within the song system.

Another issue concerns the relationship between the acquired auditory template and sensorimotor learning. One view is that template formation does not itself shape motor pathways for song, but rather during vocal practice a comparison between auditory feedback (or expected auditory feedback) and the encoded template guides incremental changes in what are initially naïve motor circuits. This view stems from the early observation that prior sensory acquisition does not alter the devastating consequences that birds experience when deafened before the onset of sensorimotor development (Konishi, 1965a, 1965b). Yet, it is apparent that AFP regions implicated in sensory acquisition also participate in the motor control of song (see Brainard, this volume). This suggests that sensory acquisition directly engages motor elements of the AFP, such that specific auditory inputs could help coordinate the same neuronal populations that will modulate vocal production at subsequent stages of vocal development. If so, then acquiring the auditory template may directly bias circuits used later in song production.

Finally, the idea that template encoding during acquisition involves basal ganglia circuitry could provide important new insights into the process of acquisition, as well as its relationship to sensorimotor learning. Models of song learning based on reinforcement theory posit that, during sensorimotor learning, birds sample "vocal space" so as to arrive at motor patterns that produce feedback matching the stored template (Doya and Sejnowski, 2000; Troyer and Doupe, 2000a; Troyer and Bottjer, 2001). Presumably, such patterns, or the feedback they produce, have reinforcement value. This, in turn, suggests that during sensory acquisition, specific acoustic elements produced by the tutor take on reward value above and beyond that of conspecific song generally. With respect to AFP circuitry, HVC projections to Area X likely convey auditory input of tutor's song during acquisition, while dopaminergic inputs from VTA/substantia nigra→X may be driven concurrently by the intrinsic reward value of conspecific song. Dopaminergic projections to the striatum from the VTA/substantia nigra have been implicated repeatedly in coding reward, prediction of reward, and reward prediction errors (Schultz, 2004; Wise, 2004), and it is conceivable that the stimulus biases evident in song learning reflect the ability of specific auditory, visual, and/or social cues to activate these inputs. The explicit pairing of dopaminergic input and tutor-driven auditory input could then drive synaptic modifications within Area X that are known to depend on concurrent glutamatergic and dopaminergic input, thereby building a specific representation of tutor's song in the AFP to serve as a template for vocal practice. During sensorimotor learning and adult song maintenance, such a representation would be critical for the generation of error signals (e.g. reward prediction errors) when expected auditory feedback fails to match the stored template.

Tests of these hypotheses will require development of methods for measuring real-time dopamine levels in awake songbirds. For example, it will be important to learn if tutoring during acquisition activates ascending dopaminergic projections, and whether such activation is essential for template encoding. The role of dopamine modulation as a substrate for "innate" learning biases could be investigated by pairing presentation of suboptimal stimuli (e.g. taped or heterospecific song) with stimulation of dopaminergic projections. During sensorimotor learning, dopamine assays could establish whether self-production of tutor-like sounds is accompanied by phasic dopamine release. Basal ganglia-thalamic-cortical loops have been implicated in many instances of implicit or procedural learning, especially the learning and execution of motor sequences (Hikosaka et al., 2002; Packard and Knowlton, 2002; Graybiel, 2005; Lehericy et al., 2005). More recently, evidence has begun to accumulate that this circuitry also participates in various aspects of human language processing (Callan et al., 2003; Dominey et al., 2003; Friederici and Kotz, 2003; Ullman, 2004). The possibility that sensory representations are also encoded within this circuitry to serve as templates for evaluating motor behavior could add an important new dimension to our understanding of basal ganglia function.

22 • Chasin' the trace: the neural substrate of birdsong memory

Johan J. Bolhuis

INTRODUCTION

Birdsong learning and imprinting are arguably the two major avian memory paradigms. For imprinting, identification of the brain substrates of memory has made possible a detailed analysis of its underlying neuronal and molecular mechanisms (Horn, 1998, 2004). In contrast, birdsong researchers have been primarily concerned with identifying brain regions involved in learning how to sing. This is not surprising, since the initial attraction of birdsong for neuroscientists was as a model system for the study of learned vocalizations, analogous to speech acquisition in humans (Doupe and Kuhl, 1999; Bolhuis and Gahr, 2006). Indeed, as Nottebohm has noted (1981, p. 1370, note 23), "Song 'learning' and song 'forgetting' are used here to refer not so much to the acquisition and loss of an auditory memory, but rather to the conversion of that memory into a motor program, with the consequent matching of an auditory model." Consequently, research on the avian "song system" (Figure 22.1) has been focused primarily upon its role as a neural substrate for song acquisition and production, i.e. upon the neural representation of the bird's own adult song (BOS), rather than as a neural substrate for the auditory memory of the tutor song. Indeed, in many of the studies which purported to demonstrate a correlation between singing behavior and some aspect of song system morphology (e.g. sex differences, repertoire size or seasonal variations in nuclear volume) it is unclear whether these correlations are related to song memory or to aspects of song production (for detailed reviews see Bolhuis and Macphail, 2001; Bolhuis, 2005; Bolhuis and Gahr, 2006). In contrast, the present review focuses specifically upon studies designed to identify the neural substrates of tutor song memory.

THE SONG SYSTEM AND TUTOR SONG MEMORY

In both anesthetized (Margoliash and Konishi, 1985; Solis *et al.*, 2000), and unanesthetized songbirds (Prather and Mooney, this volume) neurons in nuclei of the song system – in particular, LMAN, Area X, HVC and RA (Figure 22.1) – are responsive to song, particularly to conspecific song (Solis *et al.*, 2000). In adult males, neurons in these regions respond more to the bird's own song (BOS) than to the tutor song (Margoliash and Konishi, 1985; Solis *et al.*, 2000) or to the song of another conspecific (Margoliash, 1986). In LMAN and Area X, some neurons respond equally well to the BOS and tutor song, while a small proportion respond more to the tutor song (Solis *et al.*, 2000). Volman (1993) found that in juvenile white-crowned sparrow males (*Zonotrichia leucophrys*) – a species in which the memorization and sensorimotor phases are separated in time – neurons in the HVC in the memorization phase showed no preference for songs with which the birds were tutored. Later, in the sensorimotor phase when the males started to sing themselves, HVC neurons showed preferential responding to the BOS. It is important to realize that these early electrophysiological studies used anesthetized birds. Theunissen *et al.* (this volume) review evidence that shows that song system neurons are considerably less responsive to song in awake zebra finches (e.g. Schmidt and Konishi, 1998; Dave *et al.*, 1998). A recent electrophysiological analysis involving unanesthetized developing zebra finches suggests that neurons in the HVC show transient preferential responding to the song of their tutor (Nick and Konishi, 2005a). In the early sensorimotor period (35–69 days after hatching), responsiveness of HVC neurons was greater to tutor song than to the BOS, novel song, heterospecific song or white noise – in some cases the preference was

Neuroscience of Birdsong, ed. H. Philip Zeigler and Peter Marler. Published by Cambridge University Press. © Cambridge University Press 2008.

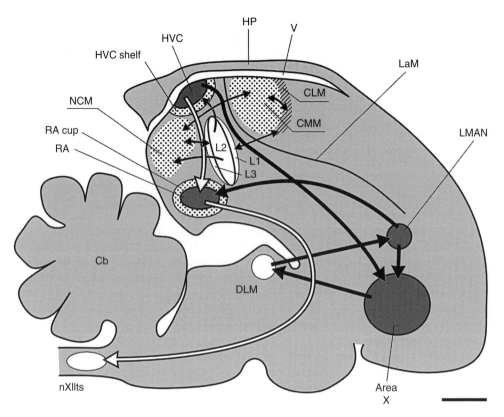

Figure 22.1 Schematic diagram of a composite view of parasagittal sections of a songbird brain, with approximate positions of nuclei and brain regions. Lesion studies in adult and young songbirds led to the distinction between a caudal pathway (white arrows), considered to be involved in the production of song, and a rostral pathway (thick black arrows), thought to play a role in song acquisition (see text for further details). Thin black arrows indicate known connections between the Field L complex, a primary auditory processing region, and some other forebrain regions. The dark gray nuclei show significantly enhanced expression of immediate early genes (IEGs) when the bird is singing. Stippled areas represent brain regions that show increased IEG expression when the bird hears song, including tutor song. Abbreviations: Cb, cerebellum; CLM, caudal lateral mesopallium; CMM, caudal medial mesopallium; DLM, nucleus dorsolateralis anterior, pars medialis; HP, hippocampus; HVC, acronym used as a proper name; formerly known as High Vocal Centre; L1, L2, L3, subdivisions of Field L; LaM, lamina mesopallialis; LMAN, lateral magnocellular nucleus of the anterior nidopallium; nXIIts, tracheosyringeal portion of the nucleus hypoglossus; RA, robust nucleus of the arcopallium; V, ventricle; Scale bar represents 1 mm. (Adapted by permission from Macmillan Publishers Ltd: *Nature Reviews Neuroscience*, Bolhuis and Gahr, 2006, copyright 2006).

significant. During the late sensorimotor phase, more than 70 days after hatching, the preference of HVC neurons for tutor song had switched to a preference for the BOS. Responsiveness of HVC neurons during the early sensorimotor period did not correlate with the similarity of the BOS to the tutor song. Taken together, these electrophysiological findings suggest that nuclei in the song system are responsive to song, particularly BOS. However, while juvenile zebra may show preferential responding of HVC neurons to tutor song early in the sensorimotor period, these findings do not support the presence of a neural representation of the tutor song in the song system of *adult* birds.

Other evidence for the possible involvement of the song system in tutor song memory was obtained in a series of studies concerning the role of N-methyl-D-aspartate (NMDA) receptors in birdsong (Aamodt *et al.*, 1995, 1996; Basham *et al.*, 1996; Livingston *et al.*, 2000; Heinrich *et al.*, 2003; see Nordeen and Nordeen, 2004; and Nordeen and Nordeen, this volume, for a detailed discussion of this work). In zebra finches, NMDA receptor binding in LMAN peaks at 30 days after hatching and subsequently declines – a decline that also occurs in socially isolated males (Aamodt et al., 1995). An early hypothesis that NMDA receptor levels might limit the sensitive period for song learning was not confirmed (Aamodt et al, 1995; Livingston et al., 2000; Heinrich et al., 2003). The hypothesis that the LMAN might nevertheless contain the neural substrate for tutor song memory was fueled by two studies involving infusion of NMDA receptor blockers, which impaired song learning in zebra finches (Aamodt et al., 1996; Basham et al., 1996; cf. Nordeen and Nordeen, this volume). Bolhuis and Gahr (2006) have argued that, because the experiments were conducted in the sensorimotor phase, it is not clear whether NMDA receptor blocker infusion affected sensorimotor integration or the formation of auditory memory. In addition, in these studies only the LMAN was targeted, not other nuclei within the song system or regions outside it. Finally, these studies were performed on the assumption that NMDA receptor-dependent synaptic plasticity underlies memory formation (Morris et al., 1986), a hypothesis that has since been questioned (e.g. Bolhuis and Reid, 1992; Saucier and Cain, 1995; Bannerman et al., 1995).

DISSOCIATING "SENSORY" AND "MOTOR" REPRESENTATIONS OF SONG

With regard to the role of the song system in song memory, it may be useful to make a distinction between a representation of the auditory memory of the tutor song and the motor memory of the bird's own song, as many authors – explicitly or implicitly – have done (e.g. Doupe and Kuhl, 1999; Nottebohm, 1981, 2000; Solis et al., 2000; Marler and Doupe, 2000; Nick and Konishi, 2005a; Adret, this volume). Bolhuis and Gahr (2006) have argued that these two aspects of song acquisition may be associated with the memorization phase and the sensorimotor phase, respectively. It then becomes important to make explicit which of these putative memory mechanisms is being investigated.

Electrophysiological studies have shown that neurons in the song system nuclei HVC, LMAN, Area X and RA are activated when the bird is singing (Margoliash, 1986; DeVoogd, 1994; Solis et al., 2000). In addition, there is increased expression of immediate early genes (IEGs, used as markers for neuronal activation) in these nuclei when the bird is singing (Jarvis and Nottebohm, 1997), but not when it hears song (Mello et al., 1992; Jarvis and Nottebohm, 1997; Mello and Jarvis, this volume), including tutor song (Bolhuis et al., 2000). These findings, combined with the absence of learning-related IEG expression in the song system (Bolhuis et al., 2000, 2001) and the electrophysiological studies discussed earlier, are inconsistent with the hypothesis that the song system of adult songbirds contains the neural substrate of tutor song memory. Rather, the results suggest that these brain nuclei are either involved in song production only, or in processing the auditory feedback of songs that occurs during the sensorimotor phase of song learning, which can also occur in adult songbirds (Brainard and Doupe, 2000a,b, 2002). In that case, it may be that there is a representation of the BOS in the song system that is being updated through continual interaction with regions beyond the song system (see below).

BEYOND THE SONG SYSTEM

One candidate region for the neural representation of the tutor song memory is the hippocampus which has been implicated in memory in both mammals (Scoville and Milner, 1957; Buckley and Gaffan, 2000) and birds (Sherry and Vaccarino, 1989; Bolhuis and Macphail, 2001; Macphail and Bolhuis, 2001; Bolhuis, 2005; Sherry, 2006) and, in songbirds, contains neurons that are responsive to auditory stimuli (M. Gahr, unpublished observations). Terpstra et al. (2004, 2006) reported that there was no significant increase in IEG expression in response to tutor song in the hippocampus of either male and female zebra finches. Overall, the expression of IEGs was very low in the hippocampus of both sexes. Similarly, there was very little expression of IEGs in the hippocampus of female budgerigars (Eda-Fujiwara et al., 2003) and male ring doves (*Streptopelia risoria*; Terpstra et al., 2005) in response to conspecific vocalizations. Bailey et al. (2002; Bailey and Wade, 2003) reported increased expression of Fos (the protein product of the IEG *c-fos*), but not of zenk (the protein product of the IEG *ZENK*), in response to conspecific song (compared with heterospecific song) in zebra finch females at 30 days after hatching and as adults. There were no such differential responses in females at 45 days after hatching (Bailey and Wade, 2005) or in males at 30 or 45 days after hatching (Bailey and Wade, 2003, 2005). Thus, the hippocampus seems an unlikely candidate as the neural substrate for tutor song memory.

In contrast, a large and growing body of evidence, much of it based upon gene expression studies (see Mello and Jarvis, this volume), suggests that regions in the caudal forebrain may contain the neural substrate for tutor song memory. Exposure to conspecific song in zebra finches or canaries led to increased IEG expression in the caudal part of the medial nidopallium (NCM) and the caudal part of the medial mesopallium (CMM, see Figure 22.1; Mello et al., 1992; Mello and Clayton, 1994; Bolhuis et al., 2000). IEG expression was greatest when birds were exposed to conspecific song, compared with heterospecific song or pure tones (Mello et al., 1992). The fact that such activation was not seen in song system structures (Jarvis and Nottebohm, 1997) suggests a dissociation between forebrain regions that are activated when the bird hears song, and those that are activated when the bird is singing itself. These findings led to the hypothesis that the regions in the caudal forebrain that are activated when the bird is exposed to conspecific song might be (part of) the neural substrate for memory of the tutor song (Bolhuis, 1994; Solis et al., 2000; Nottebohm, 2000). The hypothesis has received support from a number of studies from our laboratory, demonstrating a significant correlation between neuronal activation in the NCM of adult male zebra finches and the strength of song learning, measured as the number of song elements that had been copied from the tutor song (Bolhuis et al., 2000, 2001; Terpstra et al., 2004; see Figure 22.2A). In one study, we investigated the expression of two IEG protein products, Fos and zenk, in the forebrain of zebra finch males that had been tape-tutored with an adult zebra finch song (Bolhuis et al., 2000). When adult, the birds were re-exposed to their tutor song (or not re-exposed: controls). There was a significant increase in expression of both Fos and zenk protein in the experimental birds compared with the control birds in the NCM and in the CMM, but not in two nuclei in the conventional song system, HVC and Area X. Furthermore, in the experimental birds, but not in the controls, there was a significant positive correlation between the number of song elements copied from the tutor song and the expression of both Fos and zenk protein in the NCM, but not in the CMM. The positive correlation between the strength of song learning and IEG expression in the NCM has been replicated twice (Bolhuis et al., 2001; Terpstra et al., 2004; see Figure 22.2A, and Nordeen and Nordeen, this volume, Figure 21.2).

An electrophysiological equivalent to this learning-related neuronal activity was discovered recently by Phan et al. (2006). These authors demonstrated that neurons in the NCM of adult zebra finch males showed steeper rates of habituation to novel song than to the tutor song. A familiarity index, based on relative habituation rates, was significantly greater in tutored males than in untutored males. In addition, there was a significant positive correlation between the familiarity index of NCM neurons and the strength of song learning (Figure 22.2B).

Additional evidence suggesting a role for the NCM as (part of) the neural substrate for tutor song memory (Bolhuis et al., 2000) is provided by studies that investigated habituation-like processes. Repeated exposure to the same song leads to a decrease in expression of the immediate early gene ZENK (Mello et al., 1995) and to decreased electrophysiological responsiveness in the NCM (Chew et al., 1996a). Chew et al. (1996a) concluded that it is likely that "the NCM is specialized for remembering the calls and songs of many individual conspecifics." These results, combined with the findings of a number of other studies (e.g. Sockman et al., 2002; Eda-Fujiwara et al., 2003; for reviews see Bolhuis and Eda-Fujiwara, 2003; Bolhuis and Gahr, 2006) indicate that the NCM and CMM are involved both in processing of perceptual information concerning song complexity and in storage of song memory in songbirds and parrots.

SONG MEMORY IN FEMALES

Zebra finch females normally do not sing, but only produce calls. Nevertheless, they can develop a preference for the song of their father over novel conspecific song (Miller, 1979; Riebel, 2000), which indicates that they are able to discriminate between different conspecific songs. In operant tests, both female and male zebra finches that were reared with their father showed a significant and similar preference for their father's song (Riebel et al., 2002). Thus, these females, although they do not sing, form a memory of the song of an adult male conspecific, just as males do. It is possible, but not necessary, that the mechanisms and neural substrate of

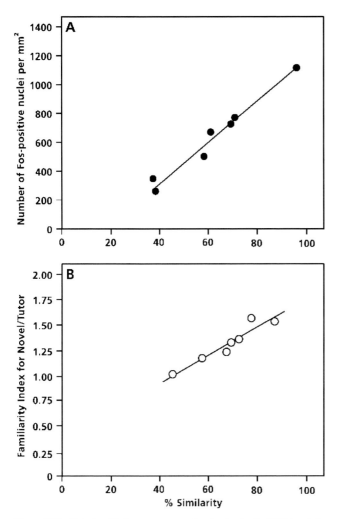

Figure 22.2 Relationship between neuronal activation in the NCM and the strength of song learning (measured as percentage of song elements copied from the tutor song) in zebra finch males. (A) Relationship between the number of cells immunopositive for Fos (the protein product of the immediate early gene *c-fos*) and the strength of song learning, in zebra finch males that had been re-exposed to the tutor song. (Adapted from Bolhuis *et al.*, 2001, *European Journal of Neuroscience*, Blackwell Publishing, with permission). (B) Relationship between the familiarity index (FI) and the strength of song learning in male zebra finches. The FI is a measure of the relative habituation rate of neurons in the NCM. (Adapted from Phan *et al.*, 2006, with permission; copyright 2006 by The National Academy of Sciences of the USA).

memory formation in the two sexes are similar, or identical. An advantage of investigating these mechanisms in nonsinging females is that learning the characteristics of the father's song is not confounded with the motor learning involved in song production.

Female songbirds – regardless of whether or not they sing – also have a song system, but with nuclei that are generally significantly smaller than in their male counterparts (MacDougall-Shackleton and Ball, 1999). There are species differences with regard to the role of the song system nucleus HVC in female song perception and discrimination. In female canaries, bilateral HVC lesions resulted in the loss of their ability to discriminate between conspecific and heterospecific songs (Brenowitz, 1991) and between different conspecific songs (Del Negro *et al.*, 1998; Halle *et al.*, 2002).

An electrophysiological study showed that sexually attractive songs of male canaries evoke different neuronal responses in the HVC of females from songs lacking these attractive syllables (Del Negro et al., 2000). Thus, in canaries the HVC appears to be involved in female perception of male song. In contrast, the results of studies in zebra finch females suggest that brain regions outside the song system play a role in song perception. Electrolytic lesions of the CMM, but not lesions of the HVC, disrupted the ability of zebra finch females to discriminate conspecific from heterospecific song (MacDougall-Shackleton et al., 1998a). Bolhuis and Gahr (2006) discussed three possible explanations for the discrepancy between the effects of lesions in the two species. First, canary and zebra finch males are open-ended and age-limited learners, respectively. The HVC in the canary would then be involved in the perception by females of new songs that are being learned by adult males. Second, unlike zebra finch females, canary females sometimes sing, and they may also learn to sing new songs, for which the HVC could be important. Third, it is possible that the HVC in female canaries is involved in another motor act – copulation solicitation display.

The suggestion that brain regions outside the song system may be involved in song perception and memory in females (MacDougall-Shackleton et al., 1998a) was supported by IEG expression studies in a number of songbird species. Song-induced IEG expression was found in the NCM in starlings *Sturnus vulgaris* (Duffy et al., 1999; Gentner et al., 2001) and canaries (Ribeiro et al., 1998), and in the NCM and the CMM in zebra finches (Bailey et al., 2002; Terpstra et al., 2006) and white-crowned sparrows *Zonotrichia leucophrys* (Maney et al., 2003). IEG expression varies with male song complexity in the NCM (and to a lesser extent in the CMM) of female budgerigars *Melopsittacus undulatus* (Eda-Fujiwara et al., 2003), a parrot species.

Zebra finch females reared with their fathers and tested as adults showed a significant preference for the song of their fathers, indicating that they had learned the characteristics of this song (Terpstra et al., 2006). Adult females that were re-exposed to their father's song showed significantly increased expression of the IEG protein product zenk in the CMM, but not in the NCM or the hippocampus, compared with females that were exposed to novel zebra finch song (Figure 22.3B). Leitner et al. (2005) reported a very similar pattern of *ZENK* expression in female canaries. There was significantly greater neuronal activation in the CMM, but not in the NCM, of female canaries that were exposed to songs they preferred compared with birds that were exposed to less attractive songs. In addition, male and female starlings trained in an operant task to recognize conspecific songs showed memory-related electrophysiological responsiveness in the CMM (Gentner and Margoliash, 2003). In the latter study, only the CMM was sampled. Taken together, these results suggest that in female songbirds, the CMM may be part of a neural substrate for song memory.

SONG-RELATED NEURONAL ACTIVATION: SPECIFIC TO MEMORY?

Is the song-induced IEG expression in the NCM and the CMM associated with a neural representation of song memory? Marler and Doupe (2000) suggested a number of alternative explanations that were explored in subsequent studies from our laboratory. We demonstrated that neuronal activation in the NCM in response to tutor song is *not* an artifact of isolation rearing, since a positive correlation between IEG expression and the strength of song learning was also found in the NCM of zebra finch males that were reared with a live tutor (Bolhuis et al., 2001; Terpstra et al., 2004). We also showed that song-related neuronal activation in the NCM does not simply reflect a predisposition in "good learners" to pay more attention to any song stimulus than "poor learners." If this were the case, one would have expected a positive correlation between the strength of learning and the extent of IEG expression in the NCM also in groups that were exposed to the BOS or to novel song, which was not the case (Terpstra et al., 2004). Finally, we showed in zebra finch females that the differential neuronal activation found in the CMM (Terpstra et al., 2006; Figure 22.3B) is not a result of the birds paying more attention to their father's song than to novel song. Such a high level of focused attention would have been expected to lead to a more consistent IEG response in the CMM in females exposed to their father's song than that in females exposed to novel song. In fact, variance in the former was significantly greater than in the latter, suggesting that neither in females nor in males is an attentional explanation likely.

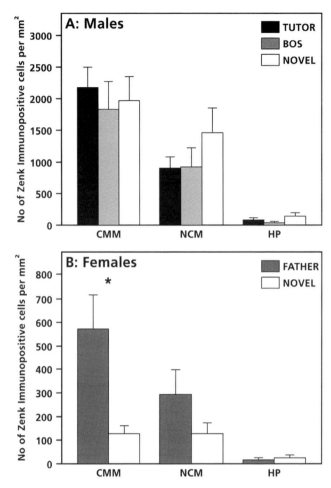

Figure 22.3 Neuronal activation (measured as the mean number of cells immunopositive for zenk, the protein product of the immediate early gene *ZENK*) in the NCM, CMM and hippocampus (HP) of zebra finch males (A) and females (B). (A) Mean (+ SEM) number of zenk-immunopositive cells per mm^2 for groups of adult male zebra finches exposed to tutor song (TUTOR), bird's own song (BOS) or novel conspecific song (NOVEL). (B) Mean number of zenk immunopositive cells (+ SEM) per mm^2 in adult female zebra finches that were re-exposed to their father's song (FATHER) and for birds that were exposed to a novel song (NOVEL). (Adapted by permission from Macmillan Publishers Ltd: *Nature Reviews Neuroscience*, Bolhuis and Gahr, 2006, copyright 2006).

FOREBRAIN MEMORY SYSTEMS IN BIRDS

Our initial results (Bolhuis et al., 2000, 2001) were insufficient to resolve the issue as to whether IEG expression in the NCM reflects a neuronal response to the tutor song, or to the BOS. The more a male has copied from the tutor song, the more this tutor song will resemble the BOS. Thus, the correlations between IEG expression and the strength of song learning could also reflect a neuronal response to songs that increasingly resemble the BOS. In a study designed to address this issue, Terpstra et al. (2004) found that in zebra finch males, IEG expression in both the NCM and the CMM does not differ in response to tutor song, the BOS or novel song (Figure 22.3A). There was a significant positive correlation between neuronal activation in the NCM and the strength of song learning *only* in response to tutor song, not to the BOS or novel song (Nordeen and Nordeen, this volume; Figure 21.2). The absence of such a correlation after exposure to the BOS is consistent with a role for the NCM in information processing for this song. However, the relatively low level of neuronal activation in the NCM in response to the BOS does not support this hypothesis. The differential responsiveness to the song of the father versus novel song in female zebra finches (Terpstra et al., 2006) cannot be a reflection of a representation of the BOS, as females do not sing.

When considering what information (tutor song or BOS) is processed in the NCM, Terpstra et al. (2004) identified at least three possible scenarios:

(1) All males have learned the complete tutor song, but they have not incorporated all of the tutor song elements into their own songs, as suggested by, for example, Marler and Peters (1982a), Nelson and Marler (1994) and Geberzahn and Hultsch (2003). Neurons in the NCM are activated by the complete tutor song. Re-exposure to the tutor song would then be expected to induce high levels of neuronal activation in the NCM of experimental males, independent of how many elements this male copied from the tutor song. Exposure to the BOS, however, would be expected to induce higher IEG levels in birds that copied many elements from the tutor song than in birds that copied only a few elements. This scenario is not likely, as both in the present study and in two previous ones (Bolhuis et al., 2000, 2001), a significant positive correlation was found between neuronal activation in the NCM and number of copied elements in males that were re-exposed to their tutor song.
(2) Neurons in the NCM are activated by the part of the tutor song that is incorporated into the bird's own song. In this case, both re-exposure to the tutor song and exposure to the BOS should induce levels of neuronal activation proportional to the strength of song learning. This hypothesis is not supported by our findings, as we did not find a significant positive correlation between neuronal activation and number of copied elements when we exposed birds to their own song.
(3) Neurons in the NCM are activated by the BOS. In this case, exposure to the BOS would be expected to lead to considerable neuronal activation in the NCM, irrespective of the strength of song learning, while the neural response to the tutor song would be expected to be proportional to the number of elements copied from the tutor song. This is consistent with the result by Terpstra et al. (2004) who reported a significant positive correlation between neuronal activation and number of copied elements when birds were exposed to their tutor song, while there was no such correlation when birds were exposed to their own song. However, this hypothesis would also predict greater neuronal activation in the NCM in the BOS group compared with the other two groups. Terpstra et al. (2004) did not find such an effect.

What then are the roles of the NCM and the CMM in auditory memory in songbirds? Studies involving zebra finch males suggest that the NCM may contain the neural substrate for tutor song memory (Bolhuis et al., 2000, 2001; Terpstra et al., 2004; Phan et al., 2006). At the same time, studies involving male (Bolhuis et al., 2000, 2001; Terpstra et al., 2004; Gentner et al., 2004) as well as female songbirds (MacDougall-Shackleton et al., 1998a; Ribeiro et al., 1998; Duffy et al., 1999; Gentner et al., 2001; Bailey et al., 2002; Maney et al., 2003; Terpstra et al., 2006) suggest that the CMM is also important for song memory. For example, male starlings showed increased neuronal activation in the CMM, but not in the NCM, when they were exposed to familiar songs (Gentner et al., 2004). In our studies with zebra finches, males and females that were reared in the same way exhibited different patterns of neural activation in response to the same song stimuli. In females, there was learning-related neuronal activation in the CMM, not in the NCM (Terpstra et al., 2006). In contrast, in males re-exposed to their tutor's song, there was a significant correlation between the strength of song learning and neuronal activation in the NCM, not in the CMM (Bolhuis et al., 2000, 2001; Terpstra et al., 2004).

Consideration of the different functions of song learning in males and females may shed more light on this issue. Recognition of the father's song is important in both sexes (and the CMM may contain the neural substrate subserving memory of that song), whereas only males produce song for which the NCM may contain the neural substrate (Terpstra et al., 2004), perhaps serving as a parallel store to the CMM. There was a significant correlation between neuronal activation in the CMM and NCM in females that were exposed to their father's song (Terpstra et al., 2006), suggesting that the two brain regions may both comprise the neural substrate for the representation of tutor song, and function in parallel. Of course, there are avian species (Gahr et al., 1998; see also Kaplan, this volume) where females also sing and learn their songs in the same way as males do. Thus, the existence of parallel stores is unlikely to be linked to sex. Rather, these data suggest that different subparts of the auditory forebrain of songbirds might be involved in different kinds of auditory memory.

The NCM and CMM are conserved widely among bird species (Jarvis et al., 2000; Reiner et al., 2004a, and this volume; Jarvis et al, 2005a; Terpstra et al., 2005). In filial imprinting in the domestic chick, memory of the visual imprinting stimulus is subserved by two parallel stores (Horn, 2004). The neural substrate of one of these stores is located in the intermediate and medial mesopallium (IMM), a brain region that partially overlaps with the CMM (Horn, 1985, 1998, 2004). In addition, the IMM is also involved in color recognition in a passive avoidance task in domestic chicks (Patterson and Rose, 1992). Song preference learning in female songbirds has been compared to sexual imprinting (Riebel, 2003). In domestic chickens, the IMM is also involved in sexual imprinting (Bolhuis et al., 1989). Thus, the medial part of the mesopallium may be (part of) a general recognition system in birds and contain representations of imprinted stimuli, conditioned stimuli and learned song (Bolhuis, 1994; Horn, 2004).

SIMILARITIES BETWEEN BIRDSONG AND HUMAN SPEECH

Can the similarities between avian song learning and human speech acquisition be extended to the neural level? In the revised interpretation of the avian brain (Reiner et al., 2004a) it is suggested by the members of the Avian Brain Nomenclature Consortium (Jarvis et al., 2005a) that the pallium, which includes the hyperpallium, mesopallium, nidopallium and arcopallium, is homologous to the mammalian neocortex. However, these researchers also concluded that it is premature to suggest one-to-one homologies between avian and mammalian pallial regions. Nevertheless, some similarities between the organization of auditory and auditory association areas in mammals and forebrain auditory-recipient areas in birds should be noted.

Within the avian forebrain, Field L2 receives auditory connections from the thalamus, and in turn projects onto Field L1 and Field L3 (Figure 22.1). These two regions project to the caudal mesopallium and caudal nidopallium, respectively. Thus, it is plausible that the Field L complex is homologous to the primary auditory cortex, in the mammalian superior temporal gyrus, which also consists of three "core" regions that receive inputs from the thalamus (Kaas and Hackett, 1999, 2000; Wise, 2003). In primates, the auditory association cortex consists of a complex network of brain regions, including the medial and lateral belt regions and the parabelt regions, that project to regions in the prefrontal cortex (Romanski et al., 1999; Kaas and Hackett, 1999, 2000; Wise, 2003). Thus, the projection regions of the Field L complex (the NCM and CMM) may be homologous to the belt and parabelt regions in the mammalian auditory association cortex. Wan et al. (2001) suggested that the rat auditory association cortex is involved in auditory recognition memory. These authors found that, compared to familiar sounds, novel sounds evoke significantly greater neuronal activation (measured as expression of the IEG protein product Fos) in the rat auditory association cortex, but not in a number of other brain regions, including the primary auditory cortex, the hippocampus and the perirhinal cortex. Similarly, in monkeys species-specific calls evoke neural activation specific to the left superior temporal sulcus (Poremba et al., 2004). In other studies with monkeys, lesions to the superior temporal cortex impaired auditory memory (Colombo et al., 1990, 1996), while lesions to the superior temporal gyrus or the temporal lobe, but not to the perirhinal cortex, impaired performance in an auditory recognition task (Fritz et al., 2005).

In humans, regions traditionally associated with speech perception, which are centred around Wernicke's area in the superior temporal lobe, are distinguished from speech motor areas including Broca's area in the frontal lobe (Wise, 2003). These regions do not function in isolation, as speech perception affects speech production from birth onwards (Kuhl, 2000; Wise, 2003). For example, there is evidence to suggest that speech perception modulates the excitability of tongue muscles (Fadiga, 2002). In their first months of life, human infants acquire sophisticated information about their native language just by listening, before they know the meaning of words (Kuhl, 2000). This early experience affects not only their discrimination ability and listening preference but also alters subsequent perception and motor performance. In addition, the experience is language specific, such that speakers that learn a second language after puberty produce it with an accent typical of the primary language (Kuhl, 2000). An fMRI study of infants (Dehaene-Lambertz et al., 2002) found that, like adults, 3-month-old babies who were exposed to speech revealed significant activity in brain regions on the left side of the brain, including the superior temporal gyrus. These findings show that precursors of adult cortical language areas are already active in infants, long before the onset of speech production. Thus, there is an interesting analogy between the mechanisms of human speech acquisition and song learning in songbirds. That is, brain regions predominantly involved in auditory learning in humans and birds are anatomically separate from those that are mainly involved in sensorimotor learning, and vocal learning involves continual interactions between them.

CONCLUSIONS AND FUTURE PERSPECTIVES

The evidence reviewed here suggests that the forebrain regions NCM and CMM may contain (part of) the neural substrate for tutor song memory, although the respective roles of the two structures remain to be determined. What role, if any, the classical song system nuclei play in tutor song memory is still unclear. What is clear is that neuronal activation related to tutor song memory (and perhaps also to memory of the bird's own song) occurs in brain regions outside the conventional song system in regions which may be homologous to mammalian auditory (association) cortex. The finding (Gobes and Bolhuis, 2007) that lesions to the NCM impaired preference for the tutor song without affecting song production in male zebra finches is consistent with the hypothesis that the NCM contains the neural substrate for tutor song memory. This finding also suggests that, as in humans, the cognitive systems of vocal production and auditory recognition memory in songbirds are subserved by distinct brain regions.

Further experiments with male and female songbirds and nonsongbirds are necessary to elucidate the roles of the nidopallium and mesopallium in the different aspects of song memory. These should include both brain imaging (Van Meir et al., 2005; Boumans et al., 2005) as well as electrophysiological analysis of NCM and CMM in awake songbirds (Schregardus et al., 2006). The neural mechanisms of female song memory also merit further investigation.

At this stage, we cannot be certain whether the NCM and CMM contain the neural substrate for song memory, or whether these brain structures are "relay stations" for a neural representation that is located elsewhere in the brain. This important problem can only be tackled in a series of experiments using different techniques, as has been done in the study of imprinting (Horn, 2004). If localization of the neural substrate of song memory in male and female songbirds can be confirmed and consolidated, the way is open for detailed cellular and molecular analyses. The role of the NCM and CMM in song acquisition during the memorization phase needs to be investigated in juveniles. In addition, localization of the neural substrate of tutor song memory is important for the analysis of the brain mechanisms of sensorimotor learning. How does information concerning the tutor song, which may be stored in the NCM and CMM, influence sensorimotor learning? Are there functional circuits linking NCM, CMM and the song system? Mooney (2004) has proposed the existence of a "comparator circuit," which he defined as "[using] auditory information generated during production of the bird's own song (BOS) to detect differences between the currently emitted song and the stored model." Mooney (2004) suggests that the HVC is a key component of the comparator circuit. Likewise, Nick and Konishi (2005a) suggested that song learning involves a "comparator," which they defined as "a behaviorally defined brain space that compares auditory feedback to the tutor song template." They speculated that such a comparator may be localized in the HVC, NCM, or CMM. At any rate, the strong and stimulus-specific responsiveness of the NCM and CMM to tutor song suggests that these brain regions continue to be important in adult songbirds. Bolhuis and Gahr (2006) have outlined a possible scenario in which the NCM and CMM are involved in song acquisition in the memorization phase, while the song system is important during the sensorimotor phase and for the production and maintenance of adult song – possibly through continual interaction with the NCM and CMM.

ACKNOWLEDGEMENTS

I wish to thank Phil Zeigler and Peter Marler for inviting me to contribute to this book, and for their valuable comments. I am extremely grateful for collaboration and stimulating discussions with Nienke Terpstra, Sharon Gobes, Guus Zijlstra, Hiroko Eda-Fujiwara, Ardie den Boer-Visser, Thijs Zandbergen, and all my other colleagues in the birdsong memory project.

23 • The template concept: crafting a song replica from memory

Patrice Adret

Template: "a gauge, pattern or mold, commonly a thin plate or board used as a guide to the form of the work to be executed."
(*Webster's International Dictionary*, 1900)

INTRODUCTION

The avian brain is endowed with a remarkable memory capacity. The birds' ability to recall the past is indispensable to the expression of various forms of recognition processes including object, place and social recognition; it also enables them to learn new skills and to act flexibly in anticipation of the future. It is reflected most impressively in the capacity of certain avian groups for vocal learning by imitation and the generation of local dialects (Marler and Slabbekoorn, 2004).

Oscine songbirds (passerines) have evolved a highly specialized song-producing system that has been well exploited in species with multiple song types. For instance, a European nightingale *Luscinia megarhynchos*, with a repertoire of 200 song types, will reproduce faithfully up to 60 song models, each heard only 20 times in youth (Hultsch and Todt, 1989c). Such a skill requires a particular kind of auditory memory – a template – to help guide the reproduction of sounds heard at an early age. Moreover, not only do oscine songbirds learn their song, but the process of avian vocal learning parallels in many respects the process of human language acquisition learning (Doupe and Kuhl, 1999, and this volume; Wilbrecht and Nottebohm, 2003), a process once thought to be an exclusive feature of humans, and one whose mechanisms remain elusive in both taxa. Despite progress in the neurobiology of birdsong learning since Konishi's (1965b) initial elaboration of the auditory template theory, the "template" construct has remained elusive and the search for its underlying mechanisms remains a central feature of birdsong research (see Adret, 2004; Bolhuis and Gahr, 2006, for recent reviews).

Understanding these mechanisms requires answers to three questions: What are the critical properties of the "template" that make possible such learning, what are their neural substrates and how are they distributed in the songbird brain? The aim of this chapter is primarily heuristic. Its concern is less with the answers to these questions than with how one goes about answering them.

SONG LEARNING AND THE TEMPLATE CONCEPT

In his pioneering studies of bird song learning W. H. Thorpe of Cambridge University showed that naïve young male chaffinches (*Fringilla coelebs*) kept in isolation developed song patterns lacking many features typical of the songs of their species (Thorpe, 1961). However, when given the opportunity to hear normal songs the deprived birds readily imitated them, eventually producing songs that conformed to those of wild chaffinches. Thorpe found that young birds were reluctant to acquire alien sounds and that they favored the songs of their own species over those of foreign species (Thorpe, 1958). He attributed this perceptual bias to an endogenous "blueprint." By blueprint he meant an inherited sensory mechanism with filtering properties that enable a young bird to recognize the sounds of his own species, and hence to select which song to learn. The blueprint idea, though a useful first step, is inadequate to convey concretely the complexity of the song development process, which involves, for example, the transformation of an internal representation of a song into a motor output (vocalization) – based, presumably, on a set of both genetic and environmental instructions.

Subsequently, Konishi (1965b) showed that song learning in oscines proceeds in two steps. A youngster first acquires his song by listening to other members of its own species; subsequently (i.e. days, weeks or months later depending on the species), guided by the

Neuroscience of Birdsong, ed. H. Philip Zeigler and Peter Marler. Published by Cambridge University Press. © Cambridge University Press 2008.

perception of its own voice, he will reproduce those sounds from memory. Konishi proposed a general principle of imitative learning known as the auditory template theory: young birds first listen to and memorize an external model and, subsequently, use the memory trace as a template with which to match their own vocalizations (Konishi, 1965b).

We can think of a template much as do carpenters and toolmakers; a model containing the essential features of an object that is used as the basis for constructing a replica of that object. Biologists are well acquainted with the term. Genes, for example, can serve as templates – as repositories of genetic information that can be passed on from one generation to the next (Squire and Kandel, 1999). A song template, by extension, can be thought of as a storehouse in the songbird brain that provides all the necessary instructions for guiding the song learning process so that, by self-perception, and with varying degrees of precision, the voice converges on those specifications, much as a toolmaker uses his templates as a guide to the final form of the tool.

Thorpe's "blueprint" evolved into the conception of song templates, first advanced by Marler (1963) who pointed out the need to distinguish between "motor programs" (instructions encoded in the vocal motor pathway) and "auditory templates" (sensory specifications converted subsequently into vocal motor commands) as potential mechanisms for the production of inherited vocalizations. Marler also envisioned a role for auditory feedback in the development and maintenance of learned vocalizations, thus stimulating experimental work on two of the central problems of birdsong research.

EXPERIMENTAL ANALYSIS OF THE TEMPLATE HYPOTHESIS: ISOLATION VERSUS DEAFENING

To examine the contribution of auditory feedback to song development, Konishi (1963; 1964a; 1965a b; 1978; 2004; and this volume) carried out deafening experiments on a variety of avian species. In nonpasserines (e.g. domestic fowl, *Gallus gallus*), bilateral extirpation of the cochlea immediately after hatching had no apparent effect on vocal development. In these birds, acquisition of a normal repertoire seems to be largely independent of auditory experience (Figure 23.1). By contrast, early deafening in oscine songbirds had a dramatic effect on song development. Konishi (1965b) demonstrated that white-crowned sparrows *Zonotrichia leucophrys* that were operated upon prior to song production produced highly abnormal song patterns while those deafened after song crystallization maintained a stable song. Moreover, songs of early-deafened birds were highly degraded regardless of the amount of prior auditory exposure to conspecific models, suggesting that (1) central motor programs are not sufficient for fully normal song development and (2) auditory feedback is essential for the conversion of memorized songs into produced songs. On this basis, Konishi conceptualized an "acquired" template, which is, simply put, a memory for guiding motor development. An analogy can be found in sexual imprinting where young animals learn the appearance of the object of sexual behavior long before they can themselves perform the appropriate reproductive behavior (ten Cate, 1994).

Subsequent deafening experiments on chaffinches showed the extent of the deterioration depended critically on the developmental stage of song development at the time of deafening: the more practiced the bird, the more structured the final song (Nottebohm, 1968). Thus it was predicted that deafening at an earlier age in these species would remove all remnants of species-identifying cues from their song (Konishi and Nottebohm, 1969).

Marler and Sherman (1983) re-examined the auditory template hypothesis in a comparative framework. Two closely related species with sharply contrasting song patterns, swamp sparrows *Melospiza georgiana* and song sparrows *Melospiza melodia* were surgically deafened prior to any sign of subsong. Surprisingly, despite the loss of many acoustic features, their highly degraded songs still retained a number of species-specific differences, mostly apparent in the degree of song segmentation (Figure 23.2). Not only did this experiment confirm the importance of auditory feedback to song acquisition (since in both species early deafening results in highly degraded songs) but it also shed some light on the contribution of central motor programs to song patterning and on species differences in the extent of that contribution. The degree of species specificity is minimal, however, and because of their lack of content the playback of such songs is without effect on the female (Searcy and Marler, 1987).

284 PATRICE ADRET

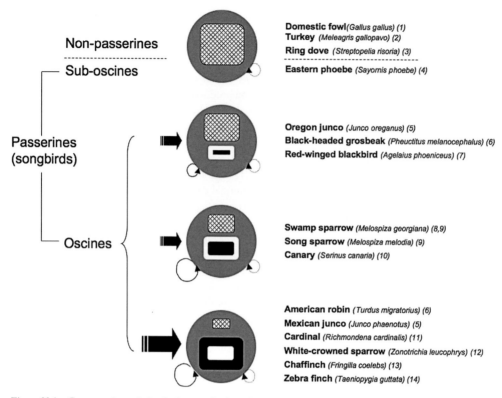

Figure 23.1 Cross-species variation in the contribution of genetic and experiential factors to avian vocal development. In nonpasserines and most suboscine songbirds, acquisition of a normal vocal repertoire is thought to be entirely under the control of a central pattern generator (CPG, crossed box) and requires no auditory feedback. An exception to the rule is the three-wattled bellbird, *Procnias tricarunculata*, which has been shown recently capable of vocal learning (see Kroodsma, 2004, 2005). The finding suggests that some suboscines do rely on auditory feedback.

In oscine songbirds, auditory feedback is necessary for normal song development; species-identifying features of song derive partly, although minimally, from the working of a CPG and partly from access by auditory feedback to an innate template (white box) modifiable by experience into an acquired template (black box). There is variation in the degree of dependence on external models (thick arrows): some birds imitate in a facultative manner (e.g. Oregon junco and song sparrow) while others strongly depend on exposure to environmental sources (e.g. white-crowned sparrow, chaffinch). Thin recursive arrows symbolize control of voice by auditory feedback. In the absence of auditory feedback, proprioceptive feedback (dashed recursive arrows) could provide a compensatory mechanism for song maintenance (Suthers *et al.*, 2002, Konishi, 1965b), as evidenced by a lack of immediate song degradation after deafening in close-ended learners.

(Adapted from Konishi, 1963 [1]; Schleidt, 1964 [2]; Nottebohm and Nottebohm, 1971 [3]; Kroodsma and Konishi, 1991 [4]; Konishi, 1964a [5]; Konishi, 1965a [6]; Marler *et al.*, 1972 [7]; Marler and Sherman, 1983 [8,9]; Mulligan, 1966 [9]; Marler and Waser, 1977 [10]; Dittus and Lemon, 1970 [11]; Konishi, 1965b [12]; Nottebohm, 1968 [13]; Price, 1979 [14].)

The effect of isolation in oscine songbirds is far less disruptive than that of early deafening. As Figure 23.3 indicates, the songs of isolates exhibit a relatively complex structure by contrast with the almost featureless songs of the deafened birds. In fact, it was the obvious difference between the relatively impoverished songs of deafened birds and the more elaborate structure of isolate songs that gave rise to the notion of "innate auditory template." Thus, comparing the effects of early deafening with those of isolation gives us access to the unmodified, innate template. As Konishi (1965b) pointed out, young isolates denied access to external models can only hear their own voice, whereas deafening, by opening the feedback loop, prevents access

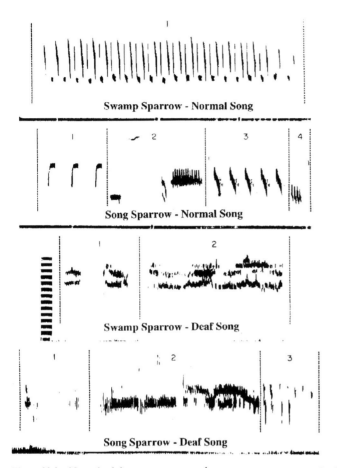

Figure 23.2 Normal, adult swamp sparrow and song sparrow songs are contrasted with the highly impoverished songs developed by a swamp sparrow and a song sparrow, both surgically deafened at an early age, prior to any sign of song production. Compared with the deaf swamp sparrow song, there is a higher degree of song segmentation in the deaf song sparrow, reminiscent of the highly segmented song of the species in the wild. Numbers indicate distinct song segments. (Adapted from Marler and Sherman, 1983.)

not only to external (acquired) sources of reference but also to inherited ones. Early-deafened birds presumably have an innate auditory template but, in the absence of auditory feedback, it remains stuck in the head, inaccessible, somewhat disconnected from the central pattern generator (Figure 23.1).

Isolation experiments also showed that the amount of genetic information encoded in the template varies across species. For instance, while many species-identifying features are lost in the isolate songs of white-crowned sparrows (Konishi, 1965b) many are retained in those of song sparrows (Mulligan, 1966; but see Kroodsma, 1977a). Thus, for normal song development, white-crowned sparrows rely more heavily on appropriate external stimulation than do song sparrows

(Figure 23.1). In both species, however, early deafening results in highly degraded songs. Furthermore, isolate songs exhibit a high degree of variation both between and within individuals, which suggests that the innate song template specifies only "crude" instructions to the vocal motor pathway. Hence, the term "crude template" that is sometimes applied in the birdsong literature (Catchpole and Slater, 1995).

ACQUISITION OF THE TEMPLATE: INNATE AND ACQUIRED FEATURES

It seems clear from the above that the acquisition of a template involves an interaction between certain

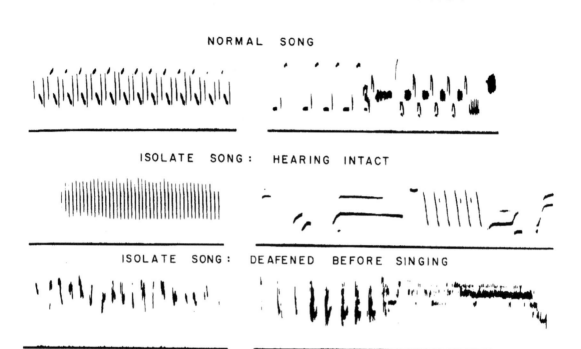

Figure 23.3 A comparison of songs produced by two sparrows, the song sparrow and the swamp sparrow, under three conditions. Note the more elaborate structure of isolate songs compared with the deaf songs. (From Marler, 1987.)

features present even in the isolate and others encountered during the bird's development. That is why, in its original interpretation, the acquired template was envisioned as a modified innate template (Konishi, 1965b) i.e. as the molding – through experience – of a crude template into an "exact" template (Catchpole and Slater, 1995). These changes in template properties take place during a "sensitive phase," a period of heightened sensitivity to song exposure. In most songbirds, this impressionable phase usually starts after fledging, but the time span during which learning occurs varies widely across different species. "Age-limited" learners acquire their song early in life, normally with no further change past the stage of song crystallization. Among them are species in which the sensory and sensory-motor phases are well separated in time (sparrows) while in others both phases may overlap substantially (finches). "Open-ended" learners such as the canary *Serinus canaria* can learn new songs as adults, varying them each season or retrieving unused song material acquired early in life (Nottebohm, 1999).

In fact, beginning in the sensitive phase of song learning, or even earlier, the songbird brain is not a *tabula rasa* (blank slate) to be written upon by the environment but exhibits distinct biases for the detection and recognition of species-specific models. For instance, newly fledged white-crowned sparrows and zebra finches *Taeniopygia guttata*, as well as nestlings of brown-headed cowbirds *Molothrus ater*, respond more strongly to own species' song than to foreign song (Dooling and Searcy, 1980; Nelson and Marler, 1993; Braaten and Reynolds, 1999; Hauber *et al.*, 2001). Moreover, 10 days after their first exposure to own species' song, not only will white-crowned sparrows chirp more to familiar than novel conspecific song but, as adults, they will also reproduce these songs from memory (Nelson *et al.*, 1997). Likewise, newly fledged sparrows do not initially distinguish between song dialects but will do so with further tutoring (Nelson, 2000). Thus, because the young bird is initially biased to certain basic song features, associated features are more likely to be attended to and hence incorporated in the developing template. Furthermore, the incorporation of other features is not haphazard but influenced by social context and selective reinforcement.

One example of such influences is a process Marler (1990) has called "action-based learning." In many songbirds there is an overproduction of syllables in the plastic song stage of song development, which is then subject to a culling process just prior to song crystallization. Marler and Peters (1981) have demonstrated the profound influence of social interactions on this attrition process. When birds engage in countersinging, for example, young males retain in their repertoire those matched song types that closely resemble the song of a rival male. Similarly, when inexperienced male cowbirds "assess" their vocal competence while courting a female, the female's response is influential in determining which of the song types is retained (West and King, 1988). This form of protracted vocal plasticity differs from more traditional (memory-based) song learning in three ways: (1) learning can occur at a time when acquisition of a novel song is no longer possible; (2) an association between experience and vocal practice provides opportunity for selective reinforcement; and (3) the learning proceeds slowly. Findings such as these indicate that song acquisition involves rapid early learning and that songbirds are attentive to many details of their own species-typical vocalizations.

But, knowing "what to sing" is clearly not sufficient for imitative learning to take place; the bird also needs to know "how to sing," i.e. how to execute the instructions encoded in the template. Such sensory-to-motor conversion relies ultimately on the operating range of the phonatory system. Indeed, recent work (Podos et al., 2004) shows that the acquired song template can be molded so as to accommodate motor constraints experienced during the sensory-motor phase. For instance, when synthetic model songs proved too hard to reproduce, young male swamp sparrows resorted to vocal experimentation. They initially deployed an unusual broken syntax but, subsequently, adopted a strategy of note deletion. Although, at maturity, the sparrows were found to produce songs at a slower tempo than normal, the species-typical syntax was preserved. The birds did not simply acquire an easy version of the target song that they would later produce. In fact, the study shows that the sparrows had memorized faithfully the challenging models but, owing to motor constraints imposed during the retrieval phase of song production, they were forced to adjust or "recalibrate" their sensory template according to their own level of proficiency (Podos et al., 2004a). A similar sort of "recalibration" has been demonstrated in canaries (Gardner et al., 2005).

Whether we focus on those template features which involve inborn predispositions to attend selectively to own species' songs (Marler, 1997) or those which reflect the consequence of auditory experience (Konishi, 1965b; Konishi and Nottebohm, 1969) the template concept serves the same basic functions: (1) to mediate in the memorization of songs heard and (2) to engage the memory trace(s) in guiding vocal development. Heuristically, the template construct provides researchers with a useful framework for investigating the physiological underpinnings of song learning. For instance, we may ask, what song features does the template store? How and where does it store them? How are they retrieved? How, in the singing bird, does the template "guide the form of the work to be executed" – the production of a species-typical song.

WHAT SONG FEATURES DOES THE TEMPLATE STORE?

The development of the bird's own song (BOS) obviously reflects the interaction of auditory feedback and information encoded in the template. But how does one distinguish between what is encoded and what is produced by the bird? Young birds spontaneously vary the frequency of begging and other calls in response to song stimulation. Using this response as an index of song preference, one can identify aspects of what is encoded prior to song production. Once singing has begun one can determine which acoustic features are selected by young birds for model imitation (Tchernichovski et al., 2000, 2004; Saar et al., this volume). In many songbirds, however, only part of the information stored in the acquired template is expressed as motor output; more experimentation is necessary if we are to access the learned, but temporarily unused song material (Hough et al., 2000; Geberzahn et al., 2002).

Experimental identification of song features selected for learning may proceed in two ways. One way is to manipulate the spectral content of the sounds produced (phonology; see Figure 23.4b) or their temporal organization (syntax; see Figure 23.4c). For instance, one can determine in the presinging bird, whether

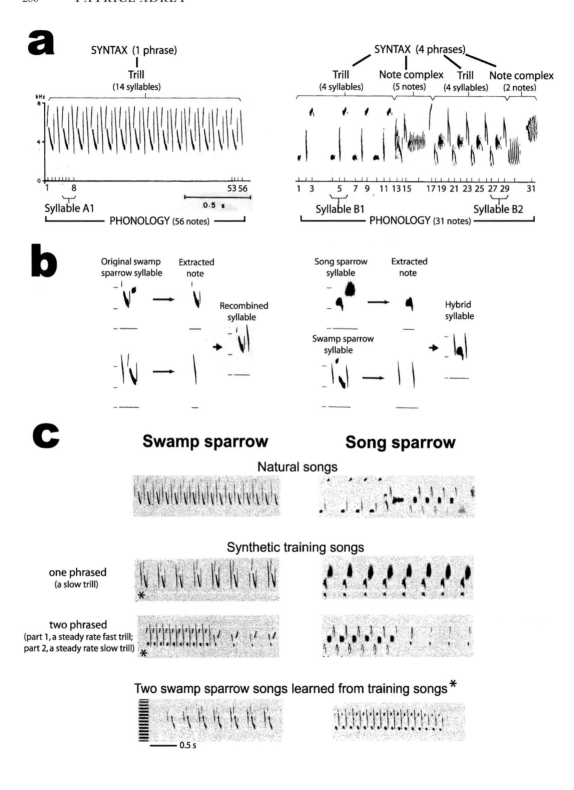

artificially modified songs are responded to and, later, whether they provide acceptable models to the singing bird; and if so, which acoustic features he tends to select for incorporation into the acquired template. Another way is to present auditory models that provide minimal information and to assess the young bird's ability to recognize critical features from such models and, later, reconstruct an entire song based on that information.

Marler and Peters (1977, 1988a) studied song development in two closely related sparrows with strikingly different song. Swamp sparrows produce a small repertoire of simple songs, each consisting of one fast-repeated syllable (trill), whereas song sparrows produce a larger repertoire of more complex songs, i.e. comprising 4–5 adjacent phrases consisting of distinct combinations of syllables (Figure 23.4a). Swamp sparrows rejected alien syllables substituted for species-typical syllables but imitated their own species' syllables regardless of the syntax in which they were embedded (Figure 23.4c). Song sparrows, on the other hand, were more ready to accept alien syllables when these were embedded in multiphrased songs compared with one-phrased songs. The authors suggest that swamp sparrows rely on phonological cues (note morphology) to select a tutor song, whereas song sparrows rely on both phonological and syntactical (phrase structure) cues. The question remains, however, whether these predispositions represent species differences operating as a sensory bias or as a production bias.

Unlike birds with a song repertoire, the white-crowned sparrow sings only one song made up of four or five ordered phrases of whistles, buzzes, note complexes and trills (Figure 23.5a). That the introductory whistle prevails in the song of all five known subspecies and emerges as the only syllable type to develop in the song of all young isolates suggests it might have a special role in selective learning. Indeed, experiments in which fledglings were tutored with an array of synthetic songs point to the whistle as a species-typical marker "facilitating the learning of phrases of other species' song when these are presented in the syntax of normal white-crowned sparrow song" (Soha and Marler, 2000). Pupils did not learn songs of their own species in which the introductory whistle had been removed but acquired foreign material (e.g. alarm calls of a Belding's ground squirrel *Spermophilus beldingi*) with a whistle added at the beginning. They even imitated sounds of unusual tonal quality from the song of a sympatric species, the hermit thrush *Catharus guttatus*, when they were manipulated to begin with a whistle (Soha and Marler, 2000).

But what about birds tested *prior* to the sensitive phase? Soha and Marler (2001a) presented fledglings with a set of modified songs in which the whistle was removed from the species-typical song or added to the beginning or at the end of a foreign song. Young sparrows responded more strongly to songs of their own species, regardless of presence and placement of a whistle. In previous work (Whaling *et al.*, 1997) fledglings were also found to respond initially indiscriminately to one-phrased (whistle, trill or buzz) songs of their own species. Thus, species-typical phonology, not syntax, is the primary cue on which young sparrows focus their attention prior to the memorization phase. Soha and Marler's (2000, 2001a) studies implicate changes in template properties. It appears that pre-encoded auditory mechanisms for the recognition of phonological cues are subsequently modified and supplemented by experience with acoustic features so as to ensure that, at a later stage, the detection of syntactical features will also help guide the learning process.

As a second research tactic, a predisposition for segmenting song was revealed by exposing young males to auditory models in which cues for segmentation had been erased (Soha and Marler, 2001b). Although birds were found to develop fewer phrases than were present in normal songs, they produced significantly more than the single phrase of model songs (Figure 23.5a). Consistent with the species' syntactical rules of song production, a whistle almost always preceded other types of syllables. Evidently, exposure to single

---

Caption for Figure 23.4
(a) Spectral and temporal organization of swamp sparrow and song sparrow songs. Lack of segmentation (one-phrased) in swamp sparrow song contrasts with the segmented (multiphrased) syntax of song sparrow song. (b) Two examples illustrating how sparrow song phonology was manipulated: recombined syllable after extraction of notes from two swamp sparrow syllables and hybrid syllable constructed by splicing two notes extracted from song sparrow and swamp sparrow syllables (from Marler, 1987). (c) Swamp sparrows exposed to a large array of synthetic songs, of which only four are shown, rejected foreign syllables but accepted their own species' syllables, irrespective of the syntactical pattern in which they were embedded. (Modified after Marler and Peters, 1977.)

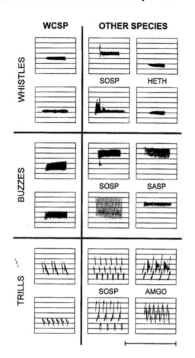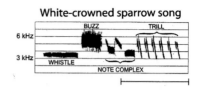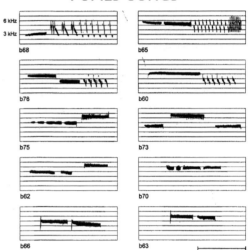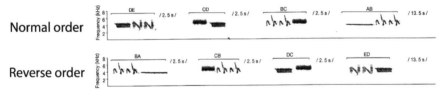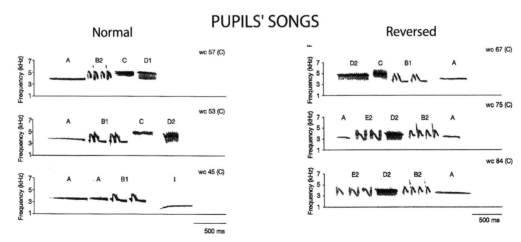

whistles, trills and buzzes does not provide sufficient information for white crowns to assemble these phrases together into a normal syntax. However, when offered pairs of normally adjacent phrases presented in that order (DE, CD, BC, etc. . . .), the sparrows were able to reconstruct the entire song (ABCDE). Remarkably, birds that had been exposed to phrase pairs presented in reverse order (BA, CB, DC, etc. . . .) sang completely reversed songs (EDCBA). In that study (Rose *et al.*, 2004) the presence of overlapping cues in the auditory fragments presumably enabled the sparrows to string the complete melody (Figure 23.5b). Phrase pairs, it seems, were encoded as a set of neural representations such that recall of ED "brought to mind" DC and so on. Thus the acquired template needs not be a representation of the full song and exhibits properties that are sufficient to override the predisposition to begin a song with a whistle (Soha and Marler, 2001b).

While the innate auditory template only specifies the production of whistles in white crowns, external stimulation in the form of isolated phrases will activate their predisposition for segmenting a song starting with an introductory whistle. The availability of syntax information results apparently in an acquired template made of fragmented auditory memories that are subsequently stitched together to generate the full song. How, then, do we go about finding "where in the brain" these representations are encoded?

IN SEARCH OF THE TEMPLATE: LOCALIZATION, RESEARCH STRATEGIES AND TACTICS

There appear to be two general strategies for the study of memory storage. One approach is to pick a mechanism of neural plasticity such as long-term potentiation (LTP) in a structure important for memory and attempt to show that it is a substrate for memories . . . The alternative strategy is to begin with a well-characterized form of learning and memory and determine the memory traces involved. In order to do this, it is first necessary to find where in the brain the memories are stored, the classical problem of localization.

(Thompson, 2005)

Contemporary studies of birdsong learning mechanisms have obviously embraced the second strategic alternative. Consequently, a major aim of that research has been a search for the template within the avian brain analogous to the mammalian "search for the engram" (Lashley, 1950). This chapter addresses the question: how will we know a song template when we see one? Put more concretely, what types of experimental designs and what kinds of data would allow one to conclude that a particular brain region functioned as a template – and to dismiss alternative interpretations of the data.

DISTINCT FORMS OF BIRDSONG MEMORY

Because songbirds learn many more songs than they produce it is useful at the outset to distinguish between two types of memory – (1) production memory, in which the song template is used to guide the motor development of song during the relatively brief developmental period and (2) recognition memory, which may operate throughout the entire life-span of an individual (Figure 23.6). My focus in this chapter is with production memory (PM), since an understanding of the mechanisms involved in this process is likely to

---

Caption for Figure 23.5

A comparison of songs developed by white-crowned sparrows tutored with isolated phrases and pairs of phrases. (a) Isolated phrases consisted of single whistles, buzzes and trills extracted both from white-crowned sparrow song (WCSP) and from the song of sympatric species, including song sparrow (SOSP), hermit thrush, *Catharus guttatus* (HETH), savannah sparrow, *Passerculus sandwichensis* (SASP) and American goldfinch, *Carduelis tristis* (AMGO). In this experiment, each isolated phrase was played many times before switching to another phrase. Despite an absence of syntactical cues, the majority of birds assembled a song incorporating two or three phrases copied from the models (modified after Soha and Marler, 2001b). (b) Normally adjacent pairs of phrases were presented either in the normal order or in the reverse order. During the plastic song stage, the sparrows were found to produce complete songs incorporating exact replicas of the models. At crystallization, their song retained appreciable complexity with a sequence of phrases based on the syntax offered during training (modified after Rose *et al.*, 2004).

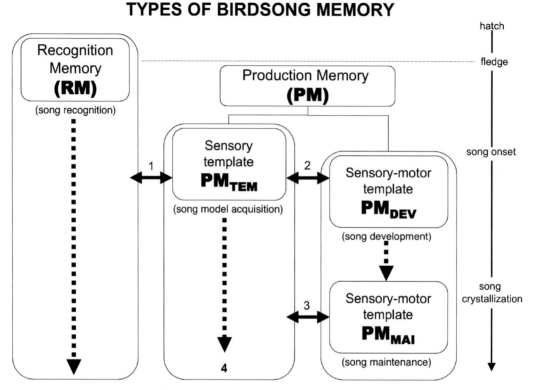

Figure 23.6 Types of birdsong memory, along with a timeline of song development. Recognition memory (RM) is in place before hatching, giving rise eventually to multiple song representations during a bird's life-span (dotted arrow). Production memory (PM) is first laid down as a sensory template (PM_{TEM}) after fledging. Subsequently, song practice generates a dynamic representation of the developing bird's own song (PM_{DEV}), best reflected in the emerging selectivity of neural responses to playback of current BOS during sensory-motor learning. I speculate that PM_{DEV} is updated on a day-to-day basis by song-related feedback such that playback of the current BOS elicits a stronger auditory response, relative to earlier versions of that motor pattern. Such process gives rise, at crystallization, to a stable production memory used for song maintenance (PM_{MAI}). Song maintenance in adulthood may be viewed as the conversion of PM_{MAI} to a stereotyped motor program, a form of procedural memory. At various points during development, some relations exist between various types of birdsong memory (double arrows): 1, Sensory templates for song model acquisition derive from innate recognition systems, as evidenced by genetically specified templates that guide learning preferences in young, naïve songbirds (Marler, 1997). 2, The onset of singing defines the time for interactions between sensory and sensory-motor templates, as occurs in action-based learning (Marler, 1990) or during song recalibration (Podos *et al.*, 2004). 3, After the stage of song crystallization, a continued link between sensory and sensory-motor templates is indicated by re-expression of songs that had been rejected during vocal learning (Hough *et al.*, 2000; Geberzahn *et al.*, 2002). 4, Persistence of sensory templates beyond the stage of song crystallization is supported by the demonstration of a long-term memory of tutor song in zebra finches (Funabiki and Konishi, 2003).

account for most of the features we subsume under the rubric of the song "template." Within the category of production memory, we may distinguish two different types: sensory templates used for vocal learning during development (PM_{TEM}), and sensory-motor templates used for song maintenance after crystallization (PM_{MAI}).

Production memory for song development

Since the template concept does not refer to a unitary process, it may not have a unitary anatomical representation. Different parts of the brain may participate in the perception, memorization, and production of different parts of the song, and may or may not be used at the same time. Similarly, one could envision a template represented by one ("grandmother") or a few cells or by a mosaic of large assemblies of distributed but interconnected neurons. In either case, their participation in template function would be indicated by a selective responsiveness to tutor songs relative to others. Both these caveats are equally applicable, by analogy, to the search for engram in mammals, and yet some progress has been made in identifying brain regions putatively involved in learning.

Identifying brain regions as participating in PM_{TEM} will require the use of a variety of procedures – anatomical, electrophysiological, genomic, pharmacological – to provide converging evidence, from both acute and chronic experiments, about the involvement of that structure in operations which characterize the template. Critical data are likely to emerge from studies carried out during the period of song acquisition and closely linked to behavioral studies of vocal development. In these experiments it will be crucial to standardize rearing and tutoring conditions across laboratories and utilize computer-assisted techniques for the acquisition and analysis of developmental data (see Saar et al., this volume).

LOCALIZING CANDIDATE BRAIN REGIONS

Over the past few decades, studies utilizing these techniques have implicated two sets of songbird brain structures as participants in "template" activity. One group includes components of the central auditory system (see Theunissen et al., this volume). The other includes structures constituting the "song system," comprised of two circuits, each originating in HVC: a basal ganglia-thalamo-pallial circuit known as the anterior forebrain pathway (AFP) – essential for song acquisition (see Brainard, this volume) – and a posterior pathway essential for song production (see Schmidt and Ashmore, this volume). These three circuits, taken together, are the territory whose exploration is most likely to yield clues as to the mechanisms mediating the behaviors from which the template construct has been inferred.

Forebrain auditory pathways: a network of reciprocal connections

Central auditory pathways in birds, like those in mammals, appear to embody a hierarchical scheme of sensory processing that conveys different streams of acoustic information in parallel pathways from low- to high-level brain regions (see Theunissen et al., 2004a, and this volume). There, auditory pathways exhibit exuberant radiation in the caudal auditory forebrain. Within Field L, a mammalian auditory cortex analog, a complex network of reciprocal connections links the recipient auditory areas to other Field L subdivisions, and from these regions to secondary auditory areas, such as the caudal mesopallium (CM) and the caudal medial nidopallium (NCM), as well as two structures (HVCshelf, RAcup) closely apposed to song system nuclei. Interestingly, the medial part of CM is reciprocally connected both with NCM and with the lateral part of CM, which in turn projects directly to the song system (see Figure 2 in Vates et al., 1996, and Figure 13.1 in Theunissen et al., this volume). As we know from primate studies, the principle of re-entry, involving reciprocal connections between different sensory maps, may prove to be an important component in encoding and decoding mechanisms of information storage (Miyashita, 2004).

In addition, robust descending projections connect the arcopallium (a motor cortex analog) to regions surrounding thalamic and midbrain auditory nuclei (monkey: Aitkin and Park, 1993; songbird: Wild, 1993a; Mello et al., 1998a). Such (re-entrant) feedback projections could in turn potentially alter the filtering properties of auditory neurons at subcortical levels. These aspects of auditory processing have been poorly studied.

The song system: a network of loops

Highly processed information in CM funnels into tertiary auditory areas, entering the song system via the interface nucleus (NIf), which is thought to be a main source of auditory input to HVC (Coleman and Mooney, 2004; Cardin et al., 2005; Wild, this volume). Each branch of the circuit consists of a chain of discrete song nuclei, suggesting serial processing of information (Williams, 1989). Moreover, at various nodes, information returns via axon collaterals in some cases, perhaps as a means to inform key structures on the nature of the signal being sent further downstream. A network of five loops involving three of these key structures (HVC, Area X and RA) have been thus far identified (see Schmidt and Ashmore, this volume). Another important attribute includes the presence of cells endowed with some of the most complex properties, including time and frequency combination sensitivity (Margoliash, 1983, 1986; Margoliash and Fortune, 1992; Volman, 1993; Doupe, 1997), properties not as yet encountered in primary and secondary auditory pathways. Lastly, a distinctive feature of the song system is the emergence of a topographic organization (to the exclusion of HVC) throughout the AFP, down to motor nuclei that innervate syringeal muscles (Vicario and Nottebohm, 1988; Vicario, 1991a; Johnson et al., 1995; Iyengar and Bottjer, 2002b). As Wilbrecht and Nottebohm (2003, p. 144) put it, "the repeated representation of the same syringeal muscles in so many nuclei may reflect the fact that those muscles and those nuclei are used for the single function of sound production and, as far as we know, nothing else."

RESEARCH STRATEGIES AND TACTICS

Knowing the circuitry

The circuitry described above is derived from a continuing series of anatomical (hodological, morphological) studies that have characterized both the connections between nuclei, and the morphology and projection targets of nuclear populations. This has provided a tentative identification of sensory, motor, and sensory-motor components of the circuits. Once the putative circuitry involved has been defined, three research methods, used singly or in converging combinations, have tended to dominate the search for template mechanisms: genomics and metabolic mapping, circuit disruption, by lesions or reversible blockade and electrophysiological recording. (Now that it is possible to image the songbird brain in vivo (e.g. Van Meir et al., 2005), we may expect such imaging to help identify potential template-related regions.)

Genomic studies

One function of such studies is to help identify candidate regions for further exploration using recording and disruption procedures. For example, gene activation markers (ZENK and c-fos expression) are *not* expressed in any of the song nuclei when birds *listen* to song, but light up to varying extents when birds *sing* (Jarvis and Nottebohm, 1997). In contrast, listening to same species' song produces robust gene activation in most telencephalic auditory areas lying outside the song system (for a review see Mello, 2002a; Mello and Jarvis, this volume). Such observations have supported a functional dissociation between auditory structures that are afferent to the song system and the classical song nuclei themselves. This distinction is consistent with the finding that deafening reduces or abolishes ZENK expression in auditory pathways afferent to the song system, but not in the song nuclei of singing deaf birds (Jarvis and Nottebohm, 1997).

Genomic studies have recently been used for template localization. Having controlled for song exposure, Bolhuis and colleagues (2000, 2001) measured neuronal activation as a function of the strength of learning. In the NCM of birds re-exposed to tutor song, but not in those kept in silence, the density of immunoreactive cells (Fos and ZENK protein products) was found to increase linearly with the fraction of crystallized song elements copied from the model (see Nordeen and Nordeen, Figure 21.2, this volume).

Marler and Doupe (2000) noted two possible problems. First, the correlation could have been imparted to a sensory bias, i.e. a predisposition of good learners to attend an auditory stimulus more persistently than poor learners. Second, higher induction in the NCM of good learners would be expected as a result of a closer match between tutor song and the BOS. Accordingly, Terpstra and colleagues (2004) went on to show that neither presentation of novel song nor presentation of the BOS produced a correlation between a ZENK response and the strength of learning, thus refuting both concerns. Remarkably, a positive correlation between gene induction and learning is obtained whether birds are tested 247 days (Bolhuis et al., 2000), 458 days (Bolhuis et al., 2001) or even 672 days (Terpstra et al., 2004) after training! These results seem congruent with long-term storage of an auditory memory (Funabiki and Konishi, 2003).

Recording studies

One criterion for identification of neuronal populations as template components would be their selective responsiveness – increased firing rates – to tutor song relative to BOS or any unfamiliar song. Such preferences have been reported in a number of studies (Solis and Doupe, 1999; Bolhuis et al., 2000, 2001; Terpstra et al., 2004; Nick and Konishi, 2005a, b; Yazaki-Sugiyama and Mooney, 2004; Phan et al., 2006).

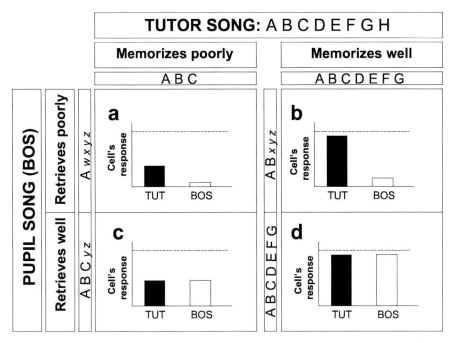

Figure 23.7 Hypothetical responses of a template cell to playback of tutor song (TUT) or the bird's own song (BOS) under four learning conditions (a–d). The cell is assumed to have highly complex properties, firing maximally to memorized sounds of tutor song. A mismemorized tutor song (left column) will give rise to weak-to-moderate response, depending on the young bird's retrieval skills. Maximal response strength is shown by horizontal dotted line. Small letters (wxyz) denote poorly imitated, abnormal or improvised song syllables. The figure shows only one of many possible outcomes. See text for more explanations.

Some caution may be appropriate since there is still a chance that the selectivity for tutor song might be confounded with a familiarity effect. Comparison of the responses to familiar, as distinct from tutor song, is necessary if we are to rule out a familiarity effect. This issue of stimulus specificity was not addressed in the studies cited above.

Searching for brain cells that exhibit tutor selectivity is further complicated by the fact that, during sensory-motor learning, the BOS comes to resemble the song model. Identifying which experience (BOS or tutor song) shapes the selectivity of the cells in a particular brain region thus becomes critical. Consider a young bird that mismemorizes the tutor song and retrieves poorly. In this case, the BOS will differ strongly from the model song. Experience that shapes the selectivity of the neurons is attributed to the stimulus that most effectively drives the neurons (Figure 23.7a). In the case of a faithful imitation, however, the BOS may prove indistinguishable from tutor song, making it impossible to determine which experience shapes the selectivity (Figure 23.7d). This is a recurring problem with which birdsong researchers – particularly electrophysiologists and molecular biologists – are confronted (for other scenarios, see Figure 23.7).

Solis and Doupe (1999; 2000) showed that the effects of tutor song versus BOS experience are indeed dissociable. They gave young birds the opportunity to first memorize a song model and subsequently made them sing a different song by denervating the syringeal muscles. Despite the fact that these birds produced abnormal songs that strongly differed from the tutor song, the authors found an overwhelming neuronal selectivity in favor of the abnormal BOS over tutor song in the AFP. Interestingly, however, many neurons responded equally to BOS *and* tutor song (see below) despite large acoustic disparities between the two songs.

The discovery of neurons especially responsive to playback of the BOS in the HVC of adult white-crowned sparrows (McCasland and Konishi, 1981;

Margoliash, 1983) provided a preparation in which to investigate how auditory feedback information interacts with the memorized song model to shape such responses during song development. Of special note, in the context of a close match between the BOS and the model song, HVC neurons fired vigorously to either one but the response to tutor song waned as a function of the acoustic dissimilarity between the two songs (Margoliash, 1986). Subsequent work reported emergence of BOS-selective responses during the plastic song stage (Volman, 1993) but found little evidence for such preference in response to tutor song, both in presinging and singing juveniles.

Auditory responses selective for tutor song were first reported in the AFP of anesthetized young male finches (Solis and Doupe, 1997). However, such neural preferences did not emerge until late in the sensory-motor phase – when young birds were in the midst of plastic song production. Moreover, only a small proportion of cells displayed the selectivity whereas a majority of neurons responded selectively to the BOS. Interestingly, however, many units exhibited a dual selectivity, discharging vigorously to either tutor song or BOS. These units may be part of a "comparator system" that evaluates acoustic differences between the template and auditory feedback (Figure 23.8a). Nevertheless, late emergence of these responses is not what one would expect of a circuit involved in early stage of song memorization. Young zebra finches given a 10-day exposure with their father, starting at 25 days post hatch, will memorize his song accurately (Roper and Zann, 2006). Yet, cells in the AFP of young males raised with their father until 30–45 days post hatch were found nonselective; they responded equally well to tutor song, adult conspecific songs or the same songs played in reverse (Doupe, 1997). More recently, however, Yazaki-Sugiyama and Mooney (2004) showed that when such limited exposure to a live male is followed by isolation until day 65 post hatch, then approximately half of LMAN neurons acquire selective responses for tutor song. Such tuning cannot be attributed to early sensory experience for reasons just mentioned (see Doupe, 1997). Rather, LMAN neurons in the Yazaki-Sugiyama and Mooney's (2004) study must have been shaped by sensory-motor experience associated with vocal practice during the isolation period, when self-produced vocalizations came to resemble tutor song. Interestingly, later exposure to a second tutor was sufficient to erase the neural signature of the first tutor in LMAN, such that neurons became tuned to the song of the most recent tutor. This result supports previous behavioral findings (Slater et al., 1991).

One approach to identifying neural substrates mediating PM_{TEM} involves recording neuronal responses to playback of the father's song in freely moving juveniles. Indeed, multicellular recordings in the HVC of chronically implanted zebra finches found tutor song an especially salient stimulus, relative to BOS and other test stimuli, during waking hours (Nick and Konishi, 2005a). When young birds fell asleep at night, the response of HVC neurons shifted in favor of the BOS, uncovering a state-dependent mechanism. The transient nature of the response to tutor song, prominent early in the sensory-motor phase and weakening in older birds (Nick and Konishi, 2005b), suggests a labile representation of the engram within HVC.

Converging operations

Having localized one or several candidate brain areas, we will need experimental paradigms that enable us to dissociate sensory, motor or plasticity functions of the region. For instance, to explore further the role of a brain structure in storing PM_{TEM}, one could reversibly inactivate that structure when song learning begins. Young birds undergoing such treatment during sensory acquisition should not be expected to memorize tutor song. After the drug wears off, song acquisition should begin again, *de novo*, as in naïve birds. Such a result would support the conclusion that PM_{TEM} is stored in that structure or in downstream targets. If the structure were needed simply to express what has been learned by other structures earlier in the flow of information then song learning would have resumed immediately after dissipation of the drug. To exclude a role for the structure in motor control rather than plasticity, the structure's output pathway should also be inactivated. In this case, learning (as indicated by the quality of the imitation) would not occur because motor performance was blocked. However, as the inactivation wears off, we would expect learning to proceed normally. This result would indicate that tutor song memorization must have occurred upstream from motor targets. One would conclude that the memory trace must have been formed

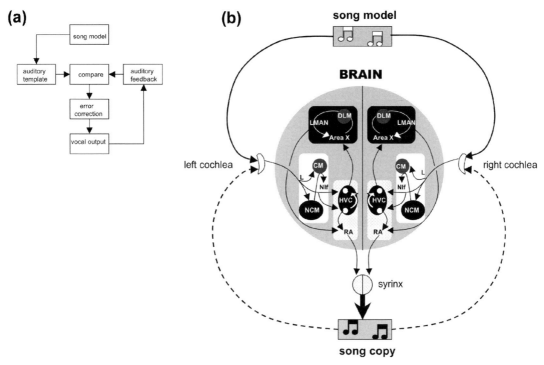

Figure 23.8 (a) Conceptual diagram showing key components of the two-step model of vocal learning, as suggested by the auditory template theory (Konishi, 1965b). (b) A sketch highlighting some of the brain regions and pathways involved. The schematic features two sets of external (acoustic) events – perception of a song model (plain arrow) and feedback (dotted arrow) resulting from self-generated vocalizations – that impinge on internal (brain) components to generate learned song (thick vertical arrow). Transformation of a percept (model song) into a vocal memory (template) during sensory learning is followed by conversion of memorized song into produced song during sensory-motor learning. This two-step process is mediated by the workings of three brain modules: ascending auditory pathways (white), the anterior forebrain pathway (a basal ganglia-thalamo-pallial circuit; black rectangle) and a posterior vocal pathway (dotted area) that generates motor commands. These descending projections activate, in coordination with respiratory mechanisms, several pairs of syringeal muscles that regulate airflow during song production. Candidate brain structures dedicated to some form of song template representation include high-level auditory area NCM, song nucleus HVC in the posterior vocal pathway, and two structures (Area X and LMAN) within the anterior forebrain pathway. A role for DLM and CM in template representation has yet to be shown. An efferent motor copy of song (white arrow within HVC) is thought to arise in RA-projecting neurons and then conveyed to X-projecting neurons via HVC interneurons (Troyer and Doupe, 2000a, Mooney and Prather, 2005). For abbreviations see text.

during the inactivation and could then be expressed in performance when the inactivation was removed. Taken together, the results of the set of experiments would support the inference that PM_{TEM} is both formed and stored in that structure. Of course any set of manipulations involving lesions or functional blockade will require its own sets of controls to allow us to move from correlation to causation. We will need to separate specific from such nonspecific effects as arousal and motivational factors that could account for changes detected in brain activity subsequent to song exposure. Similarly, in dealing with lesions or other forms of disruption, we need to control for such factors as impairments in hearing or perception (disruptions in attention, altered emotional reaction to a pleasurable stimulus) before concluding that we are dealing with template mechanisms. None of these is an easy task.

Having implicated a definite structure, one could then initiate analysis at a cellular level. For example, one could look for "hotspots," that is, regions of the neuropil contacted by axonal terminations that undergo retraction subsequent to sensory acquisition (Iyengar and Bottjer, 2002a). One might expect that afferent neurons, unselective prior to learning, would become progressively more selective during acquisition, exhibiting less divergent axon arbors (synaptic selection) after learning. Hotspots have been found at a high level of brain organization, in visual pathways mediating the transformation of a percept into a memory trace, a critical step in signal encoding (Miyashita and Hayashi, 2000; Yoshida et al., 2003). Impaired sensory acquisition following hotspot lesions or pharmacological blockade of biochemical pathways essential to learning would provide a further indication that hotspots encode critical features of a learned stimulus (tutor song). However, such disruptive effects could do no more than implicate the hotspot as a locus for memory storage. It remains a challenge to distinguish mechanisms for memorization from mechanisms for production, especially in birds in which the phases of vocal development overlap, such as zebra finches. This point cannot be overemphasized.

To function as a template component, a hotspot should be uniquely linked to both the sensory and motor sides of the circuit such that information encoded in a hotspot would be accessed by auditory feedback during sensory-motor learning, either directly in a feedforward fashion, or via re-entrant maps between highly interconnected brain areas (Naya et al., 2003). Thus, efferent projections from multiple hotspots would be expected to connect with the song system in order to retrieve template information. The retrieval process may be construed as involving three steps: (1) auditory feedback provides a mechanism for access to the template; (2) a "comparator" evaluates the match between actual (BOS) and intended (memorized tutor song) vocal output and generates an error signal; and (3) an error-correction mechanism enables the modification of vocal output toward the desired target (see Figure 23.8a).

IDENTIFYING THE BUILDING BLOCKS OF A SONG MEMORY

A large body of evidence points to synaptic processes as a neural substrate for the mechanisms mediating memory storage (but see Pytte et al., this volume, for a discussion of the possible role of neurogenesis). Long-term memory processing is presumed to trigger a cascade of biochemical events leading to the synthesis of proteins necessary for synaptic alteration. Some researchers (e.g. Nordeen and Nordeen, this volume) are seeking to identify key molecules involved in the making of auditory memories in songbirds.

For instance, by activating specific receptors on the postsynaptic membrane, the neurotransmitter glutamate plays a critical role in model systems of learning and memory. Because synaptic transmission in the anterior pathway requires the NMDA type of glutamate (Kubota and Saito, 1991; Mooney and Konishi, 1991; reviewed in Bottjer, 2002), blocking NMDA receptors during sensory learning should impair song memorization. Basham and colleagues (1996) tested this idea by infusing an NMDA receptor antagonist into the LMAN of young birds just prior to a tutoring session. These experimental birds learned significantly less than controls similarly treated but tutored on subsequent days, i.e. after dissipation of the drug. Impaired learning was not just the consequence of a perceptual defect because infusing the drug into LMAN did not prevent the birds from discriminating zebra finch from canary song. Failure of experimental birds to imitate the tutor song cannot be attributed to a deficit in memory retrieval since the controls were not impaired. Rather, the effect of treatment during tutoring sessions was sufficient to prevent memory formation, which resulted in birds producing aberrant songs. The study confirms the importance of NMDA receptors in learning and is congruent with a role of LMAN in tutor song representation (see Nordeen and Nordeen, this volume).

Other researchers are manipulating auditory and social experience of young songbirds in order to investigate the role of certain enzymes (protein kinases) known to be critical in synaptic plasticity (e.g. Sakaguchi, 2004; Singh et al., 2005). However, these biochemical studies are essentially correlative rather than causal. Similarly, the observation that a long-term potentiation (LTP) mediated by NMDA receptors was inducible in LMAN at the onset of sensory learning (Boettiger and Doupe, 2001) is potentially interesting given the suggestive role of LTP in model systems of learning and memory. The relevance of such "slice" preparations to the behaving animal is currently unclear. On the other hand, the availability of gene knockout technology, such as is available in mice, would have a formidable impact in linking learning to the action of specific molecules (Tsien et al., 1996; see Clayton and Arnold, this volume).

SONG TEMPLATE REPRESENTATION: A BIRD'S EYE VIEW

Current research suggests that the song "template," i.e. a neural representation of tutor song, is likely to be distributed among secondary and tertiary auditory areas in the songbird brain. Figure 23.8b highlights candidate regions that may function as repositories of vocal memories. (1) Gene mapping and electrophysiological studies combined with behavioral studies have identified the adult NCM, a region outside the conventional song system, as a potential locus for memory storage (Bolhuis et al., 2000; 2001; Terpstra et al., 2004; Phan et al., 2006). In light of these results, one would anticipate similar or perhaps even stronger neuronal activation in the juvenile NCM, when auditory templates first operate. (2) Fulfilling closely such a condition, chronic records in the HVC of freely moving juveniles have uncovered a state-dependent representation of tutor song, laid down early in the sensory-motor phase of learning (Nick and Konishi, 2005a, b). (3) Within the AFP, neurons have been found to acquire selectivity for tutor song somewhat later than expected (Solis and Doupe, 1997) but pharmacological inactivation of this pathway during sensory learning impairs imitative learning (Aamodt et al., 1996; Basham et al., 1996).

In this coming decade, a major research effort will be (1) to clarify the role of these regions in encoding and/or storing template information and (2) to determine how such information might be retrieved by auditory feedback. A role of high-level, secondary auditory areas – I would speculate – may be to encode spectral characteristics of tutor song (phonology). That information could then be transmitted, via HVC, to the AFP (a basal ganglia-thalamo-pallial circuit) whose main function would be to assemble these phonological units into coherent motor sequences of syllables (Doupe et al., 2005; Graybiel, 2005).

CONCLUDING REMARKS

The concept of the song template has the heuristic value of focusing our research efforts on specific issues: how and where do the auditory mechanisms underlying selective song perception operate in young, naïve birds? Where are song memory traces stored? Are they moved around in the brain, as is known to take place in chick visual imprinting (Horn, 2004; see also Rose, 2003) and, if so, is the pattern of movement predetermined or influenced by whether or not memorization has occurred? Are we dealing, as in mammals, with a multiprocess memory system (Squire and Kandel, 1999; Eichenbaum and Cohen, 2001)? At the moment templates (innate or acquired) represent constructs, rather than mechanisms. Giving them "a local habitation and a name" has been and will continue to be a daunting task.

ACKNOWLEDGEMENTS

I am greatly indebted to Phil Zeigler for his editorial comments and criticism during the preparation of this chapter. Peter Marler was influential at an early stage of my writing about birdsong templates. For his advice and suggestions, I am grateful to him. Marc Konishi commented on the manuscript and provided valuable input. I thank Richard Rosenblatt for discussions on some of the ideas formulated in an early draft.

PART V
Mechanisms of modulation and plasticity

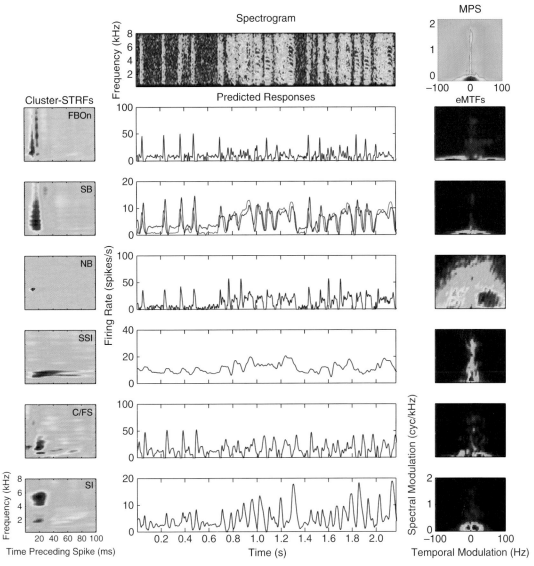

Plate 1 (Fig. 13.5) Examples of the six functional classes of neurons found in the zebra finch auditory forebrain region, Field L. Classes are defined by both spectral tuning (e.g. broadband versus narrowband) and temporal tuning (e.g. onset versus sustained responses) as determined from their spectrotemporal receptive field (STRF). FBOn, fast broadband onset; SI, sideband inhibition; SB, slow broadband; NB, narrowband; SSI, slow sideband inhibition; C/FS, constant/frequency sweep. The majority of cells (71 out of 105; 68%) fit into one of these classes. The left column shows the STRF in time-frequency space. The corresponding plots on the right column show the modulation transfer function for all neurons in that functional class in temporal-spectral modulation space. The middle column shows the response of each type of neuron to the song shown in the top row, in blue. The amplitude envelope of the song is shown in red along with the SB prediction. The plot on the right next to the spectrogram of song is the modulation spectrum of an ensemble of zebra finch song. The modulation power spectrum and the modulation transfer functions can be compared with each other to study which features of sound are represented in the different classes. (Data from Woolley *et al.*, unpublished.)

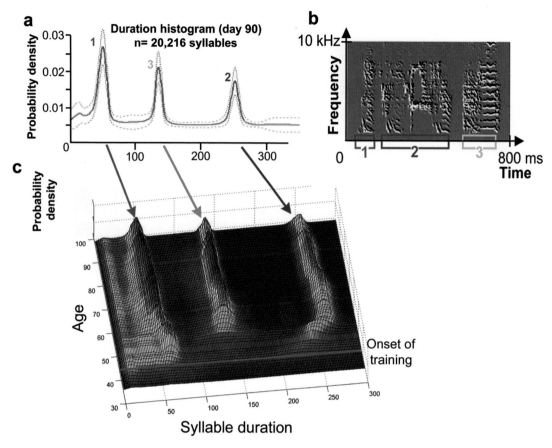

Plate 2 (Fig. 18.1) Dynamic vocal development (DVD) map of syllable durations. The map presents developmental histograms which represent the distribution of syllable durations over time – i.e. it traces the way in which syllable durations change over the course of development. DVD maps are composed of a series of frames each of which represents a cross section of the status of a feature (such as syllable duration) at a specific time (e.g. over one day) during the course of song development.
(a) A smooth histogram (locfit density plot) of the durations of all the song syllables produced in one day (20 216 syllables) by a 90-day-old zebra finch. Dotted lines represent a 95% confidence interval. The peaks in the histograms indicate that three particular durations are much more frequent than the rest.
(b) Those peaks correspond to the durations of three distinct syllable types. Each syllable type can be identified in the sound spectrogram.
(c) A DVD map was composed by plotting duration histograms for each day during development and joining the histograms together. The three 1D peaks have turned into three 3D ridges. We can easily see when each ridge emerged during development. The first ridge (corresponding to syllable 1) emerged between days 44–49. Note that training with song playbacks started on day 43. The ridge that corresponds to syllable three shows an interesting trajectory; it emerged at about day 50, and then took a turn to the right. This turn indicates a smooth increase in duration, demonstrating that this vocal change occurred mainly to this particular syllable. The less structured histograms of the days prior to training show no apparent continuity with the three ridges that appeared after the onset of training.

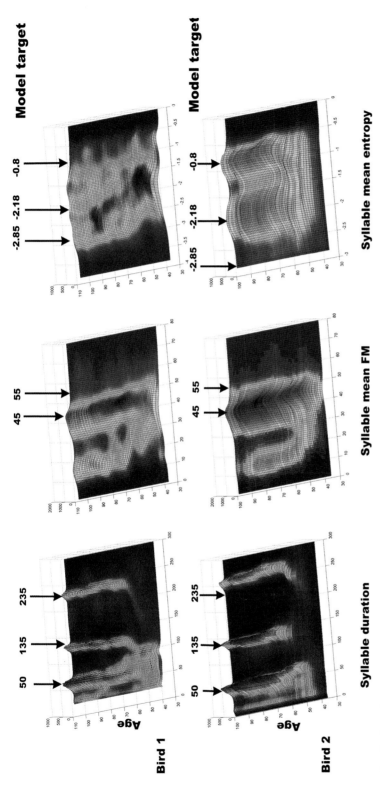

Plate 3 (Fig. 18.3) Developmental histograms of two birds (top and bottom) that successfully imitated the same syllable type in 10 birds that were trained by the same song model. We followed the development of this syllable type in 10 birds that were trained by the same song model. We found that the duration of this syllable type increased by at least 20% in 8/10 birds between days 50–70 post hatching. The DVD maps present three features: syllable duration (left panel), mean frequency modulation (middle panel) and mean Wiener entropy (right panel). Note that across features (left to right) the developmental maps are not similar in their details, namely different features that describe the same developmental data show different dynamics. For example, just as we see how the distribution of duration becomes more structured over development, we also see how the distribution of mean frequency modulation changes from a broad range of values into two distinct ridges, indicating that we now have two categories of syllables: one of high frequency modulation and one of low frequency modulation. Across the DVD maps, we see ridges appearing, bifurcating from other ridges, and smoothly shifting right or left. During development, ridges become more distinct, and they tend to move away from each other, i.e. the different syllables tend to differentiate over time. Overall, these images indicate both a differentiation of structure and a closer approximation to a model target. The model target is indicated by the arrows and values above each figure.

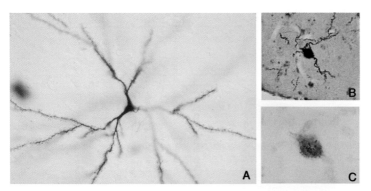

Plate 4 (Fig. 36.7) (A) HVC neuron of an adult canary that received a series of intramuscular ^{3}H-thymidine injections one month before it was anesthetized. This cell was then impaled by a hollow glass electrode and its action potentials recorded, as well as its synaptic potentials in response to sound stimulation. After the electrophysiological recordings were made, the same electrode was used to fill the cell with HRP. The bird was then killed, its brain sectioned at intervals of 100 μm and the resulting HVC sections were cleared in glycerine, so that this cell could be seen and photographed. Notice the well-developed and typically neuronal dendritic tree and dendritic spines. (B) The same cell was then sectioned at 6 μm intervals. In this section we see the cell's soma. (C) Same as in (B), but focusing on the overlying photographic emulsion. The diameter of this cell's soma is approximately 9 μm. The 30 or so exposed silver grains clustering over the cell's nucleus suggest that this cell was born a month earlier, while the bird received ^{3}H-thymidine. This was the first direct evidence that a vertebrate brain cell that looked and behaved like a neuron was born in adulthood and had become connected to existing circuits. (With changes, from Paton and Nottebohm, 1984.)

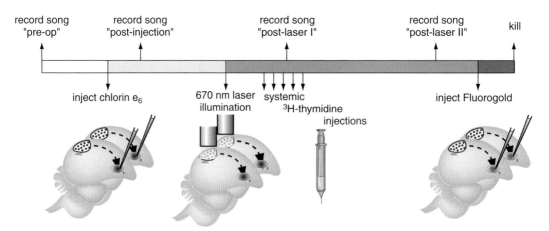

Plate 5 (Fig. 36.13) Method used for selective killing of a particular type of projection cell and evaluating the effect of this removal on song quality and new cell numbers. Song was recorded at the beginning of the experiments involving adult zebra finches. After bilateral injection of chlorin e_6 conjugated nanospheres into Area X (as depicted) or RA, song was recorded again to assure that no behavioral changes had occurred due to potential tissue damage at the injection site. After allowing sufficient time for retrograde transport of the chlorin e_6, HVC was noninvasively illuminated with long wavelength 674 μm laser light, photoactivating the production of oxygen singlets in the cells that had been backfilled by the nanospheres and thus producing their death. To monitor subsequent neurogenesis, ^{3}H-thymidine was injected intramuscularly 5 times over a 10 day period, starting with the second day after the laser illumination. During a survival period of 3 months, newly generated neurons were incorporated into HVC. During this time, short-term and long-term effects of the induced neuronal death on song were monitored through song recordings. To test whether some of the new neurons projected to the original nanosphere injection site, Fluorogold was injected 5 days before perfusion into the same target site. There were no new Area X-projecting neurons, but many new RA-projecting neurons. (From Scharff *et al.*, 2000.)

24 · Introduction

Peter Marler

The surge of endocrine research on singing behavior is relatively recent, contingent in part on the development of sensitive radioimmunoassay techniques for measuring hormone levels in the blood plasma. Soon after discovery of the song system the presence of androgen receptors was established by autoradiography, not only where anticipated, in hypothalamic and limbic areas, but more unexpectedly in parts of the forebrain that control song development, HVC and RA. Since then there have been many demonstrations of hormonal modulation of function in the song system, as well as effects on neuroanatomical development, especially in sexually dimorphic parts of the brain, and the control of neurogenesis. Most attention has been given to steroid hormones, though others are known to influence singing, including adrenal corticoids, vasoticin, GnRH, melatonin and thyroid hormones. Even with respect to steroid hormones it turns out that the gonads are only one of a number of sources, including the brain itself (see Clayton and Arnold, this volume). As Harding demonstrates, the underlying mechanisms mediating hormonal modulation of song and the location of the relevant structures have proved to be more complex and difficult to specify than originally thought. Hormonal effects on behavior can be direct or indirect via metabolites. They can be mediated by multiple receptors that differ in their thresholds and connectivity and further modulated by external factors like daylength and social stimulation. Harding argues persuasively that the focal nature of the song system's anatomy and function makes it especially tractable for addressing the many general unanswered questions about exactly how hormones come to influence behavior.

Females need more study. The propensity of females to sing varies greatly from species to species, and generally speaking the variations correlate well with sex differences in the structure of the song system. However, as the chapter by Ball, Riters, MacDougall-Shackleton and Balthazart shows, the linkage, for example between song complexity and the size of HVC, is not always perfect. Also the implications of phylogenetic relationships are not always taken into account in interpreting the functional significance of sexual differences in song across species. Contributions of female song system circuitry to variations in song complexity, stereotypy, loudness, and amount of singing are all worthy of further study. More investigation of the perceptual mechanisms underlying female responsiveness to song, already implicating parts of the song system in canaries, may throw light on the distinction between recognition and production memory that Adret (this volume) has delineated. The chapter by Ball and his colleagues suggests that to achieve a full understanding of the seasonality and amount of singing and other reproductive behaviors we have to look to parts of the brain, other than the classical song system, especially the preoptic medial nucleus (POM). Its contributions to singing are complex, complementing and interacting with those from the song system in many ways, especially with regard to the influence of social factors, perhaps providing a distinct pathway by which testosterone can influence singing behavior.

In a thoughtful overview Brenowitz makes clear that many questions remain about the functional significance of seasonal changes in singing and the associated growth of the song system. So far these have been found in every songbird studied, whether an open-ended or a closed-ended learner, including at least one tropical species. It is still not fully clear why at the end of the season regression should occur, unless it is a reflection of the expensive cost of maintaining these brain structures when they are not in use. Similarly, regarding the underlying mechanisms, a careful review of the evidence reveals differences of opinion about the relative contributions to onset of the singing season of androgens and estrogens, photoperiod, self-stimulation from song production and

Neuroscience of Birdsong, ed. H. Philip Zeigler and Peter Marler. Published by Cambridge University Press. © Cambridge University Press 2008.

other reproductive activities, and effects of social stimulation by others. Extensive comparative studies will be needed, in both laboratory and field conditions before these issues are fully clarified.

As Nottebohm and his students have shown (see Nottebohm, this volume) the periodic addition of new neurons to different parts of the adult brain, including the song system, is another manifestation of seasonality. Pytte, Wilbrecht and Kirn review the current status of songbird neurogenesis. They suggest that changes in androgen levels are implicated, perhaps under the influence of daylength. In canaries the spring period when the rate of neurogenesis peaks coincides with the phase of vocal development where plasticity is maximal, suggesting that there may be a direct link to the song learning process. However, interpreting the function of neurogenesis is complicated by the occurrence of similar processes, in many, if not all birds, including birds that do *not* learn to sing. Nevertheless the establishment of connectivity to the song systems by new neurons and their differentiation into different physiologically active neuronal types has been documented, but their functional significance remains a puzzle. They may be involved in the perceptual processing required for recognition memory systems, the need for which, especially in social species, continues throughout life. As Pytte *et al.* remind us, the apparent stereotypy of adult song in many birds has proved to be deceptive. In fact dynamic feedback controls continue to operate, only revealed when the pattern of feedback is in some way perturbed (Woolley, this volume). Pytte *et al.* make the interesting suggestion that neurogenesis may be important in providing the underpinnings of the process of song maintenance, especially during the early phases of adult life, when crystallized song is still not fully stablilized. These and many other questions about neurogenesis will only be resolved finally by experiments in which rates of neuronal recruitment are experimentally manipulated, one of many exciting prospects for the future.

25 • Hormonal modulation of singing behavior: methodology and principles of hormone action

Cheryl F. Harding

INTRODUCTION

Studies of the hormonal modulation of the songbird vocal control system have revolutionized the field of neuroscience, altering our basic understanding of how hormones modulate brain function and activate behavior. Although such studies are of recent origin, the observation that singing behavior varied seasonally has long been common knowledge. Ultimately, seasonal changes in singing were correlated with increased gonadal size and hormone secretion. Observers also noted the sexual dimorphism in singing behavior. In most species, males sing more frequently and their songs are more complex than those of females. The relation between those observations and mechanisms of hormonal action is the subject of this review.

BASIC ENDOCRINE MECHANISMS AND BEHAVIOR

The term hormone is derived from the Greek word *hormao* – to stimulate. Thus, hormones were originally defined as stimulatory chemical signals secreted by endocrine glands into the bloodstream and acting at multiple sites to integrate physiology and behavior. The definition of hormones was expanded as researchers discovered that some hormones were inhibitory, some were secreted by tissues other than endocrine glands, and some acted without being secreted into the bloodstream. In most cases, hormones are potentially available to every cell, depending on their rates of metabolism and clearance. Hormones exert their biologic actions by binding to specific receptors in either the cell membrane or inside the cell to activate receptor-coupled effector mechanisms (Kacsoh, 2000). Thus, hormones primarily affect those tissues which contain receptors (i.e. target tissues). Hydrophilic hormones (e.g. peptide/protein hormones like those of the anterior pituitary) and a few lipophilic hormones (e.g. melatonin) act through receptors in the plasma membrane. These receptors span the plasma membrane and rapidly alter cellular function by (1) modulating ion channels, membrane potentials, or intracellular electrolyte composition, or (2) stimulating protein kinases inducing phosphorylation of intracellular proteins. Most lipophilic hormones (e.g. gonadal steroids) bind to intracellular receptors which act as ligand-regulated transcription factors, binding to DNA response elements and altering the transcription of specific genes and the production of proteins, such as enzymes or receptors. Effects on gene transcription are much slower than those regulated by membrane receptors. Hormone receptors typically have a high specificity (i.e. will only bind to hormones with a certain structure) and high affinity (i.e. have a strong chemical attraction) for either a specific hormone or class of hormones. This allows low concentrations of hormone (e.g. pg/ml) to exert major behavioral and physiological changes. But when high levels of hormone are available or when synthetic hormones are administered, there can be cross-talk with other receptor systems.

Early research emphasized the differences between the endocrine system and the nervous system. Endocrine effects tend to be slower and longer lasting than those mediated by the nervous system (e.g. increased aggression during the breeding season). Endocrine effects also tend to integrate physiology and behavior by acting at multiple sites (e.g. assuring that sexual behavior occurs close enough to ovulation to maximize chances of fertilization). Today, the distinction between neural and endocrine mechanisms is becoming blurred. Neurons produce hormones. In male songbirds, the brain, not the gonads, appears to be the major source of estrogens circulating throughout the body (Schlinger and Arnold, 1992). Some neurotransmitters have also been shown to function as hormones. In mammals, dopamine has been shown to act on progesterone receptors and elicit gene transcription. This is caused by cross-talk between membrane-associated dopamine receptors and intracellular progesterone receptors, allowing dopamine to act through progesterone receptors to alter protein synthesis (Brann *et al.*, 1995). Similarly, some hormones have been shown to act via neurotransmitter receptors. Many steroids have the ability to activate or inhibit the $GABA_A$ receptor (Brann *et al.*, 1995). Hormone effects are also more rapid than traditionally thought (see discussion in Harding, 1981). Preventing changes in hormone levels during behavioral interactions has been shown to alter the outcome of the interactions (Nock and Lesher, 1976).

Behavioral endocrinology

Three methods are used to determine if hormones modulate behavior. First one looks for correlations between endocrine function and behavior. As noted earlier, the frequency of singing behavior varies seasonally, being most frequent during the breeding season and correlating with increased testicular volumes (Catchpole and Slater, 1995). In temperate-zone species, the initial increase in gonadal hormones is stimulated by the increase in daylength. Interactions with males (Harding and Follett, 1979; Harding, 1999) or females (Harding, 1999) can result in further increases in hormone production by the brain, pituitary, and gonads (Wingfield, 2005). On closer inspection, temperate zone species often show multiple peaks in singing during the breeding season, with singing rates rising when males first arrive in breeding areas and begin to defend territories. Male starlings showed a peak when they obtained a nestbox (Figure 25.1). A second peak may occur when females return to the area. In many species, male singing rates peak as females lay their first clutch, though in other species, males stop singing once they have acquired a mate (Catchpole and Slater, 1995). Singing rates typically plummet at the close of the breeding season when birds enter the postnuptial molt. It is not clear if this is merely the result of decreased activity of the hypothalamic-pituitary-gonadal axis, or if other hormones may actively inhibit singing during this period. This cyclicity is not as obvious in tropical zones, where breeding may occur year round, and the onset of breeding may be cued by factors such as rainfall or food availability. Later,

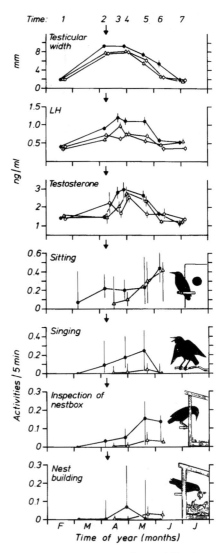

Figure 25.1 Changes in testicular width, plasma concentrations of LH and testosterone, and the frequencies of four patterns of behavior: sitting in front of a nestbox, singing, inspection of a nestbox, and nestbuilding in three groups of European starlings. Solid circles, Group 1: 12 males with access to 12 nestboxes throughout the experiment. Open triangles, Group 2: 12 males with access to 12 nestboxes starting on April 7, indicated by the arrows. Open circles, Group 3: 12 males with no access to nestboxes. In the upper three graphs, data are presented as means ± SEM. In the lower four graphs, data are presented as medians ± 95% confidence limits. (Figure reprinted from Gwinner *et al.*, 1987 with permission of VSP Publishers, an imprint of Brill Academic Publishers, Leiden, The Netherlands © 1987).

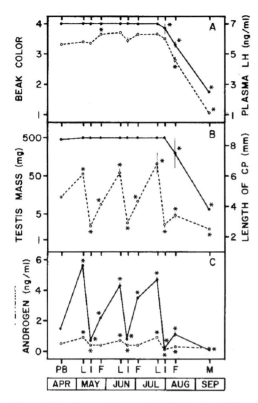

Figure 25.2 Seasonal changes in (A) beak color (solid circles) and plasma levels of luteinizing hormone (LH) (open circles), (B) mass of the left testis (solid circles) and length of the cloacal protuberance (open circles), and (C) testosterone (closed circles) and dihydrotestosterone (open circles) for male house sparrows. Testis mass is graphed on a logarithmic scale, all other axes are arithmetic. All data are means ± SEM. When error bars are not visible, it is because they are so small they fall within the symbol. Points marked with an asterisk differ significantly from the previous data point ($p < 0.05$, Dunn test). All data are organized according to a schematic calendar divided into prebreeding (PB), egg-laying (L), incubating (I), and nestling (F) stages for each brood of the season, and followed by the postnuptial moult (M). (Figure reprinted from Hegner and Wingfield, 1986, © 1986 with permission from Elsevier.)

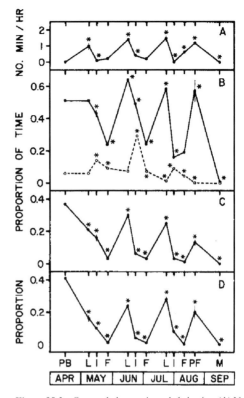

Figure 25.3 Seasonal changes in male behavior. (A) Number of minutes in which males were observed singing or courting. (B) The proportion of observation time in which males were at the nestbox (closed circles) or in the nestbox (open circles). (C) The proportion of time both members of a pair were at the nestbox simultaneously. (D) The proportion of arrivals to and departures from the nest in which the male accompanied the female. This figure follows the format of Figure 25.2. (Figure reprinted from Hegner and Wingfield, 1986, © 1986 with permission from Elsevier.)

researchers demonstrated that peaks in singing behavior were, in fact, correlated with peaks in androgen secretion (Catchpole and Slater, 1995; Whitney and Johnson, 2005). In the multiple-brooded house sparrow, each nesting effort was associated with a peak in androgen levels which was correlated with both sexual and aggressive interactions (Hegner and Wingfield, 1986; Figures 25.2, 25.3).

Second, one needs to establish that removal of the hormone source decreases behavior. Research showed that castration significantly reduced the frequency of singing and the complexity of songs sung (Catchpole and Slater, 1995). Once again there are species differences. While castration significantly decreases female-directed singing in all species examined, its effects vary across species. In some species males stop singing entirely, in others they stop singing to females but continue to sing in other contexts, while in yet other species female-directed singing is merely reduced (redwings: Harding *et al.*, 1988; starlings: Pinxten et al., 2002; zebra finches: Harding, 1983; Harding et al., 1983, respectively).

The final piece of evidence involves restoration of behavior in castrates by hormone treatment. Hormone replacement with testosterone reinstated normal levels of singing and other reproductive behaviors in castrated birds. This included both visual displays that normally accompany singing (e.g. the song-spread display of red-winged blackbirds), nest building, and aggressive displays (Harding et al., 1988; Catchpole and Slater, 1995; Wingfield, 2005).

Measurement and experimental manipulation of hormones

The development and refinement of new assay techniques have led to major insights into hormone–behavior relationships. Circulating hormone levels are typically measured by radioimmunoassay (RIA). The invention of this sensitive technique in the early seventies allowed hormone levels to be measured in microliter samples of plasma or serum, or from fecal samples, even in small birds. Precise amounts of antibody and radioactive hormone are added to extracted samples and allowed to incubate. The amount of radioactive hormone bound to the antibody is compared with a standard curve to determine the amount of hormone in the sample. The more hormone in the sample, the less radioactive hormone is bound since both are competing for a limited amount of antibody. The specificity of the assay is determined by the antibody. In recent years, very specific antibodies have become available. But some antibodies bind to multiple hormones. For example, some antibodies to testosterone bind other androgens, though with lesser affinity. If these other androgens are present in the sample, they need to be separated from testosterone prior to assay. This is typically done by running extracted samples through chromatographic columns. Chromatography allows measurement of multiple hormones in one small sample (e.g. the measurement of androgens, estrogens, and corticoids in one 100 μl sample), but it increases the difficulty and the cost of the RIA. Chromatography is also sometimes necessary to remove interfering substances. For example, high lipid levels in avian blood can cause spuriously high steroid measurements if the RIA is run on unchromatographed samples. Good studies present data on the specificity of the antibodies used, correct for losses in each sample during extraction and chromatography, and provide both intra- and interassay coefficients of variation. Intra assay variability should be ≤ 5%, while interassay variation should be ≤ 12%. Enzymeimmunoassays (EIA) and enzyme-linked immunosorbant assays (ELISA) are RIA alternatives that work on the same basic principle. However, these assays do not use radioactive tags. Antibodies are tagged with a chromogenic compound that changes color when it binds to its antigen.

Hormone receptors can be measured by a variety of techniques: (1) "grind and bind" assays, (2) autoradiography, or (3) immunocytochemistry (ICC). "Grind and bind" assays homogenize the tissue of interest, incubate it with radioactive hormone, and then quantify the amount of radioactive hormone bound by the homogenate. The assay can either measure all receptors available, or an exchange assay can measure only those receptors occupied by endogenous hormone. Autoradiography involves exposing the tissue of interest to radioactive hormone. For in vivo autoradiography, radioactive hormone is administered to the animal, which is sacrificed shortly thereafter, and the tissue of interest sectioned, and applied to slides. For in vitro autoradiography, the tissue is mounted on slides, then exposed to radioactive hormone. For both techniques, the slides are then incubated with photographic emulsion, which is later developed to quantify the level of hormone bound to receptors. Alternatively, receptors can be visualized by ICC, incubating the tissue with an antibody to the receptor and then visualizing the antibody. "Grind and bind" assays are the most sensitive, but the distribution of receptors cannot be determined. Both autoradiography and ICC show the distribution of receptors in tissue, but both techniques tend to underestimate total receptor numbers.

Receptors and other proteins can also be measured by blot tests. DNA and RNA involved in transcription and translation to produce proteins can be quantified in a similar manner. Western blots measure proteins. Southern blots measure DNA coding for protein, and northern blots measure RNA. The tissue is homogenized, and the cells lysed (i.e. broken open) with detergent. Electrophoresis is then used to separate sample constituents according to their mass. The separated sample is then transferred from the electrophoresis gel to a membrane and bound in position. The membrane is incubated with a labeled probe for the chemical of interest and the result compared with a standard curve. *In situ* hybridization (ISH) is used to localize cells actively producing mRNA coding for a specific receptor or peptide/protein hormone. It is an in vitro autoradiography technique. A labeled cDNA probe is incubated with the tissue instead of a radioactive hormone. Because steroid hormones act by altering protein synthesis, many of these techniques are used to quantify changes in DNA or RNA encoding important cellular proteins or changes in proteins themselves to monitor steroid effects.

Interestingly, the results of different techniques often do not agree. Sometimes these mismatches are relatively easy to explain. For example, studies measuring receptors by binding to radioactive estrogen suggested there were higher levels of estrogen receptors in several brain areas than seen with ICC techniques. Initially, researchers thought there was only one estrogen receptor subtype, and ICC techniques utilized antibodies generated against that receptor (alpha). Later studies ultimately demonstrated the existence of a second estrogen receptor subtype (beta) with a different affinity and distribution. One interesting recent study (Whitney and Johnson, 2005) found that singing stimulated the production of mRNA for the immediate-early gene *zenk* in RA, but translation of the mRNA to ZENK protein only occurred if the birds were housed in social groups. So protein synthesis was clearly affected by social context. Another critical issue is antibody specificity. Researchers often rely on specificity data provided by suppliers. This has proved problematic, leading to publication of erroneous results when researchers failed to validate the specificity of the antibodies they used (Rhodes and Trimmer, 2006). Some journals now require authors to demonstrate the specificity of the antibodies they use before considering manuscripts for publication.

Understanding the mechanisms of hormonal action generally involves the use of a number of complementary methods. For example, if engaging in an aggressive interaction decreased vasotocin levels in the septum, does this indicate that vasotocin synthesis has been downregulated or alternatively that synthesis cannot match current levels of utilization? Additional studies are needed to answer such questions. Similarly, studies often use ICC techniques to compare levels of enzymes available in particular tissues under different conditions. In such cases, not only must we distinguish between decreased synthesis or increased utilization, but there is also the question of the activity of the enzyme. Many antibodies recognize enzymes without regard for their activity. But activity is a crucial issue when studying enzymes,. Studies have shown that phosphorylation can increase enzyme activity a thousand fold (Cooper *et al.*, 2003). Most enzyme antibodies do not differentiate between the possible phosphorylation states. For example, suppose one quantified the amount of the androgen metabolizing enzyme aromatase available in the preoptic area of male sparrows during the spring breeding season and during the fall using an antibody generated against aromatase and found that preoptic area tissue from males captured in the spring showed ten times more staining than tissue from birds captured in the fall. One could not then conclude that there were higher levels of conversion of androgens to estrogens in the spring, when ten times more aromatase appeared in the stained tissue. Even if the amount of aromatase found in the stained tissue were an accurate indicator of the amount available in the preoptic area at that season, if aromatase were phosphorylated in the fall in a way which made it 1000 times more active, then estrogen production should be 100 times higher in the fall, despite the lower levels of aromatase. While such extreme differences in activity with phosphorylation state have not yet been demonstrated for aromatase, they have been found for other enzymes.

Hormone administration

Today hormones are typically administered via silastic implants or osmotic minipumps. These implants offer the advantage of steady hormone release for weeks or, in the case of silastics, even longer. Silastic implants are used for lipophilic hormones like steroids that diffuse easily through the silastic. The dose is determined by the surface area of the implant; the larger the surface area, the more hormone released per unit time. The more polar the hormone, the faster it diffuses through silastic. Polarity is determined by the number of double bonds in the central steroid structure and the side groups attached. For example, testosterone which has a double bond in one of its four carbon rings is more polar than dihydrotestosterone which has none. Thus, testosterone will diffuse from implants more quickly than dihydrotestosterone. This means that it is difficult to compare the effects of equivalent doses of different hormones; implants of the same size result in different doses if the hormones' polarities differ, as they typically do because of differences in structure. Osmotic minipumps are used to administer peptide hormones. These pumps release a set volume per hour, so the hormone dose depends on its concentration in the vehicle. Pumps are available in different sizes, providing hormones for different periods of time. (NB: The release rates provided by the manufacturer are for mammals, and in most cases need to be adjusted for the higher temperatures of birds.)

Pharmacological manipulations

Studies often use pharmacological manipulations like hormone receptor blockers or estrogen synthesis inhibitors to determine which class of steroids activates singing. It is relatively easy to interpret drug studies resulting in decreased behavior. Studies using metabolism blockers always require a control showing that drug treatment plus hormone (e.g. aromatization blocker + estradiol) reinstates the behavior, demonstrating that the drug-induced decrease in behavior is specific to the drug's mode of action and not caused by merely making the animal feel ill. The absence of drug effects is difficult to interpret if only one drug or one dose has been used. Unfortunately, dose-response studies are rarely done.

What can we conclude from studies treating animals with exogenous hormones?

Most researchers assume that when they administer exogenous gonadal steroids they are mimicking the effects of increased circulating hormones during the breeding season. While treatment with exogenous androgen does increase the volumes of the vocal control nuclei in males of many temperate-zone species as expected, this hormone treatment results in differential effects in other brain areas (Smulders, 2002). A meta-analysis of the literature found that brain mass, telencephalon volume and n. rotundus (a thalamic visual nucleus) volume increased from the nonbreeding season (low testosterone) to the breeding season (higher testosterone). However, treating males with exogenous testosterone had exactly opposite effects; volumes of the telencephalon, and nuclei outside the vocal control system, were lower in androgen-treated animals than in controls. These results suggest that artificial hormone manipulations do not necessarily mimic the effects of natural variations in hormone levels and that results from experiments using hormone implants to mimic natural hormonal effects should be interpreted with caution. In part, the differences between the effects of naturally occurring increases in circulating androgens and those caused by experimental manipulation may reflect differences in the photoperiods to which the birds were exposed. It should also be noted that birds in the laboratory are living in an impoverished environment relative to birds in the field. We gain experimental control, but we lose normal complexity.

HISTORICAL OVERVIEW OF RESEARCH ON THE NEUROENDOCRINE MODULATION OF SINGING

Research on singing entered a new era in 1976 with the identification of some of the brain areas involved in its control (Nottebohm et al., 1976) and the finding these brain areas contained high levels of hormone receptors. When Arnold and coworkers (1976) investigated the distribution of androgen receptors in the zebra finch brain, they found high levels of receptors in hypothalamic and limbic areas, as expected. These brain areas contain high levels of gonadal steroid receptors in every vertebrate species studied. However, they also found high levels of androgen receptors in two forebrain areas (HVC, RA) involved in the control of singing. Such high levels of gonadal hormone receptors, sufficient to delineate the vocal control areas from surrounding brain tissue, had never been found in such recently-evolved brain areas in mammals (Arnold et al., 1976). These data made it clear that the vocal control nuclei are target tissues for gonadal steroids.

Organizational hormone effects

But the most influential publication of that year documented the gross anatomical differences between the vocal control system in male and female songbirds (Nottebohm and Arnold, 1976). Within four years, researchers demonstrated that female songbirds could be induced to develop male-like brain structure and singing behavior by early hormone treatment (Gurney and Konishi, 1980). This marked a major change in our

Introduction to hormones in the song system 311

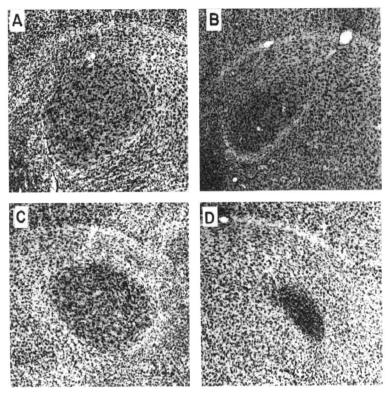

Figure 25.4 Frontal sections through the robust nucleus of the arcopallium in (A) a male and (B) a female canary and (C) a male and (D) a female zebra finch. The canary photographs are from the left hemisphere, and those of the zebra finch are from the right hemisphere. In each case, the photograph was taken at the level of the nucleus where the diameter was largest in this plane of section. The relatively unstained eyebrow-shaped structure is the lamina arcopallium dorsalis which separates the nidopallium (dorsal) from the arcopallium in which RA is located. The prominent ellipsoidal nucleus is RA. Cresyl violet-stained sections 50 μm thick (×42). (Reprinted with permission from Nottebohm and Arnold, 1976, © 1976 AAAS.)

understanding of how early hormone exposure affects the brain. In 1959, Phoenix and coworkers (Phoenix et al., 1959) proposed that hormone effects on behavior could be divided into two types, organizational and activational. According to this paradigm, exposure to particular hormones during critical periods in development "organized" the nervous system, permanently altering animals' behavioral potential. Stimulation by hormones later in life "activated" hormone-dependent behaviors as long as the activational hormones were available. By 1976 there was plenty of evidence that early hormone exposure permanently altered animals' behavioral potential. Many sexual dimorphisms in behavior had been documented. However, 40 years ago, researchers did not expect early hormone exposure to cause gross changes in neuroanatomy. Structural differences had been found previously, but they were subtle – differences in synapse number in the hypothalamus of male and female rats that could be seen only with an electron microscope (Raisman and Field, 1971). The structural differences in the sizes of the vocal control nuclei of male and female songbirds were so large that they could be seen with the naked eye (Figure 25.4). The other important revelation of this study was that the dimorphisms in brain structure correlated with clear dimorphisms in singing behavior. Since this paper was published, a plethora of sexual dimorphisms in brain structure have been documented in a wide variety of vertebrates (e.g. see reviews by Kawata, 1995; Cooke et al., 1998; Goodson and Bass, 2001) but aside from those in songbirds, few are so clearly related to dimorphisms in behavior. While there are clearly cases in which differences in brain structure do not correlate with differences in behavior, overall, the songbirds have proven a remarkable model for examining the relationships between variations in brain structure and behavior.

Activational hormone effects

Prior to 1981, studies had shown that changes in circulating hormone levels caused changes in brain neurochemistry, but scientists did not expect changes in circulating steroid levels to alter adult brain structure. In 1981, the Nottebohm lab published two studies demonstrating that changes in circulating hormones in adult birds caused striking changes in brain anatomy. Nottebohm (1981) found that there were large seasonal changes in the volumes of two vocal control nuclei in canaries. Two nuclei on the motor pathway controlling singing, HVC and RA, were 99% and 76% larger, respectively, in the spring breeding season when males sang at high rates compared with the fall when they had not sung for several months. As expected, androgen levels were significantly higher in the spring. The second study (DeVoogd and Nottebohm, 1981b) demonstrated that treating female canaries with androgen stimulated the growth of dendrites in RA. Treating females with testosterone increased dendritic fields in one type of neuron in RA 250% compared with those in ovariectomized controls. Testosterone treatment also stimulated the females to sing. Once again, these results took the research community by surprise. Since that time, researchers have demonstrated that altering brain structure is an important mechanism in hormonal activation of behavior. Changes in brain structure can occur quite rapidly, and typically precede changes in behavior. But unlike organizational changes in brain structure underlying behavioral dimorphisms, activational changes in structure are temporary and last only as long as increased hormone levels are available.

Neurogenesis in the adult brain

The third discovery which fundamentally altered our vision of brain function was finding that new neurons are born in adult brains and incorporated into functional circuits. When adult female canaries are treated with testosterone, the volume of HVC increases, and the females begin to sing. To determine whether neurogenesis was involved Goldman and Nottebohm (1983) treated adult females with ^{3}H-thymidine. When sacrificed five weeks later, the females had significant numbers of labeled neurons, glia, endothelia, and ventricular zone cells in and around HVC (Goldman and Nottebohm, 1983). Testosterone treatment did not affect the number of labeled neurons, though it significantly increased the numbers of labeled glia and endothelia. This study was greeted with scepticism, because everyone knew that new neurons were not born in adult brains. But over the next few years, the Nottebohm lab documented that these new cells were born in the ventricular zone, migrated along radial glia into HVC where they differentiated into neurons which projected to RA or became interneurons (Nottebohm and Alvarez-Buylla, 1993; see Pytte et al., this volume). The process of adding new neurons to HVC proved to be hormone sensitive, but rather than directly affecting cell birth, hormones modulated the recruitment and/or survival of neurons incorporated into HVC as well as their life-span (Alvarez-Buylla and Kirn, 1997).

Contributions of studies of the avian vocal control system to behavioral neuroendocrinology

The songbird vocal control system evolved from extant neural circuits with other functions found in many other taxa (see Farries and Perkel, this volume). Similar vocal control systems also evolved in parrots and hummingbirds. Although the general pattern of connections between telencephalic vocal nuclei are similar in budgerigars and oscine songbirds, specific connections differ. This probably reflects different evolutionary pressures (see discussion in Durand et al., 1997; Mello and Jarvis, this volume). At some point during evolution, these vocal control systems became steroid sensitive like other brain areas that control behaviors involved in reproduction. The exquisite hormone sensitivity of the vocal control system together with its extreme specialization in controlling singing have made it a critical model in elucidating both organizational and activational neuroendocrine mechanisms controlling behavior. Research clearly illustrates the integrative nature of gonadal hormone effects on singing behavior. These do not occur in a vacuum, but coordinate singing with other important reproductive behaviors such as territorial behavior, aggressive displays, nest building and nest defense (Figures 25.1, 25.2, 25.3). While studies initially focused on the hormone sensitivity of the vocal control nuclei, more recent studies are beginning to appreciate the interactions between the vocal control system and other hormone-sensitive brain areas that appear to be involved in motivating singing to females (e.g. the pre-optic area of the hypothalamus, Riters et al., 2004; see also Ball et al., this volume) and the noradrenergic innervation of the vocal control system (Barclay et al., 1992, 1996).

ISSUES IN STUDYING THE HORMONAL MODULATION OF BEHAVIOR

Understanding the hormonal modulation of singing has been complicated by a number of factors.

Hormone metabolism

Initial studies of the hormonal control of singing behavior focused on the effects of gonadal androgens such as testosterone, showing that testosterone injections or implants stimulated singing in castrated males or in intact males outside the normal breeding season. However, researchers soon demonstrated that, as in other vertebrates, many of the effects of testosterone were caused by its metabolism to other hormones (DeVoogd and Nottebohm, 1981b; Harding et al., 1983). In all songbirds investigated thus far, the activation of singing by testosterone treatment appears to involve metabolism to estrogenic metabolites as well as to other androgens. Songbird brains, like those of other vertebrates, contain significant concentrations of aromatase, the enzyme that converts androgens to estrogens (Silverin et al., 2000). Singing is stimulated by the combined actions of androgenic and estrogenic metabolites; treatments that provide only androgenic or only estrogenic metabolites do not activate singing (DeVoogd and Nottebohm, 1981b; Harding et al., 1983, 1988; see data from red-winged blackbirds in Figure 25.5). As might be expected from these data, estrogenic metabolites also stimulate some of the structural changes in the vocal control system associated with increased singing. For example, hormone treatments that provided both androgenic and estrogenic metabolites increased the dendritic fields of neurons in RA significantly more than treatments that provided only estrogens or only nonaromatizable androgens (DeVoogd and Nottebohm, 1981b). Blocking estrogen formation in androgen-treated males significantly decreased singing behavior, similar to the effects of castration (Walters and Harding, 1988). Not only is singing itself activated by the combination of androgenic and estrogenic metabolites, but also the visual displays used during singing in many species such as courtship dance displays in zebra finches and song-spread displays in red-winged blackbirds (Harding et al., 1983, 1988). Estrogens and androgens interact in many ways to modulate singing. For example, androgen levels regulate aromatase activity in the brain (Silverin et al., 2000; Fusani et al., 2001), while estrogen levels regulate androgen receptor availability in vocal control areas (Kim et al., 2004)

Estrogenic metabolites also appear to be involved in sexual differentiation of the songbird brain. Early studies on sexual differentiation in zebra finches found that estrogenic metabolites were much more effective in masculinizing female brains than androgenic metabolites (Gurney and Konishi, 1980). The production of estrogens appears to play an important role in the neurochemical cascade controlling sexual differentiation of the brain across vertebrate species, even in species in which sexual differentiation is cued by environmental factors like temperature rather than chromosomal sex (Gutzke and Bull, 1986; Elbrecht and Smith, 1992). However, the precise role of estrogen in sexual differentiation in songbirds remains in question, because while many studies have masculinized female brains and behavior by early estrogen treatment, no one has successfully blocked masculinization of genetic males in vivo by blocking estrogen's actions during early development, despite years of attempts (Wade and Arnold, 2004). However, this strategy has worked in vitro (Holloway and Clayton, 2001). Blocking estrogen synthesis in brain slices cultured from male zebra finches blocked normal differentiation of the vocal control system. However, some aspects of masculine differentiation appear to develop independent of hormone exposure (Agate et al., 2003; see Clayton and Arnold, this volume).

That patterns of hormone metabolism are labile adds yet another level of complexity to the analysis of endocrine modulation of singing. Hormone metabolism is affected by a variety of internal and environmental

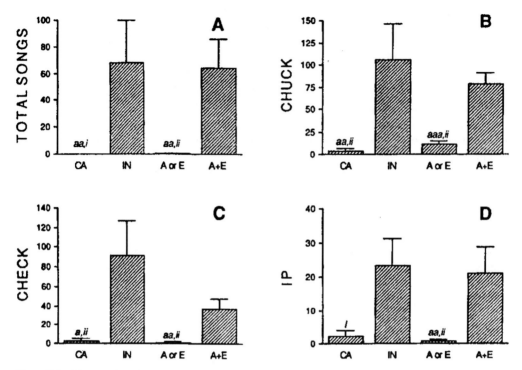

Figure 25.5 Mean (± SEM) frequencies per hour of singing and the three most common calls by castrated (CA), intact (IN), castrated male redwing-winged blackbirds given hormone treatments providing either androgenic or estrogenic metabolites (A or E, androsterone, 5 alpha- or 5 beta-dihydrotestosterone, or estradiol), and castrated redwings given hormone treatments providing both androgenic and estrogenic metabolites (A + E, testosterone, androstenedione, or estradiol + dihydrotestosterone). Surprisingly, not only singing, but the three most common calls males used were hormone dependent and only restored by treatments which provided both androgenic and estrogenic metabolites. a, differs from A + E, $p < 0.05$; aa, differs from A + E, $p < 0.01$; aaa, differs from A + E, $p < 0.001$; i, differs from intact, $p < 0.05$; ii, differs from intact, $p < 0.01$. Ns: CA = 5, IN = 5, A or E = 12, A + E = 14. (Reprinted from Harding et al., 1988, © 1988 with permission from Elsevier.)

factors including changes in photoperiod, age, social interactions, and the availability of other hormones that are metabolized by the same enzymes (Harding, 1986).

One obvious question is why singing should be activated by multiple metabolites rather than by testosterone itself. In other avian taxa, including pigeons, quail, and chickens, vocalizations used during interactions with females appeared to be purely androgen dependent, although reinstating the full repertoire of male reproductive behaviors in these groups requires both androgenic and estrogenic metabolites (Harding, 1999). Involving multiple metabolites allows finer-grained control of singing across multiple social contexts. For example, during the breeding season song is used both to attract females and repel other males. In many species, males also sing undirected songs. The need for singing in these three contexts may diverge. Modulation by multiple metabolites also allows more opportunity for interspecific variation in behavior. (See further discussion of the role of social contexts below.)

Multiple production sites

Hormones are produced at multiple sites. While the typical paradigm for determining if gonadal hormones activate a behavior is the castration/hormone replacement study, sex steroids can also be produced by other tissues, such as the adrenal glands. Researchers working with mammals have often combined adrenalectomy with gonadectomy to remove the two most likely sources of sex steroids. Because songbirds are difficult to adrenalectomize, researchers have rarely attempted this strategy. This means that if a behavior persists following castration, one cannot assume that it is not dependent on sex steroids, because these hormones may be produced by other tissues. While the adrenals are the most likely secondary source, they are not the only possibility. One study found that castrating male zebra finches significantly increased circulating estradiol (Adkins-Regan et al., 1990). This anomalous finding was ultimately explained by the unexpected discovery that the brain is the primary source of estradiol in peripheral circulation in male zebra finches (Schlinger and Arnold, 1992). Although the ability of neural targets to metabolize steroids to both more active and inactive forms was well documented, researchers had never imagined that the brain might release significant quantities of steroids into general circulation. Some studies have claimed that singing can be stimulated in castrated males by androgenic metabolites alone, but they have not demonstrated that estrogens are not being provided by extragonadal sources like the adrenals or brain.

Multiple sites of hormone action

Hormones act at multiple sites to influence singing. Hormones act directly on the various nuclei of the vocal control system, stimulating their growth and connectivity during development (Wade and Arnold, 2004). Hormones also influence vocal control nuclei indirectly. Several of the hormone effects on the structure and function of RA have been shown to be indirect effects mediated by hormone-stimulated changes in HVC (Herrmann and Arnold, 1991; Brenowitz and Lent, 2002). Many gonadal hormonal effects on singing appear to involve the catecholamine neurotransmitters, norepinephrine and dopamine. Gonadal steroids strongly modulate levels and turnover of norepinephrine and dopamine available in the vocal control system (Barclay and Harding, 1990). Some effects are estrogen dependent, some are androgen dependent, and both classes of steroids are necessary to restore normal catecholamine levels and turnover throughout the brain in castrated males. Norepinephrine and dopamine modulate the probability that males will sing to females, female responsiveness to male songs, and song learning (Barclay et al., 1992; Appeltants et al., 2002b; Harding, 2002). These effects are probably related to the effects of these neurotransmitter systems on auditory processing as well as on the vocal control system. This is just one example of hormonal modulation of other neural systems that affect singing or the response to songs.

Hormones also modulate the syrinx, the avian vocal organ. While brain areas controlling singing are sensitive to both estrogens and androgens, syringeal structure and function appear to be primarily androgen sensitive. Androgens stimulate syringeal growth, the number and the morphology of synapses, and its neurotransmitter function (e.g. acetylcholine; Luine et al., 1983; Bleisch et al., 1984; Clower et al., 1989). Thus, hormonal modulation of male singing behavior appears to follow the pattern found in most vertebrates. Brain areas controlling male behavior are modulated by the combined effects of androgenic and estrogenic metabolites, while peripheral tissues involved in the behavior are strongly androgen dependent (Harding, 1986).

Even when the same hormone affects multiple targets in the vocal control system, it often affects them differently or it may achieve the same result through different mechanisms. For example, adult male zebra finches typically have more androgen receptors in their vocal control nuclei than females. The number of androgen receptors in LMAN and HVC in females can be significantly increased by early estrogen treatment. However, the increase in androgen receptor levels is established by two different mechanisms. In LMAN, estrogen treatment preserves androgen-sensitive cells, while in HVC estrogen treatment promotes the development of additional androgen-sensitive cells (Nordeen et al., 1987).

The role of other hormones

Studies of endogenous hormones and behavior typically focus on just one or two hormones (e.g. testosterone or testosterone and its important metabolite, dihydrotestosterone, which is responsible for many of the peripheral effects of testosterone). This means that many other hormones and metabolites are rarely studied and their effects little appreciated. For example, many species of birds sing outside the breeding season. In some species, such as great tits and willow tits, the frequency of singing is correlated with maximal plasma testosterone levels in the spring and smaller testosterone peaks in the fall (Catchpole and Slater, 1995). While testosterone treatment can stimulate singing in both males and females outside of the normal breeding season, several studies have found that in some species, including mockingbirds, song sparrows, and European robins, singing in fall or winter appeared to be independent of circulating levels of gonadal androgens (Harding, 1999). Even castrated males continued to sing and defend territories. However, as mentioned previously, just because the testes have been removed does not demonstrate that the behavior is independent of sex steroids, since these hormones can be produced by other tissues, including the adrenals and brain. In male song sparrows, castration did not reduce singing and territorial defense in the fall. However, treatment with an inhibitor of estrogen synthesis blocked fall singing behavior and singing was restored by estrogen treatment (Soma et al., 2000). Thus, singing outside the breeding season in this species is not independent of sex steroids as previously concluded on the basis of castration studies. More recently, studies demonstrated that in song sparrows, singing and territorial defense in the fall appear to rely on extragonadal synthesis of dehydroepiandrosterone (DHEA). Because DHEA was considered a weak androgen, an unimportant precursor in testosterone production, it was never assayed. However, when researchers assayed DHEA in song sparrows, they found significant levels in circulation year round (Soma and Wingfield, 2001). Treating birds with DHEA outside of the breeding season increased the size of HVC and stimulated singing and territorial defense (Soma et al., 2002). Thus, endogenous DHEA appears to normally stimulate singing and territorial behavior in male song sparrows outside the breeding season, and these effects appear to be due, in part, to conversion to estrogenic metabolites (Soma and Wingfield, 2001).

The importance of hormones other than gonadal steroids

Most studies of hormonal modulation of the vocal control system and singing have focused on the effects of gonadal steroids. However, a wide variety of other hormones appear to be involved in modulating the vocal control system and singing behavior. The effects of vasotocin are probably the best documented. Voorhuis (Kiss et al., 1987; Voorhuis and de Kloet, 1992) found evidence of projections containing vasotocin innervating RA in canaries and the area around RA in both canaries and zebra finches. Despite the apparent paucity of direct vasotocin input to the vocal control system, vasotocin can both inhibit and stimulate courtship singing depending on a number of factors. Zebra finches provide an interesting example of a species in which nongonadal hormones play a critical role in modulating singing and other reproductive behaviors. This species evolved in arid areas of Australia, and their breeding is cued by water availability, which under natural conditions is unpredictable. Cheng (1993) hypothesized that the primary mechanism controlling reproduction in species relying on unpredictable cues should be inhibitory. The onset of stimulatory environmental conditions (e.g. rainfall) terminates the inhibition, allowing rapid initiation of reproduction. As the primary hormone regulating water balance in birds, vasotocin appeared a likely candidate to modulate reproduction in finches. Drought conditions cause sustained systemic vasotocin release, which in other species inhibits androgen production. Systemic infusions of nanogram levels of vasotocin inhibited singing and reproductive behavior in intact males fairly rapidly, presumably by inhibiting androgen production (Harding and Rowe, 2003). Inhibition of reproductive behavior by vasotocin may be a more general phenomenon. Vasotocin is released when animals are stressed, and high levels may inhibit reproductive behavior. Vasotocin's extremely short half life (minutes) means that once proximal factors become more favorable, the gonads should rapidly be released from its inhibitory actions. In contrast to the systemic effects of vasotocin, infusions of very low levels (0.01 ng) into the lateral ventricle in male finches stimulated courtship singing (unpublished data). Thus, vasotocin can both inhibit and stimulate singing, depending on season, dose, route of administration, and gonadal hormone status (Voorhuis et al., 1991; Maney et al., 1997; Harding and Rowe, 2003). Goodson (1998) found intracerebral vasotocin infusions directed at the septum increased agonistic singing in field sparrows, but had no effect on singing the multipurpose song type. Further studies from his lab have shown that vasotocin modulates aggressive interactions and plays a major role in birds' responses to stress (Goodson et al., 2005).

Thyroid hormones are involved in the seasonal control of reproductive behavior in songbirds (Dawson et al., 2001) and modulate apoptosis in brain areas that normally show neuron turnover (Tekumalla *et al.*, 2002). Neurons that project from HVC to RA turn over at varying rates in adult birds. In zebra finches, many of these neurons have thyroxine receptors. Acute treatment with thyroxine induced cell death in HVC and other telencephalic areas that normally show neuron turnover. Long-term thyroxine treatment reduced neuron number in HVC. Thyroxine treatment appears to modulate cell death; no effects were seen on cell birth or recruitment (Tekumalla *et al.*, 2002). Melatonin may also modulate singing. Melatonin and testosterone appear to function antagonistically in regulating the volume of vocal control areas, with testosterone increasing and melatonin decreasing volumes (Dawson *et al.*, 2001). The efficacy of melatonin appears to be modulated by thyroid hormones (Bentley, 2001).

Immunocytochemical studies have found evidence of other peptide hormones in the vocal control system, including vasoactive intestinal peptide (VIP), cholecystokinin, substance P, and corticotropin releasing hormone (Harding, 1992). Their function in the vocal control system remains to be elucidated, though evidence thus far suggests several inhibit singing. VIP stimulates prolactin release in songbirds during the breeding season (Maney *et al.*, 1999) and may be involved in decreasing singing by males that show paternal care. Corticotropin releasing factor (CRF) infusions rapidly inhibited female responsiveness to male song through a mechanism involving endogenous opioids (Maney and Wingfield, 1998). It is likely that CRF also decreases singing in males.

The role of social context in the action of reproductive hormones

During the breeding season, birds sing in many contexts. Their songs may repel other males and defend territories and/or attract females and stimulate their ovarian development. Some species, like many American warblers, have different song types to serve these two functions (Catchpole and Slater, 1995). In American warblers, unaccented songs appear to be used primarily in interactions between males and are sung primarily at dawn and dusk and during territorial interactions. Accented songs are sung more in the presence of females, during courtship, and appear to promote pair-bond maintenance. While it appears likely that endocrine factors differentially modulate these two song types, this remains to be investigated. Perhaps the balance of different hormone metabolites plays a role in this differential control, as was demonstrated for the agonistic and multipurpose songs in field sparrows (Goodson, 1998).

The effects of context have been investigated more extensively in zebra finches. Male finches sing relatively simple songs. These songs are typically classified depending on whether they are clearly directed at another bird or not. The initial classification scheme differentiated between female-directed courtship songs and undirected songs. During courtship singing, males clearly orient towards a female, and the song is often accompanied by specific patterns of feather erection and a pivoting dance display. Undirected songs are sung in a variety of social contexts, including in mixed social groups, near the nest when the female is incubating, and in complete isolation from other birds. As the term "undirected" implies, males do not orient towards other birds while singing such songs. There are subtle acoustic differences between the two song types. Although a male's courtship and undirected songs are composed of the same elements, courtship songs contain more introductory notes, more repeated sequences, are more stereotyped, and are sung slightly faster (Sossinka and Böhner, 1980). During observations, we noticed that males also sang to other males. The hormonal basis of these three song types differs (Walters *et al.*, 1991). Estrogenic metabolites stimulate males to direct songs to other birds, both males and females, but undirected song does not require estrogenic activation. Estrogenic metabolites also stimulate males to sing more songs per bout when they sing to females than when they sing to other males or when they sing undirected song. Finally, estrogenic metabolites cause males to sing more rapidly when they are courting females than when they sing to other males or sing undirected song. Thus, males subtly alter their singing behavior depending on the intended recipient, and this phenomenon is hormone dependent.

Both intact and castrated testosterone-treated canaries sang more when tested in male–male than in male–female pairs (Boseret *et al.*, 2006). Singing in male–male pairs was influenced by dominance status, with dominant males singing more than subordinates. Interestingly, the volume of HVC was larger in males housed in male–female pairs, though they sang less. In canaries, androgenic and estrogenic metabolites appeared to control different song parameters (Fusani *et al.*, 2003; Rybak and Gahr, 2004). Androgenic metabolites reduced song plasticity, decreasing the number of different phrases used in female directed songs, while estrogenic metabolites were very important in increasing syllable repetition rate, a characteristic to which female canaries appear particularly responsive (Clark *et al.*, 1995; Vallet *et al.*, 1998). During the breeding season, HVC neurons in female canaries are differentially responsive to these syllables (Del Negro *et al.*, 2000). Exposure to these high-repetition-rate "sexy syllables" has been reported to increase females' own hormone levels (Marshall *et al.*, 2005) and the levels of androgens that they deposit in their eggs (Gil *et al.*, 2004). High androgen levels in eggs stimulate chick development (Schwabl, 1996; Pilz *et al.*, 2004). Thus endocrine status and singing of the father can affect the endocrine status, development, and behavioral potential of his chicks.

Social context has also been found to modulate the effects of playbacks of male song on female endocrine status in zebra finches (Tchernichovski *et al.*, 1998). While many studies have employed estrogen-treated females to evaluate which aspects of male song stimulate copulatory displays, this study tried to stimulate female estrogen secretion by playing back male songs. The authors found that song playback could increase, decrease, or have no effect on female estrogen levels, depending on how the playback was presented. Song playbacks alone caused a decrease in female estrogen levels. Song playback only stimulated female estrogen secretion and egg laying when song was broadcast from inside a male model positioned at a distance from the female's nest and a second silent male model was positioned on the rim of the nest. When song was broadcast from the male positioned on the nest, it did not stimulate estrogen secretion. Thus, context is as important to the song recipient as to the singer.

Other studies have examined contextual effects in natural settings. One study (Galeotti *et al.*, 1997) examined the relationships between song rate and 14 variables describing song structure, plasma testosterone levels, body mass, body condition, and social context in male barn swallows. Song rate was not correlated to any of the other variables. One particular song component, which the researchers called the harsh rattle, was positively related to testosterone levels, and its peak amplitude frequency varied inversely with male body mass and body condition. Several song features varied according to social context. Males sang longer and more varied songs when they had fewer or no neighbors. Presumably males in this condition are singing primarily to females. Males in highly competitive areas sang short songs, were more likely to interrupt their songs, and were more likely to include rattles in their songs. These data are consistent with the hypothesis that intersexual selection has resulted in longer more complex songs, while male-male interactions select for shorter, simpler songs (Catchpole and Slater, 1995). Neighboring males sang more similar songs and this resulted in matched countersinging. The authors suggested that female choice may depend on context so that matched countersinging may promote female choice for dominant males providing short, harsh aggressive displays in competitive situations, but that females prefer longer, more complex songs in less competitive situations.

CONCLUSIONS

Research on the hormonal control of singing has fundamentally altered basic concepts about how hormones modulate brain function and activate behavior. Circulating hormones organize sexual dimorphisms in brain anatomy during development, activate changes in brain anatomy during adulthood, and modulate the addition of new neurons in the adult brain. In part, these discoveries resulted from the specialized nature of the songbird vocal control system. Because the vocal control nuclei are clearly delineated and because they have only one function, controlling singing behavior, it was much easier to study brain/behavior relationships in this hormone-sensitive system than in classic mammalian hormone target tissues like the limbic system that is involved in regulating so many different physiological processes and behaviors. These discoveries also resulted from the fact that the researchers involved were open to data that conflicted with current dogma about how the brain functioned. Research has now demonstrated that all of these phenomena occur in other vertebrates, including humans.

Although studies of songbird brains have revolutionized the field of neuroscience, there is still much to be learned. Most research has focused on the role of gonadal steroids in modulating singing, but we know that many other hormones also influence this behavior, including adrenal corticoids, vasotocin, gonadotropin-releasing hormone, luteinizing hormone, melatonin, and thyroid hormone. A bird's response to endocrine manipulations also depends on other factors, such as photoperiod and social interactions. We have learned that endocrinology is not as simple as once thought – the primary source of estradiol found in peripheral circulation in male finches is the brain, not the gonads or the adrenal glands. If we administer a hormone and behavior is altered, was it a direct effect of that hormone or was the administered hormone converted to other active metabolites? Which tissue was responsible for this conversion? Similarly, we need to determine where hormones act to modulate behavior. Research to date has shown that all of the vocal control nuclei respond to androgens and estrogens. In some cases, the hormone affects the nucleus directly. In other cases, the effects are indirect, either acting via effects in another vocal control nucleus or even another brain system (e.g. dopaminergic). The effects of social context on the type of song sung, the hormonal control of different song types, the hormonal and behavioral responses to songs sung in different contexts, and interactions between the vocal control system and other steroid-sensitive brain areas that comprise a "social behavior network" (Goodson et al., 2005) have just begun to be studied.

Most of the research to date has focused on relatively few species, particularly zebra finches. Zebra finches are opportunistic breeders whose endocrine control mechanisms are likely to be rather different from species whose reproduction is cued by changes in photoperiod. For example, singing in zebra finches appears to be less hormone dependent than that in photoperiod-cued species. Although castration significantly lowers singing behavior in zebra finches, it does not eliminate it. In photoperiod-cued species, castrated males rarely, if ever, sing. There is a rich diversity in songbird species that invites further exploration of the hormonal modulation not only of singing but of reproductive behavior in general.

ACKNOWLEDGEMENTS

Support from the National Center for Research Resources of the National Institutes of Health, which supports the infrastructure of the Biopsychology Program at Hunter College and from multiple grants from PSC-CUNY is gratefully acknowledged.

26 • Sex differences in brain and behavior and the neuroendocrine control of the motivation to sing

Gregory F. Ball, Lauren V. Riters, Scott A. MacDougall-Shackleton, and Jacques Balthazart

ASKING QUESTIONS ABOUT HORMONES AND THE ACTIVATION OF SONG BEHAVIOR

Introduction

The production of birdsong, especially among temperate-zone species, is often linked with reproduction (Ball, 1999). It has long been observed by field naturalists that birds tend to sing in the spring concomitant with the initiation of reproduction (White, 1789; Armstrong, 1973; Catchpole and Slater, 1995). In many avian species, but especially among temperate-zone populations, it is usually the male who sings the most (Nottebohm, 1975; Catchpole and Slater, 1995). These two observations, that song is often produced at its highest rates during a particular time of the year, and that there are often prominent sex differences in the rate of song production, raise many important questions as to the neuroendocrine mechanisms mediating these behavioral patterns. From an evolutionary perspective, these patterns of intraspecific variation may be adaptive. The sex difference in song production is hypothesized to be related to the fact that a widespread function of song is to attract mates and defend a territory. Because of the general pattern of sexual selection, males are more apt to compete for mates often via song while females are more apt to actively choose mates and thereby perceive and assess song but not necessarily produce it at a high rate (Andersson, 1994; Catchpole and Slater, 1995; Collins, 2004). This male-typical song is most apt to occur in the spring because this is the most favorable time for enhanced reproductive success in seasonally breeding songbirds in temperate regions (Perrins, 1970; Dawson *et al.*, 2001). As for proximate causation, the production of male-typical song occurs with the highest frequency in the spring, when seasonally breeding species have enlarged gonads and high plasma concentrations of gonadal steroid hormones (Wingfield and Farner, 1993; Ball, 1999; Ball and Bentley, 2000; Schlinger and Brenowitz, 2002). For this reason male song production has long been considered a hormonally regulated behavior (e.g. Beach, 1948). The hormone most often implicated is the gonadal sex steroid hormone testosterone (Schlinger and Brenowitz, 2002; Harding, 2004). Many studies have shown that high rates of singing tend to correlate with high plasma concentrations of testosterone in males and/or that exogenous testosterone can enhance song production (Ball, 1999; Schlinger and Brenowitz, 2002; Harding, 2004 for reviews). As illustrated in Figure 26.1, male song sparrows treated with testosterone when either photosensitive or photorefractory and housed in soundproof chambers greatly increase their rate of singing as compared with untreated males in either physiological state (Nowicki and Ball, 1989). However, the quality of song induced by testosterone in these two different reproductive states does not differ (Ball and Nowicki, 1990).

The reason why testosterone exerts these effects on song is also relatively clear. Gonadal sex steroid hormones orchestrate suites of traits so that they can be coordinated temporally to insure successful reproduction (Wingfield and Farner, 1993; Ball and Bentley, 2000). Like other important traits, song is a behavioral trait that must be produced at the right time of the year to be maximally effective.

Reviews of the neuroendocrine control of song production and of the song learning process are available elsewhere (e.g. Balthazart and Adkins-Regan, 2002; Schlinger and Brenowitz, 2002; see also Brenowitz, and Harding, this volume). This chapter addresses two complementary but somewhat neglected questions about the neuroendocrine control of adult song production. First, what do we know about the neural basis of the pronounced sex differences in singing behavior that have been described in many species? Studies of songbirds have been especially valuable in illustrating how species variation in the degree to which there is a sex difference in behavior is reflected by sex differences in

Neuroscience of Birdsong, ed. H. Philip Zeigler and Peter Marler. Published by Cambridge University Press. © Cambridge University Press 2008.

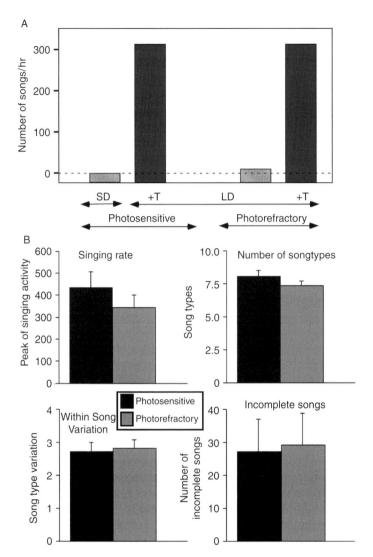

Figure 26.1 (A) Song rates (number of songs/4 hours) observed in photosensitive and photorefractory song sparrows before (gray bars) and after (dark dotted bars) treatment with exogenous testosterone. Photosensitive birds were held on short days (SD) and then transferred to long days (LD) at the beginning of the treatment with exogenous testosterone (+T); photorefractory birds were maintained on long days. (Redrawn from data in Nowicki and Ball, 1989.) (B) Assessment of measures of song production and quality in the same birds. Note that there are no discernable differences in the effectiveness of testosterone when administered to the sparrows in either the photosensitive or photorefractory state. (Based on data reported in Ball and Nowicki, 1990.)

defined neural pathways. Second, what neuroendocrine mechanisms underlie the control of the motivation to sing? Motivation in this sense refers to the seasonal change in the probability of singing in response to an appropriate stimulus that is commonly observed, especially in seasonally breeding species (Ball, 1999). Most studies of the neural and endocrine control of song have conflated two questions, namely what physiological factors influence the probability of singing and what physiological factors influence the quality of singing? We will argue here that these two aspects of the neuroendocrine basis of song need to be distinguished, and that the neural sites where testosterone acts to regulate these aspects are quite distinct.

The importance of this distinction is apparent from the study that first described the song system and its functional significance (Nottebohm et al., 1976). In this study the authors reported on the behavior of a male canary that had received bilateral lesions to the key song control nucleus HVC. They noted that the lesioned male attempted to sing in a manner similar to an intact bird but produced no apparent song and they described this behavior as "silent song" (Nottebohm et al., 1976, p. 475). Their overall conclusion was "... that even though the motivation for singing was present, there was no recruitment of output patterns controlling the song performance of bill, larynx, syrinx, and respiratory musculature" (Nottebohm et al., 1976, pp. 474–475). To insure that the canary of interest was attempting to sing at the maximal rate Nottebohm et al. (1976) subsequently administered exogenous testosterone and in response to this treatment the lesioned male engaged in singing-related behaviors more vigorously but he was still unable to produce any song. Thus the initial finding on the functional significance of the song system established the notion that hormonally modulated brain areas that regulate the attempt to sing might be quite distinct from areas regulating how well you sing. Our second goal in this chapter is therefore to discuss what is known about the neuroendocrine basis of this aspect of song behavior. A discussion of this issue leads us to make more general conclusions about the neural basis of sex differences in song behavior since males and females can differ both in the probability that they will sing in response to different stimuli and in how well they will sing.

The song system and related neural structures from a neuroendocrine perspective

The production of song and the auditory feedback needed for sensorimotor learning during ontogeny and song maintenance in adulthood is controlled by a well-defined neural circuit called the song system (Figure 26.2). Many features of this circuit appear to be unique to the songbird suborder (Passeres) and thus the song system represents a neural specialization that evolved in songbirds in association with the ability to produce a learned complex song (see also Wild, Schmidt and Ashmore, and Reiner et al., this volume). We will focus here on neuroendocrine aspects of the chemical neuroanatomy of the song circuit. Given the

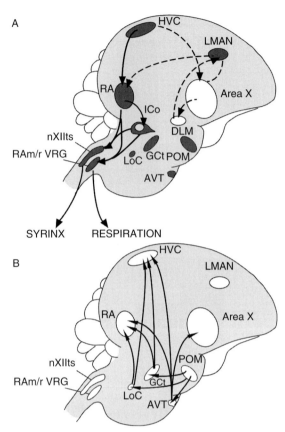

Figure 26.2 (A) A generalized view of the songbird vocal control system. Several telencephalic nuclei including HVC, robust nucleus of the arcopallium (RA) and the lateral part of the magnocellular nucleus of the anterior nidopallium (LMAN) represent neural specializations for vocal learning that are only observed in songbird species. One forebrain pathway HVC to RA to the nucleus of the tracheosyringeal division of the XIIth cranial nerve (nXIIts; black arrows) is involved in song production. Another pathway HVC to X to LMAN to RA (dotted arrows) is involved in song learning during ontogeny and song maintenance in adulthood. The preoptic medial nucleus (POM), a key area in the control of male sexual behaviors, is also illustrated. The steroid-sensitive nuclei (expressing androgen and/or estrogen receptors of the alpha subtype) are filled with dark dots. (B) A diagrammatic representation of catecholamine cell groups that have been described that project to the forebrain song control nuclei and their anatomical relationships with the POM. The locus coeruleus (LoC) projection is noradrenergic in nature while the periaqueductal gray and ventral (or substantia grisea centralis, GCt) tegmental area (AVT) are primarily dopaminergic. See text for more detail. ICo, intercollicular nucleus; RAm/r VRG, retroambiguus of ventral respiratory group.

evidence that testosterone plays a modulatory role in singing behavior, initial studies on the neuroendocrine control of song were quick to explore the possibility that testosterone is able to act directly on song control nuclei to activate song (Arnold et al., 1976). Initial autoradiographic studies utilizing [^{3}H] testosterone revealed uptake in several forebrain song control nuclei (Arnold et al., 1976; Arnold and Saltiel, 1979). Subsequent studies utilizing immunocytochemistry to identify the receptor protein or in situ hybridization to reveal the corresponding messenger RNA have confirmed the presence of steroid-sensitive neurons in these nuclei (e.g. Balthazart et al., 1992; Gahr, 2001). As summarized schematically in Figure 26.2, androgen receptors (AR) are present in several forebrain song nuclei and estrogen receptor (ER)α is observed in HVC of some species (Gahr et al., 1993; Bernard et al., 1999). The expression of AR and ERα in discrete pallial nuclei represents a neural specialization that distinguishes songbirds from other vertebrates (Kelley and Pfaff, 1978; Ball et al., 2002). Thus receptors for androgens as well as estrogenic metabolites of androgens such as 17ß estradiol are expressed in key forebrain song control areas. This anatomical observation provided a potential link between sex steroid hormone secretion in the periphery and song behavior, perhaps via modulation of the central circuit controlling song production (Nottebohm, 1980a). However, song behavior is only one component of male sexual behaviors that vary seasonally and other brain areas such as the preoptic region are important in controlling a variety of male sexual behaviors in birds and mammals (Ball et al., 2002; Hull et al., 2006). AR, ERα and ERß are also expressed in the preoptic medial nucleus (POM) a key site controlling male sexual behavior in birds (Panzica et al., 1996; Ball and Balthazart, 2004) and, as we shall see, there is evidence that testosterone acting here is important for the initiation of singing, at least in some motivational contexts.

Sex differences in brain and behavior in songbirds

Prior to the discovery of the song circuit (Nottebohm et al., 1976) there were no reports suggesting differences in the brains of male and female birds. (e.g. Brown, 1971). In 1976, Nottebohm and Arnold reported significant (male-biased) differences in the key forebrain nuclei of the song system (Arnold et al., 1976). For example, the volumes of HVC and RA are approximately 5 times larger in male than in female zebra finches and 2.5 to 3 times larger in male than in female canaries (see Figure 26.3). In the case of Area X, this nucleus was not even distinguishable in the female zebra finches and is nearly 4 times larger in volume in male as compared with female canaries (Nottebohm and Arnold, 1976). This discovery had significant implications for the way neuroscientists think about sex differences in the brain, encouraging studies in other species of relatively gross morphological differences between males and females (e.g. Arnold, 1980a; Goy and McEwen, 1980; Arnold and Gorski, 1984).

Studies of sex differences in the volume of song control nuclei

Nottebohm and colleagues subsequently conducted several studies of intraspecific variation in the volume of nuclei such as HVC to examine the hypothesis that such variations might be related to variations in such song features as repertoire size. For example, it was found that marsh wrens on the west coast of North America exhibit an HVC significantly larger in volume than males on the east coast and this difference is correlated with geographic variation in male song repertoire size (Canady et al., 1984). If there really is a link between aspects of song behavior in males and females and the volume of song control nuclei, would such differences exist in species where sex differences in song production are often quite limited or even non-existent? Tropical duetting species provided excellent subjects for such studies because males and females will sing simultaneously and intersperse notes. Although there is evidence that the song serves different functions in males and females (e.g. Levin, 1996a, b), both males and females defend territories and tend to sing at very high rates. Brenowitz et al. (1985) compared sex differences in song structure volume in males and females of three neotropical duetting species: white-browed robinchats, bay wrens and buff-breasted wrens. They found that the extent of the sex differences in nuclear volume in HVC, RA and Area X was correlated with the extent of the sex difference in repertoire size in any given species (Brenowitz et al., 1985). This and subsequent studies clearly linked species variation in the degree of sex differences in the song system to variation in parameters associated with song

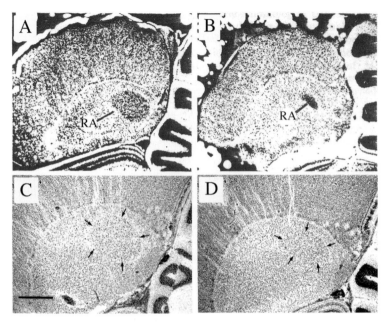

Figure 26.3 Photomicrographs illustrating the sex difference in volume of the robust nucleus of the arcopallium (RA) in zebra finches (A, B) and European starlings (C, D). Male sections are in the left column (A, C) and females are on the right (B, D). Note that the difference is more extreme in zebra finches than in starlings and this corresponds to species variation in the degree to which there is a sex difference in vocal behavior. (Modified from Arnold, 1980a, for the zebra finch and Bernard et al., 1993, for the starling.)

complexity and song production (Gurney, 1981; Brenowitz et al., 1985; Brenowitz and Arnold, 1986).

Sex differences in the cellular properties of the song system

In addition to variations in the volume of nuclei, a variety of other cellular measures were found to differ in males and females. Briefly, there are differences in the number of neurons in RA and HVC (Gurney, 1981), in the size of neuronal somata in HVC, RA and LMAN, in the length of dendrites in RA (DeVoogd and Nottebohm, 1981a) and in the number of cells serving as androgen targets (Arnold and Saltiel, 1979; Nordeen et al., 1986). Neurons immunoreactive for choline acetyltransferase in areas of the ventral pallidum that project to HVC and RA in males are larger than those in females (Sakaguchi and Taniguchi, 2000). Zebra finches or canaries served as subjects for most of these studies that have been reviewed in detail elsewhere (Arnold, 1992; Balthazart and Adkins-Regan, 2002). The basic findings about sex differences in cell number have been confirmed in recent reanalyses using more accurate stereological counting methods (Tramontin et al., 1998).

Moreover, anatomical and physiological studies have shown that the projection from HVC to RA is absent (or greatly reduced) in female zebra finches (Konishi and Akutagawa, 1985). A recent study employing real-time optical recording techniques found that HVC does make a functional connection to RA in female zebra finches, though the responses in RA to stimulation in HVC are reduced in amplitude and of a longer latency in females as compared with males (Wang et al., 1999a). One mechanism responsible for the sex difference in the HVC to RA projection is that, in female zebra finches, GABA inhibits fibers from HVC while in male zebra finches this is not the case (Wang et al., 1999a). Interestingly, sex differences in some of these cellular properties are greatly reduced in the duetting species (Arnold et al., 1986; Brenowitz and Arnold, 1986).

Comparative studies of sex differences in songbirds

These studies suggested a coevolution of sex differences in the brain and sex differences in singing

Table 26.1. *Sex differences in volume of song control nucleus HVC among songbirds*

Species	F/M HVC volume	Reference
Females normally never sing		
Marsh wren (*Cistothorus palustris*)	0.08	Canady et al., 1984
Carolina wren (*Thryothorus ludovicianus*)	0	Nealen and Perkel, 2000
Orange bishop (*Euplectes franciscanus*)	0	Arai et al., 1989
Bengalese finch (*Lonchura striata*)	0	Okanoya, 1997; Tobari et al. 2005
Zebra finch (*Taeniopygia guttata*)	0.14	Nottebohm and Arnold, 1976
Dark-eyed junco (*Junco hyemalis*)	0.26	Deviche and Gulledge, 2000
Screaming cowbird (*Molothrus rufoaxillaris*)	0.39	Hauber et al., 1999
Shiny cowbird (*Molothrus bonariensis*)	0.14	Hauber et al., 1999
Females sing, but less than males		
European starling (*Sturnus vulgaris*)	0.63	Bernard et al., 1993
Rufous-and-white wren (*Thryothorus rufalbus*)	0.46	Brenowitz et al., 1985
White-browed robin-chat (*Cossypha heuglini*)	0.34	Brenowitz et al., 1985
Red-cheeked cordon bleu (*Uraeginthus bengalus*)	0.67	Gahr and Guttinger, 1985
Black-capped chickadee (*Poecile atricapillus*)	0.64–0.73	MacDougall-Shackleton et al., 2003; Phillmore et al., 2005
Red-winged blackbird (*Agelaius phoenicus*)	0.31	Kirn et al., 1989
Domestic canary (*Serinus canaria*)	0.3	Nottebohm and Arnold, 1976; Gahr et al., 1998
Cassin's finch (*Carpodacus cassinii*)	0.32	MacDougall-Shackleton et al. 2005
White-throated sparrow (*Zonotrichia albicollis*)	0.55	DeVoogd et al., 1995
White-crowned sparrow (*Zonotrichia leucophrys*)	0.27	Baker et al., 1984
Bay-winged cowbird (*Molothrus badius*)	0.39	Hauber et al., 1999
Females duet or sing similar to males		
Slate-colored boubou (*Laniarius funebris*)	0.56	Gahr et al., 1998
Buff-breasted wren (*Thryothorus leucotis*)	0.77	Brenowitz et al., 1985
Bay wren (*Thryothorus nigricapillus*)	0.67	Brenowitz and Arnold, 1986
Northern cardinal (*Cardinalis cardinalis*)	0.60	J. M. Jawor, unpublished

behaviour (Brenowitz et al., 1985). However, conclusions from the data are constrained by methodological issues related to across-species comparisons. Because species are not independent data points but are the terminals of a branching lineage, phylogenetic history must be taken into account (Harvey and Pagel, 1991; Martins, 1996). Thus, although there may be an apparent correlation between two variables (e.g. brain and behavior) across species, the observed correlation may reflect a confound due to the inability to control for phylogeny rather than a functional relationship. For example, in the initial observation of a cross-species relationship outlined above, the duetting species with little sex dimorphism in the song-control system were two wren species (family Certhiidae) whereas the extremely dimorphic zebra finch is an estrilid finch (family Passeridae). Thus the apparent relationship between sex differences in the brain and sex differences in song could have been a product of divergence among different families of songbirds rather than reflecting a functional difference.

Table 26.1 summarizes the currently available data on the relationship between differences in song structure and in song-related behaviors in males and females. In general, these data have supported the initial idea that extreme sex differences in the brain have coevolved with extreme sex differences in behaviour (for an alternative point of view, see Gahr et al., 1998). To assess a cross-species relationship between two phenotypic variables (brain sex difference and behavior sex difference) requires some method to control for the effects of phylogeny. Fortunately, evolutionary biologists have developed a number of statistical techniques to do just that (Martins, 1996). When the phylogenetic history of songbirds, as assessed by molecular genetic data (Sibley and Ahlquist, 1990), is controlled for, the hypothesis that sex differences in the song-control system have coevolved with sex differences in behavior does seem to hold (MacDougall-Shackleton and Ball, 1999; Ball and MacDougall-Shackleton, 2001). Moreover, the development of extreme sex differences in brain and behavior seems to have evolved multiple times independently. For example, among wrens (family Certhiidae), marsh wren (*Cistothorus palustris*) and Carolina wren (*Thryothorus ludovicianus*) females never sing (Canady et al., 1984; Nealen and Perkel, 2000) whereas females of other *Thryothorus* species commonly sing in duet with males. Similarly, in the family Passeridae females of many species never sing (see Table 26.1), whereas red-cheeked cordon bleu (*Uraeginthus bengalus*) females commonly sing (Gahr and Güttinger, 1986). In both of these families, the species that have extreme sex differences in behavior have the most extreme sex differences in the song-control system. In both families this dimorphism can be as extreme as a complete lack of delineation of the nucleus (as in zebra finches). Thus, the degree to which there is a sex difference in song behavior, and in the song-control system, appear to be evolutionarily labile traits. The direction and degree of these male/female differences appear to have evolved in tandem.

Although comparative methods do support the brain-behavior correlated sex differences, several issues remain. First, any comparative method is only as good as the phylogeny on which it is based. Further work is needed to both verify phylogenetic relationships of species for which we have data on sex differences in brain and behavior and to acquire data on sex differences for species for which we have good phylogenies. Second, further work is needed to understand the functional relationship between sex differences in the brain and sex differences in behavior within species. It is not clear why male cordon bleus should have twice the HVC volume of females when they sing equally complex songs (Gahr et al., 1998). Similarly, male and female white-throated sparrows (*Zonotrichia albicollis*) sing equally complex songs, but males have much larger HVC volume (DeVoogd et al., 1995). Thus, a complete understanding of sex differences in songbird brain and behavior will require further comparisons of targeted species, as well as improved understanding of the functional significance of sex differences.

Functional interpretations of these sex differences

What aspects of song behavior are different?
The existence of substantial male–female differences in the song-control system raises the question of their functional significance. That is, how do differences in nuclei volume and cellular properties of the song control system map on to differences in singing behavior? Because song is a complex learned behavior that varies widely among songbird species it is not surprising that the nature of sex differences in singing behavior also vary widely. In the most extreme case (e.g. zebra finches, Carolina wrens) females never sing. In these species, song is a male activity only and functions to deter rivals and/or attract and court females. In other species (e.g. white-crowned sparrows) females rarely sing and when they do it is with less complexity and stereotypy than males (Baptista et al., 1993). In yet others (e.g. some white-throated sparrows) females may sing with equal complexity to, but less often than, males (Falls and Kopachena, 1994). Finally, females may duet with males (antiphonally or otherwise) with complexity and amount of singing equal to that of males (e.g. Brenowitz et al., 1985). Thus, female song varies among species from being nonexistent to being essentially identical to male song. It is also apparent that singing behavior varies between the sexes in at least three dimensions: song complexity (e.g. song repertoire size), song stereotypy (reliability of song production from rendition to rendition), and amount of singing. One approach to better understand the functional significance of sex differences in the brain would be to explore sex differences in connectivity, volume and cellular properties of the song-control system in target species in which the sexes differ along only one of these dimensions.

It is important to note that, in the most commonly studied temperate-zone songbirds, although females rarely sing, female song does not appear to be a functionless trait (Langmore, 1998). In such species female song may serve as a signal to defend territories from other females, to deter competing females from approaching one's mate, and rarely even to attract males (Langmore et al., 1996). Thus, it can be argued that female song and the underlying female song-control system is an adaptive product of sexual selection. Because female song may differ in complexity, organization and timing of production from male song, detailed examination of the female song-control system may shed light on the neural and neuroendocrine control of song. It is thus unfortunate that almost all studies of the connectivity, development, and electrophysiological properties of the song-control system have been carried out exclusively in males. Studies of the neural basis of song in females may provide more than just a comparison point with males but rather provide fundamental insights into the neural control of song behavior.

An aspect of birdsong for which females have been the subject of study is that of song perception. One of the functions of male songs is to court females, thus female songbirds should be well equipped to perceive and discriminate among songs of their own species. In fact, given the difference in fitness costs between males and females in responding to a song of the wrong species, females may even have been selected to discriminate among songs more than males (e.g. Searcy and Brenowitz, 1988). In males, the importance of song perception may be more limited to acquisition during song learning or auditory feedback during singing. Behavioral studies of sex differences in song perception are relatively rare (but see Ratcliffe and Otter, 1996). Moreover, evidence for the role played by the song-control system in perception is mixed (Brenowitz, 1991; Uno et al., 1997; MacDougall-Shackleton et al., 1998a; Scharff et al., 1998; Burt et al., 2000). However, HVC and LMAN are of some interest. Both of these regions are largest in females that respond strongly to sexually stimulating male conspecific songs, and discriminate strongly between sexually stimulating versus non-stimulating male songs in multiple species (cowbirds, canaries, and starlings; Hamilton et al., 1997; Leitner and Catchpole, 2002; Riters and Teague, 2003). Lesion and electrophysiology studies in female canaries demonstrate that HVC is involved in female sexual responses to sexually attractive elements of male canary song (Brenowitz, 1991; Del Negro et al., 1998, 2000; Burt et al., 2000), and female canaries with lesions to LMAN also showed deficits in discrimination among acoustic stimuli, including canary and heterospecific song (Burt et al., 2000).

An interesting possibility is that different kinds of song perception may be relatively independent cognitive modules, such as perception during auditory feedback to control singing, perception during the acquisition phase of song learning, perception of an aggressive signal by a territorial bird, or perception of a courtship signal from a potential mate. If these forms of song perception are modular and independent of each other, they may be mediated by different components of song-control and auditory systems in the songbird brain, as has been argued for memory systems involved in sensory and sensorimotor learning (Bolhuis and Gahr, 2006). Again, comparisons of the functional neurobiology of males and females of target species could shed light on this issue.

What is the significance of the smaller nuclei in females?

Sex differences in nuclear volume and cellular properties of the song-control system are some of the most prominent CNS sex differences present in nature. In some species, the sex difference is so great that a brain nucleus present in males is completely absent in females. For example, in female zebra finches, Area X cannot be distinguished from the surrounding medial striatum using standard Nissl stains or cytochrome oxidase histochemistry (Nottebohm and Arnold, 1976). This absence of song-control nuclei has been reported in several species (e.g. Arnold et al., 1976; Nealen and Perkel, 2000). However, it is unclear whether the absence of a distinct brain region is qualitatively different from the presence of a region much reduced in size. For example HVC can be distinguished in female zebra finches, but as noted previously, is poorly defined and does not functionally innervate RA. Does such a vestigial brain region play any functional role in the birds singing behavior? Female zebra finches never sing nor does HVC appear to mediate song perception in this species (MacDougall-Shackleton, 1997; MacDougall-Shackleton et al., 1998b). Whether HVC has any function in female zebra finches or is sexually vestigial (such as nipples in male mammals) remains an unanswered question.

From an evolutionary perspective, sex differences in the brain present an interesting problem. Sexual selection has driven the evolution of male song (reviewed in MacDougall-Shackleton, 1997) and neural structures that mediate song production (DeVoogd et al., 1993;

Székely et al., 1996; Airey et al., 2000). Why however, should females have small nuclei? Presumably there is a cost in maintaining a large and active song-control system, and for males the benefits of singing outweigh these costs. It is not clear, however, exactly what these costs are and why or how they have resulted in reduced song-control nuclei in female songbirds. What costs are males incurring by their possession of large song-control nuclei that females have avoided? There may be direct costs in maintaining metabolically active tissue, but more likely there may be indirect costs. Sex steroid hormones are able to develop and maintain large song-control nuclei. High concentrations of these hormones may be costly to maintain for females by interfering with reproductive physiology and behavior or by suppressing immune function. Thus, we may speculate that the cost may be related, not to the song-control nuclei themselves, but to the regulatory system required to maintain them (Catchpole, 1996). Alternatively, the cost incurred may be a developmental one (Nowicki et al., 1998). The song system is relatively late to develop compared with most of the brain when young songbirds have fledged the nest and are beginning to forage for themselves. Thus, smaller brain nuclei may reflect the costs of developing the brain at a time of nutritional and other stressors.

Studies of the neuroendocrine regulation of the motivation to sing

We have argued that sex differences in the *quality* of singing can be directly related to variation in the size and the morphology of the song control system. We now ask what neuroendocrine mechanisms control the *probability* of singing, i.e. the bird's motivation to sing? As discussed at the outset of this chapter, one of the most important observations about the effect of HVC lesions is that although the birds can not sing they still *try* to sing (Nottebohm et al., 1976), indicating the existence of a causal dissociation between song quality and song motivation.

A separable role for testosterone can be discerned in both processes. In wild populations the importance of testosterone in song motivation is often apparent when one compares seasonal changes in song rate and the stimulus control of song production. For example, European starlings will sing in the spring and in the fall but the ability of a female stimulus to elicit high rates of singing is apparent only in the spring and summer when the male has high concentrations of testosterone in the plasma (Eens, 1997). There is good evidence that seasonal changes in song quality can be related to testosterone regulated seasonal changes in the morphology of the song control system (Schlinger and Brenowitz, 2002). However, it is likely that the effects of testosterone on seasonal changes in song rate and the stimulus that might elicit song are mediated by indirect connections rather than by testosterone acting directly on nuclei in the song system.

The preoptic region and motivational control of singing behavior

Many lines of evidence support the idea that testosterone activity acting upon the preoptic medial nucleus (POM) plays a key role in the regulation of male sexual motivation, including song, in male songbirds (Figure 26.4). In a seasonally breeding nonsongbird, the Japanese quail (Balthazart and Ball, 1998) the POM is larger in males with high testosterone and the larger volume is associated with the expression of male sexual behavior (Panzica et al., 1991, 1996; Thompson and Adkins-Regan, 1994). Androgen and estrogen receptor (both the α and ß forms) proteins and the mRNA encoding these receptors have been located within the POM in male quail (Balthazart et al., 1989, 1992; Ball et al., 1999; Foidart et al., 1999). The POM is also rich in aromatase, the enzyme converting testosterone into estrogen (Balthazart et al., 1990; Foidart et al., 1995). Peripheral blockade of aromatase in castrated, testosterone-treated male quail abolishes the expression of behaviors associated with sexual motivation (i.e. the amount of time a male quail spends near or looking at a female located behind a window (Balthazart et al., 1997), suggesting that estrogen is the active metabolite involved in the expression of these behaviors. POM lesions in quail are followed by a significant reduction in these same behaviors (Balthazart et al., 1998). Moreover, castrated male quail with testosterone implants located directly in the POM exhibit an elevation in measures of sexual arousal (Riters et al., 1998), suggesting that the POM is the site in which the aromatization of testosterone is critical for the regulation of behaviors associated with male sexual arousal and appetitive sexual behaviors related to the pursuit of a female.

In songbirds, as in quail, the POM is rich in the enzyme aromatase (zebra finches, Balthazart et al.,

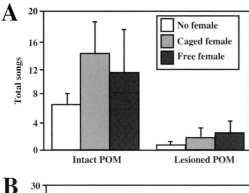
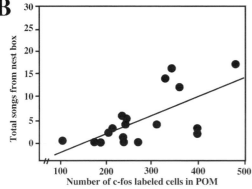
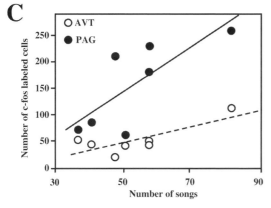

Figure 26.4 Several studies indicate that the preoptic medial nucleus (POM) and some catecholaminergic nuclei are implicated in the control of singing behavior. (A) Male starlings with lesions directed at the POM sing less frequently than control birds with an intact POM during the breeding season when presented to a caged female or to a free female or even in the absence of a female in their aviary. (B) The total number of songs produced by house sparrows correlates positively with the induction of c-fos as indicated by the number of FOS-immunoreactive cells in the POM. (C) The number of FOS-immunoreactive cells in two catecholaminergic nuclei, the ventral tegmental area (AVT) and the periaqueductal gray (PAG) also correlates positively with the number of songs produced by male song sparrows in territorial encounters. (Redrawn from data in Riters and Ball, 1999; Riters et al., 2004; Maney and Ball, 2003.)

1996a; starlings, Riters et al., 2000a), and cells within the starling POM express the mRNA for androgen receptors, and for both forms of the estrogen receptor (Bernard et al., 1999). In male zebra finches, castration results in a significant reduction in courtship behaviors, including song (Harding et al., 1983). This deficit can be completely restored by treating males with aromatizable androgens, such as androstenedione (Harding et al., 1983) whereas treatment with an aromatase antagonist completely blocks the effects of androstenedione, and estrogen treatment restores courtship behavior, including song, disrupted by the blockade of aromatase (Walters and Harding, 1988). Thus the aromatization of testosterone appears to play an important role in the regulation of behaviors reflecting male sexual arousal (i.e. courtship including song) in zebra finches as in quail.

The brain locus of aromatase effects upon courtship behaviors is not known. Because the nuclei known to participate exclusively in song learning and production do not contain either the aromatase protein (Balthazart et al., 1996a; Saldanha et al., 2000) or cells expressing aromatase mRNA (Shen et al., 1995), courtship singing may be regulated by aromatase activity outside of these brain areas, perhaps in the aromatase rich POM. Lesions of POM in reproductively active male starlings interfere with both song expression in response to the presentation of females and the gathering of green nest materials (another courtship behavior; Eens et al., 1993; Gwinner, 1997; see Figure 26.4). These observations are consistent with the hypothesis that the POM plays a role in male courtship behaviors including song (Riters and Ball, 1999).

Converging lines of evidence offer further support for a role for the POM in sexually motivated song. Outside the breeding season in fall aromatase is completely undetectable in the POM of starlings, whereas in spring this nucleus contains dense aromatase immunoreactive cells, a change that is likely steroid dependent, as has been shown in quail (Balthazart et al., 1996b). The POM is larger in male starlings when testosterone concentrations are high during, compared with outside of, the breeding season, and larger in sexually active males that sing sexually motivated song in response to a female compared with males that sing, but not in response to a female (Riters et al., 2000b). The seasonal changes in the POM correspond to seasonal changes in the primary function of male song (i.e. the POM is largest and contains dense aromatase when males use song largely to attract females and is smallest when the function of song is not related to immediate mate attraction). Furthermore, the volume of the POM related positively to song bout length (Riters et al., 2000b). Males sang longer songs in spring, when a longer song bout serves to attract mates and repel competitors (Eens et al., 1991b; Mountjoy and Lemon, 1995; Gentner and Hulse, 2000b), providing additional evidence for a critical role for the POM in song behaviors exclusively observed in spring. Additional recent data revealed positive relationships between the numbers of FOS labeled cells within POM and sexually motivated song, but not song within other less or nonsexually motivated contexts in two studies on different species (starlings: Heimovics and Riters, 2005; house sparrows: Riters et al., 2004).

Together the findings from quail, zebra finches, starlings, and house sparrows suggest the attractive hypothesis that the role of the POM in song is exclusively related to its role in sexual motivation. However, methionine-enkephalin (mENK; an opioid peptide) fiber densities in the POM were found to relate positively to song *outside*, but *not* within, the breeding season (Riters et al., 2005), suggesting that the role of the POM may not be restricted to sexually motivated song. Further support for this idea comes from a lesion study in which the effects of POM lesions on song within and outside a breeding context were examined (Alger and Riters, 2006). As expected, POM lesions suppressed song exclusively within a breeding context. Surprisingly however, unlike control males, POM lesioned males failed to display reductions in song within non- or less sexually relevant contexts. These data indicate that the role of POM in communication differs depending upon the social context, stimulating sexually relevant song, but possibly inhibiting song in other motivational contexts. These new findings suggest that the POM plays a critical role in adjusting communication to match an appropriate social context, a hypothesis consistent with past studies showing the preoptic area as an important site for the integration of sensory and hormonal information necessary for appropriate behavioral responses to environmental and conspecific social stimuli (Wood, 1998; Ball and Balthazart, 2004).

The evidence so far suggests that the POM interacts with the song control system to initiate song within the context of breeding and possibly to inhibit song within other contexts. Song has generally been thought to be regulated by seasonal changes in testosterone action and the resultant neural plasticity within the song control system (e.g. Nottebohm, 1987; Smith et al., 1997a; see also Ball, 1999; Tramontin and Brenowitz, 2000). The close associations among seasonal changes in steroid concentrations, song activity, and the volume of song control nuclei have been taken as evidence that androgens and estrogens acting directly on cells within nuclei of the song control system are critical for the activation of song (Arnold, 1980b, 1990). However, the exact function of testosterone action in the song system is not clear. One possibility is that testosterone in the song system might act to fine-tune song behavior, e.g. increase song stereotypy (Smith et al., 1997a) or song bout length in spring, while testosterone activity within brain areas known to regulate male sexual arousal, such as the POM, might be involved in the motivation to sing in a reproductive context. A simple way to think about this hypothesis is that testosterone action within POM makes songbirds want to sing while testosterone action within HVC makes them sing better.

How might the preoptic region influence song system activity?

Give that the preoptic region and more specifically the POM have effects on song production there should be some sort of anatomical connection between the POM and the song system that can implement this regulation. Brain sites outside of the song control system proper in which steroids could act to modify song production include catecholaminergic nuclei of the mesencephalon and pons that send afferent projections to the song control nuclei (Appeltants et al., 2000, 2002a; Maney and Ball, 2003). It is known in mammals and birds that the POM projects to these nuclei (e.g. Absil et al., 2001), suggesting one way that the POM might regulate the song system in songbirds. The possible neuroanatomical connections between the POM and song control nuclei have been investigated in male starlings (Riters and Alger, 2004). While no direct connections were identified between the POM and song control nuclei, labeled fibers were found to terminate in a region bordering dorsal-medial portions of RA. Additionally, several indirect routes via which the POM might communicate with the song control system were identified. Specifically, POM projected to dorsomedial nucleus intercollicularis (DM), the periaqueductal gray (PAG, or *substantia grisea centralis*; GCt), the ventral tegmental area (VTA), and the locus coeruleus (LoC), structures projecting directly to nuclei involved in song production (DM → vocal-patterning and respiratory nuclei; PAG, VTA, and LoC → RA and HVC), and the context in which song is sung (VTA → Area X). This work suggests that these projections could represent an anatomical substrate that coordinates the actions of POM and of song control nuclei on song production (Figure. 26.2). In support of a role for these areas in the control of singing Maney and Ball (2003) found that immediate early expression in PAG and VTA correlates with the production of territorial song in free-living male song sparrows (Figure. 26.4). Furthermore, in starlings densities of the opioid mENK protein within VTA related positively to male song (Riters et al., 2005), and the number of FOS labeled cells within VTA related positively to song within, but not outside of a breeding context (Heimovics and Riters, 2005). Through these pathways, the POM might provide contextual input to the song control system and this interaction could be what ensures that song occurs in an appropriate context and in response to appropriate stimuli.

Finally, studies on the motivation to sing have focused primarily on sexually motivated song production in males and the role of the POM. Questions about how the female brain regulates song within specific social contexts have not been addressed. Furthermore, little is known about the neuroendocrine control of song within a sexual context by brain regions other than POM, or about song produced within nonbreeding contexts. In addition to POM, VTA, and PAG, reviewed above, several brain regions outside of the song control system, well known for regulating sexual motivation and social behavior, have been implicated in the regulation of male song behavior. Specifically, lesion, pharmacology, and/or immediate early gene studies have identified the lateral septum, medial bed nucleus of the stria terminalis, anterior portions of the hypothalamus, and the ventromedial nucleus of the hypothalamus (VMH; Goodson, 1998; Goodson et al., 1999; Heimovics and Riters, 2006) in song produced within a variety of motivational contexts. Specifically, roles for these nuclei in the regulation of vocal behavior in songbirds have been associated with sexually motivated song (Heimovics and Riters, 2005), song outside the breeding season in large social groups (Heimovics and Riters, 2005, 2007), song in response to territorial intrusion (Goodson, 1998; Maney and Ball, 2003), or song observed in gregarious or territorial species (Goodson et al., 1999). Each of these nuclei also shares connections with the POM (Riters and Alger, 2004), and together these brain regions might act in concert with the song control system to ensure that birds sing in an appropriate social context.

Summary and conclusions

This chapter had two related goals. One was to account for sex differences in song production. Studies concerning the neural basis of sex differences in song behavior have focused on differences in the morphology of the song control system. These differences are obvious in many species and the systematic species differences correlate with species differences in behavior. Our second goal was to explain how the hormonal regulation of song, in addition to its actions on the song control system, acts upon motivational systems to influence both the probability of singing and the quality of the song. Recent studies of the hormonal control of the motivation to sing have implicated the preoptic area, a region generally involved in the control of male-typical sexual behavior, in the hormonal regulation of courtship song as well.

The two topics are related because they both concern how song can be modulated within a species. In many species the probability that a male will sing a song of a specific quality is quite different from the probability that a female will sing such a song, and similarly the probability that a temperate zone male will sing a song of a particular quality will also change seasonally. Studies of sex differences in song behavior have not sufficiently distinguished song quality from song rate (probability). Females differ from males in the quality of song they produce as well as the probability of singing and to whom song is directed. Now that it is clear that song quality is probably regulated directly by the song system while song motivation is controlled indirectly by areas such as the preoptic region, the control of sex differences in song behavior should be re-examined. Until recently the song system has been studied primarily from the perspective of cognitive neuroscience. We are suggesting that a research program which investigates the role of motivational systems in explaining sex differences in song behavior as well as seasonal differences in male behavior has the potential to link songbird neurobiology with the field of affective neuroscience, which studies how emotional and motivational events modulate cognitive processes.

27 • Plasticity of the song control system in adult birds

Eliot A. Brenowitz

INTRODUCTION

Seasonal changes of the environment that are critical to survival and reproduction have a profound effect on all animals. A fundamental feature of nervous systems is that they provide plasticity of structure and function that, in turn, allows animals to adapt to changes in their environment. It is therefore not surprising that seasonal plasticity of the brain has been observed in every vertebrate taxon (Tramontin and Brenowitz, 2000). The avian song control system provides the best model for studying the mechanisms and functional significance of seasonal plasticity in brain and behavior, with changes that are the most pronounced yet observed in any vertebrate model. Song is a critical aspect of reproduction in songbirds. In most species song is produced only by the male, and is used to defend a territory and attract females. In several tropical species, both sexes sing, sometimes in complex vocal duets. Song is a learned stereotyped behavior that can be quantitatively analyzed; it is regulated by well-identified neural circuits, and sex steroids and their metabolites exert a strong influence on the morphology and physiology of these neural circuits.

Various forms of plasticity are observed in the birdsong system, including (1) ongoing neurogenesis and neuronal recruitment in forebrain song control regions, (2) pronounced seasonal changes in the structure and physiology of song nuclei, (3) social influences on seasonal growth and immediate early gene expression, and (4) changes in adult song behavior, including the learning of new songs in some species. Here I will examine the proximate mechanisms underlying seasonal plasticity, discuss potential adaptive benefits of plasticity in the song control system, and speculate on the possible relationship between brain plasticity and the evolution of adult song learning. See the chapter by Ball *et al.*, this volume, for a discussion of social influences on brain plasticity.

SONG BEHAVIOR AND LEARNING

Song must be learned by songbirds (see Hultsch and Todt, this volume). Young birds acquire a sensory model of song by listening to adult conspecifics and subsequently convert this memory to a motor pattern of song production in the sensorimotor phase of development. There is a progressive improvement of song structure from subsong to crystallized (i.e. stereotyped) song. Species vary widely in the timing of song learning (Beecher and Brenowitz, 2005; Brenowitz and Beecher, 2005), but two general developmental patterns exist. In "closed-ended" species, such as the zebra finch (Eales, 1985) and the white-crowned sparrow (Marler, 1970b), song learning is usually restricted to the first year of life. In "open-ended" species, such as the canary (Nottebohm and Nottebohm, 1978) and the starling (Bohner *et al.*, 1990), substantial song learning continues beyond the first year. In the adult male canary, motor song learning is most pronounced in the nonbreeding season (Nottebohm, 1987).

SONG CONTROL SYSTEM AND INFLUENCES OF STEROID SEX HORMONES

Our current view of the network of interconnected discrete brain regions involved in song learning and production is summarized in Figure 27.1. The main descending motor pathway is essential for song production. The anterior forebrain pathway (AFP) is essential for song learning and for the maintenance of crystallized song in relation to auditory feedback (Adkins-Regan *et al.*, 1994; Brainard and Doupe, 2000a). Parallels have been drawn between the anterior forebrain pathway in songbirds and the basal ganglia system in mammals (Bottjer, 1993; Perkel and Farries, 2000; see Farries and Perkel, this volume).

Neuroscience of Birdsong, ed. H. Philip Zeigler and Peter Marler. Published by Cambridge University Press. © Cambridge University Press 2008.

Plasticity of the song control system in adult birds

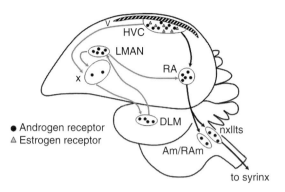

Figure 27.1 Simplified schematic sagittal view of the avian song control system showing the distribution of steroid receptors. Black arrows connect nuclei in the main descending motor circuit, and gray arrows connect nuclei in the anterior forebrain circuit. Am/RAm, ambiguus retroambiguus; DLM, dorsolateral nucleus of the medial thalamus; LMAN, lateral portion of the magnocellular nucleus of the anterior nidopallium; nXIIts, the tracheosyringeal portion of the hypoglossal nucleus; RA, the robust nucleus of the arcopallium; syrinx, vocal production organ; V, lateral ventricle; X, Area X of the medial striatum.

Steroid sex hormones have pronounced effects on song learning and production, and the juvenile development and adult plasticity of the song circuits (reviewed in Schlinger and Brenowitz, 2002; see also Ball et al., and Harding, this volume). Nuclear receptors for testosterone and/or its androgenic metabolites (androgen receptors) are present in the song nuclei HVC, RA, LMAN, ICo (intercollicular nucleus), and nXIIts in all species of songbird examined, and androgen receptor mRNA is present in Area X (Perlman et al., 2003). Estrogen receptors occur to varying degrees in different species in HVC and ICo (reviewed by Schlinger and Brenowitz, 2002). In HVC, both RA- and Area X-projecting neurons contain androgen receptors (Sohrabji et al., 1989), and Area-X projecting neurons also contain estrogen receptors (though not in the same individual cells) (Gahr, 1990). Males castrated as juveniles can acquire a sensory model of conspecific song and go through the early stages of sensorimotor song learning, but cannot develop stereotyped adult song without testosterone replacement (Arnold, 1975a; Marler et al., 1988; Bottjer and Hewer, 1992). Castration of adult males reduces or eliminates song, and treatment with testosterone or a combination of estrogenic and α-reduced androgenic steroids reinstates it (see Harding, this volume).

SEASONALITY OF BREEDING AND SONG BEHAVIOR IN BIRDS

In arctic, temperate, and subtropical birds, breeding is usually restricted to spring and early summer, and photoperiod is the most common environmental factor that influences activation of the avian reproductive system (see Ball et al., this volume). Reproduction may also be seasonal in tropical species in which there are seasonal changes in other environmental factors, such as rainfall, that influence breeding (e.g. Moore et al., 2004b). Song behavior occurs most often or only in the breeding season. In both closed-ended and open-ended species that sing throughout the year, song structure also changes seasonally (Nottebohm and Nottebohm, 1978; Smith et al., 1997a; Brenowitz et al., 1998). Open-ended learners such as canaries develop more new songs during the nonbreeding period of song instability (Nottebohm and Nottebohm, 1978; Nottebohm et al., 1986).

SEASONAL PLASTICITY IN THE BRAIN

The song control system provides the most pronounced example of seasonal plasticity in an adult brain, and remains the leading model for study of this process. Seasonal changes in song behavior are accompanied by changes in the morphology of song nuclei in essentially every seasonally breeding songbird species that has been examined, including rufous-collared sparrows (Zonotrichia capensis) that breed seasonally in the foothills of the Andes on the equator (Moore et al., 2004b) (for reviews see Brenowitz, 1997; Tramontin and Brenowitz, 2000; Ball et al., this volume) (Table 27.1). The volumes of HVC, RA, X, and nXIIts increase by up to 200% during the breeding season in both open-ended and closed-ended song learners (Figure 27.2). These seasonal changes are observed regardless of what cytological markers are used to define the borders of the nuclei (reviewed in Tramontin and Brenowitz, 2000). Cellular attributes of song regions also change (Table 27.1) (e.g. Nottebohm, 1987; Brenowitz et al., 1991; DeVoogd, 1991). The spontaneous neurophysiological activity of RA neurons is greater in breeding white-crowned sparrows (Zonotrichia leucophrys) (Park et al., 2005) and song sparrows (Melospiza melodia) (Meitzen, et al., 2007).

Table 27.1. *Attributes of the song system that change seasonally*

Volumes of HVC, RA, Area X, and nXIIts
Neuronal number in HVC
Neuronal soma size in RA, Area X, and LMAN
Neuronal density in RA and Area X
Synaptic/dendritic traits in RA
Metabolic capacity of neurons in HVC, RA, and X
Spontaneous neurophysiological activity of RA neurons[a]
Incorporation of new neurons into HVC
Song stereotypy, duration, and rate of production

---

[a] *Source:* Park *et al.* (2005). See reviews by Ball and Bentley (2000) and Tramontin and Brenowitz (2000) for other references.

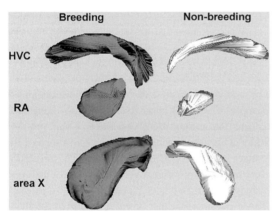

Figure 27.2 Song nuclei of spotted towhees represented in 3D, from a caudal perspective. HVC, RA, and Area X are each greater in 3-dimensional extent in breeding-condition birds (shown in gray) than in nonbreeding-condition birds (shown in white). (Figure adapted from Bentley and Brenowitz, 2002.)

An especially interesting form of seasonal plasticity is a dramatic change in neuron number in HVC. In wild-caught song sparrows (*Melospiza melodia*), for example, neuron number in HVC increases from about 150 000 in the fall to 250 000 during the breeding season (Smith *et al.*, 1997a). This change in neuron number results from seasonal patterns of cell death and ongoing neurogenesis. At the end of the breeding season, circulating testosterone levels drop and there is an increase in the death of existing HVC neurons (see Pytte *et al.*, this volume). There is a subsequent increase in the rate of incorporation of new neurons to HVC in nonbreeding birds, in both open-ended and closed-ended species (Alvarez-Buylla *et al.*, 1990a; Tramontin and Brenowitz, 1999).

COMPARISON OF FIELD AND LABORATORY STUDIES OF SEASONAL PLASTICITY

Seasonal plasticity of the song system has been studied in both wild (e.g. Smith *et al.*, 1997a; Brenowitz *et al.*, 1998; Tramontin *et al.*, 2001) and captive individuals of the same species (e.g. Brenowitz *et al.*, 1991; Bernard, 1995; Smith *et al.*, 1997a). In most cases the results of field and laboratory studies of a species are consistent with each other. There are, however, conflicting reports of seasonal changes of the song system in captive versus wild individuals of two species of seasonally breeding songbirds, the black-capped chickadee (*Poecile atricapillus*) (Phillmore *et al.*, 2006; Smulders *et al.*, 2006) and the island canary (Leitner *et al.*, 2001a, b). In both species seasonal changes of the song nuclei have been observed in birds exposed to breeding versus nonbreeding photoperiods in the laboratory (Nottebohm, 1981; MacDougall-Shackleton *et al.*, 2003). Studies of these same species in the wild, however, have reported lesser (Phillmore *et al.*, 2006) or no differences (Leitner *et al.*, 2001a, b; Smulders *et al.*, 2006) in the size of song nuclei in birds collected at different times of year. Such discrepancies should, however, be interpreted with caution.

In these field studies there is an underlying assumption that individuals collected at the same time of year experience the same physiological conditions which influence the song system. This assumption of synchrony among individuals, however, may only be justified very early in the breeding season, or in extreme environments where the time during which environmental conditions are suitable for breeding is very limited, as in the arctic, high altitudes, or arid habitats with unpredictable transient rainfall. In environments where birds can breed over longer time periods, such as those occupied by black-capped chickadees and canaries, individuals may rapidly fall out of synchrony with each other due to such factors as time of arrival from spring migration, predation of egg or young, stage of breeding (e.g. courtship, nest building, egg incubation, feeding of the young), and multiple breeding efforts. As one proceeds from high to low latitudes, the length of the breeding season generally increases and asynchrony among individuals in a population can become quite pronounced; in wild canaries, for example, the breeding season for a population may last 5 months (Leitner et al., 2001a, b).

A male's plasma testosterone levels in a given year are maximal during the first bout of courtship and copulation, but typically decrease during other phases such as incubation and feeding young, and throughout subsequent breeding efforts (Wingfield and Moore, 1987). When different individuals become asynchronous in breeding stage, birds collected at the same time can have very different circulating testosterone levels, as seen in the extreme variability of plasma testosterone in different "breeding" canaries in Leitner et al.'s study (2001a; see their Fig. 2). As will be discussed below, the song system grows in response to increased testosterone levels, and HVC in particular can change in size rapidly in response to short-term changes in circulating T. The net effect of the above considerations is that collecting different birds at the same time of year without regard to each individual's specific stage of breeding can result in pronounced individual variance in both plasma testosterone and the size of song nuclei. If the variance in testosterone and song nuclei within a season is high, then this can make it difficult to detect differences between seasons. In laboratory studies of the same species, however, the variance in testosterone and song nuclei in any seasonal sample is typically smaller because birds within different treatment groups are generally exposed to differing photoperiod and hormone manipulations more synchronously than is true of wild birds. Studies of wild birds are essential to confirm the results of laboratory manipulations, but it is necessary to ensure that individuals are at the same stage of breeding and to measure their plasma testosterone levels. Unfortunately, both chickadee studies (Phillmore et al., 2006; Smulders et al., 2006) used testes size as an indirect measure of hormonal condition, rather than measuring plasma testosterone. Testes size, however, is an unreliable measure in this context because only the first stage of testicular growth is related to increased testosterone synthesis and secretion by the Leydig cells, with most growth related to spermatogenesis in the seminiferous tubules. Also, testosterone levels in the testes and the plasma need not be the same (Donham et al., 1982). Plasma testosterone therefore provides a clearer indication of the steroid levels transported from the testes to the brain by the blood circulation.

SEX STEROID INFLUENCES ON SEASONAL PLASTICITY

As described above, sex steroid hormones and their metabolites influence the development of song behavior and the song control circuits. The secretion and metabolism of gonadal steroids vary with season; plasma testosterone levels in males are high during breeding, and low after breeding. These seasonal changes in circulating hormone levels modulate song production (birds sing frequently in the breeding season, and less or not at all outside of breeding), and are correlated with the morphological changes in the song control regions (which are fully grown when testosterone levels are high, and regressed when testosterone is low) (e.g. Smith et al., 1997a).

The seasonal changes in the song nuclei are primarily regulated by changes in gonadal testosterone and its metabolites. Smith et al. (1997a) showed that testosterone induced volumetric and neuronal growth in HVC, RA, X, and nXIIts in castrated Gambels' white-crowned sparrows housed on either long days or short days. A long day photoperiod in the absence of a testosterone implant induced only small increases in neuron size and spacing in RA. Other studies are consistent with the conclusion that growth of the song nuclei is due primarily to increased levels of testosterone and its metabolites in breeding birds (e.g. Bernard et al., 1997; Gulledge and Deviche, 1997).

The sensitivity of song nuclei to sex steroids varies seasonally. Immunostaining for the androgen receptors in the HVC of male Gambel's white-crowned sparrows is more intense and labels more cells during the breeding season when plasma testosterone levels are high, than in the fall when plasma testosterone drops to basal levels (Soma et al., 1998). The expression of androgen receptor and estrogen receptor mRNA by cells in HVC is greater in breeding canaries (Gahr and Metzdorf, 1997) and white-crowned sparrows (E. A. Brenowitz, R. Steiner, and G. Fraley, unpublished observations). In starlings that are photorefractory, treatment with exogenous testosterone does not increase HVC volume (Bernard and Ball, 1997). Testosterone does induce growth of HVC in photorefractory white-crowned sparrows (Wennstrom et al., 2001), however, suggesting that there are species differences in seasonal patterns of sensitivity to the trophic effects of testosterone.

Testosterone secreted by the gonads may be metabolized in the brain to either physiologically active androgens, such as 5α-dihydrotestosterone (DHT), or estrogens, such as 17–β estradiol (E_2). This observation raises the possibility that the effects of testosterone on seasonal growth of the song circuits may be mediated by androgens, estrogens, or both. In adult female canaries exogenous dihydrotestosterone and estradiol delivered together stimulated greater dendritic growth in RA than did either metabolite alone (DeVoogd and Nottebohm, 1981b). Estradiol promoted the survival of new neurons and decreased neuronal turnover in the HVC of adult male canaries (Hidalgo et al., 1995). To measure the relative contributions of androgens and estrogens to seasonal growth, castrated Gambel's white-crowned sparrows with regressed song nuclei were implanted with either testosterone, dihydrotestosterone, estradiol, or a combination of dihydrotestosterone and estradiol (Tramontin et al., 2003). All four steroid treatments increased the volumes of HVC, RA, and X when compared with controls (Figure 27.3). These data demonstrate that androgen and estrogen receptor binding are each sufficient to trigger seasonal growth of the song circuits, and that testosterone's effects may depend, in part, upon enzymatic conversion of testosterone to active metabolites. Song production was highly variable within these treatment groups. Interestingly, only one of seven birds treated with estradiol alone was observed to sing, whereas a majority of birds with testosterone or dihyrdotestosterone sang. Despite the low level of song in estradiol-treated birds, however, this hormone induced substantial growth of the song nuclei. This observation suggests that increased song behavior is not the primary factor driving changes in the morphology of the song circuits.

Additional evidence in support of an estrogenic contribution to seasonal growth of the song circuits comes from a study of wild male song sparrows (Soma et al., 2004). Territorial males were implanted in both the breeding and nonbreeding seasons with osmotic pumps that released fadrazole, an inhibitor of the aromatase enzyme which catalyzes the conversion of testosterone to estradiol. In breeding males, aromatase inhibition caused the volume of HVC to decrease,

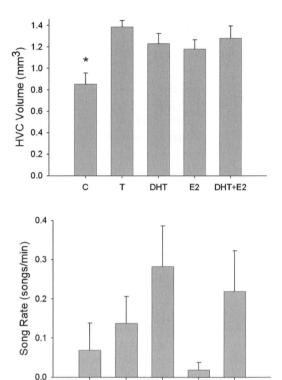

Figure 27.3 Effects of different sex steroids on the size of HVC (top panel) and on song behavior (bottom panel) in castrated male Gambel's white-crowned sparrows. Androgens and estrogens significantly increased the mean volume of HVC. * indicates that HVC in controls (C) was significantly smaller than in all steroid-treated groups. No significant differences were detected among groups that received steroid implants. Bars in lower panel represent means of maximum song rate. Numbers beneath the x-axis represent the ratio of singing birds to all birds in each treatment group. While maximum song rate did not differ significantly among treatment groups, only one of seven estradiol-treated birds sang, and that bird sang very little. Error bars represent the standard error of the mean. (Figure adapted from Tramontin et al., 2003.) See text for abbreviations.

and this effect was partially rescued by concurrent estrogen replacement (Figure 27.4). It is noteworthy that fadrazole decreased HVC volume even though plasma testosterone increased in breeding males due to disrupted neuroendocrine feedback. Fadrazole-treated males sang less than controls in response to song playback and a live decoy. In nonbreeding males, estradiol treatment caused HVC and X to grow to maximal breeding size within two weeks. Fadrazole decreased

Plasticity of the song control system in adult birds 337

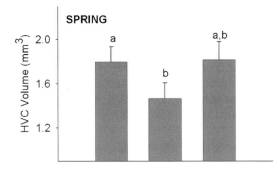

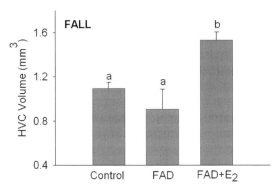

Figure 27.4 Effects of estrogen manipulations on the volume of HVC. Subjects were wild adult male song sparrows in the spring and fall. Birds were treated with vehicle (control), the aromatase inhibitor fadrozole (FAD), or fadrozole and concurrent estrogen replacement (FAD + E$_2$). Letters above bars denote groups that are significantly different (ANOVA followed by *post hoc* pairwise comparisons). In the spring, FAD-treatment reduced HVC volume, and estradiol blocked this effect. In the fall treatment with estradiol stimulated growth of HVC. (Figure adapted from Soma *et al.*, 2004.)

singing behavior and estradiol-treatment rescued song in the nonbreeding birds. Taken together the results of Soma *et al.* (2002) and Tramontin *et al.* (2003) strongly suggest that both estrogenic and androgenic metabolites of gonadal testosterone contribute to seasonal growth of the song circuits and song production.

NONSTEROIDAL CUES CONTRIBUTE TO SEASONAL PLASTICITY

The seasonal growth of the song nuclei is primarily regulated by the secretion of gonadal testosterone and its metabolism in the brain, but nonsteroidal factors such as melatonin and thyroid hormone may modulate this growth (see Ball *et al.*, this volume; also Bentley *et al.*, 1997; Tekumalla *et al.*, 2002). In the laboratory, social cues from sexually receptive female Gambel's white-crowned sparrows enhance the photo-induced growth of two song nuclei in their male cage-mates (Tramontin *et al.*, 1999).

SEASONAL GROWTH AND REGRESSION OF THE SONG CONTROL CIRCUITS OCCUR RAPIDLY AND SEQUENTIALLY

Seasonal changes in the song circuits seem to occur quickly following changes in photoperiod and circulating testosterone levels. Smith *et al.* (1997a) measured changes in plasma testosterone and the morphology of the song nuclei in a population of wild song sparrows at four times of the year: early spring, at the onset of breeding; late spring, at the peak of breeding; early fall, immediately after the prebasic feather molt; and late fall (December). HVC was already fully grown by early spring, and RA was intermediate in size between its regressed late fall and fully grown late spring states. These results suggested that the growth of the song system occurs rapidly and may precede the onset of breeding. To get a better sense of the time course of growth of the song system relative to the onset of breeding, Tramontin *et al.* (2001) measured changes in plasma T, gonadal size, and the size of song nuclei in wild song sparrows from November to April. Testosterone levels began to increase in early February. Less than a month later, HVC and RA were fully grown, even though plasma testosterone levels had reached only about 20% of the breeding maximum at that time, and the testes were less than 10% of typical breeding size (Figure 27.5). These results indicate that growth of the song system occurs rapidly once plasma testosterone levels first start to rise as daylength increases in late winter, and precedes full seasonal reproductive development.

The fact that the song nuclei of these sparrows grew fully when testosterone levels were still below peak breeding levels has been interpreted by some authors as evidence that the song system can grow in response to photoperiodic cues independent of the action of testosterone (Ball *et al.*, 2002; Caro *et al.*, 2005). We interpret this finding instead as evidence that the song system grows in response to testosterone levels lower than those required for breeding. Testosterone levels in the study of Tramontin *et al.* (2001) were just over 1 ng/ml

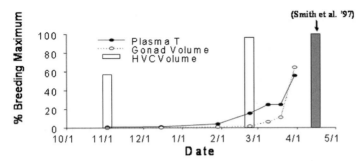

Figure 27.5 The seasonal growth of the song control system in western song sparrows precedes seasonal reproductive development. Data are expressed as a percentage of the typical level observed during the height of the breeding season in April. Mean breeding gonad volume (open circles) is 350 mm^3 in late April (data from Tramontin and Brenowitz, 1999). Mean breeding plasma testosterone (filled circles) is approximately 5.5 ng/ml in late April (data from Wingfield, 1984). HVC volume during April was measured by Smith *et al.* (1997a) and was approximately 1.60 mm^3. In this study, HVC (white bars) was fully developed by late February and was similar in volume to that reported for males during April. These results show that growth of the song system occurs at lower testosterone levels than does reproduction. (Adapted from Tramontin *et al.*, 2001.)

in late February, which is consistent with the mean testosterone level of 1.5 ng/ml observed by Smith *et al.* (1997a) in early spring, when HVC was fully grown. These field studies and observations from several laboratory studies suggest that a plasma testosterone level of about 1 ng/ml may be a threshold for stimulating growth of the song circuits in white-crowned sparrows and song sparrows. Also, as noted above, increased photoperiod in castrated birds has only minimal effects on growth of the song nuclei, independent of testosterone (Smith *et al.*, 1997a). These and other studies are consistent with the conclusion that the song system is stimulated to grow by the initial increases in circulating testosterone that occur as daylength increases in late winter.

The early growth of HVC and RA observed by Tramontin *et al.* (2001) raised the question of just how rapidly the song circuits can grow in response to exposure to breeding photoperiod and testosterone levels. This question was addressed in a laboratory study of Gambel's white-crowned sparrows (Tramontin *et al.*, 2000). Two groups of male white-crowns were shifted overnight from short day to long day photoperiods and implanted with testosterone at a dose sufficient to produce breeding-typical plasma levels. One group was sacrificed 7 days after the initial exposure to long days plus testosterone, and the other group was sacrificed after 20 days. HVC increased its volume to breeding size within 7 days of exposure to long days plus testosterone. Neuron number in HVC increased from about 90 000 in short day control birds to 140 000 in long days plus testosterone birds by 7 days. This addition of 50 000 neurons within such a short time period represents a striking example of plasticity. HVC's efferent targets, RA and Area X, grew more slowly and were not significantly larger until day 20. In the field study of Soma *et al.* (2002), estradiol treatment of nonbreeding song sparrows caused X (as well as HVC) to grow to maximal breeding size within two weeks. These studies taken together indicate that the song control circuitry grows rapidly and sequentially in response to seasonal cues.

Regression of the song system at the end of the breeding season also occurs rapidly and sequentially. HVC and RA in the wild song sparrows studied by Smith *et al.* (1997a) were already fully regressed in the early fall, just after the prebasic feather molt. We have investigated the time course of seasonal regression of the song circuits in Gambel's white-crowned sparrows in the laboratory (Thompson *et al.*, 2007). Male white-crowns were exposed to long days plus testosterone for 20 days to induce full growth of the song nuclei. The birds were castrated, their testosterone pellets were removed and they were shifted overnight to short days, and they were sacrificed at different times 0.5–20 days later. With great rapidity, HVC regressed fully to a nonbreeding volume by 12 hours. HVC neuron number decreased within four days, again pointing to pronounced plasticity of this nucleus to changes in circulating hormone levels. Area X was regressed by day 7. RA did not regress until day 20. These efferent targets therefore regress more slowly than does HVC.

SITE OF HORMONE ACTION IN SEASONAL GROWTH OF SONG CIRCUITS

Given the pronounced and rapid effects of steroid hormones on seasonal growth of the song circuits, we may ask which nuclei within this circuit are directly targeted to initiate such growth. Some insights may be provided by the observation that HVC grows rapidly and its efferent targets RA and X grow more slowly (Tramontin et al., 2000). Two hypotheses can be proposed to explain this pattern of growth. (1) As described above, steroid receptors or their mRNA are present in all of the major song nuclei (Figure 27.1). It is therefore possible that steroids act directly and independently on each nucleus in the song circuits, perhaps with different time courses which could account for the differences in the time required for growth of the various nuclei. (2) Alternatively, the rapid growth of HVC and slower growth of its efferent targets may indicate that testosterone and its metabolites initially act directly on HVC, which subsequently stimulates growth of RA and X transynaptically.

We have conducted several types of studies to test these hypotheses. In one study we lesioned HVC on one side of the brain in Gambel's white-crowned sparrows housed on a short day photoperiod, and then exposed these birds to long days plus systemic testosterone implants for 30 days (Brenowitz and Lent, 2001). The unlesioned side of the brain acted as an internal control. The unilateral HVC lesions completely blocked the growth of the ipsilateral RA and X, but the contralateral RA and X, as well as the contralateral HVC, showed normal growth. In a second study we placed small testosterone implants unilaterally in the brain near HVC or RA in male Gambel's white-crowned sparrows housed on short days for 30 days (Brenowitz and Lent, 2002). The testosterone implant near HVC produced significant growth of the ipsilateral (but not contralateral) HVC, RA, and X, and increased neuronal number in the ipsilateral HVC. The testosterone implant near RA, however, did not produce growth of the ipsilateral RA, HVC, or X. The failure of this latter implant to stimulate growth of RA is noteworthy given that this nucleus has abundant androgen receptors (Schlinger and Brenowitz, 2002). In a third study unilateral intracerebral infusion of dihydrotestosterone plus estradiol adjacent to HVC produced significant growth of neurons in ipsilateral but not contralateral RA (Meitzen et al., 2007). These results together indicate that testosterone and its metabolites induce seasonal growth by acting directly on HVC, which in turn provides some permissive or trophic support of growth to its efferent targets RA and X.

Testosterone-induced trophic support provided by HVC to its efferent targets may require activity-dependent release of neurotrophins. Presynaptic terminals from HVC release brain-derived neurotrophic factor (BDNF) and perhaps other neurotrophins that stimulate growth of postsynaptic neurons. Neurotrophins can be transported anterogradely and taken up by postsynaptic neurons (von Bartheld et al., 1996; Kohara et al., 2001). BDNF and other neurotrophins are present in the song system and influence the development of the song circuits in juveniles and their T-induced growth in adult females (Johnson et al., 1997; Akutagawa and Konishi, 1998; Dittrich et al., 1999; Rasika et al., 1999). Singing increases BDNF mRNA in HVC, and BDNF increases neuronal recruitment in adult male canaries (Li et al., 2000). BDNF mRNA expression in HVC increased in male Gambel's white-crowned sparrows exposed to a long day photoperiod plus testosterone for 7 days compared with short day controls (Wissman and Brenowitz, 2003). Infusion of recombinant BDNF into RA stimulates growth of RA neurons (Wissman and Brenowitz, 2004). Inhibiting electrical activity of RA neurons with infusions of tetrodotoxin blocks the neurotrophic effects of a systemic testosterone implant (E. A. Brenowitz and K. Lent, unpublished observation). The steroid-stimulated seasonal growth of HVC and its efferent targets may thus be mediated by the action of neurotrophins and require electrical activity in HVC axon terminals and in RA neurons.

SEASONAL CHANGES IN SONG BEHAVIOR

Seasonal changes in various aspects of song behavior accompany plasticity of the song circuits. In some

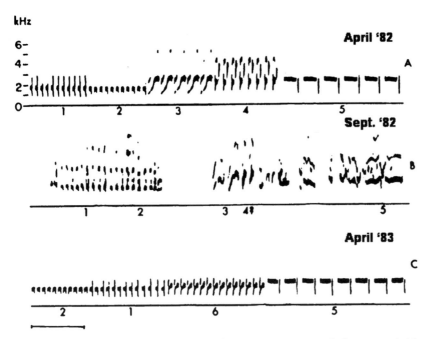

Figure 27.6 Song stereotypy changes seasonally. Spectrograms are presented of songs recorded from a single adult male canary during two successive breeding seasons (April 1982 and 1983) and the intervening nonbreeding season (September 1982). During April 1982, this bird's repertoire contained highly stereotyped syllables. In September, song stereotypy was markedly decreased and syllables were poorly structured. By the following April (1983), this bird's song production was once again stable and stereotyped. Bar = 0.5 s. (Modified, with permission, from Nottebohm et al., 1986.)

species of birds, such as the spotted towhee (*Pipilo maculatus*) and sedge warbler (*Acrocephalus schoenobaenus*), song is produced only during the breeding season and is absent at other times of year. Other species, such as song sparrows, white-crowned sparrows, and canaries, sing throughout most of the year. Even in these year-round singers, however, song is produced at much higher rates during the breeding season, as a result of the higher circulating testosterone levels then. As described above, sex steroids are necessary for the activation of song behavior.

Song structure also changes seasonally in year-round singers. Songs typically become more variable after the breeding season. The morphology of individual song syllables becomes less stereotyped in species including canaries, song sparrows, and white-crowned sparrows (Nottebohm et al., 1986; Smith et al., 1997a; Brenowitz et al., 1998; Tramontin et al., 2000) (Figure 27.6). Song sparrows sing a greater number of variations of specific song types outside the breeding season (Smith et al., 1997a). Songs also are shorter in nonbreeding song sparrows, white-crowned sparrows, and wild island canaries (Smith et al., 1997a; Brenowitz et al., 1998; Leitner et al., 2001a, b) (Figure 27.7). The stereotypy of song duration and of the "fee" note of black-capped chickadees (*Poecile atricapillus*) is greater during the breeding season (Smulders et al., 2006). Stable song produced during the breeding season, however, does not change in structure from year to year in closed-ended learners like song sparrows and white-crowned sparrows (Hough et al., 2000; Nordby et al., 2002).

Two hypotheses, which need not be mutually exclusive, can be proposed to explain why regression of the song circuits leads to increased variability of song structure outside the breeding season.

(1) As the number of neurons in HVC decreases by one third or more at the end of the breeding season, the accuracy of timing for song production is decreased. In discussing the evolutionary enlargement of the human cortex as a possible adaptation

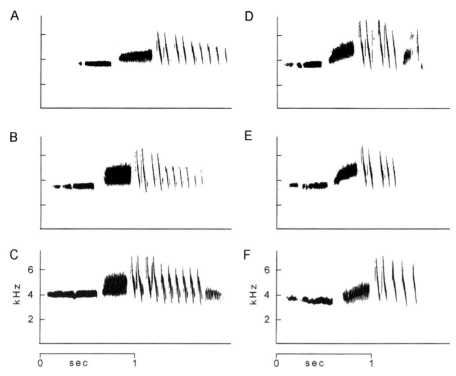

Figure 27.7 Song duration changes seasonally. Songs of male Nuttall's white-crowned sparrows recorded in the spring breeding season (A–C), and from sparrows at the same locality in the autumn after the breeding season (D–F). Note the greater duration of songs in the spring, as well as the quavering quality of the introductory whistles and the second phrases in the autumn songs. (Reprinted from Brenowitz et al., 1998.)

to improve the accuracy of throwing weapons at prey, Calvin (1993) pointed out that due to the inherent jitter or noise introduced by ion channels, individual muscle or brain cells cannot time events with great accuracy. He suggested that a way of decreasing this cellular jitter is to assign a large number of cells to the same timing task; with an increase in the number of cells involved in timing, the timing signals can be averaged and a more accurate estimate obtained. An analagous argument can be made for the song system. Some evidence suggests that neurons in HVC generate the timing signal for the motor production of song (Yu and Margoliash, 1996; Hahnloser et al., 2002, see Fee et al., 2004). As neuron number in HVC decreases after the breeding season, the effective population size generating the timing signal for song is decreased and the averaged signal therefore has greater variablility, which may be manifested as decreased song stereotypy.

(2) The aerobic capacity of the syringeal muscles involved in generating sound may decrease outside the breeding season, which might in turn decrease their ability to produce sustained, stereotyped song. The muscle fibers of the syrinx contain androgen receptors (Wade and Buhlman, 2000), and the mass of the syrinx decreases when testosterone levels drop after the breeding season (Tramontin et al., 2001). In another hormone-sensitive vocal motor system, the sonic muscle of plainfin midshipman fish (*Porcithys notatus*), there is a similar seasonal change in muscle mass with changes in circulating androgens, and this is coupled with changes in the aerobic capacity of the muscle (Mommsen and Bass, cited in Walsh et al., 1995). If comparable changes in aerobic capacity of the syringeal muscles occur seasonally, then the regressed muscles in nonbreeding birds may not be able to sustain prolonged production of stereotyped song. If these muscles do fatigue more

rapidly when they are regressed, then this could introduce variability in syllabic structure between successive renditions of song. Such reduced muscular aerobic capacity could also help explain the decreased duration of song in nonbreeding birds.

CAUSAL RELATIONSHIPS BETWEEN SEASONAL CHANGES IN SONG BEHAVIOR AND SONG NUCLEI

Are changes in song behavior a consequence or a cause of seasonal changes in the song circuits? (see Adkins-Regan, 2005). Although it is generally assumed that causation works forward from hormonally induced changes in the song circuit to song, it has also been suggested that changes in the circuit are a consequence of the stimulating effects of song production itself (see Ball et al., this volume). One account posits that increased circulating testosterone in the breeding season stimulates song behavior by acting on hormone-sensitive brain regions related to sexual and aggressive behavior outside the song system, rather than acting directly on the cells of the song nuclei (Ball et al., 2002). The increased song production, in turn, is seen as inducing seasonal growth of the song nuclei by increasing the expression of the gene for BDNF, which increases neuronal recruitment in HVC (Rasika et al., 1999; Li et al., 2000). In this scenario, the growth of the song nuclei in breeding birds is viewed as the consequence of high rates of singing at this time of year (Sartor et al., 2002).

Because it can be difficult to separate causation and correlation, the hypothesis, though appealing, is challenging to test experimentally. A fundamental limitation of this hypothesis stems from the idea that song behavior can be meaningfully dissociated from its physical substrates. Song production, for example, is the result of a constellation of events, including neuronal activity in the brain, contraction of the syringeal and respiratory muscles, pressure in the air sacs, the flow of air through the bronchotracheal pathway, vibration of the sound-generating membranes in the syrinx, changes in the configuration of the vocal tract, and beak movement (reviewed by Suthers, 1997; see chapters by Goller and Cooper, and Suthers and Zollinger, this volume). We cannot divorce a behavior such as song production from this underlying ensemble of events. When we measure song behavior, we are indirectly measuring the outcome of this series of actions. Thus, a phenomenon such as enhanced gene expression with increased song production may reflect changes in one or more of these underlying processes, rather than being a result of changes in song behavior per se.

One way to test the idea that enhanced song production increases processes such as neuronal recruitment to HVC or seasonal growth of the song system is to devocalize birds, by cutting the tracheosyringeal nerve or lesioning the syrinx (e.g. Sartor et al., 2002; Wilbrecht et al., 2002a). With such manipulations, however, the independent variable is the nerve cut or syringeal lesion, not the desired changes in song output. Identifying the direct causal basis of changes in a dependent variable such as growth of the song nuclei following manipulations therefore may be problematic. Such experimental manipulations may affect several systems, only some of which are being measured. Tracheosyringeal nerve cuts, for example, will alter syringeal function, which in turn might have unintended consequences in terms of altered feedback to brain nuclei (Bottjer and Arnold, 1982), and/or altered retrograde trophic support from syringeal muscles to nXIIts and its afferent nuclei. Syringeal lesions could alter respiration as well. Such considerations constrain our ability to attribute changes in neuronal recruitment or song nuclear volume after "devocalization" to changes in "song behavior" itself.

Similarly, the issue of seasonal growth of the song system presents particular challenges for experimental manipulations. In gonadally intact birds, many factors covary with each other, including circulating testosterone and its metabolites in the brain, the morphology and physiology of the song circuits, expression of genes for BDNF and other neurotrophins, and song behavior. The level of causal complexity increases still further given that steroid hormones likely motivate song behavior by acting on areas of the brain outside the song system (Ball et al., 2004a, b). With this diverse pattern of interactions, manipulating testosterone levels may simultaneously influence song production and neural circuits in and out of the song system. It thus becomes easy to confound correlation and causation, and difficult to determine the direction of causal relationships.

There is also practical difficulty in trying to dissociate the direct effects of sex steroids on the song circuits from such nonsteroidal cues as those associated with photoperiod and social factors. Typically birds are castrated to remove the primary endogenous source of the sex steroids (e.g. Bernard et al., 1997; Smith et al., 1997a; Gulledge and Deviche, 1998). When partial or complete growth of the song system occurs in castrated birds exposed to long day photoperiods or social cues, one might conclude that such growth is mediated by factors other than sex steroids (Bernard et al., 1997; Gulledge and Deviche, 1998). Castration, however, even when complete, does not eliminate all sex steroids. For example, the avian adrenal gland may release the androgen precursor dehydroepiandrosterone (DHEA), which in turn may be converted in the brain to active sex steroids such as testosterone and estradiol (Vanson et al., 1996; Schlinger et al., 1999). Castrated male song sparrows and swamp sparrows (*Melospiza georgiana*), for example, had elevated plasma levels of estradiol (Marler et al., 1988). Treatment of nonbreeding wild song sparrows with DHEA stimulated growth of HVC (Soma et al., 2002). Furthermore, all of the essential enzymes for the *de novo* synthesis of androgens and estrogens from cholesterol have been identified in the songbird brain (Freking and Schlinger, cited in Schlinger and Brenowitz, 2002). It is therefore possible that physiologically active levels of sex steroid hormones may be present in the brain but undetectable in the peripheral circulation. For these reasons, it is necessary to use caution in interpreting growth of the song nuclei in castrated birds exposed to long day photoperiods and social cues.

There is also specific evidence in some species which suggests either that behavioral changes follow rather than precede changes in the song nuclei, or implies that behavioral contributions to changes in the song circuits are minor compared with the effects of testosterone and its metabolites.

(1) Wild song sparrows sing at higher rates in the early fall, when juvenile males establish territories, than in the late fall. The song nuclei, however, are fully and equally regressed at both times of year (Smith et al., 1997a). This result shows that enhanced song production need not stimulate growth of the song nuclei.

(2) As noted earlier, HVC grows to its full size by 7 days of exposure to long days plus testosterone in Gambel's white-crowned sparrows. Song stereotypy, however, did not increase until 20 days of treatment (Tramontin et al., 2000). This observation suggests that significant growth of the song circuit must occur before song becomes more stereotyped.

(3) Treatment of castrated Gambel's white-crowned sparrows with estradiol and dihyrdrotestosterone stimulated HVC and RA to grow equally. Only one of seven estradiol-treated birds was observed to sing, however, and that one sang only at a very low rate, whereas most dihyrdrotestosterone-treated birds sang and did so at a high rate (Figure 27.3). Song production in dihyrdrotestosterone-treated birds was considerably greater but this did not, however, produce increased growth of the song nuclei compared with the estradiol-treated birds, which essentially did not sing.

(4) Intracerebral testosterone pellets implanted unilaterally near HVC in Gambel's white-crowned sparrows housed on a short day photoperiod are sufficient to induce growth of the ipsilateral HVC, RA, and X, but these birds were not observed to sing (Brenowitz and Lent, 2002). This result demonstrates again that sex steroids can induce growth in the absence of song production.

(5) The maximum song rate observed in deafened white-crowned sparrows is less than 10% of that seen in hearing birds. Despite this pronounced difference in song production, however, HVC, RA, and X grew equally in deafened and hearing sparrows exposed to long days plus testosterone. This observation shows that the trophic effects of breeding testosterone levels and photoperiod on the song system are not mediated primarily by increases in song production in this species, nor is auditory feedback necessary (Brenowitz et al., 2007).

(6) Alvarez-Borda and Nottebohm (2002) compared neuronal recruitment to HVC in gonadally intact and castrated male canaries that produced comparable amounts of song in the autumn. When matched for song behavior, the intact birds had about 4 times as many new RA-projecting neurons, and about 2.6 times as many total new HVC neurons as did the castrated birds. This result suggests that the contribution of testosterone to the incorporation of HVC neurons is considerably greater than that of song production per se.

Such observations are inconsistent with the hypothesis that seasonal change in song behavior is the primary factor driving changes in the morphology and physiology of the song circuits in all species. Instead, the burden of evidence is consistent with the idea that at least in most species examined, seasonal changes in the song nuclei are predominantly regulated by changes in gonadal testosterone and its metabolites, and that subsequent changes in song behavior may play only a secondary or modulatory role in reinforcing the neural changes by mechanisms such as song-induced expression of BDNF. It is possible, and perhaps even likely, that there are species differences in the relative contributions to seasonal plasticity of the song circuits of sex steroids and physical actions associated with song behavior. The relative contibutions of different factors, such as sex steroids, social cues, and photoperiod, may differ among species. But caution should be used in invoking species differences, however, given the complex nature of interactions between sex steroids, neural and muscular function, gene expression, and behavior.

ADAPTIVE VALUE OF SEASONAL PLASTICITY

What is the adaptive value of the extensive seasonal changes observed in the song circuits? Regrowing the song system each spring, which includes the addition of 50 000 or more neurons to HVC, must impose an energetic cost on birds. Is the cost of such yearly growth outweighed by some advantage that is gained? After all, other hormone sensitive regions of the avian brain, such as the hippocampus, do not undergo the seasonal regression and growth characteristic of the song system (Lee et al., 2001).

One hypothesis of the benefit of seasonal plasticity was presented by Nottebohm (1981) in the original study of this phenomenon in canaries. Male canaries develop new song patterns as adults, and do so in a seasonal manner. Nottebohm reasonably proposed that the synaptic plasticity associated with the seasonal changes in the song nuclei provides a neural substrate for this adult song learning. He predicted that seasonal plasticity would be restricted to open-ended song learning species. Comparable patterns of seasonal changes in the song system, however, have been observed in every species of seasonally breeding songbird examined thus far, regardless of whether they are open-ended or closed-ended song learners. Thus, while seasonal plasticity may be necessary for adult song learning, it is not sufficient for such learning to occur and is therefore unlikely to have evolved explicitly as a mechanism for learning.

An alternative hypothesis about the adaptive value of seasonal plasticity is that rather than having evolved specifically as a mechanism for adult song learning, it is a form of performance-associated hypertrophy (Tramontin and Brenowitz, 2000). The sustained peak performance of a seasonally predictable behavioral or physiological task is often preceded by hypertrophy of the organs and/or tissues involved in performance of the task (Piersma and Lindstrom, 1997). For example, the size of the gonads and other reproductive structures increases dramatically in preparation for the annual breeding season, and these organs regress when the breeding season is terminated. Anticipatory changes such as these are stimulated by seasonal environmental cues and mediated by neural and endocrine signaling mechanisms. The maintenance of hypertrophied organ systems and tissues is thought to be energetically expensive and so these systems regress when peak performance is not required (Gaunt et al., 1990).

The principle of performance-associated hypertrophy might also pertain to seasonal plasticity of the song circuitry. This hypothesis predicts that song performance should be enhanced during the breeding season. This prediction is supported by data from canaries, white-crowned sparrows and song sparrows. As described above, males of these species sing more stereotyped songs, and sing more frequently, during the breeding season.

Another prediction of the performance-associated hypertrophy hypothesis is that the growth of the song nuclei should precede behavioral changes. In the study of Tramontin et al. (2000), HVC in male Gambel's white-crowned sparrows grew fully by 7 days of exposure to long days plus testosterone, but song stereotypy only increased between 7 and 20 days.

A third prediction of the performance-associated hypertrophy hypothesis is that the energetic costs of maintaining a fully developed song system throughout the nonbreeding season outweigh those associated with regrowing the song system each spring. The relative metabolic costs of maintaining and regrowing the song system each year are not yet known. Some insight into this issue, however, comes from the study of

Wennstrom et al. (2001). They measured the activity of cytochrome oxidase (CO), an enzymatic marker of cellular metabolic capacity, in the song nuclei of Gambel's white-crowned sparrows with regressed song systems in comparison with birds who were treated with testosterone to induce growth of the song system. T-induced growth of the song system increased CO activity considerably in HVC, RA, and X. This result suggests that the song system does impose a greater metabolic cost in its fully grown state than when regressed.

Although additional research is necessary, the existing evidence is consistent with the performance-associated hypertrophy hypothesis, and suggests that seasonal plasticity of the song system may be an adaptation to reduce the energetic costs imposed by the song system in the fall and winter. Outside the breeding season, there is no need to produce frequent, stereotyped song for mate attraction and territorial defense. At the same time, birds may experience the energetic stress of migration, increased thermoregulatory demands, and decreased food availability. Songbirds are relatively small animals with large surface area to volume ratios, and are therefore particularly subject to energetic constraints (Calder and King, 1974). Given that the brain requires large amounts of energy to maintain signaling activities (Ames, 2000), regression of the song system outside the breeding season reduces the energetic costs imposed by the song nuclei and this is probably advantageous for birds. On balance, the reduced energy required by a regressed song system throughout the fall and winter may outweigh the energy required to regrow it the following spring.

EVOLUTIONARY RELATIONSHIP BETWEEN SEASONAL PLASTICITY AND ADULT SONG LEARNING

Plasticity of the song system may provide insights into the evolution of adult song learning. Birdsong researchers have traditionally distinguished between "closed-ended" species, in which sensorimotor song learning is restricted to the first year of life, and "open-ended" or "age-independent" species, in which song learning continues into adulthood (e.g. Immelmann, 1969; Slater, 1983; Nottebohm, 1984; Marler and Peters, 1987). In the song learning literature there has been a largely unstated assumption that these represent two distinct strategies, perhaps reflecting a dichotomous evolutionary divergence from a common ancestral pattern (but see Kroodsma and Pickert, 1984; Marler and Peters, 1987). Various types of evidence suggest, however, that adult song plasticity may be more common than was originally thought, particularly in closed-ended species, and that closed-ended and open-ended song learning are not dichotomous but lie on a continuum (Figure 27.8) (Beecher and Brenowitz, 2005; Brenowitz and Beecher, 2005; see also Williams, this volume). In reviewing this evidence, I will consider adulthood to begin when birds first become

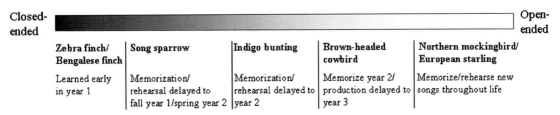

Figure 27.8 The diversity of song learning programs observed across species forms a continuum between closed-ended and open-ended learning strategies. "Year" refers to calendar years. Zebra and Bengalese finches are extreme examples of closed-ended learners; once song is crystallized in the first year, males sing the same song for the rest of their life. A young song sparrow learns song in the summer and fall of his first year, and perhaps in the following spring, when he memorizes songs of adult males in the area where he will subsequently establish his own breeding territory (Nordby et al., 2001). Indigo buntings typically delay sensory acquisition of song models until their first breeding season. In brown-headed cowbirds (*Molothrus ater*) males memorize new songs in the first breeding season but do not sing them until the next breeding season (Beecher and Brenowitz, 2005). At the opposite end of the spectrum from strictly closed-ended learners like zebra finches are species that modify their songs throughout life, such as mockingbirds and starlings. These species are presented to illustrate the concept of a continuum of temporal song learning programs, with no implication that they are closely related.

reproductively competent. Due to space constraints, only selected examples in support of each point can be cited.

(1) The timing of song memorization and sensorimotor development may be strongly influenced by photoperiod and the amount of song a male hears. Kroodsma and Pickert (1980) showed that male marsh wrens (*Cistothorus palustris*) hatched early in the breeding season and exposed to long day photoperiods in the laboratory complete their acquisition of song models in the hatching year. Males born late in the breeding season and exposed only to shorter photoperiods, however, retained the ability to learn songs the following spring. Due to short breeding seasons at high altitudes or northern latitudes, and a parasitic reproductive pattern where young are raised in the nest of other species, wild migratory brown-headed cowbirds (*Molothrus ater*) also show evidence of prolonged sensitive periods for development of local dialect songs (O'Loghlen and Rothstein, 1995, 2002a, b). Young males hatched late in the breeding season have restricted opportunities to hear and acquire a sensory memory of local songs, and the memorization phase of development continues during the next breeding season, a year after they were hatched. These sexually mature yearling males, however, do not produce their newly acquired local songs until the start of their second breeding season.

(2) The timing of song development may be influenced by social interactions. White-crowned sparrows can be induced to learn heterospecific songs if housed with other species, and to learn the songs of live tutors first encountered beyond the close of the sensitive period for birds tutored by tape recording (Baptista and Petrinovich, 1986). In migratory indigo buntings (*Passerina cyanea*), males may defer memorizing songs until the first breeding season the year after hatching, when they learn the song of a local territorial neighbor with whom they will interact aggressively (reviewed in Payne and Payne, 1993). These year-old bunting males appear to selectively copy the songs of adult males that either arrived early in the breeding season, whose song would thus be heard over the greatest time period, or that had relatively more blue plumage, which is typical of older birds. Similarly, young saddleback (*Philesturnus carunculatus*) males seem to delay song learning until their first breeding season, when they learn songs produced by adult males in that area, rather than in the area where the young males hatched (Jenkins, 1978). A potential advantage of such social copying is that learning the songs of adjacent territorial males allows a young bird to engage in aggressive song-matching interactions with them. It is possible that some young males could have memorized these songs from local adults in their hatching year (Margoliash *et al.*, 1994; Hough *et al.*, 2000). This suggestion does not seem likely, however, for males that do not return to their natal area to breed, and thus would not have opportunities to learn songs that specifically match those of neighbors in the previously unfamiliar breeding area.

(3) The selective retention and "discarding" of crystallized songs to form the adult song repertoire is influenced by social interactions. Some species produce more syllable types in the plastic song phase of sensorimotor learning than are retained in the stable adult song (e.g. Marler and Peters, 1982a). In field sparrows (*Spizella pusilla*), song sparrows, and other species, males seem to selectively retain previously learned song or note types that are most similar to those of conspecifics on neighboring territories, which facilitates song matching (Nelson, 1992; Marler, 1997; Beecher *et al.*, 2000; Nordby *et al.*, 2000).

(4) Some species retain the ability to memorize or improvise new song patterns well into adulthood. Adult European starlings and nightingales (*Luscinia megarhynchos*) can memorize and produce entirely new songs (Chaiken *et al.*, 1994; Geberzahn and Hultsch, 2003). Adult song plasticity is perhaps most pronounced in species that mimic a wide range of sounds in their environment and consequently have very large repertoires, such as northern mockingbirds (*Mimus polyglottos*), brown thrashers (*Toxostoma rufum*), and lyrebirds (*Menura superba*). In mockingbirds, adult repertoires may include 400 or more song types, and increase in size with age (Derrickson, 1987). An adult brown thrasher may produce in excess of 1800 different types of sounds in his song, and continually improvise new sounds (Kroodsma,

1977b). In such species it is more parsimonious to hypothesize that pronounced age-related increases in song complexity are the outcome of improvisation or new auditory memorization than to assume that all of these signals were learned as juveniles.

(5) Disrupting auditory feedback can result in increased song variability in adult birds. Traditionally it was thought that closed-ended song learners do not require ongoing auditory feedback to maintain their crystallized songs, whereas adult open-ended learners do need to hear themselves to maintain song (Nottebohm, 1968; Konishi and Nottebohm, 1969; Nottebohm *et al.*, 1976). More recent studies, however, have shown that some closed-ended learning species do have to hear themselves in order to continue producing stereotyped songs as adults. Song begins to deteriorate within one week of deafening in Bengalese finches (*Lonchura striata domestica*), and after 2–8 weeks in zebra finches, depending upon their age (Nordeen and Nordeen, 1992; Woolley and Rubel, 1997; Lombardino and Nottebohm, 2000; see Woolley, this volume). Leonardo and Konishi (1999) showed that exposing adult zebra finches to delayed auditory feedback similarly leads to song deterioration after several weeks.

(6) Castration prevents the normal development of stable motor patterns of song. Nottebohm (1969) castrated a male chaffinch (*Fringilla coelebs*) at six months of age, before it developed crystallized song. When he treated this bird with testosterone at two years of age, it learned to copy a conspecific tutor song presented then. Song learning in gonadally intact chaffinches does not occur past the first year of life. Castration therefore extended the sensitive period for song development. Male swamp sparrows and song sparrows castrated by four weeks post hatching acquired a sensory model of conspecific song and went through the subsong and early plastic song stages of sensorimotor learning. These castrated birds, however, continued to produce variable song until they were implanted with testosterone, at which point they rapidly crystallized song (Marler *et al.*, 1988). When testosterone pellets were removed from these castrates, they quickly reverted to producing more variable song.

(7) Seasonal plasticity of the song system induces plasticity of song behavior. As discussed above, even in closed-ended learning species like song sparrows and white-crowned sparrows, there are pronounced seasonal changes in song. Song becomes more variable in structure and shorter in duration in nonbreeding birds. Also, male white-crowned sparrows can temporarily sing song types early in adult breeding seasons that had been produced as juveniles during the plastic phase of song learning but were deleted at the time of crystallization (Hough *et al.*, 2000). By the peak of the breeding season, however, the adult birds revert to producing only the one song type on which they crystallized as a juvenile.

Taken together, these observations indicate that plasticity is a common feature of adult songs in both closed-ended and open-ended species. Rather than representing two distinct adaptations, juvenile and adult song learning may represent a continuum of plasticity (Figure 27.8) (Brenowitz and Beecher, 2005). From such a perspective, closed-ended and open-ended song learning species can be regarded as differing in the degree of plasticity present in adult song, rather than in the presence or absence of plasticity. In closed-ended species, song becomes variable in structure outside the breeding season but then reverts to its previous stereotyped form in the following breeding season. In open-ended species, the plasticity of song structure outside the breeding season is more pronounced and can be exploited to develop entirely new song patterns. The observation that canaries develop more new songs outside the breeding season (Nottebohm and Nottebohm, 1978) is consistent with this suggestion.

An extension of this scenario is that seasonal plasticity of the song control circuits, which appears to be widespread among the songbirds and is therefore perhaps an ancestral trait, may have served as a preadaptation that enabled the evolution of adult song learning in some species. In this view, the synaptic plasticity that occurs with seasonal changes of the song system is necessary but not sufficient for adults to develop new songs. A broad comparative study of seasonal plasticity of song behavior and the song control system of a wider range of closed-ended and open-ended taxa would help to test the generality of this model, and will foster a better understanding of the adaptive value(s) conferred by adult song learning.

SUMMARY AND CONCLUSIONS

Seasonal plasticity of the song control system occurs in every species of seasonally breeding songbird examined thus far, including a tropical species that breeds on the equator. Figure 27.9 presents a summary of some of the many interactions between photoperiod, hormones, brain, and behavior that characterize seasonal growth. As daylength increases beyond a threshold level in late winter, the hypothalamic-pituitary-gonadal axis is stimulated. The testes begin to recrudesce and secrete increased levels of testosterone into the blood. Testosterone is transported to the brain, where it acts directly on cells in HVC via the androgen receptor. In addition, testosterone is metabolized to estradiol in adjacent areas of the caudal telencephalon, and the estradiol may also act on cells in HVC. Both androgenic and estrogenic hormones act on HVC predominantly to increase neuron number and the overall volume. Growth of HVC induces the growth of its efferent targets RA and X transynaptically, by the activity-dependent release of chemical neurotransmitters and/or neurotrophins. Volume increases of both RA and X largely result from increases in neuron size and spacing, but not in neuron number. As the song circuits grow and the syringeal muscles hypertrophy, songs become longer and more stereotyped in structure. There is also an increase in the rate of song production, perhaps due to the action of sex steroids on areas of the brain outside the song system that are associated with sexual and aggressive motivation. The neural activity associated with increased singing may reinforce or consolidate the growth of the song circuits by increased expression of the gene for BDNF in HVC neurons.

Regression of the song system after the breeding season may have evolved as an adaptation to reduce the energetic demands imposed by these regions of the brain at a time of year when song production is greatly reduced or absent. The neural and behavioral plasticity that occurs with seasonal changes of the song circuits may in turn have served as preadaptations for the evolution of adult song learning in some species of birds.

Many new questions have been raised for future research. Do androgenic and estrogenic metabolites of testosterone act in a *complementary* manner directly on HVC to stimulate growth of the song circuits, or is growth mediated predominantly by one of these two classes of metabolites? What is the nature of the transynaptic support provided by HVC to stimulate the growth of its efferent targets RA and X? To what extent does the neural activity underlying increased song production contribute to growth of the song circuits, and can behavioral contributions to neural plasticity even be addressed in an interpretable way given the complex interactions between hormones, neural circuits, gene

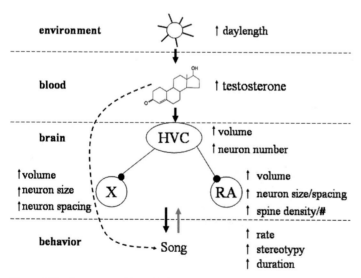

Figure 27.9 A summary of the main interactions between photoperiod, hormones, brain, and behavior that characterize seasonal growth of the song system. See text for explanation. (Modified from a figure by A. M. Wissman.)

expression, and song behavior? Why does the song system undergo extensive seasonal changes whereas other hormone sensitive brain systems remain stable throughout the year? Is there a functional relationship between seasonal plasticity of the song circuits and the ability of some species to develop new songs as adults? The pursuit of questions such as these will ensure that the song system continues to be a highly productive model for studying relationships between behavioral flexibility and plasticity of the adult brain.

ACKNOWLEDGEMENTS

This research was supported by NIH MH53032, MH 66939, and the Virginia Merrill Bloedel Hearing Research Center of the University of Washington. My thanks to Arthur Arnold, Michael Beecher, Adrian O'Loghlen, and David Perkel for comments on sections of the manuscript, and to Phil Zeigler and Peter Marler for helpful editorial comments.

28 • Regulation and function of neuronal replacement in the avian song system

Carolyn Pytte, Linda Wilbrecht, and John R. Kirn

INTRODUCTION

The remarkable complexity of the adult vertebrate brain requires that appropriate numbers of neurons are produced, that they find their way to the correct locations and that they adopt the appropriate phenotype – the latter including morphology, afferent and efferent connections, and neurotransmitter type. Most neurons are formed some distance from where they will ultimately reside and are often guided to their final destinations by other cells. During the latter stages of migration or, more typically, once young neurons have reached their final destination, they form elaborate dendritic and axonal arbors and establish connections. For the majority of neurons in birds and mammals, all of these steps are completed before or shortly after birth. However, it is now well established that in some cases neurons accomplish the same feats in the adult brain. In humans and other mammals, adult neurogenesis appears to be normally limited to the hippocampus and olfactory bulb (reviewed by Gage, 2002; Alvarez-Buylla and Lim, 2004; Abrous et al., 2005; Benedetta et al., 2006; Lledo et al., 2006; but see Gould and Gross, 2002). In songbirds, however, neurons formed in the juvenile and adult brain are incorporated throughout much of the telencephalon (Figure 28.1) where they replace other neurons that have died (reviewed by Alvarez-Buylla and Kirn, 1997; Wilbrecht and Kirn, 2004; Nottebohm, 2004; see also Nottebohm, this volume).

The mechanisms which control posthatching neurogenesis and the functions that it serves are not well understood. While new neurons are probably added to the telencephalon of all birds (Nottebohm, 1985b; Ling et al., 1997), songbirds possess the rare ability to learn vocalizations, and we can trace the formation and insertion of neurons formed in juveniles and adults into discrete and easily identifiable nuclei that are dedicated to the learning, production and perception of song. Moreover, a major cell type that undergoes renewal forms part of the efferent pathway for song motor control, implicating neuron addition in the overt expression of a behavior. This unprecedented neural plasticity in a warm-blooded vertebrate that is linked to a well-defined and quantifiable behavior makes the avian brain a valuable model that both challenges existing concepts of brain function and may provide insights relevant for advancing methods of brain repair. In this chapter we review recent work on the posthatching addition and replacement of neurons in the song system and discuss experiments designed to elucidate the mechanisms that control neuronal replacement and its possible functions. With respect to function, we speculate unapologetically, in part because that is the most exciting aspect of doing science and also with the hope of stimulating further thinking and research. For additional perspectives see reviews by Banta-Lavenex et al., 2001; Gahr et al., 2002; Nottebohm, 2002; see also Brenowitz, and Nottebohm, this volume).

A BRIEF DESCRIPTION OF SONG DEVELOPMENT AND MAINTENANCE

Although all oscine songbirds studied to date rely on auditory feedback to develop normal song, there are thousands of different songbird species and they vary in several ways, including the number of songs they sing, song complexity, the age at which they acquire songs, and the frequency and context in which they sing. The subjects in our studies, the zebra finch and the canary, both copy songs they have heard as juveniles from adult conspecifics, though they learn and produce songs on very different schedules. The learning process first involves the formation of an auditory memory or "template" of the song or songs that they will imitate, and this takes place during a sensitive period for song learning (for review, see Marler, 1997; and Adret, this

---

Neuroscience of Birdsong, ed. H. Philip Zeigler and Peter Marler. Published by Cambridge University Press. © Cambridge University Press 2008.

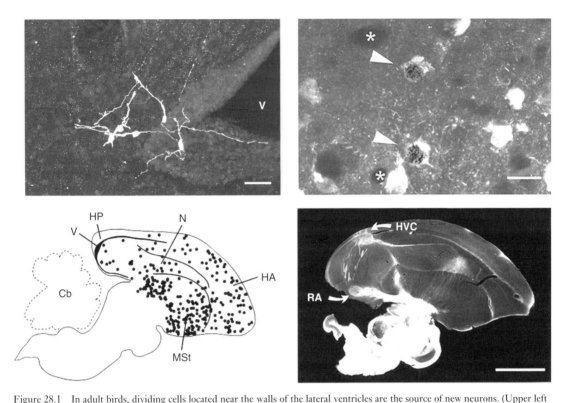

Figure 28.1 In adult birds, dividing cells located near the walls of the lateral ventricles are the source of new neurons. (Upper left panel) Dividing cells near the ventral tip of the lateral ventricle (V) in an adult zebra finch male were infected with a replication-incompetent retrovirus expressing green fluorescent protein (GFP). GFP-expressing daughter cells, some with a migratory phenotype, can be seen. (Photo courtesy of Kate Stellitano.) From the ventricular zone, young neurons migrate long distances and are incorporated throughout the telencephalon. (Lower left panel) Reconstruction of incorporation sites for new neurons in adult canary brain in sagittal view. Rostral is to the right. Each dot represents a single [^{3}H]- labeled neuron in a canary that had received two injections of [^{3}H]-thymidine/day for 14 days 1 month prior to sacrifice. (Modified from Alvarez-Buylla et al., 1994.) Many neurons incorporated into HVC become RA-projecting cells. (Upper right panel) Two HVC-RA neurons formed in adulthood (arrows) are shown. These neurons have autoradiographically developed silver grains overlying their nucleus following systemic injections of [^{3}H]-thymidine, and Fluoro-Gold in their cytoplasm following injection of this retrograde tracer into RA. Asterisks label capillaries. (Lower right panel) Darkfield sagittal view of an unstained section from canary brain showing motor control regions HVC and RA. White streaks represent bundles of axons coursing from HVC to RA, a distance of 2–3 mm. Rostral is to the right. (From Kirn et al., 1991.)

Abbreviations: HA, hyperpallium accessorium; HP, hippocampus; HVC, used as a proper name; RA, robust nucleus of the arcopallium; MSt, medial striatum; N, nidopallium; V, ventricle; CB, cerebellum. CB and other subtelencephalic regions do not receive new neurons in adulthood. (Nomenclature based on Reiner et al., a, c.) Scale bars: upper left 50 μm, upper right 10 μm, lower right 3 mm.

volume). In the zebra finch, attempts at imitation can occur in month-old juveniles as soon as a few hours after song model presentation (Tchernichovski et al., 2001). In juvenile canaries, songs heard in the first summer are imitated in the following fall and spring (Waser and Marler, 1977; Nottebohm et al., 1986; Weichel et al., 1986). The sensory-motor phase is thought to involve a comparison between expected and received auditory feedback during singing and this comparison is used to shape vocal output until a good match is made between the bird's song and auditory memories of a tutor's song.

The sensitive period for song learning is thought to end when songs become stereotyped or "crystallized." Despite the fact that zebra finches can imitate some sounds a few hours after presentation, the natural

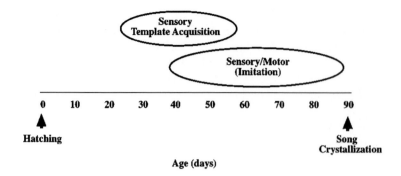

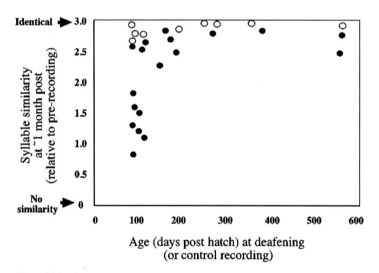

Figure 28.2 Approximate time-course for song learning in the zebra finch (upper panel). The sensory phase (template acquisition), when males memorize the songs of adult conspecifics, begins early and overlaps with the sensory/motor phase, where the template is used to guide vocal imitation. The sensitive phase for song learning is completed by 90 days of age, and song remains largely unchanged thereafter (Immelmann, 1969: Eales, 1985). However, even song maintenance may be an active process relying on auditory feedback and this reliance wanes with increasing age or practice (lower panel). Shown are changes in song syllable morphology between songs recorded predeafening and 1 month postdeafening (filled circles) or over a similar interval in hearing male zebra finches (open circles) as a function of bird age. Note that in young adults (100–200 days of post hatch) deafening has a much more dramatic impact on song compared with birds deafened at older ages. (From Brainard and Doupe, 2001.) It should be noted that deafening at any age probably leads to the eventual deterioration in song structure but the latency increases and the magnitude of deterioration decreases as adults age (Lombardino and Nottebohm, 2000). (From Wilbrecht and Kirn, 2004, reprinted with permission from Blackwell Publishing.)

course of song learning in this species takes approximately two months, and song becomes stereotyped approximately 90–100 days after hatching (Figure 28.2). In the canary, song becomes stereotyped during the first breeding season, six months to a year after hatching (Nottebohm et al., 1986; Weichel et al., 1986). Zebra finches learn a song once during this initial 90 days after hatching and then continue to sing it in a highly stereotyped manner throughout their lifetime (Immelmann, 1969; Eales, 1985), while canaries modify their song patterns seasonally as adults (Nottebohm et al., 1986; Leitner et al., 2001a). Thus, the zebra finch represents an example of an "age-limited" learner while the canary is an example of an "open-ended" learner.

Adult male canaries can have 30–40 or more distinct syllables in their repertoire and they modify their songs on a seasonal basis (Güttinger, 1985; Nottebohm et al., 1986). Between year 1 and 2 after hatching in laboratory-housed birds, approximately one third of the syllables are dropped and replaced by new ones, one third are modified, and the remainder are unchanged (Nottebohm et al., 1986). Although there is clear evidence that adult canaries change their songs, and can copy the songs of other males (Güttinger, 1979), the extent to which seasonal song modifications reflect the copying of new syllables from conspecifics, improvisation, or the selection of different subsets of syllables from a "library" of songs learned during development (Hough et al., 2000; Leitner et al., 2001a) is unknown (reviewed by Marler, 1997). This is an important issue because the neural mechanisms underlying these different scenarios may vary in significant ways. While all of these scenarios are likely to share a need for auditory feedback and, therefore, a sensory-motor mapping process in the brain, this may be where commonalities end. For example, if birds are selecting from a subset of previously learned songs, then perhaps what is being learned in such cases is an association between a particular song and a particular social context (Marler, 1997). Moreover, it seems unlikely in this case that entirely new auditory memories are established to guide vocal change as would be expected during new song learning.

In some species, the maintenance of a stereotyped song may be as active a process as song learning. Song learning requires intact hearing, which is used not only to acquire a model but also to monitor and guide vocal output (Price, 1979; Burek et al., 1991; Scharff and Nottebohm, 1991). However, even after song crystallization and throughout adulthood, deafening or distorting auditory feedback results in deterioration of song structure (Nordeen and Nordeen, 1992; Okanoya and Yamaguchi, 1997; Woolley and Rubel, 1997; Leonardo and Konishi, 1999; Wang et al., 1999b; Cynx and von Rad, 2001; for review, see Konishi, 2004, and Woolley, and Brainard, this volume). This suggests that, as with the learning of new song elements, song *maintenance* relies on comparisons between expected and received auditory feedback during singing. Mismatches between expected and received auditory feedback may create error signals (or the withholding of reinforcement signals) (for review, see Margoliash, 2002) used to make adjustments to motor commands in order to maintain a stereotyped song. Recent work has shown that as in human speech (Waldstein, 1990; also see Doupe and Kuhl, this volume) the reliance of adult song on auditory feedback wanes with increasing age, suggesting that the motor program for song becomes increasingly stable with age, singing experience, or both (Lombardino and Nottebohm, 2000; Brainard and Doupe, 2001) (Figure 28.2). The possibility that postcrystallization singing experience contributes to age-related increases in motor program stability is intriguing because it would indicate that even in so-called 'age-limited' learners, song development is a lifelong process (Lombardino and Nottebohm, 2000).

Although song stereotypy after crystallization is high even in young adults, there are subtle changes in song with age or cumulative singing experience. For example, with increasing age, individual song syllables become less "noisy" and variable in fine structural detail and are delivered more rapidly by zebra finches (Arnold, 1975a; Brainard and Doupe, 2001; Pytte et al., 2007). Perhaps reproducing notes rapidly and with precisely the same duration is a difficult skill that improves with protracted (lifelong?) practice.

NEUROGENESIS AND THE AVIAN SONG SYSTEM

The dynamic nature of song learning and song maintenance suggests a high degree of plasticity in the neurons that make up the song system (see Figure 5.2. in Reiner et al., this volume). However, no one expected how dramatic neural plasticity in the songbird brain might be until Steve Goldman and Fernando Nottebohm (Goldman and Nottebohm, 1983) discovered that neuron addition and replacement continues in the song system well into adulthood (see Nottebohm, 2004, and this volume).

Vocal control regions that receive neurons after hatching and the neuronal life cycle

The most detailed information about neuron addition and loss has been obtained from studies of zebra finches and canaries. Area X and HVC are the only vocal control regions known to exhibit large-scale neuron addition both after hatching and throughout adulthood (reviewed by Alvarez-Buylla and Kirn, 1997). HVC, part of the posterior (sensory-motor) pathway, is necessary for song production at all ages, while Area X and the anterior forebrain pathway are critical for the acquisition of song in juveniles (Nottebohm et al., 1976; Bottjer et al., 1984; Sohrabji et al., 1990; Scharff and Nottebohm, 1991), and may also play a role in song maintenance in adults (Williams and Mehta, 1999; Brainard and Doupe, 2000a; Kao et al., 2005), also see Brainard, this volume). A third region, the caudomedial nidopallium (NCM), an auditory region that has indirect projections to the vocal control system (reviewed by Mello, 2002a; also see Mello and Jarvis, this volume), also receives new neurons in adulthood (Alvarez-Borda, 2002; Lipkind et al., 2002). NCM appears to store song-specific auditory information of potential use by juveniles during song acquisition and, perhaps, song maintenance by adults (Bolhuis and Gahr, 2006; Phan et al., 2006). Relatively little is known about neuron addition to Area X or NCM (see Sohrabji et al., 1993; Alvarez-Borda, 2002; Lipkind et al., 2002). However, because HVC has been implicated in song production and perception (Nottebohm et al., 1976; Brenowitz, 1991; Gentner et al., 2000; Gentner, 2004; reviewed by Alvarez-Buylla and Kirn, 1997) neuronal recruitment in this nucleus has been extensively studied and will be the main focus of this chapter.

In both canaries and zebra finches, HVC is not fully developed at hatching but adds substantial numbers of new neurons during juvenile life as shown in Figure 28.3 for the zebra finch. Neurons formed post hatching in songbirds arise exclusively from cells in the ventricular zone (VZ) lining the walls of the lateral ventricles (Alvarez-Buylla and Nottebohm, 1988; Figure 28.1). From their birth site, young neurons migrate considerable distances through mature brain tissue to reach their final destination. Migration is aided by the retention into adulthood of a matrix or scaffolding comprised of the long processes of radial glia, upon which young neurons move (Alvarez-Buylla and Nottebohm, 1988; Alvarez-Buylla et al., 1988a). Neuronal migration also follows a path perpendicular to that associated with radial glia, known as tangential migration. In this mode, cells migrate near or within the walls of the lateral ventricles for varying distances before dispersing along radial glial fibers (Doetsch and Scharff, 2001). Cell proliferation can be measured by sacrificing birds less than 24 hours after injection of [^{3}H]-thymidine or bromodeoxyuridine (BrdU). Both of these markers are taken up by dividing cells and can be used to identify a cell's birth date. Survival times of less than 24 hours provide enough time for the cell birth dating molecule to become incorporated into dividing cells and, at the same time, this interval is too short for most labeled daughter cells to have migrated away from the VZ. In zebra finches, proliferation in the VZ is higher in juveniles than in adults, although it is not known whether this is true specifically for the production of neurons destined for HVC (DeWulf and Bottjer, 2002). Therefore, it is unclear whether the high incorporation rates in HVC during this time are due to cell production or survival. Regardless, in both species, high rates of neuron addition to HVC occur at the very time song is initially learned.

Once HVC has reached adult size in both species, total HVC neuron number does not increase further with increasing age, yet new HVC neurons continue to be produced and added throughout life (Alvarez-Buylla et al., 1988b, 1990; Kirn et al., 1991; Wang et al., 2002), indicating that neuron addition is coupled with neuron loss. Pyknotic, degenerating cells have been found in HVC at all ages examined from 10 days after hatching to adulthood (Kirn and DeVoogd, 1989; Kirn et al., 1994; Burek et al., 1997; Tekumalla et al., 2002). Thus, neuron addition and loss occur throughout posthatching life, but during the juvenile growth phase neuron addition surpasses loss and once adult neuron numbers have been attained, these two processes are more closely matched. The balancing act between net addition and loss may also be regulated seasonally in adults of some species. In western song sparrows total HVC neuron number in the fall is only about two thirds the number found in spring (Tramontin and Brenowitz, 1999). Therefore, in some birds, the balance appears to alternate between net cell addition and loss at various times throughout life.

Canaries also show seasonal changes in neuron addition (Alvarez-Buylla et al., 1990a; Kirn et al., 1994), which may reflect differential survival, rather than production, of new neurons. Though roughly similar numbers of new HVC neurons are initially incorporated in the spring and fall, considerably more are preserved within the fall-born cohort compared with neurons formed in the spring (Alvarez-Borda et al., 2004). Cell death probably plays a pivotal role in the regulation of neuronal replacement. In canaries,

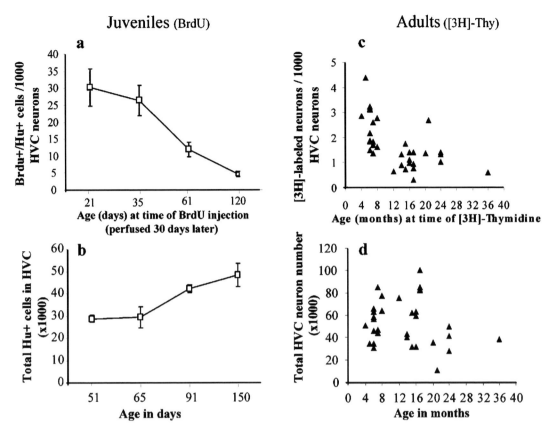

Figure 28.3 Neurons labeled with the cell birth marker BrdU are added to HVC in juvenile zebra finches (a) as the total number of neurons, Hu + cells, grows to adult levels (b). After day 150 the number of new neurons (identified using [^{3}H]-thymidine and retrograde labeling from RA (not shown) or Nissl stain) added to HVC declines with age (c) and total HVC neuron number remains stable (d). Neuronal replacement probably occurs throughout life; however, the juvenile stage is dominated by net addition whereas addition and loss are more closely matched in adulthood. Values in left panels are means ± SEM and the data are from Wilbrecht et al. (2002b). In panels to the right each symbol represents 1 bird and the data are from Wang et al. (2002). The use of different methods of quantification in the two studies prevents direct numerical comparisons. Nevertheless, it is clear from these data that HVC neuron recruitment is higher during song acquisition than after song crystallization and that in adulthood, neuron recruitment continues to decrease with increasing age. (From Wilbrecht and Kirn, 2004, reprinted with permission from Blackwell Publishing.)

seasonal peaks in HVC neuron recruitment are preceded by peaks in cell death (Kirn et al., 1994). Moreover, the selective killing of HVC neurons leads to a subsequent increase in the incorporation of new neurons of the same kind (Scharff et al., 2000). It follows by extension that for replaceable neuron types, the number of healthy neurons present can, at least under some conditions, regulate and even constrain the number of new cells that are added. In this model, new neuron addition is contingent on available synaptic space in HVC and, perhaps, RA (for the axon terminals of HVC-RA projection neurons). Periods of high cell death (such as late summer–fall in some seasonal breeders) may make room for a wave of new neuron addition. This wave of addition continues until available circuit space is occupied once again (Kirn et al., 1994; Scharff et al., 2000; also see Nottebohm, this volume).

Neuron types formed after hatching

Area X neurons produced after hatching are exclusively interneurons (Sohrabji et al., 1993). At least two neuron types are replaced in HVC. HVC has two types of projection neurons. HVC neurons projecting to Area X (HVC-X) are produced before hatching and are not replaced in adulthood (Alvarez-Buylla et al., 1988b; Gahr, 1990b; Kirn et al., 1999; Scharff et al., 2000). In contrast, the majority of adult-formed HVC neurons send projections to RA (Figure 28.1) and the rest are interneurons (Alvarez-Buylla et al., 1988b, 1990a; Nordeen and Nordeen, 1988; also see Nottebohm, this volume). Thus, many adult-formed HVC neurons perform the remarkable feat of migrating from their birth place in the walls of the lateral ventricles to HVC and then sending an axon 2–3 mm away to their target cells in RA. Perhaps even more remarkable is the highly specific temporal firing patterns of HVC-RA projection neurons. These neurons exhibit discrete and selective bursts of premotor activity during singing (Hahnloser et al., 2002). In the zebra finch, as few as 20–40 HVC-RA projecting neurons participate in each 6 ms portion of song (Hahnloser et al., 2002). How loss, replacement, and "training" of these neurons are orchestrated is a fascinating question.

What fraction of HVC neurons is replaced and how long do replaceable neurons live?

In the canary, 50–60% of all HVC neurons project to RA, roughly 20% project to Area X (HVC-X) and the remainder are interneurons (Alvarez-Buylla et al., 1988b; Nottebohm et al., 1990; Kirn et al., 1991, 1999). Replaceable RA-projecting neurons represent a large fraction of the pathway connecting HVC to RA. Roughly 50% of this pathway is replaced over 6 months from spring to fall in adult canaries (Kirn and Nottebohm, 1993). We do not know whether there is complete replacement in 12 months, nor do we know when these dying cells were formed after hatching. However, these results suggest that a large fraction of the dominant cell type in HVC undergoes loss and replacement.

The life-span of neurons formed after hatching can vary from days to months. This is known from studies in which adult birds were first injected with the cell birth marker [^{3}H]-thymidine and then sacrificed at different intervals in order to follow the development of a specific labeled cohort of neurons. The death of new neurons can be inferred indirectly by changes in their numbers, or more directly by counting degenerating cells, which have a characteristic appearance that distinguishes them from healthy cells using several conventional stains. With this approach it has been shown that some cells die even while migrating (Alvarez-Buylla and Nottebohm, 1988). The journey from the ventricular zone to HVC can take as little as one week and among those cells that reach HVC, many more die between the ages of 2 and 3 weeks, at the time when their axons can first be retrogradely labeled from RA (Burek et al., 1994; Barami et al., 1995; Kirn et al., 1999). New HVC neurons that survive to 4–5 weeks of age have life-spans of 4 to 8 months or longer (Kirn et al., 1991; Nottebohm et al., 1994; Wang et al., 1999b; 2002). These results suggest that adult-formed neurons can have substantial life-spans and, therefore, opportunities to contribute to brain function, but that their survival is challenged at many stages during their lifetime.

The functions of vocal control neuron replacement

While much is known about the basic kinetics of neuronal replacement, the most intriguing and poorly understood puzzle may be why neuronal replacement occurs at all. Among vertebrates, there are many instances in which neuron addition is accompanied by substantial overall postnatal body growth (Zupanc and Horschke, 1995), which is not surprising if the ratio of central neurons to sensory receptors and muscle fibers needs to be conserved. However, the function of ongoing neuron addition and replacement long after animals have reached maximum size is not so obvious.

It is possible that adult neurogenesis, particularly in birds like the zebra finch that normally do not learn new songs in adulthood, has little adaptive significance. The recruitment of new neurons to the telencephalon (including the caudal nidopallium and medial striatum) could simply be a vestigial characteristic enduring from an early vertebrate ancestor in which adult neurogenesis did serve a function (such as one related to continual growth of the animal). Similarly, perhaps HVC and Area X evolved in these particular neurogenic brain regions for reasons other than their capacity to incorporate new neurons, and thus do not require or use this phenomenon for any aspect of song learning or song production. A related argument is that adult neurogenesis may be an epiphenomenal holdover from brain development. The genesis of some vocal control regions is delayed relative to surrounding regions; however, in a broader sense there is a general tendency for most vocal control nuclei to follow the same developmental timetable as surrounding regions. Arcopallium develops relatively early as does RA and neither receive neurons in adulthood, whereas HVC and Area X exhibit extensive adult neuron incorporation as do surrounding nidopallium and medial striatum, respectively (Kirn and DeVoogd, 1989; Konishi and Akutagawa, 1990; Alvarez-Buylla et al., 1994). The recruitment of new neurons to HVC in juveniles may serve only to populate the brain with sufficient neurons to control the later expression of adult song. Once adult neuron numbers have been achieved in zebra finches, relatively few new neurons are added per day compared with the case in juveniles, and these numbers may be low enough that they have no functional consequences. Within either an evolutionary or developmental scenario, the idea of continual neuron production in the absence of an adaptive function implies that there simply has not been sufficient selective pressure to curtail adult neurogenesis.

Perhaps then, the first question to be answered is whether there is sufficient evidence to entertain the notion that new neurons play *any* functional role in song. If they do serve some song-related function(s), then (a) new neurons ought to become active during singing or in response to song playbacks, (b) if the specific neuron types produced in adulthood are destroyed, song structure and/or perception should be disrupted and (c) if these cells are replaced with newly formed neurons of the same type, this disruption should be reversed.

The first of these predictions is supported by the finding that adult-formed HVC neurons respond with action potentials to acoustic stimulation (Paton and Nottebohm, 1984). The now classic experiment that showed this was exceedingly difficult for the following reasons. In adult canaries, as is likely true of most songbirds, relatively few new HVC neurons are formed per day (0.3–0.5%) (Alvarez-Buylla et al., 1990a; Kirn et al., 1991) and with the cell birth dating techniques available at that time, these cells could only be identified post mortem. Therefore, to determine whether new neurons are synaptically connected and physiologically active, these researchers had to do random intracellular recordings from HVC neurons in birds preinjected with $[^3H]$-thymidine, with the hope that a few of the cells recorded from would be identified as new neurons after sacrifice. The researchers also had to be able to identify the cells they had recorded from and so after recordings, these cells were filled with horseradish peroxidase (HRP) so that they could be identified in fixed tissue. As might be expected, there were many cells from which detailed recordings were made but that were not labeled by $[^3H]$-thymidine and, therefore, their age was unknown. However, in a few cases, electrophysiological recordings were made from cells that were also shown to be labeled by HRP and $[^3H]$-thymidine. Most importantly, these cells responded to sound presentation with excitatory postsynaptic potentials and action potentials and this provided the necessary evidence for functional integration of adult-formed neurons.

What about the other two predictions supporting a functional role for adult neuron replacement? That is, if one selectively kills adult-formed neurons, is there a disruption in song and, if so, does song recover following replacement of lost cells? This too is a difficult question to answer for two reasons. One major hurdle is the technical problem of how to kill these cells. HVC lesions would accomplish this but would also kill cells that project to Area X. A laser ablation technique (Macklis and Madison, 1985) provided the answer. This method employs a conventional retrograde tracer that has attached to it the molecule chlorin e_6. When this tracer is injected into the brain, only neurons with axon terminals in the injection site will retrogradely transport the tracer along with the chlorin e_6. The chlorin e_6 is relatively benign on its own. However, when excited by 670 nm wavelength laser light, a chemical reaction is activated, the result of which is the production of singlet oxygen and free radicals that kill the cell (Macklis and Madison, 1985; Madison et al., 1990; Sheen et al., 1992). This method permits the selective targeting of HVC-RA neurons for death. However, we don't know whether the HVC-RA neurons formed specifically during adulthood are a specialized subclass of this projection neuron type, and thus whether laser ablation also kills neurons that are not normally replaced. We cannot rule this out, although work showing that 50–60% of all HVC-RA neurons are lost and replaced over 6 months (Kirn and Nottebohm, 1993) suggests that most, if not all, of these neurons are replaced at least once over a bird's lifetime. When HVC-RA neurons were selectively killed by laser ablation in adult zebra finch males (Scharff et al., 2000), this procedure resulted in a dramatic disruption in song acoustic structure, followed by varying amounts of recovery. These birds were injected with [^{3}H]-thymidine shortly after laser treatment, followed by retrograde labeling of HVC-RA neurons with Fluoro-Gold (FL), a tracer recognizably distinct from the one used to deliver the chlorin e_6, shortly before sacrifice. The number of [^{3}H]-FL cells provided an index of HVC-RA neuron replacement post laser. It was found that the death of HVC-RA neurons was followed by a dramatic compensatory replacement by neurons of the same type and the timetable for neuronal replacement paralleled that for behavioral recovery. This work strongly suggests that neuronal loss and replacement can have important functional consequences. However, it does not help us understand why new cells should need to be added to existing, presumably competent, circuits.

Studies of the function of new neurons in the song system initially focused on their potential role in song learning. However, more recent work has provided correlational evidence suggesting a role for new neurons in lifelong song production and song maintenance. Here we discuss some hypotheses regarding the potential functions of postnatal neurogenesis and neuronal replacement. These include the possibilities that neurogenesis and neuronal replacement serve functions related to (a) perceptual learning, (b) motor activity and the replacement of damaged or damage-prone, premotor neurons, (c) sensory-motor control of song learning, and (d) the establishment of song motor stereotypy.

Neuronal replacement and perception
Three kinds of evidence suggest a role for new neurons in song-related perceptual function. First, new neurons are found in the auditory association area NCM (Alvarez-Borda, 2002; Lipkind et al., 2002). Moreover, neurons in Area X and HVC show both auditory and motor related activity (Katz and Gurney, 1981; Bankes and Margoliash, 1993; Solis et al., 2000; Mooney et al., 2002), and lesions to Area X, HVC and surrounding tissue disrupt auditory discrimination (Brenowitz, 1991; Scharff et al., 1998; Gentner et al., 2000). All of these findings raise the possibility that new neurons are involved in perceptual functions, including perceptual learning. In addition, as noted earlier, physiological recordings from adult-formed HVC neurons show that they respond to sound (Paton and Nottebohm, 1984). Finally, rates of neuronal recruitment appear to covary with perceptual demands. For example, neuronal replacement is correlated with the richness of the auditory environment. When compared with birds housed singly or in pairs, birds housed in complex social groups recruit more neurons to HVC, Area X and NCM (Lipkind et al., 2002). Moreover, deafening alters new neuron recruitment to HVC (Wang et al., 1999a), perhaps by interfering with the learning or perception of the songs of other birds.

These findings are suggestive of a link between neuron addition and perceptual functions, including the formation of perceptual memories of the songs of other birds. This is a particularly attractive hypothesis for zebra finches, which live in large groups whose membership changes frequently (Zann, 1996). Neurogenesis could provide the substrate for updating perceptual memories of the songs of conspecifics (Nottebohm et al., 1990). One prediction of this model would be that deafening would block the stimulatory effects of complex social housing on neuron addition. However, social enrichment may affect many variables including, but not limited, to perceptual learning and so this hypothesis requires further study.

Neuronal replacement is a mechanism for renewal of damaged or damage-prone neurons
Physiological recording experiments in singing birds show that RA-projecting HVC neurons fire at high rates for brief intervals during singing (Hahnloser et al., 2002), and many species are renowned for their high song production rates. Perhaps the high metabolic demands associated with repeated use result in short neuronal life-spans. Thus one possible function for neurogenesis and neural replacement may be to replace premotor HVC neurons that become damaged by use. If neurogenesis serves to replenish damage-prone neurons, then one would expect a correlation between rates of neuron incorporation and amount of singing. Singing rate in zebra finches peaks during the plastic phase of song learning in juveniles, declining to lower levels in adulthood and this correlates with changes in HVC neuron incorporation rates (Nordeen and Nordeen, 1988; Johnson et al., 2002; Wilbrecht et al., 2002b). Naturally occurring individual differences in amount of singing are also positively correlated with

differences in HVC neuron incorporation (Alvarez-Borda and Nottebohm, 2002). Moreover, in the only direct test of this association, it was found that suppression of singing (by distracting birds every time they began to sing) resulted in decreased incorporation of new HVC neurons in adult canaries (Li *et al.*, 2000). It is difficult to manipulate singing in a way that does not also alter other variables of potential importance in the regulation of neuron recruitment. Nevertheless, these results are all consistent with the proposal that singing regulates new neuron incorporation.

However, work with birds that sing seasonally suggest that this hypothesis alone is insufficient. Singing in canaries and western song sparrows is highest in the spring mating season, a time when HVC neuron incorporation is much lower than in the fall, when amount of song is reduced (Alvarez-Buylla *et al.*, 1990a; Tramontin and Brenowitz, 1999). If, however, one focuses on the *survivorship* of new neurons once they are incorporated into HVC, life-spans of neurons formed in spring, at least in canaries, are shorter than those of neurons born in the fall (Nottebohm *et al.*, 1994), which would be consistent with this model. A more serious problem for this hypothesis is that RA neurons, which receive direct projections from HVC (as well as neurons downstream from RA in the motor pathway) are likely to experience similar, singing-related metabolic challenges, yet they are *not* replaced in adulthood.

Neuronal replacement is necessary for song learning
A hypothesis first advanced and later refined by Nottebohm (Nottebohm, 1989, 2002) proposes that as a cell ages it becomes progressively less plastic in a manner analogous to cellular differentiation and commitment during development. If true, it follows that an evolutionarily adaptive strategy would be to build increasingly larger brains, thus providing a surplus of cells for adaptive plasticity in long-lived animals. He reasoned further, however, that an alternative strategy to achieve the same goal, particularly for animals that place a premium on minimizing body weight such as birds, would be to have a smaller brain and to discard and replace old neurons that have limited potential for acquiring new information and that encode information that is no longer relevant to the animal. Thus, one would predict that neuronal replacement should be highest at times when birds are learning new song, and the recruitment of new HVC neurons should be different in species with different learning trajectories. Indeed, in zebra finches, more new HVC neurons are added during the sensitive phase for song learning than after song is crystalized (Nordeen and Nordeen, 1988b; Wilbrecht *et al.*, 2002b) (Figure 28.4). Moreover, in zebra finches individual differences in the total number of HVC neurons by the time song has fully developed are positively correlated with the degree to which birds had imitated their tutor's song (Ward *et al.*, 1998). After song crystallization, new neurons continue to be added to the zebra finch HVC but at lower levels. In adult canaries, new HVC neurons are added at the highest rate during the fall when song modifications are greatest (Kirn *et al.*, 1994) (Figure 28.4). Another peak in new neuron recruitment is seen in the early spring, when there is a second, more modest peak in song modification. New neuron recruitment is notably lower in the summer and mid-winter months when song is modified little.

If new neuron recruitment is closely tied to and regulated by song learning, then blocking imitation during the sensitive period for song learning should disrupt the recruitment of new neurons. When imitation was prevented in juvenile zebra finches either by deafening, bilateral lesions of nucleus LMAN, or bilateral denervation of the syrinx, they were unable to imitate their tutor's song at day 90. Nevertheless, however poor their ability to imitate, they recruited a normal number of HVC neurons between day 61 and day 91 (Wilbrecht *et al.*, 2002a, b; Figure 28.5, left panel). This suggested that there was little relation between the song imitation process and new neuron recruitment to HVC. However, one group that sustained only a unilateral denervation of the syrinx at day 26 recruited nearly twice the number of new HVC neurons on the intact side as controls over the same day 61 to day 91 period (Wilbrecht *et al.*, 2002a, b) (Figure 28.5, right panel). Interestingly, these birds *were* capable of making an imitation of their father's song. When the unilateral denervation was combined with bilateral LMAN lesions or deafening, song learning and augmented neuron recruitment were blocked (Figure 28.5, right panel), suggesting that the unilateral increase in new neurons was caused not only by unilateral denervation, but by learning a song with input to only one side of the syrinx. These data suggest that the process of imitating song under the abnormal conditions of unilateral

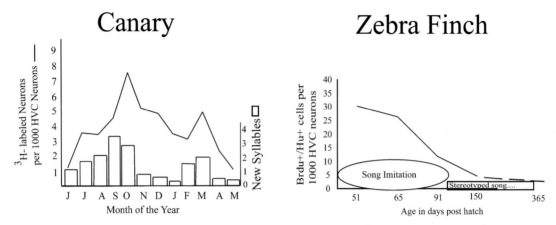

Figure 28.4 Schematic of new neuron recruitment in the canary and zebra finch. More new neurons are found in HVC during the periods when song is being learned or modified. This is age-limited in zebra finches, and open-ended and seasonal in canaries. Data for canaries are from Kirn *et al.* (1994). Data for zebra finches day 51–150 are from Wilbrecht *et al.* (2002b) and days 150–365 are roughly extrapolated from Wang *et al.* (2002) taking into account differences in number of injection days. The use of different methods in the experiments precludes direct species comparisons of neuron recruitment rates. However, in adult canaries between 1–2 years old, HVC neuron recruitment rates at times of year when song is relatively stereotyped are similar to those for 8–13 month old zebra finch males (Alvarez-Buylla *et al.*, 1990a). (From Wilbrecht and Kirn, 2004, reproduced with permission from Blackwell Publishing.)

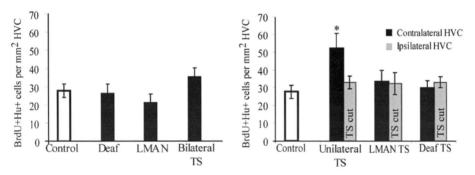

Figure 28.5 (Left panel) New neuron recruitment into the zebra finch HVC after imitation was blocked by various surgeries at day 26. BrdU was given on days 61–65 and birds were killed on day 91. All of the experimental birds represented in black made poor imitations of their tutor's song at day 90, yet they had normal levels of new neuron recruitment. (Right panel) Under the same conditions as in the left panel, the intact side HVC of unilateral tracheosyringeal (TS) nerve cut birds had nearly twice the number of new neurons as control HVCs, and they made successful imitation of their father's song. When unilateral TS cuts were combined with bilateral LMAN lesions or deafening, also at day 26, this effect at day 90 disappeared. Error bars = SEM. Data are from Wilbrecht *et al.* (2002a, b). (From Wilbrecht and Kirn, 2004, reprinted with permission from Blackwell Publishing.)

syringeal control *can* affect neuronal recruitment to HVC. As the effect of unilateral denervation was only seen at the latest stages of song learning (Wilbrecht *et al.*, 2002b; data not shown), a time normally associated with the transition from high to much lower rates of HVC neuron recruitment, it could be that imitation under these difficult circumstances prolonged the song learning process, as well as the normal period of high cell recruitment. It is known that the sensitive period for song learning can be extended by isolating zebra finches from other birds (Eales, 1985; Morrison and Nottebohm, 1993). Interestingly, the normal rate of decline in new neuron recruitment to HVC roughly between 60 and 90 days after hatching is attenuated in isolated zebra finches until day 150 and there is a correlation between the number of neurons incorporated in HVC and variability in song 25–30 days after the neurons were labeled (Wilbrecht et al., 2006).

In summary, neuron addition in juvenile zebra finches can occur at normal rates under some conditions that block song learning. However, manipulations that *do* alter neuron recruitment also affect song learning. These results suggest that new neuron addition to HVC may largely be permissive for song plasticity, while the process of song imitation has conditional effects on neuronal recruitment and replacement that we do not yet fully understand. Studies of neuronal replacement in both juveniles and adults may ultimately show that neuronal replacement is necessary but not sufficient for song learning; however, current data cast doubt on the simple equation between new neurons and new song learning.

Neuronal replacement provides motor flexibility for the achievement and maintenance of song stereotypy

The relationship between new neurons and new song learning is weakened further by the observation that new neurons continue to be added in zebra finches after song crystallization. It also does not explain data from field studies of the western song sparrow, a species which, like the zebra finch, learns songs only in its first year of life, but which, like the canary, annually recruits more HVC neurons in the fall than in the spring (Tramontin and Brenowitz, 1999; see Brenowitz, this volume). Song sparrows sing more variable song in the fall than in the spring suggesting that in this species, and perhaps all species, song instability, rather than song learning, may be a better correlate of high HVC neuron recruitment levels (Tramontin and Brenowitz, 1999). Even in the adult canary, it has been shown that some syllables are retained from one year to the next, although they go through a period of low stereotypy in the fall (Nottebohm et al., 1986; Tramontin and Brenowitz, 2000; Leitner et al., 2001a). Of course, adult song should become less stereotyped as it goes through changes directed by learning. In this sense new neurons may provide the variability upon which a new song can be sculpted. Perhaps the magnitude of seasonal changes in neuron recruitment varies with the extent to which song is modified. Seasonal changes in neuron recruitment appear to be substantially smaller in song sparrows than in canaries (Alvarez-Buylla et al., 1990a; Kirn et al., 1994; Tramontin and Brenowitz, 1999); however, a better test of this hypothesis would involve species comparisons within the same study using the same protocols.

When viewed collectively, the data across species and ages suggest that the best correlate of high rates of neuron incorporation is a period spanning the transition between song instability and high stereotypy. A modified version of the learning hypothesis that is consistent with these data is that neuronal replacement provides flexibility to the motor pathway permitting the achievement of stereotypy, regardless of when initial song learning takes place. This flexibility could permit adaptive responses to mismatches between expected and received feedback from singing. Perhaps neurons receive selective feedback signals just after they have been active, indicating their participation in an error or a more optimized sound. These signals could promote the death or long-term survival of a neuron (respectively), and thereby act as a mechanism that sculpts the motor output pathway and song. In this model new neurons are essential raw material for motor plasticity that could not be achieved by more conservative changes in synapse weight or modifications to the dendrites and axonal arbor of perennial neurons.

As previously mentioned, error correction systems may be important for both song learning *and* song maintenance. If new neuron addition and survival are regulated by song error correction, then one would predict that neuron recruitment rates would be highest at times when error signals are likely to be the strongest or most abundant and song stereotypy is low (Figure 28.6). This prediction is borne out by the overall reduction in neuron incorporation between the initial song-learning phase in juvenile zebra finches and the song maintenance phase of adult zebra finches (the latter being a time when song stereotypy is high and error signals are presumably low). If adult song maintenance is an active process driven by error correction, then this could also explain why neuron recruitment persists, albeit at low rates, after song crystallization. Moreover, this scenario is consistent with the finding that in adult zebra finches, an age-related decline in neuron incorporation occurs specifically over the period when reliance of song on auditory feedback wanes (Lombardino and Nottebohm, 2000;

Based on comparative studies, the best correlate of high rates of neuron recruitment is low song stereotypy

This instability may be associated with the initial learning of song in development

Seasonal learning in adulthood

Seasonal re-expression of a previously learned song

Or adult, age-related changes in song stereotypy.

Figure 28.6 A summary integrating results from studies on behavioral correlates of high rates of neuron incorporation across species and ages. Upper row of photographs, from left to right: song sparrows, canaries and zebra finches. These three species have different song learning schedules and vary in the extent and nature of adult song modifications. Despite these differences, all three species exhibit highest rates of HVC neuron incorporation at times when song variability is high. See text for specific citations.

Brainard and Doupe, 2001; Figure 28.3, right panel; compare with Figure 28.2) and song note stereotypy increases (Brainard and Doupe, 2001; Pytte *et al*, 2007). It could also explain why neuronal recruitment is high during the seasonal decrease in adult stereotypy both for new song (canary) as well as for a previously learned song (western song sparrow) since it is likely that the achievement of stereotypy in both cases relies on error correction.

Perhaps the building of optimal (error free) circuitry for song motor control is a gradual process that is regulated by singing history (Figure 28.7). A bird that has sung more in his life will have had a greater opportunity to build a long-lived and optimal collection of neurons in HVC. Thus with time and practice a greater cohort of "correctly" firing neurons would be assembled, leading to more reliable output and thus decreased reliance on auditory "instruction" for appropriate activity patterns. As these cells accumulate with age, motor program stability would increase, and the neuronal replacement process would decrease. Singing might up-regulate the pool of new neurons, from which a subset are kept based on the extent to which they contribute to a desired motor pattern. That is, we could have some sort of loop.

"Optimal" neurons would be those that ultimately contribute to the production of a target song. However, the most immediate role new neurons play could even be to destabilize song. Maintaining close correspondence between vocal output and a target song might be best achieved by introducing jitter or noise in the motor output that can then serve as the substrate for vocal improvement (much like detuning a guitar string first in order to achieve a target pitch). So long as the end result is the same, this proposal is consistent with the previously described hypothesis that ongoing neuron addition provides a reservoir of neurons used to achieve song stereotypy.

There are also ways in which song stereotypy and neuron recruitment rates could be functionally related but where neither relies on auditory-based error correction. Repeated use of song motor circuits could lead to a progressively stronger covariance in premotor neuron firing patterns and the emergence of a single song variant. RA-projecting HVC neurons send axon collaterals to other neurons of the same type (Mooney and Prather, 2005). There could be a direct motor-to-motor matching process that, with repeated execution of motor commands, leads to the accumulation and strengthening of neurons with similar firing patterns and these cells could have a competitive survival advantage over incoming neurons. Such a process could account for the decreased reliance of song on auditory feedback with age as well as the decrease in neuron recruitment. However, this scenario still requires a process in which cells are evaluated based on firing patterns.

Instructional signals derived from an auditory or nonauditory based evaluation process might be provided by modulatory neurotransmitters such as dopamine, which has been shown to increase under novel or unexpected outcomes (Waelti *et al*., 2001; Schultz, 2002). HVC is rich in tyrosine hydroxylase (TH), an enzyme involved in the synthesis of dopamine, and TH immuno-staining intensifies in zebra finches at ages when song learning occurs (Soha *et al*., 1996) and neuronal recruitment is high. Moreover, singing influences BDNF expression (Li *et al*., 2000) which, in turn,

Figure 28.7 A metaphor illustrating the potential role of neuronal replacement in error correction. Songbirds may audition new neurons based on a cell's ability to contribute to song development and maintenance. Neurons with response properties consistent with optimal song structure are cast while those that do not fit the part are rejected, prompting more neurons to be auditioned. The assembly of the entire cast may be a lifelong process but one that diminishes with age and experience as ever fewer roles in the musical remain unfilled. (From Wilbrecht and Kirn, 2004, reprinted with permission from Blackwell Publishing.)

has been shown to promote new HVC neuron survival (Rasika et al., 1999; Alvarez-Borda et al., 2004; see Nottebohm, this volume). It is entirely possible that a key feature of singing that is involved in BDNF expression is the prevalence of error signals. It would be interesting to compare BDNF expression in birds that sing at similar rates but which differ with respect to song stereotypy.

While many data fit well with this model, some do not. When a deaf bird sings there are likely to be gross error signals related to the absence of song feedback and so one might expect neuronal replacement to increase in deaf birds. However, deafening apparently does not affect HVC neuron incorporation in juveniles except when they have unilateral section of the tracho-syringeal nerve, and even in this situation, deafening attenuates, rather than augments neuron addition (Wilbrecht et al., 2002b). Deafening in adulthood also leads to a decrease, rather than an increase in neuron incorporation (Wang et al., 1999b).

Dissociating the role of neuronal replacement in motor activity, perceptual or sensory-motor learning, and the establishment of stereotypy

We have reviewed a substantial body of data suggesting that HVC neuronal recruitment and replacement (a) serve a perceptual function, (b) serve a singing-related motor function associated with replacement of damage-prone neurons, (c) are specifically linked with song learning, and (d) enable adaptive adjustments to song by error correction with the goal of motor stereotypy in a more general sense. Yet for each of these hypotheses the data are inconsistent. At present, the best correlate of high rates of HVC neuron recruitment across species and age appears to be a transition period from low to high song stereotypy.

Many of the inconsistencies in the data are likely to result from the complexity of the variables that might regulate neuron addition and a failure to standardize experimental paradigms in a way that enables us to explore one variable while holding others constant. First, social housing conditions are known to influence HVC neuron addition (Lipkind et al., 2002). In group-housed birds, manipulations such as deafening may alter social dynamics that influence amount of singing or other relevant behaviors in ways that would not be readily observed when birds are housed in more impoverished social environments. Indeed, recent work suggests that in certain housing conditions, zebra finches deafened in adulthood have *higher*, rather than lower rates of HVC neuron recruitment compared with their hearing cage mates (A. P. Hurley et al., unpublished data). The critical variables affected by social housing are not known but could relate to perceptual learning, changes in singing rate, or both. More extensive examinations of how social complexity affects neuron recruitment are needed.

Second, song learning, the maintenance of stereotypy by error correction, and amount of singing may not be independent factors. For example, one could argue that so long as a bird relies on auditory feedback for song maintenance, every time it sings it is learning something by a template matching process. Manipulations intended to alter one song attribute may inadvertently alter others, making distinctions between hypotheses (b), (c) and (d) difficult. Third, the more a bird has sung in its life the more resistant song may be to perturbations (perhaps due to a highly stable HVC-RA cell population), potentially resulting in a weaker or absent correlation between experimental manipulations and cell recruitment. However, with appropriate experimental design, these variables can be systematically explored. For example, in order to fully appraise the relative contributions of error signals and song-related motor activity to neuronal replacement, hearing and deaf birds could be matched for lifelong singing rate and age and then compared with respect to neuron recruitment. Finally, changes in neuron recruitment can be transient. Thus, the timing of cell birth dating relative to experimental manipulations, as well as the length and number of survival times, can determine whether or not one captures the window during which experimentally induced changes occur. This can be a major concern when making inferences on the functions of adult neurogenesis (Wang et al., 1999b; Alvarez-Borda, 2002; Wilbrecht et al., 2002b). Controlling for these variables is likely to provide a better appraisal of the potential functions of neuronal replacement both during song learning by juveniles as well as song maintenance by adults.

SUMMARY AND FUTURE DIRECTIONS

It is now more than 20 years since the discovery that new neurons are generated and incorporated into some of the vocal control nuclei of songbirds (see Nottebohm, this volume). During this period behavioral and neurobiological work has revised and broadened the list of potential functions of vocal control neuron addition and replacement. There is evidence that neurogenesis may be necessary, but not sufficient for song learning. New neuron recruitment levels in HVC can be correlated with song learning, low song stereotypy, and the amount of singing, which may not be independent variables. Even in age-limited learners such as zebra finches, replacement may play a role in song maintenance. This latter possibility is especially intriguing because it would suggest the rather radical idea that neuronal replacement is actually necessary for behavioral stability. Future work should systematically examine the relative contributions of each of these song attributes in the control of cell replacement.

Does neurogenesis serve a single function in the adult brain? This review has focused mainly on HVC because relatively little is known about neuron recruitment to Area X and other regions. However, even within the song system, there is evidence that the regulation and, perhaps, function of neuron addition to HVC and Area X are different. Seasonal increases in the volumes of HVC and Area X are associated with increases in neuron addition in the former but only increases in cell spacing in the latter in sparrows (Thompson and Brenowitz, 2007). Moreover, age-related decreases in HVC neuron recruitment in the zebra finch are not accompanied by similar decreases in Area X (Pytte et al., 2007). It might advance our understanding of adult neurogenesis more generally, and clarify the range of functions this process serves, if future work examined both of these song system regions in the same experiments.

Correlational studies exploring natural variation in song attributes and neuronal replacement have provided valuable contributions in formulating hypotheses about the behavioral relevance of neuronal turnover. These types of studies are especially valuable when they demonstrate dissociations between neuronal turnover rates and proposed functions (e.g. Tramontin and Brenowitz, 1999). However, the ultimate test of the utility of any hypothesis must involve experimental manipulation of the variables of interest. A preliminary step in this direction has been to determine the outcome of targeted killing of HVC cells, which augments replacement. A complementary approach could be to examine the behavioral effects of attenuating cell recruitment by antimitotic treatments, as has been done in mice (Shors et al., 2001). This approach could provide insights regarding the necessity of neuron recruitment for any of the proposed functions we have discussed.

Future work characterizing the morphology and physiological response properties of adult-formed neurons with retroviral techniques (Goldman et al., 1996; Cepko et al., 1998; van Praag et al., 2002; Carleton et al., 2003) is likely to also improve our understanding of neuronal replacement. Retroviruses carrying the reporter gene for green fluorescent protein (GFP) (Okada et al., 1999) could be especially useful (see Figure 28.1). These new methods have major advantages over the laborious process of combining [^{3}H]-thymidine with electrophysiological recordings described earlier. Retroviruses only infect mitotically active cells and daughter cells express GFP in vivo. The fluorescence is bright enough to be used to target these cells in vitro (and in some cases, in vivo) for electrophysiological recordings and reconstruction of cell morphology. There are many important questions that could be addressed with this method. Are there morphological and physiological subtypes of adult-formed HVC interneurons and RA-projection neurons that differ with respect to their life-span, time of year when they are born, or sensitivity to environmental and behavioral change? How do adult-formed neurons change as they age and as a function of experience? Are young neurons actually more plastic than older ones?

Another major technical innovation, gene microarray analyses, although still in their infancy, have already provided potentially valuable data for molecular explanations of why some neurons are replaced and others are not. A gene implicated in the regulation of protein degradation, ubiquitin carboxyl-terminal hydrolase (UCHL1) is underexpressed in replaceable HVC-RA neurons compared with nonreplaceable HVC-X neurons. Interestingly, this same gene is also underexpressed in replaceable neurons in the rodent hippocampus and olfactory bulb (Lombardino et al., 2005). There are many potential reasons for this common pattern of gene expression that have nothing to do with cell turnover; however, this gene has been implicated in neurodegenerative disease and is upregulated by singing, which promotes neuron survival (Lombardino et al., 2005).

While such experimental approaches will continue to provide information on mechanisms of neuronal replacement, behavioral studies will continue to provide valuable information on the processes of song learning and maintenance essential for understanding the functions of posthatching neurogenesis. Careful studies of song development in some age-limited learners have revealed that young birds overproduce notes that are then culled during the establishment of crystallized song and yet, some of the deleted material can re-emerge in adulthood (Hough et al., 2000; reviewed by Marler, 1997). Similar longitudinal studies in open-ended learners would begin to address the question of whether new songs are learned each year or whether adult changes in song reflect a process where a different subset of sounds is selected each year from a larger reservoir learned early in life. This important work would not only address issues regarding assumptions about when song is learned but also the question of when and even whether syllables are ever truly forgotten. Perhaps the latter could be addressed with operant techniques (Cynx and Nottebohm, 1992; Gentner et al., 2000) where rates of acquisition of a song discrimination task using apparently forgotten song elements could be compared with acquisition rates using unfamiliar song. Even if a new song is learned yearly and then forgotten, questions still remain about whether the new song material arises by improvisation or by copying song from conspecifics. The answers to these questions could have considerable impact on current models of the functions of neuronal replacement in relation to seasonal changes in song.

While direct manipulation of cell recruitment is the most powerful approach to functional questions, this method is useless without first knowing what to look for as possible behavioral outcomes. The notion that song stereotypy requires flexibility and error correction – a form of plasticity that continues long after song crystallization – raises new possibilities for the functions of adult neurogenesis. At the very least, there is now both behavioral and neurobiological evidence that even age-limited learners may continue to perfect their songs in subtle ways throughout life (Lombardino and Nottebohm, 2000; Brainard and Doupe, 2001; Pytte et al., 2007). Should these refinements be shown to rely on neurogenesis, an ultimate test of the functional significance of neuron turnover would be to see whether such subtle improvements in song structure influence female mate choice (see Nowicki and Searcy, 2004). Perhaps the most important answer to the question of *why* adult neurogenesis occurs will address the ways in which it impacts reproductive success and is transmitted from parent to offspring.

PART VI
The genomic revolution and birdsong neurobiology

29 • Introduction

H. P. Zeigler

Clayton and Arnold explore some of the implications of the genomic revolution for birdsong neurobiology, noting that the sequencing of the zebra finch genome is long overdue. They suggest that while all organisms are equal, in the sense that they share a common evolutionary history, some organisms, like songbirds, are more equal than others, because of the light they may throw upon a broad range of physiological processes and behavior. They outline some of the tools that were formerly only available to workers studying mice, fish, flies, and worms and discuss their potential application to such current research issues as the evolution of social behavior, mechanisms of sexual differentiation, and the development of vocal behaviors.

They also suggest that the impact of the genomic revolution will be conceptual as well as empirical. As they note, the early development of animal behavior studies was dominated by the question of the relative contributions to behavior of genes and environment. Experimental resolution of this issue requires the manipulation of the critical variables involved. However, while the environment could be manipulated with relative ease genetic effects could not. Yet one of the things that songbird research has taught us is that the learning process itself is under significant genetic constraint, e.g. with respect to filtering and timing aspects of song learning. Until recently, the genetic mechanisms underlying those constraints have remained elusive. Despite current difficulties in the application of some genomic methods to songbirds (e.g. gene deletions and insertions) the two variables in the dichotomous formulation (nature versus nurture) are now almost equally accessible to experimentation. Clayton and Arnold conclude that perhaps the most important outcome of songbird genomic research will not be the new answers it provides but the way in which the results will change the nature of the questions we ask.

As the Mello and Jarvis chapter indicates, one genomic method, the measurement of gene expression as a tool for the study of birdsong, has been in widespread use for at least a decade. Studies of immediate early genes (IEG) are based on the fact that activation of brain neurons during exposure to specific stimuli or performance of specific behaviors results in an initial transient change in gene expression whose products may be identified with good spatial resolution, though without the degree of temporal resolution provided by neurophysiological recording techniques. While the exact relationship between neuronal activation and IEG expression is not entirely clear, the method has been successful in implicating specific brain regions as components of behaviorally relevant neural circuits related to song.

One important result of such studies has been the dissociation of those brain regions implicated in listening to or recognizing song and those regions mediating song production. IEG studies demonstrating singing-related gene expression in the classical vocal-motor and anterior forebrain pathways have been important in complementing lesion and recording studies that have implicated these areas in the higher-order control of song production. IEG studies have also led to the identification of brain areas outside the classical "song system" which seem to process auditory stimuli involved in song memory and song recognition and discrimination. Finally, the method has been used with success in comparative studies of vocal learning to identify common structures in the brains of all three groups of vocal learners (oscines, parrots, and hummingbirds) that are responsive during singing or listening to song. Such data can help generate hypotheses about evolutionary relationships among existing song-learning and vocal nonlearning birds, and, more cautiously, about links between song learning in birds and speech in humans.

The Haesler and Scharff chapter explores such links more directly. It represents an important first step in assessing the extent to which songbirds and

Neuroscience of Birdsong, ed. H. Philip Zeigler and Peter Marler. Published by Cambridge University Press. © Cambridge University Press 2008.

humans share common genetic mechanisms associated with vocal learning. In humans disruptions in the function of the FoxP2 gene are associated both with relatively selective disruption of speech production and with structural and functional abnormalities of the basal ganglia. In songbirds FoxP2 is expressed at high levels during phases of song plasticity in a basal ganglia region (Area X) critical for song acquisition. Because it is not yet possible to create a "knockout" songbird, these investigators produced a "knockdown" preparation using the molecular technology of RNA interference (RNAi). They used a lentivirus injected into Area X as a vector to deliver short interfering RNA (siRNA) with a sequence complementary to mRNA transcript in order to decrease expression of the FoxP2 gene in vivo (hence the term "knockdown"). They describe the disruptive effects of this manipulation on song acquisition and offer accounts of the possible neurobehavioral mechanisms involved. Their conclusion, "that similarities between human speech and bird song may well extend to the genetic level" is perhaps premature, given the uncertainty of FoxP2's cellular role in birds and the uncertainties about the neural substrate of the deficit in the human FoxP2 cases. What is clear, however, is that these and similar genomic-based methods are likely to make a very significant contribution to our understanding of the links between birdsong learning and human language development.

30 • Studies of songbirds in the age of genetics: what to expect from genomic approaches in the next 20 years

David F. Clayton and Arthur P. Arnold

INTRODUCTION

In 2005, the U.S. National Institutes of Health selected the zebra finch for whole genome sequencing, recognizing the unique potential of songbirds to illuminate many aspects of basic biology relevant to human health and disease. The first draft sequence should be released by the time this book is in print (www.ncbi.nlm.nih.gov/genome/guide/finch/). The zebra finch genome sequence will enable studies never before considered, much less undertaken, in songbirds. With it will come descriptions of global gene content, predictions of all the proteins in the organism, and insight into the structure and function of noncoding regions of the genome. It will spawn new tools for analyzing the diversity of zebra finch populations, the inheritance of specific traits, and the evolutionary history of songbirds. Perched as we are on the edge of a new field of discovery, we can only just begin to discern the contours of the land ahead. We anticipate that the genome will provide new solutions to old questions – but it will also change the questions asked. In this chapter we introduce some of the basic elements of genomic science as it will be applied to study of songbirds. We predict, very tentatively, some of the insights that might come from combining the new genomic information with the fascinating unsolved questions of songbird biology.

WHY STUDY SONGBIRDS IN THE GENOMIC ERA?

To provide a mechanistic understanding of function and development of any biological system, the current state of the art requires that the genes controlling the relevant traits be identified, and their impact be understood at the level of molecules, cells, systems, and individuals. Often this means manipulating candidate genes to demonstrate the function of the genes underlying emergent phenotypes. Because genetic manipulation is performed relatively easily only in a few model organisms such as mice, zebra fish, fruit flies, and *C. elegans*, research on those species now represents the vast majority of all research on mechanisms controlling animal phenotypes. Overinvestigation of a small number of life forms could ultimately choke the growth of biological science. A complete understanding of the mouse, even if that were possible, would leave us ignorant of many critical biological phenomena simply because mice provide only one window onto biological processes.

Songbirds have already repeatedly provided valuable perspectives that have not emerged from the study of more popular model organisms. As is documented extensively in other chapters of this book, important discoveries *initially described in songbirds* have subsequently been found to apply also to mammals.

Songbirds are also, of course, interesting and worthwhile subjects for research, all on their own. They represent one of the most successful and robust radiations in the evolution of vertebrates, and are a common part of the natural world that every human experiences. One need not look beyond those motivations to justify a significant investment in further research. It is important, however, to appreciate that investment in songbird research has a history of paying dividends that enrich the study of all animal life and have application to, for example, human medicine. Because a modern approach to understanding the biology of any species of birdsong requires a comprehensive knowledge of the genes encoding proteins that make up the species phenotype, the sequencing of the genome is a critical step for future work on songbirds.

WHAT GENOMICS BRINGS TO THE STUDY OF BIRDSONG

The availability of a complete genome sequence forces one to think about genetics in large, integrative terms.

Neuroscience of Birdsong, ed. H. Philip Zeigler and Peter Marler. Published by Cambridge University Press. © Cambridge University Press 2008.

The number of known genes in the avian genome exceeds 20 000 (Hillier et al., 2004). Considering alternative splicing events and posttranslational modifications, the number of different protein forms could easily be ten times as large. This is a jungle of molecular diversity. Yet somehow, from this jungle emerges the orderly and predictable development of every young songbird. The parts specified by the genome work together smoothly and efficiently to determine the physiology and behavior of the singing animal and all its functional subsystems. And, on a much different timescale, these components have slowly varied and evolved to generate the 4000 different species of songbirds currently known on Earth. Thus the genome unites the perspectives of development, physiology, and evolution.

The parts list

One immediate outcome of genome sequencing is the generation of a complete "parts list" for the species (e.g. http://parts.mit.edu). The notion that the list is "complete" is of course fanciful, not only because it is only an approximation of the complete genome, but also because only some of the molecules in an organism (i.e. proteins and functional RNAs) are directly encoded in the genome. Nevertheless, those genome-encoded elements are responsible for synthesizing most of the remaining macromolecular components (e.g. lipids and carbohydrates), and thus a deep knowledge of the genome can provide comprehensive insight into the molecular makeup of the organism.

The parts list is also a virtual toolbox. Every sequenced gene contains the information needed to measure the expression of that gene (as mRNA), using DNA/RNA hybridization. Antibodies can be raised to the predicted protein and used to map expression of a protein, inhibit its function, or purify it. DNA microarrays are being used (http://songbirdgenome.org) to measure simultaneously the expression of many genes – indeed it seems reasonable to expect the imminent development of a microarray which makes possible simultaneous measurements of expression for essentially all the genes in the songbird genome. Comparative analyses will soon reveal how different patterns of gene expression are associated with different developmental and physiological states in songbirds. These results in turn will drive the development of experiments to identify the regulatory signals and pathways that determine these gene expression patterns.

Rules of regulation

The genome sequence will provide much more than just a list of molecular parts for the organism (Pearson, 2006). It will also contain information about non-protein-coding (regulatory) regions responsible for when, where and how each part is produced. Many of the driving questions in songbird biology are about regulation and dynamic change. Answers to many of these questions may lie at the level of the regulatory sequences and networks that govern specific gene expression. Having the genome sequence does not automatically reveal these sequences and networks, but through comparative genomics progress is being made in learning how to recognize and infer regulatory information from genome sequence.

The most familiar regulatory elements are the DNA sequences (promoters and enhancers) within each gene that interact with the proteins (transcription factors) responsible for transcribing the gene into RNA. Genome sequencing can directly reveal the presence of some of the better-defined and conserved promoter and enhancer elements in a specific gene. Whole genome sequencing is also catalyzing appreciation of microRNAs (miRNAs), short pieces of RNA that do not code for proteins, but instead bind to their complementary sequences in specific protein-coding mRNAs (He and Hannon, 2004; Miska et al., 2004). This typically results in either degradation of the mRNA or suppression of its translation. Thus the production of a particular miRNA sequence in a cell can modulate the effective expression of specific targeted genes, in a way independent of the enhancers and promoters that govern the transcription of the target gene. At this writing, miRNAs have not yet been described in songbirds but they are certainly present. MiRNAs are highly conserved among vertebrate genomes, and the chicken genome has been shown to contain the instructions for producing at least 152 of these (http://microrna.sanger.ac.uk/cgi-bin/sequences/browse.pl). Future research is sure to probe the possible involvement of miRNA-based regulatory mechanisms in the development and physiology of the song control system.

Window on evolution and diversity

Why do some songbirds learn their song only once, and others continue to modify it throughout life? Why do some species sing with monotonous repetition whereas others warble long and changing melodies? Fundamentally, these distinctions must rest on differences in DNA. All songbirds share a common ancestral genome that dates back roughly 100 million years. But their genomes have not been static, and comparison of songbird genomes should ultimately reveal how particular traits and properties are associated with different gene structures and organizations.

To date, the genetic structure of songbird populations has been studied by use of microsatellite sequences (Hansson et al., 2005; Backstrom et al., 2006; Dawson et al., 2006). The genome sequencing project will greatly increase the sensitivity and resolution of comparative genomic analyses, first by providing a collection of mapped single nucleotide differences (single nucleotide polymorphisms, SNPs) that can be used to track different alleles, or forms of the same gene. The SNPs can be used to determine the presence or absence of specific alleles in an individual, the distribution of alleles in a population, and the statistical association of a particular allele with a particular phenotype of interest (using linkage mapping of quantitative trait loci, QTLs; Crabbe, 1996). At a much larger scale of action, different songbird genomes may also be distinguished by wholesale gene losses, duplications and inversions. Comparative analyses of other animals are revealing that gene copy number varies significantly between species (Proulx and Phillips, 2006; Wang et al., 2006) and even between individuals of the same species (Aitman et al., 2006). Inversion of whole subregions of a chromosome (Stefansson et al., 2005) may effectively isolate different versions of that chromosome, suppressing recombination between them and allowing them to drift independently (Dobzhansky and Epling, 1948).

Emerging techniques

Although songbirds have been bred in captivity for centuries, the techniques of genetic analysis and manipulation have so far had little impact on formal scientific research using songbirds. The genome sequence will catalyze a change in this. Most immediately, availability of a genome sequence virtually mandates the development of a formal genetic linkage map for the species. As of yet there is no genetic linkage map for any songbird species. A selective breeding experiment for corticosteroid response in the zebra finch was recently reported (Evans et al., 2006), and a potential base for genetic linkage analysis may be available in the work of Birkhead, Slate and colleagues on the population genetics of sperm competition in the zebra finch (Birkhead et al., 2005). A genetic linkage map will be an important complement to the whole genome sequence because it helps localize gene sequences to specific chromosomes and provides a bridge between the abstract level of sequence data and the observable properties of individual organisms and populations. The linkage map will form a foundation for future studies to correlate allelic variation of many genes with behavioral and other phenotypes.

The genome sequence will also propel efforts to develop methods for production of genetically modified (transgenic) songbirds. In theory, the relative accessibility of the avian embryo should offer some advantages for transgenic manipulation. However, progress with avian transgenesis has lagged behind the dramatic advance of mammalian transgenesis, perhaps in part due to a smaller scale of effort and investment but also due to the unique challenges presented by birds. Transgenic mice are produced by injecting DNA into the pronuclei of the zygote in mice, after which the embryo is transferred into a female mouse whose uterus easily accepts the embryo for implantation into the uterus. In birds, the zygote cannot be easily harvested in quantity, nor can it easily be placed back into the female prior to formation of the egg shell. Once the egg is laid, the avian embryo already has about 40 000 cells, so that manipulation at that stage is not easy. Moreover, it is challenging to support full development of manipulated embryos in shell-less culture (Brown et al., 2003). Recent developments using chickens include the successful isolation, culture and genetic modification of primordial germ cells (van de Lavoir et al., 2006) and the successful application of lentiviral vectors to insert DNA and produce transgenic founder lines (McGrew et al., 2004). Optimized vectors have also recently been developed for generating and delivering interfering RNAs (RNAi, synthetic relatives of natural microRNAs reviewed earlier) in avian systems (Das et al., 2006a). It has been possible to electroporate DNA constructs into the nervous systems of avian embryos in ovo, a technique that might be modified for introduction of transgenes into songbird tissues (Itasaki *et al.*, 1999).

OLD THEMES, NEW PERSPECTIVES

To illustrate how songbird neurobehavioral biology will be changed by the imminent exponential increase in the amount and quality of genomic information, we next review four traditional foci of interest involving songbirds, and consider how the new genomic information is likely to influence future research in these areas. Our list below is not exhaustive and we can only consider a few of the many points of intersection between the traditions of songbird research and the potential impact of the new genomics.

Nature versus nurture, revisited

The dominant issue in ethology and comparative psychology in the 1950s and 1960s was the nature–nurture debate. To what extent are specific behaviors controlled by genetics versus the environment? The debate surrounded attempts to reconcile the writings of European ethologists such as Lorenz (1970) and Tinbergen (1951) with those of, for example, American behaviorists (Skinner, 1938). The ethologists emphasized that behavior, like bone structure, had evolved. For evolution to sculpt behavioral processes, however, the behaviors had to be innate. The graylag gosling, Lorenz maintained, was genetically programmed to imprint on its mother in the first hours after hatching, so that it would adaptively follow the mother in the early weeks of its life. The herring gull chick, Tinbergen claimed, was born with a sensory image in its brain of the red dot on the end of its parent's long bill, so that it was genetically pre programmed to peck at the bill to obtain food. The pecking was mediated by genetically hard-wired motor mechanisms called fixed action patterns (Marler and Hamilton, 1966; Hinde, 1970). In contrast, the American behaviorists, fresh from decades of research that demonstrated the potent and subtle effects of conditioning on behaviors, naturally tended to view many complex behaviors as learned chains of responses. Indeed, it was claimed that any behavior could be conditioned to occur in response to any arbitrary stimulus. The animal could be seen as a *tabula rasa* on which stimulus-response connections were programmed by its experiences.

The resolution, or perhaps cease-fire, of the debate, came with the realization that all behaviors have inherited and learned components (Brown, 1975; Gould and Marler, 1987). The herring gull chick, born with certain behavioral tendencies, very quickly refines its bill-pecking behavior as a result of whether the pecking leads to food reward or not (Hailman, 1969). Studies of conditioning showed that a wide variety of animals learn certain associations much more easily than others (Rozin and Kalat, 1971; Gould and Marler, 1987). The kinds of associations that are easy to learn appear to be species-specific, and to make sense in terms of the animal's evolved adaptations to its world. The ability to learn – like other abilities – has evolved and is under genetic constraint.

The study of birdsong played an important role in the nature–nuture controversy. Many male oscine passerine birds learn the form of their songs by imitating other males. However, the learning process itself is under significant (genetic) constraint. The white-crowned sparrow male easily learns to copy other birds of its own species, but neatly filters out and does not copy similar sounds sung by song sparrows in its natal district (Marler and Tamura, 1964; also see Adret, this volume). The reproductive advantage seems clear – to attract and breed with white-crowned sparrow females, the male needs to sing the song she appreciates. Moreover, the timing of the learning process is species-specific, and therefore appears also to be under genetic constraint. Thus, the study of birdsong offered numerous examples of the fascinating interplay of genetic and learned factors to produce the fully differentiated adult behavioral phenotype, and helped refine the nature–nurture debate.

To demonstrate environmental effects on a behavior, one manipulates the environment and measures the effect on the behavior. For example, by varying the developmental stage at which a white-crowned sparrow hears its species song, one can establish what can be copied, and when. To demonstrate a genetic influence on the behavior, one manipulates the composition or expression of the genome, and measures the behavioral effects. Manipulation of the genome has been accomplished classically (ever since Mendel) by breeding experiments. Parents with different traits are crossed, and the distribution of traits in the offspring are compared with a genetic model to test the model. If the data fit the model, then variation in the trait is found to be influenced by genetics. More recently, other techniques show how genetic differences cause phenotypic differences. If two inbred strains of mice, for example, differ in a trait when their environment is held relatively constant, the difference is attributed to genetic differences. Moreover, specific behavioral or other phenotype differences can be convincingly correlated with (linked to) measured differences in the genome at specific locations (loci) in the genome, an experiment that establishes the genomic regions controlling the trait. Most convincing is evidence that manipulation of the gene alters the trait – for example, by removing a gene from the genome (null mutation) – or increases its expression (by insertion of a transgene).

It is interesting that the nature–nurture issue of 50 years ago was debated in the near absence of rigorous genetic evidence of the type just described. Because it was relatively easy to vary the environment, the effects of experience on behaviors were convincingly demonstrated. The effect of genetics on complex behaviors, however, was often inferred (at least in slowly reproducing vertebrate species like songbirds) from the properties of the behavior when the effect of the environment was assumed to be small, or when manipulations of the environment had little effect. For example, the song of the white-crowned sparrow that developed when the bird was deprived of hearing his conspecifics was sometimes called the "innate song," based on the idea that this song was what occurred in the absence of the normally rich acoustic environment. However, the untutored bird clearly had an environment, since genes cannot give rise to traits in the complete absence of an environment. It was not obvious how depauperate environments had contributed to the innate song. Similarly, the herring gull chick that pecked at its parent's bill within minutes of hatching was assumed not to have learned much before hatching, although it clearly had a sensory environment then, and had engaged in ("practiced") movements before hatch (Lehrman, 1953). Indeed, the difficulty in making conclusions about innateness by restricting the environment led to numerous discussions that illustrated the problems with the approach (see Lehrman, 1953; Lorenz, 1965). The insurmountable obstacle in the mid twentieth century was the inability to manipulate the genome. If one was restricted to performing experiments by manipulating the environment, it was impossible to estimate the effects of genetic variation on traits as clearly as it was to measure the effects of environmental variation.

Although a small number of studies in the 1950s and 1960s did, in fact, show that interbreeding bird species produced offspring with intermediate behaviors (Dilger, 1962; Lade and Thorpe, 1964) in general the genetic experiments made only a minor contribution to the discussions of nature versus nurture in avian behaviors. A major point was that many behaviors are species-specific, and the specificity was often inferred to be caused by genetic differences between the species, even though the species often also differ in their environments (Lorenz, 1965, 1970). In contrast, the field of mouse behavioral genetics progressively utilized all of the sophisticated genetic methods outlined in the preceding paragraph and at the beginning of this chapter, and the study of mice still defines what is possible in the study of genetics of complex vertebrate traits. The availability of many highly inbred strains of mice provided the genetic material suitable for the application of these genetic techniques. Well-controlled genetic experiments were difficult in birds, and were not performed.

We can predict that this situation will change somewhat in the next decade or two. With the new information on the sequence of the zebra finch genome, it will be possible to examine in detail (1) how specific genetic variations among different songbird individuals or populations (in captivity or in the field) correlate with complex behavioral traits, and (2) how specific regions of the genome in different songbird species correlate with species differences in behavioral traits. For example, a large list of specific single-nucleotide polymorphisms (SNPs) has already emerged from the large-scale sequencing of the zebra finch genome and expressed sequences. The SNP maps will allow sophisticated linkage mapping, for example among zebra finches with known familial relationships, to determine where in the genome genetic variation is associated with (causes) variation in phenotype. When in vivo methods for manipulation of gene expression are developed (transgenic insertion of genes, knock down of gene expression using RNAi techniques), it will be possible to test directly how genes influence behaviors, how genes influence which experiences influence behavior, and how specific variations in the environment influence which genes impact the behavior.

Will the new genetic information lead to a reanalysis of the nature–nurture debate? Yes, and no. Yes, the knowledge of gene and intergene sequences will offer new power for true genetic experiments (linkage, breeding, gene manipulations), much more than what was available in the 1950s and 1960s. But no, the question of gene-experience interaction will be framed more specifically, and differently than 50 years ago. That debate taught us that we can never fully disentangle the genetic and environmental factors.

Neurogenomic mechanisms of song learning

As the nature–nurture debate was waning, neuroscientists began making progress mapping the structures in the songbird brain that mediate a young bird's abilities to memorize a song model, learn to reproduce it himself by trial and error, and discriminate among the songs of his (or her) neighbors the rest of its life. Cellular and biochemical studies of learning in various model systems increasingly implicated the genome as an active participant in memory storage. Thus an abstract, purely behavioral concept – "learning" – came to be grounded in the real tissue of the brain. This tissue is made of cells, and at the center of every cell is a genome specifying the proteins and processes that make the cell work.

The neuroanatomical basis for learned singing began to emerge in the mid 1970s through application of lesion and pathway mapping techniques, which demonstrated a set of discrete, interconnected brain nuclei necessary for song production (Nottebohm et al., 1976) and juvenile song learning (Bottjer et al., 1984; Scharff and Nottebohm, 1991; see Figure 5.2 in Reiner et al., this volume). These nuclei comprise the vocal control pathway and are found as discrete entities so far only in songbirds (Gahr et al., 1993). Early application of cDNA cloning and hybridization techniques began to reveal the molecular complexity of the songbird brain (Clayton et al., 1988; Clayton, 1997, 2006). Measurement of gene responses to song stimulation (Mello et al., 1992) opened the door to study of the processing network in the auditory forebrain responsible for representing complex song patterns and participating in lifelong song discrimination (Chew et al., 1995, 1996a; Mello et al., 1995; Stripling et al., 1997, 2001; Kruse et al., 2004) and juvenile song memorization (Bolhuis et al., 2000, 2001; Terpstra et al., 2004, 2006; Phan et al., 2006; see also Bolhuis, this volume).

Meanwhile, research into the biochemistry of learning in rodents, flies and other model organisms showed that disruption of RNA or protein synthesis would interfere with development of a stable memory in various learning tasks (Davis and Squire, 1984). Studies using transgenic or targeted antisense manipulations demonstrated essential requirements for a number of specific genes (Bourtchuladze et al., 1994; Yin et al., 1994; Guzowski and McGaugh, 1997; Jones et al., 2001; Taubenfeld et al., 2001; Guzowski, 2002; Pittenger et al., 2002; Ressler et al., 2002; Wei et al., 2002; Bozon et al., 2003; Lee et al., 2004a). Signaling pathways that dynamically modulate gene expression were implicated in various models of learning and memory including long-term potentiation (English and Sweatt, 1996, 1997; Adams and Sweatt, 2002), fear conditioning (Atkins et al., 1998; Schafe et al., 2000; Athos et al., 2002), spatial memory (Blum et al., 1999), conditioned taste aversion (Berman et al., 1998), and classical conditioning and sensitization (Crow et al., 1998; Sharma et al., 2003). Some of these gene products and pathways are now being implicated in aspects of song learning as well (Cheng and Clayton, 2004; Singh et al., 2005; Huesmann and Clayton, 2006). Dynamic changes in gene expression have been hypothesized to perform an essential role in integrating information over time and across modalities of experience, functioning as a "genomic action potential" (Clayton, 2000).

We anticipate that the next 20 years will see an explosion of information about the role of specific genes in building the song control circuit in the developing brain, constraining what a bird learns, determining when a bird learns and from whom, and executing the different phases of birdsong learning. Even as we write, numerous studies using DNA microarray technologies are under way, exploring changes in gene expression in specific parts of the songbird brain associated with formation of the song control system, the response to early song exposure, activities during juvenile song rehearsal, and adult song recognition learning (e.g. see http://songbirdgenome.org).

Sexual differentiation of the brain

Mechanisms of sexual differentiation are of broad and fundamental importance in biology. Songbirds are a classic model for studying sex differences in brain and behavior. Male songbirds sing a courtship and/or territorial song that females usually lack, although female song is intriguing and a subject for further genetic study (see Kaplan, this volume). Zebra finches in particular show striking sexual differentiation of the primary telencephalic pathway controlling song production, as the central connection between nuclei HVC and RA forms only in males not in females. Recent research suggests that genes may play a surprisingly direct role in sexual differentiation of the zebra finch brain, not anticipated from the dogma of mammalian sexual differentiation.

Sex differences in passerine birdsong have been obvious to any observer of birds since pre-history. By the early twentieth century, evidence began to accumulate that sex differences in body phenotypes were controlled by gonadal secretions (Lillie, 1916). By the 1930s, when relatively pure preparations of gonadal steroids became available, it was found that injecting female canaries with androgens would make them sing like a male (Shoemaker, 1939). By 1960, solid experimental evidence from studies in mammals suggested that sex differences in body phenotype and behavioral capacity were controlled by gonadal secretions during early stages of development (Jost, 1947; Phoenix et al., 1959). Research defined two basic modes of gonadal hormone action, organizational and activational. Permanent organizational actions occur when the testes secrete testosterone early in life, before and after birth, which causes the male's body to differentiate into the male form (penis, scrotum, sperm ducts). Testosterone also acts on the developing brain to cause the formation of circuits typical of the male, and to prevent formation of circuits typical of the female (Arnold, 2002). Later in life, however, the activational effects of gonadal steroids contribute further to sex differences in the brain, body, and behavior. The ovaries and testes secrete different gonadal hormones, in different temporal patterns, that act on numerous target sites, including the brain, to make them function differently in the two sexes.

Thus, when large morphological sex differences in the brain were discovered in discrete brain regions controlling song (Nottebohm and Arnold, 1976), the dominant expectation was that the sex differences would be controlled by the organizational and activational effects of hormones produced by the gonads. It was quickly determined that castrating adult male zebra finches and canaries did not eliminate the sex difference in the song control circuit (Arnold, 1980c). In contrast, when female zebra finches were treated with estradiol at hatching, they developed a dramatically more masculine neural circuit (in terms of volumes of brain regions, cell size and neuron number) and they sang as adults (Gurney and Konishi, 1980). Estradiol is a major metabolite of testosterone in the brain, and is known to be critically involved in permanent organizational effects of testosterone on the neonatal rodent brain. In the zebra finch brain, high levels of the enzyme that converts testosterone to estrogen (aromatase) were found in the telencephalon near the song circuit. Collectively, these findings seemed consistent with the mammalian model of gonadally driven sexual differentiation: males secrete testosterone from their testes during development, which then enters the brain and is converted to estradiol, which causes masculinization of neural circuits.

Other observations, however, did not support the classical theory that sex differences in this brain circuit were controlled by gonadal hormones (Arnold and Schlinger, 1993). Males were not consistently found to have higher levels of androgens or estrogens in their blood after hatch. All attempts to manipulate the effect of estrogens and androgens in young males failed to prevent masculine development of the song circuit (Wade and Arnold, 2004). Other evidence suggested that that estradiol normally plays a role in brain masculinization of males, but that the estradiol is actually produced *de novo* in the brain rather than deriving from testicular testosterone. When sections of the forebrain of adolescent zebra finches were grown in tissue culture, the male sections underwent some steps of the masculinization process in vitro (Holloway and Clayton, 2001). The female sections did not. Moreover, the masculine developmental steps in vitro required estradiol, which was synthesized *de novo* from the male but not female brain sections. These results, while consistent with a critical role for estrogen in guiding brain masculinization, left open the question of what sex-specific processes cause the male brain to make more estradiol? Other factors need to be invoked, especially because treating females with estradiol has never fully sex-reversed the brain (they are about half as masculine as males), and blocking steroid hormone action in males has little demasculinizing effect (Wade and Arnold, 2004).

The most obvious other set of sex-specific factors are the genes encoded on the sex chromosomes, which are differentially represented in the male (ZZ) and female (ZW) genome. Thus, the double dose of Z genes in male brain cells (Chen et al., 2005) might directly masculinize these cells. Conversely, masculine development might be prevented in females because of the expression of W chromosome gene(s) in female brain cells. This hypothesis of direct genetic control of brain sexual differentiation has been generally supported in several cases in which the sexual phenotype of the brain correlates not with the type of gonad (ovary versus testes), but with the genetic sex of brain cells. Female zebra finches have been produced that have substantial amounts of testicular tissue that secrete androgens, but their genetically female brain remains feminine and they do not sing (Wade and Arnold, 1996, 2004). A mutant gynandromorphic zebra finch was described in which the right half of the body was genetically male and the left half genetically female. The plumage and gonads followed genetic sex, so the left half had female feathers and an ovary, but the right half had male feathers and a testis. In the brain, too, the right half was more masculine than the left, correlating with the genetic sex of the cells comprising those halves of the brain (Agate et al., 2003). In another mutant finch, the plumage and brain and song behavior were all masculine, correlating with the genetic sex (ZZ) of the bird. However, there appeared to be a mutation in the gonad-determining path, so that the bird had an ovary but not testes. In this case the genetic sex of the bird and brain were concordant, and the brain seems to have been masculinized in the absence of testes (Itoh et al., 2005).

Thus, although research to understand sexual differentiation originally focused on hormonal effects on the brain, the current aim is to determine which sex chromosome genes are expressed in the brain, and how might they initiate sex-specific patterns of development of the song system. The good news is that the sex-specific signals must be encoded on the sex chromosomes, so the search is narrowed to about 5% of the genome. One goal therefore has been simply to identify the Z and W genes, and compare their expression patterns in male and females. Now that the question is genetic rather than hormonal, however, a new set of experimental tools needs to be utilized. In some cases these techniques have never previously been applied in birds. The situation in 2008 in this field is similar to that in the field of mammalian sex determination in the 1980s. At that time it was realized that the gene(s) responsible for testis differentiation in male mammals reside on the Y chromosome (Arnold and Burgoyne, 2004). Solving which gene was responsible was possible through the careful study of numerous XX humans possessing a small region of the Y chromosome, who had testes. By determining which region of the Y chromosome was present in each case, a small segment of the Y chromosome was recognized as containing the testis-determining gene. Detailed study of that region led to the discovery of a novel gene *Sry* (Sinclair et al., 1990), which was subsequently proven to control testis determination when transgenic insertion of the gene into an XX genome caused testis development, and deletion of the gene in XY mice prevented testis development.

To prepare for the same sort of experiments to identify the brain-masculinizing (or antimasculinizing) gene(s) in zebra finches, we must identify the genes on the sex chromosomes; develop cDNA and genomic (BAC) probes to detect these genes in normal and mutant individuals; study their expression in the brain; study which Z and W genes are present in the genomes of mutant ZW birds that have masculine song, or which genes are absent in ZZ males that do not sing; manipulate their expression in vitro cell systems to understand their function; and compare the characteristics of Z and W genes in brains of species that do and do not have large sex differences in the song system. When transgenic and knockout methods become available in birds, the Z and W genes must be manipulated to test their importance for sex differences in song system development.

This new focus on the biology of the avian and passerine sex chromosomes has led to the first large-scale analysis of sex differences in gene expression in avian sex chromosome genes (Itoh et al., 2007). Using microarrays to measure hundreds to thousands of mRNAs simultaneously, we discovered that many or most Z genes are expressed constitutively higher in male zebra finches and chickens than in females. In contrast, autosomal genes were expressed on average at the same level in the two sexes. Z genes were expressed about 30% higher in males than females. This widespread sexual disparity of gene expression is surprising, because it has not been found in any other species. In mammals, for example, the sex difference in genomic dose (number of copies in genome) of X chromosome genes is efficiently eliminated by transcriptional inactivation of one X chromosome in each female cell. The sexual disparity found in birds has been thought to be disadvantageous, because a sex-specific imbalance would lead to suboptimal doses of gene expression in one or both sexes. Birds, apparently, have evolved mechanisms to reduce the disadvantages.

Study of birds therefore offers a fresh perspective on the problem of sex chromosome gene dosage compensation. Because the evolution of sex differences in any phenotype requires that a sex-specific signal (hormone, genes, etc.) evolves control over development of the trait, the generally higher expression of Z genes in male birds may make Z genes more available as sex-specific signals than, for example, sexually equivalent X genes are in mammals. Accordingly, sex differences in phenotypes may more often be driven by sex chromosome genes in birds than in mammals. Nevertheless, even in mammals it is now emerging that sex chromosomes may directly cause sex differences in neural phenotypes (Arnold and Burgoyne, 2004; Dewing et al., 2006). Indeed, the precedent from songbirds was a major stimulus to efforts to look for direct genetic control of sexual differentiation in mammals (Carruth et al., 2002; De Vries et al., 2002).

Evolution of social behaviors

The genome participates in behavior on multiple timescales (Ben-Shahar et al., 2002; Robinson et al., 2005). As touched on earlier, it has an active role in each individual organism, organizing the nervous system in development and mediating the ongoing storage of information acquired from experience. Every individual has a slightly different genome and, partly as a result, some individuals are more successful in their environment than others. Over the vast scale of evolutionary time, gene variants correlated with success are propagated and those with failure are eliminated. Thus the genome is the repository for the verdicts of natural selection.

Songbirds provide an extraordinary opportunity to explore the role of the genome in the evolution of social behaviors. Songbirds comprise ~4000 species, representing nearly half of the bird species in the world today (Sibley and Monroe, 1990). They represent one of the most explosive and successful of evolutionary radiations (Barker et al., 2004). Social behavior is at the center of this evolution: a principal difference between songbirds (oscines) and their evolutionary sister group, the suboscines, is the ability to produce and modify vocal signals based on social experience. Songbird species also vary significantly in other social behaviors. Some are territorial, some are colonial. Mating systems are diverse, including monogamy, polygamy, polyandry and polygyny. Daily activities (e.g. foraging) may be done individually or in the context of a structured flock. All this diversity evolved over the past 77 million years (van Tuinen and Hedges, 2001).

We anticipate that comparative genomics will begin to reveal specific sequence variations that are correlated with species differences in some of these behaviors. An intriguing specific example is provided by the white-throated sparrow, *Zonotrichia albicollis*, found in northeast North America. A half-century ago, Lowther (1961) documented the presence of two plumage morphs in the natural population, and Thorneycroft (1966, 1975) demonstrated association of this polymorphism with a specific chromosomal inversion evident by karyotyping. The two morphs also show different behavioral phenotypes. Birds of the "tan" morph show more parental care and less aggression than do birds of the "white" morph (Tuttle, 2003). The two morphs mate with each other, which maintains the chromosomal polymorphism in the population. Analysis of the chromosomal inversion may lead to identification of gene sequences that differ in the two morphs. Obviously, the genome sequence of the zebra finch will not reveal these differences in another species, but because of the close relationships of all songbirds, the zebra finch genome will be invaluable as a staging and reference point for targeted analysis of specific chromosomal domains in the white-throated sparrow, and other passerines.

FROM SONGBIRDS TO HUMANS (AND BACK)

A primary social justification for committing national resources to sequence the zebra finch genome is the potential to inform human biology and ultimately improve human health. Study of the songbird genome may yield fundamental insight into broad phenomena already considered in this review: nature/nurture, mechanisms of learning, sexual differentiation, evolution of social behavior. The zebra finch also has potential as a new laboratory model for study of specific genes involved in human disease and recovery. The striking conservation of genes involved in visual system development – from flies to man, despite their very different eye structures (Gehring and Ikeo, 1999) – lends hope to the notion that strong and explicit parallels will connect humans and songbirds when viewed at a genomic level.

Of particular interest are the many parallels between birdsong and human speech. Songbirds are the best and perhaps the only viable animal model of complex human-like vocal communication (Marler, 1967; Doupe and Kuhl, 1999; Kuhl, 2003). About 13% of the American population (roughly 36.5 million people) struggle with a communicative disorder. Among these are spasmodic dysphonia, stuttering, and familial language impairment (Gallagher and Watkin, 1997). Stuttering, for example, affects Americans with a prevalence of 1% and a lifetime incidence of 5%, and has a strong genetic component (Andrews and Harris, 1964; Bloodstein, 1995; Mansson, 2000). A number of other heritable syndromes have major if not exclusive involvement with vocal communication systems including autism, Parkinson's disease and Fragile X syndrome (Roberts *et al.*, 2005).

FoxP2 is the first gene associated with a human communicative disorder that has been studied in some detail in the songbird. The gene encodes a transcription factor protein, and it is believed to act in the development of cerebellar and corticostriatal circuits that control voluntary orofacial movements. In humans, defined FoxP2 mutations cause orofacial dyspraxia. Intriguingly, the human mutations are associated with deficits not only in speech production but also in language comprehension. Taxonomic comparisons have failed to reveal any simple association of the protein's sequence and vocal learning ability across species (Haesler *et al.*, 2004; Scharff and Haesler, 2005). However, FoxP2 is expressed in Area X, the striatal nucleus of the song learning circuit, where it appears to be regulated during developmental and seasonal periods of song system plasticity (Haesler *et al.*, 2004; Teramitsu *et al.*, 2004), and also in adults during undirected singing (Teramitsu and White, 2006). Hence it seems likely that FoxP2 contributes to the development and function of vocal communication systems in both songbirds and humans (see Haesler and Scharff, this volume). A second example of a human disease-associated gene under study in songbirds is the gene first called "synelfin" in canaries (George *et al.*, 1995). Synelfin is the first gene formally shown to be regulated in the song control system at the time of song learning (George *et al.*, 1995; Jin and Clayton, 1997b; Hartman *et al.*, 2001). It was later recognized to be the ortholog of the alpha-synuclein gene (SNCA) implicated in Parkinson's disease (Clayton and George, 1998; George, 2001). The normal function of SNCA is obscure, despite numerous gene knockout studies in laboratory mice. For both FoxP2 and SNCA, studies in songbirds may complement research in more conventional model systems (such as the mouse) by allowing tests of function in the context of vocal system development and learning.

The complete annotation of the whole genome sequence of the zebra finch is likely to create more new opportunities for study of human disease gene orthologs in songbirds, much as it has in *Drosophila* (Reiter *et al.*, 2001) and rodents (Peltonen and McKusick, 2001; Huang *et al.*, 2004). The combination of active adult neurogenesis, rich behavior and accessible neuroanatomy in songbirds may inspire new approaches to therapy that we have yet to imagine (Nottebohm, 1985a).

CONCLUSION

We close with confidence that the coming zebra finch genome will transform the study of songbirds. We are optimistic that it will have even broader consequences in 20 years, strengthening the links between songbird research and many other domains of knowledge and investigation.

31 • Behavior-dependent expression of inducible genes in vocal learning birds

Claudio V. Mello and Erich D. Jarvis

INTRODUCTION: ACTIVITY-DEPENDENT GENE EXPRESSION AS A MAPPING TOOL

Brain cells communicate rapidly with each other using action potentials. These fast traveling electric pulses are generated within milliseconds of neuronal activation and are essential for normal brain function and expression of behavior. On a longer timescale, i.e. minutes to hours, neuronal cells also undergo changes in gene expression in response to their activation (Figure 31.1). This gene response consists of a transient increase in the transcription of specific genes that have low expression in the absence of stimulation, resulting in the accumulation of their mRNAs and encoded protein products in the cytoplasm of activated neurons (Goelet *et al.*, 1986; Morgan and Curran, 1989; Sheng and Greenberg, 1990; Clayton, 2000). The analysis of expression of these activity-dependent genes in brain sections allows the identification of recently activated neurons (Hunt *et al.*, 1987; Morgan *et al.*, 1987; Rusak *et al.*, 1990; Worley *et al.*, 1991; Mello *et al.*, 1992; Chaudhuri, 1997; Jarvis and Nottebohm, 1997; Guzowski *et al.*, 1999; Tischmeyer and Grimm, 1999). This approach provides very useful insights into the functional organization of brain pathways (reviewed in Herrera and Robertson, 1996; Kaczmarek and Robertson, 2002; Pinaud and Tremere, 2006).

By analogy with the early stages of viral infections, where a rapid induction of "immediate-early" viral genes occurs, genes displaying a rapid response to neuronal activation have also been called immediate early genes (IEGs). Many IEGs encode transcription factors that bind to specific motifs in the promoters of downstream target genes and subsequently modify the transcription of these targets (Morgan and Curran, 1989; Sheng and Greenberg, 1990). The most extensively studied IEGs encoding transcription factors are *c-fos*, *c-jun*, and *zif-268/egr-1* (Morgan and Curran, 1991; Knapska and Kaczmarek, 2004). Other IEGs such as *arc*, *homer*, and *actin* encode proteins linked to various aspects of neuronal cell physiology. These early or direct effectors, e.g. *arc*, *homer*, and *actin* are thought to exert their effects soon after neuronal cell activation (Guzowski *et al.*, 2000). In contrast to early effectors, the action of transcription factors on neuronal cells is indirect and delayed. Altogether, early effectors and transcription factors represent the early stages of a regulatory cascade that couples cell activation to long-lasting modulation of the neuronal properties (Goelet *et al.*, 1986; Clayton, 2000). Activity-dependent regulatory cascades have been postulated as a basic contributing mechanism for long-term memory formation in the nervous system of a wide range of organisms, from sea mollusks to mammals (Goelet *et al.*, 1986; Kaczmarek and Robertson, 2002).

Although the exact causal relationship between neuronal activation and the ensuing gene expression response is unknown, it is thought that depolarization-induced presynaptic neurotransmitter release leads to postsynaptic neurotransmitter receptor activation, which in turn induces calcium entry through voltage-dependent gates, activation of calcium-dependent signaling pathways (e.g. kinases), and activation of already present early transcription factor proteins (ETFs; Figure 31.1) such as CREB (Sheng and Greenberg, 1990; Treisman, 1996). CREB and other factors then regulate the mRNA expression of IEGs encoding transcription factors and effectors. The IEG response to neural signaling has been referred to as the *genomic action potential*, in reference to the parallel electrophysiological action potential (Clayton, 2000). However, the extent to which the electrophysiological and gene expression responses are coupled varies among different areas and neuronal populations (Mello and Clayton, 1995; Jarvis and Mello, 2000; Kaczmarek and Robertson, 2002).

Neuroscience of Birdsong, ed. H. Philip Zeigler and Peter Marler. Published by Cambridge University Press. © Cambridge University Press 2008.

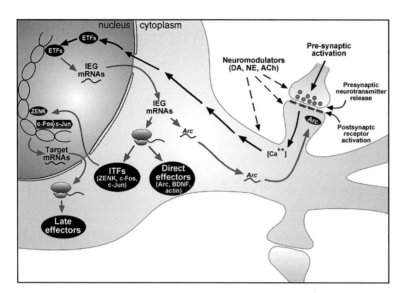

Figure 31.1 Hypothesis of synaptic mechanism for IEG induction associated with neuronal cell activation. Shown is a schematic representation of the cascade (arrows) triggered upon synaptic activation and calcium entry into a postsynaptic neuron. DNA is drawn as a double-stranded helix inside the nucleus, mRNAs are drawn as squiggly lines, which are translated into proteins by ribosomes (gray ovals). This cascade of events occurs simultaneously while the animals perceive and perform behaviors, and is thought to have long-term consequences on subsequent perception and behavior. Abbreviations: Ach, acetylcholine; DA, dopamine; ETF, early transcription factors; ITF, inducible transcription factors; NE, norepinephrine. (Modified from Velho et al., 2005.)

The expression of IEGs can be detected using *in situ* hybridization or immunocytochemistry for the localization of mRNA and protein products, respectively. In this manner, one can identify and analyze, with cellular resolution, brain areas activated in response to a specific stimulus or in association with a specific behavior. The mapping of activity-dependent gene expression can be performed in serial sections through the entire brain of freely moving animals without interfering with normal behaviors (e.g. Jarvis and Nottebohm, 1997). The standard paradigm involves repeated presentation of a stimulus or repeated production of a behavior within a set time window (30–60 min). The continued activation of responsive neurons results in the cellular accumulation of IEG mRNAs and proteins.

The identification of activity-dependent IEG expression does not imply a causal link between the gene(s) and other effects of the specific stimulus or behavior under study; determining such a link requires experiments involving manipulation of gene expression. Conversely, the absence of IEG expression in a given experiment does not necessarily imply a lack of neuronal activation, as some areas do not express specific IEGs. Moreover, although the temporal resolution of the method is low, its spatial resolution is high by comparison with electrophysiology recording methods. When used judiciously and in combination with behavioral manipulations and electrophysiological recordings, the analysis of IEG expression is a powerful tool for understanding information coding by brain circuits.

The mapping of IEG expression has yielded significant advances in birdsong neurobiology, including the identification and analysis of brain areas that participate in song perceptual processing, the functional organization of the circuitry that controls the production and acquisition of learned vocalizations, and a comparative analysis of perceptual and vocal control pathways across vocal learning avian groups (reviewed in Mello, 2002a, b; Jarvis, 2004a, b; Mello et al., 2004; Bolhuis and Gahr, 2006). This chapter discusses these three major areas of study in songbirds and other vocal learning birds (hummingbirds and parrots). We use "song-induced" to refer to gene expression triggered by auditory stimulation with birdsong, and "singing-induced" to refer to expression triggered by the vocal act of singing. Italicized lowercase refers to genes and uppercase refers to proteins – e.g. *zenk* and

Table 31.1. *Anatomical abbreviations*

AAC, central nucleus of the anterior arcopallium
Ai, intermediate arcopallium
ACM, caudal medial arcopallium
Area X, area X of the striatum
Av, avalanche
CLM, caudal lateral mesopallium
CMM, caudal medial mesopallium
CM, caudal mesopallium
CN, cochlear nucleus
CSt, caudal striatum
DLM, medial nucleus of dorsolateral thalamus
DM, dorsal medial nucleus of the midbrain
DMM, magnocellular nucleus of the dorsomedial thalamus
HVC, (a letter based name)
MOc, oval nucleus of the mesopallium complex
L2, Field L2
MMSt, magnocellular nucleus of the anterior striatum
MAN, magnocellular nucleus of anterior nidopallium
MLd, mesencephalic lateral dorsal nucleus
NAOc, oval nucleus of the anterior nidopallium complex
NCM, caudal medial nidopallium
NDC, caudal dorsal nidopallium
NIDL, intermediate dorsal lateral nidopallium
NIf, interfacial nucleus of the nidopallium
NLC, central nucleus of the lateral nidopallium
nXIIts, tracheosyringeal subdivision of the hypoglossal nucleus
Ov, nucleus ovoidalis
PAm, paraambiguus
RAm, retroambiguus
RA, robust nucleus of the arcopallium
Uva, nucleus uvaeformis
VA, vocal nucleus of the arcopallium
VAM, vocal nucleus of the anterior mesopallium
VAN, vocal nucleus of the anterior nidopallium
VAS, vocal nucleus of the anterior striatum
VLN, vocal nucleus of the lateral nidopallium
VMM, vocal nucleus of the medial mesopallium
VMN, vocal nucleus of the medial nidopallium

ZENK. We use the revised avian brain nomenclature (Reiner *et al.*, 2004a, b; Jarvis *et al.*, 2005a); anatomical abbreviations are in Table 31.1.

SONG-INDUCED AUDITORY EXPRESSION OF GENES IN SONGBIRDS

When songbirds hear song from other birds of their species, several IEGs are rapidly, markedly and transiently induced in discrete areas of the brain. This phenomenon was first described for *zenk* (Mello *et al.*, 1992), the avian homologue of the IEG *zif-268/egr-1/ngf1A/krox24*, which encodes a zinc-finger transcription factor (Milbrandt, 1987; Christy *et al.*, 1988; Lemaire *et al.*, 1988; Sukhatme *et al.*, 1988). This was one of the first clear demonstrations that an IEG can be induced in the brain by a natural stimulus of behavioral relevance, rather than by direct electrical stimulation, seizure activity, or stress (reviewed in Kaczmarek and Robertson, 2002). Song-induced *zenk* expression is of relevance to auditory learning, considering that this gene is involved in long-term synaptic plasticity and memory formation in mammals (Jones *et al.*, 2001). Other song-induced IEGs linked to neuronal plasticity and memory include the transcription factors *c-fos* and *c-jun* and the early effector *arc* (Nastiuk *et al.*, 1994; Velho *et al.*, 2005). All four genes are expressed in similar patterns and can colocalize in the same neuronal cells, indicating that they participate in an orchestrated cascade triggered by song stimulation (Velho *et al.*, 2005). The song-induced auditory expression of *zenk* requires MAP-kinase signaling (Cheng and Clayton, 2004; Velho *et al.*, 2005), but other components of the signaling cascade remain to be identified.

The brain areas that show song-induced IEG expression (Figure 31.2A) include the midbrain's auditory nucleus mesencephalic lateralis pars dorsale (MLd; equivalent to the central nucleus of the mammalian inferior colliculus), pallial forebrain regions (subfields L1 and L3 of the telencephalic auditory Field L), the caudomedial nidopallium – NCM, the caudomedial mesopallium – CMM, the shelf adjacent to vocal nucleus HVC (used as a letter-based abbreviation), the cup region adjacent to vocal nucleus RA (robust nucleus of the arcopallium), and a subpallial forebrain region (the caudal medial part of the striatum – CSt) (Mello and Clayton, 1994; Mello and

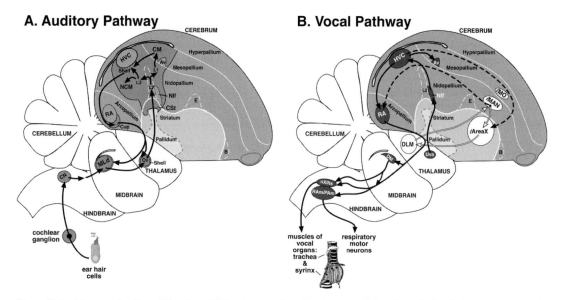

Figure 31.2 Diagram of auditory (A) and vocal (B) pathway regions that show song-induced and singing-induced gene expression. Only the most prominent or most studied projections are indicated. For the auditory pathway, NCM actually lies in a parasagittal plane medial to that depicted and reciprocal connections between pallial areas are not indicated. For the vocal pathways, black arrows show connections of the nuclei (in dark gray) of the posterior vocal pathway, white arrows show connections of the nuclei (in white) of the anterior forebrain pathway, and dashed lines indicate connections between the two pathways. For clarity, only the lateral part of the anterior vocal pathway is shown, and the connection from Uva to HVC is not depicted. Abbreviations are in Table 31.1. (Modified from Jarvis et al., 2005a.)

Ribeiro, 1998). Song-induced *zenk* expression has not yet been examined systematically in auditory nuclei caudal to the midbrain.

Forebrain areas showing song-induced IEG expression are part of a network that constitutes the central auditory processing pathways in songbirds (Figure 31.2A) (Kelley and Nottebohm, 1979; Vates et al., 1996; Mello et al., 1998a; reviewed in Mello, 2002a, b; Mello et al., 2004; for a discussion of auditory pathways and song selectivity, see Theunissen et al., this volume). This neuronal circuitry resembles closely the auditory pathways of nonoscine avian species such as pigeon, chicken, and budgerigar, indicating that the central auditory pathways are conserved among avian species, regardless of whether they evolved vocal learning (Karten, 1968; Brauth et al., 1987; Brauth and McHale, 1988; Wild et al., 1993; Metzger et al., 1998). This avian auditory system also shares important features with auditory pathways in mammals, such as thalamo-telencephalic reciprocity and the occurrence of descending sensory projections. For example, the auditory thalamic nucleus ovoidalis (Ov) both projects to and receives input from auditory telencephalic areas (Kelley and Nottebohm, 1979; Wild et al., 1993; Vates et al., 1996; Mello et al., 1998a), while MLd receives input from a descending projection originating in an auditory telencephalic area (Wild et al., 1993; Mello et al., 1998a).

Gene expression induced in auditory areas upon hearing song occurs whether or not a bird sings, and it also occurs in females, who normally do not sing (Mello et al., 1992). Furthermore, deafening abolishes *zenk* induction in these areas, whether or not a bird sings (Jarvis and Nottebohm, 1997). Thus, song-induced expression of *zenk* and other IEGs in auditory areas correlates with the experience of hearing birdsong, rather than with singing behavior. In fact, while early studies focused on the nuclei involved in song production, the use of IEG expression mapping has helped to shift the analysis to areas more fundamentally involved in the auditory processing of birdsong (reviewed in Mello, 2002a, b; Mello et al., 2004; Bolhuis and Gahr, 2006). Several songbird telencephalic areas have been analogized by diverse criteria to regions of the mammalian auditory system beyond the auditory primary projection area. Two of these regions (NCM and CM) exhibit robust IEG expression whose properties are suggestive of a role in song perception.

NCM

NCM shows the most robust *zenk* response to hearing song (Figures 31.2A, 31.3A) (Mello *et al.*, 1992; Mello and Clayton, 1994). It is a large caudomedial nidopallial expanse surrounded dorsally, ventrally, medially, and caudally by the lateral ventricle, but lacking a distinct lateral boundary (Mello and Clayton, 1994). Rostrally, NCM is contiguous with the medial part of Field L and separated from CMM by the mesopallial lamina. NCM's main inputs are from Field L and CMM, and it is reciprocally connected with the latter (Vates *et al.*, 1996). Based on connectivity, NCM occupies a position comparable to that of superficial (supragranular) layers of the mammalian auditory cortex (Karten and Shimizu, 1989; Vates *et al.*, 1996; Mello *et al.*, 1998a). Together with medial Field L, NCM and CMM comprise a caudomedial auditory lobule involved in various aspects of song perceptual processing.

In awake birds, NCM neurons have electrophysiological responses to several auditory stimuli, and these responses differ from those in L2 (Chew *et al.*, 1995, 1996a, b; Ang, 2001). Recordings in the canary have shown that NCM responses have wider tuning, lower amplitude, and rarely show inhibitory effects, in comparison to those in L2, which are typically vigorous and phasic, and with a tuning width that narrows over the course of the response, terminating rapidly at stimulus offset (Terleph *et al.*, 2006). In addition, tuning curves in NCM are often multipeaked and flatter than those in Field L2. Tuning width in NCM increases as the response unfolds, and poststimulus excitation is often sustained. Interestingly, neither NCM nor L2 responses seem to be selective for the bird's own song (Chew *et al.*, 1995; Ang, 2001), in contrast to the responses that can be recorded in the song control nuclei of different songbird species under anesthesia (Margoliash, 1983, 1986; Doupe and Konishi, 1991). This suggests that at least regarding response selectivity, NCM properties are intermediate between those of the primary auditory area L2 and the song control nuclei.

IEG studies have helped reveal the internal organization of NCM and its functional role in song processing. The expression of both *zenk* and *arc* in NCM is most robust for conspecific song, followed by heterospecific song, and nonsong auditory stimuli (Figure 31.3B, right panel) (Mello *et al.*, 1992; Velho *et al.*, 2005). Since this analysis is based on average densitometric measurements from autoradiograms, it reflects the overall activity of song processing circuits in NCM and not necessarily the tuning properties of individual neurons. The *zenk* response decreases upon repeated presentations of the same conspecific song and is reinstated upon presentation of a novel song (Figure 31.3B, middle panel) (Mello *et al.*, 1995). The decreased IEG response parallels the rapid and long-lasting decrease (habituation) of the electrophysiological responses of NCM neurons in awake birds (Figure 31.3C) (Chew *et al.*, 1995, 1996b; Stripling *et al.*, 1997). This habituation is song-specific, since the responses in habituated birds are reinstated upon presentation of a novel conspecific song. The habituation rates derived from habituation curves can be interpreted as an indication of whether a song is "remembered" or not by neuronal cells or circuits. Thus, NCM neurons, either individually or acting as ensembles of units, are able to discriminate between different conspecific songs, a property required for song perception and discrimination.

Stimulation with whole songs of most species induces widespread *zenk* expression in NCM, making it difficult to use such stimuli to infer the rules of auditory representation in this area. One approach to this problem has been to use canary song, because it can be dissected into phrases containing individual component syllables that are much simpler in their acoustic structure than a whole song (Güttinger, 1985; Vallet *et al.*, 1998) and a semi-automated mapping method that takes into account ZENK expression levels and the location of ZENK-expressing cells (Cecchi *et al.*, 1999). This approach revealed that the topographical patterns of ZENK expression in canary NCM correlate with acoustic features of the stimulus (Ribeiro *et al.*, 1998). Presentation of the prominent whistle component of canary song results in clusters of ZENK-expressing cells whose position along the dorsoventral axis of rostral NCM varies as a function of frequency, but yields very little ZENK expression in caudal NCM (Figure 31.3D, left panel). Combinations of whistles elicit ZENK expression patterns that are not the sum of the patterns resulting from the individual component

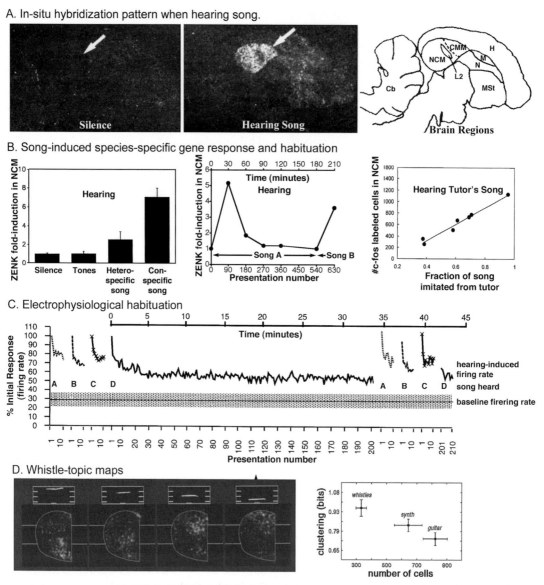

Figure 31.3 Song-induced IEG expression and related neuronal firing in songbird NCM. (A) Accumulated *zenk* mRNA in adult male zebra finches, comparing a silent control and a bird presented with playbacks of other male songs for 30 minutes. Arrow points to regions of high hearing-induced *zenk* expression on the autoradiogram; as detailed by the anatomical diagram on the right, this includes NCM and CMM. (B) Accumulated *zenk* mRNA in zebra finch NCM, relative to silence, when songbirds hear tones, conspecific or heterospecific songs (left panel); when birds hear novel versus familiar songs (middle panel, high induction occurs upon hearing novel song A for 30 minutes [90 presentations], but not after hearing many more presentations [up to 540], followed by high induction again upon hearing novel song B [90 presentations]); and when birds hear their tutor's song as adults (right panel, showing relationship of amount of song imitated and amount of *c-fos* induced in NCM). (C) Electrophysiological activity in zebra finch NCM recorded in response to hearing novel songs for the number of presentations indicated. (D) Maps of ZENK protein induction in canary NCM after hearing whistles of different frequencies (left panel), and the relative degree of clustering of activated neurons in NCM when hearing whistles, pure synthetic whistles, or guitar notes (right panel). (Modified from Mello *et al.*, 1992 for A, B left panel, 1995 for B middle panel; Bolhuis *et al.*, 2000 for B, right panel; Stripling *et al.*, 1997 for C; Ribeiro *et al.*, 1998 for D.)

whistles, but are instead distinct patterns of their own including ZENK-positive cells in caudal NCM. Similarly, the response to natural whistle stimuli cannot be predicted from the response to equivalent, but modified synthetic stimuli (Figure 31.3D, right panel). This observation suggests that NCM neurons may be tuned to subtle features present in the natural stimuli taken directly from the song. Finally, fast frequency modulations elicit low levels of ZENK expression across broad regions of NCM, including the regions activated by individual whistles. Overall, the analysis of ZENK induction in canaries indicates that the activation patterns in NCM contain enough information to discriminate among the various components of the song stimulus (Ribeiro et al., 1998).

Further evidence that *zenk* expression patterns in NCM correlate with acoustic features of the song stimulus comes from experiments in starlings and white-crowned sparrows. In starlings, greater ZENK expression occurs within the dorsal subregion of NCM when females hear their preferred songs, i.e. long song stimuli, as compared with short ones (Gentner et al., 2001). Long songs are likely to contain more complex acoustic features due to the substantially larger motif repertoire size compared with short songs. Thus, local variations in ZENK expression in starling NCM depend on features present in the song stimulus. In white-crowned sparrows, greater ZENK expression occurs within the ventral subregion of NCM when females hear their home dialects of song relative to foreign dialects (Maney et al., 2003). Taken together with the work in canaries, these observations are consistent with a role for NCM in the perceptual processing of various features of song.

Although the evoked electrophysiological activity and the ensuing gene expression response in NCM are primarily driven by auditory stimulation, mounting evidence indicates that such activity can be modulated by other sensory modalities and/or behavioral state. For instance, song-induced *zenk* expression in NCM can be enhanced by associatively coupling the song stimulus with a foot shock during active avoidance learning, even though the shock by itself has no effect on NCM (Jarvis et al., 1995). Modulation of the *zenk* response also occurs when combining song presentation with behaviorally relevant visual stimulation, such as the image of another bird (Avey et al., 2005), by manipulations of the spatial location and intensity of the song stimulus, or by coupling song with light stimuli (Kruse et al., 2004), but not by the simultaneous presentation of song and background noise (Vignal et al., 2004a). A modulatory effect has also been observed when comparing restrained versus unrestrained birds (Park and Clayton, 2002). The song-induced *zenk* expression is abolished by anesthetic agents (Mello, 1993), indicating the strong influence of behavioral state. Other studies show that the quality and behavioral relevance of the song stimulus can affect the expression levels of *zenk* (Gentner et al., 2001; Eda-Fujiwara et al., 2003; Maney et al., 2003; Phillmore et al., 2003), suggesting that the bird's degree of attentiveness to the song is an important factor in determining *zenk* expression in NCM. Finally, the social context, i.e. the presence of another quiet bird, causes higher song-induced *zenk* expression in NCM (Vignal et al., 2005). There is no anatomical evidence for direct inputs from other sensory modalities to NCM, nor is there any evidence that nonauditory modalities can directly affect neuronal activity in NCM. Therefore, these indirect effects on song-inducible gene expression are more likely to be associated with changes in the bird's attention and/or alertness, and potentially mediated by neuromodulators such as noradrenaline or acetylcholine (Ribeiro and Mello, 2000), as observed for stimulation-driven IEG expression in the mammalian brain (Cirelli et al., 1996; Pinaud et al., 2000). Based on these precedents, neuromodulatory systems probably do not act by themselves in NCM, but rather act by modulating NCM's activation by auditory stimuli.

Several studies have examined whether song-induced gene expression in NCM is linked to song memorization and/or the storage of song-related memories. For example, like the findings on song habituation in zebra finches, recent auditory experience with long and short bouts of songs influences the number of neurons with ZENK and c-Fos responses in the NCM of adult starlings (Sockman et al., 2002, 2005). In zebra finches, at early posthatch ages, *zenk* is expressed at relatively high levels in NCM and other brain areas, but lacks inducibility by song (Jin and Clayton, 1997a). Around posthatch day 30, in conjunction with the beginning of the sensitive period for vocal learning, *zenk* expression (as well as the levels of electrophysiological activation in NCM) becomes responsive to song auditory stimulation, suggesting a temporal correlation between IEG inducibility and song learning (Jin and Clayton, 1997a; Stripling et al., 1997; Stripling et al., 2001). Furthermore, the number of cells that express *zenk* in adult males stimulated with tutor songs correlates positively with the extent to which these males copied their songs from the tutors during the song learning period (Figure 31.3B, right panel) (Bolhuis et al., 2000, 2001; see also Bolhuis, this volume). This effect does not appear to simply reflect individual differences in attention, as a comparable correlation is absent in birds exposed to novel conspecific songs, which they are likely to attend to during adulthood (Terpstra et al., 2004).

These observations suggest that the induction of an IEG like *zenk* in NCM may be involved in some aspect of the acquisition and/or retention of song auditory memories. Interestingly, other developmental studies have provided evidence that factors such as sex, age and early auditory experience are modulators of the ZENK and c-Fos protein responses to song in the NCM of zebra finches and other songbird species (Bailey and Wade, 2003, 2005; Phillmore et al., 2003; Hernandez and MacDougall-Shackleton, 2004). These findings suggest that the modulation of the expression of plasticity-associated IEGs such as *zenk* could be a mechanism through which such factors influence the processing of song during the sensitive period for song learning.

The habituation of NCM's electrophysiological responses to song has been suggested as a possible cellular mechanism contributing to the perceptual memorization of song. The degree to which a bird has copied the tutor's song during the learning period correlates positively with the habituation rate of NCM units upon presentation of the tutor's song during adulthood (Phan et al., 2006). Such findings parallel those of gene expression studies and provide further evidence for a role of NCM circuits in the formation or storage of song auditory memories. Importantly, the long-term maintenance of the habituation of NCM responses depends on local song-induced gene expression. This was shown by placing injections of protein or RNA synthesis blockers in NCM during the time window that follows song stimulation, and testing whether the habituation was still present after various poststimulation intervals (Chew et al., 1995). *Zenk* and other song-induced IEGs are regulated in NCM within this same time window and, therefore, could be involved in long-term habituation, but testing this hypothesis will require manipulating the expression of specific genes.

CM

Pronounced ZENK induction by song also occurs in the caudal mesopallium, CM (Figure 31.2A), triggering a renewed interest in this area. Tract-tracing in zebra finches (Vates et al., 1996) has established that CM can be divided into medial and lateral subdomains, respectively CMM and CLM. CMM is reciprocally connected with NCM and CLM, and CLM is reciprocally connected with several Field L subdivisions. In addition, CLM sends projections to the shelf region under vocal nucleus HVC and to vocal nucleus NIf (interfacial nucleus). The latter projection is particularly important, as it represents an anatomical substrate for a direct entry of auditory information into the song control system.

In nonoscine species CM is reciprocally connected with Field L, indicating a role in the processing of auditory signals (Bonke et al., 1979a; Brauth and McHale, 1988; Wild et al., 1993). This is consistent with electrophysiological studies, which established that responses in CM have a longer latency than those in Field L and are tonotopically organized (Muller and Leppelsack, 1985; Heil and Scheich, 1991). In addition, CM contains a high density of neurons that have selectivity for complex stimuli, suggesting that it represents a higher-order auditory station than Field L (Muller and Leppelsack, 1985).

In songbirds, lesions to CLM result in a disruption of normal song preferences in female zebra finches (MacDougall-Shackleton et al., 1998a), indicating that this area participates in the processing and/or discrimination of behaviorally relevant stimuli. Electrophysiological recordings have shown that the selectivity properties of CM neurons are strongly dependent on the bird's experience and can be modified by perceptual learning (Gentner and Margoliash, 2003), providing further evidence for the participation of CM in the processing of behaviorally relevant song stimuli. Interestingly, several studies in various species have found significant differences in the ZENK activation patterns between CMM and NCM (Avey et al., 2005; Bailey and Wade, 2005; Leitner et al., 2005; McKenzie et al., 2006; Terpstra et al., 2006), but further studies are needed to uncover the functional significance of such differences.

Other areas with song-induced gene expression

As described above, song presentation induces *zenk* in the shelf under HVC and the cup rostroventral to RA (Figure 31.2A) (Mello and Clayton, 1994). The Nissl delineation of the respective vocal nuclei provides a clear-cut boundary for these two areas of *zenk* expression. Their intermediate location between auditory areas and song control nuclei suggested that the shelf and/or cup might constitute sites of entry for auditory information into the song system, and thereby play a role in sensorimotor integration (Nottebohm, 1996). Morphological and tract-tracing studies revealed that this may be partly the case, as well as other important features of the shelf and cup regions (Kelley and Nottebohm, 1979; Fortune and Margoliash, 1995; Vates *et al.*, 1996; Foster and Bottjer, 1998; Mello *et al.*, 1998a):

(1) Both the shelf and cup are termination zones of projections from subfields L1 and L3, and from CLM.
(2) Neurons in the shelf send fibers into HVC, where they branch out and terminate.
(3) Dendrites from HVC neurons located close to its ventral border reach out into the shelf, where they could receive terminations from Field L projections.
(4) No conclusive evidence has been uncovered for a direct projection from Field L subfields to HVC.
(5) The shelf projects in a spatially organized manner to the cup.
(6) The cup originates a descending projection that terminates onto nuclei of the ascending auditory pathway, similar to descending projections that originate in infragranular layers of the mammalian auditory cortex.
(7) There is no evidence for a direct projection from the cup into RA.

In summary, the shelf and cup regions are integral components of the central auditory pathway. The former provides potential access for auditory information to the song system whereas the latter is in a unique position to influence the flow of auditory information along the ascending auditory pathway (for further discussion of these regions, see chapters by Wild, Prather and Mooney, and Theunissen *et al.*, this volume).

Areas lacking song-induced gene expression

Neither *zenk*, *c-fos*, *c-jun* or *arc* are induced in the thalamic auditory nucleus Ov or in the primary thalamo-recipient telencephalic zone Field L2 (Mello and Clayton, 1994; Nastiuk *et al.*, 1994; Mello and Ribeiro, 1998; Velho *et al.*, 2005), even though these areas are major stations of the ascending auditory pathway (Karten, 1968; Kelley and Nottebohm, 1979; Vates *et al.*, 1996), and the neuronal cells in L2 are known to be activated during song auditory stimulation (Chew *et al.*, 1995; Sen *et al.*, 2001; Terleph *et al.*, 2006). Even the use of metrazole, a strong depolarizing agent that causes seizures and widespread gene induction (Mello and Clayton, 1995), fails to induce *zenk* expression in Field L2. Metrazole is also ineffective in inducing *zenk* in other thalamo-recipient sensory regions of the telencephalon, namely the visual entopallium (E) and the somatosensory basorostralis (B; Figure 31.2A). Therefore, it seems that induction of these IEGs has been uncoupled from neuronal activation in Ov, L2, and other primary thalamo-recipient zones. The functional significance of such uncoupling could be that Ov and L2 serve primarily as relays of auditory inputs to higher-order areas in the telencephalon, instead of representing sites that undergo active experience-dependent plasticity. Alternatively, it is possible that a different set of IEGs not yet identified are involved in IEG expression in these primary "sensory" structures.

Importantly, auditory stimulation with conspecific song or even the bird's own song does not induce IEG expression in the nuclei of the song control system (Mello *et al.*, 1992; Mello and Clayton, 1994; Nastiuk *et al.*, 1994; Jarvis and Nottebohm, 1997; Mello and Ribeiro, 1998; Velho *et al.*, 2005). This finding seemed paradoxical at first, given that robust electrophysiological responses (recorded in anesthetized birds) occur in song control nuclei upon auditory stimulation with conspecific birdsong, particularly with playbacks of the bird's own song (Katz and Gurney, 1981; McCasland and Konishi, 1981; Williams and Nottebohm, 1985; Margoliash, 1986; Doupe and Konishi, 1991; Vicario and Yohay, 1993). Possible explanations were: (1) neuronal activation might be uncoupled from regulation of IEGs in vocal nuclei, as appears to occur in auditory Ov and L2; (2) hearing song might induce IEGs in vocal nuclei only during an as yet undefined developmental stage, and (3) song stimulation might not lead to the activation of vocal nuclei in awake birds. The subsequent finding that the act of singing causes a robust induction of *zenk* and c-Fos in vocal nuclei (see next section) excluded an absolute uncoupling between electrophysiological activation and IEG expression within the song system. In addition, the electrophysiological activation of vocal nuclei in response to hearing song was found to be robust during sleep but weak and sometimes absent in awake zebra finches (Dave *et al.*, 1998; Schmidt and Konishi, 1998; Nick and Konishi, 2001; Rauske *et al.*, 2003; Cardin and Schmidt, 2004a), the same state as in gene expression studies. Although it is not clear what role auditory responses recorded during sleep might play in vocal communication, the fact that such responses are much weaker in awake birds argues against a prominent role of vocal nuclei in the perceptual processing of birdsong, at least in zebra finches. Interestingly, recent electrophysiological recordings reveal significant song-evoked auditory responses in the HVC of awake swamp sparrows (see chapter by Prather and Mooney, this volume), pointing to important species differences in the processing of song.

SINGING-REGULATED EXPRESSION OF GENES IN SONGBIRDS

When songbirds sing, the expression of IEGs is markedly increased in nuclei of both the direct vocal-motor pathway and the anterior forebrain pathway of the song control system (Figure 31.2B). Specifically, singing-induced IEG expression is seen in seven telencephalic nuclei: HVC, NIf, and the magnocellular nucleus of the nidopallium (MAN; lateral and medial parts, respectively LMAN and MMAN), RA, nucleus avalanche (Av) and an oval-like nucleus of the mesopallium (MO-like), and Area X of the striatum, as well as in the midbrain's dorsomedial nucleus of the intercollicular complex (DM; Figure 31.4) (Jarvis and Nottebohm, 1997; Kimpo and Doupe, 1997; Jarvis *et al.*, 1998; Wada *et al.*, 2006).

IEG activation in HVC and RA was consistent with the observation that these nuclei show premotor neuronal firing during singing (McCasland and Konishi, 1981; Yu and Margoliash, 1996). Analysis of the other nuclei, however, yielded several additional insights. Av is the target of a projection from HVC (Nottebohm *et al.*, 1982) about which very little was known and the *zenk* experiments demonstrated that this nucleus is active during singing. MO-like, a small nucleus named for its similar location to vocal nucleus MO in parrots (Jarvis and Mello, 2000; Reiner *et al.*, 2004a), had not been previously described in songbirds. Although knowledge about songbird MO is still scarce, other IEGs (*c-fos*, *c-jun*, and *arc*) are also activated in this nucleus during singing (Figure 31.4) (Wada *et al.*, 2006). Within HVC, experiments combining neuronal tract-tracing and IEG expression revealed that both of its projection neuron types (RA- and X-projecting neurons) have singing-activated *zenk* expression (Jarvis *et al.*, 1998). This finding was confirmed with electrophysiological recordings of identified RA- and X-projecting neurons within HVC during singing (Hahnloser *et al.*, 2002). *Zenk* is not induced by singing in the thalamic vocal nucleus DLM (Jarvis *et al.*, 1998).

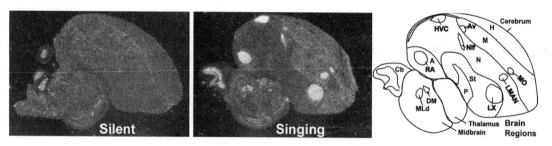

Figure 31.4 Singing-induced gene expression in vocal pathways. Accumulated *c-fos* mRNA in adult male zebra finches comparing a silent bird and one that sang while alone for 30 minutes. A *c-fos* pattern is depicted because it shows robust expression in all vocal nuclei, including those in the mesopallium (M). All seven cerebral vocal nuclei as well as DM of the midbrain can be seen to have singing-induced gene expression. MLd shows expression that is thought to be due to the bird hearing itself sing. (Figure courtesy of Haruhito Horita and Kazuhiro Wada, modified from Wada *et al.*, 2006.) A, arcopallium; Cb, cerebellum; H, hyperpallium; M, mesopallium; N, nidopallium; P, pallidum; St, striatum. Other abbreviations as in Table 31.1.

DLM is within the anterior vocal pathway and is expected to show changes in neuronal firing during singing (Luo et al., 2001; Person and Perkel, 2005). Thus, it is possible that electrophysiological and transcriptional activation are uncoupled in DLM, as they appear to be in auditory thalamic nucleus Ov, or that a different set of IEGs are regulated by activity in this nucleus. These findings of IEG studies during singing have consolidated the notion that there is a distinct separation of brain areas activated by song perception and song production (Jarvis and Nottebohm, 1997), as detailed below.

Singing-induced IEG expression is motor driven

Because singing activity results in auditory feedback of the bird's own vocalizations, which may reach the song control system under some conditions, it was important to determine whether such feedback contributed to singing-induced IEG expression. However, singing-induced expression was present in deafened birds at the same levels as in intact animals producing similar amounts of song (Jarvis and Nottebohm, 1997; Kimpo and Doupe, 1997). Furthermore, birds made mute by section of the tracheosyringeal branch of nXII to the syringeal muscles still showed induced gene expression in all seven telencephalic vocal nuclei when they attempted to sing (Jarvis and Nottebohm, 1997). These findings suggest that neither auditory nor muscle proprioceptive feedback is necessary for singing-induced expression. Other experiments revealed that the levels of accumulated IEG mRNA in vocal nuclei are linearly proportional to the amount of singing for a 30 minute period. The IEG mRNA expression in song control nuclei is not present before a bird that has been silent for over 1 hour sings, but appears within 5–10 min after singing starts. As a bird continues to sing past 30 min, the mRNA levels for *zenk* do not increase further, but also do not habituate as is the case in auditory areas; instead, the mRNA levels are maintained at lower post 30-min singing levels. This suggests that, as singing continues, either *zenk* mRNA is continuously synthesized and degraded at a steady state level, or it is stabilized. In either case, the mRNA levels reflect the moment-to-moment changes in the birds' singing behavior. Taken together, these results confirm that the singing-induced gene expression in vocal nuclei is motor driven.

Area X and LMAN

One of the most intriguing results of the singing-induced IEG experiments was the observation that Area X and MAN (both MMAN and LMAN) showed singing-induced *zenk* expression, with Area X showing the highest levels of activation of all vocal nuclei (Jarvis and Nottebohm, 1997). Both Area X and LMAN, nuclei of the anterior vocal pathway, were previously considered not to have a functional role in adults, since lesions to these nuclei in adult zebra finches had no noticeable effects on song, while lesions in juveniles caused profound deficits in vocal learning (Bottjer et al., 1984; Sohrabji et al., 1990; Scharff and Nottebohm, 1991). Although one early electrophysiological study found no singing-associated neuronal firing in Area X (McCasland, 1987), subsequent electrophysiological findings showed that Area X and LMAN show premotor neuronal firing during singing (Hessler and Doupe, 1999a), confirming that the singing-associated activation of these nuclei is motor-driven (Jarvis and Nottebohm, 1997).

A possible role of Area X and LMAN in adult song production emerged with the finding that singing-induced IEG expression in these nuclei (as well as in RA) varies dramatically with the social context in which song is produced (Jarvis et al., 1998). Zebra finches produce two song types, undirected and directed. The two differ little in the song pattern but greatly in its mode of delivery. Undirected song is sung to no particular bird and is thought to be used for advertisement or vocal practice, whereas directed song is produced with a dance while facing another bird, usually a female, and is thought to be used for courtship (chapter 10 in Zann, 1996). Undirected singing results in high levels of *zenk* expression in all seven telencephalic vocal nuclei, whereas directed singing results in high expression only in HVC, the dorsal cap of RA, NIf, MO, MMAN, and the medial part of Area X, and moderate to no induction in the core of RA, LMAN and the lateral part of Area X (Figure 31.5A). These observations suggest the existence of two parallel anterior forebrain pathways (a medial and a lateral one) that are differentially activated according to the context of singing; the former, through the MMAN-to-HVC projection, would modulate singing-related activation inside HVC, and the latter, through the LMAN-to-RA projection, would modulate activation in RA (Jarvis

A. Social Context-Dependent Gene Regulation

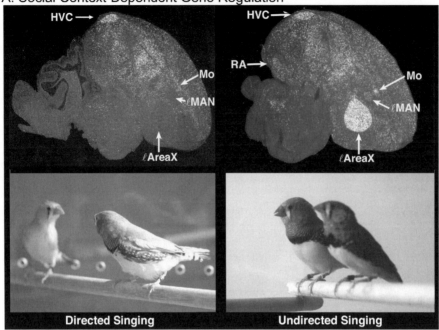

B. Social Context-Dependent Electrophysiological Activity

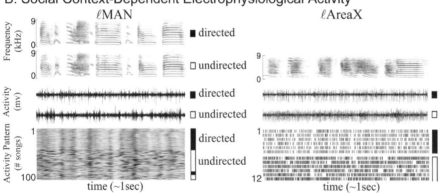

C. Activation Model

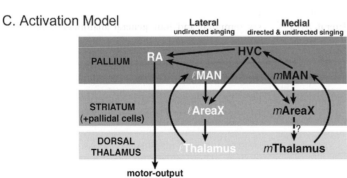

and Nottebohm, 1997). This hypothesis is partly supported by connectivity data (indicated by arrows in Figure 31.5C), but further analysis is required to fully verify it.

The finding that Area X and LMAN in zebra finches show greater neuronal activation during undirected than directed singing (Hessler and Doupe, 1999b) is consistent with the IEG findings. Interestingly, the firing pattern is quite variable from song bout to song bout during undirected singing, but is closely aligned to the song syllables during directed singing (Figure 31.5B). Furthermore, LMAN activity is required for the small but detectable amount of variability in song syllable structure seen during undirected singing, but not for the highly stereotyped song produced during directed singing (Kao et al., 2005). Chipping sparrows also sing a more variable song to interact with males and a more stereotyped song to attract females. In this species, the amount of variability in the song being produced is the strongest predictor of zenk expression in lateral Area X and LMAN (Liu and Nottebohm, 2005). Taken together with the result that Area X lesions can cause stuttering in Bengalese finches (Kobayashi et al., 2001), these findings indicate that the anterior forebrain nuclei exert some online control over singing, and that the activation of the lateral portion of these nuclei helps introduce variability to the song structure.

Investigations into the mechanism of social context differences have yielded strong indications for a modulatory role of catecholamines. For example, the dopaminergic ventral tegmental area (VTA) in the midbrain shows singing-associated gene expression in song sparrows and starlings, with the expression levels dependent upon context and breeding season, indicating that the VTA could influence the song system (Maney and Ball, 2003; Heimovics and Riters, 2005; see also Ball et al., this volume). Dopamine levels within lateral Area X increase during directed but not undirected singing, an effect that requires the dopamine reuptake transporter at the terminals of the VTA-to-X projection (Sasaki et al., 2006). This selective rise in dopamine levels in Area X might be associated with the lack of zenk expression during directed singing, a hypothesis that requires further experimentation. An involvement of noradrenaline is suggested by the finding that systemic removal of noradrenergic innervation causes high zenk expression in lateral Area X during directed singing, thus abolishing the context difference in this nucleus (Castelino and Ball, 2005). Currently available data indicate that the brain's main noradrenergic nucleus, the locus coeruleus (LoC), does not provide direct input to Area X (Mello et al., 1998b). Therefore, further studies are needed to determine whether noradrenaline exerts a direct or indirect effect in Area X.

A cascade of singing-regulated genes

Although early studies on singing-regulated gene expression were focused on *zenk* and to a lesser extent on *c-fos*, subsequent studies have shown that singing behavior involves regulation of a substantially larger number of genes (Li et al., 2000; Lombardino et al., 2005, 2006; Velho et al., 2005; Lombardino et al., Teramitsu and White, 2006; Wada et al., 2006). The most extensive study (Wada et al., 2006) used high-throughput microarray techniques and revealed the regulation of 33 genes (including *zenk* and *c-fos*) whose functions span a wide range of molecular and cellular categories, from structural to signal transduction, trafficking, and synapse-related molecules (for details, see Figures 3 and 4 in Wada et al., 2006). These genes are up- or down regulated in different patterns within 1–3 hr after singing. There are at least six temporal and four general anatomical patterns of

---

Caption for Figure 31.5

Social context-dependent gene expression and neural activity in the song system of adult male zebra finches. (A) Top panels: Brain sections showing high singing-driven *zenk* mRNA expression (white) in all vocal nuclei during undirected singing (right) and low expression in lateral Area X (LAreaX), LMAN, and RA during directed singing (left). Bottom panels: Examples of birds singing to a female (left) or singing undirected song (right). (B) Sonograms (top), electrophysiological activity recordings (middle), and activity patterns across repeated song motifs (bottom) from LMAN and LAreaX during directed and undirected singing. Each tick mark represents one burst of action potentials. (C) Proposed connectivity model of the functional medial and lateral components of the anterior vocal pathway and their respective influence on the posterior vocal pathway nuclei HVC and RA. Dashed arrow depicts a connection demonstrated only in the anterograde direction; dashed arrow with question mark depicts a hypothesized connection. (Modified from Jarvis et al., 1998, for A and C; and Hessler and Doupe., 1999b for B.)

expression, with most showing the highest regulation in Area X, followed by HVC. In general, the data suggest that singing-regulated genes are integrated into regulatory networks, and that overlapping, but distinct, signal transduction pathways are activated in different vocal nuclei by singing. Several genes that peak after 1 hr, such as beta-actin and actin-associated molecules, glyceraldehyde-3-phosphate dehydrogenase, and NADH dehydrogenase, can be considered "housekeeping" genes, suggesting that the late gene response could have a general function such as replenishing protein products and/or other cellular components that are used up during neuronal cell activation. Further insight into singing-induced regulatory networks comes from studies of some individual genes.

The singing upregulation of *arc* in all telencephalic vocal nuclei (Velho et al., 2005; Wada et al., 2006) is intriguing in that *arc* is localized to recently activated sites along dendrites (Link et al., 1995; Lyford et al., 1995; Steward et al., 1998) and is associated with synaptic plasticity (Guzowski et al., 2000). This suggests that functional and/or structural changes at dendrites may occur as a result of singing behavior. Synaptotagmin IV, a gene involved in neurotransmitter release from synaptic vesicles, is upregulated in pallial vocal nuclei during undirected but not directed singing (Poopatanapong et al., 2006; Wada et al., 2006). This suggests that HVC has differential modulation across different social contexts, and that synaptotagmin IV is sensitive to this modulation but not *zenk*, since the latter is upregulated in HVC by both directed and undirected singing. In contrast, the *foxP2* (forkhead box 2) gene, which encodes a known transcriptional suppressor, is expressed in Area X and downregulated after 2 hours of undirected but not directed singing (Teramitsu and White, 2006). Thus, genes that are suppressed by FoxP2 in Area X are likely turned on after a long period of singing. FoxP2 basal levels increase during the late critical period of vocal learning in zebra finches and during the seasonal period of new song acquisition in canaries (Haesler et al., 2004). These results are exciting because *foxP2* is the first gene described that when mutated in humans causes apraxia (Lai et al., 2001). Further studies are required, however, to establish a role for this gene in vocal learning in songbirds (see Haesler and Scharff, this volume).

Another singing regulated gene, brain derived neurotrophic factor (BDNF), is a synaptically released protein involved in cell survival and neurogenesis (Li et al., 2000). In mammals, BDNF mRNA levels are regulated by neural activity in pallial/cortical areas and the protein is transported anterogradely to connected striatal neurons (Altar et al., 1997). In songbirds, singing increases BDNF mRNA in pallial vocal nuclei (Li et al., 2000; Wada et al., 2006). HVC neurons that show BDNF regulation by singing are those that project to RA, which are continually replaced throughout adulthood (Nottebohm, 2002). Experimentally enhancing HVC's BDNF expression in the absence of singing enhances the local survival of new neurons that enter HVC (Rasika et al., 1999; Alvarez-Borda et al., 2004). Likewise, singing enhances the survival of new neurons in HVC; as with BNDF expression, this enhancement is linearly proportional to the amount of singing (Li et al., 2000; Alvarez-Borda and Nottebohm, 2002). A blockade of BDNF expression during singing is still needed to determine whether this gene is required for singing-enhanced neuronal survival. Another interesting gene is ubiquitin carboxy-terminal hydrolase L1 (UCHL1), which is expressed at low levels in replaceable neuronal populations (of both birds and mammals) and upregulated by singing in HVC's RA-projecting neurons (Lombardino et al., 2005). UCHL-1 dysfunction has been implicated in neuronal cell loss in diseases such as Parkinson's, Alzheimer's, and Huntington's (Das et al., 2006b). An intriguing possibility that requires testing is that the upregulation of this gene by singing or exercise might lead to increased neuronal survival (see Pytte et al., this volume).

Potential mechanisms of singing-regulated gene expression

Like song-induced gene expression in auditory areas, singing-induced gene expression may depend on signal transduction pathways activated after synaptic neurotransmitter receptors bind their specific ligands (Figure 31.1). However, gene regulatory pathways in vocal nuclei may be more specialized than in other parts of the brain. For example, all four major vocal nuclei (HVC, RA, MAN, and Area X) have differential expression of particular glutamate receptors (Wada et al., 2004). AMPA receptors, involved in fast synaptic transmission, tend to be low in vocal nuclei. NMDA receptors, involved in slower synaptic transmission and in synaptic plasticity, have a more complex distribution: the NR2A subtype, involved in synapse stabilization (Dumas, 2005), is higher in vocal nuclei relative to the surrounding regions, whereas the NR2B subtype, involved in synapse plasticity, is lower relative to the surroundings (Wada et al., 2004). These receptor levels change during development, such that NR2A is lower and NR2B higher during the early stages of vocal learning (Basham et al., 1999; Heinrich et al., 2002). Interestingly, zenk expression in vocal nuclei during singing is higher in juveniles than in adults (Jarvis and Nottebohm, 1997; Jin and Clayton, 1997a; Jarvis et al., 1998). Because NMDA receptor activation induces zenk and c-fos expression in mammalian neurons (Platenik et al., 2000), the developmental changes in glutamate receptor and singing-regulated gene expression may be linked.

COMPARATIVE ANALYSIS OF VOCAL LEARNERS AND NONLEARNERS

The discovery of regulated gene expression in the songbird brain in the contexts of hearing song and singing prompted similar investigations in the other vocal learning avian groups, namely parrots and hummingbirds. The goals were twofold: to determine whether behavior-dependent gene regulation was unique to the song system of songbirds or a general feature of vocal learners, and to use the findings to gain insights into the evolution of vocal learning and associated brain structures. Phylogenetic studies place parrots, hummingbirds, and songbirds as distant relatives, with each having more closely related vocal nonlearning orders (Figure 31.6A) (Sibley and Ahlquist, 1990). This led to the hypothesis that vocal learning evolved independently in the three vocal learning avian lineages (Nottebohm, 1972a). However, not all avian orders have been experimentally tested for vocal learning, and an isolated species within a given order, such as bellbirds within the suboscines (Kroodsma, 2005) could be a vocal learner. Nevertheless, comparative studies have revealed some remarkable similarities among songbirds, parrots, and hummingbirds with respect to both vocal-related brain structures and inducible gene expression.

Auditory pathways in parrots, hummingbirds, and vocal nonlearning birds

In both budgerigars (a representative small parrot species) and hummingbirds (sombre and rufous-breasted), playbacks of species-specific songs induces zenk mRNA in caudal telencephalic areas (Figure 31.6B) (Jarvis and Mello, 2000; Jarvis et al., 2000). Although the exact shape of these auditory regions differs among the three vocal learning groups, likely reflecting broad differences in the relative sizes and shapes of telencephalic subdivisions across avian groups, their relative locations are quite similar. These include a cluster of medial regions (NCM, CMM, and CSt), a lateral region in the nidopallium (parrot NIDL, hummingbird NDC) and a region in the arcopallium (parrot ACM, hummingbird Ai). Zenk upregulation is also seen in the midbrain's MLd, but not in Ov or L2 (although Brauth et al., 2002, report low level induction in L2). IEG expression in parrots, as in songbirds, habituates with repeated presentation of learned calls, and is greater with more complex warble song stimuli (Brauth et al., 2002, 2003; Eda-Fujiwara et al., 2003).

Song-induced expression of zenk or arc has also been found in auditory regions of vocal nonlearning species, namely quail, chicken, and ring dove, in experiments that often use imprinting and other developmental/learning paradigms (Long et al., 2002; Bock et al., 2005; Terpstra et al., 2005; Thode et al., 2005). IEG expression in response to species-specific calls or other sounds is found in CMM, sometimes in NCM and in the arcopallium, but, as in songbirds, it is absent in L2. Together with data showing that the connectivity of auditory regions is similar in vocal learners (songbirds and budgerigars) and nonlearners (pigeon and chicken), the findings of IEG studies suggest that the auditory pathway in vocal learning birds was inherited from a common ancestor of birds (Mello et al., 1998a; Jarvis, 2004a).

Vocal pathways in parrots and hummingbirds

As in songbirds, singing in these orders induces IEG expression in seven well-defined telencephalic nuclei (Figure 31.6B) (Jarvis and Mello, 2000; Jarvis et al., 2000). Three of these nuclei are in similar anterior locations with respect to three overlying brain subdivisions of the telencephalon (mesopallium, nidopallium,

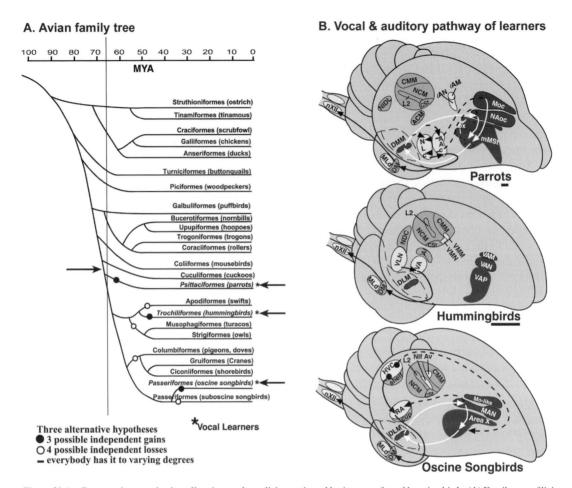

Figure 31.6 Comparative organization of hearing- and vocalizing-activated brain areas of vocal learning birds. (A) Family tree of living avian orders based upon DNA–DNA hybridization analyses of Sibley and Ahlquist (1990). Passeriformes are divided into its two suborders: suboscine and oscine songbirds. The vertical line indicates the Cretaceous–Tertiary boundary, the time of the dinosaur extinction; MYA, millions of years ago. Open and closed circles show the minimal ancestral nodes where vocal learning could have either evolved or been lost independently. Independent losses would have required at least one common vocal learning ancestor, located by the right-pointing arrow. (B) Proposed corresponding vocal and auditory brain areas among vocal learning birds. White regions and black arrows indicate proposed posterior vocal pathways; black regions and white arrows indicate proposed anterior vocal pathways; dashed lines show connections between the two vocal pathways; gray indicates auditory regions. For simplification, not all connections are shown; connectivity data in hummingbirds is based on tract-tracing using DiI in post-fixed brains (Gahr, 2000). The vocal nuclei of each vocal learning group have been given different names, due to the possibility that they evolved independently. Abbreviations are in Table 31.1. Scale bar: ∼7 mm. (Modified from Jarvis et al. 2000, and Jarvis 2004.)

and striatum). In songbirds, these are the nuclei that are part of the anterior vocal pathway involved in vocal learning and context-dependent singing. Other nuclei are more posterior within comparable brain subdivisions (mesopallium, nidopallium, and arcopallium) respectively as their songbird counterparts. In songbirds, these nuclei are part of the posterior pathway involved in vocal production. These posterior vocal nuclei vary in their medial-to-lateral location across orders, but their positions relative to each other is well conserved. While in songbirds the posterior nuclei are adjacent to or embedded within the caudomedial auditory regions, in hummingbirds and parrots respectively these vocal regions are situated laterally to, and far away from, auditory areas, which have a medial position similar to that of their songbird counterparts.

One important similarity between songbirds and parrots is that their anterior nuclei form a pallial-basal-ganglia-thalamic-pallial loop (Figures 31.5C, 31.6B). This loop resembles circuitry found in areas outside of the song system, and is akin to cortical-basal-ganglia-thalamic loops in mammals, which are thought to be involved in coordination and learning of motor behaviors (Bottjer et al., 1989; Vates and Nottebohm, 1995; Brauth et al., 1997; Durand et al., 1997; Luo et al., 2001; Jarvis, 2004a; see Farries and Perkel, this volume). Another similarity between these two groups and possibly with hummingbirds as well (Gahr, 2000) is that the posterior vocal nuclei of the nidopallium and arcopallium originate a series of projections to the brainstem vocal motor neurons in nXIIts. The differences across vocal learning groups are in the connections between the anterior and posterior vocal pathways (reviewed in Jarvis, 2004a). In songbirds, pallial nucleus HVC projects to the striatal vocal nucleus Area X within the anterior vocal pathway, and the anterior vocal nuclei MMAN and LMAN project to the posterior nuclei HVC and RA respectively; in budgerigars, the arcopallial vocal nucleus (AAC) projects to the anterior vocal nucleus NAO, which in turn has an intermingling of cells that project to both the posterior nidopallial and arcopallial vocal nuclei (NLC and AAC, respectively; Figure 31.6B). Thus, although the organization of the two pathways may be similar, the manner in which these two pathways influence each other is likely to differ across orders. It is not yet clear how auditory signals reach the vocal pathways in all these avian groups. Finally, although they have been sought for by a number of investigators (Nottebohm, 1980a; Kroodsma and Konishi, 1991; Gahr, 2000; Wada et al., 2004), there are as yet no reports on the existence of telencephalic vocal nuclei in vocal nonlearning birds, strengthening the correlation between these nuclei and the presence of vocal learning.

Evolution of vocal learning pathways

The marked similarities among vocal learners (Figure 31.6B) suggest three alternative possibilities (Figure 31.6A): (1) vocal learning evolved independently at least three times among birds, under genetic and/or epigenetic constraints that led to similar, but not identical, brain structures and pathways; (2) vocal learning was present in a common ancestor of vocal learning birds, but this trait was lost at least four independent times in the closely related descendants, suggesting strong negative selective pressures were operating; or (3) vocal learning and associated brain structures are present in a rudimentary form among all birds and was independently amplified in the three established vocal learning groups (Jarvis et al., 2000). The fact that vocal pathways in all species appear to have connectivity similar to brain regions outside of the vocal circuitry suggests that this circuitry may have evolved independently across vocal learners, but were constrained by pre-existing patterns of connectivity (Farries, 2004; Jarvis, 2004a). Some preliminary evidence (Jarvis et al., 2005b) suggests that vocal structures in both songbirds and parrots are closely linked to brain areas generally involved in motor behaviors.

Summary and conclusions

Activity-dependent IEG expression studies have helped researchers to identify brain areas involved in perceptual and production aspects of birdsong, to characterize some basic aspects of the function of these areas, and to provide comparative information of relevance to the evolution of avian vocal learning. We suggest that the application of this approach to a wider variety of groups such as fish, frogs, reptiles, and mammals (Wan et al., 2001; Hoke et al., 2004; Burmeister and Fernald, 2005) is likely to extend our understanding of the functional and evolutionary aspects of vocal communication and learning in vertebrates in general.

32 • Genes for tuning up the vocal brain: FoxP2 in human speech and birdsong

Sebastian Haesler and Constance Scharff

INTRODUCTION

Although all animals communicate in some fashion, the faculty of language distinguishes us from all other animals. We learn the sounds used for communication through imitation as do some ocean mammals (Payne and McVay, 1971; Ralls *et al.*, 1985; Janik, 2000), bats (Esser, 1994), many birds (Kroodsma and Baylis, 1982) and apparently also elephants (Poole *et al.*, 2005). But unlike animal communication, all languages allow the expression of infinite and symbolic meaning by combining a finite repertoire of sounds or gestures through the combinatorial rules of grammar.

Already Charles Darwin postulated an innate disposition towards language learning. The linguist Noam Chomsky extended and formalized the idea that learning language is not solely based upon experience. He developed the concept of a "universal grammar," which posits the existence of a universal set of core rules common to all languages (Chomsky, 1957). However, recent theoretical work also suggests that syntactic rules can emerge as a by-product when language is computationally modeled as a scale-free network. In such networks words occur according to their frequency of use (e.g. "the" is common, "adoxography" is rare) and are linked to each other via shared reference to objects (e.g. car is linked to driving, wheel, etc.) (Nowak *et al.*, 2001; i Cancho *et al.*, 2005). It is likewise not settled whether language learning requires an intrinsic, genetically defined structural and functional specialized neural tool kit, or whether a generalized capacity for learning to segment and group complex sensory inputs suffices (Fitch *et al.*, 2005; Pinker and Jackendoff, 2005). The fact that individuals with severe mental retardation can have superior language skills, as in Williams Beuren syndrome (OMIM#194050) could be interpreted in favor of Chomsky's position. Regardless of whether language learning uses general-purpose or special machinery, genes must contribute to the process at all levels, e.g. the formation of brain regions involved in language processing, the establishment of appropriate neural connections and the function of those circuits.

The first gene known to contribute specifically to language function is FoxP2[*] (Vargha-Khadem *et al.*, 1998; Alcock *et al.*, 2000; Watkins *et al.*, 2002b). Mutations cause developmental verbal dyspraxia (DVD), a condition that impairs the fluent production of words and also the correct use and comprehension of grammar, but largely spares other motor or cognitive functions (Lai *et al.*, 2001; MacDermot *et al.*, 2005; Feuk *et al.*, 2006). Brain imaging studies of FoxP2 patients reveal structural and functional abnormalities in the striatum of the basal ganglia, which also expresses high levels of FoxP2 (Ferland *et al.*, 2003; Lai *et al.*, 2003). These clinical findings raise the question whether FoxP2 influences the initial formation and/or subsequent function of neural circuits vital for speech.

Because songbirds like humans acquire their vocal repertoire by auditory-guided imitation (Doupe and Kuhl, 1999), they provide a genuine model for unraveling the basic principles of speech development, its neural substrates and its pathologies. In both humans and songbirds, learning "to play the instrument" proceeds in characteristic stages and relies on the interaction of auditory and motor centers in the brain. Optimal learning occurs during a "critical period" before puberty. Similarities also extend to social influences on vocal learning (Goldstein *et al.*, 2003) a fact that may help in developing novel experimental paradigms for

---

[*] In order to improve the readability of this chapter we use a unified nomenclature for the FoxP2 gene and protein from all species, which is based on Kaestner *et al.* (2000), with the exception of human FOXP2 and mouse Foxp2 being all written as FoxP2.

studying aspects of human communication disorders like autism (Moldin *et al.*, 2006). Bird species that do not need to learn their communication sounds through imitation of vocalizations of adult conspecifics provide ready-made control subjects for comparative studies.

In this chapter we summarize evidence for a genetic influence on speech and birdsong and discuss how genes underlying vocal communication behaviors can be identified. We then describe recent advances in the development of techniques for gene manipulation in songbirds. Using the example of FoxP2, a gene involved in human speech, we review work implicating FoxP2 in vocal learning of birds. Finally, we suggest some ways in which songbird studies may inform our understanding of the FoxP2-dependant speech disorder and speech in general.

GENES AND VOCAL COMMUNICATION: A GENETIC BASIS FOR SPEECH AND BIRDSONG

Innate mechanisms govern some aspects of how vocal learning proceeds. Birds can discriminate between homo- and heterospecific song early in life (Dooling and Searcy, 1980; Nelson and Marler, 1993). They also show an initial predisposition for species-typical signals (Marler and Peters, 1977). When two canary strains with distinct vocal repertoires are crossed, offspring develops mixed repertoires when tutored with song types from both repertoires. Pure-bred birds tutored accordingly learned songs of their own genetic type (Mundinger, 1995). In the absence of any auditory input, deaf canaries still produce some of the species-typical syntactical rules (Marler and Waser, 1977). Similarly, deaf human infants start babbling normally, but their vocalizations soon become distinguishable from those of hearing babies. When exposed to sign language, deaf babies babble using their hands (Petitto and Marentette, 1991).

The spontaneous development of sign language among deaf people from different cultures further suggests that the gestures of naturally evolving sign languages are assembled according to grammatical rules (Goldin-Meadow and Mylander, 1998; Senghas *et al.*, 2004; Sandler *et al.*, 2005). Moreover, simple "pidgin" languages can develop from a crude mixture of different languages into discrete, more complex languages, as children naturally "infuse" grammar without explicitly being taught to do so (Pinker, 1994). Ample evidence also exists for perceptual biases for human speech sounds, among them the demonstration that babies can discriminate different phonetic units as early as in their first weeks of life (Kuhl, 2004).

HUNTING FOR GENES INVOLVED IN LEARNED VOCALIZATION

To identify candidate genes for vocal learning and vocal production, two avenues can be followed. In comparative approaches differences in gene expression between species, sexes, brain regions, or behavioral conditions can be identified by a number of techniques, e.g. differential display (Denisenko-Nehrbass *et al.*, 2000) and more recently through powerful genomic tools such as microarrays (see Clayton and Arnold, this volume). This method does not require prior knowledge about particular genes or molecular pathways and has already been used with tremendous success to identify a suite of genes regulated by singing behavior (Wada *et al.*, 2006) and genes expressed differentially in females and males (Wade *et al.*, 2004; Ping Tang and Wade, 2006). The second avenue is the "candidate gene approach" (Fitzpatrick *et al.*, 2005). Here, information from other systems guides the search. Following up on some obvious candidates, i.e. genes that play a role in vocal learning in other species, e.g. FoxP2 (Haesler *et al.*, 2004), or genes that are involved in synaptic plasticity underlying learning and memory, e.g. glutamate receptor subtypes (Wada *et al.*, 2004), has been fruitful. Not surprisingly, many genes involved in neural transmission, hormonal control, or development are also central players in the song circuitry (Scharff and White, 2004; Velho *et al.*, 2006; also see Mello and Jarvis, this volume). Predicting additional candidate genes relevant for vocal learning will be aided considerably by combining comparative genome analyses of pylogenetic lineages that differ in vocal learning ability (e.g. human versus nonhuman primates, chicken versus zebra finch) with information about tissue- and cell-type-specific gene expression in the mouse (Liss and Roeper, 2004; Lobo *et al.*, 2006; www.brainatlas.org/).

GENE FUNCTION ANALYSIS IN SONGBIRDS

Genes could contribute to learned vocalizations at many levels, e.g. to acoustic recognition, to memory, to motor learning and production. Some genes will contribute more than others, and for each candidate it needs to be experimentally determined whether it is necessary for a particular process. Only then can we say that a gene is specifically involved in learned vocalization. However, this by no means indicates that the gene is exclusively involved in learned vocal communications. In fact, gene knockout experiments in mice have revealed that many genes have pleiotropic effects, i.e. that the same gene has diverse functions in different contexts.

To test the functional significance of candidate genes involved in learned vocalizations, methods for experimental gene manipulation in songbirds are needed. In mice and other genetic model organisms like Drosophila, both transgene overexpression and disruption ("knockout") of a gene of interest has been used extensively for studying gene function. Recent success in the generation of transgenic chicken (McGrew et al., 2004) and quails (Scott and Lois, 2005) using lentiviral vectors has brought the production of transgenic songbirds closer to reality. Lentiviral vectors are derived from human retroviruses and are well suited for transferring genetic information because they integrate stably into the host genome (Lois et al., 2002). Lentiviral vectors can be targeted to particular brain regions by stereotactic injection. Since lab-made lentiviruses are designed to be replication incompetent, the virus does not spread beyond the originally infected cells. This approach has successfully been used to overexpress genes in the zebra finch brain (Wada et al., 2006).

In contrast to transgene overexpression, the generation of "knockout" birds remains elusive (Zajchowski and Etches, 2000). But the method of RNA interference (RNAi) now allows experimental blocking of the expression of genes in vivo. In contrast to the physical disruption of the gene in a "knockout" animal, with RNAi the gene products are targeted. RNAi is a mechanism of posttranscriptional gene silencing through sequence specific degradation of mRNA (Dykxhoorn et al., 2003) (Figure 32.1). It is induced by 21–23 nucleotide-long double-stranded RNAs (short interfering RNA, siRNA), designed to match endogenous mRNA sequences (Pekarik et al., 2003).

SiRNA can be expressed from vectors using promoters derived from mammalian small nuclear RNA genes. The "double-strandedness" of the expressed siRNA is achieved by encoding first a sense siRNA sequence, followed by a "loop" sequence and the antisense siRNA sequence. This design leads to a backfolding of the linear transcript into a hairpin structure, because the bases of the sense siRNA are exactly complementary to those of antisense sequence. Accordingly, expressed siRNA are called short hairpin RNA (shRNA). ShRNA induces RNAi efficiently in vitro and in vivo (Krichevsky and Kosik, 2002; Rubinson et al., 2003). The exact efficiency of gene silencing induced by RNAi has to be determined experimentally. Usually it is around 60–80%, but up to 95% knockdown can be achieved. When expressed from a lentiviral vector injected into specific brain regions shRNA can alter neural gene expression, electrophysiology (Komai et al., 2006) and behavior in mice (Hovatta et al., 2005).

The induction of RNAi by expression of shRNA can affect mRNAs with sequences similar to the target mRNA (Jackson and Linsley, 2004). Therefore, at least two shRNA should be designed against different sections of the mRNA sequence of the gene of interest. All hairpins should have equivalent knockdown efficiency and cause a similar phenotype. Moreover, the activation of the molecular machinery required for RNAi can have side effects (Jackson and Linsley, 2004; Grimm et al., 2006). In fact, the RNA-induced silencing complex (RISC), which is essential for knockdown is also involved in the formation of long-term memory in Drosophila (Ashraf et al., 2006). To address this issue, additional genes, such as a virally expressed reporter gene (e.g. green fluorescent protein, GFP) should be targeted as control. Taken together, lentiviral expression systems and RNAi add new possibilities for studying the influence of genes on behavior in songbirds.

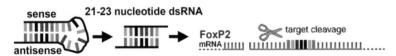

Figure 32.1 Induction of RNA interference (RNAi) by short hairpin RNA (shRNA). ShRNA is processed by a protein complex including the RNAse III nuclease DICER. This generates double-stranded RNA (dsRNA) molecules. The dsRNA mediates the recognition of the homologous mRNA target by the RNA-induced silencing complex (RISC). The catalytic component of the RISC then degrades the target mRNA by endonucleolytic cleaveage (Dykxhoorn, 2005).

FoxP2, a gene involved in human speech and language

One example of a gene involved in learned vocalizations is the transcription factor FoxP2. This gene belongs to the large family of winged helix transcription factors that are characterized by a conserved Forkhead box (Fox) DNA-binding domain. The forkhead box binds to distinct sequences in promoter regions of a specific set of target genes, allowing their transcriptional regulation. In vitro FoxP2 mostly acts as a transcriptional repressor, i.e. it reduces target gene expression when bound to promoter regions.

A causative link between FoxP2 and the speech and language disorder DVD was established when genomic alterations were identified in all 16 affected members of the so-called KE-family and an unrelated individual with a similar pathophenotype. Affected KE family members across three generations carry a point mutation in one FoxP2 allele, which prevents the protein from binding to DNA. This mutation is dominantly inherited. In the unrelated individual FoxP2 is disrupted by a balanced translocation (Lai et al., 2001). The search for FoxP2 mutations in DVD patient panels meanwhile revealed more individuals with one disrupted FoxP2 allele (MacDermot et al., 2005; Feuk et al., 2006). Thus FoxP2-dependent DVD is likely to result in all cases from a reduction of functional FoxP2 protein by 50%.

What is the behavioral phenotype of individuals with DVD? Affected KE family members have severe problems with articulation since they are impaired in performing fine movements of the mouth, tongue and lips (Watkins et al., 2002a) a feature called orofacial dyspraxia. Importantly, these difficulties do not reflect a general impairment of motor control, as limb movements are not affected (Watkins et al., 2002a). Neither is hearing. KE family patients also show deficits in receptive and grammatical language tests. They sometimes inflect words incorrectly (i.e. change tense or number) and are slower than their unaffected relatives in matching syntactically complex sentences describing relationships between objects to corresponding images. This is not a function of IQ, since IQs overlap considerably among affected and unaffected siblings (Vargha-Khadem et al., 1998; Alcock et al., 2000; Watkins et al., 2002a). Thus, the primary deficit in KE patients is likely a disruption of the sensorimotor mechanisms mediating the selection, control, and sequencing of learned fine movements of the mouth and face. An open question remains if the receptive cognitive problems result from the primary articulation problem or if they constitute a second independent core deficit of the disorder. The first possibility is consistent with the motor theory of speech perception (Liberman and Mattingly, 1985), positing that decoding speech requires the activation of the brain circuitry involved in its production. Recent studies support this concept (Fadiga et al., 2002; Watkins et al., 2003).

Structural brain imaging of affected and unaffected KE family members show reduced gray matter in the caudate nucleus in the basal ganglia (Vargha-Khadem et al., 1998; Watkins et al., 2002b), the ventral cerebellum (Belton et al., 2003), and Broca's area in KE patients, but an increase in the putamen and Wernicke's area. Caudate volume correlates well with the performance in oral praxis (Watkins et al., 2002b), supporting caudate involvement in the pathology. Given the well-established role of the basal ganglia in motor planning and sequencing, these structural changes are consistent with impaired motor control. However, it is not clear why specifically orofacial movements are compromised.

Functional imaging during a verb generation test revealed that in contrast to the typical left-dominant activation of Broca's area in unaffected KE family members, the signal in affected individuals is more bilaterally distributed. Bilateral activation is also observed when DVD subjects are asked to repeat words they hear. Consistent with the morphological findings, Broca's area and the putamen are less active in the affected KE family members (Liegeois et al., 2003). The overactivation of areas normally not involved in language might result from compensatory recruitment of additional brain areas, increased attention or a higher cognitive effort to solve the task. Taken together, the imaging work points to the frontostriatal and frontocerebellar networks as key circuitry affected in KE family members carrying FoxP2 mutations.

FoxP2 expression in the vertebrate brain

Since not all genes are expressed in all cells simultaneously, the temporal and spatial expression pattern of a gene provides information about its possible function. Many forkhead transcription factors are crucial for embryonic development (Carlsson and Mahlapuu, 2002).

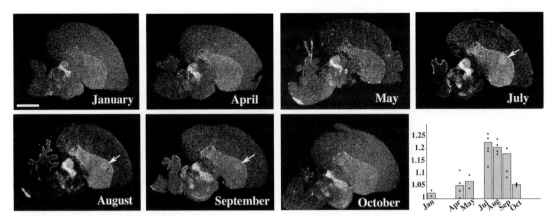

Figure 32.2 FoxP2 expression in canaries varies seasonally. Area X expressed noticeably more FoxP2 than the surrounding striatum only during the months of July, August, and September, resulting in higher ratios of Area X to striatum expression. Bar graphs show mean ratios for each month; superimposed points represent the values for individual birds.

Early FoxP2 expression in regions of the avian embryo where inductive signals organize adjacent neuroepithelium and neuronal migration (Haesler et al., 2004) is consistent with this notion. However, strong postnatal FoxP2 expression also suggests that FoxP2 is important after brain development has been completed. FoxP2 has a highly similar pattern in all species studied, from crocodiles to humans. Across species, the strongest signal is always observed in the striatum of the basal ganglia, nuclei of the dorsal thalamus and midbrain, the inferior olive, and the Purkinje cells of the cerebellum (Ferland et al., 2003; Lai et al., 2003; Takahashi et al., 2003; Haesler et al., 2004; Teramitsu et al., 2004). In pallial/cortical regions FoxP2 is expressed at very low levels, with the exception of the mesopallium in birds and cortical layer 6 in mammals (Ferland et al., 2003).

Songbirds possess specialized brain areas for song recognition, song production and song learning, the so-called song system. FoxP2 is absent in nuclei of the motor pathway, but it is strongly expressed in Area X, a striatal nucleus essential for the acquisition of song (Scharff and Nottebohm, 1991). The analysis of FoxP2 expression during different stages of song ontogeny revealed an upregulation in Area X at the time when young zebra finches learn to imitate song (Haesler et al., 2004). Moreover, during the months of the year when adult canaries remodel their songs, FoxP2 expression in Area X is also elevated (Figure 32.2). Both observations were made in birds that had not sung in the hours prior to analysis. Thus, FoxP2 expression correlates with phases of vocal plasticity in a forebrain structure uniquely associated with learned vocal communication.

Postnatal expression of FoxP2 might also be related to singing activity, similar to the immediate-early gene *zenk* (see Mello and Jarvis, this volume). A recent report shows downregulation of FoxP2 in Area X of adult zebra finches after undirected song ((Teramitsu and White, 2006), which was not observed in a previous study (Haesler et al., 2004). These conflicting results might in part be explained by different experimental conditions (Haesler et al., 2004; Teramitsu and White, 2006) and will be resolved by further studies. Taken together, the expression pattern of FoxP2 suggests its involvement in the formation and/or function of brain pathways including, but not limited to, those essential for learned vocal communication.

ANALYSIS OF FOXP2 FUNCTION IN VIVO

FoxP2 knockout mouse

Mice with disruption of both FoxP2 alleles only live for about three weeks (Shu et al., 2005). They are developmentally delayed, and have impaired motor function. Heterozygous mice perform moderately worse than wild-types and catch up by their second week of life. Consistent with the cerebellar FoxP2 expression (Ferland et al., 2003; Lai et al., 2003), FoxP2 knockout (KO) mice display cerebellar abnormalities suggestive of impaired cell migration. The cerebellum may be particularly vulnerable to lack of FoxP2, because it does not coexpress FoxP1 (Tamura *et al.*, 2004), which can interact with FoxP2 to regulate gene expression (Li *et al.*, 2004). FoxP1 might compensate for the absence of FoxP2 in the basal ganglia and the thalamus, which express both. In fact, the basal ganglia of KO mice are histologically gross normal, but detailed quantification of the different basal ganglia nuclei, as done in KE patients (Watkins *et al.*, 2002b) is missing.

Given the speech pathophysiology of FoxP2 patients, it is interesting that vocal behavior in the KO mice is impaired. Homozygous FoxP2 knockout pups vocalize less in the sonic range than heterozygous and wild-type animals when separated from their mothers. In the ultrasonic range, both homo- and heterozygous knockout animals utter fewer whistles. The acoustic structure of the vocalizations is preserved in FoxP2 KO pups indicating that the motor areas controlling acoustic features of sound production are intact. Ultrasound communication in adult homozygotes could not be tested because they die too early (Shu *et al.*, 2005). The recent finding that adult male mice vocalize with a previously unrecognized complexity reminiscent of birdsong (Holy and Guo, 2005), begs for a more detailed study of FoxP2 KO mouse vocalizations. In light of the relative ease of genetic manipulation in mice this seems a particularly promising area for future research. However, whether mouse vocalizations, like human speech and birdsong, are learned has yet to be determined.

FUNCTIONAL ANALYSIS OF FOXP2 IN SONGBIRDS

Because FoxP2 expression in Area X correlates with vocal plasticity (Haesler *et al.*, 2004), lentivirus-mediated RNA-interference (see above) was used in a recent study to experimentally reduce FoxP2 expression levels in Area X in vivo (Haesler *et al.*, 2007). Lentivirus targeting FoxP2 or control viruses, were injected stereotactically into Area X at the beginning of the song learning period of young zebra finches. Injected animals were then tutored by adult male zebra finches and the pupils' songs were analyzed to evaluate the behavioral consequence of FoxP2 knockdown (Figure 32.3). Histological analysis and real-time PCR confirmed localized knockdown of FoxP2 in Area X. As in FoxP2 patients, knockdown birds experienced *reduced* levels of FoxP2, not a complete absence of the protein (Lai *et al.*, 2001; MacDermot *et al.*, 2005).

The most striking behavioral consequence of FoxP2 knockdown was an incomplete and inaccurate imitation of the tutor song (Figures 32.4, 32.5). Moreover, syllable production in knockdown birds was more variable than in controls. The reduced accuracy of FoxP2 knockdown animals in copying tutor syllables raises the question whether the results reflect a motor deficit, i.e. whether knockdown animals were simply unable to generate particular sounds. Three arguments speak against this. Syllables with similar acoustic features could either be learned or omitted by the same animal. Second, acoustic features of omitted and imitated syllables did not differ. Third, the distribution of acoustic features across the syllable repertoire was indistinguishable between knockdown and control birds. Furthermore, knockdown of FoxP2 apparently did not delay song development, because the sequence stereotypy of motif delivery was indistinguishable between groups.

The limited imitation success of FoxP2 knockdown birds could also result from an imprecise neural representation of the tutor model. Although there is biochemical evidence for a role of Area X in sensory learning (Singh *et al.*, 2005) pharmacological inactivation of Area X during song tutoring does not prevent song imitation (Andalman and Fee, 2006). If sensory

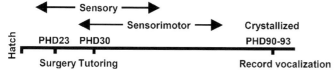

Figure 32.3 Timeline of knockdown experiments. By PHD20, fathers and older male siblings were removed from family cages to prevent experimental zebra finches from instructive auditory experience prior to the onset of tutoring. At the beginning of the sensory learning period on PHD23, virus was injected into Area X. From PHD30 on, injected birds were housed in sound-recording chambers together with an adult male zebra finch as tutor. We recorded the song of adult pupils on PHD90 to 93 using an automated recording system.

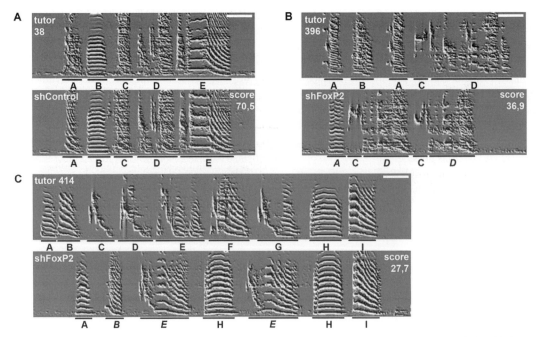

Figure 32.4 Representative sonograms from FoxP2 knockdown and control birds. Each sonogram depicts a typical motif of one animal (scale bars 100 ms, frequency range 0–8600 Hz). Tutor syllables are underlined with black bars and identified by letters. The identity of pupil syllables was determined by similarity comparison to tutor syllables using SAP software. Imprecisely copied pupil syllables are designated with italic letters. (A) tutor #38 and shControl-injected zebra finch; (B) tutor #396 and shFoxP2-injected animal (C) tutor #414 and shFoxP2-injected animal. The motif imitation scores from each pupil to the respective tutor are indicated in the right upper corner of the sonograms.

acquisition of the tutor model does not require Area X, it seems more likely that FoxP2 is involved in sensory-motor integration, a hypothesis that is also consistent with the increased expression of FoxP2 in Area X during times of vocal plasticity in juvenile zebra finches and adult canaries (Haesler et al., 2004). FoxP2 seems to be particularly important when the birds are adjusting their songs through vocal motor practice. We therefore propose that knockdown of FoxP2 results in poor tutor imitation because animals failed to match their own vocalization to the memorized tutor model.

This hypothesis is supported by the phenotypic overlap of song deficits observed in FoxP2 knockdown birds and birds otherwise prevented from matching vocal output with memorized tutor song. For instance, perturbed auditory feedback and juvenile Area X lesions cause song abnormalities similar to those of knockdown birds (Scharff and Nottebohm, 1991; Leonardo and Konishi, 1999). Given that specific gene knockdown in a particular cell type is a much more precise experimental approach than electrolytic lesions, it is not surprising that song deficits of FoxP2 knockdown birds were not identical to those of birds with early Area X lesions. Different from FoxP2 knockdown, juvenile Area X lesions result in reduced song sequence stereotypy and longer syllables and intervals (Scharff and Nottebohm, 1991), song features not typical in FoxP2 knockdown finches.

How might FoxP2 affect vocal variability and song imitation? In Area X, pallial auditory and song motor efference information converges with nigral ascending reinforcement signals in the spiny neurons, which express FoxP2 (Haesler et al., 2004; Reiner et al., 2004a). By analogy with the mammalian system these cells might be involved in reward learning (Doupe et al., 2005). The integration of these signals provides a candidate mechanism for tuning the motor output to the tutor model during learning. The increase of FoxP2 expression in Area X of zebra finches during times of vocal plasticity could be functionally related to this process. FoxP2 might mediate adaptive structural and functional changes of the spiny neurons while the song is learned. In canaries, increased

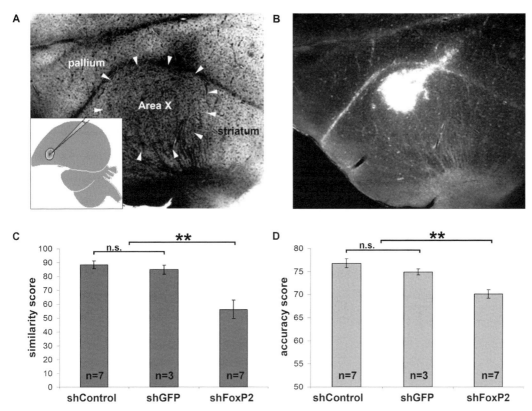

Figure 32.5 Lentivirus-mediated knockdown of FoxP2 in zebra finch Area X results in poor tutor imitation. (A) Phase contrast image of a sagittal 50 μm thick brain section from a male zebra finch. Area X is outlined by white arrowheads (scale bar 1 mm). The microinjection into Area X and its position within the brain are schematized in the inset. (B) Fluorescent microscopy image of (A). Virus-infected cells expressed GFP (white). (C) Tutor song imitation was measured using Sound Analysis Pro software (Tchernichovski, 2001) in FoxP2 knockdown birds (shFoxP2) and two different control virus-injected animals (shControl and shGFP). ShControl expressed an shRNA that has no target gene and shGFP expressed a shRNA targeting the viral reporter gene GFP. The mean similarity between pupil and tutor motifs was significantly lower in shFoxP2 injected animals than in control virus injected birds, indicating that knockdown animals copied less acoustic material from their tutors (± standard error of the mean, SEM; two-tailed t-test, **$p < 0.001$, Bonferroni-corrected α-level). There was no significant difference between shGFP and shControl injected animals (not significant, n.s., $p > 0.5$). (B) Average motif accuracy was significantly reduced in shFoxP2 knockdown animals compared with control animals, indicating that they imitated their tutors less exactly (± SEM; two-tailed t-test, **$p < 0.001$, Bonferroni-corrected α-level). shControl and shGFP-injected birds copied their tutors with similar precision (n.s., $p > 0.3$).

FoxP2 expression in the fall months might similarly be involved in seasonal song modifications.

Since FoxP2 is a transcription factor, it could affect vocal plasticity by positively or negatively regulating neural plasticity-related genes. If FoxP2 functions as a plasticity-promoting factor, knockdown animals should have been less plastic during learning, resulting in impoverished imitation and abnormally invariant song. Syllable omissions of FoxP2 knockdown birds are consistent with this notion, but more variable syllable production is clearly not. Alternatively, if FoxP2 restricts neuronal plasticity, knockdown birds should sing more variably. In fact, this is the case, but then the observed syllable omissions are not easily explained.

In adult zebra finches, FoxP2 is downregulated after they sing slightly variable, undirected song, but not after they sing more stereotyped female-directed song (Teramitsu and White, 2006). The vocal variability during undirected singing likely results from some form of underlying neural plasticity, as suggested by strong induction of the immediate early gene *zenk* (Jarvis *et al.*, 1998). If FoxP2 promoted plasticity it should therefore be upregulated under this condition like *zenk*, but the opposite is the case. In view of the complementary expression patterns of FoxP2 and *zenk*, it seems thus plausible that the transcriptional repressor FoxP2 restricts neural plasticity by repressing genes induced by recurrent neuronal activity. The identification of the downstream target genes of FoxP2 and the electrophysiological characterization of medium spiny neurons with reduced FoxP2 levels will shed further light on the function of FoxP2 in vocal learning.

CONCLUSION AND FUTURE PROSPECTS

The observation that the brain regions with strongest FoxP2 expression are also the site of structural and functional abnormalities in DVD patients has led to the proposition of a FoxP2-dependent speech and language circuitry (Vargha-Khadem *et al.*, 2005). This circuitry comprises a cortico-striatal-thalamic (CS) and a cortico-cerebellar (CC) loop, both of which are known to be involved in the acquisition and execution of temporally sequenced muscle movements (Doyon and Benali, 2005) including those required for speech (Lieberman, 2001; Marien *et al.*, 2001). Whereas the CS circuit exhibits striking parallels to the AFP in songbirds, the CC loop has not yet been implicated in learning and production of song. Recent work highlights the importance of cortical-subcortical networks also for speech and language. During speech production an increase of both syllable and sequence complexity increases the activation of the basal ganglia, the anterior thalamus, and the cerebellum (Bohland and Guenther, 2006), the three major sites of FoxP2 expression. During perceptive tasks, the speed and accuracy of phonological processing are associated with dopaminergic release in the caudate and putamen (Tettamanti *et al.*, 2005). Moreover, disorders that affect basal ganglia function such as Huntington's and Parkinson's disease, specifically interfere with speech and language (Grossman, 1999; Hochstadt *et al.*, 2006; Teichmann *et al.*, 2006). The basal ganglia-thalamocortical pathway seems particularly important for fluency aspects of speech, and is involved in neurogenic stuttering (Alm, 2004). Interestingly, zebra finches can naturally develop abnormal song syllable repetitions, suggesting that the neural basis of stuttering might also be studied in songbirds (Helekar *et al.*, 2003).

In view of the functional similarities of the basal ganglia in human speech and birdsong, the vocal behavior of FoxP2 knockdown zebra finches offers an interesting interpretation of the speech abnormality in individuals with genetic aberrations of FoxP2 (Watkins *et al.*, 2002b). An involvement of FoxP2 not only in brain development but also in brain function during speech acquisition should be considered for the pathophysiology of individuals with genetic aberrations of FoxP2. In analogy to the interpretation of the poor tutor imitation observed in FoxP2 knockdown birds, reduced levels of FoxP2 in humans may interfere with adjusting the vocal output to articulation rules during speech learning. The articulation problems of DVD patients might thus rather represent an inability to acquire the correct motor programs needed for fluent speech production, than a hard-wired motor deficit.

The apparent discrepancy between the requirement of FoxP2 for human speech and birdsong and its overall sequence conservation among all vertebrates, most of which are not capable of auditory-guided vocal imitation, leads to the question why the FoxP2 protein changed so little during vertebrate evolution. Under the assumption of a neutral model of evolution (Kimura, 1968) in which random genetic drift provides a large source for genetic variation, this strongly suggests that the FoxP2 gene was under selection pressure unrelated to vocal learning. Consistent with this, homozygous knockout of FoxP2 in mice causes perinatal death (Shu *et al.*, 2005), although it remains open why lack of FoxP2 is lethal. Most likely, multiple selective constraints act on the FoxP2 gene, indicating that FoxP2 function is important in several biological processes. A more detailed analysis of the different functional domains of the FoxP2 protein and the identification of the genes that interact with it in both vocally learning and non-learning species is needed to disentangle the different evolutionary constraints from a functional perspective.

ACKNOWLEDGEMENTS

The authors regret that, owing to space constraints, many key pieces of research are cited only in reviews. We gratefully acknowledge our collaborators Erich Jarvis, Kazuhiro Wada, Wolfgang Enard, Pawel Licznerski and Pavel Osten for contributing to the work presented here. Thanks to Anna Zychlinsky for language editing.

PART VII
On a personal note

33 • Introduction

H. P. Zeigler

> And Thorpe begat Marler and Marler begat Konishi and Nottebohm. Of the number of their offspring there is no end.
>
> (Zeigler and Marler, 2004, p. xiv)

The biblical overtones are not inappropriate because, to an extraordinary extent, the development of a "neuroscience of birdsong" is the product of a closely knit (almost inbred) community of workers, many of whom are the intellectual descendants of the four men whose histories are to be found in this section. Most of them can trace their lineage back to W. H. Thorpe (Figure 33.1) via the intermediary of his student Peter Marler and of Marler's students.

Hinde's obituary memoir of Thorpe, which opens this section, provides one view of the man and his contribution. Marler's autobiographical memoir, which closes it, provides another, as that contribution figured in his own intellectual development. In the intervening chapters, Mark Konishi and Fernando Nottebohm outline their respective intellectual and personal odysseys (Figure 33.2).

My own experience is marginally relevant because I was, in a sense, "present at the creation." When I arrived at Cambridge in 1958 on a postdoctoral fellowship it was, ostensibly, to work with Thorpe on song learning. I found myself assigned to Hinde, who was at that point working on his studies on the habituation of great tit alarm calls. (Hinde has one of the most incisive minds I have ever encountered and provided a quality of mentoring for which I shall always be grateful. I was privileged to be an interested bystander to his role in the integration of what then appeared to be the irreconcilable views of the American behaviorists and the European ethologists). On my very first day at the Madingley field station, I was introduced by Thorpe to Peter Marler, who was leaving to take up a position at Berkely. The rest is history. Peter and I were not to meet again until almost 50 years later when he attended the Hunter College Conference on Birdsong Neurobiology, the forerunner of the present volume. Without his contribution as coeditor, this book would not have been possible.

Figure 33.1 William Homan Thorpe (1902–1986). Reprinted from *Biographical Memoirs of Fellows of the Royal Society*, Vol. 33 (Dec. 1987), pp. 620–639. Photo courtesy of Professor R. A. Hinde and the Fellows of the Royal Society.

Neuroscience of Birdsong, ed. H. Philip Zeigler and Peter Marler. Published by Cambridge University Press. © Cambridge University Press 2008.

Figure 33.2   Three birdsong researchers: (left) Masakazu Konishi, (center) Peter Marler, (right) Fernando Nottebohm. Photo taken at Brandeis University in May 2004, during ceremonies attendant upon their receipt of the 33rd Lewis T. Rosenstiel Award for Distinguished Work in the Basic Medical Sciences. ©A.Brilliant/BrilliantPictures.Inc. 2004.

34 • William Homan Thorpe
(1 April 1902–7 April 1986, elected F.R.S. 1951)

R. A. Hinde

William Homan Thorpe came from a Sussex family. Born in 1902 William Thorpe was effectively an only child, his half-brother and half-sister being so much older that they provided little in the way of companionship for him. Bill's parents – and it is easier to write of him as Bill, though it carries here none of the implications of jovial sociability that are associated with the abbreviation of William – both liked the countryside, and his mother was fond of identifying the flowers she found on her walks and cycle rides. By the age of eleven he had made up his mind to be a naturalist.

Although Bill's interest in natural history was fostered by his parents' love of the countryside, he had a clear view that its source lay elsewhere. At school Bill was rather delicate and had long periods of ill-health, which kept him out of team games. When he was fourteen his father decided that he was living too abnormal and solitary a life, and that he should go to a boarding school. At Mill Hill he found no real enthusiasm for chemistry and physics. Instead his energies went into natural history and music. Convinced that he would not be up to the standard for a university course, Bill set his sights on being a naturalist, or failing that a farmer. However, a report on the need for economic entomologists led his mother to arrange an interview for him in the Royal College of Science in London. Professor H. Maxwell Lefroy suggested an agricultural degree at Cambridge followed by postgraduate work at Imperial College. After cramming for Little-Go Latin with a private tutor, in 1921 Bill went up to Jesus College, Cambridge.

As an undergraduate he was one of the founder members of the Cambridge Bird Club, and published a number of notes and papers on field ornithology. He spent much of his time with two men who were to become lifelong friends, Edward Armstrong and John Clarke. Cambridge became a centre for a number of like-minded colleagues. He also regularly attended meetings of the British Ornithologists' Club in London.

Obtaining a second-class Ordinary degree in Agriculture, Bill then read for a Diploma in Agricultural Science (Plant Pathology), writing a thesis on biological control. It was apparently in the diploma year that Bill started to come into his own. Stanley Gardiner asked him to give some lectures in advanced entomology. In 1925 he had started research in the Department of Agriculture, with F. R. Petherbridge, studying the currant capsid bug (*Lygus pabulinus*) and "biological races" in moths of the genus *Hyponomeuta*. He then took a two-year Rockefeller Fellowship (1927–29) at the University of California, studying the biology of the parasitic dipterous *Cryptochaetum* and other insect parasites. He also made a detailed study of the petroleum fly (*Psilopa petrolei*), which lives in oil pools in California. During this period he made the most of opportunities to travel widely in the U.S.A. and to visit Hawaii. He obtained his Cambridge Ph.D. in 1929.

On returning to England he took a post at the Farnham Royal Parasite Laboratory of the Imperial Institute of Entomology, working on problems of biological control and the biology of parasitic insects. In 1932 Professor Stanley Gardiner and Dr. A. D. Imms invited him to return to Cambridge as Lecturer in Entomology. Simultaneously Jesus College appointed him as Fellow and Tutor.

His early work on the physiology and behavior of insect parasites resulted in a large number of papers and led him along a number of divergent paths. One concerned the chemical orientation of insects. Working with the highly variable ermine moth *Hyponomeuta padella*, he assessed the rigidity of food choice in the larvae, and obtained evidence that there were distinct races laying on apple and hawthorn.

Subsequently he demonstrated "pre-imaginal conditioning" in the endoparasitic *Nemeritis canescens*, which normally parasitizes *Ephestia kühniella* but will also lay on *Meliphora grisella*. Parasites reared on *Ephestia* showed a strong oviposition preference for that species over

Meliphora, but that preference could be significantly reduced by rearing on *Meliphora*. He subsequently showed that this was not due to simple conditioning of oviposition to the new host by the newly emerged parasite, but was rather an association "with a favourable environment as a whole." The work was extended also to *Drosophila*. These results were interpreted as potentially important for understanding the formation of biological races and thus speciation. This view was followed up in 1945 by an important review on "The evolutionary significance of habitat selection." It argued that geographical, topographical and ecological isolation differed only in degree, and that learnt habitat preferences could give rise to ecological isolating mechanisms and lead to speciation. This view was not in accord with the views expressed around that time by Mayr, Lack and others, but is now widely accepted as a likely possibility in other fields, such as marine biology. His emphasis on the existence of biological races influenced the development of modern approaches to systematics. From 1938 Bill Thorpe was chairman of the Pacifists' Service Bureau, which attempted to direct the efforts of pacifists into constructive social service, and when war was declared he applied for registration as a conscientious objector on religious grounds. Nevertheless, he found it consistent with his pacifist stand to put his entomology at the service of his country's war effort towards self-sufficiency in food production. He contributed to a project on wireworms, which were causing serious agricultural damage. He expressed pleasure that this work prevented estrangement from colleagues who supported the war. During this time he devoted his major energies to a review of insect behaviour in general.

In the meanwhile Bill's enthusiasm for ornithology had not waned since his schoolboy days, and his entomological publications had been interlaced with notes and longer publications on birds. His studies of olfactory conditioning in the 1930s focused his interest on the relation between "instinct" and "learning" and extensive reviews of animal learning resulted, on insects in 1943 and on birds in 1951. He was impressed by the writings of Konrad Lorenz, feeling that Lorenz had managed to pinpoint the key issues on which further observation and experiment would be profitable. As a result Bill played a unique role in establishing ethology in the English-speaking world after the war. He was responsible for publicizing Lorenz's views (as well as the work of von Frisch on bees) in special numbers of the *Bulletin of Animal Behaviour*, and managed to visit Lorenz in Soviet-occupied Austria soon after Lorenz was released from a Soviet prisoner-of-war camp. Indeed he played a leading role in attempting to find a post for Lorenz in the U.K., an enterprise that would have been successful had Lorenz not received an offer from the Austrian government.

He was a founding editor of the journal *Behaviour*. He was also largely responsible for a meeting of the Society of Experimental Biology and the Association for the Study of Animal Behaviour in Cambridge in 1949. This was an important forum for the presentation of Lorenz's and Tinbergen's postwar views, and Bill's own paper, "The concepts of learning and their relation to those of instinct" was a foretaste of his later book. He also established a group that attempted to define many of the terms used in the blossoming subdiscipline of ethology. The 1949 meeting also permitted overt presentation of opposing views in the "central versus peripheral" controversy: Lorenz (leaning in part on the work of Erich von Holst, who was absent), postulating the endogenous generation of motivation, while James Gray (Professor of Zoology at Cambridge) and Hans Lissmann attempted to account for locomotion in lower tetrapods in terms of successive reflexes.

Seeing that birds offered an ideal combination of stereotyped and readily identifiable movement patterns and considerable learning ability, he initiated the establishment of an Ornithological Field Station at Madingley. The station was established in 1950; a wire fence was erected around the site, and 60 aviaries, made partly from scrap left over from wartime beach defences, were purchased. James Gray, who had initially been supportive, became much less so, in part because of his antipathy to the Lorenzian views that Bill shared. Bill therefore had many difficulties to surmount. The first additions were G. V. T. Matthews, who worked on bird navigation, and Peter Marler, who carried out a field study of the chaffinch. Over the years, under Bill's direction, the Ornithological Field Station developed into a Field Station for the Study of Animal Behaviour and then into the Sub-Department of Animal Behaviour. Help came from the Josiah Macy Foundation and the number of students increased. In 1960, by the generosity of the Nuffield Foundation and the Rockefeller Foundation, a substantial laboratory building was erected. Bill was made a Reader in 1959 and elected into a personal chair in 1966. He was greatly helped by the appointment of Carl Pantin as Professor of Zoology: Carl was both a friend and an enthusiastic supporter of Bill's academic plans, and also had a quiet tolerance for Bill's somewhat erratic way of dealing with the financial side of his enterprise.

Once the Madingley Field Station was established, Bill's main research concerned the development of chaffinch (*Fringilla coelebs*) song. By raising chaffinches in varying degrees of social isolation, and by "tutoring" them by exposing them to other individuals or to tape-recorded songs, Bill was able to show that (1) chaffinches reared in auditory isolation from other chaffinches produced only a simple, unpatterned song; (2) isolate-reared chaffinches exposed to chaffinch song in their first winter produced near normal song; (3) further learning occurs as the full song "crystallizes out" from the amorphous rambling of the subsong produced early in the bird's first spring; (4) chaffinches learn preferentially chaffinch song, and not that of alien species.

The findings indicated that song learning involved two stages: the chaffinch learns what to sing, and then subsequently how to sing it. This work led to the template theory of song learning – the bird supposedly first acquires a template (or modifies a pre-existing one) of what the song "ought to be like," and then modifies its vocal output to agree with the template.

Peter Marler, after completing his Ph.D. on chaffinch behavior in the field, worked with Bill on chaffinch song development and then started his own program on song development in the USA. It was taken up by his students (Fernando Nottebohm and Mark Konishi) and subsequently by many others. There is no doubt that the work initiated by Bill in Madingley has led to one of the most important enterprises in ethology. Not only has it achieved his initial aim of throwing light on the relations between "instinct" and "learning," but it has paved the way for detailed neurophysiological investigation of the mechanisms involved. In addition, Bill pioneered the use of the sound spectrograph for studies of birdsong, and subsequently used it for studies of vocal imitation, duetting and individual recognition in birds.

Two other lines of empirical investigation initiated by Bill in the 1950s deserve mention. Lorenz had shown that the young of many nidifugous birds would follow the first moving object they saw, and direct to it many of the responses that they would normally direct to their parents. He regarded the learning process involved as unique, and termed it "imprinting." Bill extended this work to moorhens (*Gallinula chloropus*) and coots (*Fulica atra*), with somewhat more carefully controlled experiments than Lorenz had done. Although confirming many of the previous findings, these experiments showed that the effects of imprinting were more reversible than had previously been thought. This work was subsequently taken up and greatly extended by Patrick Bateson, who was later to follow in Bill's footsteps as Director of the Madingley laboratory.

The other line of work initiated by Bill at this time concerned the analysis of "insight learning" in birds. The principal test used concerned the bird's ability to pull up food suspended on a piece of string. This work was subsequently continued by Margaret Vince, who made considerable progress in analyzing the apparently complex learning into simpler processes.

During the 1950s Bill attempted a synthesis of Lorenz's ideas on "instinctive behavior" with the work of American (and other) psychologists on learning in his book *Learning and Instinct in Animals* (Thorpe, 1956). In this he classified learning processes phenomenologically into a number of distinct types, discussed their relations with each other and related them to Lorenz's concept of instinctive behavior. In some ways it was a book ahead of its time. The emphasis on the importance of processes of perception, and on the goal-directedness of many aspects of behavior, was less in tune with current concepts then than it would be today. Nevertheless, the immense scholarship displayed earned widespread admiration.

In the late 1950s and 1960s he organized, with Oliver Zangwill, the Professor of Psychology, a discussion group for psychologists and biologists interested in behavior. It started with a prolonged discussion of Hebb's "Organization of behaviour," and continued to meet regularly for many years. The Thorpe–Zangwill group had a profound influence on many of those who attended.

For Bill scientific and religious thought were different but parallel. He once quoted Sir William Bragg as saying that "Religion and science are opposed ... as my finger and thumb are opposed – and between the two one can grasp everything." His studies of animal behavior were guided by the view that some element of self-consciousness could be traced in the animal kingdom, and he therefore laid himself open to the charge of preferring complex to simple explanations, though more recently the climate of opinion has moved more in his direction. At the same time he held that man had evolved powers of abstract thought, of moral choice and of religious feeling involving a difference in kind from other species.

The Gifford Lectures (Thorpe, 1974) devoted much space to a discussion of the light thrown by the study of behavior on animal nature. In it Bill went on to discuss human nature in the same terms, but stressing man's uniqueness and the emergence of the human spirit. He took these issues further in his book *Purpose in a World of Chance* (Thorpe, 1978), which was an attack on Monod's *Chance and Necessity* (1971) and emphasized the role of purpose in animal and human behaviour.

Bill Thorpe was awarded the Godman-Salvin medal of the British Ornithologists' Union in 1968, the Frink medal of the Zoological Society of London in 1980, and in 1982 the Fyssen prize for contributions to knowledge about the origin and nature of man's cognitive abilities.

Although so much of Bill's life can be seen as an attempt to bring together nature, science, esthetics, philosophy and religion, and although so much of his writing involved attempts to convey to others his perceptions of the relations between them, this was not apparent in his everyday life. He seemed to keep the various domains of his life separate. There were some with whom he shared his love of music, others with whom he went bird-watching or worked for nature conservation, scientific colleagues and students, philosophers and Quakers – many who knew one side of his complicated nature but few who penetrated further. Outside his family, and with the possible exception of the ornithologist Edward Armstrong, he had few, if any, really close friends. Even his capacity for total involvement appeared on the surface only rarely. Niko Tinbergen once said that he had only once felt he could get close to this rather distant, apparently self-sufficient man, and that was when they were staying on Scolt Head together. With a cold sea mist blowing in across the dunes Bill stripped off all his clothes and ran stark naked into the sea. He loved swimming, riding, listening to music, watching birds, but only rarely, and then mostly when he was intensely involved, did his capacity for intense enjoyment become apparent to others. On most occasions, and for many people, his reputation for intellectual brilliance and his natural shyness formed an impenetrable barrier. But he was capable of being utterly kind-hearted, and those among his juniors who took the plunge necessary to ask him for advice were amply rewarded.

35 • My journey with birdsong

Masakazu Konishi

The first section describes my childhood experiences that might have led me to ethology in later years. I grew up in Kyoto, the ancient capital of Japan. I became familiar with animals from an early age, because, as the only child, I spent more time with animals than with other children. My parents did not dictate what animals I could bring home except for cats, which my mother disliked. Also, we did not have space for big animals like goats and donkeys, which I badly wanted. My pets included dogs, rabbits, chickens, Bengalese finches, turtles, singing frogs (kajika or *Buergeria japonica*), goldfish, medaka fish (*Oryzias latipes*), and insects such as bell crickets (*Homoeogryllus japonica*), fireflies, rhinoceros beetles. These frogs and insects were and are still sold in the open-air animal section of department stores. As I grew older, I collected them in the wild. I learned not only how to take care of animals but also how to play with some of them. I let rhinoceros beetles pull a toy wagon. They also fought like sumo wrestlers! The domesticated pet animals in the above list were my regular playmates. One day I took my dog and pet rabbit, which often played hide and seek under the floor of our house, to an island in the shallow river that ran through the middle of the city. I chose the island, because I did not want to lose the rabbit. My idea was to play a hunter. I freed the rabbit and waited until it disappeared, then I told my dog to find it. I saw the rabbit run back and forth followed by the dog. Suddenly, the dog lost his prey. I could not find it either. Then, to my great surprise, the rabbit was on a small rock in the middle of the river! I played this game, because I had assumed that rabbits could not swim. Rabbits and chicken were important sources of meat during the worst period of the war. I raised chicks and bunnies for the above purpose and sometimes for sale to other children. Chickens turned out to be more interesting than rabbits. I found that my flock of chickens was not just a bunch of individual birds but had a pecking order. The king was always a big shiny proud-looking rooster. He would even try to attack me. To give him a lesson, I took a low ranking rooster in my hands and forced it to assume fighting postures and movements in front of the king. As the boss tried to retaliate, I grabbed the young cock's neck to land his beak on the boss's head. When I repeated these procedures, the boss would walk away, although I did not succeed in dethroning him.

As I graduated from raising puppies, bunnies, and chicks, I tried something unusual, wild songbirds. The most common passerine bird was the sparrow (*Passer montanus*), which farmers hated, because sparrows raided rice fields just before the harvest season. Sparrow nestlings were easily available from nests built in spaces between roofs and rain gutters. A friend of mine and I checked nest sites at our school to take a few nestlings for hand rearing. However, we could never raise young nestlings with cooked rice or rice bran on which adult sparrows thrived. This puzzled me a bit, because my Bengalese finches, seedeaters, regurgitated food into the mouth of their young. I also saw every summer swallows bring insects to their young. So, the rule appeared to be that nestlings are fed the same food that their parents eat. How about the seed-eating sparrow? Watching sparrows carry food to their nests, I discovered that they too brought insects to their young. I successfully raised a nestling sparrow by feeding spiders and insects. The bird became so tame that I could leave it free in our house. In order to catch spiders I had to learn their habitat and behavior. I would find a spider web and tested how I could fool the animal by tossing small objects into the web. I also enjoyed seeing two spiders fight. The web owner always won over the ones I introduced. When my grade school teacher showed our class how big spiders fought upon meeting in the middle of a stick he held horizontally, I was absolutely thrilled, thinking that even teachers, our gods, played like me.

I loved the zoo in our city and my big dream was to become an animal caretaker there. My parents liked the idea, because they wanted me to become a "salary man" so that I could lead a stable life. This made sense, for I was born in the middle of the great depression (1933). My fun with pets had to come to an end around the second year of high school, because I had to prepare for college entrance examinations. The idea of going to college had never occurred to me until I started to associate with high school friends from salary men's families. They were mostly members of the biology club that I led.

SAPPORO

I chose to receive my college education in Sapporo, which is the capital of Hokkaido, because this large northern island fit my frontier image. Also, the school was well known for an American missionary professor William Clark who said, "Boys be ambitious" as he was leaving after his tour of duty. Japan adopted the U.S. system of college education at the end of the war; two years of general education followed by two years of majoring in specific subject areas. I chose zoology in my junior year (1953) against my original plan, which was ranching. I got the taste of pure science after reading and seeing what and how zoologists studied animals.

The zoology department had some good professors. Shoichi Sakagami who was known for his behavioral and comparative studies of bees, taught a course in ethology. He recommended that we (about 12 students) read *The Study of Instinct* by N. Tinbergen (1951). I bought a copy and read it from cover to cover, although I could barely afford it when the currency exchange rate was very bad for Japan. I thought I found my natural niche in which I can fool animals with dummies. As I was starting my Master's thesis work in 1956 under Sakagami, he personally told me to read the classic work of Peter Marler entitled "Characteristics of some animal vocalizations" which appeared in *Nature* (1955a). This was the first paper that discussed the functional significance of acoustic properties of animal vocal signals. I began to study the territorial behavior of great reed warblers (*Acrocephalus arundinaceus*), which bred in reed beds on the campus. I got Sakagami interested in these birds. We took turns to watch a single male in the reed swamp nearest to our laboratories. Sakagami was obsessed with quantification. While we were mapping its territory, we found multiple females nesting in it. We measured the length of time the male was singing by marking in our notebook every minute during which the bird was singing. I was particularly interested in the role of song in territorial defense. Our department had only one tape recorder. Since this machine did not operate on batteries, I had to use a long electric cord for this work. Nevertheless, vigorous responses of the bird to playback songs were very encouraging to me. I also combined the above field study with a project in which I investigated the number and distribution of reed warbler nests containing cuckoo eggs or nestlings. Each female cuckoo, as judged by the markings on her eggs, appeared to have her own domain in which she parasitized warbler nests. Sometimes, I found two different cuckoo eggs in a reed warbler nest, indicating overlaps in the parasitizing areas of two cuckoos. I watched newly hatched cuckoos show stereotyped movements to carry on their back host eggs out of the nest. The same species of cuckoos (*Cuculus canorus*) occurs in Europe where the marking on their eggs matches that of host eggs. I did not find this phenomenon in Hokkaido. I wrote to an eastern German fellow named Wolfgang Makatsch who published a nice book on cuckoos, offering him cuckoo eggs from Hokkaido. He was gracious enough to send me a copy of his book in exchange. I heard later that he was so well known that he could freely travel to the then West Berlin during the communist regime.

BERKELEY

After completing the two-year Master's period in Sapporo, I decided to go abroad. I applied to several U.S. schools that were known for vertebrate zoology. I got offers of admission from University of California Berkeley, University of Michigan, and Yale. Although I gathered information about their graduate schools at the U.S. cultural center in Sapporo, my decision to go to Berkeley was based on two simple facts; it was the first to send me the letter of admission and my roommate's statement that his father liked Berkeley. I applied for and got a Fulbright travel grant to cover my travel expenses. I left Japan with only 50 borrowed dollars. I knew that I wanted to work under Alden H. Miller who was a well-known avian systematist and ecologist.

I read his review paper about the role of behavior in avian speciation, which was the subject I wanted to pursue. However, just before I sailed from Yokohama, I received a letter from Cadet Hand, who was the acting chair of the zoology department, informing me that Miller was on a sabbatical leave and that a new ethologist named Peter Marler just arrived from Cambridge. What a miracle! Also, I could change my mind without betraying Miller, because he did not know my original plan to join him.

I arrived in Berkeley on September 9, 1958. I was busy for a while taking courses and earning money as a teaching assistant. I received a very good graduate fellowship in my third year, enabling me to work full time for my thesis research. My first project in the Marler laboratory was to determine the acoustic properties of song that birds use for species recognition. I chose birds with simple songs like the Oregon junco (*Junco oreganus*) in order to be able to modify song with ease. The laboratory had a portable tape recorder (Magnemite 610) in which turning of the reels was by a coiled spring as in old phonographs. I had to spin a heavy flywheel by hand to start the machine turning and crank up the spring every so often. The machine basically worked flawlessly and I recorded many songs on the Berkeley hills. I was also curious to see if there was anything like song dialects in juncos. Although I did not detect any local differences in songs, I found that syllables were not infinitely variable but fell into a few distinct types (Konishi, 1964b). The meaning of this phenomenon was not discussed until a similar trend was discovered by Marler and Pickert (1984) in swamp sparrows (*Melospiza georgiana*). I used my first Kay sonagraph in the Marler laboratory to look at the acoustic properties of songs. How do I change recorded song though? Computers were not available then. I recall asking people at the Haskins Laboratory whether they could synthesize birdsongs with their Visible Speech machine, which scanned and converted cutout patterns (holes in paper) into sounds. Years later I got to know Alvin Liberman, who was one of the designers of the machine. He did not remember any letter from me. He said that he would have helped me, if he had read my letter. If I could see sounds on magnetic tapes, I might be able to cut and paste recordings. I either figured it out by myself or learned from someone that the magnetized parts of the tape might pick up fine iron particles. I got hold of iron powder and passed tape through a mound of it. I was delighted to see patches of iron powder corresponding to song syllables in the trill type song of Oregon juncos. Assuming that the constant silent interval between syllables was important for species recognition, I cut and pasted tape to make the interval highly variable. When I played back this type of song in the field, wild juncos responded to it. This result was a great disappointment for me, because I had expected no response. Also, this project turned out to be very time consuming; I could do field experiments only during the spring breeding season. I had to become realistic, because I was in my third year of graduate study.

Although I was very comfortable with behavioral experiments, I also had a strong inclination towards physiological studies. I had a very good teacher named Mitsuo Tamashige who taught neurophysiology in my junior and senior years in Sapporo. He covered everything from biophysics to behavior. Some of his on-campus and marine laboratory exercises included neuroethological studies of invertebrate animals. One of them was about the central coordination of locomotion in a snail. The professor allowed me to use his experimental setup in his own office. No student could expect this treatment from a Japanese professor. When I showed him rhythmic neural potentials in one of the nerves, he said "the rhythm seems centrally generated." The idea of central coordination goes back to the turn of the last century when people like Friedländer (1888) and Biedermann (1904) carried out simple but clever experiments to prove or disprove the theory. Later people like von Holst and Gray performed sophisticated behavioral experiments to obtain evidence for or against central coordination. Peter Marler covered central coordination and endogenous rhythms quite extensively in his animal behavior course, because central coordination was at the core of the Lorenz–Tinbergen model of instinctive behavior. It was Donald M. Wilson who used neurophysiological methods to provide the most convincing evidence for central control of wing beating in the locust (Wilson, 1961). While I was in Berkeley, Wilson joined our department and he served on my thesis committee, which also included the famous Frank Beach of sexual behavior from the Department of Psychology. Wilson later moved to Stanford and invited me from Princeton to give a couple of lectures in the course he and Donald Kennedy were teaching. Shortly after this visit Wilson died in a rafting accident, trying to save his girl friend who fell from the fast moving raft. She managed to come ashore, while he got dragged down to the bottom by the long rope tied to his body. I remember him telling me how he and she swam in a flooded river in Palo Alto all the way to the ocean. I always wonder what Wilson would be doing if he were alive today. His work triggered a bandwagon effect in which other people tried to replicate his finding in every possible preparation. In retrospect, it is interesting to realize that the idea of central coordination did not affect the students of birdsong at that time, because few of them were interested in neurophysiology. Also, mechanistic views of birdsong simply did not exist.

I thought that the relationship between vocalization and hearing resembles that between locomotory coordination and sensory feedback. It was already known that humans could not speak normally when auditory feedback was removed or delayed. I thought that similar experiments in birds had to be done. I was also aware that I could not fail in this project, because either positive (deafening affects song) or negative results were worth publishing.

I checked the literature on the subject and found Johann Schwartzkopff (1949) who developed a method for removing the avian cochlea. He also reported that the flute-like quality of a learned social call in adult bullfinches (*Pyrrhula pyrrhula*) gradually became shrill after deafening, although this operation did not affect other vocalizations. Similarly, Messmer and Messmer (1956) for whom Schwartzkopff deafened blackbirds (*Turdus merula*) heard some abnormal sounds from these birds. However, I could not check the accuracy of their impressions, because they had no pictorial way to visualize birdsongs before the age of the sonagraph, which apparently did not reach German zoology laboratories until after 1956.

I read Schwartzkopff's paper in German to learn his methods. This was not a big problem, because I had learned enough German in my undergraduate years in Sapporo. His illustrations of relevant anatomical structures and head holding devices were very helpful. The main problems were the tools that I needed for his methods. The Marler laboratory was not equipped to do surgeries. The most advanced surgical technique the laboratory used was laparotomy, i.e. making a hole on the bird's body wall to see the gonads. I learned this method from Alden H. Miller and introduced it to the Marler laboratory. Both deafening and laparotomy methods required a dissecting microscope and a light source that could illuminate the bottom of a small hole. The question was how to direct a light beam into a small hole without obstructing the view with the light source itself. Today, we can buy a dissecting microscope like the Zeiss Operating Microscope that comes with a vertical illuminator. Another graduate student who knew the method of vertical illumination told me how to solve the problem. According to his idea, I should use a mirror, which is coated only on one side, and place it 45 degrees relative to the optical axis of the dissecting scope. So, what this arrangement did was to allow me to see the bottom of the hole through the mirror, while this was directing some light into the hole. Where do I get such a mirror? He told me how to make one by exposing one side of a large cover glass to smoke from a candle.

The zoology department in Berkeley had a good machine shop with helpful technicians. One of them modified a pair of tweezers for operating in small and deep holes. My next problem was to find material for making fine fish hooks. Schwartzkopff put a small wire hook at the end of a probe like a thin chopstick. He inserted the hook through a hole made in the bony cavity containing the cochlea. I looked for fine but relatively stiff wires without success. Then, it occurred to me that light bulbs contained wires holding the filaments. I collected a few burned out light bulbs. They did contain fine tungsten wires that were just perfect for making those hooks. Actually, I have never found a better material. As soon as I discovered this fact, I asked everyone around to save burned-out light bulbs. The remaining problem was how to adjust the angle of the bird's head relative to the optical axis of the dissecting microscope, because the scope was on a fixed stand. I needed a small table that could be tilted around a pivotal point. I went to a junkyard and found a material suitable for the above purpose and that was an automobile rear-viewing mirror. I replaced the mirror with a plastic plate and constructed a simple device for holding the bird's head. The plastic table could be moved up and down around the ball joint that came with the mirror. When I was almost finished with my research, the Marler laboratory got an NSF grant that included a dissecting microscope with a vertical illuminator! I gave my operating table to Fred (Fernando Nottebohm) who used it for his thesis on chaffinches in Cambridge, England. I recall seeing the table in one of Fred's laboratories at the Rockefeller University years later.

I operated on several species of songbirds using the Schwartzkopff methods. Most of the data in my thesis came from these species. As I worked on larger birds and also more abundant species like the domestic chicken, I found that I could remove the cochlea through the ear canal instead of a hole made in the skull. I described this in my chicken paper (Konishi, 1963), although few people seem to know the origin of this method. In recent years, I have taught several people to deafen zebra finches (*Taeniopygia guttata*) with this method using Zeiss Operating Microscopes.

The Marler laboratory had a menagerie of animals ranging from fish to unusual mammals like kinkajous and a badger. No one seemed to be bothered by the crowing of my roosters, which I kept in an old greenhouse in the central courtyard of the building. When the Animal Behavior Field Station was built up on the Berkley hills, I moved some of the chickens there to make clean recordings. When Fred and I met there, we would return to his apartment for lunch with steaks and red wine as in his country, Argentina. Since we did not have enough soundproof boxes to house a large number of birds individually, I put all my deaf passerines in a small pent-house on the roof of the Life Science Building. I lined the pent-house walls with cheap sound-absorbing materials. I spent most of my day time sitting there listening and recording, because nothing was automated as it is today. I made about 3000 sonagrams for my thesis. I still have them in my office. To make one sonagram took a few minutes. I sometimes read a book while I was making sonagrams.

The first set of data came from the chickens, because they matured much faster than wild birds. I knew enough about the vocalizations of chickens from my childhood experience. It was particularly interesting to see how deaf chickens failed to respond to vocal signals such as cackling and aerial alarm calls of their flock mates. Sonagrams of several vocalizations showed no systematic differences between normal and deaf roosters. Around this time a German named Erich Bäumer published a paper on chicken vocalizations. I could guess what vocalizations he was referring to from his German descriptions. He was kind enough to send me his tape recordings upon my request. I made sonagrams of his recordings and compared them with my own recordings. He and I agreed on all identifiable adult vocalizations. When I published my results in the same journal, I wrote his German terms next to my English ones (Konishi, 1963). Pictorial catalogues of animal voices with their functional significance were rare at that time except for the one for chaffinches (Marler, 1956). I also noted that there were graded and discrete signals (I used the terms analogue versus digital; D. Wilson did not like the term digital, because the signal was not digital in the true sense of the word). Peter had already pointed out this distinction in his theoretical essay (Marler, 1961). Later he also found examples in the voices of several primate species (Marler, 1965, 1976b). The chicken results were neither discouraging nor encouraging. Had I just worked on chickens, what would I have concluded? Auditory feedback is not necessary for avian vocalizations? I already knew from the work of Schelderup-Ebbe (1922) that isolated chickens could develop normal vocalizations. Of course, his finding alone tells nothing about the role of auditory feedback.

I had reasons to expect that deaf songbirds would develop abnormal songs. It was already known from the work of Thorpe in the chaffinch and also from the work in progress in the Marler laboratory with the white-crowned sparrow (*Zonotrichia leucophrys*) that young birds memorize tutor songs before they can sing. This fact suggested to me that auditory feedback should be indispensable for vocal reproduction of tutor song. I thought that the only way this expectation could be shown wrong would be to have a situation in which vocal memory somehow directly controlled vocal motor centers of the brain. This possibility was inconceivable, because birds have to know how their song sounds in order to know the degree of match between the memorized and vocalized songs. I was, therefore, delighted to see the dramatic effects of deafening on the development of song in the white-crowned sparrow (Konishi, 1965b). All other songbirds I used also developed abnormal songs (Konishi, 1964a, 1965a). In my thesis, I summarized my thoughts above in a model in which birds use auditory feedback to match their vocal output with a stored song template. I also reported that deafness did not affect the song of adult white-crowned sparrows. Although recent studies appear to contradict this conclusion (Nordeen and Nordeen, 1992; Woolley and Rubel, 1997; Okanoya and Yamaguchi, 1997; Leonardo and Konishi, 1999), a systematic study of the relationship between age and the effects of deafening in zebra finches by Lombardino and Nottebohm (2000) showed that the song of birds 5–6 years of age remained unchanged after the operation for a much longer time than that of younger birds.

Of many memorable events in the Marler laboratory, trips to Inverness (a coastal area north of San Francisco) were my favorites. White-crowns nest in coastal chaparrals. We would arrive there the night before and camp out on the meadows. Peter always brought the whole family including his wife Judith, their young son Christopher, and a basenji dog. In the evening, we would talk around the campfire. Chris would babble before going to sleep. His babbles appeared to contain some elements of English to my ears, which were used to the babbling of Japanese babies. Peter was very interested to hear my impression. Another story that must be told is about Fred. When Peter took us to the Chiricahua Mountains in New Mexico to collect nestlings of slate-colored juncos (*Junco hyemalis*), Fred got lost and spent all night wandering the mountains. He was carrying a nest with young birds until they died. Fortunately, a passing ranger truck picked him up as he finally hit a road in the morning. As he came back to our campsite, he gobbled a breakfast and threw it up right away before he went to his tent to sleep half a day. Had he disappeared, would we know of the existence of the song control system today?

GERMANY

After I finished my thesis work in Berkeley, I had to leave the country, because I had an exchange visitor visa. Instead of going back to Japan where I had no place to return to anyway, I chose to go to Schwartzkopff's laboratory to learn more about the auditory system of birds. On my way to Germany I attended my first international congress of ethology in Leiden (1963). Peter Marler managed to send me and another student of his as speakers, which were more like the plenary speakers of today. I could see famous people like Konrad Lorenz, Niko Tinbergen, William Thorpe, and Otto Koehler sitting quite close to the podium. After my talk Don Wilson congratulated me and Koehler came to me to ask if I would publish my results in "his" journal, which was then called *Zeitschrift für Tierpsychologie*. I was so flattered that I simply said yes and kept my promise. John Emlen of the University of Wisconsin approached me to ask if I would be interested in a position in his department. Later the job came through and I accepted it. This is what Sidney Brenner calls the American way in contrast to the British way. Your future is decided upon not by inheritance but by the people you meet.

After the congress I went to Seewiesen where Konrad Lorenz was. I thought that Seewiesen was a heaven for ethologists. Jürgen Aschoff, who was famous for his study of circadian rhythm, invited me to stay at his "castle" in a nearby village. He had a lot of Japanese art objects, which his father (medical professor) received from some 40 Japanese medical students he trained. There was no dull moment in the castle, because Aschoff knew how to use time for useful purposes. He asked me to give a talk for him alone. When I used the term "template" in this talk, he proposed an equivalent German term "Sollmuster" I really liked this term, because it is so expressive. Soll means "should or must" and Muster "pattern" I used it in the German summary of my white-crown paper in Koehler's journal (Konishi, 1965b). Koehler liked the word and asked me how I got this nice term. I enjoyed my two years in Germany, because I met so many people doing excellent ethological studies including Walter Heiligenberg and Klaus Immelmann. A visit to the Immelmann laboratory was my first introduction to zebra finches (*Taenopygea guttata*). Also, neuroethology, particularly the neuroethology of insects, was in full bloom there. People like Franz Huber (cricket brain and song) and Dietrich Schneider (silkworm moth pheromone) were doing very interesting work. Later I lived in Munich for a year and half. I would drive to Seewiesen almost every week to hear a seminar or play ping-pong with Walter. Lorenz told me that I should feel like a member of his institute.

One time I left a reprint of my white-crown paper on Lorenz's desk in his office. Later, he thanked me and said he seldom read anything, because he did not want to change his ideas. I know he read the paper I left on his desk, because he mentioned my name and work in an interview with Joseph Alsop of the *New Yorker*. Either Lorenz or Alsop mixed the species when quoting my work. I wrote to Alsop that I enjoyed his interesting article except for a small error in the story. Alsop wrote back "glad the error was small." My chairman at Princeton, John Bonner, was very excited about the article and told me that he would put a copy in my file for future reference. I did not know that the *New Yorker* was such an influential journal.

The primary goal of my sojourns in Germany was to study the avian auditory system under Schwartzkopff. This plan did not work, because his laboratories were not ready after a move from Munich. His group was housed in an old mill. Since my time was limited, I decided to move to the Max Planck group led by Otto Creutzfeld in Munich to map the receptive fields of neurons in the cat's visual cortex using intracellular recording methods. This work did not go far, because we could not hold neurons long enough to map their receptive fields. Nevertheless, I developed a taste for intracellular methods.

PRINCETON

After I returned from Germany to the U.S.A., I settled in Princeton via Madison, Wisconsin, where I taught for about a year. My laboratories in Princeton were in the basement of the former psychology building where Wever and Bray (1930) discovered "cochlear microphonics." Their hand-made wooden soundproof room was still there for my use. Wever's group had moved to a new set of buildings outside the main campus. Wever was conducting comparative studies of reptilian and amphibian ears. It was no accident that my first graduate student there, Geoff Manley, currently professor at the Technical University of Munich, conducted a comparative physiological study of the reptilian auditory system.

I started to study song development in white-crowned sparrows. My aim was to test whether or not white-crowns raised in complete isolation from the egg could distinguish the song of their own species from that of multiple other species sharing the same habitat (Konishi, 1985). I had this plan despite the fact that Peter had shown not only the inability of nestlings under 10 days of age to learn even the song of their own species but also the ability of older nestlings to choose the song of their own species over the song of another species (Marler, 1970b). This work shows that nestlings younger than 10 days of age could not or did not reproduce the tutor song in adulthood. If such birds had been given a second chance of choosing between the original tutor song and a new song during the normal critical period, which song would they have chosen? The question is whether an early exposure to song affects a later choice of song. This was the rationale for raising white-crowned sparrows in complete isolation from the earliest stage of embryogenesis. I wanted to answer this question by collecting newly laid white-crown eggs and incubating them and raising chicks without exposing them to any birdsong before tutoring. This project required a logistic planning. For example, I took a graduate student with me to the Inverness area where we collected a few eggs. We wrapped each of these eggs in cotton and slid it into a test tube. We connected the test tubes side by side with strings into a belt, which we wore across our belly. Our own body heat kept the eggs alive. We brought back the eggs to Princeton and next morning the student drove up to Millbrook, NY, where Peter put the eggs in canaries' nests. Incredibly, most eggs hatched. Despite this success, I began to think that the number of nestlings we could raise per year severely limited our progress. I also thought that we had to raise nestlings entirely artificially without the help of canaries. So, I decided to suspend the white-crown work.

While the introduction of the sonagraph revolutionized research on birdsong, little was known about what songbirds could hear. I chose neurophysiological methods to answer this question. My research strategy was simple; I collected or bought birds whose songs differed clearly in the frequency domain. I recorded single neurons in one of the cochlear nuclei and determined threshold sensitivities of these neurons. The results were clear-cut; birds that produced high frequencies in their song had auditory neurons that responded to these frequencies (Konishi, 1969, 1970). However, all birds could hear low frequencies whether their song contained these sounds or not. Some birdsongs contain rapid transients in frequency. These transients can exceed the highest frequency to which auditory neurons can respond. I found that the threshold of the most sensitive neuron in a given frequency range was close to the sensitivity measured by behavioral methods for that frequency range. Fortunately, Bob Dooling, who was doing his thesis work under my friend Jim Mulligan from my Berkeley days, had a behavioral audibility curve for canaries. I compared it with my neurophysiological results from canaries to find a very good match between the two sets of data. So, if one draws a curve connecting the most sensitive neurons in all frequency bands, one gets a curve similar to the bird's audibility curve (Konishi, 1969). This relationship has been established not only in birds but also in other species including cats (e.g. Kiang, 1965).

Having learned the usefulness of single unit recording, I addressed another issue that occupied my mind. Recall how Lehrman (1953) used Kuo's interpretation of behavioral development of chicks to argue that we had to know more about behavioral development before birth instead of assuming the inborn nature of behavior. This line of argument spread fast to make ethologists apprehensive. For example, according to Gottlieb (1971), mallard duck embryos, which were prevented from vocalizing, discriminated poorly between the maternal call of their own species and that of chickens. He also reported that duck embryos responded to the maternal call a week before hatching. These studies got me interested in hearing in avian embryos. I checked if and when duck embryos began to hear in the egg. I showed that neurons of the cochlear nucleus in embryos became sensitive to low frequency sound about a week before hatching. As embryos developed further, neurons became more sensitive and responded to higher frequencies. Both the sensitivity and the range of frequency became adult-like two days before hatching (Konishi, 1973).

CALTECH

My fond memories of Berkeley pulled me towards California. This force moved me west from Princeton to Caltech in 1975 against the advice of some of my Princeton colleagues who reminded me of the smog problem. I resumed my work on the white-crowned sparrow mentioned earlier. We had to solve a few important practical problems in bird rearing. First, we had to develop a new method of holding and transporting eggs. Our electrical engineer Mike Walsh built a battery-operated portable incubator. This allowed us to stay days at collecting sites at elevations up to 8000 ft in areas in the Sierra Nevada where mountain white-crowns breed. Also, we did not have to drive many hours nonstop to rush the eggs to the laboratory incubator. He also built an incubator, which periodically changed the orientation of eggs as in chicken egg incubators. It was generally thought that passerines could not be raised from birth on the so-called 'steak food.' My able assistant Gene Akutagawa found that liquid from the crop of canaries raising nestlings contained something that enabled chicks (of other passerines) to consume the steak food. We raised white-crowned sparrows from birth in complete individual isolation with this method. Gene further found that the "liquid" was not necessary, if he fed predigested food for human babies to newborn white-crowns. He even figured out how to raise newborn zebra finches, which normally receive semi-digested seeds from their parents. The trick was to feed babies dehusked millet, which is available in health food stores. We showed that young white-crowns isolated as eggs preferred the song of their own species to alien songs sung by other inhabitants in the same area (Konishi, 1985). However, some of these white-crowns initially developed both a copy of the white-crown tutor song and a copy of one of the alien songs. As the season progressed, these birds dropped the alien song.

In my early days at Caltech, all postdoctoral fellows wanted to work with owls, but I began advising graduate students to work in the field of songbird research, which was to become very attractive to neurobiologists because of the discovery of the song control system by Nottebohm *et al.* (1976). I always liked and encouraged graduate students starting new things in my laboratory. I had some adventurous students who would do anything. Larry Katz, who tragically passed away a short while ago, was the most adventurous and skilful. I was so charmed by him that I allowed him to rent a small airplane to fly to Stanford to get a new histological tracer. Next, he suggested that we introduce brain slice techniques. So, he and I drove down to UC Irvine to see slice setups. On our way home, we bought a couple of components, which Larry assembled into a functioning system within a few days. He developed a powerful new method to study the anatomical organization of neural tissues. He would inject a fluorescent tracer into the target area of neurons residing some distance away. He would then make slices of the tissues containing the somata of the neurons. He discovered that neurons projecting to different targets had different soma and dendritic morphologies. I told him that he could make big contributions if he would apply these methods to the cat visual cortex, which was the darling of the time. So, Larry wrote his thesis on the cat's visual cortex in my laboratory. A big loss to the birdsong field, but a big future for Larry. Rich Mooney later inherited Larry's setup to do his very original thesis work on the nature of synaptic inputs to RA, which receives signals from LMAN by NMDA receptors and from HVC by non-NMDA glutamate receptors (Mooney and Konishi, 1991; Mooney, 1992). His project was the first extensive in vitro and intracellular developmental study of the song system in my laboratory and in the birdsong field. This was his idea, because I did not know what NMDA was. His work started a new NMDA cottage industry in the birdsong community (see Nordeen and Nordeen, this volume).

Mark Gurney was another adventurous student. The Nottebohm laboratory and we independently discovered sexual dimorphism in the song system of the zebra finch. I had this conversation with Mark Gurney: he said, "These gender differences may be genetic." I responded, "Genetics is molecular biology." He said, "You are right." He did the simplest experiment by injecting sex hormones into developing zebra finch eggs and newly hatched chicks. Mark found that estrogen masculinized the female song system. These birds sang when treated with testosterone in adulthood (Gurney and Konishi, 1980). Why estrogen instead of androgen? The brain (of rodents and birds) contains an enzyme that converts testosterone from the gonads into estrogen, which induces masculine differentiation in some areas of the brain. After Mark left, Gene Akutagawa took over the hormone project. Using radioactive markers to identify neurons, he showed that the neurons that migrated into RA were born on the seventh day of incubation. These neurons are large and equal in size in both sexes on the first day of hatching. However, they undergo gradual atrophy and ultimate death in the female RA, whereas they grow in size in the male. Gene further showed that exogenous estrogen could prevent the atrophy and death of these marked neurons. There was a gradient of estrogen action; the earlier it was injected, the more effective it was in preventing cell atrophy and death (Akutagawa and Konishi, 1985; Konishi and Akutagawa, 1990). Today, estrogen is thought to be good for postmenopausal women not only for the maintenance of normal physiological conditions but also for preventing the death of their brain cells, although some experts disagree on this point. Who would have thought of a link between women's health and songbirds?

Mark Gurney and Larry Katz were good buddies when they were exploring something new. One day they set up the necessary gear to do intracellular recordings in HVC of a zebra finch. I told them to clap hands to see if neurons responded to sound. To their great surprise, they saw responses in HVC. I showed them how to use auditory instruments and measure sound level. At any rate, they wrote up a simple report. It was flatly rejected twice as an artifact, although it was eventually published (Katz and Gurney, 1981). Another student, Jim McCasland, was recording multiunits in the HVC of behaving canaries. I told him to play canary song. He found that neurons responded much better to the song of a bird of the same breed than to the song of another breed of canary (McCasland and Konishi, 1981). Jim also showed that HVC neurons did not respond to playback of the bird's own song, while the canary was singing and immediately after the end of song. These preliminary findings were exciting, because the presence of auditory responses within the vocal control pathway suggested a possible link between the auditory and vocal control systems. The discovery of neurons selective for the bird's own song was exciting, because they might represent the song template. I wanted to know what features of song these neurons were detecting. This study required analysis and synthesis of sounds. The song of zebra finches was too complex for both analysis and synthesis at that time. I suggested to Dan Margoliash to undertake this project with white-crowned sparrows with simple tonal song, because Dan was the only student who could use computers. His results clearly showed the importance of both syllable structure and sequence (Margoliash, 1983, 1986). In the meantime, the Nottebohm laboratory also reported that they saw auditory responses in the hypoglossal nucleus (William and Nottebohm, 1985). Separate groups of HVC neurons project to RA and X. When Allison Doupe joined my group as a postdoctoral fellow, she decided to check for auditory responses in the anterior forebrain pathway. She found selectivity for the bird's own song (BOS) in LMAN and X. Furthermore, she showed that injections of a local anesthetic to HVC abolished auditory responses in both X and RA, suggesting that these nuclei received their song selective property from HVC (Doupe and Konishi, 1991).

Caltech has a graduate program called Computation and Neural System (CNS). CNS students are bright. When these students appreciate biological problems, they can do excellent research. I was telling my group in one of our lunchtime gatherings that I had heard about new methods of recording from neurons in vitro called "whole cell clamp." I told my group that it would be interesting to try the methods in vivo. No one said anything at that time, but Mike Lewicki, a former mathematic student from Carnegie Mellon, came to my office to ask if the methods would work in vivo. I said why not? He started right away. He read that he could count the number of bubbles to measure the tip diameter of a capillary electrode. Since this method was too crude for him, he took electrodes to a scanning electron microscope on campus. When he plotted the tip diameters measured with this method and those with the bubble method, he got a straight diagonal line. This episode impressed me very much, because I like students who go beyond my knowledge and ability. Then, we heard that an assistant professor elsewhere was doing in vivo whole cell clamping. Mike went to see the person and came back to tell me that their methods were similar. Mike turned out to be a very good neurophysiologist. He showed that the sensitivity of HVC neurons to syllable sequences involved inhibition; for example, a neuron responded preferentially to syllable A followed by syllable B. When the sequence was reversed, the neuron responded with an inhibitory potential. He developed a simple circuit model that detected specific syllable orders (Lewicki and Konishi, 1995; Lewicki, 1996).

From my time in the Marler laboratory, one topic stuck in my mind. It is about designing experiments to test whether or not delayed auditory feedback affects song. I got some people interested in the subject at Caltech. I read that someone designed a theater in which the audience wore wireless headsets and listened to music. This alone is not new, but coils surrounding the theater transmitted the electrical signals. When I told Dan Margoliash about this story, he got interested and built a small version of this setup. He wore magnetic earphones and stuck his head in the coils he made. This was a short-lived project, because Dan heard no sounds! More recently another ambitious student took this topic seriously and got excellent results by different methods. Anthony Leonardo, another CNS student from Carnegie Mellon, was not only smart but also technically skilled. He built a computer-based system to detect song and play back its delayed version. Although birds heard both natural and delayed feedback, they gradually changed the probability of syllable sequences and also syllable structure in some cases (Leonardo and Konishi, 1999). Remarkably, the original song gradually returned after normal feedback was restored. One summer, he went to the Bell Telephone Laboratory to work with Michale Fee in designing and testing the now well-known microdrive for zebra finches. He assembled two microdrives for his use in our laboratory. He quickly figured out how to place electrodes in LMAN. Recording single neurons in the LMAN of singing birds would answer the most important question about its role in the feedback control of song. Anthony did not find any effects of delayed feedback on the firing patterns of LMAN neurons (Leonardo, 2004). Despite these advances the control of song by auditory feedback remains one of the most important issues in birdsong research.

EPILOGUE

I have tried to tell the reader how ideas, methods, and people have shaped my journey in birdsong research. This is not the whole story, because I selected events and people to highlight some of the memorable and exciting things in my career. I have been lucky to have good graduate students and postdoctoral fellows. I mention mostly graduate students, because I could closely observe how they grew as scientists. Students and postdoctoral fellows worked not for me but for themselves. This is the best way to discover and encourage good students. I enjoyed my freedom in the Marler laboratory and I tried to create a similar environment for my people. Nothing pleases me more than hearing about the accomplishments of former members of my laboratory. When they thank me, I say to them that I did not do anything except giving them spiritual support and freedom.

ACKNOWLEDGEMENTS

I thank Peter Marler for his ever-lasting spiritual support since my first day in Berkeley and for reviewing this paper.

36 • The discovery of replaceable neurons

Fernando Nottebohm

INTRODUCTION

Early seafarers often set forth without a clear idea of where their next landfall would be. So with journeys of scientific discovery whose destination, at the onset, is unknown, but there is another, perplexing, difference between these two kinds of travel. While seafarers start from a known port, so there is a clear beginning, journeys of scientific discovery often have a less clear starting point – instead of a beginning, there is a gradual emergence. The result is that sometimes the origin of ideas is hard to trace, because visibility during that early stage is poor and the aspiring seafarer of the mind meanders, palpates the cottony fog and is not sure where he is going.

From Buenos Aires to Berkeley, via Nebraska

– Why birds? What made you choose birds for your work?
– Oh, so many things.

E. O. Wilson says that he chose to study ants because he was very shortsighted, yet with his glasses on and on all fours he could carefully observe, as a child, what ants were doing. He liked that. I was attracted to birds because I could not take my eyes from them. I grew up in a residential area of Buenos Aires where all the houses had gardens in the back. There was much birdlife. I loved the regulars, but even more the exotics, as when a canary turned up. I had to get my hands on it, and so I designed crude traps. Once caught, I would spend hours marveling at the beauty of the new arrival, which I now could inspect at close quarters. I also spent long holidays at my family's ranch in central Argentina (Southeast Cordoba), usually with a pair of binoculars in hand. My hero in my early teens was W. H. Hudson, who had grown up in Argentina. Many years later (1917), after he migrated to England, he wrote, '*Far Away and Long Ago*', his adventures as a child in the pampas, surrounded by the birds he loved. Hudson was a wonderful naturalist. I wanted to be like him, but was also intrigued by other matters.

I grew up Catholic. Sundays I went to church shepherded by my sister, but I did not warm up to the Christian faith. Animals seemed to have been created to satisfy the needs of people. Not one of the ten commandments spoke about kindness to animals and I was distraught that pets were not allowed in heaven. But those were the complaints of a child. During my high-school years, I read about physics, biology and philosophy and gradually my dissatisfaction with religion took a deeper turn. No one disputed that life was made of matter and that matter had narrowly defined properties. If so, then under identical circumstances, matter must always perform in the same way. How did this relate to free will and to guilt? Uncertainty principles did not trouble me since surely freedom of choice was not about statistical flukes. I must have shared some of these thoughts with my zoology teacher, for during my penultimate year of high-school he asked me to explain them to the rest of the class. I wrote on the blackboard "A identical to A" and explained how everything else followed. I do not remember the reaction of the other boys. Religion had introduced me to the paradox of a free spirit that was part and parcel of a material body. I liked the poetry of this, but did not see how it could work. Ever since, I have had a soft spot for paradoxes, I like them better than knowledge. I was now an agnostic.

During the summer of my seventeenth year I read *The Story of Philosophy* by Will Durant (1926). That writer chronicled the way in which different civilizations and philosophers had met their need for existential meaning and for an afterlife. That book gave me a level of comfort. It was not as if I had renounced the one and only and natural God. I had just given up one of the many Gods created by successive civilizations. As a result, my Sundays were now simpler. I was able to get earlier on the train to the Parana Delta, there to spend a day in a waterworld of boats, half sunken islands and interesting birds. I also repaired to that world when, with my mother's approval, I played hooky. School helped me develop my mind, but on a daily basis taught me little that was of interest. I had by then started to read about Charles Darwin and asked my parents to send me to Cambridge University so that I could pick up on Darwin's tradition and line of work. During my last year in Argentina I was also lucky to see a film about the dance language of bees by the great Austrian biologist Karl von Frisch.

My mother encouraged my interest in biology, but my father, more practical, was opposed. He pointed out that as a biologist I would be unable to support a family because there was no interest for this kind of work in Argentina, where the only animals that received attention were the ones one could eat. He recommended that I become a rancher and that I watch birds on weekends. It was not a bad solution and so the next question was where to study ranching? Argentina's universities were, at that time, in self-destructive turmoil. Misinformation led me to choose the School of Agriculture at the University of Nebraska in Lincoln. I was told that ranching in Nebraska was the closest to the conditions encountered in the pampas. Yet when I got there I found that Nebraska's cattle country was nothing like Argentina's, certainly not like the pampas. It had a surfeit of snow, blizzards, tornadoes and huge sandhills.

I arrived at Lincoln in January of 1959. I was 19 years old and spent the summer after my first undergraduate semester working as a hired hand at the Bar Eleven ranch in Valentine, northwest Nebraska, owned by Frank and Edna Reece. There I enjoyed the wild animals in the sandhill country, riding in rodeos and waterskiing in the evenings. During the days I milked cows, painted barns, herded cattle and put up endless haystacks. My nickname as a hired hand was "Freddie." By summer's end, Freddie's haystacks had a very good reputation. That opportunity to live with an American family and share with them ranch life was magnificent. Entering the United States through its very center helped me understand the values that had made the country great and gave me a glimpse of its history. I met in Valentine the father of my boss's wife, Mr. Highlander Philander Young. He was in his 80s and as a young lad had gotten to Valentine on a prairie wagon. That was when Nebraska was first settled by Europeans. Despite the richness of that summer and the graciousness of the Reece family, I could not get into the mood of becoming a farmer or rancher. Rather, I decided that there were three big issues that needed answering: time, space, and consciousness. Time and space, I felt, were unsolvable, since a universe bounded in time and space was incomprehensible, and infinite time and space were incomprehensible too. Consciousness, on the other hand, showed more promise since it happened inside our head and did not seem to ask impossible questions. Moreover, it fell within the realm of biology, where I had a bit more confidence.

Back at school, my fellow students had other interests, usually related to more immediate matters. Buenos Aires had many flaws, but it was a cosmopolitan town. I found no one in the circles I moved in Nebraska with whom I could talk about the things that intrigued me and so one year after my arrival I drove west to the University of California at Berkeley and it was there, during my first semester, as a student in Zoology IB, that I met Peter Marler, who had a russet beard and talked about animals and their lives as I imagined Darwin would. It was intellectual love at first sight. The fact that this great man devoted his research to the song of birds made him even more irresistible. Marler was very hospitable and invited students to his home, where he lived with his family and basenji dog. I did my undergraduate honors thesis in Marler's lab. He was a clear thinker, had a knack for blending field and laboratory techniques and observations, was a very resourceful experimentalist and helped me with my honors project. I placed a rooster's syrinx in a pressure chamber that mimicked the role of the interclavicular air sac that surrounds the syrinx, then ran air through the bronchi and trachea generating vocal sounds. I also could make a dead rooster vocalize by clamping a pressurized air hose to the bird's humerus, thereby accessing the air sac system. The final touch – Marler's idea – was to mount a stroboscopic light outside the pressure chamber that held the syrinx so that intermittent illumination was reflected from the syringeal membranes. When the frequency of the strobe light and the frequency of oscillation of the external tympaniform membranes recorded on film were the same, these membranes seemed to cease oscillating. That frequency, I was pleased to find, was the same as the fundamental frequency of the sound produced.

As my last undergraduate semester neared its end and I was preparing to return to Argentina, Marler sent word through his doctoral student and good friend of mine Mark Konishi, that I would be welcome to stay for doctoral studies. I had not gone to Berkeley because Marler was there, but it was my good fortune to find him and Mark Konishi there. Marler was a splendid role model. I learned from him that it was possible to study birds – even the song of birds – and support a family. I was by then also enamored of the United States, of Berkeley, of California. Life was very good, and I was just getting started. I realized that by studying birds I was taking a detour, away from my deeper philosophical interests, but it promised to be a happy detour, working with the birds I loved. I could only hope that the road I traveled would, eventually, bring me back to the big questions that interested me. Had I been asked then, as one is often asked in grant proposals, to map the road ahead, I would have never imagined the twists and turns it would take. As I journeyed, I often thought about "A is identical to A," about choice and freedom. There must be a great biological advantage that leads us to believe that we are in charge.

A key, life-long partner

I met my future wife Marta when she was 16 and finishing high school in San Francisco. I married her when she was barely 18. I found in her the best mind that I encountered in my entire career. Not only was she the person that, for many years, kept my laboratory and research well organized, but she also provided ideas, suggestions and perceptive criticism and the style and joy and laughter that made me the happiest man on earth. She recorded and analyzed the song of the birds with which I worked, at a time when analysis was done by looking at long streams of film of the visual representation of sound on an oscilloscope screen. Analyzing the song of canaries, with their large and changing repertoires, took many days of work per bird; each new syllable when first encountered had to be drawn on a sheet of paper, given a number and memorized. This work was tiring on the eyes and a labor of love, yet indispensable before the times of computer-aided analysis of sounds. Marta also read and edited every publication that came from the laboratory and she did all this as she created and presided over our home and raised two lovely children, Lawrence and Olivia. I will never be able to thank her enough. Without her understanding, encouragement, support and hard work, my career would have dried up early on, like a river that disappears in the sand.

The "instinct to learn"

The discovery that some brain cells are constantly replaced in an adult vertebrate brain came as a surprise because it was widely believed that central nervous system neurogenesis stopped soon after birth and that the same complement of neurons, perhaps with some losses, was present throughout life. An account of this discovery belongs in this book because it was made in the song system of birds and may help explain some aspects of vocal learning.

I indulged earlier in reminiscing how I became a student at the University of California in Berkeley, how I met Peter Marler and Mark Konishi and became a zoologist. I will now describe how the study of vocal learning in birds led to the discovery of neuronal replacement. In 1958 W. H. Thorpe of the University of Cambridge published "The learning of song patterns by birds, with special reference to the song of the chaffinch, *Fringilla coelebs*" (Thorpe, 1958). In that study, Thorpe showed that young chaffinches acquired their song by imitating the song of adults of their own kind, for which they seemed to have a preference. Imitation of the model occurred during a *sensitive period* that normally ended at sexual maturity. However, if conspecific models were withheld, song could still be altered in future years. Thorpe interpreted his observations as evidence for the existence of an innate "blueprint" that guided song learning.

Peter Marler was involved as a doctoral student and research assistant with much of Thorpe's work on song learning. Marler became very interested in the natural history of song dialects in birds and he continued that work when he settled in the late fifties at the University of California in Berkeley. The laboratory that Marler established focused on vocal learning in birds. Many of us that have worked on this topic – Baker, Dooling, Konishi, Kroodsma, Nowicki, Searcy and myself, among others – apprenticed in Marler's laboratory.

Marler and his colleagues used California's white-crowned sparrows, *Zonotrichia leucophrys*, and other songbirds to confirm and extend Thorpe's observations, adding new experimental refinements, better quantification and controls and exploring new conceptual issues (Marler and Tamura, 1962, 1964; Marler, 1970b; Marler and Peters, 1977, 1981, 1982a, b, c, d, 1987, 1988a; Marler and Sherman, 1983, 1985. Marler 1991b) concluded from all these studies that oscine songbirds had an inherited program typical of each species that could be thought of as an "instinct to learn." This instinct defined the "when", "what" and "how" of learning so only the details of the imitated pattern were left to external sources of information. There are similarities between Thorpe's "blue-print" for song learning and Marler's "instinct to learn" (see Adret, this volume) but Marler put more work into fleshing out this concept and provided many more details.

Auditory feedback is necessary for vocal learning

Mark Konishi did his doctoral work in Marler's laboratory. Konishi (1963, 1965b) investigated how auditory experience exerted its effect on song development (see Konishi, this volume). Using the technique of cochlear removal developed by Schwartzkopff (1949), Konishi (1965b) noticed that the marked effect of early deafening on song learning in white-crowned sparrows was the same whether the operation occurred before or after exposure to a model song, provided that the model had not yet been imitated. This result suggested that the auditory model did not have a direct effect on vocal output but, rather, guided learning through a two-step process: (1) first the auditory memory of the model was acquired; (2) as vocal development proceeded, vocal output was modified until the auditory feedback generated matched the auditory model, which acted as a *template*. Konishi's insight went well with the observation, later emphasized by Marler (Marler and Peters, 1982d), that several months can elapse from the time a model is heard until it is imitated – i.e. acquisition of the template and its imitation are two separate processes that can also be separate in time.

Konishi (1965b) made another tantalizing observation. The song of intact white-crowned sparrows reared as isolates included tonal whistles also present in wild-type song but typically absent from the song of early-deafened individuals. Thus, access to auditory feedback helped a young bird steer song development in the direction of wild-type song even when this bird had never heard an external model. This type of vocal ontogeny differs from that of non vocal learners such as domestic fowl, *Gallus domesticus* (Konishi, 1963) ring-doves, *Streptopelia risoria* (Nottebohm and Nottebohm, 1971) and eastern phoebes, *Sayornis phoebe* (Kroodsma and Konishi, 1991), that produce normal adult vocalizations even when deafened very early in ontogeny.

Left hypoglossal dominance

Many of the seminal insights of Thorpe, Marler and Konishi were already in place or being finalized by the time I started my doctorate in Marler's laboratory in 1962. In the fall of 1963 I left for Cambridge, U.K. to spend a year in Thorpe's laboratory as part of my doctoral training, while Marler was on a sabbatical in East Africa studying monkeys. It was during that year that I started to experiment with syringeal denervation. Konishi (1965b) thought that the stability of learned song that he observed in white-crowned sparrows deafened as adults might be maintained by proprioceptive feedback, though it was not known whether the syrinx had sensory nerves. I expected that if the syrinx were denervated in an adult that had been previously deafened, the bird would still have a voice but that many of the learned sounds would disappear. I speculated that under these conditions proprioceptive information that normally accompanied song would be altered. If proprioception played a role in the song stability of deafened adults, then denervation of the syrinx (Figure 36.1) should lead not just to song that was aberrant in its motor execution, but also that was unstable. As I set to test this prediction, I encountered a roadblock. When both halves of the syrinx of an adult male chaffinch were denervated by section of the right and left tracheosyringeal nerves, the birds, when alarmed, had difficulty breathing – in fact, if the disturbance continued, they suffocated. I then denervated the syrinx unilaterally. Breathing now was unimpeded, but in addition I encountered a paradox, and of course, that set me aquiver. Though both halves of the syrinx were anatomically similar, denervation of the left half eliminated most of the sounds of a chaffinch's learned song while denervation of the right half had little or no effect (Figure 36.2). I called

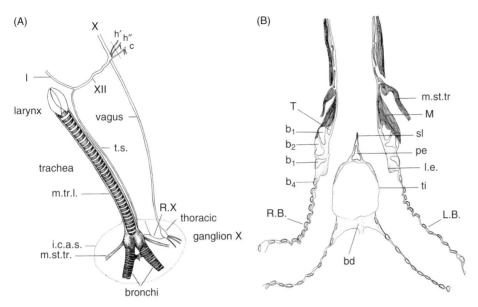

Figure 36.1 (A) Schematic ventral view of the syrinx, trachea and laryinx of a male chaffinch, *Fringilla coelebs*, and attendant musculature and innervation. All anatomical details are very similar in canaries and zebra finches. The syrinx is the vocal organ of birds, at the confluence of both bronchi and the trachea. It is surrounded by an air space, the interclavicular air sac (i.c.a.s.). Each syringeal half has its own air flow, sound source, muscular control and innervation. Each tracheosyringeal branch of the hypoglossal nerve innervates just the muscles of the ipsilateral, syringeal half; this drawing shows just the left side innervation; a very slender vagal branch reaches the syrinx, but its role is unknown. Unilateral syringeal denervation can be achieved by cutting all three roots (h', h'' and c) of one hypoglossal (XII) nerve or just by cutting the tracheosyringeal (ts) branch on one side. The anatomy and pattern of innervation of the songbird syrinx allow it to function as two separate sound sources (Greenewalt, 1968; Nottebohm, 1971). The involvement of the larynx in phonation has not been described; the extent of its opening probably affects sound loudness and the emphasis or suppression of different harmonics, as described for bill movements (Hoese et al., 2000). (B) A longitudinal section of the syrinx of an adult male canary shows the point of constriction, on each side, where the external labium (l.e.) extends towards the rostral end of the internal tympaniform membrane (t.i.); sound is thought to be generated by the periodic oscillations of the external labium (Goller and Larsen, 1997a; Fee et al., 1998). The bronchidesmus, a membrane, anchors the syrinx to the dorsal wall of the inter clavicular air sac. The drawings, with modifications, come from Nottebohm (1971). Notice the somewhat heavier muscle mass on the left syringeal half.

Abbreviations: R.B and L.B., right and left bronchi; X, tenth cranial nerve; RX, recurrens branch of the vagus, that reaches the syrinx; m. st.tr., sternotrachealis muscle; m.tr.l., tracheolateralis muscle; L, lingual branch of the hypoglossus nerve that innervates the tongue (not shown). The syrinx also has several bony components: the tympanum (T), bronchial half-rings 1–4, and the pessulus (pe); the pessulus is crested by the semi-lunar membrane (sl).

this phenomenon *left hypoglossal dominance* or "the vocal left" (Nottebohm 1970, 1971a, 1972b). Now I had something to work with!

Left hypoglossal dominance is also present in canaries and white-crowned sparrows (Nottebohm, 1976; Hartley and Suthers, 1990), so it is not a one-species anomaly. As I returned from England, where wild-born chaffinches had been kindly provided by Thorpe's laboratory, I switched my research on laterality to canaries, which were easy to breed in captivity and therefore better suited for laboratory work. Peter Marler kindly gave me the birds I needed, which came from his colony of Belgian Waterslagers.

The discovery of left hypoglossal dominance in chaffinches, canaries and white-crowned sparrows delighted me because it suggested that the song of at least some oscine songbirds might blend three very human traits: vocal learning, handedness and hemispheric dominance. If some songbirds had left hypoglossal dominance for the production of learned song, might they also show left hemispheric dominance for this same behavior and if so, which parts of the brain

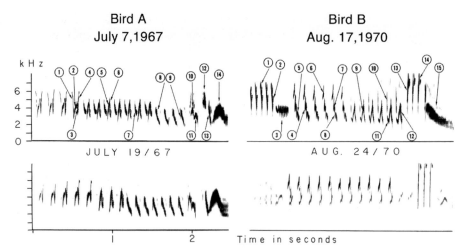

Figure 36.2 The song of two intact adult male chaffinches, A and B. Both birds were recorded again after denervation of the left (Bird A) or right (Bird B) syringeal halves. The elements of preoperative song, defined as frequency upsweeps, downsweeps or buzzes, were numbered. In bird A, only three out of fourteen elements (numbers 1, 7 and 12) were lost after section of the right tracheosyringeal nerve. In bird B, only three elements out of fifteen (numbers 6, 13 and 14) survived section of the left tracheosyringeal nerve. These results suggest that a majority of the elements of adult chaffinch song is under left hypoglossal control, while the remainder of elements (usually simple in structure and high pitched) is under control of the left syringeal innervation. Lost elements were replaced by silence or by altered sounds. The horizontal axis indicates time in seconds, the vertical one frequency in kilohertz. Dates stand for when the preoperative and postoperative recordings were made. (From Nottebohm, 1971.)

were responsible for the acquisition and production of learned song? Philosophers and evolutionary biologists have struggled for a long time with the origins of vocal learning in humans. Could it be that birds would shed independent light on this key evolutionary step? I now had a strong incentive to delve into the songbird brain.

The song system comes of age

In January of 1967 I moved, at Peter Marler's invitation, to his laboratory at Rockefeller University, where I started as an assistant professor. Soon thereafter, I teamed up with a brilliant comparative neuroanatomist, Christiana Leonard, who I knew from Berkeley, and a very gifted technician, Tegner Stokes. Our first task was to produce a stereotaxic atlas of the canary brain (Stokes et al., 1974). We were guided in this work by the recently published pigeon atlas of Karten and Hodos (1967). Our next task was to identify the motor neurons that innervated the muscles of the syrinx. We found them in the hypoglossal nucleus of the medulla. Then we placed electrolytic lesions in the vicinity of the forebrain's known auditory projection, Field L of Rose (1914), and noticed which of these lesions had an effect on song and how these regions were connected, directly or indirectly, to the hypoglossal motor neurons that innervated the syrinx. That is how the discrete nuclei that we now call the high vocal center (HVC), the robust nucleus of the Arcopallium (RA) and Area X of the striatum were discovered. We saw that HVC projected to RA which, in turn projected to nXIIts; in addition, HVC projected to Area X. All these projections were uncrossed (Nottebohm et al., 1976). That was our first glimpse of the *song system*. Studies by many different laboratories later added more anatomical details (Figure 36.3).

The reversal of left hemispheric dominance

Lesions of the left HVC and of the left RA of adult male canaries had a greater effect on the quality of postoperative song than similar lesions on the right side; complete lesions of the left HVC were particularly devastating in that, without muting the bird, they erased all traces of learned song. If HVC was destroyed bilaterally the bird still adopted a singing posture and tried to sing, but with exception of a few clicking sounds the song was silent; there was no recovery from this

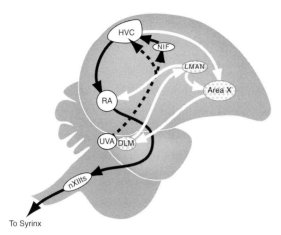

Figure 36.3 Schematic view of the song system in a parasaggital section of the adult songbird brain. The anatomical relations shown apply equally to adult male canaries and zebra finches, the two species that have received the greatest anatomical and neurophysiological attention. The details shown here are not complete. A posterior pathway, responsible for the production of learned song, leads from the nucleus interface (NIf) to the high vocal center (HVC), from HVC to the robust nucleus of the archistriatum (RA), and from there to a number of midbrain and medullary nuclei that control structures involved with phonation and respiration – including the caudal half of the hypoglossal nucleus, which houses the motor neurons that innervate the muscles of the syrinx. An anterior pathway necessary for song learning but not song production goes from HVC to Area X, then to the medial portion of the dorsolateral thalamic nucleus (DLM) and from there to the lateral magnocellular nucleus of the anterior neostriatum (LMAN); LMAN sends a projection back to Area X and another one to RA; other feedback loops are not shown. Interestingly, nucleus uvaeformis (Uva) of the thalamus, which projects to NIf and HVC, is capable of driving the firing of HVC neurons. All the projections shown here are ipsilateral. (This drawing is based on anatomical information from Nottebohm *et al.*, 1976 and 1982; Okuhata and Saito, 1987; Bottjer *et al.*, 1989; Vates and Nottebohm, 1995; Nixdorf-Bergweiler *et al.*, 1995a; and Vates *et al.*, 1997; and on neurophysiological observations from McCasland, 1987; Williams, 1989; Yu and Margoliash, 1996; and Hahnloser *et al.*, 2002.)

condition (Nottebohm, 1977). It seemed from the above results that the HVC of adult canaries did not initiate song, yet the left HVC played a dominant role in the production of learned sounds. To my surprise, this dominance could be reversed. Following the elimination of the left HVC of an adult male canary, song quality was back to normal some seven months later, this time under right side control (Nottebohm, 1977) (Figure 36.4). We now had three useful handles on the neurobiology of learned canary song: (1) a well-defined pathway whose principal role was the production of learned vocal patterns; (2) functional lateralization; and (3) adult plasticity that allowed for laterality reversal and recovery of function. There is even now, at the time of this writing, no other known brain system that shows this combination of attributes. It seems likely that the ability of the right HVC to redevelop normal song in adult canaries and the time it takes to do this results from the replacement of many of its neurons, but this idea has not been tested and was not in our mind at the time.

There are in canaries no obvious, systematic differences in the size of the right and left HVC and RA (Nottebohm *et al.*, 1981), so the functional asymmetry that normally occurs is probably not dictated by asymmetrical pathway limitations. Indeed, the very fact that following lesion of the dominant side the previously subordinate one takes over suggests that in terms of their potential both sides are similar (Nottebohm, 1977; Nottebohm *et al.*, 1979).

Gross sexual dimorphism in brain structure and its relation to gonadal hormones

The next surprise occurred when Art Arnold and I noticed – he in zebra finches, *Taeniopygia guttata*, I in canaries – that the nuclei of the song system were much larger in males than in females (Nottebohm and Arnold, 1976). Gross sexual dimorphism in a specific CNS pathway had not been observed before. We already knew that several of the nuclei of the adult song system concentrated testosterone and its metabolites (Zigmond *et al.*, 1973; Arnold *et al.*, 1976). Given this ability to concentrate hormones, the song system's sexual dimorphism could result from sexual differences in the hormone levels present at any one time or from differences in development that occurred earlier in ontogeny. Both these possibilities proved to be correct.

Adult female canaries sing relatively little and when they sing their song tends to be simpler and less stereotyped than that of males. However, after systemic treatment with testosterone, adult female canaries sing much more (Leonard, 1939; Shoemaker, 1939; Baldwin *et al.*, 1940; Herrick and Harris, 1957) and their song syllables become more stereotyped, though their diversity does not increase. This change in behavior is accompanied by gross anatomical changes. Song nuclei HVC and RA are

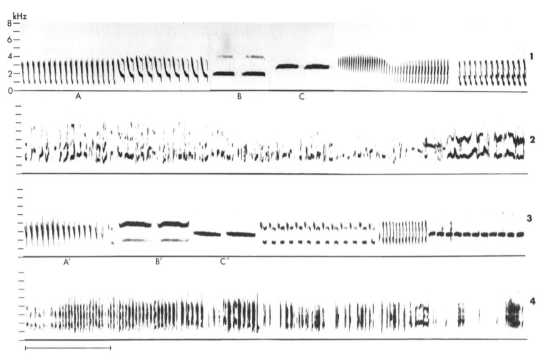

Figure 36.4 Effects on the song of an adult male canary of an electrolytic lesion that fully destroyed the left HVC. (1) Fragment of preoperative song. (2) Song recorded 7 days after lesion of left HVC. (3) Song recorded 7 months after left HVC lesion. (4) Song recorded 4 days after section of right tracheosyringeal nerve. Letters A, B and C identify preoperative syllables; A′, B′ and C′ identify the somewhat altered postoperative version of these same syllables. No syllable survived section of the right tracheosyringeal nerve. (From Nottebohm, 1977.)

90% and 50% larger, respectively, one month after onset of hormone treatment than in untreated controls. At that time, though, these two song nuclei remain in females still half as large as those of males (Nottebohm, 1980b). Full masculinization of the song system can only be achieved by hormone treatment early in juvenile life, as shown by Mark Gurney in Konishi's laboratory (Gurney and Konishi, 1980; Gurney, 1981).

Seasonal changes in brain structure

The observation of hormone-induced changes in the size of the female song control nuclei led to the next question: did the size of the song control nuclei of adult males change seasonally, as the birds went in and out of reproductive state? The answer was yes. The size of HVC and RA became considerably smaller after the end of the breeding season, then bounced back. This was the first report of seasonal changes in brain structure (Nottebohm, 1981).

Brain space for a learned task

The marked sexual dimorphism in the size of song nuclei can be related to sexual differences in singing behavior. Male canaries and zebra finches sing a lot, but female canaries sing little and female zebra finches never sing. In addition, the song of male canaries has a greater diversity of syllable types than that of females. Perhaps our observations of song system sexual dimorphism could be explained as a simple principle of brain economy: the amount of brain space allotted by each sex to a learned behavior is proportional to the relative complexity of that behavior in males and females. This, of course, raises the question of whether male canaries, which show marked individual variability in their number of song syllable types, might show a corresponding difference in the size of song nuclei. The answer to this question is intriguing. There is a significant relation between the size of HVC and the number of different song syllable types, but this relation seems to be a permissive one. Adult canaries

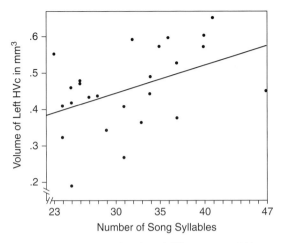

Figure 36.5 Regression of number of different song syllables on the volume of left HVC for 25 adult male canaries, aged 1–3 years. Song recordings were made during March and April when the birds, in full breeding condition, sang adult stereotyped song; samples of songs of 600 or more seconds were used for the syllable counts. Brains were obtained in mid summer, after the birds had bred. There was no significant age difference in HVC volume. Though at that time the testes of some of the birds had already regressed, there was no significant correlation between testis volume and HVC volume. Only left HVC volumes are plotted here; there was no consistent right–left asymmetry in the volume of HVC. (From Nottebohm et al., 1981.)

that have a large syllable repertoire also tend to have a large HVC and birds with a small HVC tend to have small song repertoires. However, a bird can have a large HVC and a small repertoire, a case, perhaps, of unfulfilled learning potential (Figure 36.5). I called the relation between complexity of learned song and amount of space devoted to it "*brain space for a learned task*" (Nottebohm et al., 1981) and, as described above, this brain space can change seasonally. Subsequent studies have found this relation between song complexity and HVC volume very robust. It holds when comparing different populations of a same species (Canady et al., 1984) or different congeners (Székely et al., 1996), or when comparing a large group of different oscine songbirds (DeVoogd et al., 1993).

Hormone-induced changes in dendritic length

How do hormones and season affect the size of adult HVC and RA? Tim DeVoogd and I used Golgi stained material to show that the dendrites of the most numerous type of RA neuron are significantly longer in adult male canaries than in adult females of the same age (DeVoogd and Nottebohm, 1981a). These differences, though, are less dependent on genetic sex than on circulating hormone levels. When adult female canaries were treated with testosterone levels as found in spring-time males they started, as expected, to sing like males. In these birds the dendrites of the cells we studied grew and many new synapses were formed (DeVoogd and Nottebohm, 1981b; DeVoogd et al., 1985; Canady et al., 1988) and for a while this seemed like a sufficient explanation for the seasonal and hormone-induced changes in the volume of song nuclei, because as dendrites grew cells would be spaced further apart. However, it puzzled me that whereas the RA of testosterone-treated adult female canaries grew by 50%, the HVC of these birds grew by 90%. Might there be in HVC still another process at work? Could it be that new cells were added? Might neurons come and go as hormone levels changed?

Neurogenesis in adult brain

I did not know how to test for the occurrence of neurogenesis in adult brain, but a doctoral student in my laboratory, Steve Goldman, knew how to do it. He and my technician Sue Kasparian (Goldman and Nottebohm, 1983) injected testosterone-treated adult female canaries with the birth-date marker ^{3}H-thymidine. As a cell prepares to divide, ^{3}H-thymidine becomes part of newly synthesized DNA. To our delight, there were many HVC neuron-like cells autoradiographically labeled with tritium (^{3}H) when the birds were killed 30 days later (Figure 36.6). Interestingly, too, these cells were present regardless of whether or not the birds had been treated with testosterone, but were not present if the birds were killed one or two days after ^{3}H-thymidine injection. At those short survivals, though, we could see many labeled cells on the wall of the lateral ventricle overlying HVC. We inferred that, as during embryogeny, new neurons were born on the walls of the lateral ventricle and, from there, migrated into HVC, where they settled and differentiated. As a double check on the neuronal identity of these cells, Steve Goldman and later Gail Burd looked at some of these cells electron microscopically. What they saw was compatible with our earlier identification of them as neurons (Goldman and Nottebohm, 1983; Burd and Nottebohm, 1985).

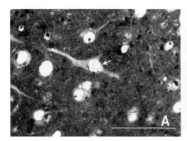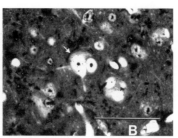

Figure 36.6 Two examples of Cresyl Violet stained, araldite embedded, 1 μm thick sections of the HVC of an adult canary killed one month after the last of a series of systemic injections of ³H-thymidine. As some of the unstable tritium (³H) atoms give up their excess energy (beta rays), it exposes silver grains in the overlying photographic emulsion, which then look like pepper grains. Two of the cells shown (arrows) have a clear neuronal profile and 10 or more exposed grains overlying their nucleus. The thinness of the sections made it unlikely that the radiation came from pieces of other, unseen, adjacent cells. Electron microscopy analysis of these same sections yielded further evidence of the neuronal identity of the labeled cells. (From Goldman and Nottebohm, 1983.)

We had learned, by then, about Joseph Altman's earlier reports, in the 1960s, of postnatal neurogenesis in the brain of rats and cats (Altman, 1962, 1963, 1970; Altman and Das, 1965) and of the resistance these reports had met. The Altman studies, like ours, used ³H-thymidine to document the presence of newly born cells. Subsequent studies by Kaplan had also used electron microscopy to show that at least some new cells born in adulthood received synapses and therefore were, presumably, neurons (Kaplan and Hinds, 1977). Yet questions remained as to what was reliable proof of neuronal identity and adult birth – could synapses be formed on glia? was a neuron-like cell necessarily a neuron? Was the nuclear presence of ³H-thymidine definitive proof of time of birth? Were the new neurons working neurons and, if the answer was yes, where could they possibly come from and how did they reach their destination? The resistance to adult neurogenesis was, perhaps, understandable, in view of the long-held dogma that neurogenesis stopped at or soon after birth.

Proof of new, working neurons

Given the persistent scepticism that continued to surround claims of adult neurogenesis in warm-blooded vertebrates, our first priority now was to test whether the cells we called new HVC neurons were in fact working neurons and were connected to existing circuits. John Paton, a wonderful neurophysiologist who at the time was a postdoctoral fellow in my laboratory, undertook this task. We knew that if we injected a male or female canary with ³H-thymidine twice a day for 2 weeks, approximately 10% of the neurons in HVC would be labeled 30 days after the last injection. At that time the bird was anesthetized and Paton entered HVC with a hollow glass electrode filled with the marker horseradish peroxidase (HRP). As the electrode advanced, Paton knew when a neuron was penetrated because of the change in electrical potential. Once the electrode was inside a cell, the bird was exposed to sound stimulation. Katz and Gurney (1981), working in the Konishi laboratory at Caltech had shown earlier that neurons in HVC fired in response to auditory stimulation. The same electrode that Paton used to impale the cells and record changes in potential in response to sound was then used to fill the cells with HRP. After the bird was killed, the brains were sliced at 120 μm intervals. Immunocytochemistry was used to visualize the HRP-filled cells. Typically, one to three HRP-filled cells were recovered per HVC. These cells showed as much detail – including dendritic spines – as if they had been stained with the Golgi method (Figure 36.7). After the HRP-filled cells were drawn, the 120 thick sections were thin sliced at 6 μm intervals. These thinner slices were used for autoradiography. Seventy-four HRP-filled cells with neuronal neurophysiological profiles and clear neuronal anatomy were recovered from HVC. Seven of these, the expected 10%, were well labeled with ³H-thymidine. Four of the latter seven had given vigorous electrophysiological responses to sound. The simplest interpretation of these results was that the cells we called new HVC neurons were, indeed, neurons and that at least some of them were connected to existing circuits (Paton

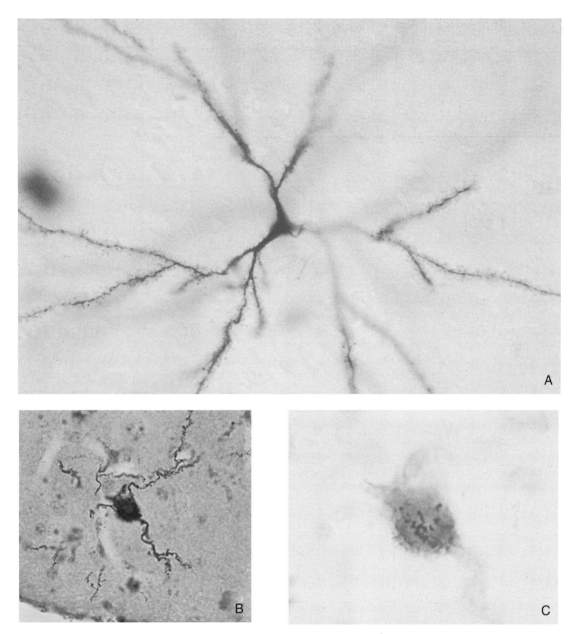

Figure 36.7 (A) HVC neuron of an adult canary that received a series of intramuscular ^{3}H-thymidine injections one month before it was anesthetized. This cell was then impaled by a hollow glass electrode and its action potentials recorded, as well as its synaptic potentials in response to sound stimulation. After the electrophysiological recordings were made, the same electrode was used to fill the cell with HRP. The bird was then killed, its brain sectioned at intervals of 100 μm and the resulting HVC sections were cleared in glycerine, so that this cell could be seen and photographed. Notice the well-developed and typically neuronal dendritic tree and dendritic spines. (B) The same cell was then sectioned at 6 μm intervals. In this section we see the cell's soma. (C) Same as in (B), but focusing on the overlying photographic emulsion. The diameter of this cell's soma is approximately 9 μm. The 30 or so exposed silver grains clustering over the cell's nucleus suggest that this cell was born a month earlier, while the bird received ^{3}H-thymidine. This was the first direct evidence that a vertebrate brain cell that looked and behaved like a neuron was born in adulthood and had become connected to existing circuits. (With changes, from Paton and Nottebohm, 1984.) (See also color plate section.)

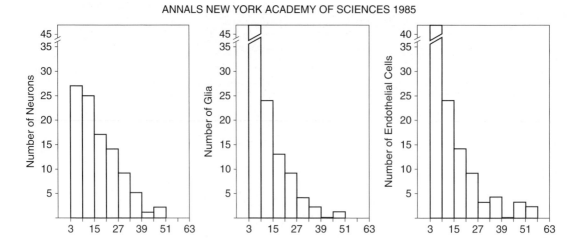

Figure 36.8 Histogram showing the relative frequency with which various numbers of exposed silver grains occurred over the nuclei of 100 labeled neurons (left), 100 labeled glia (middle) and 100 labeled endothelial cells (right). The sample of labeled cells was obtained from the HVC and adjacent neostriatum of an adult canary treated with ^{3}H-thymidine one month earlier; the tissue was sectioned at 6 μm intervals, stained with Cresyl Violet and exposed for autoradiography for 10 days. The criterion for labelling was three silver grains, which corresponds here to 10 × background. However, in all three cell classes a majority of labeled cells had nine or more exposed silver grains. The short exposure time was chosen to reduce the instances in which the number of exposed silver grains over the nucleus was too high to count or too high to allow for identification of the underlying cell. (From Nottebohm, 1985b).

and Nottebohm, 1984). I have dwelt on this particular experiment because it provided the first direct evidence that some cells born in adulthood became working neurons. This was the evidence that had not been available in Altman's work and whose absence accounted, probably, for the resistance it had encountered. Direct proof of the recruitment of new functional neurons in rodents came 16 years later (van Praag et al., 2002).

Another important question was whether the amount of ^{3}H label we saw in neurons of ^{3}H-thymidine treated birds warranted the conclusion that these cells had been born at the time the birth marker was given. It was known that thymidine was incorporated into existing cells not just during the stage of DNA synthesis that precedes mitosis but also during DNA repair, which is an ongoing process in living tissues. Might we be dealing with a situation in which, for some reason, in some cells DNA incorporation was exacerbated without this being part of a mitotic event? Our answer to this question was simple. We showed that the extent of autoradiographic label in the ^{3}H-labeled neurons we saw in the adult canary HVC was similar to that in other kinds of ^{3}H-labeled cells, such as glia and endothelia, also found in the same animals (Nottebohm, 1985b) (Figure 36.8). Since it was generally accepted that glia and endothelia continue to be produced in adulthood, the most parsimonious interpretation was that in all three kinds of cells ^{3}H-thymidine had been incorporated into DNA during the S-phase that precedes mitosis and that this was the main source of autoradiographic labeling. The same conclusion has been drawn more recently from studies in rodents (Palmer et al., 2000).

The new neurons added to the adult HVC are projection neurons

Now that we were confident that new neurons were added to nucleus HVC, what kind of neuron were they? Initially, we thought that the new neurons were predominantly interneurons because very few were backfilled when HRP was injected into either of HVC's target nuclei, RA or Area X (Paton et al., 1985). However, this result was misleading and probably resulted from the fact that HRP is not a universal agent for retrograde transport. When, in a later study, we used fluorogold (FG) instead of HRP we found, to our surprise, that a majority of the new neurons added to the adult HVC projected from HVC to RA, i.e. were projection neurons;

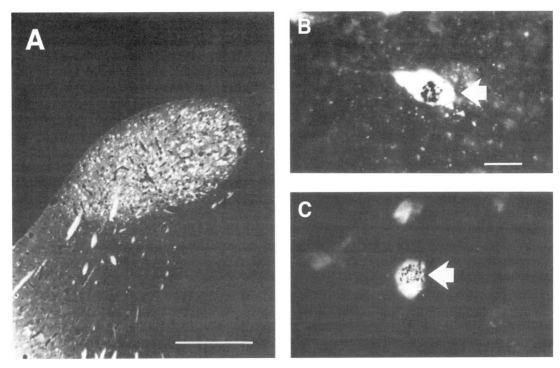

Figure 36.9 (A) Nucleus HVC of an adult male canary backfilled by Fluorogold injected into RA. Backfilled axons and cell bodies are clearly visible. (B, C) Two examples of HVC neurons backfilled with Fluorogold from RA and also labeled with 3H-thymidine (black grains over soma). These observations provided evidence that a majority of the neurons added to the adult HVC projected to RA. (From Kirn et al., 1991.)

these cells were backfilled by FG injected into RA (Figure 36.9), but not into Area X (Kirn et al., 1991); we thought, at the time, that the remainder of new HVC neurons were interneurons because they could not be backfilled by tracer injections into RA or Area X. Recent work shows that while interneurons and HVC→RA projection neurons are added to the HVC of juvenile zebra finches at the time they first learn their song (Scott and Lois, 2007), only HVC→RA neurons are added when these birds become adults (Scotto-Omassese et al., 2007). Another study has shown that new neurons recruited into the HVC of adult zebra finches and that cannot be backfilled from RA have other properties that set them apart from the new HVC→RA neurons that can be backfilled (Alvarez-Borda and Nottebohm, 2002), but maybe this difference has to do with some of the HVC→RA neurons connecting faster with their target nucleus than others, a phenomenon for which there is some preliminary, indirect evidence (Fig. 3B in Kirn et al., 1991).

The new neurons are replaceable neurons and their comings and goings coincide with the life of replaceable memories

Next came evidence that the neurons added to adult HVC persisted for a number of months and then disappeared. Addition and disappearance occurs even at times of year when total HVC numbers remain constant. Apparently the neurons added to adult HVC hold their job for a limited time and are then replaced by new ones. Evidence for this "replacement" is numerical. We do not know if the new neuron is employed in the same position and capacity as the one it replaces. The notion of "throw-away" neurons, i.e. neurons used for a while and then discarded, is still a novel one (Kirn and Nottebohm, 1993).

Soon thereafter we put together seasonal data collected over a number of years and seasons. Adult male canaries change their song every year by adding new syllable types and discarding others (Nottebohm

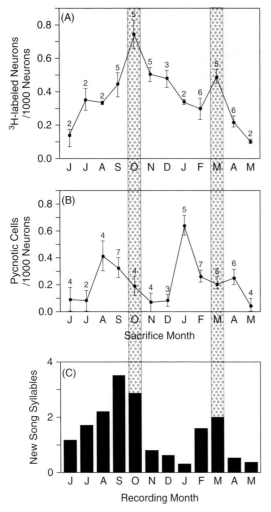

and Nottebohm, 1978). This process of recruiting new syllable types, which occurs during all months of the year, shows a major peak in September, after the end of the breeding season, and a less pronounced one in February–March (Nottebohm et al., 1987). Recruitment of new HVC neurons also occurs spontaneously throughout the year, but shows seasonal peaks in October and March. These peaks are preceded, two months earlier, by peaks of cell death (Kirn et al., 1994) (Figure 36.10). The peaks in syllable addition are sandwiched between the peaks in cell death and cell recruitment. This temporal relation suggests that the learning of new song syllables is enabled by cell-death-related weakening in the motor control of the pre-existing syllables, which as a result become more variable, thus providing raw material and opportunity for new learning. As new syllables emerge from this variability, the peak in new neuron recruitment is associated with a restoration of stereotypy; thereafter new syllable learning is attenuated. Presumably the new neurons receive instruction in song after or while they connect with the existing circuitry. More on this later.

Interestingly, whereas a majority of new HVC neurons born during the spring, when canary song is relatively stable, have disappeared by the end of the summer, when song becomes more variable, those born at the end of the summer and early fall, when the birds are acquiring a new repertoire, are still present eight months later, when that repertoire is used during the following breeding season (Nottebohm et al., 1994). The life of many of the new neurons in nucleus HVC seems commensurate with the life of memories that underlie the behavior controlled by that nucleus.

The migration of young neurons in adult brain and evidence for overproduction

Soon after neuronal recruitment was discovered in adult HVC (Goldman and Nottebohm, 1983), it became clear that it also occurred throughout most of the songbird telencephalon (Nottebohm, 1985b). If, as evidence suggested, new neurons were part of a choreography

Figure 36.10 (A) Mean number of ^{3}H-labeled neurons ($\pm$ SEM) per 1000 HVC neurons and per day of ^{3}H-thymidine treatment in adult male canaries that received 8 successive IM injections of ^{3}H-thymidine at 12 h intervals and were killed 27 days after the last injection. The number of birds in each monthly sample is indicated above the SEM bar. The letter in the horizontal axis stands for the month of the year when the birds were killed, starting with June (J) on the extreme left. (B) Pycnotic cells per 1000 labeled neurons were counted in the same birds as in (A) plus in 10 more added to make up for the fact that pycnotic cells occur at very low densities and thus require more individuals to yield a reliable mean. (C) Mean number of new syllable types that appeared every month in the song of six male canaries in their second year of life (modified from Nottebohm et al., 1986). For a syllable to be counted it had to be different from all others produced by the bird at that time and recur with sufficient stereotypy as to make its identification unambiguous, though these syllables are not, at that time, fully

Caption for Figure 36.10 (cont.)

stereotyped. The shaded bars linking A–C are meant to make it easier to compare the temporal relation between peaks of cell death (pycnotic cells), new cell recruitment and first appearance of new song syllables. (From Kirn et al., 1994.)

of constant replacement, we needed a mechanism to show how it occurred and that was applicable to much of the forebrain. Dan Buskirk, doing postdoctoral work in my laboratory, discovered a crucial component of this mechanism by producing a monoclonal antibody that revealed cells with a small body lodged in the wall of the lateral ventricle. These cells had a long, unbranching process that penetrated the adjacent parenchyma (Figure 36.11A) and at times could be seen reaching the pial surface. This typical anatomy was well known to embryologists. These cells were called by them *radial glia* and they had not been observed before in adult brain. They were very plentiful in the brain of our adult canaries. Arturo Alvarez-Buylla, then doing doctoral work in my laboratory, characterized the antibody that had revealed these cells and showed it to be positive to vimentin, an intermediate filament protein also found in astrocytes (Alvarez-Buylla et al., 1987).

Soon after radial glia were first visualized in our adult birds, we noticed that small elongated cells of uncertain identity, closely apposed to the radial fibers and showing their same orientation, were particularly common close to the lateral ventricle and in parts of the telencephalon rich in radial glia fibers. It was tempting to imagine that these young elongated cells (Figure 36.11B), which had not been previously described in adult brain, were young migrating neurons (Alvarez-Buylla et al., 1988a). To test this possibility, adult canaries received two systemic injections of ^{3}H-thymidine 12 hours apart and were then killed after 1 h or 1, 3, 6, 15, 20, 30 or 40 days later. At 1 h and 1 d survivals there were many ^{3}H-labeled cells on the walls of the lateral ventricle but none of the small elongated cells that we suspected of being young migrating neurons was labeled nor were there any labeled differentiated neurons. A small number of the small, elongated cells was labeled, however, 3 d after injection and all of them were close to the lateral ventricle. On subsequent days these small, elongated, ^{3}H-labeled cells were found further and further away from the lateral ventricle, giving the impression of a migrating wave. Some of these cells had reached the farthest corners of the telencephalon by day 20, which was, too, when we saw in that material the first, differentiated (postmigratory) ^{3}H-labeled neuron. Between days 20 and 40 the number of labeled elongated cells decreased, while that of labeled neurons increased (Figure 36.12). We inferred from these results that our time series showed the migration and eventual settling and differentiation of a wave of young

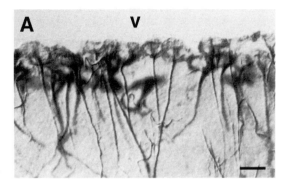

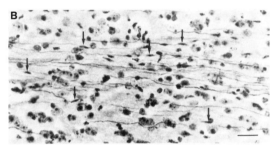

Figure 36.11 (A) Radial glia in adult canary brain stained with an antivimentin antibody. The body of the cells, with clear nuclei, is in the ventricular zone that forms the wall of the lateral ventricle (V). These cells project a single, unbranching process into the underlying brain parenchyma. Calibration bar: 10 μm. (From Alvarez-Buylla et al., 1987.) (B) Section of forebrain showing radial fibers stained brown by an antivimentin antibody in association with small elongated cells closely apposed to the fibers and showing the same orientation. These elongated cells were suspected, at the time, of being young migrating neurons. (From Alvarez-Buylla et al., 1988a).

neurons that had been born in the adult canary brain on a particular day. We noticed, too, that the maximal number of cells thought to be young migrating neurons seen on a particular day (day 20 after ^{3}H-thymidine injection) was three times higher than the number of labeled neurons seen on day 40. Apparently, as during embryogeny, a majority of the migrating neurons or newly differentiated neurons were culled, so that by day 40 only one third of the initial cohort survived (Alvarez-Buylla and Nottebohm, 1988).

The identity of neuronal stem cells

The next step was to identify the cells that gave birth to new neurons. We noticed that labeled ventricular zone cells were not evenly distributed throughout the walls

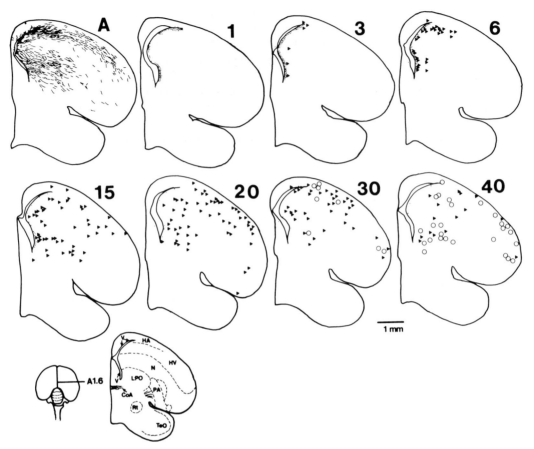

Figure 36.12 (Section A) Distribution of radial glia fibers in a transverse section of adult canary brain cut in the plane of the canary atlas (Stokes et al., 1974) at the level of the anterior comissure (CoA, inset). The number on the other sections indicates survival time, in days, after the second of two IM doses, 12 h apart, of ^{3}H-thymidine. At the earliest survivals there are many ^{3}H-labeled cells (indicated as dots) in the wall of the lateral ventricle, where new neurons are thought to be born. On day 3 one sees the first small, elongated, ^{3}H-labeled cells (filled triangles) close to the wall of the lateral ventricle, presumed to be young migratory neurons born after label injection and starting to move away from their birth-site. Thereafter the number of these cells increases. The first ^{3}H-labeled cells with a postmigratory phenotype (open circles) were seen on day 20 and their number much increased thereafter. (From Alvarez-Buylla and Nottebohm, 1988.)

of the lateral ventricle, but were particularly abundant in the dorsal and ventral reaches of the lateral ventricular wall, which we called "hot spots." We dissociated the cells in these hot spots one hour after a systemic ^{3}H-thymidine injection. At that time, the majority of the individually dissociated ^{3}H-labeled cells was positive to vimentin and had long, undividing processes, i.e. a majority of the dividing cells were radial glia. We suggested, from this observation, that new neurons were formed when radial glia divided and, presumably, one of the daughter cells assumed the identity of young migrating neuron (Alvarez-Buylla et al., 1990b). This view was confirmed later by more detailed, ultrastructural work in adult birds (Alvarez-Buylla et al., 1998) and in mouse embryos (Noctor et al., 2001). More recently Alvarez-Buylla and coworkers found that in adult mammals too the brain's neuronal progenitors are glia – astrocytes in the case of the subventricular zone cells that give rise to new olfactory bulb neurons (Doetsch et al., 1999) and radial glia in the case of the cells that give rise to hippocampal granule cells (Alvarez-Buylla et al., 2002). The accepted view at this time is that there is in the adult vertebrate brain a subset of glial cells – as defined by ultrastructure and molecular markers – that gives rise to neurons and that these cells are, therefore, neuronal stem cells.

Depending on where in the brain one looks, the process from birth of a young neuron to its postmigratory differentiation can take from eight days, which is the case for new cells in HVC (Kirn et al., 1999), to 20–40 days elsewhere in telencephalon (Alvarez-Buylla and Nottebohm, 1988). This timing is of general interest when looking for evidence of new neuron production in adult brain. It is possible that in larger brains, such as those of primates, where the distance from point of origin to final destination might be greater, the time spent in migration is longer than in songbirds, so comparisons between birds and mammals should take this possible timing difference into account. We do not know, yet, how often a given stem cell can divide and give rise to a new neuron. This information is important when tracking a cohort of cells labeled with a birth-date marker.

The natural orchestration of neuronal replacement

Now that all the cellular players are in place, I will review the experiments that show how neuronal replacement is orchestrated. We know that canary song is a seasonal behavior (Nottebohm et al., 1986, 1987), that HVC size in adult male canaries changes seasonally (Nottebohm, 1981) and that testosterone can double HVC size in adult female canaries (Nottebohm, 1980b). It seemed likely, from this, that testosterone levels would play a role in neuronal replacement and Rasika, a doctoral student in my laboratory set out to investigate this possibility using ^{3}H-thymidine as the birth-date marker. She noticed that the number of cells dividing in the ventricular wall above the HVC – the presumptive (Goldman and Nottebohm, 1983) and recently confirmed (Scott and Lois, 2007) origin of new neurons added to the adult HVC – of adult female canaries two days after onset of testosterone treatment was the same as in controls. However, if the females that received ^{3}H-thymidine were killed 60 days later and treated with physiological doses of testosterone during their last 40 days, then the number of ^{3}H-labeled neurons tripled compared with controls (Rasika et al., 1994). Apparently testosterone or its metabolites promoted new neuron survival.

Some years later Rasika found evidence that the testosterone effect on new neuron survival was mediated by brain derived neurotrophic factor, BDNF, as suggested by the following four kinds of evidence. (1) The HVC of male and female canaries has the TrkB receptor, which is specific to BDNF. (2) Systemic testosterone treatment of adult female canaries increases the levels of BDNF protein in their HVC. (3) In the absence of testosterone, BDNF infusion into the HVC of adult females triples new neuron survival. (4) Infusion of an antibody that binds to BDNF into the HVC of testosterone-treated females, thus preventing BDNF from binding to its receptor, blocks testosterone's effect on new neuron survival (Rasika et al., 1999).

Rasika's experiments left a question unanswered. Did testosterone exert its effect on HVC's BDNF levels directly, or by inducing higher singing levels? We knew that the act of singing could induce the expression of selected genes in HVC (Jarvis and Nottebohm, 1997); perhaps this applied, also, to BDNF expression. Xiao-Ching Li provided a first answer. She showed that BDNF expression in HVC is proportional to amount of singing. Moreover, males treated with the birth-date marker BrdU and allowed to sing as much as they wanted had twice as many new neurons in HVC as males prevented from singing during their last eight days (days 31–38 post BrdU injection) of life (Li et al., 2000). Interesting as these results were, however, they did not tell us whether singing, in the absence of testosterone, might affect BDNF production and neuronal survival, an issue addressed by the next experiment.

We knew that testosterone levels were very low in early fall in adult male canaries, though these birds sang a lot at that time (Nottebohm et al., 1987). We presumed that castration of those males would not affect their singing and so these birds might be able to tell us whether under those circumstances singing, in the absence of gonadal hormones, affected new neuron survival. Doctoral student Benjamin Alvarez-Borda did these experiments and showed that the number of new HVC neurons was lower in the castrates than in the noncastrates, even though both groups of birds sang, as expected, similar amounts of song. However, when the castrates were ranked by their amount of singing, then again there was a relation between amount of singing and new neuron survival. These results and the earlier work by Rasika et al. (1994) suggest that singing and blood testosterone levels affect new neuron survival in an additive manner (Alvarez-Borda and Nottebohm, 2002). It is possible that both these effects are mediated by BDNF, but direct evidence for this is not yet available for singing. Intriguingly, a positive effect of BDNF on the survival of new HVC neurons occurs only during a restricted time window 14–20 days after the new cells are born (Alvarez-Borda et al., 2004).

Evidence from a doctoral thesis suggests that the great majority, if not all, of the HVC cells that project to RA in the canary brain, regardless of whether they are born in nestlings or in adults, is eventually replaced (Alvarez-Borda, 2002). This is a population of cells that, over the ages sampled (2–9 years), seems to have a rather stable yearly turnover (Alvarez-Borda, 2002). The situation is very different in zebra finches, whose adult song changes little from year to year; in them the recruitment of new HVC neurons is drastically reduced with age (Wang et al., 2002). From this comparison of the two species, it would seem that age of the cells involved is not the primary reason for their replacement, but that the stability of the behavior they serve is.

It was mentioned earlier that in adult canaries the rate of turnover differs between times of year. For example, cells born in late summer/early fall and that presumably partake in song learning at that time, are still around next spring, when the song learned 8 months earlier is used during the breeding season. By contrast, half of the neurons added to HVC in the spring have disappeared by 4 months later, and by that time the spring song has also been much modified (Nottebohm et al., 1994). In this case, too, we must conclude that cell age is not the primary reason for replacement. Presumably seasonal changes in testosterone levels and in amount of singing account, in part at least, for this effect of season of birth on new neuron survival.

The role of vacancies in neuronal replacement

There is one more variable that affects the choreography of neuronal replacement and that I have not discussed yet. That variable is cell death and its consequence, *vacancies*.

We knew that some brain cells continued to be produced and replaced in the adult songbird brain, but we did not know if the death of some cells promoted the survival of others. If adult neurogenesis was, as we believed, part of a process of replacement, might it be that the type of cell that continued to be produced and recruited into adult circuits depended on the type of cells that died? If that were so, then the fact that RA-projecting but not Area X-projecting neurons continued to be added to the adult HVC could result from the fact that in our birds cells of one type, but not of the other, died and thus created a specific type of vacancy.

Scharff et al. (2000) explored the relation between death and replacement. Using technology recently developed by Jeffrey Macklis at Harvard, Scharff et al. selectively removed RA-projecting neurons from the HVC of adult male zebra finches (Figure 36.13) without leaving behind scar tissue. Elimination of RA-projecting cells resulted in a marked increase in the recruitment of new cells of this kind. However, elimination of HVC's Area X-projecting cells in adult male zebra finches was not followed by increased recruitment of new cells. Apparently, the brain of songbirds has a program for the constant replacement of some cells, but not of all cells and the effect of "vacancies" in upregulating new neuron survival may be specific to a relatively small class of replaceable neurons. If this is true of neuronal replacement in general, then we will have to learn a good deal more before it becomes possible to orchestrate, at will, the replacement of any central nervous system neuron that dies. We do not know how the effect of vacancies on new neuron survival comes about. Maybe this effect is mediated by "empty space" (e.g. synaptic space) left behind by a neuron that dies, or by an excess of unused neurotrophins, or by chemical signals produced by dying cells, to name a few possibilities.

The new neurons must be trained

The experiment of Scharff et al. (2000) had another, remarkable consequence. Some of the male zebra finches that lost bilaterally many of their RA-projecting neurons, also lost their learned song. In two of these birds, over a period of weeks, as the lost cells were replaced, the learned song was reinstated. This was of great interest to us. Zebra finches, like chaffinches and white-crowned sparrows, learn the motor pattern of their song during a sensitive period that precedes full sexual maturity and change it little if at all thereafter (Immelmann, 1969). Was reinstatement of the learned song after elimination of RA-projecting neurons a case of relearning of the motor program or a case of reconstituting the motor pathway so that a program that had not been affected by the cell loss could be expressed? In the first of these two instances, auditory feedback would be expected to play a role, and this has proven to be the case.

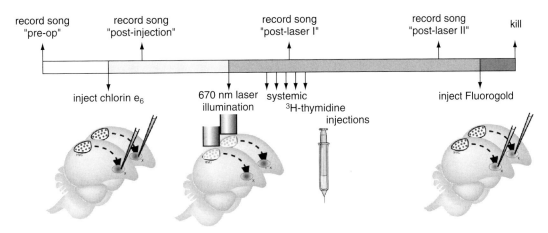

Figure 36.13 Method used for selective killing of a particular type of projection cell and evaluating the effect of this removal on song quality and new cell numbers. Song was recorded at the beginning of the experiments involving adult zebra finches. After bilateral injection of chlorin e_6 conjugated nanospheres into Area X (as depicted) or RA, song was recorded again to assure that no behavioral changes had occurred due to potential tissue damage at the injection site. After allowing sufficient time for retrograde transport of the chlorin e_6, HVC was noninvasively illuminated with long wavelength 674 μm laser light, photoactivating the production of oxygen singlets in the cells that had been backfilled by the nanospheres and thus producing their death. To monitor subsequent neurogenesis, ^{3}H-thymidine was injected intramuscularly 5 times over a 10 day period, starting with the second day after the laser illumination. During a survival period of 3 months, newly generated neurons were incorporated into HVC. During this time, short-term and long-term effects of the induced neuronal death on song were monitored through song recordings. To test whether some of the new neurons projected to the original nanosphere injection site, Fluorogold was injected 5 days before perfusion into the same target site. There were no new Area X-projecting neurons, but many new RA-projecting neurons. (From Scharff et al., 2000.) (See also color plate section.)

Selective elimination of HVC's RA-projecting neurons (HVC→RA neurons) is followed, over a period of 1–4 months, by a reconstitution of the original number of cells of this class. This return to the original number of neurons occurs in hearing as well as in deaf birds. However, recovery of the preoperative song occurs only in the hearing birds (B. Alvarez-Borda and F. Nottebohm, unpublished data). One possibility is that these cells need auditory guidance to assume their role in the patterning of learned song. This interpretation is in line with electrophysiological evidence that the HVC→RA neurons carry the tune of learned song (Hahnloser et al., 2002). Of course, it is possible that the targeted killing of HVC→RA neurons also requires auditorily guided adjustments in other parts of the song system.

Taken together, the observations of Kirn et al., (1994), Rasika et al., (1994, 1999), Li et al., (2000), Scharff et al., (2000), Alvarez-Borda and Nottebohm (2002), and Alvarez-Borda et al. (2004) suggest the following choreography, which may apply, in general, to temperate zone birds whose song is seasonal. A drop in testosterone levels and singing, such as occurs in midsummer towards the end of the breeding season, induces a drop in the levels of BDNF present in HVC, which leads to the death of many of HVC's replaceable neurons. The death of these cells creates vacancies. The new neurons that are constantly added to HVC sense the existence of these vacancies – indeed may home on them – and this increases the likelihood of their survival. However, for this survival to occur, the new cells or others nearby must engage in moderate to high levels of circuit activity (singing), thereby increasing BDNF expression and the amount of BDNF protein present in HVC. A rise in the level of testosterone or of its metabolites also enhances new cell survival. This cycle of events is summarized in Figure 36.14. It is possible that trophic substances other than BDNF are also involved and so our list of factors promoting new neuron survival could grow longer. Regulation of new neuron survival by extent of circuit use may be a general mechanism for ensuring that neuronal replacement is closely attuned to environmental change and hence to the brain's and the individual's needs. We infer that the demise of replaceable neurons results from the underuse of these cells, not from their overuse. Put differently,

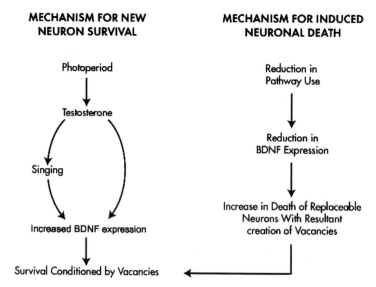

Figure 36.14 Chain of events that promote (left panel) or suppress (right panel) new neuron survival in the HVC of adult songbirds. Although the role of BDNF in promoting new neuron survival has been shown, other neurotrophins may also act in this manner. Arrows indicate the temporal order in which events are driven. Notice that the events that induce neuronal death, with the consequent creation of "vacancies," will also affect the survival of new neurons. (From Nottebohm, 2002a).

neuronal replacement does not seem to result from "wear and tear" associated with normal pathway use. If the above choreography is correct, then it may not be possible to have at the same time and in the same part of the brain a winnowing and recruitment of neurons, but the two processes may occur side by side if the signals that regulate them are contained within cells.

It should be noted that though we know that intact hearing is necessary for the new HVC cells to engage in the production of learned song, we do not yet know if, in the absence of hearing, the new cells turn over faster. If the answer were yes, then hormone levels, circuit activity, trophic substances, vacancies, *and learning* would configure the chain of events that made for a successful replacement of cells, renewal of function and long-term survival.

Silastic implants filled with testosterone can be used to maintain the typically high, spring testosterone levels throughout the summer and early fall. When this treatment is applied to adult male canaries, relatively few HVC neurons are added in late summer/early fall, perhaps because high testosterone levels prevent the death of existing neurons that normally occurs at the end of the breeding season and so there are fewer vacancies to promote the survival of new neurons (Rasika, 1998).

Replaceable neurons are in a minority

Direct evidence of neuronal replacement in adult avian brain is available only for the HVC neurons that project to RA. If we assume that 15 or so nuclei are involved with the acquisition and production of song (including nuclei also involved with respiration) and if we assume that each of these nuclei has two neuron types and that neurons produced in adulthood are of the replaceable kind, then only 2 neuron types out of an estimated 30 (6–7%) are in the replaceable category. We do not know, at this time, whether this percentage is representative of other functional systems and whether, therefore, it would apply to the whole brain. Evidence from mammals suggests that the percentage of replaceable neurons in them is smaller than in birds. We do not yet know why a subset of neurons falls into the replaceable class, nor what is their special function, but this is now a key research question.

Why replace neurons in the healthy adult brain?

Early on it became clear that while some birds learn new songs only once, before sexual maturity, others continue to learn new songs as adults and can change their song from one year to the next. I thought that eventually our understanding of how the song system works must meet a stringent test: that we be able to turn a sensitive period learner such as the chaffinch (Thorpe, 1958), white-crowned sparrow (Marler, 1970b; Marler and Tamura, 1964) or zebra finch (Immelmann, 1969), into an open-ended learner such as the canary (Nottebohm and Nottebohm, 1978; Nottebohm et al., 1986, 1987). This goal has not yet been achieved. However, it has been possible to delay in the chaffinch and zebra finch the closure of the sensitive period for song learning. The behavioral studies of W. H. Thorpe (1958) and Lucy Eales (1985, 1987) suggest that this can be achieved by withholding exposure to a wild-type model. Linda Wilbrecht may have uncovered the mechanism that maintains the sensitive period open. In socially reared male zebra finches the number of new HVC neurons born 30 days earlier and still present at the time the animal is killed decreases from day 50 until day 120 after hatching (Wilbrecht et al., 2002b). These birds produce stable adult song by day 90. However, as indicated above, young zebra finches kept singly are still able to imitate the song of a live model presented after day 90 (Eales, 1985, 1987). In these isolates the number of new neurons labeled with a birth-date marker and still present in HVC 30 days later when the bird is killed at the age of 90, 120 or 150 days is 50% higher than in the colony-reared individuals (Wilbrecht et al., 2006). We speculate that though isolates develop a song by day 90, *they are not committed to this song*; it seems as if this holding back of commitment is reflected in the turnover of existing neurons that continues at relatively high levels. Perhaps not surprisingly, this higher turnover is reflected in greater syllable variability. However, the higher turnover does not affect the total number of HVC neurons, which is, at those three ages, similar in the socially reared and isolate individuals (Wilbrecht et al., 2006).

The above study fits well with observations from two other laboratories. When adult Bengalese finches (*Lonchura domestica*) and zebra finches are deafened at 6 months, after the imitated song has been mastered, song deterioration in the former proceeds much more rapidly than in the latter. This outcome might be related to the fact that new neuron addition is twice as high, at that time, in the HVC of the Bengalese finches as in that of the zebra finches (Scott et al., 2000). Presumably, the neurons added after deafening are "uneducated" while the ones they replace are "educated," leading to an erosion of the learned skill. In zebra finches, song deterioration following deafening occurs faster in young adults than in older adults (Lombardino and Nottebohm, 2000) and in the former more new HVC neurons are added than in the latter (Wang et al., 2002).

Taken together, several of the song system studies reviewed here (Alvarez-Buylla et al., 1990a; Kirn et al., 1994; Nottebohm et al., 1994; Scott et al., 2000; Wilbrecht et al., 2002b, 2006; see also Pytte et al., this volume) suggest that the addition of new HVC neurons is related to the acquisition or forgetting of learned song. This relation between memory and neuronal replacement is reinforced by three other studies that focus on parts of the brain other than the song system. A study of free-ranging black-capped chickadees, *Parus atricapillus* showed that in these food-storing birds seasonal peaks in hippocampal new neuron addition occur at times of year when these birds learn much new spatial information (Barnea and Nottebohm, 1994, 1996). In the second study, which used zebra finches, the number of new neurons in caudal telencephalon was shown to increase when adults were moved from cages in which they lived as male/female pairs into others where they encountered a more complex social group, where they presumably had to learn to identify many new individuals and forge new social relations (Lipkind et al., 2002). In the third study, focusing on the same region of caudal telencephalon as the previous one, the number of new neurons rose in breeding zebra finches during the time the parents fed their young, when parents of this colonial breeder presumably memorized the vocal identity of their own offspring (Barkan et al., 2007). The sum of these studies (see also Nottebohm, 2002a) suggests that the recruitment and replacement of neurons is widespread in the adult songbird brain and that its occurrence is related to turnover of memorized information. This may be a basic and important aspect of how the brain functions.

However, correlations should be treated with care. The song sparrow, *Melospiza melodia*, learns its song during its first year of life and retains it thereafter, with no addition of new songs (Marler and Peters, 1987), yet shows, as adult, seasonal changes in HVC volume and in the recruitment of new HVC neurons. As in canaries, HVC becomes smaller at the end of summer and song at that time becomes more variable (Smith et al., 1997a). Again, as in canaries, more new neurons are added in the fall than in the spring. This remarkable study by Tramontin and Brenowitz (1999) was conducted with free-ranging individuals. It is not known whether these birds would keep their learned song if they were deafened. If they did not, then, as in adult zebra finches whose HVC→RA neurons were selectively killed (B. Alvarez-Borda and F. Nottebohm, unpublished data), the replacement cells might have to relearn the pattern the bird had already mastered. The logic or selective advantage of such an arrangement is not obvious at this time. Perhaps the songs of all song learners, whether of the sensitive period or open-ended kind, become better with age in ways that we do not yet appreciate and this subtle improvement depends on neuronal replacement. If substantial neuronal replacement were to occur in adult song sparrows after deafening, yet fail to interfere with the retention of learned song, then the argument that memory resides in the replaceable neurons would have to be re-examined. It is not clear, though, that the level of neuronal replacement observed in the HVC of adult canaries and song sparrows is similar, because the manner of delivery of the birth-date marker used in both studies was very different and so the counts of labeled neurons cannot be compared. It is possible, also, that some neurons are more sensitive to aging than others, and thus benefit more from replacement, or that they play in the adult brain roles that we do not yet suspect and that would necessitate periodic replacement even when this offers no learning advantages and poses an extra learning burden. If so, then it would be fascinating to know why any of these conditions is present in HVC's replaceable RA-projecting cells, but not in the X-projecting cells that are not replaced (Scharff et al., 2000).

Searching for molecular cues

As the birdsong saga unfolded and became part of the neurosciences, it became clear that insights from the various levels of analysis must eventually be related to the world of molecules. I was particularly interested in using molecular tools to explore the extent to which replaceable neurons might differ from nonreplaceable ones. The idea was that the lifestyles of these two kinds of cells – the former shorter lived than the latter – might result in profiles of gene expression that differed systematically between the two even as one moved from birds to mammals and from one part of the brain to another. The first test of this hypothesis, using a microarray, produced by Xiao-Ching Li, of 800 genes expressed in the zebra finch brain and using laser capture microdissection (Lombardino et al., 2006) yielded positive results. That study revealed, to our surprise, that one of the commonest proteins in nerve cells, ubiquitin carboxyl-terminal hydrolase (UCHL1) was under expressed in the replaceable neurons of both zebra finches and mice (Lombardino et al., 2005). Since UCHL1 is involved in protein degradation, perhaps the periodic demise of HVC→RA neurons results from a buildup of unwanted proteins. Singing behavior, known to increase the survival of adult-born HVC neurons, also results in an upregulation of UCHL1 expression in the replaceable neurons, with no such change in the nonreplaceable ones. Interestingly, UCHL1 dysfunction has been associated with neurodegeneration in Parkinson's, Alzheimer's and Huntington's disease patients. In all these instances, reduced UCHL1 function may jeopardize the survival of CNS neurons. Molecular tools may reveal why some kinds of neurons are replaced in healthy brain. Might there be trade-offs between what shorter-lived and longer-lived neurons can do and how they must, accordingly, organize their molecular machinery? Might there be burdens to cellular upkeep past a given cellular age, and advantages in learning flexibility when young neurons are used? Might it be possible to use the song system to revisit the possibility of an anatomical/molecular memory engram? (Lashley, 1950). There are grounds to believe that by querying the profile of gene expression one cell at a time, using cells of known age and experience, answers to these questions might become possible and that the song system of birds will enter a new sphere of usefulness, as a convenient place in which to study the molecular biology of learning, aging and neuronal degeneration. That is as far as this trail, that started in the pampas far away and long ago, has come.

Once, a secretary, Emme, in the days when manuscripts were typed, asked me, "Don't you ever get tired of writing about the same things." No, because the same thing kept changing.

Overview

My account stops here, but the journey that started in the pampas has not ended. I have described the many stops in a long trail and it will be apparent that this trail did not really start in the pampas. The beginning of modern day studies of the biology of vocal learning in birds started in the 1950s with the work of Thorpe and Marler. The part that it was my lot to play came later. I have highlighted in this account, how it became apparent, quite unexpectedly, that new neurons were constantly added to some circuits of the adult brain, where they replaced older ones that had died. This insight should not cloud the fact that a majority of the neurons of the adult central nervous system are born during embryonic development or very early in postnatal life and are probably never replaced. Yet even if the class of *ephemeral neurons* is in the minority, its very existence suggests that our understanding of the neurobiology of learning and of the brain's potential for self-rejuvenation and self-repair is still very incomplete. If synapses were able to acquire and maintain – through their modification in number, specificity, and efficacy – all the changes needed to learn and forget, then neuronal replacement would seem unnecessary. Since neuronal replacement occurs and does not seem to be linked in any obvious way to circuit wear and tear, it may serve a unique function in the healthy brain that cannot be easily achieved by synaptic plasticity alone. I have speculated that this function is the acquisition of long-term memory through irreversible changes in cell configuration and connectivity, possibly mediated by irreversible changes in gene expression (Nottebohm, 1985b, 2002a, b), as are thought to occur during cell differentiation. In this scenario, the role of synaptic plasticity remains undiminished but perhaps regulated by each cell's profile of gene expression. If a learning-related change in gene expression is responsible for the acquisition and maintenance of long-term memories, then this secure repository for long-term memories creates a problem, because as more and more neurons become learned entities, there may not be enough pupils left to acquire new knowledge. Neuronal replacement may take care of this dilemma as a process that balances the need to learn and the need to remember. I suspect that if and when neuronal replacement slows down and comes to a halt, then the aging of the whole brain reaches an irreversible point in which the acquisition of new long-term memories becomes very difficult, if not impossible. The song system of birds alerted us to this phenomenon and it may also provide clues on how to bring this aging process under control.

The journey described started with an abundance of ignorance and the hope that, in some mysterious and fortunate manner, a love for birds and a passion for the deeper questions of philosophy might come together. In my mind, this has happened. I no longer think of the mind as a whole, but of the component cells, each of them with its potential, its biases, and its memories. Physicists and cosmologists have long yearned for a unified theory of the basic physical forces, capable of explaining all matter and energy as found in atoms and the universe and they refer to this as the Theory of Everything. There is, however, another theory of everything that focuses on the brain and on its way of acquiring, retaining, and integrating information. The world of physicists and of their Theory of Everything is but one manifestation of the workings of our brain and therefore subordinate to a theory of brain function and of its limitations. To the extent that our brain evolved to meet the needs of a changing lifestyle, we must assume that our understanding of the world will continue to change as the keyhole through which we look – our brain and senses – continues to change. A full understanding of how the brain does its work may help us appreciate the tentative nature of our cosmic theories. In these many ways, the song of birds and the circuitry evolved to acquire it and maintain it has much to tell us.

ACKNOWLEDGEMENTS

The work I review was funded by NIMH grants 18343 and 63132. Three private individuals also supported my work: Herbert Singer, who became a close friend and made very generous provisions in his will; Howard Phipps, whose love for birds and nature includes a strong curiosity for how things work; and Tim Collins, who gave his help in memory of my son Lawrence, whom he knew well. My laboratory's research was supported, too, by the Mary Flagler Cary Charitable Trust, under the guidance of my good friend Ned Ames. The trail I followed would have been very different had I not met Peter Marler and Mark Konishi, two great scientists and treasured friends. Over the years I benefited, too, from the expertise and wisdom of many wonderful doctoral and post-doctoral students whose names appear in the text and references and with whom I published much of my laboratory's work. They taught me much more than I ever gave back. I was much helped, as well, by my gifted secretary, Patricia Tellerday, by wonderful technicians such as Sue Kasparian, Gus Pandazis, Barbara O'Loughlin and Sattie Haripal, and by a devoted and skilled group of animal caretakers that included Daun Jackson, Sharon Sepe and Helen Ecklund. To all of these people, my unlimited gratitude. Yet all I did would not have been done had it not been for the love, company, patience, encouragement and hands-on help and ideas and hard work of my wonderful wife, Marta Elena Seeber de Nottebohm, who also edited most of the things I wrote. My final thanks go to Rockefeller University, which has supported me in innumerable and wonderful ways since 1967 and provided a secure intellectual and financial haven during the many years of this adventure.

37 • Birdsong and monkey talk: an ethological journey

Peter Marler

I have been fascinated by birds and their songs as long as I can remember. As an eight-year-old, I was already an avid birdwatcher. The first explicit announcement of ornithology as an avocation seems to have been made at age eleven. I was invited to vacation with fond relatives as a reward for passing the elitist rite of passage from elementary school to grammar school, known as the eleven-plus exam. The year was 1939. Everyone was preoccupied with the threat of war which cast a shadow over my youth, much as the First World War had done over my father's.

I was still in Somerset when war was declared, and we began excavating an abandoned root cellar as an air-raid shelter. Back home my parents were already digging a pit in our suburban garden, to be covered with sod, after lining it with the sheets of corrugated steel provided by the government. As an industrial town, Slough was a potential bombing target, and by the time of the Battle of Britain it was bristling with anti-aircraft guns. We three children spent many nights underground with our parents listening to the drone of German planes and the booming guns. Slough was bombed several times. Once, at the height of the bombing my father took me up to London. We were awed by the wasteland around St. Paul's Cathedral, the tragedy brought home to us by a man we saw searching for his belongings in still-smoldering rubble that had been an apartment building the day before.

Despite frequent air-raid warnings the entire family still assembled most weekends for the Sunday excursion, embarking on foot from the town suburbs into the countryside. Often we walked out to Stoke Poges where everyone would sit in the churchyard to read the elegy inscribed in the wall of Thomas Gray's tomb. I was usually more interested in a pair of bullfinches that lived in the yew trees nearby. I owe more of my love of natural history than I ever appreciated at the time to these nature walks, which provided my father with a much-needed change of pace after laboring all week in a factory as a toolmaker. A shed in the garden was fixed up as a workshop with a jeweler's lathe and all kinds of beautifully kept hand tools. He gave me free access and even allowed the invasion of a collection of noxious and in some cases corrosive chemicals that I began to experiment with as a teenager.

CHEMISTRY OR BOTANY?

In grammar school I enjoyed all subjects although there was never any doubt that I was to be a scientist. But I could not imagine anyone would employ me as an ornithologist. Instead it seemed more practical to take up chemistry, a decision that created new problems for the family. Reluctant aunts were coaxed into investing in homemade cosmetics, often more depilatory than emollient. Some relatives were generous enough to order a regular supply, thus unwittingly underwriting my less well publicized diversions into pyrotechnics. The air-raid shelter was ideal for detonating homemade fireworks. My parents' patience and understanding must have been limitless, for I cannot recall once having been reprimanded by them for explosive and inflammatory experimentation, although neighbors were less inhibited. Despite my enthusiasm for chemistry, I gradually realized that my mathematical aptitudes were not up to the abstractions of physical chemistry. Instead I came under the influence of an outstanding biology teacher and with his collusion I decided that a botany degree was probably my most appropriate preparation for employment. Meanwhile I had begun to keep birds at home, and to breed them, first domesticated species and later wild birds taken as nestlings and reared by hand, a pursuit I find fascinating and instructive to this day. I registered as a day student in botany at University College London, with minors in zoology and chemistry. After my early morning newspaper round, I sprinted each day for the morning train to attend lectures by such luminaries as J. B. S. Haldane,

Neuroscience of Birdsong, ed. H. Philip Zeigler and Peter Marler. Published by Cambridge University Press. © Cambridge University Press 2008.

J. Z. Young, and above all, W. H. Pearsall, my botany professor. My first publications on plant distribution and ecology derived from expeditions as a botany undergraduate.

As a day student much of my life was still centered at home. In school I joined forces with friends to establish a local Natural History Society, and weekends became a ferment of feverish activity, surveying migrant birds at local reservoirs and sewage farms, organizing fungus forays and botanical excursions. These Society excursions enriched my experience as a naturalist but nothing could match the inspiration of a field trip with W. H. Pearsall. He had an almost mystical appreciation of the relationships between plants and animals, rocks and soil, and the debt of the present to the past. I used to return from field courses in plant ecology on the Norfolk coast, with its bird colonies and salt marshes, deeply envious of those who were lucky enough to devote their lives to the study of plants and animals in wild places. This emotive feeling for natural history persisted even as I became more and more intellectually preoccupied with the logic of science. An early formative experience was participation in the annual Christmas bird conference organized by the great ecologist David Lack at Oxford. These remarkable gatherings brought together 30 or so students, mostly undergraduates from a wide range of disciplines, to present papers on some ornithological theme. Despite our garrulous inexperience, he managed to imbue these occasions with high scientific standards. There were guest lectures and at one of those I first savored the delights of a lecture by Niko Tinbergen. He had recently arrived from Holland to take up a position in animal behavior at Oxford University, effectively launching ethology in Britain. My first serious bird research on song variation and niche expansion in the Azores was conducted while I was a botanist. It became my first publication on bird behavior (Marler and Boatman, 1951). We found significant changes in the songs of birds living on those remote islands resulting from the reduced number of species there.

SONG DIALECTS

On graduating with my botany degree I would probably have joined a forest survey team in East Africa if Pearsall had not offered me a graduate studentship at UCL. I enjoyed working as a teaching assistant especially on field trips. Discipline was lax enough to permit a good deal of birdwatching on the side; one of the zoology undergraduates, destined to become a leading ethologist, Aubrey Manning, was as eager for the diversions as I was. My research assignment was to study the ecology of plant succession in the Lake District. I spent long hours drilling mud cores in the postglacial deposits around Esthwaite Water and subjecting them to chemical analysis. Looking back I was less inspired by the chemistry of mud than by the discovery of dialects in the songs of chaffinches in the surrounding valleys, which reinforced my determination to develop a notation for transcribing birdsongs and to begin exploring geographic variation in song.

After I finished my London Ph.D. I took a job with the newly formed Nature Conservancy surveying possible conservation sites in Scotland. Although my twenty-first birthday had come and gone I was still in some sense drifting. I enthusiastically scoured Scotland for the dwindling examples of primal plant communities, while becoming increasingly preoccupied with bird behavior, especially song dialects and my growing conviction that birdsongs must be learned (cf. Baker and Cunningham, 1985). Beginning in 1949, while playing hooky from my graduate studies in botany and my duties as a plant ecologist for the Nature Conservancy, I hiked around in England, Scotland, France, and the Azores chasing chaffinches. Altogether I transcribed by ear more than 500 chaffinch songs, learning much about their behavior and ecology in the process. Looking back, my interpretation of the mosaic structure of chaffinch song dialects was not far off target (Marler, 1952). The variation was on such a local scale that a genetic basis seemed improbable. Could it be that the dialects are learned and culturally transmitted? Before long I began pressuring the Nature Conservancy to let me launch a research program on birdsong and the possible role of learning in its development.

A significant event in my education as an ethologist was the publication by the Society of Experimental Biology of the proceedings of a conference held at Cambridge in July 1949 on "Physiological Mechanisms in Animal Behaviour." It included contributions by Konrad Lorenz and Niko Tinbergen arguing persuasively that, contrary to the prevailing view, research on the behavior of animals can be just as objective as studies of their anatomy and morphology. For the first time

I became truly engaged intellectually in science as a creative enterprise. It turned out that my proposal for birdsong research found a sympathetic ear in one of the organizers of the Cambridge conference, William Homan Thorpe. By a heaven-sent coincidence he was in the process of establishing the Ornithological Field Station at the University of Cambridge, to support research on bird behavior focusing especially on birdsong and its development (Thorpe, 1954). The Nature Conservancy arranged for me to transfer to the University of Cambridge in 1951 complete with a research fellowship.

CAMBRIDGE, THORPE, AND HINDE

Thorpe welcomed me warmly in his laboratory, which had just been equipped with the first Kay Electric Company sound spectrograph available in Britain for the analysis of animal sounds. Although we now see it as a relatively primitive machine, at the time it was ideal for launching the new science of bioacoustics. Although basically simple in conception the sound spectrograph was temperamental to operate and obtaining reproducible results was an art. Once we had figured out how to use it there seemed no limit to its revelations, especially since Bill Thorpe had free access to the rapidly growing birdsong library at the BBC. We also had high ambitions for our own recordings but in those days equipment problems were serious. To begin with Thorpe used to record songs by laboriously cutting original phonograph discs. Recordings from those days are sadly deficient by modern standards and I still get twinges of embarrassment listening to them.

As a new student in Cambridge I was invited by Thorpe to be a member of Jesus College where he was senior tutor. In my first encounter with Oxbridge chauvinism I found my London Ph.D. unacknowledged, so I decided to reregister for a Cambridge Ph.D., this time in animal behavior. My true initiation into ethology fell largely into the hands of Robert Hinde, recently appointed to direct Thorpe's new Field Station in the village of Madingley, outside Cambridge. Hinde had begun research as a graduate student at Oxford, but David Lack's primary focus was in ecology and when a year or so after he began, biology at Oxford was enriched by Tinbergen's arrival from Holland, Robert Hinde struck an immediate chord with him, transferred, and completed his Oxford Ph.D. under Tinbergen's guidance. In Cambridge I had the double benefit of Hinde the Tinbergen disciple and Hinde as critic and innovator. Every new idea and each word I wrote was subject to the closest analytical scrutiny. He was generous in sharing his own ideas, and several of the dominant themes in the field study of the behavior of chaffinches that became my zoology Ph.D. thesis are direct reflections of his current enthusiasms, especially ethological drive theory and the idea that much display behavior reflects the underlying behavioral conflicts to which animals are subject during social interactions.

My conversion to ethology was completed by the Second International Ethological Congress held in Oxford in 1952. It was intellectually inspiring and as with all good conferences, it served to cement what were to become many lifelong friendships, especially the coterie of Tinbergen students who ran the conference. The high point of the meeting was the entertainment on the last evening. As nostalgically described by Desmond Morris (1980), the front desk in the main lecture hall was covered with a Rube Goldberg apparatus illustrating Lorenz's "water closet" principle of action-specific potential. At the climactic moment bulging water-filled balloons exploded, assaulted by sperm launched on wires from the back of the lecture hall, drenching the wildly applauding audience.

By this time, based on many hours of study in Madingley Wood, I had a complete picture of the chaffinch's vocal repertoire including not just song, but also all of its distinct call types (Marler, 1956a). I began to question Lorenz's assertion that animal signals are arbitrary with no necessary linkage between meaning and physical structure. I became convinced that problems of sound localization held the key to understanding the peculiar acoustic structure of some animal calls (Marler, 1955a). In the breeding season male chaffinches have a thin, high-pitched whistle that they use in extreme danger, especially when a hawk is nearby. I found the "seet" call to be strangely ventriloquial so that the calling male was hard to track down. Moreover the search for the vocalizer sometimes revealed, not a chaffinch, but a great tit. Not only do these two species have virtually identical hawk alarm calls but they also engaged in mutual exchanges with each other and several other species living in Madingley Wood. Unlike many vocal signals with purely private functions, these alarm calls also provide a public interspecific service, hence their lack of species-specificity. Several songbirds appear to have sacrificed specific distinctiveness in the interest of the mutual communication of extreme danger. The varying selection pressures for divergence and convergence in the course of vocal signal evolution were the topic of my first paper in the journal *Behaviour* in 1957, at that time the major outlet for ethological publications (Marler, 1957).

To learn more about how sound sources are localized, I sought out Donald Broadbent, working on human sound localization at the Medical Research Council Unit on Applied Psychology. As he explained to me the intricacies of human sound localization, it became obvious that chaffinches and great tits had designed an alarm call that is actually difficult to locate, by using a narrow bandwidth and a call with a gradual onset and termination. Conversely, most bird calls are adapted to maximize localizability, with a wide bandwidth and abundant sharp discontinuities, rich in transients. It seemed that certain aspects of the physical structure of many animal sound signals could be understood in terms of selection either to maximize or to minimize the cues for localization. The relative emphasis will depend on the caller's vulnerability to sound-oriented predators and the balance of benefits and disadvantages of easy localization for the signaler and its companions. More than 20 years elapsed before tests on the ability of hawks to localize sounds of various types proved that my hypothesis was valid (Brown, 1982). Mark Konishi also pointed out that while the high pitch of the "seet" call does not actually discourage localization by owls, it hinders detection by predators due to rapid attenuation with distance. I was especially pleased when the dean of German scientific ornithology Erwin von Stresemann expressed interest in my 1955 paper on alarm call structure and published a translation of it in the *Journal für Ornithologie* (Marler, 1956b).

Since the possibilities for winter field work in Cambridgeshire were limited I turned to the extensive aviary facilities at Madingley and began studying the behavior of captive chaffinches. I focused especially on aggression. In one interesting finding I demonstrated that the critical distance in the winter flock at which one bird must approach another before aggression is triggered can be measured by adjusting the distance between feeders. I found that the red breast of a male chaffinch, absent in females, is a classic example of a Lorenz/Tinbergen releaser, modulating the distance at which another bird would be tolerated. Males always dominated females in the winter hierarchy, but within the female hierarchy faking male coloration on a female's breast feathers with red ink guaranteed her high rank when placed with females she had not previously met (Marler, 1955b). We all participated in the annual program for rearing chaffinches by hand for Bill Thorpe's song development studies and I used some of them to explore the developmental basis of female chaffinch responsiveness to the male red breast: it was evident on their very first exposure to a disguised female.

I had expected song learning to be a primary focus for me at Cambridge, but as things turned out I never achieved that aim until I moved to California. Thorpe made it clear that song learning was his domain. When I came to submit my 1958 paper on "The voice of the chaffinch and its function as a language," illustrated by sound spectrograms, there was some question as to whether songs should be illustrated at all. The problem was solved diplomatically by keeping the figure small.

In 1955 I joined the competition for a much-coveted Research Fellowship at Jesus College and to my delight I was successful. As junior fellow my main duty was to serve the port after dinner, doing my best to serve the assembled fellows equally from the bottles provided without leaving any dregs. Thorpe's rooms in college were the rendezvous for a seminar that he organized with the chairman of the psychology department, neurolinguist Oliver Zangwill. The small group encompassed a wide range of interests, including Horace Barlow and Richard Gregory on visual perception and physiology, Larry Weiskrantz and Zangwill on neuropsychology, Donald Broadbent on learning and perception, and Hinde and Thorpe on ethology. A book called *Current Problems in Animal Behaviour* published in 1961, after I left Cambridge, summarized many of our discussions, including my own contribution on the filtering of external stimuli during instinctive behavior (Marler, 1961). The experience helped to broaden my still rather narrow horizons, especially in neurobiology and psychology.

In 1954 I married Judith Golda Gallen, an alumnus of the Slough High School for Girls, strategically placed to be as far from the Grammar School for Boys as it could be while still inside the town boundaries. In those days, women were virtually nonpersons in Cambridge college life, so we took the opportunity to travel whenever we could. Konrad Lorenz's student Jürgen Nicolai, one of the world's great aviculturalists, was studying cardueline finches and we visited him and his wife for several weeks at the old castle in Westphalia where Lorenz's institute was housed. This was before the Max Planck Institute moved Lorenz with Wolfgang Schleidt, to more luxurious quarters at Seewiesen in Bavaria. Lorenz was a cordial and generous host. We assembled regularly for afternoon tea, and had endless discussions about ethology, especially with Irenäus Eibl-Eibesfeldt. Deeply immersed in his own studies of mammalian behavior, he wrote many books and became a major spokesman for classical Lorenzian ethology.

Nicolai and I had planned a survey of the ethology of finches and we dreamed of an expedition to the Himalayas where many cardueline finches are to be found. We planned to drive a Land Rover out from Britain with our wives, by way of the Middle East and Afghanistan. We had already applied to the Royal Society for funds when I got wind of a job possibility at the University of California and the trip was never made. The fact that we even contemplated such a journey is a reminder of how much the world has changed. Instead we moved to California, to take up my new position in Berkeley as a faculty member in the Department of Zoology.

BERKELEY

The process of appointment to the Berkeley position was somewhat unorthodox. It had been offered first to psychologist David Vowles, but his wife did not want to leave Britain. I was next on the list, but because time and funds were short, the zoology department decided to hire me sight unseen. Ecologist Gordon Orians who was a graduate student at Berkeley at the time, on leave for a year at Oxford, stood in as a job interviewer and we got on famously. My three letters of recommendation were from Haldane, Lorenz, and Thorpe. Chairman at the time was Richard Eakin the embryologist, famous for his work on the pineal eye. He was a superb teacher, renowned for disguising himself for lectures as famous figures like Charles Darwin. He began the process of acculturation to Californian informality by insisting that our correspondence should be on first name terms. It took several letters and some anxious consultation before we succumbed. Cambridge traditions were quite otherwise. Bill Thorpe and I had known each other for four years before he invited me to use his first name when he came personally to tell me about my Jesus fellowship.

We traveled on a freighter from Glasgow to San Francisco by way of the Panama Canal and Los Angeles. Our baggage included a large cage containing a dozen jackdaws we had reared by hand. Crossing London we had to strap the cage on the taxi roof and as it began to move, the breeze brought the jackdaws into full preflight display, calling loudly all the way. Once on the boat the crew encouraged us to exercise the birds, which was fine as long as we released them one at a time. On more than one occasion the captain had to slow the boat down in mid-Atlantic so that a desperately flapping jackdaw could catch up.

Beforehand, thinking about how I would structure my animal behavior course at Berkeley, I had discussed the alternatives with several people on another trip to Germany. One was cyberneticist Bernhard Hassenstein, then an assistant under Franz Möhres at Tübingen, who was in favor of a physiological approach, using a homeostatic model of interactions between the behaving organism and the environment. Otto Koehler, with a professorship in Freiburg that Hassenstein was eventually to occupy, entertained us in the garden of their home with his beautiful young wife Amelie, daughter of world famous ornithologist Erwin von Stresemann. He was more inclined to favor a psychological approach, in line with his own research on animal cognition and communication. We discussed the alternatives over tea at which my wife and I were regaled for the first time with *waldhonig*, the strong flavored honey gathered by bees from the honeydew of aphids living in pine trees in the Black Forest.

As I prepared lectures for the first time I was acutely conscious of the many gaps in my own education. I learned a great deal from the Berkeley zoology department. Its major strengths were in cellular and organismal biology. The great human geneticist Curt Stern immediately made my wife and me welcome and always introduced us when distinguished visitors such as Dobzhansky or Goldschmidt were around. The major focus of interest in ethology was in the Museum of Vertebrate Zoology, closely integrated with the zoology department under the broadminded directorship of ornithologist Alden Miller. The curator of birds, Frank Pitelka, had an outstanding program in ecology and population biology in full spate and his many brilliant students were eager to learn more about ethology and ready to serve as teaching assistants in my laboratory course, often teaching me as much as I taught them. Subjects ranged from fruit flies and honey bees to fighting fish to testosterone-treated day-old chicks, to red tree-mice and of course various birds, including zebra finches and ring doves (Stokes, 1968).

In the psychology department, the physiological psychologist Mark Rosenzweig had soundproof chambers in the basement for work on psychoacoustics that I inherited. The behavioral geneticist, Gerald McClearn, was a close friend, and we taught courses together in comparative psychology. Once the world authority on sexual behavior, Frank Beach arrived from Yale, I found myself teaching animal behavior to many of his students, including Irven De Vore. More often than not, half of the students in my class were nonzoologists, including anthropologists, entomologists, geneticists, as well as psychologists. Hybrid vigor was rampant in the weekly seminar that my wife Judith and I hosted in our house in the Berkeley hills.

Anthropology students like Irven DeVore were urged by Sherwood Washburn to extend their education into the biological domain. Kroeber was still active and much interested in evidence of protoculture in animals. With NSF funding the three departments led by a committee of Beach, Washburn, and myself persuaded the University to give us some precious land in Strawberry Canyon for construction of an animal behavior station.

Once in California I was eager to launch my own research program on song learning. I immediately began recording different songbirds, searching for potential subjects for the studies of song development that I was finally free to begin. I had to design my own soundproof chambers to control the acoustic environment in which young birds were raised. After several experiments, I chose the white-crowned sparrow, common enough around Berkeley that we could easily find nests and bring newly hatched babies into the laboratory to raise by hand under controlled conditions (Marler and Tamura, 1962, 1964). It proved to have a system of song dialects as elaborate as that of the chaffinch in Europe, but with much less within-dialect variation. This was valuable because we could pose questions not only about the role of learning in songs of the white-crowned sparrow as a species, but also about the developmental basis of the local dialects of the various locations where birds were taken. Using tape recordings for tutoring we were able over a period of several years to work out the month-long sensitive period for song learning. We established the structure of the rather simple innate song of the white-crowned sparrow, quite variable, but with no trace of the local dialect. We demonstrated its ability in the song learning process to reject sounds of another cohabiting species to which it was exposed in the normal course of development in the wild. These were all themes that Thorpe had begun to deal with in Cambridge, now better secured because of our much more rigorous control of early experience, and the larger numbers of subjects. It became clear that, although the song dialects were unquestionably learned, the learning process was guided by innate learning preferences providing a unique example of the way in which nature and nurture can be entwined in behavioral development (Marler, 1970b).

My graduate student group in Berkeley grew into an independent organism with its own way of life. The greenhouses in the Life Science Building courtyard were soon filled with an endless procession of birds and small mammals, some local, some exotic. There was a ferment of work on birdsong. One student, a Jesuit, worked out the basis of song development in the song sparrow (Mulligan, 1966). Another persuaded white-crowned sparrows to sing in helium air, suggesting that, as Steve Nowicki proved later, songbirds make sounds according to rather different principles than the human vocalization, relying less heavily on tract resonances and more on the primary vibrators. When Masakazu Konishi joined the group from Japan he immediately impressed everyone with his good humor and brilliance. His demonstrations of the impact of deafening on vocal development in songbirds provided an elegant complement to my own ontogenetic studies, changing the way we think about song learning (Konishi, 1965a, b, and this volume). His standing with the Berkeley faculty was high by the time he left for postdoctoral work in Germany. He was the first of my students to face the physiological implications of ethology squarely and he set out to acquire the necessary skills for what was to become a new beachhead in neuroethology in his laboratory at Caltech.

The interplay between behavior and physiology in Berkeley was enriched by a new faculty member, Donald Wilson. His demonstrations that the central nervous system of the locust can generate flight rhythms endogenously revolutionized thinking about the origins of patterned motor activity, in vertebrates as well as invertebrates (Wilson, 1961). He found himself sympathetic to the emphasis Lorenz and Tinbergen placed on endogeneity. We joined forces to coordinate teaching in neuroethology and collaborated on a training grant. Our students exchanged ideas and facilities freely. Ingrid Waldron (1964) discovered *Drosophila* songs in my laboratory before transferring to Wilson's group to study insect flight. Later Wilson moved to Stanford for another burst of creativity before a tragic drowning accident in the Snake River in Idaho ended his meteoric career. He was a political radical, deeply involved in the student unrest in the mid sixties, peaking while I was on sabbatical in Africa in 1964–65. I came back to find one of my students, Mildred Eley, who had been working on the behavioral significance of the pecten in the eye of birds, now in court and threatened with jail. Like other students from that time, her studies were never completed.

Like everyone else teaching animal behavior we were in desperate need of a textbook. I set to work on *Mechanisms of Animal Behavior*. I asked William J. Hamilton, whose death in 2006 we are still mourning, for help with the chapters on orientation (Marler and Hamilton, 1966). We wrote the book more from the viewpoint of organismal biologists, with a physiological flavor, than as evolutionists. Its dominant theme, like that of early ethology, was the interplay of forces arising both within and outside the individual organism in the control of behavior. The preoccupation with proximate factors was paramount and the disciplines with which we sought to link ethology in this book were sensory physiology, endocrinology, and neurobiology. We made little or no connection with population and evolutionary biology, which were obviously crucial to understanding the origins of social behavior but called for a different treatment. I revisited this theme a few years later, when I was invited to speak at the annual meeting of the American Society of Zoologists on "The Interface between Population and Organismal Biology." I argued for renewed attention to the relationship between behavior and population genetics. In particular I singled out William D. Hamilton's two papers on the genetics of social behavior as "containing the seeds of a revolution in our thinking about social organization" (Marler, 1969b). Within five years the sociobiological revolution was upon us and the plethora of population-genetic theorizing that ensued made it possible for the first time to complement the ethological emphasis on proximate mechanisms with an organized and coherent consideration of the ultimate factors bearing on the evolution of animal behavior. One result of this revolution was that teaching and research in ethology gradually shifted its focus to function, leaving the study of underlying physiological mechanisms to neuroethologists, whose affinities were closer to neuroscience than to evolutionary biology. Simultaneously there were relevant and rapidly emerging developments in endocrinology, and after a few years a new confluence emerged between behavior, hormones, and neuroscience, creating the new subdiscipline of behavioral endocrinology, while maintaining some of the traditions of classical ethology (Marler, 2005).

PRIMATE COMMUNICATION

In the midst of the Berkeley years I found intellectual inspiration from an unexpected source; research on the behavior of monkeys and apes. Such is the power of long-standing academic traditions; primatology has always tended to be the province of physical anthropologists and psychologists. Except for a few courageous mavericks such as John Emlen and Robert Hinde, most zoologists have been hesitant to sponsor research on primate behavior. In Berkeley modern primatology arrived, in the person of Sherwood Washburn. He was an intellectual innovator from the outset. In the face of departmental opposition, he forcefully introduced animal behavior to the curriculum of his graduate students and I found myself closely involved, both as a teacher and as a thesis examiner of his students. In 1962, for a conference on primate behavior at the Institute for Advanced Study in the Behavioral Sciences in Palo Alto I was asked to review problems of communication in relation to birds and other animals. The conference gave rise to a book that effectively launched a new science of behavioral primatology, shifting the emphasis in physical anthropology from the laboratory to the field, stressing the importance of the structure and ecology of the natural environment in understanding the behavioral subtleties of monkeys and apes. Preparation of my chapter on "Communication in monkeys and apes" prefaced a new phase in my ethological career (Marler, 1965).

The deluge of new facts was an education in itself although there was more common ground with the work on communication in birds than I had expected, not so much to do with song as with the calls that are an important part of the signal repertoires of birds and, as it turned out, of monkeys. Thinking about primate social signals I became more conscious of the logical and practical problems of defining signal categories, especially difficult when there is continuous intergradation of signal features, as often occurs in higher primates. I was struck by the implications for the evolution of signal structure of using auditory and visual displays not separately, but in close concert, differing in the details of their content, but also generating a high degree of redundancy. This occurs commonly in the complex social groupings of primates, a far cry from the more dispersed social organization of most songbirds. With a sabbatical coming up, I took a year off from bird studies for a trip to Africa to look at the vocal behavior of forest monkeys, about which almost nothing was known at the time. Aided by a Guggenheim Fellowship, Judith and I and two children embarked for Uganda. We lived in Kampala, where an appointment was arranged at Makerere University and made periodic expeditions out into the rain forest. Like every biologist visiting a tropical forest for the first time I was mesmerized by the richness and diversity of the flora and fauna. The primates were abundant, but under forest conditions they were difficult to habituate to human presence. I came back with hours of tape recordings, but often with little clear idea of the context in which vocalizations were used. But still the basic lineaments of several vocal repertoires became clear. Parallels with some of my findings with birds were immediately evident. For example, I found the relationships I had described in birds between call function and rates of evolutionary divergence of call structure could be generalized to primates. The alarm calls of blue and red-tailed monkeys, living in close proximity in the forest, were less divergent than other calls in their repertoires concerned with more intraspecific functions such as intertroop spacing and rallying of the social group prior to movement (Marler, 1973).

From the viewpoint of human comparisons chimpanzees have a special place. When the opportunity arose some years later to visit Jane Goodall's research site at the Gombe Stream in Tanzania I jumped at the chance. I spent a summer in 1967 documenting the vocal repertoires of Jane Goodall's study population of chimpanzees at the Gombe Stream in Tanzania (Marler, 1969a; Marler and van Lawick-Goodall, 1970). We made a sound film of chimpanzee calls and displays which is still one of the best introductions to the signal repertoire of a nonhuman primate. Had I found dialects or other signs of vocal learning in the chimpanzee, I probably would have changed the main course of my research and concentrated on laboratory experimentation with primates, but no clear evidence was forthcoming. I remain skeptical about whether any nonhuman primates display truly learned local dialects in their vocal behavior, like songbirds, although other colleagues have found some indications that I might be wrong (Mitani et al., 1992).

For his thesis research, Steve Green (1975) analyzed the vocalizations of Japanese monkeys, grappling with the problem of their graded vocal signals and their meaning. He showed that even the subtle variations in vocal morphology they display are in fact orderly, with different call variants produced in distinct circumstances. Green stayed on to work in India on the endangered lion-tailed macaque. Meanwhile, the Japanese monkey story unfolded further. At a meeting of the advisory board of the Primate Research Center at Duke University, William Stebbins, the leading expert on primate psychoacoustical research, complained about the sterility of tones and clicks as the traditional stimuli in psychometric studies. It did not take much to convince him that new revelations would ensue if natural vocal stimuli could be substituted. Before the site visit was over, plans emerged for a collaborative project.

Green edited out samples of field recordings of some "coo" calls of the Japanese macaque, where subtle variation in the position of a frequency inflection appeared to have meaning to the monkeys. Stebbins showed that Japanese monkeys do in fact respond to the position of this inflection. It emerged that they process their vocal signals in a different way than other species do when confronted with exactly the same sounds, implying that they are perceptually specialized for the processing of conspecific signals – precisely the kind of result that a classical ethologist would have forecast. Remarkably, this study also yielded evidence of cerebral lateralization, paralleling the right ear dominance that we ourselves display in the perception of speech. One of Stebbins' students had the inspiration to present test stimuli to the Japanese monkeys monaurally, changing randomly from ear to ear. The distribution of errors in call type judgments revealed a strong right-ear advantage when subjects were required to classify calls by a feature that was thought to be "linguistically" relevant (position of the frequency inflection). But the lateralization disappeared when the task was switched to a "linguistically irrelevant" cue. Other monkey species, for whom these calls were meaningless, showed no such effect (Petersen et al., 1984). The parallels with the processing of speech sounds remind us that the roots of language may reach further back into our primate ancestry than we suppose.

Students continued to be fascinated with the meaning of monkey calls. In Africa Peter Waser conducted pioneering studies of reactions of mangabeys to vocal playbacks in Uganda (Waser, 1977), actively advised by Tom Struhsaker. Tom himself had previously described a fascinating system of alarm calls, each one apparently specific to a particular predator (Struhsaker, 1967). In 1975 I invited postdocs Dorothy Cheney and Robert Seyfarth to come to Rockefeller and, taking a leaf from Peter Waser's book, see if wild vervet monkeys would respond to alarm call playbacks. The experiments were successful beyond our wildest expectations. The monkeys responded differently and adaptively to eagle, leopard, and snake alarm calls, and did so in the absence of any predator (Seyfarth et al., 1980; Cheney and Seyfarth, 1990). We found ourselves thinking about vocal communication in a new way, with some signals so tightly linked to particular situations and eliciting such specific reactions that they seemed almost symbolic in nature.

We wondered if this could be true of signals with a social function, like rhesus monkey calls that carry information about dominance relationships. Sarah and Harold Gouzoules found five different "scream" calls, used to recruit allies during agonistic encounters. It appeared from playback experiments that the calls given by sparring youngsters label different classes of opponent as defined by dominance rank and matrilineal relatedness, helping parents to decide at a distance whether to intervene in their offspring's quarrels (Gouzoules et al., 1984). Later work on those same rhesus monkeys grappled with a range of important theoretical issues. Multimodal signaling was a special focus for Sarah Partan and deception was another among many summarized in Marc Hauser's synthetic overview of animal communication (Hauser, 1996; Partan and Marler, 1999).

Our investigations of what animal signals mean also embraced alarm and food calls of birds. We used both live hawks and video techniques to study alarm calls in chickens, making a compelling case for what we came to call "functional reference" (Marler et al., 1992). This is a concept epitomized by signals that seem to mentally represent objects in the environment, like a predator or food, doing so both to the caller and to others listening. Since we coined this term it has influenced a good many others studying social communication in mammals and other animals.

THE ROCKEFELLER YEARS

While we were in Africa studying monkeys, Berkeley was in turmoil and letters from home gave us an inkling of how pressured students felt. We were further unsettled by two job offers, different enough from Berkeley to be intriguing. The New York Zoological Society decided to establish a research institute for work on the ethology of animals in the zoo collections. Then in midyear, Ernst Mayr wrote to ask me to consider an Agassiz Professorship in the Museum of Comparative Zoology at Harvard. Several new developments followed in short order. Unbeknownst to us, Don Griffin, chairman of Biology at Harvard and discoverer of echolocation in bats, was advising the NYZS on its new Institute. We now learned that he was tempted to join himself, but the lack of an academic affiliation was a serious deterrent. Don solved this problem by persuading the Rockefeller University to cosponsor the project. Both he and I were now possible candidates for professorships at the Rockefeller, with joint appointments at the NYZS, and research facilities in both places. Such a conjunction of attractions was irresistible.

As director of the new Institute, Don Griffin was generous to everyone involved. Work by him and his students flourished on the radar tracking of night-migrating birds, and the echolocation of bats. One of his first students, Jack Bradbury, was destined eventually to author with his wife a path-breaking book on animal communication (Bradbury and Vehrencamp, 1998). In those early halcyon days at Rockefeller, Don and I each had funds for two junior faculty members. He invited Richard Penny, an expert on penguin behavior and navigation, and Roger Payne, studying prey catching in owls. Roger's life underwent a literal sea change when he and his wife Katy first described the song of the humpback whale, ultimately devoting themselves to the behavior and conservation of whales.

I invited two Berkeley students, Fernando Nottebohm and Tom Struhsaker, to join me as assistant professors at Rockefeller. After launching new projects on West African monkeys, Tom established an immensely fruitful long-term study site in the Kibale Forest in Uganda, where I had made a fleeting visit in my sabbatical year. He went on to become a world authority on the behavior, ecology, and conservation of forest primates (Struhsaker, 1997).

I offered the second junior faculty position to Fernando Nottebohm. After graduating with the Departmental Citation in zoology at Berkeley in 1963, Fernando had approached me, in a discussion over a bottle of Argentinian wine at the animal behavior station about graduate work. Konishi's work, which he had heard about in my behavior course, interested him and he was especially intrigued by the contrast Mark had found between the effects of early and late deafening on sparrow song. Early deafening resulted in the kind of highly degraded song that Mark had found in several species, whereas postponement of deafening until song had matured seemed to have less effect or none at all (Konishi, this volume). Had proprioceptive control taken over, or an endogenous motor tape such as Donald Wilson, working down the hall, had invoked in his studies of locust flight? Under Konishi's tutelage, Fernando quickly gained the necessary surgical skills, for which both had a natural talent.

Fernando's first subjects were ring doves, but auditory feedback proved unimportant in the development of their vocal repertoire (Nottebohm and Nottebohm, 1971). While I was in Africa on sabbatical, Fernando got a fellowship to work with Bill Thorpe for a year in Cambridge. Deafening chaffinches at various stages of song development, he showed that they can retain much of the song structure already achieved once a certain point is reached, thus explaining some of the variation Konishi had found in the effects of deafening in different individuals (Nottebohm, 1968). At Rockefeller, after a spell of field research in Trinidad on vocalizations of Amazon parrots, Fernando returned to the question that had preoccupied him in Berkeley, about proprioception and its role in vocal control. As the fates would have it, a focused attack on that question was delayed (see Wild, Suthers and Zollinger, and Goller and Cooper, this volume) by an exciting discovery.

Proprioceptive afferents from the syrinx pass along the hypoglossal nerve, and Fernando's first step was to see what effect cutting this nerve had on mature song. Again chaffinches and white-crowned sparrows were the first subjects. Study of effects of bilateral section was difficult because of interference with respiration, but he noticed that even with one side cut there was some song disruption, but the effects were variable. It gradually became clear that the effects of left hypoglossectomy were more drastic than on the right (Nottebohm, this volume). This unexpected discovery, that song production is lateralized, suggested that the time was ripe to launch a broader research effort on the neural basis of singing behavior. Fernando decided to use the inbred strain of German Wasserschlager canaries we had been working with for some time. Song control again proved to be lateralized, and once more the left side was dominant. Analogies with human vocal control, always lurking in the back of Fernando's mind, now came into the forefront. Would the lateralization extend up into the brain? But before this question could be addressed, the basic connectivity of circuits for song control had to be worked out. This was a daunting undertaking, given the lack of information about the songbird brain. But Fernando rose to the challenge and with the aid of skilled Rockefeller colleagues made the remarkable discovery of a previously unsuspected subsystem in the canary brain, specialized for the control of song.

THE SONG SYSTEM

A new page was turned in understanding the neural basis of song learning. Nottebohm was still working in my laboratory at the time, so I had a ringside seat; I can vouch that this new undertaking was nothing like as easy as it sounds. In the first place, it took a good deal of courage. A neuroanatomical laboratory had to be set up, with all of the necessary techniques. Not unexpectedly, with neuroanatomical neophytes making such a radical discovery, there was skepticism about whether the results were credible. I seem to recall some difficulty at first in getting the findings published. Editorial incredulity was followed by an acute case of study section indigestion during the review of a grant application at the National Institutes of Health. Eventually it was realized that the anatomical work was indeed sound, and that this represented a remarkable, even revolutionary development in the study of functional vertebrate neuroanatomy.

Many discoveries followed in quick succession, as related in detail in Nottebohm's chapter in this book. I still recall the day when Fernando and Art Arnold, then a graduate student in Nottebohm's laboratory, rushed into my office, with stained brain sections in hand, Art with zebra finches, Fernando with canaries. They were thrilled to the point of near incoherence by the discovery that, even with the naked eye, you could see the sex differences in the song systems of both species, flatly contradicting the current dogma that there is no sexual dimorphism in the brains of higher vertebrates. Shortly afterwards, Fernando was promoted to head his own laboratory at Rockefeller. Fifteen years would pass before the 1976 Nottebohm/Arnold paper in *Science* was acknowledged as a scientific classic. Meanwhile the word had spread to other laboratories about the unique value of the song system for tackling questions about the neural basis of behavioral plasticity.

The brain nuclei that make up the song system proved to have well-defined boundaries, making it relatively easy to measure their volume. This led to another remarkable discovery that their size can vary, not only between the sexes, but in the same bird. Female canaries sing much more than usual if they are treated with testosterone. It turned out that certain song system nuclei also responded to the androgen treatment, growing significantly in size within three or four weeks. Exploring the size issue further, Nottebohm then discovered seasonal changes, reaching a maximum in the breeding season, when singing behavior is at its peak, a pattern that has since been described in other songbirds.

Gradually a convincing picture emerged of a relationship between the complexity of oscine birdsongs and the size of some of these song system nuclei, captured by Nottebohm's felicitous phrase, "brain space for a learned task." The relationship proved to be valid at many levels, from species differences in the size and complexity of the song repertoire at one extreme down to individual and population differences in the size of the syllable repertoire from which songs are constructed. As far as we know there are no such relationships between brain structure and song behavior in birds with unlearned songs that lack a song system. So the process of song learning, with all its complexities, appears to be dependent on the evolution of a particular type of brain structure, seemingly designed for the purpose. Intriguingly, the two other bird groups with learned vocalizations, hummingbirds and parrots, possess structures within their brain that although different from the classical song system, have a number of features in common with songbirds. The commonalities hint at a basic set of brain circuits required for vocal learning to take place, providing us with a unique illustration of the necessary interdependence of complex behavior and their underlying neural circuitry, usually taken for granted, but rarely analyzed with the attention to detail that the song system has received (see Mello and Jarvis, this volume).

The seasonal changes in the song system left another major question unanswered. Does the number of neurons in a nucleus like HVC actually increase in the springtime, and if so, where do they come from? Nottebohm and a student in his laboratory, Steve Goldman, had the inspired notion of injecting canaries with radioactive thymidine and then searching the brain for radioactively labeled neurons. Lo and behold they were indeed present, newly created in the 30-day period since the injections were made. This remarkable discovery of neurogenesis in the adult songbird brain launched yet another saga that reverberates to this day. Much has been learned about the dynamics of the process by which they originate, and new questions are still emerging. What is the precise function of these new neurons? They migrate, not only into the song system, but to other parts of the forebrain as well, not necessarily directly involved in song learning. Yet it seems inconceivable that the temporal correlation between neurogenesis and the dynamics of song learning and development is just coincidental (see Pytte *et al.*, this volume).

In addition to involvement in song learning, neurogenesis could make other kinds of contributions to the physiology of singing behavior, even when song structure is quite stereotyped. We probably misinterpret stereotyped behavior if we assume that the underlying mechanisms are necessarily fixed and immutable. Once a young bird's songs are fully developed they frequently become quite stereotyped, sometimes for a season, often for life. But we know that if you disrupt the auditory feedback controls, by deafening or by delayed auditory feedback, the song of an adult zebra finch can be completely destabilized (see Woolley, this volume). The bird begins to stutter and repeat itself and syllable quality degenerates. If normal feedback is restored the original song is gradually reinstated, though often with some potential for change. So, far from being set in stone, the stereotypy of adult song has a dynamic basis. As the geneticist Waddington suggested years ago, if the emergence and maintenance of a particular, narrowly prescribed phenotype is to be guaranteed, developmental mechanisms for detecting and correcting for deviations from a preferred target may be a necessity, even in maturity. This is another challenge to which neurogenesis could make a contribution. And of course motor patterns of other kinds vary in frequency and structure from season to season. One wonders to what extent new neurons born in adulthood can play multiple roles, or whether their fate is already predetermined. In the neuroscientific study of birdsong new vistas are always emerging to challenge the next generation.

THE MILLBROOK FIELD STATION

In New York we had begun to experience growing pains from budgets and space limitations, and eventually the Rockefeller offered to take over the Institute entirely. After an extended search, we found a new home 90 miles north of New York City in Dutchess County, on several hundred acres of land. We began moving house and building new laboratories, with the final move in spring 1972. Among many other things it was the perfect place for John Wingfield and his students to launch his special brand of behavioral endocrinology when I invited him to join my group. His laboratory for hormone assays was a stone's throw from the places where he trapped his birds; he could repeatedly capture them on their territories, take a blood sample, and release them. This was the first time anyone had mapped the changes in hormone levels in different species as the seasons progressed, and males engaged in territorial encounters. In some remarkable field experiments, he demonstrated how the behavior of birds changes when you inject them with various hormones (Wingfield and Marler, 1988; Wingfield et al., 1999).

Meanwhile the student group grew dramatically, always with a strong emphasis on field work, some in Millbrook, some elsewhere. One postdoc worked on manakin leks in the rain forests of Trinidad, defining for the first time the female role in these unusual and complex mating systems (Lill, 1976). Another project on the evolution of polygyny in birds focused on the sage grouse in Wyoming, confirming that a tiny proportion of males do most of the mating (Wiley, 1973). Another project focused on the evolution of the mating systems of dragonflies. Switching from biophysics to work on social communication in electric fish, Carl Hopkins worked in a field laboratory in Guyana where they breed at the height of the rainy season. He found that the electric sense serves as a highly sophisticated communicative system (Hopkins, 1999). Electric fish have since become a favored neuroethological preparation, demonstrating many sensory specializations for the processing of species-specific communicative signals.

Throughout this period of more than 20 years at Rockefeller there were many song learning studies. We had superb laboratory facilities, including more than 50 soundproofed chambers. Canaries continued to be favorite subjects. By yoking together chambers with and without an adult singer we showed that canaries learn well both from a live tutor and from a loudspeaker; not unexpectedly, a live tutor teaches them more (Marler and Waser, 1977; Waser and Marler, 1977; see Saar et al., this volume). We persuaded them to breed with white noise in the chamber so loud that it masked their hearing of their own voice. As Konishi predicted, canaries reared from the egg in white noise and immediately deafened had highly degraded songs with two or three syllable types instead of the usual 30 or so. However, if, instead of deafening them, their residual hearing was left intact, they eventually developed a degree of normal phrase structuring in their songs even in isolation (Marler et al., 1973). They behaved as though they possess motor programs that can define the gross features of song endogenously. One wonders whether something similar might be true of our own speech behavior.

Above all, Millbrook had outstanding potential for studying song development in local birds, whose nests we could find in large numbers as a source of naïve nestlings that we could raise by hand in the laboratory. I set my sights on locating two birds with different songs that we could rear under identical conditions. We figured that any species differences that emerged under those conditions might provide a window on innate contributions to song development that would lend themselves to genetic investigation in the future (see Clayton and Arnold, this volume). In particular any abilities they displayed as fledglings to respond selectively to songs of others as models for the learning process would tell us something novel about the learning process. Song and swamp sparrows, with very different songs, one simple, the other much more complex, turned out to be ideal (Marler and Sherman, 1983, 1985); both favored tutor songs of their own species. By using synthesized songs with different features systematically varied we identified the very different properties they were using as naïve learners to identify conspecific song. Sparrows taught us many other things about song development. One of the most remarkable was the overproduction of song material in subsong and plastic song, winnowed down to size as birds mature, by a process reminiscent of the emergence of speech from the babbling of young children (Marler 1970a; Marler and Peters, 1980; see Hultsch and Todt, this volume).

Red-winged blackbirds were common in Millbrook and nested in marshes nearby. Their mating behavior was studied intensively, with many new insights into the dynamics of breeding in a colony, written up eventually in book form (Searcy and Yasukawa, 1995). Bill Searcy also studied mate selection in our sparrows, including an imaginative series of playback studies to males and to estradiol-treated females. He used both natural and synthetic songs as well as songs of deaf and isolated birds, showing that the buzzy deaf songs are functionally ineffective: natural songs on the other hand are potent species-specific signals, and isolate songs somewhere between (Searcy and Marler, 1981, 1987). Later Searcy brought these and a plethora of other new facts and ideas in a book he coauthored with Steve Nowicki, *The Evolution of Animal Communication* (Searcy and Nowicki, 2005). Steve himself had clarified in some elegant experiments how the songbird syrinx works as an instrument before going on to make the significant discovery that the nutritional state of a young bird affects its learning ability and thus the way in which it sings, with functional implications that are still unfolding (Nowicki, 1987; Nowicki *et al.*, 2002).

To understand song learning you have to take account of what a bird can hear. The field studies at Millbrook were complemented in the laboratory by Bob Dooling, joining us from the Central Institute for the Deaf in St. Louis. He brought the special insights of the psychoacoustician to bear on the perception by birds of songs and other complex sounds (Dooling, 1980). With Jeff Baylis, Bob initiated us into the mysteries of synthesizing birdsongs, a valuable experimental tool. He has an extraordinary knack for getting wild birds to submit happily to a variety of audiometric conditioning procedures, including tests in which birds were half asleep so that reliable heart rate measures could be taken. As a result he has learned more than anyone about hearing in birds and the special predispositions they bring to bear on the perception of song.

In 1989 I became an emeritus professor at Rockefeller and we decided to pull up stakes and return to California. In retrospect the years at Rockefeller were more productive than I could have hoped for. We had established some of the rules governing the process of learning to sing. In the hands of Fernando Nottebohm the neuroscientific study of birdsong was launched. We had trained an outstanding cohort of students, postdocs and colleagues working on all aspects of animal communication, many of whom pioneered as teachers themselves.

For research on mechanisms of developmental plasticity, the virtuoso quality of song learning makes it an ideal subject. Its consequences are measurable with precision, both directly, by its effects on responsiveness to song stimulation, and indirectly by the effects on song production. On the other hand we need to keep in mind that singing is emotional in nature, with few of the cognitive nuances that other bird and primate signals display. Indeed even in birds, the complex and varied ways in which social galliforms use their innate alarm and food calls in communication have taught us more about animal semantics than birdsong (Gyger *et al.*, 1987; Gyger and Marler, 1988; Evans *et al.*, 1993; Marler and Slabbekoorn, 2004). I make this point as a reminder that unlearned vocal signals can also be a valuable source of insights of a different kind, also with neurobiological implications. I am thinking of the cognitive functions which they may serve, with profound implications for the ways in which they are used as social signals. In this regard, a comparison with what we have found out about primate communication is instructive. Apart from our own speech, the developmental plasticity of the actual acoustic morphology of nonhuman primate signals is limited. But when we explore the ways in which monkey calls are used in communication we find that they easily outclass birdsongs as a source of new insights into their cognitive implications. The neurobiology of the song system, so productive for the study of learning, has told us virtually nothing about the meaning of vocal signals. For that we will have to look elsewhere in the brain. How does production of an innate call become linked in ontogeny to the external referents that govern its use, such as a food object or a predator or certain social companions? Where are these links to be found in the brain and when are they established? We have the methodology to answer these questions, with the unusual potential to illuminate otherwise difficult issues like where predator recognition occurs in the brain, where it connects with centers for escape and aggressive behavior, and how a particular call is elicited and responded to by others. There are endless new frontiers in the neuroscience of vocal behavior still awaiting our attention.

BACK TO CALIFORNIA

In Davis I was blessed with yet another group of first class students. As I have mentioned, in the study of primate communication new discoveries were made on referential signaling and the use of multiple sensory modalities. Making use of our renewed proximity in California to white-crowned sparrows, we took a new look at the predispositions they bring to the song learning process. When we compared birds living in the benign environment on the California coast on the one hand, and in the rigorous, short season in the high Sierras, we uncovered a number of innate differences between these local populations in the timing and specificity of song learning (Nelson et al., 1995). A valuable new assay for song recognition in very young birds was developed, based on their readiness to call in response, making it possible for the first time to map changes with time and experience from infancy: we showed that development proceeds differently in males and females (Nelson et al., 1997). Cleverly designed experiments by students revealed how innate responsiveness to certain song components shapes the process of song learning in sparrows, and these paradigms also proved useful in some novel physiological and neurobiological studies (Whaling et al., 1997; Soha and Marler, 2000). In another bird, the American cardinal, the details of song learning in females were explicated (Yamaguchi, 1998). Meanwhile during the ten-year period in Davis I worked hard as director of the Center for Animal Behavior and as principal investigator on a big training grant. Most recently I have been engaged in two stimulating editorial collaborations with Phil Zeigler. Publication of the proceedings of the first international conference in New York on the behavioral neurobiology of birdsong that he organized was a major event (Zeigler and Marler, 2004). Now the present book is bringing together the work of an entire generation of birdsong researchers, a few of whom I am tempted to regard, perhaps more sentimentally than substantively, as scientific grandchildren. I like to think that Bill Thorpe, to whom the book is dedicated, would have been happy to give this explosion of knowledge about the biology of birdsong his own paternal blessing.

References

Aamodt, S. M., Kozlowski, M. R., Nordeen, E. J., Nordeen, K. W. (1992) Distribution and developmental change in [3H]MK-801 binding within zebra finch song nuclei. *J Neurobiol*, **23**:997–1005.

Aamodt, S. M., Nordeen, E. J., Nordeen, K. W. (1995) Early isolation from conspecific song does not affect the normal developmental decline of N-methyl-D-aspartate receptor binding in an avian song nucleus. *J Neurobiol*, **27**:76–84.

Aamodt, S. M., Nordeen, E. J., Nordeen, K. W. (1996) Blockade of NMDA receptors during song model exposure impairs song development in juvenile zebra finches. *Neurobiol Learn Mem*, **65**:91–98.

Abraham, W. C., Bear, M. F. (1996) Metaplasticity: the plasticity of synaptic plasticity. *Trends Neurosci*, **19**:126–130.

Abramson, A. S., Lisker, L. (1970) Discriminability along the voicing continuum: cross-language tests. In *Proceedings of the Sixth International Congress of Phonetic Sciences Prague 1967*, pp. 569–573. Prague: Academia.

Abrous, D. N., Koehl, M., Le Moal, M. (2005) Adult neurogenesis: from precursors to network and physiology. *Physiol Rev*, **85**:523–569.

Absil, P., Riters, L. V., Balthazart, J. (2001) Preoptic aromatase cells project to the mesencephalic central gray in the male Japanese quail (*Coturnix japonica*). *Horm Behav*, **40**.

Adams, J. P., Sweatt, J. D. (2002) Molecular psychology: roles for the ERK MAP kinase cascade in memory. *Annu Rev Pharmacol Toxicol*, **42**:135–163.

Adkins-Regan, E. (2005) Activity dependent brain plasticity: does singing increase the volume of a song system nucleus? Theoretical comment on Sartor and Ball (2005). *Behav Neurosci*, **119**:346–348.

Adkins-Regan, E., Ascenzi, M. (1987) Social and sexual behavior of male and female zebra finches treated with testosterone. *Anim Behav*, **35**:1100–1112.

Adkins-Regan, E., Abdelnabi, M., Mobarak, M., Ottinger, M. A. (1990) Sex steroid levels in developing and adult male and female zebra finches (*Poephila guttata*). *Gen Comp Endocrinol*, **78**:93–109.

Adkins-Regan, E., Mansukhani, V., Seiwert, C., Thompson, R. (1994) Sexual differentiation of brain and behavior in the zebra finch: critical periods for effects of early estrogen treatment. *J Neurobiol*, **25**:865–877.

Adret, P. (1993) Operant conditioning, song learning and imprinting to taped song in the zebra finch. *Anim Behav*, **46**:149–159.

Adret, P. (2004) In search of the song template. *Ann NY Acad Sci*, **1016**:303–324.

Adret-Hausberger, M., Jenkins, P. F. (1988) Complex organization of the warbling song in starlings. *Behaviour*, **107**:138–156.

Agate, R. J., Grisham, W., Wade, J., et al. (2003) Neural, not gonadal, origin of brain sex differences in a gynandromorphic finch. *Proc Natl Acad Sci USA*, **100**:4873–4878.

Ahumada, J. A. (2001) Comparison of the reproductive biology of two neotropical wrens in an unpredictable environment in northeastern Colombia. *Auk*, **118**:191–210.

Airey, D. C., DeVoogd, T. J. (2000) Greater song complexity is associated with augmented song system anatomy in zebra finches. *NeuroReport*, **11**:2339–2344.

Airey, D. C., Buchanan, K. L., Székely, T., Catchpole, C. K., DeVoogd, T. J. (2000) Song, sexual selection, and a song control nucleus (HVc) in the brains of European sedge warblers. *J Neurobiol*, **44**:1–6.

Aitkin, L., Park, V. (1993) Audition and the auditory pathway of a vocal new world primate, the common marmoset. *Prog Neurobiol*, **41**:345–367.

Aitman, T. J., Dong, R., Vyse, T. J., et al. (2006) Copy number polymorphism in Fcgr3 predisposes to glomerulonephritis in rats and humans. *Nature*, **439**:851–855.

Akutagawa, E., Konishi, M. (1985) Neuronal growth, atrophy and death in a sexually dimorphic song nucleus in the zebra finch. *Nature*, **315**:145–147.

Akutagawa, E., Konishi, M. (1994) Two separate areas of the brain differentially guide the development of a song control nucleus in the zebra finch. *Proc Natl Acad Sci USA*, **91**:12413–12417.

Akutagawa, E., Konishi, M. (1998) Transient expression and transport of brain-derived neurotrophic factor in male zebra finch's song system during vocal development. *Proc Natl Acad Sci USA*, **95**:11429–11434.

Alcock, K. J., Passingham, R. E., Watkins, K. E., Vargha-Khadem, F. (2000) Oral dyspraxia in inherited speech and language impairment and acquired dysphasia. *Brain Lang*, **75**:17–33.

Alger, S. J., Riters, L. V. (2006) Lesions to the medial preoptic nucleus differentially affect singing and nest box-directed behaviors within and outside of the breeding season in European starlings (*Sturnus vulgaris*). *Behav Neurosci*, **120**:1326–1336.

Allan, S. E., Suthers, R. A. (1994) Lateralization and motor stereotypy of song production in the brown-headed cowbird. *J Neurobiol*, **25**:1154–1166.

Alm, P. A. (2004) Stuttering and the basal ganglia circuits: a critical review of possible relations. *J Commun Disord*, **37**:325–369.

Altar, C. A., Cai, N., Bliven, T., et al. (1997) Anterograde transport of brain-derived neurotrophic factor and its role in the brain. *Nature*, **389**:856–860.

Altman, J. (1962) Are new neurons formed in the brains of adult mammals? *Science*, **135**:1127–1128.

Altman, J. (1963) Autoradiographic investigation of cell proliferation in the brains of rats and cats. *Anat Rec*, **145**:573–591.

Altman, J. (1970) Post-natal neurogenesis and the problem of neural plasticity. In *Developmental Neurobiology* (Himwich, W. A., ed.), pp. 192–237. Springfield, IL: Thomas.

Altman, J., Das, G. D. (1965) Autoradiographic and histological evidence of postnatal hippocampal neurogenesis in rats. *J Comp Neurol*, **124**:319–335.

Alvarez-Borda, B. (2002) *On New Neurons in Canary Brains*. Thesis dissertation. New York: The Rockefeller University.

Alvarez-Borda, B., Nottebohm, F. (2002) Gonads and singing play separate, additive roles in new neuron recruitment in adult canary brain. *J Neurosci*, **22**:8684–8690.

Alvarez-Borda, B., Haripal, B., Nottebohm, F. (2004) Timing of brain-derived neurotrophic factor exposure affects life expectancy of new neurons. *Proc Natl Acad Sci USA*, **101**:3957–3961.

Alvarez-Buylla, A., Nottebohm, F. (1988) Migration of young neurons in adult avian brain. *Nature*, **335**:353–354.

Alvarez-Buylla, A., Kirn, J. R. (1997) Birth, migration, incorporation, and death of vocal control neurons in adult songbirds. *J Neurobiol*, **33**:585–601.

Alvarez-Buylla, A., Lim, D. A. (2004) For the long run: maintaining germinal niches in the adult brain. *Neuron*, **41**:683–686.

Alvarez-Buylla, A., Buskirk, D. R., Nottebohm, F. (1987) Monoclonal antibody reveals radial glia in adult avian brain. *J Comp Neurol*, **264**:159–170.

Alvarez-Buylla, A., Theelen, M., Nottebohm, F. (1988a) Mapping of radial glia and of a new cell type in adult canary brain. *J Neurosci*, **8**:2707–2712.

Alvarez-Buylla, A., Theelen, M., Nottebohm, F. (1988b) Birth of projection neurons in the higher vocal center of the canary forebrain before, during, and after song learning. *Proc Natl Acad Sci USA*, **85**:8722–8726.

Alvarez-Buylla, A., Kirn, J. R., Nottebohm, F. (1990a) Birth of projection neurons in adult avian brain may be related to perceptual or motor learning [published erratum appears in Science 1990 Oct 19;250(4979):360]. *Science*, **249**:1444–1446.

Alvarez-Buylla, A., Theelen, M., Nottebohm, F. (1990b) Proliferation "hot spots" in adult avian ventricular zone reveal radial cell division. *Neuron*, **5**:101–109.

Alvarez-Buylla, A., Ling, C.-Y., Yu, W. S. (1994) Contribution of neurons born during embryonic juvenile and adult life to the brain of adult canaries: regional specificity and delayed birth of neurons in the song control nuclei. *J Comp Neurol*, **347**:233–248.

Alvarez-Buylla, A., Garcia-Verdugo, J. M., Mateo, A. S., Merchant-Larios, H. (1998) Primary neural precursors and intermitotic nuclear migration in the ventricular zone of adult canaries. *J Neurosci*, **18**:1020–1037.

Alvarez-Buylla, A., Seri, B., Doetsch, F. (2002) Identification of neural stem cells in the adult vertebrate brain. *Brain Res Bull*, **57**:751–758.

Ames, A., 3rd (2000) CNS energy metabolism as related to function. *Brain Res Rev*, **34**:42–68.

Amin, N., Grace, J. A., Theunissen, F. E. (2004) Neural response to bird's own song and tutor song in the zebra finch field L and caudal mesopallium. *J Comp Physiol A Neuroethol Sens Neural Behav Physiol*, **190**:469–489.

Andalman, A., Fee, M. S. (2006) Spiking activity in Area X is not necessary for sensory acquisition during tutor exposure. Program No 448, 2006, Neuroscience Meeting Planner. Atlanta, GA: Society for Neuroscience, 2006.

Andersson, M. (1994) *Sexual Selection*. Princeton, NJ: Princeton University Press.

Andrews, G., Harris, M. (1964) *The Syndrome of Stuttering*. Clinics in Developmental Medicine, No. 17. London: William Heineman Medical Books Ltd.

Ang, C. W.-Y. (2001) Emerging auditory selectivity in the caudomedial neostriatum of the zebra finch songbird. Ph.D. dissertation. New York: The Rockefeller University.

Appeltants, D., Absil, P., Balthazart, J., Ball, G. F. (2000) Identification of the origin of catecholaminergic inputs to HVc in canaries by retrograde tract tracing combined with tyrosine hydroxylase immunocytochemistry. *J Chem Neuroanat*, **18**:117–133.

Appeltants, D., Ball, G. F., Balthazart, J. (2002a) The origin of catecholaminergic inputs to the song control nucleus RA in canaries. *NeuroReport*, **13**:649–653.

Appeltants, D., Del Negro, C., Balthazart, J. (2002b) Noradrenergic control of auditory information processing in female canaries. *Behav Brain Res*, **133**:221–235.

Appeltants, D., Gentner, T. Q., Hulse, S. H., Balthazart, J., Ball, G. F. (2005) The effect of auditory distractors on song discrimination in male canaries (*Serinus canaria*). *Behav Processes*, **69**:331–341. Epub 2005 Feb 2024.

Arai, A., Taniguchi, I., Saito, N. (1989) Correlation between the size of song control nuclei and plumage color change in orange bishop birds. *Neurosci Lett*, **98**:144–148.

Arends, J. J., Wild, J. M., Zeigler, H. P. (1988) Projections of the nucleus of the tractus solitarius in the pigeon (*Columba livia*). *J Comp Neurol*, **278**:405–429.

Ariëns Kappers, C. U. (1909) The phylogenesis of the paleocortex and archi-cortex compared with the evolution of the visual neo-cortex. *Arch Neurol Psychiat*, **4**:161–173.

Ariëns Kappers, C. U. (1922) The ontogenetic development of the corpus striatum in birds and a comparison with mammals and man. *Proc Kon Akad v Wetens te Amsterdam*, **26**:135–158.

Ariëns Kappers, C. U. (1928) *Three Lectures on Neurobiotaxis and Other Subjects*. Copenhagen: Levin & Munksgaard.

Ariëns Kappers, C. U., Huber, G. C., Crosby, E. C. (1936) *The Comparative Anatomy of the Nervous System of Vertebrates, Including Man*. (2 vols.) Oxford: Macmillan.

Armstrong, E. A. (1973) *A Study of Bird Song*. New York: Dover.

Arnold, A. P. (1975a) The effects of castration on song development in zebra finches (*Poephila guttata*). *J Exp Zool*, **191**:261–278.

Arnold, A. P. (1975b) The effects of castration and androgen replacement on song, courtship, and aggression in zebra finches (*Poephila guttata*). *J Exp Zool*, **191**:309–326.

Arnold, A. P. (1980a) Sexual differences in the brain. *Am Sci*, **68**:165–173.

Arnold, A. P. (1980b) Logical levels of hormone action in the control of vertebrate behavior. *Am Zool*, **21**:233–242.

Arnold, A. P. (1980c) Effects of androgens on volumes of sexually dimorphic brain regions in the zebra finch. *Brain Res*, **185**:441–444.

Arnold, A. P. (1990) The passerine bird song system as a model in neuroendocrine research. *J Exp Zool*, **4** (suppl):22–30.

Arnold, A. P. (1992) Developmental plasticity in neural circuits controlling birdsong: sexual differentiation and the neural basis of learning. *J Neurobiol*, **23**:1506–1528.

Arnold, A. P. (2002) Concepts of genetic and hormonal induction of vertebrate sexual differentiation in the twentieth century, with special reference to the brain. In *Hormones, Brain, and Behavior* (Pfaff, D. W., Arnold, A. P., Etgen, A., Fahrbach, S., Rubin, R., eds.), pp. 105–135. Vol. 4. San Diego: Academic Press.

Arnold, A., Burgoyne, P. (2004) Are XX and XY brain cells intrinsically different? *Trends Endocrinol Metab*, **15**:6–11.

Arnold, A. P., Gorski, R. A. (1984) Gonadal steroid induction of structural sex differences in the central nervous system. *Ann Rev Neurosci*, **7**:413–442.

Arnold, A. P., Saltiel, A. (1979) Sexual difference in pattern of hormone accumulation in the brain of a song bird. *Science*, **205**:702–705.

Arnold, A. P., Schlinger, B. A. (1993) Sexual differentiation of brain and behavior: the zebra finch is not just a flying rat. *Brain Behav Evol*, **42**:231–241.

Arnold, A. P., Nottebohm, F., Pfaff, D. W. (1976) Hormone concentrating cells in vocal control and other areas of the brain of the zebra finch (*Poephila guttata*). *J Comp Neurol*, **165**:487–511.

Arnold, A. P., Bottjer, S. W., Brenowitz, E. A., Nordeen, E. J., Nordeen, K. W. (1986) Sexual dimorphisms in the neural vocal control system in song birds: ontogeny and phylogeny. *Brain Behav Evol*, **28**:22–31.

Asher, J. J., Garcia, R. (1969) The optimal age to learn a foreign language. *Mod Lang J*, **53**(5):334–341.

Ashmore, R. C., Wild, J. M., Schmidt, M. F. (2005) Brainstem and forebrain contributions to the generation of learned motor behaviors for song. *J Neurosci*, **25**:8543–8554.

Ashmore, R. C., Bourjaily, M. A., Schmidt, M. F. (2007) Hemispheric coordination is necessary for song production in adult birds: implications for a dual role for forebrain nuclei in vocal motor control. *J Neurophysiol*, **10**:1152.

Ashmore, R. C., Renk, J., Schmidt, M. F. (2008) Bottom-up activation of the vocal motor forebrain by the respiratory brainstem. *J Neurosci*, in press.

Ashraf, S. I., McLoon, A. L., Sclarsic, S. M., Kunes, S. (2006) Synaptic protein synthesis associated with memory is regulated by the RISC pathway in *Drosophila*. *Cell*, **124**:191–205.

Aston-Jones, G., Cohen, J. D. (2005) An integrative theory of locus coeruleus-norepinephrine function: adaptive gain and optimal performance. *Annu Rev Neurosci*, **28**:403–450.

Athos, J., Impey, S., Pineda, V. V., Chen, X., Storm, D. R. (2002) Hippocampal CRE-mediated gene expression is required for contextual memory formation. *Nat Neurosci*, **5**:1119–1120.

Atkins, C. M., Selcher, J, C., Petraitis, J. J., Trzaskos, J. M., Sweatt, J. D. (1998) The MAPK cascade is required for mammalian associative learning. *Nat Neurosci*, **1**:602.

Aubin, T., Jouventin, P., Hildebrand, C. (2000) Penguins use the two-voice system to recognize each other. *Proc Biol Sci*, **267**:1081–1087.

Avey, M. T., Phillmore, L. S., MacDougall-Shackleton, S. A. (2005) Immediate early gene expression following exposure to acoustic and visual components of courtship in zebra finches. *Behav Brain Res*, **165**:247–253.

Backstrom, N., Brandstrom, M., Gustafsson, L., Qvarnstrom, A., Cheng, H. H., Ellegren, H. (2006) Genetic mapping in a natural population of collared flycatchers (*Ficedula albicollis*): conserved synteny but gene order rearrangements on the avian Z chromosome. *Genetics* (online).

Bailey, D. J., Rosebush, J. C., Wade, J. (2002) The hippocampus and caudomedial neostriatum show selective responsiveness to conspecific song in the female zebra finch. *J Neurobiol*, **52**:43–51.

Bailey, D. J., Wade, J. (2003) Differential expression of the immediate early genes FOS and ZENK following auditory stimulation in the juvenile male and female zebra finch. *Brain Res Mol Brain Res*, **116**:147–154.

Bailey, D. J., Wade, J. (2005) FOS and ZENK responses in 45-day-old zebra finches vary with auditory stimulus and brain region, but not sex. *Behav Brain Res*, **162**:108–115.

Baker, A. M., Mather, P. B., Hughes, J. M. (2000) Population genetic structure of Australian magpies: evidence for regional differences in juvenile dispersal behaviour. *Heredity*, **85** (Pt 2):167–176.

Baker, M. C., Bottjer, S. W., Arnold, A. P. (1984) Sexual dimorphism and lack of seasonal changes in vocal control regions of the white-crowned sparrow brain. *Brain Res*, **295**:85–89.

Baker, M. C., Cunningham, M. A. (1985) The biology of bird-song dialects. *Behav Brain Sci* **8**:85–133.

Bakin, J. S., Weinberger, N. M. (1990) Classical conditioning induces CS-specific receptive field plasticity in the auditory cortex of the guinea pig. *Brain Res*, **536**:271–286.

Baldwin, F. M., Goldin, H. S., Metfessel, M. (1940) Effects of testosterone proprionate on female roller canaries under complete song isolation. *Proc Soc Exp Biol Med*, **44**:373–375.

Ball, G. F. (1999) Neuroendocrine basis of seasonal changes in vocal behavior among songbirds. In *The Design of Animal Communication*, (Hauser, M., Konishi, M., eds.), pp. 213–253. Cambridge, MA: MIT Press.

Ball, G. F., Balthazart, J. (2004) Hormonal regulation of brain circuits mediating male sexual behavior in birds. *Physiol Behav*, **83**:329–346.

Ball, G. F., Bentley, G. E. (2000) Neuroendocrine mechanisms mediating the photoperiodic and social regulation of seasonal reproduction in birds. In *Reproduction in Context: Social and Environmental Influences on Reproductive Physiology and Behavior* (Wallen, K., Schneider, J. E., eds.), pp. 129–158. Cambridge: MIT Press.

Ball, G. F., MacDougall-Shackleton, S. A. (2001) Sex differences in songbirds 25 years later: what have we learned and where do we go? *Microsc Res Tech*, **54**:327–334.

Ball, G. F., Nowicki, S. (1990) Differences in song quality in photorefractory and photosensitive song sparrows. *Anim Behav*, **40**:986–987.

Ball, G. F., Casto, J. M., Bernard, D. J. (1994) Sex differences in the volume of avian song control nuclei: comparative studies and the issue of brain nucleus delineation. *Psychoneuroendocrinology*, **19**:485–504.

Ball, G. F., Bernard, D. J., Foidart, A., Lakaye, B., Balthazart, J. (1999) Steroid sensitive sites in the avian brain: does the distribution of the estrogen receptor alpha and beta types provide insight into their function? *Brain Behav Evol*, **54**:28–40.

Ball, G. F., Riters, L. V., Balthazart, J. (2002) Neuroendocrinology of song behavior and avian brain plasticity: multiple sites of action of sex steroid hormones. *Front Neuroendocrin*, **23**:137–178.

Ball, G. F., Auger, C. J., Bernard, D. J., *et al.* (2004a) Seasonal plasticity in the song control system: multiple brain sites of steroid hormone action and the importance of variation in song behavior. *Ann NY Acad Sci*, **1016**:586–610.

Ball, G., Auger, C., Bernard, D., *et al.* (2004b) Seasonal plasticity in the song control system: steroid metabolism, brain sites, and mechanisms of hormone action. In *Behavioral Neurobiology of Birdsong* (Ziegler, H., Marler, P., eds.), pp. 586–610. New York: Ann NY Acad Sci.

Ballam, G. O., Clanton, T. L., Kunz, A. L. (1982) Ventilatory phase duration in the chicken: role of mechanical and CO_2 feedback. *J Appl Physiol*, **53**:1378–1385.

Ballentine, B. (2006) Morphological adaptation influences the evolution of a mating signal. *Evolution*, **60**:1936–1944.

Ballentine, B., Hyman, J., Nowicki, S. (2004) Vocal performance influences female response to male bird song: an experimental test. *Behav Ecol*, **15**:163–168.

Ballintijn, M. R., ten Cate, C. (1998) Sound production in the collared dove: a test of the 'whistle' hypothesis. *J Exper Biol*, **201**:1637–1649.

Ballintijn, M. R., ten Cate, C., Nuijens, F. W., Berkhoudt, H. (1995) The syrinx of the collared dove (*Streptopelia decaocto*): structure, inter-individual variation and development. *Netherlands J Zool*, **45**:455–479.

Balthazart, J., Gahr, M., Surlemont, C. (1989) Distribution of estrogen receptors in the brain of the Japanese quail: an immunocytochemical study. *Brain Res*, **501**:205–214.

Balthazart, J., Adkins-Regan, E. (2002) Sexual differentiation of brain and behavior in birds. In *Hormones, Brain and Behavior* (Pfaff, D. W., Arnold, A. P., Etgen, A. M., Fahrbach, S. E., Rubin, R. T., eds.), pp. 223–301. San Diego, CA: Academic Press.

Balthazart, J., Ball, G. F. (1998) The Japanese quail as a model system for the investigation of steroid-catecholamine interactions mediating appetitive and consummatory aspects of male sexual behavior. *Ann Rev Sex Res*, **9**:96–176.

Balthazart, J., Foidart, A., Surlemont, C., Vockel, A., Harada, N. (1990) Distribution of aromatase in the brain of the Japanese quail, ring dove and zebra finch: an immunocytochemical study. *J Comp Neurol*, **301**:276–288.

Balthazart, J., Foidart, A., Wilson, E. M., Ball, G. F. (1992) Immunocytochemical localization of androgen receptors in the male songbird and quail brain. *J Comp Neurol*, **317**:407–420.

Balthazart, J., Absil, P., Foidart, A., Houbart, M., Harada, N., Ball, G. F. (1996a) Distribution of aromatase-immunoreactive cells in the forebrain of zebra finches (*Taeniopygia guttata*): implications for the neural action of steroids and nuclear definition in the avian hypothalamus. *J Neurobiol*, **31**:129–148.

Balthazart, J., Tlemçani, O., Harada, N. (1996b) Localization of testosterone-sensitive and sexually dimorphic aromatase-immunoreactive cells in the quail preoptic area. *J Chem Neuroanat*, **11**:147–171.

Balthazart, J., Castagna, C., Ball, G. F. (1997) Aromatase inhibition blocks the activation and sexual differentiation of appetitive male sexual behavior in Japanese quail. *Behav Neurosci*, **111**:381–397.

Balthazart, J., Absil, P., Gérard, M., Appeltants, D., Ball, G. F. (1998) Appetitive and consummatory male sexual behavior in Japanese quail are differentially regulated by subregions of the preoptic medial nucleus. *J Neurosci*, **18**:6512–6527.

Bankes, S. C., Margoliash, D. (1993) Parametric modeling of the temporal dynamics of neuronal responses using connectionist architectures. *J Neurophysiol*, **69**:980–991.

Bannerman, D. M., Good, M. A., Butcher, S. P., Ramsay, M., Morris, R. G. (1995) Distinct components of spatial learning revealed by prior training and NMDA receptor blockade. *Nature*, **378**:182–186.

Banta-Lavenex, P., Lavenex, P., Clayton, N. S. (2001) Comparative studies of postnatal neurogenesis and learning: a critical review. *Avian Poult Biol Rev*, **12**:103–125.

Baptista, L. F. (1996) Nature and its nurturing in avian vocal development. In *Ecology and Evolution of Acoustic Communication in Birds* (Kroodsma, D. E., Miller, E. H., eds.), pp. 39–60. Ithaca & London: Cornell University Press.

Baptista, L. F. (1975) Song dialects and demes in sedentary populations of the white-crowned sparrow (*Zonotrichia leucophrys nuttalli*). *U Calif Publ Zool*, **105**:1–52.

Baptista, L. F., Morton, M. L. (1981) Interspecific song acquisition by a white-crowned sparrow. *Auk*, **98**:383–385.

Baptista, L. F., Petrinovich, L. (1984) Social interaction, sensitive phases and the song template hypothesis in the white-crowned sparrow. *Anim Behav*, **32**:172–181.

Baptista, L. F., Petrinovich, L. (1986) Song development in the white-crowned sparrow: social factors and sex differences. *Anim Behav*, **34**:1359–1371.

Baptista, L. F., Schuchmann, K. L. (1990) Song learning in the anna hummingbird (Calypte anna). *Ethology*, **84**:15–26.

Baptista, L. F., Trail, P. W., DeWolfe, B. B., Morton, M. L. (1993) Singing and its functions in female white-crowned sparrows. *Anim Behav*, **46**:511–524.

Barami, K., Iversen, K., Furneaux, H., Goldman, S. A. (1995) Hu protein as an early marker of neuronal phenotypic differentiation by subependymal zone cells of the adult songbird forebrain. *J Neurobiol*, **28**:82–101.

Barclay, S. R., Harding, C. F. (1990) Differential modulation of monoamine levels and turnover rates by estrogen and/or androgen in hypothalamic and vocal control nuclei of male zebra finches. *Brain Res*, **523**:251–262.

Barclay, S. R., Harding, C. F., Waterman, S. A. (1992) Correlations between catecholamine levels and sexual behavior in male zebra finches. *Pharmacol Biochem Behav*, **41**:195–201.

Barclay, S. R., Harding, C. F., Waterman, S. A. (1996) Central DSP-4 treatment decreases norepinephrine levels and courtship behavior in male zebra finches. *Pharmacol Biochem Behav*, **53**:213–220.

Barkan, S., A. Ayali, Nottebohm, F. Barnea, A. (2007) Neuronal recruitment in adult zebra finch brain during a reproductive cycle. *Dev Neurobiol*, **67**:687–701.

Barker, F. K., Cibois, A., Schikler, P., Feinstein, J., Cracraft, J. (2004) Phylogeny and diversification of the largest avian radiation. *Proc Natl Acad Sci USA*, **101**:11040–11045.

Barnea, A., Nottebohm, F. (1994) Seasonal recruitment of hippocampal neurons in adult free-ranging black-capped chickadees. *Proc Natl Acad Sci USA*, **91**:11217–11221.

Barnea, A., Nottebohm, F. (1996) Recruitment and replacement of hippocampal neurons in young and adult chickadees: an addition to the theory of hippocampal learning. *Proc Natl Acad Sci USA*, **93**:714–718.

Barrett, G., Silcocks, A., Barry, S., Cunningham, R., Poulter, R. (2003) *The New Atlas of Australian Birds*. Hawthorn-East, Victoria, Australia: Royal Australasian Ornithologists' Union.

Barria, A., Malinow, R. (2005) NMDA receptor subunit composition controls synaptic plasticity by regulating binding to CaMKII. *Neuron*, **48**:289–301.

Basham, M. E., Nordeen, E. J., Nordeen, K. W. (1996) Blockade of NMDA receptors in the anterior forebrain impairs sensory acquisition in the zebra finch (*Poephila guttata*). *Neurobiol Learn Mem*, **66**:295–304.

Basham, M. E., Sohrabji, F., Singh, T. D., Nordeen, E. J., Nordeen, K. W. (1999) Developmental regulation of NMDA receptor 2B subunit mRNA and ifenprodil binding in the zebra finch anterior forebrain. *J Neurobiol*, **39**:155–167.

Bates, E. (1992) Language development. *Curr Opin Neurobiol*, **2**:180–185.

Bates, E., Elman, J., Johnson, M., Karmiloff-Smith, A., Parisi, D., Plunkett, K. (1998) Innateness and emergentism. In *A Companion to Cognitive Science* (Bechtel, W., Graham, G., eds.). Oxford: Basil Blackwell.

Bayer, K. U., LeBel, E., McDonald, G. L., O'Leary, H., Schulman, H., De Koninck, P. (2006) Transition from reversible to persistent binding of CaMKII to postsynaptic sites and NR2B. *J Neurosci*, **26**:1164–1174.

Beach, F. A. (1948) *Hormones and Behavior*. New York: Paul B. Hoeber, Inc.

Beckers, G. J. L., ten Cate, C. (2006) Nonlinear phenomena and song evolution in *Streptopelia* doves. *Acta Zool Sinica*, **52** (Supplement):482–485.

Beckers, G. J. L., Suthers, R. A., ten Cate, C. (2003) Pure-tone birdsong by resonance filtering of harmonic overtones. *Proc Natl Acad Sci USA*, **100**:7372–7376.

Beckers, G. J. L., Nelson, B. S., Suthers, R. A. (2004) Vocal-tract filtering by lingual articulation in a parrot. *Curr Biol*, **14**:1592–1597.

Beecher, M. D. (1991) Success and failures of parent-offspring recognition in animals. In *Kin Recognition* (Hepper, P. G., ed.). Cambridge: Cambridge University Press.

Beecher, M. D. (1996) Birdsong learning in the laboratory and field. In *Ecology and Evolution of Acoustic Communication in Birds* (Kroodsma, D. E., Miller, E. H., eds.), pp. 61–78. Ithaca & London: Cornell University Press.

Beecher, M. D., Brenowitz, E. A. (2005) Functional aspects of song learning in the songbirds. *Trends Ecol Evol*, **20**: 143–149.

Belton, E., Salmond, C. H., Watkins, K. E., Vargha-Khadem, F., Gadian, D. G. (2003) Bilateral brain abnormalities associated with dominantly inherited verbal and orofacial dyspraxia. *Hum Brain Mapp*, **18**:194–200.

Bendix, J., Rafiqpoor, M. D. (2001) Studies on the thermal conditions of soils at the upper tree line in the Páramo of Papallacta (eastern cordillera of Ecuador). *Erdkunde*, **55**:257–276.

Benedetta, L., Gould, E., Shors, T. J. (2006) Is there a link between adult neurogenesis and learning? *Hippocampus*, **16**:216–224.

Benoit, T. C., Jocelyn, L. J., Moddemann, D. M., Embree, J. E. (1996) Romanian adoption. The Manitoba experience. *Arch Pediatr Adolesc Med*, **150**:1278–1282.

Ben-Shahar, Y., Robichon, A., Sokolowski, M., Robinson, G. (2002) Influence of gene action across different time scales on behavior. *Science*, **296**:741–744.

Bentley, G. E. (2001) Unraveling the enigma: the role of melatonin in seasonal processes in birds. *Microsc Res Tech*, **53**:63–71.

Bentley, G. E., Brenowitz, E. A. (2002) Three-dimensional analysis of avian song control nuclei. *J Neurosci Methods*, **121**:75–80.

Bentley, G. E., Goldsmith, A. R., Dawson, A., Glennie, L. M., Talbot, R. T., Sharp, P. J. (1997) Photorefractoriness in European starlings (*Sturnus vulgaris*) is not dependent upon the long-day-induced rise in plasma thyroxine. *Gen Comp Endocrinol*, **107**:428–438.

Benton, S., Cardin, J. A., DeVoogd, T. J. (1998) Lucifer Yellow filling of area X-projecting neurons in the high vocal center of female canaries. *Brain Res*, **799**: 38–147.

Bergman, A. (1990) *Auditory Scene Analysis*. Cambridge, MA: MIT Press.

Berman, D. E., Hazvi, S., Rosenblum, K., Seger, R., Dudai, Y. (1998) Specific and differential activation of mitogen-activated protein kinase cascades by unfamiliar taste in the insular cortex of the behaving rat. *J Neurosci*, **18**:10037–10044.

Bernard, D. J. (1995) The effects of testosterone, photoperiod, and season on plasticity in the song control system of European starlings. Ph.D. dissertation. Baltimore, MD: Johns Hopkins University.

Bernard, D. J., Ball, G. F. (1997) Photoperiodic condition modulates the effects of testosterone on song control nuclei volumes in male European starlings. *Gen Comp Endocrinol*, **105**:276–283.

Bernard, D. J., Casto, J. M., Ball, G. F. (1993) Sexual dimorphism in the volume of song control nuclei in European starlings: assessment by a Nissl stain and autoradiography for muscarinic cholinergic receptors. *J Comp Neurol*, **334**:559–570.

Bernard, D. J., Wilson, F. E., Ball, G. F. (1997) Testis-dependent and -independent effects of photoperiod on volumes of song control nuclei in American tree sparrows (*Spizella arborea*). *Brain Res*, **760**:163–169.

Bernard, D. J., Bentley, G. E., Balthazart, J., Turek, F. W., Ball, G. F. (1999) Androgen receptor, estrogen receptor alpha, and estrogen receptor beta show distinct patterns of expression in forebrain song control nuclei of European starlings. *Endocrinology*, **140**:4633–4643.

Berridge, C. W., Waterhouse, B. D. (2003) The locus coeruleus-noradrenergic system: modulation of behavioral state and state-dependent cognitive processes. *Brain Res Rev*, **42**:33–84.

Bickerton, D. (1990) *Language and Species*. Chicago, IL: University of Chicago Press.

Biedermann, W. (1904) Studien zur vergleichenden Physiologie der peristaltischen Bewegungen Plüg. *Arch ges Physiol*, **102**:475–542.

Bigalke-Kunz, B., Rubsamen, R., Dorrscheidt, G. J. (1987) Tonotopic organization and functional characterization of the auditory thalamus in a songbird, the European starling. *J Comp Physiol [A]*, **161**:255–265.

Birkhead, T. R., Pellatt, E. J., Brekke, P., Yeates, R., Castillo-Juarez, H. (2005) Genetic effects on sperm design in the zebra finch. *Nature*, **434**:383–387.

Bischof, H. J. (1998) Song learning, filial imprinting, and sexual imprinting: three variations of a common theme? *Biomed Res*, **18**:133–146.

Bischof, H. J., Geissler, E., Rollenhagen, A. (2002) Limitations of the sensitive period for sexual imprinting: neuroanatomical and behavioural experiments in the zebra finch (*Taeniopygia guttata*). *Behav Brain Res*, **133**: 317–322.

Bleisch, W., Luine, V. N., Nottebohm, F. (1984) Modification of synapses in androgen-sensitive muscle. I. Hormonal regulation of acetylcholine receptor number in the songbird syrinx. *J Neurosci*, **4**:786–792.

Bliss, T. V., Collingridge, G. L. (1993) A synaptic model of memory: long-term potentiation in the hippocampus. *Nature*, **361**:31–39.

Bloodstein, O. (1995) *A Handbook on Stuttering*. 5th edn. San Diego, CA: Singular Pub. Group.

Blum, S., Moore, A. N., Adams, F., Dash, P. K. (1999) A mitogen-activated protein kinase cascade in the CA1/CA2 subfield of the dorsal hippocampus is essential for long-term spatial memory. *J Neurosci*, **19**:3535–3344.

Bock, J., Braun, K. (1999a) Filial imprinting in domestic chicks is associated with spine pruning in the associative area, dorsocaudal neostriatum. *Eur J Neurosci*, **11**: 2566–2570.

Bock, J., Braun, K. (1999b) Blockade of N-methyl-D-aspartate receptor activation suppresses learning-induced synaptic elimination. *Proc Natl Acad Sci USA*, **96**: 2485–2490.

Bock, J., Thode, C., Hannemann, O., Braun, K., Darlison, M. G. (2005) Early socio-emotional experience induces expression of the immediate-early gene Arc/arg3.1 (activity-regulated cytoskeleton-associated protein/activity-regulated gene) in learning-relevant brain regions of the newborn chick. *Neuroscience*, **133**:625–633.

Boettiger, C. A., Doupe, A. J. (1998) Intrinsic and thalamic excitatory inputs onto songbird LMAN neurons differ in their pharmacological and temporal properties. *J Neurophysiol*, **79**:2615–2628.

Boettiger, C. A., Doupe, A. J. (2001) Developmentally restricted synaptic plasticity in a songbird nucleus required for song learning. *Neuron*, **31**:809–818.

Bohland, J. W., Guenther, F. H. (2006) An fMRI investigation of syllable sequence production. *Neuroimage*, **32**:821–841.

Böhner, J. (1983) Song learning in the zebra finch (*Taeniopygia guttata*): selectivity in the choice of a tutor and accuracy of song copies. *Anim Behav*, **31**:231–237.

Böhner, J. (1990) Early acquisition of song in the zebra finch, *Taeniopygia guttata*. *Anim Behav*, **39**:369–374.

Böhner, J., Todt, D. (1996) Influence of auditory stimulation on the development of syntactical and temporal features in European starling song. *Auk*, **113**:450–456.

Bohner, J., Chaiken, M. L., Ball, G. F., Marler, P. (1990) Song acquisition in photosensitive and photorefractory male European starlings. *Horm Behav*, **24**:582–594.

Boland, C. R. J. (1998) Helpers improve nest defence in cooperatively breeding white-winged choughs. *Emu*, **98**:320–324.

Boland, C. R. J., Heinsohn, R., Cockburn, A. (1997) Deception by helpers in cooperatively-breeding white-winged choughs and its experimental manipulation. *Behav Ecol Sociobiol*, **88**:295–302.

Boles, W. E. (1995) The world's oldest songbird. *Nature*, **374**:21–22.

Bolhuis, J. J. (1994) Neurobiological analyses of behavioural mechanisms in development. In *Causal Mechanisms of Behavioural Development*. (Hogan, J. A., Bolhuis, J. J., eds.), pp. 16–46. New York, NY: Cambridge University Press.

Bolhuis, J. J. (2005) Function and mechanism in neuroecology: looking for clues. *Anim Biol*, **55**:457–490.

Bolhuis, J. J., Eda-Fujiwara, H. (2003) Bird brains and songs: neural mechanisms of birdsong perception and memory. *Anim Biol*, **53**:129–145.

Bolhuis, J. J., Gahr, M. (2006) Neural mechanisms of birdsong memory. *Nat Rev Neurosci*, **7**:347–357.

Bolhuis, J. J., Macphail, E. M. (2001) A critique of the neuroecology of learning and memory. *Trends Cogn Sci*, **5**:426–433.

Bolhuis, J. J., Reid, I. C. (1992) Effects of intraventricular infusion of the N-methyl-D-aspartate (NMDA) receptor antagonist AP5 on spatial memory of rats in a radial arm maze. *Behav Brain Res*, **47**:151–157.

Bolhuis, J. J., Johnson, M. H., Horn, G., Bateson, P. (1989) Long-lasting effects of IMHV lesions on social preferences in domestic fowl. *Behav Neurosci*, **103**:438–441.

Bolhuis, J. J., Van Mil, D. P., Houx, B. B. (1999) Song learning with audiovisual compound stimuli in zebra finches. *Anim Behav*, **58**:1285–1292.

Bolhuis, J. J., Zijlstra, G. G., den Boer-Visser, A. M., Van Der Zee, E. A. (2000) Localized neuronal activation in the zebra finch brain is related to the strength of song learning. *Proc Natl Acad Sci USA*, **97**:2282–2285.

Bolhuis, J. J., Hetebrij, E., Den Boer-Visser, A. M., De Groot, J. H., Zijlstra, G. G. (2001) Localized immediate early gene expression related to the strength of song learning in socially reared zebra finches. *Eur J Neurosci*, **13**:2165–2170.

Bonke, B. A., Bonke, D., Scheich, H. (1979a) Connectivity of the auditory forebrain nuclei in the guinea fowl (*Numida meleagris*). *Cell Tissue Res*, **200**:101–121.

Bonke, D., Scheich, H., Langner, G. (1979b) Responsiveness of units in the auditory neostriatum of the guinea fowl (*Numida meleagris*) to species-specific calls and synthetic stimuli. I. Tonotopy and functional zones of field L. *J Comp Physiol*, **132**:243–255.

Borror, D. J., Reese, C. R. (1956) Vocal gymnastics in wood thrush songs. *Ohio J Sci*, **56**:177–182.

Boseret, G., Carere, C., Ball, G. F., Balthazart, J. (2006) Social context affects testosterone-induced singing and the volume of song control nuclei in male canaries (*Serinus canaria*). *J Neurobiol*, **66**:1044–1060.

Bottjer, S. W. (1993) The distribution of tyrosine hydroxylase immunoreactivity in the brains of male and female zebra finches. *J Neurobiol*, **24**:51–69.

Bottjer, S. W. (2002) Neural strategies for learning during sensitive periods of development. *J Comp Physiol A*, **188**:917–928.

Bottjer, S. W. (2005) Silent synapses in a thalamo-cortical circuit necessary for song learning in zebra finches. *J Neurophysiol*, **94**:3698–3707.

Bottjer, S. W., Arnold, A. P. (1982) Afferent neurons in the hypoglossal nerve of the zebra finch (*Poephila guttata*): localization with horseradish peroxidase. *J Comp Neurol*, **210**:190–197.

Bottjer, S. W., Arnold, A. P. (1984) The role of feedback from the vocal organ. I: Maintenance of stereotypical vocalizations by adult zebra finches. *J Neurosci*, **4**:2387–2396.

Bottjer, S. W., Hewer, S. J. (1992) Castration and antisteroid treatment impair vocal learning in male zebra finches. *J Neurobiol*, **23**:337–353.

Bottjer, S. W., Johnson, F. (1997) Circuits, hormones, and learning: vocal behavior in songbirds. *J Neurobiol*, **33**:602–618.

Bottjer, S. W., Miesner, E. A., Arnold, A. P. (1984) Forebrain lesions disrupt development but not maintenance of song in passerine birds. *Science*, **224**:901–903.

Bottjer, S. W., Halsema, K. A., Brown, S. A., Miesner, E. A. (1989) Axonal connections of a forebrain nucleus involved with vocal learning in zebra finches. *J Comp Neurol*, **279**:312–326.

Bottjer, S. W., Brady, J. D., Walsh, J. P. (1998) Intrinsic and synaptic properties of neurons in the vocal-control nucleus lMAN from in vitro slice preparations of juvenile and adult zebra finches. *J Neurobiol*, **37**:642–658.

Bottjer, S. W., Brady, J. D., Cribbs, B. (2000) Connections of a motor cortical region in zebra finches: relation to pathways for vocal learning. *J Comp Neurol*, **420**:244–260.

Boughman, J. W. (1998) Vocal learning by greater spear-nosed bats. *Proc Biol Sci*, **265**:227–233.

Boumans, T., Vandersmissen, L., Gobes, S. M. H., et al. (2005) Functional magnetic resonance imaging of the zebra finch auditory forebrain during exposure to original and altered versions of the bird's own song. *Soc Neurosci Abstr*, 1002.16.

Bourtchuladze, R., Frenguelli, B., Blendy, J., Cioffi, D., Schutz, G., Silva, A. (1994) Deficient long-term memory in mice with a targeted mutation of the cAMP-responsive element-binding protein. *Cell*, **79**:59–68.

Bout, R. G., Dubbeldam, J. L. (1991) Functional morphological interpretation of the distribution of muscle spindles in the jaw muscles of the mallard (*Anas platyrhynchos*). *J Morphol*, **210**:215–226.

Bout, R. G., Tellegen, A. J., Dubbeldam, J. L. (1997) Central connections of the nucleus mesencephalicus nervi trigemini in the mallard (*Anas platyrhynchos* L.). *Anat Rec*, **248**:554–565.

Bower, G. (1970) Organizational factors in memory. *Cogn Psychol*, **1**:18–46.

Bozon, B., Davis, S., Laroche, S. (2003) A requirement for the immediate early gene zif 268 in reconsolidation of recognition memory after retrieval. *Neuron*, **40**:695–701.

Braaten, R. F., Reynolds, K. (1999) Auditory preference for conspecific song in isolation-reared zebra finches. *Anim Behav*, **58**:105–111.

Bradbury, J. W., Vehrencamp, S. L. (1998) *Principles of Animal Communication*. Sunderland, MA: Sinauer Assoc.

Brainard, M. S., Knudsen, E. I. (1998) Sensitive periods for visual calibration of the auditory space map in the barn owl optic tectum. *J Neurosci*, **18**:3929–3942.

Brainard, M. S., Doupe, A. J. (2000a) Interruption of a basal ganglia-forebrain circuit prevents plasticity of learned vocalizations. *Nature*, **404**:762–766.

Brainard, M. S., Doupe, A. J. (2000b) Auditory feedback in learning and maintenance of vocal behaviour. *Nat Rev Neurosci*, **1**:31–40.

Brainard, M. S., Doupe, A. J. (2001) Postlearning consolidation of birdsong: stabilizing effects of age and anterior forebrain lesions. *J Neurosci*, **21**:2501–2517.

Brainard, M. S., Doupe, A. J. (2002) What songbirds teach us about learning. *Nature*, **417**:351–358.

Brann, D. W., Hendry, L. B., Mahesh, V. B. (1995) Emerging diversities in the mechanism of action of steroid hormones. *J Steroid Biochem Mol Biol*, **52**:113–133.

Brannon, E. M., Terrace, H. S. (1998) Ordering of the numerosities 1 to 9 by monkeys. *Science*, **282**:746–749.

Braun, K., Scheich, H., Schachner, M., Heizman, C. (1985) Distribution of parvalbumin, cytochrome oxidase activity and 14C-2-deoxyglucose uptake in the brain of the zebra finch. I. Auditory and vocal motor systems. *Cell Tissue Res*, **240**:101–115.

Brauth, S. E., McHale, C. M. (1988) Auditory pathways in the budgerigar. II. Intratelencephalic pathways. *Brain Behav Evol*, **32**:193–207.

Brauth, S. E., McHale, C. M., Brasher, C. A., Dooling, R. J. (1987) Auditory pathways in the budgerigar. I. Thalamo-telencephalic projections. *Brain Behav Evol*, **30**:174–199.

Brauth, S. E., Heaton, J. T., Shea, S. D., Durand, S. E., Hall, W. S. (1997) Functional anatomy of forebrain vocal control pathways in the budgerigar (*Melopsittacus undulatus*). *Ann N Y Acad Sci*, **807**:368–385.

Brauth, S., Liang, W., Roberts, T. F., Scott, L. L., Quinlan, E. M. (2002) Contact call-driven Zenk protein induction and habituation in telencephalic auditory pathways in the budgerigar (*Melopsittacus undulatus*): implications for understanding vocal learning processes. *Learn Mem*, **9**:76–88.

Brauth, S. E., Tang, Y. Z., Liang, W., Roberts, T. F. (2003) Contact call-driven zenk mRNA expression in the brain of the budgerigar (*Melopsittacus undulatus*). *Brain Res Mol Brain Res*, **117**:97–103.

Brennan, P. A. (1994) The effects of local inhibition of N-methyl-D-aspartate and AMPA/kainate receptors in the accessory olfactory bulb on the formation of an olfactory memory in mice. *Neuroscience*, **60**:701–708.

Brenowitz, E. A. (1991) Altered perception of species-specific song by female birds after lesions of a forebrain nucleus. *Science*, **251**:303–305.

Brenowitz, E. A. (1997) Comparative approaches to the avian song system. *J Neurobiol*, **33**:517–531.

Brenowitz, E. A., Arnold, A. P. (1986) Interspecific comparisons of the size of neural song control regions and song complexity in duetting birds: evolutionary implications. *J Neurosci*, **6**:2875–2879.

Brenowitz, E. A., Beecher, M. D. (2005) Song learning in birds: diversity and plasticity, opportunities and challenges. *Trends Neurosci*, **28**:127–132.

Brenowitz, E. A., Lent, K. (2001) Afferent input is necessary for seasonal growth and maintenance of adult avian song control circuits. *J Neurosci*, **21**:2320–2329.

Brenowitz, E. A., Lent, K. (2002) Act locally and think globally: intracerebral testosterone implants induce seasonal-like growth of adult avian song control circuits. *Proc Natl Acad Sci USA*, **99**:12421–12426.

Brenowitz, E. A., Arnold, A. P., Levin, R. N. (1985) Neural correlates of female song in tropical duetting birds. *Brain Res*, **343**:104–112.

Brenowitz, E. A., Nalls, B., Wingfield, J. C., Kroodsma, D. E. (1991) Seasonal changes in avian song nuclei without seasonal changes in song repertoire. *J Neurosci*, **11**:1367–1374.

Brenowitz, E. A., Baptista, L. F., Lent, K., Wingfield, J. C. (1998) Seasonal plasticity of the song control system in wild Nuttall's white-crowned sparrows. *J Neurobiol*, **34**:69–82.

Brenowitz, E. A., Lent, K., Rubel, E. W. (2007) Auditory feedback and song production do not regulate seasonal growth of song control circuits in adult White-crowned Sparrows. *J Neurosci*, **27**:6810–6814.

Bricker, P. D., Pruzansky, S. (1976) Speaker recognition. In *Contemporary Issues in Experimental Phonetics* (Lass, N. J., ed.), pp. 295–326. New York: Academic Press.

Broca, P. (1861) Remarques sur le siège de la faculté du langage articulé, suivies d'une observation d'aphémie. Trans. J. Kann (1950). *J Speech Hear Disord*, **15**:16–20.

Brown, C. H. (1982) Ventroloquial and locatable vocalizations in birds. *Z Tierpsychol*, **59**:338–350.

Brown, E. D., Farabaugh, S. M. (1991) Song sharing in a group-living songbird, the Australian magpie, *Gymnorhina tibicen*. III. Sex specificity and individual specificity of vocal parts in communal chorus and duet songs. *Behaviour*, **118**:244–274.

Brown, E. D., Farabaugh, S. M., Veltman, C. J. (1988) Song sharing in a group-living songbird, the Australian magpie, *Gymnorhina tibicen*. I. Vocal sharing within and among social groups. *Behaviour*, **104**:1–28.

Brown, J. L. (1971) An exploratory study of vocalization areas in the brain of the redwinged blackbird (*Agelaius phoeniceus*). *Behaviour*, **39**:91–127.

Brown, J. L. (1975) *The Evolution of Behavior*. New York: WW Norton & Co.

Brown, J. L. (1987) *Helping and Communal Breeding in Birds*. Princeton, NJ: Princeton University Press.

Brown, W. R., Hubbard, S. J., Tickle, C., Wilson, S. A. (2003) The chicken as a model for large-scale analysis of vertebrate gene function. *Nat Rev Genet*, **4**:87–98.

Brumm, H., Hultsch, H. (2001) Pattern amplitude is related to pattern imitation during the song development of nightingales. *Anim Behav*, **61**:747–754.

Brumm, H., Todt, D. (2004) Male-male vocal interactions and the adjustment of song amplitude in a territorial bird. *Anim Behav*, **67**:281–286.

Brummett, R. E., Fox, K. E., Kempton, J. B. (1992) Quantitative relationships of the interaction between sound and kanamycin. *Arch Otolaryngol Head Neck Surg*, **118**:498–500.

Buckley, M. J., Gaffan, D. (2000) The hippocampus, perirhinal cortex and memory in the monkey. In *Brain, Perception, Memory: Advances in Cognitive Neuroscience* (Bolhuis, J. J., ed.), pp. 279–298. Oxford, UK: Oxford University Press.

Buller, A. L., Larson, H. C., Schneider, B. E., Beaton, J. A., Morrisett, R. A., Monaghan, D. T. (1994) The molecular basis of NMDA receptor subtypes: native receptor diversity is predicted by subunit composition. *J Neurosci*, **14**:5471–5484.

Burchuladze, R., Rose, S. P. R. (1992) Memory formation in day-old chicks requires NMDA but not non-NMDA glutamate receptors. *Eur J Neurosci*, **4**:535–538.

Burd, G. D., Nottebohm, F. (1985) Ultrastructural characterization of synaptic terminals formed on newly generated neurons in a song control nucleus of the adult canary forebrain. *J Comp Neurol*, **240**:143–152.

Burek, M. J., Nordeen, K. W., Nordeen, E. J. (1991) Neuron loss and addition in developing zebra finch song nuclei are independent of auditory experience during song learning. *J Neurobiol*, **22**:215–223.

Burek, M. J., Nordeen, K. W., Nordeen, E. J. (1994) Ontogeny of sex differences among newly-generated neurons of the juvenile avian brain. *Dev Brain Res*, **78**:57–64.

Burek, M. J., Nordeen, K. W., Nordeen, E. J. (1997) Sexually dimorphic neuron addition to an avian song-control region is not accounted for by sex differences in cell death. *J Neurobiol*, **33**:61–71.

Burgoon, J. K., Saine, T. (1978) *The Unspoken Dialogue: An Introduction to Nonverbal Communication*. Boston, MD: Houghton Mifflin.

Burmeister, S. S., Fernald, R. D. (2005) Evolutionary conservation of the egr-1 immediate-early gene response in a teleost. *J Comp Neurol*, **481**:220–232.

Burnham, D. K., Earnshaw, L. J., Quinn, M. C. (1987) The development of the categorical identification of speech. In *Perceptual Development in Early Infancy: Problems and Issues*. (McKenzie, B. E., Day, R. H., eds.), pp. 237–275. Hillsdale, NJ: Erlbaum.

Burt, J. M., Lent, K. L., Beecher, M. D., Brenowitz, E. A. (2000) Lesions of the anterior forebrain song control pathway in female canaries affect song perception in an operant task. *J Neurobiol*, **42**:1–13.

Butler, A. B., Hodos, W. (1996) *Comparative Vertebrate Neuroanatomy: Evolution and Adaptation*. New York: Wiley-Liss.

Butler, P. J. (1991) Exercise in birds. *J Exp Boil*, **160**:233–262.

Calder, W. A. (1970) Respiration during song in the canary (*Serinus canaria*). *Comp Biochem Physiol*, **32**:251–258.

Calder, W. A., King, J. R. (1974) Thermal and caloric relations in birds. In *Avian Biology* (Farner, D. S., King, J. R., Parkes, K. C., eds.), pp. 259–413. New York: Academic Press.

Callan, D. E., Tajima, K., Callan, A. M., Kubo, R., Masaki, S., Akahane-Yamada, R. (2003) Learning-induced neural plasticity associated with improved identification performance after training of a difficult second-language phonetic contrast. *Neuroimage*, **19**:113–124.

Calvin, W. H. (1993) The unitary hypothesis: a common neural circuitry for novel manipulations, language, plan-ahead, and throwing? In *Tools, Language, and Cognition in Human Evolution* (Gibson, K. R., Ingold, T., eds.), pp. 230–250. Cambridge, UK: Cambridge University Press.

Campbell, C. B., Hodos, W. (1970) The concept of homology and the evolution of the nervous system. *Brain Behav Evol*, **3**:353–367.

Canady, R. A., Kroodsma, D. E., Nottebohm, F. (1984) Population differences in complexity of a learned skill are correlated with the brain space involved. *Proc Natl Acad Sci USA*, **81**:6232–6234.

Canady, R. A., Burd, G. D., DeVoogd, T. J., Nottebohm, F. (1988) Effect of testosterone on input received by an identified neuron type of the canary song system: a Golgi/electron microscopy/degeneration study. *J Neurosci*, **8**:3770–3784.

Capsius, B., Leppelsack, H. J. (1996) Influence of urethane anesthesia on neural processing in the auditory cortex analogue of a songbird. *Hearing Res*, **96**:59–70.

Carcraft, J. (1967) Comments on homology and analogy. *Syst Zool*, **16**:356–359.

Cardin, J. A., Schmidt, M. F. (2003) Song system auditory responses are stable and highly tuned during sedation, rapidly modulated and unselective during wakefulness, and suppressed by arousal. *J Neurophysiol*, **90**:2884–2899.

Cardin, J. A., Schmidt, M. F. (2004a) Auditory responses in multiple sensorimotor song system nuclei are co-modulated by behavioral state. *J Neurophysiol*, **91**:2148–2163.

Cardin, J. A., Schmidt, M. F. (2004b) Noradrenergic inputs mediate state dependence of auditory responses in the avian song system. *J Neurosci*, **24**:7745–7753.

Cardin, J. A., Raksin, J. N., Schmidt, M. F. (2005) Sensorimotor nucleus NIf is necessary for auditory processing but not vocal motor output in the avian song system. *J Neurophysiol*, **93**:2157–2166.

Carleton, A., Petreanu, L. T., Lansford, R., Alvarez-Buylla, A., Lledo, P. M. (2003) Becoming a new neuron in the adult olfactory bulb. *Nat Neurosci*, **6**:507–518.

Carlsson, P., Mahlapuu, M. (2002) Forkhead transcription factors: key players in development and metabolism. *Dev Biol*, **250**:1–23.

Carmignoto, G., Vicini, S. (1992) Activity-dependent decrease in NMDA receptor responses during development of the visual cortex. *Science*, **258**:1007–1011.

Caro, S. P., Lambrechts, M. M., Balthazart, J. (2005) Early seasonal development of brain song control nucleu in male blue tits. *Neurosci Lett*, **386**:139–144.

Carr, C. E. (1992) Evolution of the central auditory system in reptiles and birds. In *The Evolutionary Biology of Hearing* (Webster, D. B., Fay, R. R., Popper, A. N., eds.), pp. 511–544. New York: Springer-Verlag.

Carrick, R. (1972) Population ecology of the Australian black-backed magpie, royal penguin and silver gull. In *Population Ecology of Migratory Birds: A Symposium*. Wildlife Research Report, Vol. 2, pp. 41–99. Washington, DC: U.S. Department of the Interior.

Carrillo, G. D., Doupe, A. J. (2004) Is the songbird Area X striatal, pallidal, or both? An anatomical study. *J Comp Neurol*, **473**:415–437.

Carruth, L., Reisert, I., Arnold, A. (2002) Sex chromosome genes directly affect brain sexual differentiation. *Nat Neurosci*, **5**:933–934.

Casey, R. M., Gaunt, A. S. (1985) Theoretical models of the avian syrinx. *J Theor Biol*, **116**:45–64.

Castelino, C. B., Ball, G. F. (2005) A role for norepinephrine in the regulation of context-dependent ZENK expression in male zebra finches (*Taeniopygia guttata*). *Eur J Neurosci*, **21**:1962–1972.

Castellani, G. C., Quinlan, E. M., Cooper, L. N., Shouval, H. Z. (2001) A biophysical model of bidirectional synaptic plasticity: dependence on AMPA and NMDA receptors. *Proc Natl Acad Sci USA*, **98**:12772–12777.

Catchpole, C. K. (1982) The evolution of bird sounds in relation to mating and spacing behavior. In *Acoustic Communication in Birds* (Kroodsma, D., Miller, E. H., eds.), pp. 297–319. New York: Academic Press.

Catchpole, C. K. (1983) Variation in the song of the great reed warbler *Acrocephalus arundinaceus* in relation to mate attraction and territorial defence. *Anim Behav*, **31**:1217–1225.

Catchpole, C. K. (1996) Song and female choice: good genes and big brains? *Trends Ecol Evol*, **11**:358–360.

Catchpole, C. K., Slater, P. J. B. (1995) *Bird Song: Biological Themes and Variations*. Cambridge, UK: Cambridge University Press.

Cecchi, G. A., Ribeiro, S., Mello, C. V., Magnasco, M. O. (1999) An automated system for the mapping and quantitative analysis of immunocytochemistry of an inducible nuclear protein. *J Neurosci Methods*, **87**:147–158.

Cepko, C. L., Ryder, E., Austin, C., Golden, J., Fields-Berry, S., Lin, J. (1998) Lineage analysis using retroviral vectors. *Methods*, **14**:393–406.

Chaiken, M., Bohner, J., Marler, P. (1993) Song acquisition in European starlings, *Sturnus vulgaris*: a comparison of the songs of live-tutored, tape-tutored, untutored, and wild-caught males. *Anim Behav*, **46**:1079–1090.

Chaiken, M., Bohner, J., Marler, P. (1994) Repertoire turnover and the timing of song acquisition in European starlings. *Behaviour*, **128**:25–39.

Chaudhuri, A. (1997) Neural activity mapping with inducible transcription factors. *NeuroReport*, **8**:iii–vii.

Chen, X., Agate, R, J., Itoh, Y., Arnold, A, P. (2005) Sexually dimorphic expression of trkB, a Z-linked gene, in early posthatch zebra finch brain. *Proc Natl Acad Sci USA*, **102**:7730–7735.

Cheney, D. L., Seyfarth, R. M. (1990) *How Monkeys See the World: Inside the Mind of Another Species*. Chicago, IL: University of Chicago Press.

Cheng, H. Y., Clayton, D. F. (2004) Activation and habituation of extracellular signal-regulated kinase phosphorylation in zebra finch auditory forebrain during song presentation. *J Neurosci*, **24**:7503–7513.

Cheng, M.-F. (1993) Vocal, auditory, and endocrine systems: three-way connectivities and implications. *Poult Sci Rev*, **5**:37–47.

Cheng, M.-F., Durand, S. (2004) A new function for the bird's own song. In *Behavioral Neurobiology of Birdsong* (Ziegler, H., Marler, P., eds.), pp. 611–627. New York: Ann NY Acad Sci, **1016**, Special Issue.

Chew, S. J., Mello, C., Nottebohm, F., Jarvis, E., Vicario, D. S. (1995) Decrements in auditory responses to a repeated conspecific song are long-lasting and require two periods of protein synthesis in the songbird forebrain. *Proc Natl Acad Sci USA*, **92**:3406–3410.

Chew, S. J., Vicario, D. S., Nottebohm, F. (1996a) A large-capacity memory system that recognizes the calls and songs of individual birds. *Proc Natl Acad Sci USA*, **93**:1950–1955.

Chew, S. J., Vicario, D. S., Nottebohm, F. (1996b) Quantal duration of auditory memories. *Science*, **274**:1909–1914.

Chi, Z., Margoliash, D. (2001) Temporal precision and temporal drift in brain and behavior of zebra finch song. *Neuron*, **32**:899–910.

Chomsky, N. A. (1957) *Syntactic Structures*. New York: Mouton.

Chomsky, N. (1980) *Rules and Representations*. New York: Columbia University Press.

Christy, B. A., Lau, L. F., Nathans, D. (1988) A gene activated in mouse 3T3 cells by serum growth factors encodes a protein with "zinc finger" sequences. *Proc Natl Acad Sci USA*, **85**:7857–7861.

Cirelli, C., Pompeiano, M., Tononi, G. (1996) Neuronal gene expression in the waking state: a role for the locus coeruleus. *Science*, **274**:1211–1215.

Clark, C. W., Marler, P., Beeman, K. (1987) Quantitative analysis of animal vocal phonology. *Ethology*, **76**:101–115.

Clark, J. T., Micevych, P. E., Panossian, V., Keaton, A. K. (1995) Testosterone-induced copulatory behavior is affected by the postcastration interval. *Neurosci Biobehav Rev*, **19**:369–376.

Clayton, D. F. (1997) Role of gene regulation in song circuit development and song learning. *J Neurobiol*, **33**:549–571.

Clayton, D. F., George, J. M. (1998) The synucleins: a family of proteins involved in synaptic function, plasticity, neurodegeneration and disease. *Trends Neurosci*, **21**:249–254.

Clayton, D. F. (2000) The genomic action potential. *Neurobiol Learn Mem*, **74**:185–216.

Clayton, D. F. (2004) Songbird genomics: methods, mechanisms, opportunities, and pitfalls. In *Behavioral Neurobiology of Birdsong* (Ziegler, H., Marler, P., eds.), pp. 45–60. New York: Ann NY Acad Sci, **1061**. Special Issue.

Clayton, D. F. (2006) Molecular neurobiology of birdsong. In *Behavioral Neurochemistry and Neuroendocrinology*. 3rd edn. (Blaustein, J. D., ed.). New York: Kluwer.

Clayton, D. F., Huecas, M. E., Sinclair-Thompson, E. Y., Nastiuk, K. L., Nottebohm, F. (1988) Probes for rare mRNAs reveal distributed cell subsets in canary brain. *Neuron*, **1**:249–261.

Clayton, N. S. (1987a) Song tutor choice in zebra finches. *Anim Behav*, **35**:714–722.

Clayton, N. S. (1987b) Song learning in Bengalese finches: a comparison with zebra finches. *Ethology*, **76**:247–255.

Clayton, N. S. (1988) Song learning and mate choice in estrildid finches raised by two species. *Anim Behav*, **36**:1589–1600.

Clayton, N. S. (1989) The effects of cross-fostering on selective song learning in estrildid finches. *Behaviour*, **109**:163–175.

Clower, R. P., Nixdorf, B. E., DeVoogd, T. J. (1989) Synaptic plasticity in the hypoglossal nucleus of female canaries: structural correlates of season, hemisphere, and testosterone treatment. *Behav Neural Biol*, **52**:63–77.

Clucas, B. A., Freeberg, T. M., Lucas, J. R. (2004) Chick-a-dee call syntax, social context, and season affect vocal responses of Carolina chickadees (*Poecile carolinensis*). *Behav Ecol Sociobiol*, **57**:187–196.

Cockburn, A. (1996) Why do so many Australian birds cooperate: social evolution in the Corvidae. In *Frontiers in Population Ecology* (Floyd, A., de Barro, P. J., eds.), pp. 451–472. Melbourne: CSIRO.

Cohen, D., Karten, H. (1974) The structural organization of the avian brain: an overview. In *Birds, Brain and Behavior* (Goodman, I., Schein, M., eds.), pp. 29–73. New York: Academic Press.

Colbran, R. J., Brown, A. M. (2004) Calcium/calmodulin-dependent protein kinase II and synaptic plasticity. *Curr Opin Neurobiol*, **14**:318–327.

Coleman, M. J., Mooney, R. (2004) Synaptic transformations underlying highly selective auditory representations of learned birdsong. *J Neurosci*, **24**:7251–7265.

Coleman, M. J., Vu, E. T. (2000) Neural activity in HVc of adult zebra finches during the song recovery following unilateral lesions of nucleus uvaeformis. *Soc Neurosci Abstract*, **26**:2031.

Coleman, M. J., Vu, E. T. (2005) Recovery of impaired songs following unilateral but not bilateral lesions of nucleus uvaeformis of adult zebra finches. *J Neurobiol*, **63**:70–89.

Coleman, M., Roy, A., Wild, J. M., Mooney, R. (2007) Thalamic gating of auditory responses in the telencephalic song control nuclei. *J Neurosci*, **27**:10024–10036.

Collins, P. W. (1988) Synergistic interactions of gentamicin and pure tones causing cochlear hair cell loss in pigmented guinea pigs. *Hear Res*, **36**:249–259.

Collins, S. (2004) Vocal fighting and flirting: the functions of birdsong. In *Nature's Music: The Science of Birdsong* (Marler, P., Slabbedorn, H., eds.), pp. 39–79. San Diego, CA: Academic Press.

Colombo, M., D'Amato, M. R., Rodman, H. R., Gross, C. G. (1990) Auditory association cortex lesions impair auditory short-term memory in monkeys. *Science*, **247**:336–338.

Colombo, M., Rodman, H. R., Gross, C. G. (1996) The effects of superior temporal cortex lesions on the processing and retention of auditory information in monkeys (*Cebus apella*). *J Neurosci*, **16**:4501–4517.

Cooke, B., Hegstrom, C. D., Villeneuve, L. S., Breedlove, S. M. (1998) Sexual differentiation of the vertebrate brain: principles and mechanisms. *Front Neuroendocrinol*, **19**:323–362.

Cooper, B. G., Goller, F. (2004) Multimodal signals: enhancement and constraint of song motor patterns by visual display. *Science*, **303**:544–546.

Cooper, B. G., Goller, F. (2006) Physiological insights into the social-context-dependent changes in the rhythm of the song motor program. *J Neurophysiol*, **95**:3798–3809.

Cooper, J. R., Bloom, F. E., Roth, R. H. (2003) *The Biochemical Basis of Neuropharmacology*. 8th edn. New York: Oxford University Press.

Corwin, J. T., Cotanche, D. A. (1988) Regeneration of sensory hair cells after acoustic trauma. *Science*, **240**:1772–1774.

Cotanche, D. A., Saunders, J. C., Tilney, L. G. (1987) Hair cell damage produced by acoustic trauma in the chick cochlea. *Hear Res*, **25**:267–286.

Cotanche, D. A., Lee, K. H., Stone, J. S., Picard, D. A. (1994) Hair cell regeneration in the bird cochlea following noise damage or ototoxic drug damage. *Anat Embryol (Berl)*, 189:1–18.

Cotman, C. W., Monaghan, D. T., Ganong, A. H. (1988) Excitatory amino acid neurotransmission: NMDA receptors and Hebb-type synaptic plasticity. *Annu Rev Neurosci*, 11:61–80.

Cousillas, H., Leppelsack, H. J., Leppelsack, E., Richard, J. P., Mathelier, M., Hausberger, M. (2005) Functional organization of the forebrain auditory centres of the European starling: a study based on natural sounds. *Hear Res*, 207:10–21.

Cowan, N. (2001) The magical number 4 in short-term memory: a reconsideration of mental storage capacity. *Behav Brain Sci*, 24:87–114; discussion 114–185.

Cowie, R., Douglas-Cowie, E. (1992) *Postlingually Acquired Deafness: Speech Deterioration and the Wider Consequences*. Berlin: Mouton de Gruyter.

Crabbe, J. (1996) Quantitative trait locus gene mapping: a new method for locating alcohol response genes. *Addict Biol*, 1:229–235.

Craigie, E. H. (1932) The cell structure of the cerebral hemisphere of the humming bird. *J Comp Neurol*, 56:135–168.

Crair, M. C., Malenka, R. C. (1995) A critical period for long-term potentiation at thalamocortical synapses. *Nature*, 375:325–328.

Crow, T., Xue-Bian, J. J., Siddiqi, V., Kang, Y., Neary, J. T. (1998) Phosphorylation of mitogen-activated protein kinase by one-trial and multi-trial classical conditioning. *J Neurosci*, 18:3480–3487.

Crowder, R. G. (1976) *Principles of Learning and Memory*. Hillsdale, NJ: Erlbaum.

Cruz, R. M., Lambert, P. R., Rubel, E. W. (1987) Light microscopic evidence of hair cell regeneration after gentamicin toxicity in chick cochlea. *Arch Otolaryngol Head Neck Surg*, 113:1058–1062.

Cull-Candy, S., Brickley, S., Farrant, M. (2001) NMDA receptor subunits: diversity, development and disease. *Curr Opin Neurobiol*, 11:327–335.

Cynx, J. (1990) Experimental determination of a unit of song production in the zebra finch (*Taeniopygia guttata*). *J Comp Psychol*, 104:3–10.

Cynx, J., Gell, C. (2004) Social mediation of vocal amplitude in a songbird, *Taeniopygia guttata*. *Anim Behav*, 67:451–455.

Cynx, J., Nottebohm, F. (1992) Role of gender, season, and familiarity in discrimination of conspecific song by zebra finches (*Taeniopygia guttata*). *Proc Natl Acad Sci USA*, 89:1368–1371.

Cynx, J., von Rad, U. (2001) Immediate and transitory effects of delayed auditory feedback on bird song production. *Anim Behav*, 62:305–312.

Cynx, J., Williams, H., Nottebohm, F. (1990) Timbre discrimination in zebra finch (*Taeniopygia guttata*) song syllables. *J Comp Physchol*, 104:303–308.

Cynx, J., Bean, N., Rossman, I. (2004) Testosterone implants alter the frequency range of zebra finch songs. *Hormones Behav*, 47.

Dabelsteen, T., Larsen, O. N., Pedersen, S. B. (1993) Habitat induced degradation of sound signals: quantifying the effects of communication sounds and bird location on blur ratio, excess attenuation, and signal-to-noise ratio in blackbird song. *J Acoust Soc Am*, 93:2206–2220.

Daley, M., Goller, F. (2004) Tracheal length changes during zebra finch song and their possible role in upper vocal tract filtering. *J Neurobiol*, 59:319–330.

Dantzker, M. S., Grant, D. B., Bradbury, J. W. (1999) Directional acoustic radiation in the strut display of male sage grouse *Centrocercus urophasianus*. *J Exp Biol*, 202:2893–2909.

Dantzker, M. S., Bradbury, J. W. (2006) Vocal sacs and their role in avian acoustic display. *Acta Zool Sinica*, 52 (Supplement): 486–488.

Das, R. M., Van, H, N. J., Howell, G. R., et al. (2006a) A robust system for RNA interference in the chicken using a modified microRNA operon. *Dev Biol*, 294:554–563.

Das, C., Hoang, Q. Q., Kreinbring, C. A., et al. (2006b) Structural basis for conformational plasticity of the Parkinson's disease-associated ubiquitin hydrolase UCH-L1. *Proc Natl Acad Sci USA*, 103:4675–4680.

Dave, A. S., Margoliash, D. (2000) Song replay during sleep and computational rules for sensorimotor vocal learning. *Science*, 290:812–816.

Dave, A. S., Yu, A. C., Margoliash, D. (1998) Behavioral state modulation of auditory activity in a vocal motor system. *Science*, 282:2250–2254.

Davidson, W. R., Langmore, N. E. (1991) Variation in the male whip-crack of the eastern whipbird *Psophodes olivaceus*. *Austral Bird Watcher*, 14:82–84.

Davies, D. C., Csillag, A., Székely, A. D., Kabai, P. (1997) Efferent connections of the domestic chick archistriatum: a Phaseolus lectin anterograde tracing study. *J Comp Neurol*, 389:679–693.

Davis, H. P., Squire, L. R. (1984) Protein synthesis and memory: a review. *Psychol Bull*, **96**:518–559.

Davis, S., Butcher, S. P., Morris, R. G. (1992) The NMDA receptor antagonist D-2-amino-5-phosphonopentanoate (D-AP5) impairs spatial learning and LTP in vivo at intracerebral concentrations comparable to those that block LTP in vitro. *J Neurosci*, **12**:21–34.

Daw, N. W., Stein, P. S., Fox, K. (1993) The role of NMDA receptors in information processing. *Annu Rev Neurosci*, **16**:207–222.

Dawson, A., King, V. M., Bentley, G. E., Ball, G. F. (2001) Photoperiodic control of seasonality in birds. *J Biol Rhythms*, **16**:365–380.

Dawson, D. A., Burke, T., Hansson, B., et al.(2006) A predicted microsatellite map of the passerine genome based on chicken-passerine sequence similarity. *Mol Ecol*, **15**:1299–1320.

de Boysson-Bardies, B. (1993) Ontogeny of language-specific syllabic productions. In *Developmental Neurocognition: Speech and Face Processing in the First Year of Life* (de Boysson-Bardies, B., de Schonen, S., Jusczyk, P. W., McNeilage, P., Morton, J., eds.), pp. 353–363. Dordrecht/New York: Kluwer Academic/Plenum Publishers.

De Vries, G., Rissman, E., Simerly, R., et al. (2002) A model system for study of sex chromosome effects on sexually dimorphic neural and behavioral traits. *J Neurosci*, **22**:9005–9014.

DeCasper, A. J., Fifer, W. P. (1980) Of human bonding: newborns prefer their mothers' voices. *Science*, **208**:1174–1176.

DeCasper, A. J., Spence, M. J. (1986) Prenatal maternal speech influences newborns' perception of speech sounds. *Infant Behav Devel*, **9**:133–150.

Dehaene-Lambertz, G., Dehaene, S., Hertz-Pannier, L. (2002) Functional neuroimaging of speech perception in infants. *Science*, **298**:2013–2015.

Del Negro, C., Gahr, M., Leboucher, G., Kreutzer, M. (1998) The selectivity of sexual responses to song displays: effects of partial chemical lesion of the HVC in female canaries. *Behav Brain Res*, **96**:151–159.

Del Negro, C., Kreutzer, M., Gahr, M. (2000) Sexually stimulating signals of canary (*Serinus canaria*) songs: evidence for a female-specific auditory representation in the HVc nucleus during the breeding season. *Behav Neurosci*, **114**:526–542.

Del Negro, C., Lehongre, K., Edeline, J. M. (2005) Selectivity of canary HVC neurons for the bird's own song: modulation by photoperiodic conditions. *J Neurosci*, **25**:4952–4963.

den Boer, P. J., Bout, R. G., Dubbeldam, L. J. (1986) Topographical representation of the jaw muscles within the trigeminal motor nucleus: an HRP study in the mallard, *Anas platyrhynchos*. *Acta Morphol Neerl Scand*, **24**:1–17.

Deng, C., Kaplan, G., Rogers, L. J. (2001) Similarity of the song nuclei of male and female Australian magpies (*Gymnorhina tibicen*). *Behav Brain Res*, **123**:89–102.

Denisenko-Nehrbass, N. I., Jarvis, E., Scharff, C., Nottebohm, F., Mello, C. V. (2000) Site-specific retinoic acid production in the brain of adult songbirds. *Neuron*, **27**:359–370.

Dennis, M., Whitaker, H. A. (1976) Language acquisition following hemidecortication: linguistic superiority of the left over the right hemisphere. *Brain Lang*, **3**:404–433.

Deregnaucourt, S., Mitra, P. P., Feher, O., Maul, K. K., Lints, T. J., Tchernichovski, O. (2004) Song development in search of the error-signal. *Ann NY Acad Sci*, **1016**:364–376.

Deregnaucourt, S., Mitra, P. P., Fehér, O., Pytte, C., Tchernichovski, O. (2005) How sleep affects the developmental learning of bird song. *Nature*, **433**:710–716.

Derrickson, K. C. (1987) Yearly and situational changes in the estimate of repertoire size in northern mockingbirds. *Auk*, **104**:198–207.

Deviche, P., Gulledge, C. C. (2000) Vocal control region sizes of an adult female songbird change seasonally in the absence of detectable circulating testosterone concentrations. *J Neurobiol*, **42**:202–211.

DeVoogd, T. J. (1991) Endocrine modulation of the development and adult function of the avian song system. *Psychoneuroendocrinology*, **16**:41–66.

DeVoogd, T. J. (1994) The neural basis for the acquisition and production of bird song. In *Causal Mechanisms of Behavioural Development* (Hogan, J. A., Bolhuis, J. J., eds.), pp. 49–81. New York, NY: Cambridge University Press.

DeVoogd, T. J., Nottebohm, F. (1981a) Sex differences in dendritic morphology of a song control nucleus in the canary: a quantitative Golgi study. *J Comp Neurol*, **196**:309–316.

DeVoogd, T. J., Nottebohm, F. (1981b) Gonadal hormones induce dendritic growth in the adult avian brain. *Science*, **214**:202–204.

DeVoogd, T. J., Nixdorf, B., Nottebohm, F. (1985) Synaptogenesis and changes in synaptic morphology related to acquisition of a new behavior. *Brain Res*, **329**:304–308.

DeVoogd, T. J., Krebs, J. R., Healy, S. D., Purvis, A. (1993) Relations between song repertoire size and the volume of brain nuclei related to song: comparative evolutionary analyses amongst oscine birds. *Proc Biol Sci*, 254:75–82.

DeVoogd, T. J., Houtman, A. M., Falls, J. B. (1995) White-throated sparrow morphs that differ in song production rate also differ in the anatomy of some song-related brain areas. *J Neurobiol*, 28:202–213.

Dewing, P., Chiang, C., Sinchak, K., et al. (2006) Direct regulation of adult brain function by the male-specific factor SRY. *Curr Biol*, 16:415–420.

DeWulf, V., Bottjer, S. W. (2002) Age and sex differences in mitotic activity within the zebra finch telencephalon. *J Neurosci*, 22:4080–4094.

Dietrich, K. (1980) Model choice in the song development of young male Bengalese finches. *Z Tierpsychol*, 52:57–76.

Dilger, W. C. (1962) The behavior of lovebirds. *Sci Amer*, 206:88–98.

Ding, L., Perkel, D. J. (2002) Dopamine modulates excitability of spiny neurons in the avian basal ganglia. *J Neurosci*, 22:5210–5218.

Ding, L., Perkel, D. J. (2004) Long-term potentiation in an avian basal ganglia nucleus essential for vocal learning. *J Neurosci*, 24:488–494.

Ding, L., Perkel, D. J., Farries, M. A. (2003) Presynaptic depression of glutamatergic synaptic transmission by D1-like dopamine receptor activation in the avian basal ganglia. *J Neurosci*, 23:6086–6095.

Dittrich, F., Feng, Y., Metzdorf, R., Gahr, M. (1999) Estrogen-inducible, sex-specific expression of brain-derived neurotrophic factor mRNA in a forebrain song control nucleus of the juvenile zebra finch. *Proc Natl Acad Sci USA*, 96:8241–8246.

Dittus, W. P. J., Lemon, R. E. (1970) Auditory feedback in the singing of cardinals. *Ibis*, 112:544–548.

Dobzhansky, T., Epling, C. (1948) The suppression of crossing over in inversion heterozygotes of *Drosophila pseudoobscura*. *Proc Nat Acad Sci USA*, 34:137–141.

Doetsch, F., Caille, I., Lim, D. A., Garcia-Verdugo, J. M., Alvarez-Buylla, A. (1999) Subventricular zone astrocytes are neural stem cells in the adult mammalian brain. *Cell*, 97:703–716.

Doetsch, F., Scharff, C. (2001) Challenges for brain repair: insights from adult neurogenesis in birds and mammals. *Brain Behav Evol*, 58:306–322.

Dominey, P. F., Hoen, M., Blanc, J. M., Lelekov-Boissard, T. (2003) Neurological basis of language and sequential cognition: evidence from simulation, aphasia, and ERP studies. *Brain Lang*, 86:207–225.

Donham, R. S., Wingfield, J. C., Mattocks, P. W., Jr., Farner, D. S. (1982) Changes in testicular and plasma androgens with photoperiodically induced increase in plasma LH in the house sparrow. *Gen Comp Endocrinol*, 48:342–347.

Dooling, R. J. (1980) Behavior and psychophysics of hearing in birds. In *Comparative Studies of Learning in Vertebrates* (Popper, A. N., Fay, R. R., eds.), pp. 261–288. New York: Springer-Verlag.

Dooling, R. J. (1992) Perception of speech sounds by birds. *Adv Biosci*, 83:407–413.

Dooling, R. J., Brown, S. D. (1990) Speech perception by budgerigars (*Melopsittacus undulatus*): spoken vowels. *Percept Psychophys*, 47:568–574.

Dooling, R., Searcy, M. (1980) Early perceptual selectivity in the swamp sparrow. *Dev Psychobiol*, 13:499–506.

Doupe, A. J. (1997) Song- and order-selective neurons in the songbird anterior forebrain and their emergence during vocal development. *J Neurosci*, 17:1147–1167.

Doupe, A. J., Konishi, M. (1991) Song-selective auditory circuits in the vocal control system of the zebra finch. *Proc Natl Acad Sci USA*, 88:11339–11343.

Doupe, A. J., Kuhl, P. K. (1999) Birdsong and human speech: common themes and mechanisms. *Annu Rev Neurosci*, 22:567–631.

Doupe, A. J., Solis, M. M. (1997) Song- and order-selective neurons develop in the songbird anterior forebrain during vocal learning. *J Neurobiol*, 33:694–709.

Doupe, A. J., Solis, M. M., Kimpo, R. (2004) Cellular, circuit, and synaptic mechanisms in song learning. *Ann NY Acad Sci*, 1016:495–523.

Doupe, A. J., Perkel, D. J., Reiner, A., Stern, E. A. (2005) Birdbrains could teach basal ganglia research a new song. *Trends Neurosci*, 28:353–363.

Doya, K., Sejnowski, T. (2000) A computational model of avian song learning. In *The New Cognitive Neurosciences* (Gazzaniga, M. S., ed.), pp. 469–482. Cambridge, MA: MIT Press.

Doyon, J., Benali, H. (2005) Reorganization and plasticity in the adult brain during learning of motor skills. *Curr Opin Neurobiol*, 15:161–167.

Dubbeldam, J. L., Bout, R. G. (1990) The identification of the motor nuclei innervating the tongue muscles in the mallard (*Anas platyrhynchos*); an HRP study. *Neurosci Lett*, 119:223–227.

Dubbeldam, J. L., den Boer-Visser, A. M., Bout, R. G. (1997) Organization and efferent connections of the archistriatum of the mallard, *Anas platyrhynchos* L.: an anterograde and retrograde tracing study. *J Comp Neurol*, **388**:632–657.

Duchowny, M., Jayakar, P., Harvey, A. S., Resnick, T., Alvarez, L., Dean, P., Levin, B. (1996) Language cortex representation: effects of developmental versus acquired pathology. *Ann Neurol*, **40**:31–38.

Duffy, D. L., Bentley, G. E., Ball, G. F. (1999) Does sex or photoperiodic condition influence ZENK induction in response to song in European starlings? *Brain Res*, **844**:78–82.

Dugatkin, L. A. (1997) *Cooperation Among Animals: An Evolutionary Perspective*. New York: Oxford University Press.

Dumas, T. C. (2005) Developmental regulation of cognitive abilities: modified composition of a molecular switch turns on associative learning. *Prog Neurobiol*, **76**:189–211.

Dunham, M. L., Warner, R. R., Lawson, J. W. (1995) The dynamics of territory acquisition: a model of two coexisting strategies. *Theor Popul Biol*, **47**:347–364.

Dunn, P. O., Cockburn, A. (1999) Extrapair mate choice and honest signalling in cooperatively breeding superb fairy-wrens. *Evol*, **53**:938–946.

Durand, S. E., Heaton, J. T., Amateau, S. K., Brauth, S. E. (1997) Vocal control pathways through the anterior forebrain of a parrot (*Melopsittacus undulatus*). *J Comp Neurol*, **377**:179–206.

Durant, W. (1926) *The Story of Philosophy*. New York: Simon & Schuster.

Dutar, P., Vu, H. M., Perkel, D. J. (1998) Multiple cell types distinguished by physiological, pharmacological, and anatomic properties in nucleus HVc of the adult zebra finch. *J Neurophysiol*, **80**:1828–1838.

Dutar, P., Vu, H. M., Perkel, D. J. (1999) Pharmacological characterization of an unusual mGluR-evoked neuronal hyperpolarization mediated by activation of GIRK channels. *Neuropharmacology*, **38**:467–475.

Dutar, P., Petrozzino, J. J., Vu, H. M., Schmidt, M. F., Perkel, D. J. (2000) Slow synaptic inhibition mediated by metabotropic glutamate receptor activation of GIRK channels. *J Neurophysiol*, **84**:2284–2290.

Dykxhoorn, D. M., Lieberman, J. (2006) The silent revolution: RNA interference as basic biology, research tool, and therapeutic. *Annu Rev Med*, **56**:401–423. Review.

Dykxhoorn, D. M., Novina, C. D., Sharp, P. A. (2003) Killing the messenger: short RNAs that silence gene expression. *Nat Rev Mol Cell Biol*, **4**:457–467.

Eales, L. A. (1985) Song learning in zebra finches: some effects of song model availability on what is learnt and when. *Anim Behav*, **33**:1293–1300.

Eales, L. A. (1987) Song learning in female-raised zebra finches: another look at the sensitive phase. *Anim Behav*, **35**:1356–1365.

Eales, L. A. (1989) The influences of visual and vocal interaction on song learning in zebra finches. *Anim Behav*, **37**:507–508.

Eda-Fujiwara, H., Satoh, R., Bolhuis, J. J., Kimura, T. (2003) Neuronal activation in female budgerigars is localized and related to male song complexity. *Eur J Neurosci*, **17**:149–154.

Edelman, G. M. (1993) Neural Darwinism: selection and reentrant signaling in higher brain function. *Neuron*, **10**:115–125.

Edinger, L. (1885) *The Anatomy of the Central Nervous System of Man and of Vertebrates in General*. 1896 5th German edn.; 1899 English published edn. Philadelphia: F. A. Davis Company.

Edinger, L. (1908) The relations of comparative anatomy to comparative psychology. *J Comp Neurol Psychol*, **8**:437–458.

Edinger, L., Wallenberg, A., Holmes, G. M. (1903) Untersuchungen über die vergleichende Anatomie des Gehirns. Das Vorderhirn der Vögel. *Abh Senckenb nat Gesellsch*, **20**:343–426.

Edwards, C. J., Alder, T. B., Rose, G. J. (2002) Auditory midbrain neurons that count. *Nat Neurosci*, **5**:934–936.

Eens, M. (1997) Understanding the complex song of the European starling: an integrated ethological approach. *Adv Study Anim Behav*, **26**:355–434.

Eens, M., Pinxten, M., Verheyen, R. F. (1989) Temporal and sequential organization of song bouts in the European starling. *Ardea*, **77**:75–86.

Eens, M., Pinxten, R., Verheyen, R. F. (1991a) Organization of song in the European starling: species-specificity and individual differences. *Belgian J Zool*, **121**:257–278.

Eens, M., Pinxten, R., Verheyen, R. F. (1991b) Male song as a cue for mate choice in the European starling. *Behaviour*, **116**:210–238.

Eens, M., Pinxten, R., Verheyen, F. R. (1993) Function of the song and song repertoire in the European starling (*Sturnus vulgaris*): an aviary experiment. *Behaviour*, **125**:51–66.

Ehrlich, P., Dobkin, D., Wheye, D. (1988) *The Birders Handbook: A Field Guide to the Natural History of North American Birds*. New York: Simon and Schuster Inc.

Eichenbaum, H., Cohen, N. J. (2001) *From Conditioning to Conscious Recollection: Memory Systems of the Brain.* Oxford, UK: Oxford University Press.

Eimas, P. D., Siqueland, E. R., Jusczyk, P., Vigorito, J. (1971) Speech perception in infants. *Science,* **171**:303–306.

Eimas, P. D. (1975a) Auditory and phonetic coding of the cues for speech: discrimination of the (r-l) distinction by young infants. *Percept Psychophys,* **18**:341–347.

Eimas, P. D. (1975b) Speech perception in early infancy. In *Infant Perception* (Cohen, L. B., Salapatek, P., eds.), pp. 193–321. New York: Academic Press.

Elbrecht, A., Smith, R. G. (1992) Aromatase enzyme activity and sex determination in chickens. *Science,* **255**:467–470.

Elemans, C. P. H., Larsen, O. N., Hoffmann, M. R., Van Leeuwen, J. L. (2003) Quantitative modelling of the biomechanics of the avian syrinx. *Anim Biol,* **53**:183–193.

Elemans, C. P. H., Spierts, I. L. Y., Muller, U. K., van Leeuwen, J. L., Goller, F. (2004) Superfast muscles control dove's trill. *Nature,* **431**:146.

Elemans, C. P. H., Spierts, I. L. Y., Hendriks, M., Schipper, H., Muller, U. K., van Leeuwen, J. L. (2006) Syringeal muscles fit the trill in ring doves (*Streptopelia risoria* L.). *Journal of Experimental Biology,* **209**:965–977.

Elman, J. L., Bates, E. A., Johnson, M. H., Karmiloff-Smith, A. (1996) *Rethinking Innateness: A Connectionist Perspective on Development.* Cambridge, MA: MIT Press.

English, J. D., Sweatt, J. D. (1996) Activation of p42 mitogen-activated protein kinase in hippocampal long term potentiation. *J Biol Chem,* **271**:24329–22432.

English, J. D., Sweatt, J. D. (1997) A requirement for the mitogen-activated protein kinase cascade in hippocampal long term potentiation. *J Biol Chem,* **272**:19103–11916.

Esser, K. H. (1994) Audio-vocal learning in a non-human mammal: the lesser spear-nosed bat *Phyllostomus discolor. Neuroreport,* **5**:1718–1720.

Evans, C. S., Evans, L., Marler, P. (1993) On the meaning of alarm calls: functional reference in an avian vocal system. *Anim Behav,* **46**:23–38.

Evans, M. R., Roberts, M. L., Buchanan, K. L., Goldsmith, A. R. (2006) Heritability of corticosterone response and changes in life history traits during selection in the zebra finch. *J Evol Biol,* **19**:343–352.

Fadiga, L., Craighero, L., Buccino, G., Rizzolatti, G. (2002) Speech listening specifically modulates the excitability of tongue muscles: a TMS study. *Eur J Neurosci,* **15**:399–402.

Falls, J. B. (1982) Individual recognition by sound in birds. In *Acoustic Communication in Birds* (Kroodsma, D. E., Miller, E. H., eds.), pp. 237–278. New York: Academic Press.

Falls, J. B. (1985) Song matching in western meadowlarks. *Canadian J Zool,* **63**:2520–2524.

Falls, J. B., Brooks, R. J. (1975) Individual recognition by song in white-throated sparrows. II: Effects of location. *Can J Zool,* **53**:1412–1420.

Falls, J. B., Kopachena, J. G. (1994) White-throated sparrow. In *The Birds of North America* (Poole, A., Gill, F., eds.). Philadelphia, PA: The Academy of Natural Sciences/The American Ornithologists' Union.

Fant, G. (1960) *Acoustic Theory of Speech Production.* The Hague: Mouton.

Farabaugh, S. M. (1982) The ecological and social significance of duetting. In *Acoustic Communication in Birds.* Vol. 2 (Kroodsma, D., Miller, E. H., eds.), pp. 85–124. New York: Academic Press.

Farabaugh, S. M., Brown, E. D., Hughes, J. M. (1992) Cooperative territorial defence in the Australian magpie, *Gymnorhina tibicen* (Passeriformes, Cracticidae), a group-living songbird. *Ethol,* **92**:283–292.

Faraci, F. M., Fedde, M. R. (1986) Regional circulatory responses to hypocapnia and hypercapnia in bar-headed geese. *Am J Physiol,* **250**:R499–R504.

Farner, D. S. (1964) The photoperiodic control of reproductive cycles in birds. *Amer Sci,* **52**:137–156.

Farries, M. A. (2001) The oscine song system considered in the context of the avian brain: lessons learned from comparative neurobiology. *Brain Behav Evol,* **58**:80–100.

Farries, M. A. (2004) The avian song system in comparative perspective. *Ann NY Acad Sci,* **1016**:61–76.

Farries, M. A., Perkel, D. J. (2002) A telencephalic nucleus essential for song learning contains neurons with physiological characteristics of both striatum and globus pallidus. *J Neurosci,* **22**:3776–3787.

Farries, M. A., Meitzen, J., Perkel, D. J. (2005a) Electrophysiological properties of neurons in the basal ganglia of the domestic chick: conservation and divergence in the evolution of the avian basal ganglia. *J Neurophysiol,* **94**:454–467.

Farries, M. A., Ding, L., Perkel, D. J. (2005b) Evidence for "direct" and "indirect" pathways through the song system basal ganglia. *J Comp Neurol,* **484**:93–104.

Fedde, M. R., R.N. Gatz, H. Slama and P. Scheid (1974) Intrapulmonary CO_2 receptors in the duck. II. Comparison with mechanoreceptors. *Resp Physiol,* **22**:115–121.

Fedde, M. R. (1987) Respiratory muscles. In *Bird Respiration* (Seller, T. J., ed.), pp. 3–37. Boca Raton: CRC Press.

Fee, M. S. (2002) Measurement of the linear and nonlinear mechanical properties of the oscine syrinx. *J Comp Physiol A*, **188**:829–839.

Fee, M. S., Leonardo, A. (2001) Miniature motorized microdrive and commutator system for chronic neural recording in small animals. *J Neurosci Methods*, **112**:83–94.

Fee, M. S., Shraiman, B., Pesaran, B., Mitra, P. P. (1998) The role of nonlinear dynamics of the syrinx in the vocalizations of a songbird. *Nature*, **395**:67–71.

Fee, M. S., Kozhevnikov, A. A., Hahnloser, R. H. (2004) Neural mechanisms of vocal sequence generation in the songbird. *Ann NY Acad Sci*, **1016**:153–170.

Feldman, J. L., Del Negro, C. A. (2006) Looking for inspiration: new perspectives on respiratory rhythm. *Nat Rev Neurosci*, **7**:232–242.

Ferland, R. J., Cherry, T. J., Preware, P. O., Morrisey, E. E., Walsh, C. A. (2003) Characterization of Foxp2 and Foxp1 mRNA and protein in the developing and mature brain. *J Comp Neurol*, **460**:266–279.

Fernald, A. (1985) Four-month-old infants prefer to listen to motherese. *Infant Behav Dev*, **8**:181–195.

Fernald, A., Kuhl, P. K. (1987) Acoustic determinants of infant preference for motherese speech. *Infant Behav Dev*, **10**:279–293.

Feuk, L., Kalervo, A., Lipsanen-Nyman, M., *et al.* (2006) Absence of a paternally inherited FOXP2 gene in developmental verbal dyspraxia. *Am J Hum Genet*, **79**:965–972.

Fields, R. D., Yu, C., Nelson, P. G. (1991) Calcium, network activity, and the role of NMDA channels in synaptic plasticity in vitro. *J Neurosci*, **11**:134–146.

Fiete, I. R., Hahnloser, R. H. R., Fee, M. S., Seung, H. S. (2004) Temporal sparseness of the premotor drive is important for rapid learning in a neural network model of birdsong. *J Neurophysiol*, **92**:2274–2282.

Fitch, W. T. (1999) Acoustic exaggeration of size in birds via tracheal elongation: comparative and theoretical analyses. *J Zool Lond*, **248**:31–48.

Fitch, W. T., Hauser, M. D. (2004) Computational constraints on syntactic processing in a nonhuman primate. *Science*, **303**:377–380.

Fitch, W. T., Kelley, J. P. (2000) Perception of vocal tract resonances by whooping cranes *Grus americana*. *Ethology*, **106**:559–574.

Fitch, W. T., Hauser, M. D., Chomsky, N. (2005) The evolution of the language faculty: clarifications and implications. *Cognition*, **97**:179–210.

Fitzpatrick, M. J., Ben-Shahar, Y., Smid, H. M., Vet, L. E., Robinson, G. E., Sokolowski, M. B. (2005) Candidate genes for behavioural ecology. *Trends Ecol Evol*, **20**:96–104.

Flege, J. E. (1991) Age of learning affects the authenticity of voice-onset time (VOT) in stop consonants produced in a second language. *J Acoust Soc Am*, **89**:395–411.

Fletcher, N. H. (1988) Bird song: a quantitative acoustic model. *J Theor Biol*, **135**:455–481.

Fletcher, N. H. (1992) *Acoustic Systems in Biology*. New York: Oxford University Press.

Fletcher, N. H. (2000) A class of chaotic bird calls? *J Acoust Soc Am*, **108**:821–826.

Fletcher, N. H., Tarnopolsky, A. (1999) Acoustics of the avian vocal tract. *J Acoust Soc Am*, **105**:35–49.

Fletcher, N. H., Riede, T., Beckers, G. J. L., Suthers, R. A. (2004) Vocal tract filtering and the 'coo' of doves. *J Acoust Soc Am*, **116**:3750–3756.

Fletcher, N. H., Riede, T., Suthers, R. A. (2006) Model for vocalization by a bird with distensible vocal cavity and open beak. *J Acoust Soc Am*, **I119**:1005–1011.

Flint, A. C., Maisch, U. S., Weishaupt, J. H., Kriegstein, A. R., Monyer, H. (1997) NR2 A subunit expression shortens NMDA receptor synaptic currents in developing neocortex. *J Neurosci*, **17**:2469–2476.

Floody, O. R. (1997) Song lateralization in the zebra finch. *Horm Behav*, **31**:25–34.

Floody, O. R., Arnold, A. P. (1997) Song lateralization in the zebra finch. *Horm Behav*, **31**:25–34.

Fodor, J. A. (1983) *Modularity of Mind: An Essay on Faculty Psychology*. Cambridge, MA: MIT Press.

Foidart, A., Reid, J., Absil, P., Yoshimura, N., Harada, N., Balthazart, J. (1995) Critical re-examination of the distribution of aromatase-immunoreactive cells in the quail forebrain using antibodies raised against human placental aromatase and against the recombinant quail, mouse or human enzyme. *J Chem Neuroanat*, **8**:267–282.

Foidart, A., Lakaye, B., Grisar, T., Ball, G. F., Balthazart, J. (1999) Estrogen receptor-beta in quail: cloning, tissue expression and neuroanatomical distribution. *J Neurobiol*, **40**:327–342.

Ford, H. A. (1989) *Ecology of Birds: An Australian Perspective*. Chipping Norton, New South Wales: Surrey Beatty & Sons.

Fortin, G., Foutz, A. S., Champagnat, J. (1994) Respiratory rhythm generation in chick hindbrain: effects of MK-801 and vagotomy. *Neuroreport*, **5**:1137–1140.

Fortune, E. S., Margoliash, D. (1992) Cytoarchitectonic organization and morphology of cells of the field L complex in male zebra finches (*Taenopygia guttata*). *J Comp Neurol*, **325**:388–404.

Fortune, E. S., Margoliash, D. (1995) Parallel pathways and convergence onto HVc and adjacent neostriatum of adult zebra finches (*Taeniopygia guttata*). *J Comp Neurol*, **360**:413–441.

Foster, E. F., Bottjer, S. W. (1998) Axonal connections of the high vocal center and surrounding cortical regions in juvenile and adult male zebra finches. *J Comp Neurol*, **397**:118–138.

Foster, E. F., Bottjer, S. W. (2001) Lesions of a telencephalic nucleus in male zebra finches: influences on vocal behavior in juveniles and adults. *J Neurobiol*, **46**:142–165.

Foster, E. F., Mehta, R. P., Bottjer, S. W. (1997) Axonal connections of the medial magnocellular nucleus of the anterior neostriatum in zebra finches. *J Comp Neurol*, **382**:364–381.

Fox, K., Daw, N. W. (1993) Do NMDA receptors have a critical function in visual cortical plasticity? *Trends Neurosci*, **16**:116–122.

Franz, M., Goller, F. (2002) Respiratory units of motor production and song imitation in the zebra finch. *J Neurobiol*, **51**:129–141.

Franz, M., Goller, F. (2003) Respiratory patterns and oxygen consumption in singing zebra finches. *J Exp Biol*, **206**:967–978.

Freeberg, T. M., Lucas, J. R. (2002) Receivers respond differently to chick-a-dee calls varying in note composition in Carolina chickadees, *Poecile carolinensis*. *Anim Behav*, **63**:837–845.

Friederici, A. D., Kotz, S. A. (2003) The brain basis of syntactic processes: functional imaging and lesion studies. *Neuroimage*, **20** Suppl 1:S8–17.

Friedlander, B. (1888) Über das Kriechen der Regenwürmer. *Biol Zbl*, **8**:362–366.

Fritz, J., Mishkin, M., Saunders, R. C. (2005) In search of an auditory engram. *Proc Natl Acad Sci USA*, **102**: 9359–9364.

Fromkin, V., S. K., S. C., D. R., M. R. (1974) The development of language in Genie: a case of language acquisition beyond the 'critical period'. *Brain Lang*, **1**:81–107.

Fry, C. H. (1972) The social organisation of bee-eaters (Meropidae) and co-operative breeding in hot climate birds. *Ibis*, **114**:114.

Funabiki, Y., Konishi, M. (2003) Long memory in song learning by zebra finches. *J Neurosci*, **23**:6928–6935.

Fusani, L., Hutchison, R. E., Hutchison, J. B. (1997) Vocal postural coordination of a sexually dimorphic display in a monomorphic species: the Barbary dove. *Behaviour*, **134**:321–335.

Fusani, L., Hutchison, J. B., Gahr, M. (2001) Testosterone regulates the activity and expression of aromatase in the canary neostriatum. *J Neurobiol*, **49**:1–8.

Fusani, L., Metzdorf, R., Hutchison, J. B., Gahr, M. (2003) Aromatase inhibition affects testosterone-induced masculinization of song and the neural song system in female canaries. *J Neurobiol*, **54**:370–379.

Gage, F. H. (2002) Neurogenesis in the adult brain. *J Neurosci*, **22**:612–613.

Gahr, M. (1990a) Localization of androgen receptors and estrogen receptors in the same cells of the songbird brain. *Proc Natl Acad Sci USA*, **87**:9445–9448.

Gahr, M. (1990b) Delineation of a brain nucleus: comparisons of cytochemical, hodological, and cytoarchitectural views of the song control nucleus HVC of the adult canary. *J Comp Neurol*, **294**:30–36.

Gahr, M. (2000) Neural song control system of hummingbirds: comparison to swifts, vocal learning (songbirds) and nonlearning (suboscines) passerines, and vocal learning (budgerigars) and nonlearning (dove, owl, gull, quail, chicken) nonpasserines. *J Comp Neurol*, **426**:182–196.

Gahr, M. (2001) Distribution of sex steroid hormone receptors in the avian brain: functional implications for neural sex differences and sexual behaviors. *Microsc Res Tech*, **55**:1–11.

Gahr, M., Güttinger, H-R (1985) Korrelation zwischen der sexualdimorphen Gehirndifferenzierung und der Verhaltensausprägung bei Prachtfinken (Estrildidae). *J Ornithologie*, **126**:310.

Gahr, M., Güttinger, H-R (1986) Functional aspects of singing in male and female *Uraeginthus bengalus* (Estrildidae). *Ethology*, **72**:123–131.

Gahr, M., Metzdorf, R. (1997) Distribution and dynamics in the expression of androgen and estrogen receptors in vocal control systems of songbirds. *Brain Res Bull*, **44**:509–517.

Gahr, M., Flugge, G., Güttinger, H. R. (1987) Immunocytochemical localization of estrogen-binding neurons in the songbird brain. *Brain Res*, **402**:173–177.

Gahr, M., Guttinger H- R., Kroodsma, D. E. (1993) Estrogen receptors in the avian brain: survey reveals general distribution and forebrain areas unique to songbirds. *J Comp Neurol*, **327**:112–122.

Gahr, M., Sonnenschein, E., Wickler, W. (1998) Sex difference in the size of the neural song control regions in a dueting songbird with similar song repertoire size of males and females. *J Neurosci*, **18**:1124–1131.

Gahr, M., Leitner, S., Fusani, L., Rybak, F. (2002) What is the adaptive role of neurogenesis in adult birds? *Prog Brain Res*, **138**:233–254.

Gale, S. D., Perkel, D. J. (2005) Properties of dopamine release and uptake in the songbird basal ganglia. *J Neurophysiol*, **93**:1871–1879.

Galeotti, P., Saino, N., Sacchi, R., Moller, A. P. (1997) Song correlates with social context, testosterone and body condition in male barn swallows. *Anim Behav*, **53**:687–700.

Gallagher, T. M., Watkin, K. L. (1997) 3D ultrasonic fetal neuroimaging and familial language disorders: in utero brain development. *J Neuroling*, **10**:187–201.

Galoch, Z., Bischof, H-J (2007) Behavioural responses to video playbacks by zebra finch males. *Behav Processes*, **74**:21–26.

Gardner, T., Cecchi, G., Magnasco, M., Laje, R., Mindlin, G. B. (2001) Simple motor gestures for birdsongs. *Phys Rev Lett*, **87**:208101.

Gardner, T. J., Naef, F., Nottebohm, F. (2005) Freedom and rules: the acquisition and reprogramming of a bird's learned song. *Science*, **308**:1046–1049.

Gaunt, A. S. (1987) Phonation. In *Bird Respiration* (Seller, T. J., ed.), pp. 71–94. Boca Raton, FL: CRC Press.

Gaunt, A. S., Gaunt, S. L. L. (1985) Syringeal structure and avian phonation. In *Current Ornithology* (Johnston, R. F., ed.), pp. 213–245. New York: Plenum Press.

Gaunt, A. S., Gaunt, S. L. L., Casey, R. M. (1982) Syringeal mechanics reassessed: evidence from Streptopelia. *Auk*, **99**:474–494.

Gaunt, A. S., Hikida, R. S., Jehl, J. R., Fenbert, L. (1990) Rapid atrophy and hypertrophy of an avian flight muscle. *Auk*, **107**:649–659.

Geberzahn, N., Hultsch, H. (2003) Long-time storage of song types in birds: evidence from interactive playbacks. *Proc Biol Sci*, **270**:1085–1090.

Geberzahn, N., Hultsch, H. (2004) Rules of song development and their use in vocal interactions by birds with large repertoires. *An Acad Bras Cienc*, **76**:209–218.

Geberzahn, N., Hultsch, H., Todt, D. (2002) Latent song type memories are accessible through auditory stimulation in a hand-reared songbird. *Anim Behav*, **64**:783–790.

Gehr, D. D., Capsius, B., Grabner, P., Gahr, M., Leppelsack, H. J. (1999) Functional organisation of the field-L-complex of adult male zebra finches. *Neuroreport*, **10**:375–380.

Gehring, W., Ikeo, K. (1999) Pax 6: mastering eye morphogenesis and eye evolution. *Trends Genet*, **15**:371–377.

Gentner, T. Q. (2004) Neural systems for individual song recognition in adult birds. *Ann N Y Acad Sci*, **1016**:282–302.

Gentner, T. Q., Hulse, S. H. (1998) Perceptual mechanisms for individual vocal recognition in European starlings, *Sturnus vulgaris*. *Anim Behav*, **56**:579–594.

Gentner, T. Q., Hulse, S. H. (2000a) Perceptual classification based on the component structure of song in European starlings. *J Acoust Soc Am*, **107**:3369–3381.

Gentner, T. Q., Hulse, S. H. (2000b) Female European starling preference and choice for variation in conspecific male song. *Anim Behav*, **59**:443–458.

Gentner, T. Q., Margoliash, D. (2002) The neuroethology of vocal communication: perception and cognition. In *Springer Handbook of Auditory Research* (Simmons, A., Popper, A. N., Fay, R. R., eds.), pp. 324–386. Dordrecht: Springer-Verlag.

Gentner, T. Q., Margoliash, D. (2003) Neuronal populations and single cells representing learned auditory objects. *Nature*, **424**:669–674.

Gentner, T. Q., Duffy, D. L., Kaloudis, K., Ellis, E., Ball, G. F. (1998) Behaviorally relevant variation in male song induces differential expression of the IEG ZENK in a sub-region of NCM in female starlings. *Soc Neurosci Abstr*, **24**:700.

Gentner, T. Q., Hulse, S. H., Bentley, G. E., Ball, G. F. (2000) Individual vocal recognition and the effect of partial lesions to HVc on discrimination, learning, and categorization of conspecific song in adult songbirds. *J Neurobiol*, **42**:117–133.

Gentner, T. Q., Hulse, S. H., Duffy, D., Ball, G. F. (2001) Response biases in auditory forebrain regions of female songbirds following exposure to sexually relevant variation in male song. *J Neurobiol*, **46**:48–58.

Gentner, T. Q., Hulse, S. H., Ball, G. F. (2004) Functional differences in forebrain auditory regions during learned vocal recognition in songbirds. *J Comp Physiol A Neuroethol Sens Neural Behav Physiol*, **190**:1001–1010.

Gentner, T. Q., Fenn, K. M., Margoliash, D., Nusbaum, H. C. (2006) Recursive syntactic pattern learning by songbirds. *Nature*, **440**:1204–1207.

George, J. G., (2001) The synucleins. *Genome Biology*, 3:Reviews 3002.3001–3002.3006.

George, J. M., Jin, H., Woods, W. S., Clayton, D. F. (1995) Characterization of a novel protein regulated during the critical period for song learning in the zebra finch. *Neuron*, **15**:361–372.

Gerrits, P. O., Holstege, G. (1996) Pontine and medullary projections to the nucleus retroambiguus: a wheat germ agglutinin-horseradish peroxidase and autoradiographic tracing study in the cat. *J Comp Neurol*, **373**:173–185.

Geschwind, N. (1970) The organization of language and the brain. *Science*, **170**:940–944.

Geschwind, N., Galaburda, A. M. (1985) Cerebral lateralization: biological mechanisms, associations, and pathology. I. A hypothesis and a program for research. *Arch Neurol*, **42**:428–459.

Ghazanfar, A. A., Nicolelis, M. A. (2001) Feature article: the structure and function of dynamic cortical and thalamic receptive fields. *Cereb Cortex*, **11**:183–193.

Gil, D., Leboucher, G., Lacroix, A., Cue, R., Kreutzer, M. (2004) Female canaries produce eggs with greater amounts of testosterone when exposed to preferred male song. *Horm Behav*, **45**:64–70.

Gilbert, C. D., Sigman, M., Crist, R. E. (2001) The neural basis of perceptual learning. *Neuron*, **31**:681–697.

Gill, P., Zhang, J., Woolley, S. M., Fremouw, T., Theunissen, F. E. (2006) Sound representation methods for spectro-temporal receptive field estimation. *J Comput Neurosci*, **22**:22.

Glaze, C. M., Troyer, T. W. (2006) Temporal structure in zebra finch song: implications for motor coding. *J Neurosci*, **26**:991–1005.

Gleeson, M. (1987) The role of vagal afferents in the peripheral control of respiration in birds. In *The Neurology of the Cardiorespiratory System* (Taylor, E., ed.), pp. 51–79. Manchester, UK: Manchester University Press.

Gleeson, M., Molony, V. (1989) Control of breathing. In *Form and Function in Birds* (King, A. S., McLelland, J., eds.), pp. 439–484. London: Academic Press.

Gobes, S. M. H., Bolhuis, J. J. (2007) Bird song memory: a neural dissociation between song recognition and production. *Curr Biol*, **17**: 789–793.

Gobes, S. M. H., Bolhuis, J. J., Roosemalen, K., Pots, J. M., Zandbergen, M. A. (2005) Effects of neurotoxic lesions to the caudomedial nidopallium on song production and tutor song preference in male zebra finches. *Soc Neurosci Abstr*, **31**:1002–1015.

Godard, R. (1991) Long-term memory for individual neighbors in a migratory songbird. *Nature*, **350**: 228–229.

Goelet, P., Castellucci, V. F., Schacher, S., Kandel, E. R. (1986) The long and the short of long-term memory – a molecular framework. *Nature*, **322**:419–422.

Goldin-Meadow, S., Mylander, C. (1998) Spontaneous sign systems created by deaf children in two cultures. *Nature*, **391**:279–281.

Goldman, S. A., Nottebohm, F. (1983) Neuronal production, migration, and differentiation in a vocal control nucleus of the adult female canary brain. *Proc Natl Acad Sci USA*, **80**:2390–2394.

Goldman, S. A., Zukhar, A., Barami, K., Mikawa, T., Niedzwiecki, D. (1996) Ependymal/subependymal zone cells of postnatal and adult songbird brain generate both neurons and nonneuronal siblings in vitro and in vivo. *J Neurobiol*, **30**:505–520.

Goldstein, M. H., King, A. P., West, M. J. (2003) Social interaction shapes babbling: testing parallels between birdsong and speech. *Proc Natl Acad Sci USA*, **100**:8030–8035.

Goller, F., Cooper, B. G. (2004) Peripheral motor dynamics of song production in the zebra finch. *Ann N Y Acad Sci*, **1016**:130–152.

Goller, F., Cooper, B. G. (2005) Altered acoustic feedback does not affect patterns of beak movement during song in the zebra finch. *Soc Neurosci Abstr*, **1002**:13.

Goller, F., Daley, M. A. (2001) Novel motor gestures for phonation during inspiration enhance the acoustic complexity of birdsong. *Proc R Soc Lond B Biol Sci*, **268**:2301–2305.

Goller, F., Larsen, O. N. (1997a) A new mechanism of sound generation in songbirds. *Proc Natl Acad Sci USA*, **94**:14787–14791.

Goller, F., Larsen, O. N. (1997b) In situ biomechanics of the syrinx and sound generation in pigeons. *J Exp Biol*, **200**:2165–2176.

Goller, F., Suthers, R. A. (1995) Implications for lateralization of bird song from unilateral gating of bilateral motor patterns. *Nature*, **373**:63–66.

Goller, F., Suthers, R. A. (1996a) Role of syringeal muscles in gating airflow and sound production in singing brown thrashers. *J Neurophysiol*, **75**:867–876.

Goller, F., Suthers, R. A. (1996b) Role of syringeal muscles in controlling the phonology of bird song. *J Neurophysiol*, **76**:287–300.

Goller, F., Suthers, R. A. (1999) Bilaterally symmetrical respiratory activity during lateralized birdsong. *J Neurobiol*, **41**:513–523.

Goller, F., Mallinckrodt, M. J., Torti, S. D. (2004) Beak gape dynamics during song in the zebra finch. *J Neurobiol*, **59**:289–303.

Goller, F., Cooper, B. G., Suthers, R. A. (2006) Respiratory dynamics and syllable morphology in songbirds. *Acta Zool Sinica*, **52**(Suppl.):471–474.

Goodson, J. L. (1998) Territorial aggression and dawn song are modulated by septal vasotocin and vasoactive intestinal polypeptide in male field sparrows (*Spizella pusilla*). *Horm Behav*, **34**:67–77.

Goodson, J. L., Bass, A. H. (2001) Social behavior functions and related anatomical characteristics of vasotocin/vasopressin systems in vertebrates. *Brain Res Brain Res Rev*, **35**:246–265.

Goodson, J. L., Eibach, R., Sakata, J., Adkins-Regan, E. (1999) Effect of septal lesions on male song and aggression in the colonial zebra finch (*Taeniopygia guttata*) and the territorial field sparrow (*Spizella pusilla*). *Behav Brain Res*, **98**:167–180.

Goodson, J. L., Saldanha, C. J., Hahn, T. P., Soma, K. K. (2005) Recent advances in behavioral neuroendocrinology: insights from studies on birds. *Horm Behav*, **48**:461–473.

Gottlieb, G. (1971) *Development of Species Identification in Birds*. Chicago, IL: University of Chicago Press.

Gould, S. J., Lewontin, R. C. (1979) The spandrels of San Marco and the Panglossian paradigm: a critique of the adaptationist programme. *Proc R Soc Lond B*, **205**:581–598.

Gould, J. L., Marler, P. (1987) Learning by instinct. *Sci Am*, **256**:74–85.

Gould, E., Gross, C. G. (2002) Neurogenesis in adult mammals: some progress and problems. *J Neurosci*, **22**:619–623.

Gouzoules, S., Gouzoules, H., Marler, P. (1984) Rhesus monkey (*Macaca mulatta*) screams: representational signalling in the recruitment of agonistic aid. *Anim Behav*, **32**:182–193.

Goy, R. W., McEwen, B. S. (1980) *Sexual Differentiation of the Brain*. Cambridge, MA: MIT Press.

Grabatin, O., Abs, M. (1986) (Efferent innervation of the larynx in domestic pigeons *Columba livia domestica* L.). *Anat Anz*, **162**:101–108.

Grace, J. A., Amin, N., Singh, N. C., Theunissen, F. E. (2003) Selectivity for conspecific song in the zebra finch auditory forebrain. *J Neurophysiol*, **89**:472–487.

Graybiel, A. M. (2005) The basal ganglia: learning new tricks and loving it. *Curr Opin Neurobiol*, **15**:638–644.

Graybiel, A. M., Aosaki, T., Flaherty, A. W., Kimura, M. (1994) The basal ganglia and adaptive motor control. *Science*, **265**:1826–1831.

Green, S. (1975) Communication by a graded communication system in Japanese monkeys. In *Primate Behavior: Developments in Field and Laboratory Research* (Rosenblum, L. A., ed.), pp. 1–102. New York: Academic Press.

Greenewalt, C. H. (1968) *Bird Song: Acoustics and Physiology*. Washington, DC: Smithsonian Institution Press.

Grieser, D. L., Kuhl, P. K. (1988) Maternal speech to infants in a tonal language: support for universal prosodic features in motherese. *Dev Psychol*, **24**:14–20.

Grimm, D., Streetz, K. L., Jopling, C. L., et al. (2006) Fatality in mice due to oversaturation of cellular microRNA/short hairpin RNA pathways. *Nature*, **441**:537–541.

Grisham, W., Arnold, A. P. (1994) Distribution of GABA-like immunoreactivity in the song system of the zebra finch. *Brain Res*, **651**:115–122.

Gross, C. G. (2002) Genealogy of the "grandmother cell". *Neuroscientist*, **8**:512–518.

Grossman, M. (1999) Sentence processing in Parkinson's disease. *Brain Cogn*, **40**:387–413.

Grubb, B., Mills, C. D., Colacino, J. M., Schmidt-Nielsen, K. (1977) Effect of arterial carbon dioxide on cerebral blood flow in ducks. *Am J Physiol*, **232**:H596–601.

Gulledge, C. C., Deviche, P. (1997) Androgen control of vocal control region volumes in a wild migratory songbird (*Junco hyemalis*) is region and possibly age dependent. *J Neurobiol*, **32**:391–402.

Gurney, M. E., Konishi, M. (1980) Hormone induced sexual differentiation of brain and behavior in zebra finches. *Science*, **208**:1380–1382.

Gurney, M. E. (1981) Hormonal control of cell form and number in the zebra finch song system. *J Neurosci*, **1**:658–673.

Güttinger, H. R. (1979) The integration of learned and genetically programmed behavior: a study of hierarchical organization in songs of canaries, greenfinches and their hybrids. *Z Tierpsychol*, **49**:285–303.

Güttinger, H.-R., Nicolai, J. (1973) Struktur und Funktion der Rufe bei Prachtfinken (Estrildidae). *Z Tierpsychol*, **33**:319–334.

Güttinger, H. R. (1985) Consequences of domestication on the song structures in the canary. *Behaviour*, **94**:254–278.

Gutzke, W. H., Bull, J. J. (1986) Steroid hormones reverse sex in turtles. *Gen Comp Endocrinol*, **64**:368–372.

Guzowski, J. F. (2002) Insights into immediate-early gene function in hippocampal memory consolidation using antisense oligonucleotide and fluorescent imaging approaches. *Hippocampus*, **12**:86–104.

Guzowski, J. F., McGaugh, J. L. (1997) Antisense oligodeoxynucleotide-mediated disruption of hippocampal cAMP response element binding protein levels impairs consolidation of memory for water maze training. *Proc Nat Acad Sci USA*, **94**:2693–2698.

Guzowski, J. F., McNaughton, B. L., Barnes, C. A., Worley, P. F. (1999) Environment-specific expression of the immediate-early gene Arc in hippocampal neuronal ensembles. *Nat Neurosci*, **2**:1120–1124.

Guzowski, J. F., Lyford, G. L., Stevenson, G. D., *et al.* (2000) Inhibition of activity-dependent arc protein expression in the rat hippocampus impairs the maintenance of long-term potentiation and the consolidation of long-term memory. *J Neurosci*, **20**:3993–4001.

Gwinner, H. (1997) The function of green plants in nests of European starlings (*Sturnus vulgaris*). *Behaviour*, **134**:337–351.

Gwinner, H., Gwinner, E., Dittami, J. (1987) Effects of nestboxes on LH, testosterone, testicular size, and the reproductive behaviour of male European starlings in spring. *Behaviour*, **103**:68–82.

Gyger, M., Marler, P. (1988) Food calling in the domestic fowl, *Gallus gallus*: the role of external referents and deception. *Anim Behav*, **36**:358–365.

Gyger, M., Marler, P., Pickert, R. (1987) Semantics of an avian alarm call system: the male domestic fowl, *Gallus domesticus*. *Behaviour*, **102**:15–40.

Haesler, S., Wada, K., Nshdejan, A., *et al.* (2004) FoxP2 expression in avian vocal learners and non-learners. *J Neurosci*, **24**:3164–3175.

Haesler, S., Rochefort, C., Georgi, B., *et al.* (2007) Incomplete and inaccurate vocal immitation after knockdown of FoxP2 in songbird basal ganglia nucleus Area X. *PLoS Biol*, **5**:e321.

Hahnloser, R. H., Kozhevnikov, A. A., Fee, M. S. (2002) An ultra-sparse code underlies the generation of neural sequences in a songbird. *Nature*, **419**:65–70.

Hahnloser, R. H., Kozhevnikov, A. A., Fee, M. S. (2006) Sleep-related neural activity in a premotor and a basal-ganglia pathway of the songbird. *J Neurophysiol*, **96**:794–812.

Hailman, J. P. (1969) How an instinct is learned. *Sci Am*, **221**:6.

Hailman, J. P., Ficken, M. S. (1986) Combinational animal communication with computable syntax: chick-a-dee calling qualifies as 'language' by structural linguistics. *Anim Behav*, **34**:1899–1901.

Hailman, J. P., Ficken, M. S., Ficken, R. S. (1987) Constraints on the structure of combinatorial chick-a-dee calls. *Ethology*, **75**:62–80.

Hall, M. L. (2000) The function of duetting in magpie-larks: conflict, cooperation, or commitment? *Anim Behav*, **60**:667–677.

Hall, M. L. (2004) A review of hypotheses for the functions of avian duetting. *Behav Ecol Sociobiol*, **55**:415–430.

Hall, M. L., Magrath, R. D. (2000) Duetting and mate-guarding in Australian magpie-larks (*Grallina cyanoleuca*). *Behav Ecol Sociobiol*, **47**:180–187.

Halle, F., Gahr, M., Pieneman, A. W., Kreutzer, M. (2002) Recovery of song preferences after excitotoxic HVC lesion in female canaries. *J Neurobiol*, **52**:1–13.

Halle, F., Gahr, M., Kreutzer, M. (2003) Effects of unilateral lesions of HVC on song patterns of male domesticated canaries. *J Neurobiol*, **56**:303–314.

Hamilton, K. S., King, A. P., Sengelaub, D. R., West, M. J. (1997) A brain of her own: a neural correlate of song assessment in a female songbird. *Neurobiol Learn Mem*, **68**:325–332.

Hansson, B., Akesson, M., Slate, J., Pemberton, J. M. (2005) Linkage mapping reveals sex-dimorphic map distances in a passerine bird. *Proc Biol Sci*, **272**:2289–2298.

Harding, C. F. (1981) Social modulation of circulating hormone levels in the male. *Am Zool*, **21**:225–231.

Harding, C. F. (1983) Hormonal specificity and activation of social behaviour in the male zebra finch. In: *Hormones and Behaviour in Higher Vertebrates* (Balthazart, J., Pröve, E., Gilles, R., eds.), pp. 275–289. Berlin: Springer Verlag.

Harding, C. F. (1986) The role of androgen metabolism in the activation of male behavior. *Ann N Y Acad Sci*, **474**:371–378.

Harding, C. F. (1992) Hormonal modulation of neurotransmitter function and behavior in male songbirds. *Poult Sci Rev*, **4**:261–273.

Harding, C. F. (1999) Androgens, effects in birds. In *Encyclopedia of Reproduction* (Knobil, E., Neill, J. D., eds.), pp. 188–196. New York: Academic Press.

Harding, C. F. (2002) The effects of manipulating catecholamines on song learning. In *Neurobiology of Birdsong*, Annual Symposium of Research Centers in Minority Institutions. New York: Hunter College.

Harding, C. F. (2004) Hormonal modulation of singing: hormonal modulation of the songbird brain and singing behavior. In *Behavioral Neurobiology of Birdsong* (Zeigler, H. P., Marler, P., eds.), pp. 524–539. New York: New York Academy of Sciences.

Harding, C. F., Follett, B. K. (1979) Hormone changes triggered by aggression in a natural population of blackbirds. *Science*, 203:918–920.

Harding, C. F., Rowe, S. A. (2003) Vasotocin treatment inhibits courtship in male zebra finches; concomitant androgen treatment inhibits this effect. *Horm Behav*, 44:413–418.

Harding, C. F., Sheridan, K., Walters, M. J. (1983) Hormonal specificity and activation of sexual behavior in male zebra finches. *Horm Behav*, 17:111–133.

Harding, C. F., Walters, M. J., Collado, D., Sheridan, K. (1988) Hormonal specificity and activation of social behavior in male red-winged blackbirds. *Horm Behav*, 22:402–418.

Hartley, R. S. (1990) Expiratory muscle activity during song production in the canary. *Respir Physiol*, 81:177–187.

Hartley, R. S., Suthers, R. A. (1989) Airflow and pressure during canary song: direct evidence for mini-breaths. *J Comp Physiol A*, 165:15–26.

Hartley, R. S., Suthers, R. A. (1990) Lateralization of syringeal function during song production in the canary. *J Neurobiol*, 21:1236–1248.

Hartman, V. N., Miller, M. A., Clayton, D. F., Liu, W.-C., Kroodsma, D. E., Brenowitz, E. A. (2001) Testosterone regulates alpha-synuclein mRNA in the avian song system. *Neuroreport*, 12:943–946.

Harvey, P. H., Pagel, M. D. (1991) *The Comparative Method in Evolutionary Biology*. Oxford, UK: Oxford University Press.

Hau, M. (2001) Timing of breeding in variable environments: tropical birds as model systems. *Horm Behav*, 40:281–290.

Hauber, M. E., Clayton, N. S., Kacelnik, A., Reboreda, J. C., DeVoogd, T. J. (1999) Sexual dimorphism and species differences in HVC volumes of cowbirds. *Behav Neurosci*, 113:1095–1099.

Hauber, M. E., Russo, S. A., Sherman, P. W. (2001) A password for species recognition in a brood-parasitic bird. *Proc R Soc Lond B*, 268:1041–1048.

Hausberger, M. (1997) Social influences on song acquisition and sharing in the European starling (*Sturnus vulgaris*). In *Social Influences on Vocal Development* (Snowden, C., Hausberger, M., eds.). Cambridge: Cambridge University Press.

Hausberger, M., Cousillas, H. (1995) Categorization in birdsong: from behavioural to neuronal responses. *Behav Processes*, 35:83–91.

Hausberger, M., Black, J. M., Richard, J.-P. (1991) Bill opening and sound spectrum in barnacle goose loud calls: individuals with 'wide mouths' have higher pitched voices. *Anim Behav*, 42:319–322.

Hauser, M. D. (1996) *The Evolution of Communication*. Cambridge, MA: MIT Press.

Hauser, M. D., Newport, E. L., Aslin, R. N. (2001) Segmentation of the speech stream in a non-human primate: statistical learning in cotton-top tamarins. *Cognition*, 78:B53–64.

Hauser, M. D., Chomsky, N., Fitch, W. T. (2002a) The faculty of language: what is it, who has it, and how did it evolve? *Science*, 298:1569–1579.

Hauser, M. D., Weiss, D., Marcus, G. (2002b) Rule learning by cotton-top tamarins. *Cognition*, 86:B15–22.

Haüsler, U., ed. (1996) *Measurement of Short-time Spatial Activity Patterns During Auditory Stimulation in the Starling*. New York: Plenum Press.

He, L., Hannon, G. (2004) MicroRNAs: small RNAs with a big role in gene regulation. *Nat Rev Genet*, 5:522–531.

Hegner, R. E., Wingfield, J. C. (1986) Behavioral and endocrine correlates of multiple brooding in the semicolonial house sparrow *Passer domesticus*. I. Males. *Horm Behav*, 20:294–312.

Heil, P., Scheich, H. (1985) Quantitative analysis and two-dimensional reconstruction of the tonotopic organization of the auditory field L in the chick from 2-deoxyglucose data. *Exp Brain Res*, 58:532–543.

Heil, P., Scheich, H. (1991) Functional organization of the avian auditory cortex analogue. II. Topographic distribution of latency. *Brain Res*, 539:121–125.

Heimovics, S. A., Riters, L. V. (2005) Immediate early gene activity in song control nuclei and brain areas regulating motivation relates positively to singing behavior during, but not outside of, a breeding context. *J Neurobiol*, 65:207–224.

Heimovics, S. A., Riters, L. V. (2006) Breeding-context-dependent relationships between song and cFOS labeling within social behavior brain regions in male European starlings (*Sturnus vulgaris*). *Horm Behav*, **50**:726–735.

Heimovics, S. A., Riters, L. V. (2007) ZENK labeling within social brain regions reveals breeding context-dependent patterns of neural activity associated with song in male European starlings (*Sturnus vulgaris*). *Behav Brain Res*, **176**: 333–343.

Hein, A. M., Hart, A. S., Nordeen, K. W., Nordeen, E. J. (2005) Dopamine signaling may contribute to tutoring-induced CaMKII activation within the avian basal ganglia. *Soc Neurosci Abstr*.

Hein, A. M., Sridharan, A., Nordeen, K. W., Nordeen, E. J. (2007) Characterization of CaMKII-expressing neurons within a striatal region implicated in avain vocal learning. *Brain Res*, **1155**:125–133.

Heinrich, J. E., Singh, T. D., Sohrabji, F., Nordeen, K. W., Nordeen, E. J. (2002) Developmental and hormonal regulation of NR2 A mRNA in forebrain regions controlling avian vocal learning. *J Neurobiol*, **51**:149–159.

Heinrich, J. E., Singh, T. D., Nordeen, K. W., Nordeen, E. J. (2003) NR2B downregulation in a forebrain region required for avian vocal learning is not sufficient to close the sensitive period for song learning. *Neurobiol Learn Mem*, **79**:99–108.

Heinrich, J. E., Nordeen, K. W., Nordeen, E. J. (2005) Dissociation between extension of the sensitive period for avian vocal learning and dendritic spine loss in the song nucleus lMAN. *Neurobiol Learn Mem*, **83**:143–150.

Heinz, R. D., Sachs, M. B., Sinnott, J. M. (1981) Discrimination of steady-state vowels by blackbirds and pigeons. *J Acoust Soc Am*, **70**:699–706.

Helekar, S. A., Espino, G. G., Botas, A., Rosenfield, D. B. (2003) Development and adult phase plasticity of syllable repetitions in the birdsong of captive zebra finches (*Taeniopygia guttata*). *Behav Neurosci*, **117**:939–951.

Hermann, K., Arnold, A. P. (1991) Lesions of HVc block the developmental masculinizing effects of estradiol in the female zebra finch song system. *J Neurobiol*, **22**:29–39.

Hernandez, A. M., MacDougall-Shackleton, S. A. (2004) Effects of early song experience on song preferences and song control and auditory brain regions in female house finches (*Carpodacus mexicanus*). *J Neurobiol*, **59**:247–258.

Herrera, D. G., Robertson, H. A. (1996) Activation of c-fos in the brain. *Prog Neurobiol*, **50**:83–107.

Herrick, C. J. (1948) *The Brain of the Tiger Salamander, Ambystoma tigrinum*. Chicago, IL: University of Chicago Press.

Herrick, C. J. (1956) *The Evolution of Human Nature*. Austin, TX: University of Texas Press.

Herrick, E. H., Harris, J. O. (1957) Singing female canaries. *Science*, **125**:1299–1300.

Herrmann, K., Arnold, A. P. (1991) The development of afferent projections to the robust archistriatal nucleus in male zebra finches: a quantitative electron microscopic study. *J Neurosci*, **11**:2063–2074.

Hersch, G. L. (1966) Bird voices and resonant tuning in helium air mixtures. Ph.D. dissertation, University of California, Berkeley.

Hessler, N. A., Doupe, A. J. (1999a) Singing-related neural activity in a dorsal forebrain-basal ganglia circuit of adult zebra finches. *J Neurosci*, **19**:10461–10481.

Hessler, N. A., Doupe, A. J. (1999b) Social context modulates singing-related neural activity in the songbird forebrain. *Nat Neurosci*, **2**:209–211.

Hestrin, S. (1992) Developmental regulation of NMDA receptor-mediated synaptic currents at a central synapse. *Nature*, **357**:686–689.

Hidalgo, A., Barami, K., Iversen, K., Goldman, S. A. (1995) Estrogens and non-estrogenic ovarian influences combine to promote the recruitment and decrease the turnover of new neurons in the adult female canary brain. *J Neurobiol*, **27**:470–487.

Hikosaka, O., Nakamura, K., Sakai, K., Nakahara, H. (2002) Central mechanisms of motor skill learning. *Curr Opin Neurobiol*, **12**:217–222.

Hile, A. G., Plummer, T. K., Striedter, G. F. (2000) Male vocal imitation produces call convergence during pair bonding in budgerigars, *Melopsittacus undulatus*. *Anim Behav*, **59**:1209–1218.

Hillier, L. W., Miller, W., Birney, E., et al. (2004) Sequence and comparative analysis of the chicken genome provide unique perspectives on vertebrate evolution. *Nature*, **432**:695–716.

Hinde, R. A. (1956) The biological significance of territories in birds. *Ibis*, **98**:340–369.

Hinde, R. A. (1970) *Animal Behaviour: A Synthesis of Ethology and Comparative Psychology*. New York: McGraw-Hill.

Hirano, S., Kojima, H., Naito, Y., et al. (1996) Cortical speech processing mechanisms while vocalizing visually presented languages. *Neuroreport*, **8**:363–367.

Hirano, S., Kojima, H., Naito, Y., et al. (1997) Cortical processing mechanism for vocalization with auditory verbal feedback. *Neuroreport*, **8**:2379–2382.

Ho, C. E., Pesaran, B., Fee, M. S., Mitra, P. P. (1998) Characterization of the structure and variability of zebra finch song elements. *Proc Joint Symp Neural Comput*, 5:76–83.

Hochstadt, J., Nakano, H., Lieberman, P., Friedman, J. (2006) The roles of sequencing and verbal working memory in sentence comprehension deficits in Parkinson's disease. *Brain Lang*, 97:243–257.

Hodos, W. (1974) The comparative study of brain–behavior relationships. In *Birds: Brain and Behavior* (Goodman, I. J., Schein, M. W., eds.), pp. 15–25. New York: Academic Press.

Hodos, W. (1976) The concept of homology and the evolution of behavior. In *Evolution, Brain and Behavior: Persistent Problems* (Masterton, R. B., Hodos, W., Jerison, H., eds.), pp. 153–167. Hillsdale, NJ: Erlbaum Associates.

Hoese, W. J., Podos, J., Boetticher, N. C., Nowicki, S. (2000) Vocal tract function in birdsong production: experimental manipulation of beak movements. *J Exp Biol*, 203:1845–1855.

Hofer, M., Prusky, G. T., Constantine-Paton, M. (1994) Regulation of NMDA receptor mRNA during visual map formation and after receptor blockade. *J Neurochem*, 62:2300–2307.

Hoffmann, H., Gremme, T., Hatt, H., Gottmann, K. (2000) Synaptic activity-dependent developmental regulation of NMDA receptor subunit expression in cultured neocortical neurons. *J Neurochem*, 75:1590–1599.

Hoke, K. L., Burmeister, S. S., Fernald, R. D., Rand, A. S., Ryan, M. J., Wilczynski, W. (2004) Functional mapping of the auditory midbrain during mate call reception. *J Neurosci*, 24:11264–11272.

Holland, J., Dabelsteen, T., Paris, A. L. (2000) Coding in the song of the wren: importance of rhythmicity, syntax and element structure. *Anim Behav*, 60:463–470.

Holloway, C. C., Clayton, D. F. (2001) Estrogen synthesis in the male brain triggers development of the avian song control pathway in vitro. *Nat Neurosci*, 4:170–175.

Holstege, G. (1989) Anatomical study of the final common pathway for vocalization in the cat. *J Comp Neurol*, 284:242–252.

Holy, T. E., Guo, Z. (2005) Ultrasonic songs of male mice. *PLoS Biol*, 3:e386.

Homberger, D. G. (1999) The avian tongue and larynx: multiple functions in nutrition and vocalisation. In *Proceedings of the 22nd International Ornithological Congress* (Adams, N. J., Slotow, R. H., eds.), pp. 94–113. Durban: BirdLife South Africa.

Hopcroft, J., Ullman, J. (1979) *Introduction to Automata Theory, Languages, and Computation*. Reading, MA: Addison-Westley.

Hopkins, C. D. (1999) Signal evolution in electric communication. In *The Design of Animal Communication* (Konishi, M., Hauser, M. D., eds.), pp. 461–491. Cambridge, MA: MIT Press.

Horn, G. (1985) *Memory, Imprinting, and the Brain: An Inquiry into Mechanisms*. Oxford/New York: Clarendon Press/Oxford University Press.

Horn, G. (1998) Visual imprinting and the neural mechanisms of recognition memory. *Trends Neurosci*, 21:300–305.

Horn, G. (2004) Pathways of the past: the imprint of memory. *Nat Rev Neurosci*, 5:108–120.

Houde, J. F., Jordan, M. I. (1998) Sensorimotor adaptation in speech production. *Science*, 279:1213–1216.

Hough, G. E., Volman, S. F. (2002) Short-term and long-term effects of vocal distortion on song maintenance in zebra finches. *J Neurosci*, 22:1177–1186.

Hough, G. E., 2nd, Nelson, D. A., Volman, S. F. (2000) Re-expression of songs deleted during vocal development in white-crowned sparrows, *Zonotrichia leucophrys*. *Anim Behav*, 60:279–287.

Houk, J. C., Davis, J. L., Beiser, D. G. (1994) *Models of Information Processing in the Basal Ganglia*. Cambridge, MA: MIT Press.

Houx, B. B., ten Cate, C. (1999) Song learning from playback in zebra finches: is there an effect of operant contingency. *Anim Behav*, 57:837–845.

Hovatta, I., Tennant, R. S., Helton, R., et al. (2005) Glyoxalase 1 and glutathione reductase 1 regulate anxiety in mice. *Nature*, 438:662–666.

Howell, P., Archer, A. (1984) Susceptibility to the effects of delayed auditory feedback. *Percept Psychophys*, 36:296–302.

Hsu, A., Woolley, S. M., Fremouw, T. E., Theunissen, F. E. (2004a) Modulation power and phase spectrum of natural sounds enhance neural encoding performed by single auditory neurons. *J Neurosci*, 24:9201–9211.

Hsu, A., Borst, A., Theunissen, F. E. (2004b) Quantifying variability in neural responses and its application for the validation of model predictions. *Network*, 15:91–109.

Huang, H., Winter, E., Wang, H., et al. (2004) Evolutionary conservation and selection of human disease gene orthologs in the rat and mouse genomes. *Genome Biol*, 5:R47.

Hudson, W. H. (1917) *Far Away and Long Ago*. New York: E. P. Dutton & Co.

Huesmann, G., Clayton, D. F. (2006) Dynamic role of postsynaptic caspase-3 and BIRC4 in zebra finch song response habituation. *Neuron*, **52**:1061–1072.

Hughes, M., Hultsch, H., Todt, D. (2002) Imitation and invention in song learning in nightingales (*Luscinia megarhynchos* B., Turdidae). *Ethology*, **108**:97–113.

Hull, E. M., Wood, R. I., McKenna, K. E. (2006) The neurobiology of male sexual behavior. In *Knobil and Neil's Physiology of Reproduction*. 3rd edn. (Neil, J. D., ed.), pp. 1729–1824. San Diego, CA: Academic Press.

Hulse, S. H. (1995) The discrimination-transfer procedure for studying auditory perception and perceptual invariance in animals. In *Methods in Comparative Psychoacoustics* (Klump, G. M., Dooling, R. J., Fay, R. R., Stebbins, W. C., eds.), pp. 319–330. Basel, Switzerland: Birkhäuser Verlag.

Hultsch, H. (1980) Beziehungen zwischen Struktur, zeitlicher Variabilität und sozialem Einsatz des Gesangs der Nachtigall (*Luscinia megarhynchos*). Ph.D. dissertation. Berlin: Freie Universität.

Hultsch, H. (1985) Sub-Repertoire-Bildung: ein Organisationsprinzip bei Erwerb und Einsatz großer Repertoires. *Verh Dtsch Zool Ges*, **78**:229.

Hultsch, H. (1991a) Song ontogeny in birds: closed or open developmental programs? In *Synapse-transmission, Modulation* (Elsner, N., Penzlin, H., eds.). Stuttgart: Thieme Verlag.

Hultsch, H. (1991b) Early experience can modify singing styles: evidence from experiments with nightingales (*Luscinia megarhynchos*). *Anim Behav*, **42**:883–889.

Hultsch, H. (1992) Time window and unit capacity: dual constraints on the acquisition of serial information in songbirds. *J Comp Physiol A*, **170**:275–280.

Hultsch, H., Kopp, M. L. (1989) Early auditory learning and song improvisation in nightingales, *Luscinia megarhynchos*. *Anim Behav*, **37**:510–512.

Hultsch, H. (1993a) Tracing the memory mechanisms in the song acquisition of birds. *Netherlands J Zool*, **43**:155–171.

Hultsch, H. (1993b) Ecological versus psychobiological aspects of song learning in birds. *Etologia*, **3**:309–323.

Hultsch, H., Todt, D. (1982) Temporal performance roles during vocal interactions in nightingales (*Luscinia megarhynchos* B.). *Behav Ecol Sociobiol*, **11**:253–260.

Hultsch, H., Todt, D. (1989a) Song acquisition and acquisition constraints in the nightingale (*Luscinia megarhynchos*). *Naturwissenschaften*, **76**:83–86.

Hultsch, H., Todt, D. (1989b) Context memorization in the song-learning of birds. *Naturwissenschaften*, **76**:584–586.

Hultsch, H., Todt, D. (1989c) Memorization and reproduction of songs in nightingales (*Luscinia megarhynchos*): evidence for package formation. *J Comp Physiol A*, **165**:197–203.

Hultsch, H., Todt, D. (1996) Discontinuous and incremental processes in the song learning of birds: evidence for a primer effect. *J Comp Physiol A*, **179**:291–299.

Hultsch, H., Todt, D. (2004) Approaches to the mechanisms of song memorization and singing provide evidence for a procedural memory. *An Acad Bras Cienc*, **76**:219–230.

Hultsch, H., Lange, R., Todt, D. (1984) Mustertyp-markierte Lernangebote: eine Methode zur Untersuchung auditorisch/vokaler Strophentypgedächtnisse. *Verh Dtsch Zool Ges*, **77**:249.

Hultsch, H., Geberzahn, N., Schleuss, F. (1998) Song invention in nightingales – cues from song development. *Ostrich*, **69**:254.

Hultsch, H., Mundry, R., Todt, D. (1999a) Learning, representation and retrieval of rule-related knowledge in the song system of birds. In *Learning: Rule Extraction and Representation* (Friederici, A. D., Menzel, R., eds.), pp. 89–115. Berlin: De Gruyter.

Hultsch, H., Schleuss, F., Todt, D. (1999b) Auditory-visual stimulus pairing enhances perceptual learning in a songbird. *Anim Behav*, **58**:143–149.

Hunt, G. R., Gray, R. D. (2003) Diversification and cumulative evolution in New Caledonian crow tool manufacture. *Proc Biol Sci*, **270**:867–874.

Hunt, S. P., Pini, A., Evan, G. (1987) Induction of c-fos-like protein in spinal cord neurons following sensory stimulation. *Nature*, **328**:632–634.

Hutt, S. J., Hutt, C., Lenard, H. G., Bernuth, H., Munt Jewerff, W. J. (1968) Auditory responsivity in the human neonate. *Nature*, **218**(5144):888–890.

i Cancho, R. F., Riordan, O., Bollobas, B. (2005) The consequences of Zipf's law for syntax and symbolic reference. *Proc Biol Sci*, **272**:561–565.

Immelmann, K. (1969) Song development in the zebra finch and other estrildid finches. In *Bird Vocalizations* (Hinde, R. A., ed.), pp. 61–77. Cambridge: Cambridge University Press.

Immelmann, K., Suomi, S. J. (1981) Behavioral development: the Bielefeld interdisciplinary project. In *Behavioral*

Development: The Bielefeld Interdisciplinary Project (Immelmann, K., Barlow, G. W., Petrinovich, L., Main, M., eds.), pp. 395–431. Cambridge: Cambridge University Press.

Ince, S. A., Slater, P. J. B. (1985) Versatility and continuity in the songs of thrushes, *Turdus* spp. *Ibis*, **127**:355–364.

Ingvar, D. H., Schwartz, M. S. (1974) Blood flow patterns induced in the dominant hemisphere by speech and reading. *Brain*, **97**:273–278.

Itasaki, N., Bel-Vialar, S., Krumlauf, R. (1999) "Shocking" developments in chick embryology: electroporation and *in ovo* gene expression. *Nat Cell Biol*, **1**:E203–207.

Itoh, Y., Chen, X., Kim, Y., et al. (2005) Brain masculinization in the absence of testes in a ZZ zebra finch. *Soc Neurosci Abstr*, **152.4**:P16.

Itoh, Y., Melamed, E., Yang, X., et al. (2007) Dosage compensation is less effective in birds than in mammals. *J Biology*, **6**:2.

Iyengar, S., Viswanathan, S. S., Bottjer, S. W. (1999) Development of topography within song control circuitry of zebra finches during the sensitive period for song learning. *J Neurosci*, **19**:6037–6057.

Iyengar, S., Bottjer, S. W. (2002a) Development of individual axon arbors in a thalamocortical circuit necessary for song learning in zebra finches. *J Neurosci*, **22**:901–911.

Iyengar, S., Bottjer, S. W. (2002b) The role of auditory experience in the formation of neural circuits underlying vocal learning in zebra finches. *J Neurosci*, **22**:946–958.

Jackson, A. L., Linsley, P. S. (2004) Noise amidst the silence: off-target effects of siRNAs? *Trends Genet*, **20**:521–524.

Jakobson, R., Fant, C. G. M., Halle, M. (1969) *Preliminaries to Speech Analysis: The Distinctive Features and Their Correlates*. Cambridge, MA: MIT Press.

Janata, P., Margoliash, D. (1999) Gradual emergence of song selectivity in sensorimotor structures of the male zebra finch song system. *J Neurosci*, **19**:5108–5118.

Janik, V. M. (2000) Whistle matching in wild bottlenose dolphins (*Tursiops truncatus*). *Science*, **289**:1355–1357.

Jarvis, E. D. (2004a) Learned birdsong and the neurobiology of human language. *Ann N Y Acad Sci*, **1016**:749–777.

Jarvis, E. D. (2004b) Brains and Birdsong. In *Nature's Music: The Science of Birdsong* (Marler, P., Slabberkoorn, H., eds), pp 239–275. New York: Elsevier.

Jarvis, E. D., Mello, C. V. (2000) Molecular mapping of brain areas involved in parrot vocal communication. *J Comp Neurol* **419**:1–31.

Jarvis, E. D., Nottebohm, F. (1997) Motor-driven gene expression. *Proc Natl Acad Sci USA*, **94**:4097–4102.

Jarvis, E. D., Mello, C. V., Nottebohm, F. (1995) Associative learning and stimulus novelty influence the song-induced expression of an immediate early gene in the canary forebrain. *Learn Mem*, **2**:62–80.

Jarvis, E. D., Scharff, C., Grossman, M. R., Ramos, J. A., Nottebohm, F. (1998) For whom the bird sings: context-dependent gene expression. *Neuron*, **21**:775–788.

Jarvis, E. D., Ribeiro, S., da Silva, M. L., Ventura, D., Vielliard, J., Mello, C. V. (2000) Behaviourally driven gene expression reveals song nuclei in hummingbird brain. *Nature*, **406**:628–632.

Jarvis, E. D., Gunturkun, O., Bruce, L., et al. (2005a) Avian brains and a new understanding of vertebrate brain evolution. *Nat Rev Neurosci*, **6**:151–159.

Jarvis, E. D., Feenders, G., Liedvogel, M., Wada, K., Mouritsen, H. (2005b) Movement-driven gene expression patterns implicate origin of brain areas for vocal learning. *Soc Neurosci Abstr*, **10027**.

Jenkins, P. F. (1978) Cultural transmission of song patterns and dialect development in a free-living bird population. *Anim Behav*, **26**:50–78.

Jin, H., Clayton, D. F., (1997a) Localized changes in immediate-early gene regulation during sensory and motor learning in zebra finches. *Neuron*, **19**:1049–1059.

Jin, H., Clayton, D. F. (1997b) Synelfin regulation during the critical period for song learning in normal and isolated juvenile zebra finches. *Neurobiology of Learning and Memory*, **68**:271–284.

Johnson, F., Bottjer, S. W. (1992) Growth and regression of thalamic efferents in the song-control system of male zebra finches. *J Comp Neurol*, **326**:442–450.

Johnson, F., Bottjer, S. W. (1994) Afferent influences on cell death and birth during development of a cortical nucleus necessary for learned vocal behavior in zebra finches. *Development*, **120**:13–24.

Johnson, F., Sablan, M. M., Bottjer, S. W. (1995) Topographic organization of a forebrain pathway involved with vocal learning in zebra finches. *J Comp Neurol*, **358**:260–278.

Johnson, F., Hohmann, S. E., DiStefano, P. S., Bottjer, S. W. (1997) Neurotrophins suppress apoptosis induced by deafferentation of an avian motor-cortical region. *J Neurosci*, **17**:2101–2111.

Johnson, F., Soderstrom, K., Whitney, O. (2002) Quantifying song bout production during zebra finch sensory-motor learning suggests a sensitive period for vocal practice. *Behav Brain Res*, **131**:57–65.

Johnson, J. S., Newport, E. L. (1991) Critical period effects on universal properties of language: the status of subjacency in the acquisition of a second language. *Cognition*, **39**:215–258.

Johnston, J. B. (1923) Further contributions to the study of the evolution of the forebrain. *J Comp Neurol*, **35**:337–481.

Joliveau, E., Smith, J., Wolfe, J. (2004) Tuning of vocal tract resonance by sopranos. *Nature*, **427**:116.

Jones, A. E., ten Cate, C., Slater, P. J. B. (1996) Early experience and plasticity of song in adult male zebra finches (*Taeniopygia guttata*). *J Comp Psychol*, **110**:354–69.

Jones, M. W., Errington, M. L., French, P. J., *et al.* (2001) A requirement for the immediate early gene Zif268 in the expression of late LTP and long-term memories. *Nat Neurosci*, **4**:289–296.

Jordan, M. I. (1995) Computational motor control. In *The Cognitive Neurosciences* (Gazzaniga, M., ed.), pp. 567–610. Cambridge, MA: MIT Press.

Jost, A. (1947) Recherches sur la différenciation sexuelle de l'embryon de lapin. *Arch Anat Microsc Morphol*, **36**:271–315.

Jurgens, U. (2002) Neural pathways underlying vocal control. *Neurosci Biobehav Rev*, **26**:235–258.

Jurisevic, M. A., Sanderson, K. J. (1994) Alarm vocalisations in Australian birds: convergent characteristics and phylogenetic differences. *Emu*, **94**:69–77.

Jurisevic, M. A., Sanderson, K. J. (1998) A comparative analysis of distress call structure in Australian passerine and non-passerine species: influence of size and phylogeny. *J Avian Biol*, **29**:61–71.

Jusczyk, P. W. (1981) Infant speech perception: a critical appraisal. In *Perspectives on the Study of Speech* (Eimas, P. D., Miller, J. L., eds.), pp. 113–164. Hillsdale, NJ: Erlbaum.

Jusczyk, P. W., Rosner, B. S., Cutting, J. E., Foard, C. F., Smith, L. B. (1977) Categorical perception of nonspeech sounds by 2-month-old infants. *Percep Psychophys*, **21**:50–54.

Jusczyk, P. W., Hirsh-Pasek, K., Nelson, D. G., Kennedy, L. J., Woodward, A., Piwoz, J. (1992) Perception of acoustic correlates of major phrasal units by young infants. *Cognit Psychol*, **24**:252–293.

Jusczyk, P. W., Friederici, A. D., Wessels, J. M., Svenkerud, V. Y., Jusczyk, A. M. (1993) Infants' sensitivity to the sound patterns of native language words. *J Mem Lang*, **32**:402–420.

Kaas, J. H., Hackett, T. A. (1999) 'What' and 'where' processing in auditory cortex. *Nat Neurosci*, **2**:1045–1047.

Kaas, J. H., Hackett, T. A. (2000) Subdivisions of auditory cortex and processing streams in primates. *Proc Natl Acad Sci USA*, **97**:11793–11799.

Kacsoh, B. (2000) *Endocrine Physiology*. New York: McGraw-Hill, Health Professions Division.

Kaczmarek, L., Robertson, H. A. (2002) *Immediate Early Genes and Inducible Transcription Factors in Mapping of the Central Nervous System Function and Dysfunction*. Amsterdam: Elsevier.

Kaestner, K. H., Knochel, W., Martinez, D. E. (2000) Unified nomenclature for the winged helix/forkhead transcription factors. *Genes Dev*, **14**:142–146.

Källén, B. (1953) On the nuclear differentiation during the embryogenesis in the avian forebrain and some notes on the amniote strio-amygdaloid complex. *Avata Anat (Basel)*, **17**:72–84.

Kamen, R. S., Watson, B. C. (1991) Effects of long-term tracheostomy on spectral characteristics of vowel production. *J Speech Hear Res*, **34**:1057–1065.

Kao, M. H., Brainard, M. S. (2006) Lesions of an avian basal ganglia circuit prevent context-dependent changes to song variability. *J Neurophysiol*, **96**:1441–1455.

Kao, M. H., Doupe, A. J., Brainard, M. S. (2005) Contributions of an avian basal ganglia-forebrain circuit to real-time modulation of song. *Nature*, **433**:638–643.

Kaplan, G. (1999) Song structure and function of mimicry in the Australian magpie (*Gymnorhina tibicen*) compared with the lyrebird (*Menura*). *Int J Comp Psychol*, **4**:219–241.

Kaplan, G. (2003) Magpie mimicry. *Nature Australia*, **27**:60–67.

Kaplan, G. (2004) *Australian Magpie: Biology and Behaviour of an Unusual Songbird*. Sydney/Melbourne: University of New South Wales Press/CSIRO.

Kaplan, G. (2005a) The vocal behaviour of Australian magpies (*Gymnorhina tibicen*): a study of vocal development, song learning, communication and mimicry in the Australian magpie. PhD thesis. St. Lucia, Brisbane: School of Veterinary Sciences, University of Queensland.

Kaplan, G. (2005b) Vocal development, vocal learning and babbling in an avian species: parallels to human development? In *Proceedings of the International Society for Developmental Psychobiology*. Washington, DC.

Kaplan, G. (2006a) Alarm calls, communication and cognition in Australian magpies (*Gymnorhina tibicen*). *Acta Zool Sinica*, **52**:614–617.

Kaplan, G. (2006b) Australian magpie: voice. In *The Handbook of Australian, New Zealand, and Antarctic Birds*, Vol. 7. (Higgins, P. J., Peter, J., Cowling, S. J., eds.), pp. 605–608, 613–616. Melbourne: Oxford University Press.

Kaplan, G. (2007) Alarm calls and referentiality in Australian magpies: between midbrain and forebrain, can a case be made for complex cognition? *Brain Res Bull*, Special Issue (in press).

Kaplan, G., Rogers, L. J. (2001) *Birds: Their Habits and Skills*. Sydney: Allen & Unwin.

Kaplan, M. S., Hinds, J. W. (1977) Neurogenesis in the adult rat: electron microscopic analysis of light radioautographs. *Science*, **197**:1092–1094.

Karten, H. J. (1968) The ascending auditory pathway in the pigeon (*Columba livia*). II. Telencephalic projections of the nucleus ovoidalis thalami. *Brain Res*, **11**:134–153.

Karten, H. J. (1969) The organization of the avian telencephalon and some speculations on the phylogeny of the amniote telencephalon. In *Comparative and Evolutionary Aspects of the Vertebrate Central Nervous System* (Pertras, J., ed.), pp. 164–179. New York: NY Academy of Science.

Karten, H. J. (1991) Homology and evolutionary origins of the 'neocortex'. *Brain Behav Evol*, **38**:264–272.

Karten, H. J., Hodos, W. (1967) *A Stereotaxic Atlas of the Brain of the Pigeon (Columba livia)*. Baltimore, MD: Johns Hopkins University Press.

Karten, H. J., Shimizu, T. (1989) The origins of neocortex: connections and lamination as distinct events in evolution. *J Cogn Neurosci*, **1**:291–301.

Katz, L. C., Gurney, M. E. (1981) Auditory responses in the zebra finch's motor system for song. *Brain Res*, **221**:192–197.

Katz, D. M., Karten, H. J. (1983) Visceral representation within the nucleus of the tractus solitarius in the pigeon, *Columba livia*. *J Comp Neurol*, **218**:42–73.

Kawaguchi, Y. (1993) Physiological, morphological, and histochemical characterization of three classes of interneurons in rat neostriatum. *J Neurosci*, **13**:4908–4923.

Kawaguchi, Y., Katsumaru, H., Kosaka, T., Heizmann, C. W., Hama, K. (1987) Fast spiking cells in rat hippocampus (CA1 region) contain the calcium-binding protein parvalbumin. *Brain Res*, **416**:369–374.

Kawata, M. (1995) Roles of steroid hormones and their receptors in structural organization in the nervous system. *Neurosci Res*, **24**:1–46.

Kay, L. M., Laurent, G. (1999) Odor- and context-dependent modulation of mitral cell activity in behaving rats. *Nat Neurosci*, **2**:1003–1009.

Kelley, D. B., Nottebohm, F. (1979) Projections of a telencephalic auditory nucleus-field L in the canary. *J Comp Neurol*, **183**:455–469.

Kelley, D. B., Pfaff, D. W. (1978) Generalizations from comparative studies on neuroanatomical and endocrine mechanisms of sexual behaviour. In *Biological Determinants of Sexual Behaviour* (Hutchison, J. B., ed.), pp. 225–254. Chichester: John Wiley & Sons.

Kent, R. D. (1997) *The Speech Sciences*. San Diego, CA: Singular Publishing Group, Inc.

Kiang, N. Y. (1965) *Discharge Patterns of Single Fibers in the Cat's Auditory Nerve*. Cambridge, MA: MIT Press.

Kilgard, M. P., Merzenich, M. M. (1998) Cortical map reorganization enabled by nucleus basalis activity. *Science*, **279**:1714–1718.

Kilgard, M. P. (2003) Cholinergic modulation of skill learning and plasticity. *Neuron*, **38**:678–680.

Kim, M., McGaugh, J. L. (1992) Effects of intra-amygdala injections of NMDA receptor antagonists on acquisition and retention of inhibitory avoidance. *Brain Res*, **585**:35–48.

Kim, K. H., Relkin, N. R., Lee, K. M., Hirsch, J. (1997) Distinct cortical areas associated with native and second languages. *Nature*, **388**:171–174.

Kim, Y. H., Perlman, W. R., Arnold, A. P. (2004) Expression of androgen receptor mRNA in zebra finch song system: developmental regulation by estrogen. *J Comp Neurol*, **469**:535–547.

Kimpo, R. R., Doupe, A. J. (1997) FOS is induced by singing in distinct neuronal populations in a motor network. *Neuron*, **18**:315–325.

Kimura, M. (1968) Evolutionary rate at the molecular level. *Nature*, **217**:624–626.

King, A. P., West, M. J. (1983) Epigenesis of cowbird song: a joint endeavour of males and females. *Nature*, **305**:704–706.

King, A. P., West, M. J. (1989) The effect of female cowbirds on vocal imitation and improvisation in males. *J Comp Psychol*, **103**:39–44.

King, A. S. (1989) Functional anatomy of the syrinx. In *Form and Function in Birds* (King, A. S., McLelland, J., eds.), pp. 105–192. London: Academic Press.

King, A. S., Molony, V. (1971) The anatomy of respiration. In *Physiology and Biochemistry of the Domestic Fowl*

(Bell, D. J., Freeman, B. M., eds.), pp. 93–169. London: Academic Press.

Kipper, S., Mundry, R., Hultsch, H., Todt, D. (2004) Long term persistence of song performance rules in nightingales: a longitudinal field study on repertoire size and composition. *Behaviour*, **141**:371–390.

Kirn, J. R., DeVoogd, T. J. (1989) Genesis and death of vocal control neurons during sexual differentiation in the zebra finch. *J Neurosci*, **9**:3176–3187.

Kirn, J. R., Nottebohm, F. (1993) Direct evidence for loss and replacement of projection neurons in adult canary brain. *J Neurosci*, **13**:1654–1663.

Kirn, J. R., Clower, R. P., Kroodsma, D. E., Devoogd, T. J. (1989) Song-related brain regions in the red-winged blackbird are affected by sex and season but not repertoire size. *J Neurobiol*, **20**:139–163.

Kirn, J. R., Alvarez-Buylla A., Nottebohm, F. (1991) Production and survival of projection neurons in a forebrain vocal center of adult male canaries. *J Neurosci*, **11**:1756–1762.

Kirn, J., O'Loughlin, B., Kasparian, S., Nottebohm, F. (1994) Cell death and neuronal recruitment in the high vocal center of adult male canaries are temporally related to changes in song. *Proc Natl Acad Sci USA*, **91**:7844–7848.

Kirn, J. R., Fishman, Y., Sasportas, K., Alvarez-Buylla, A., Nottebohm, F. (1999) The fate of new neurons in adult canary high vocal center during the first 30 days after their formation. *J Comp Neurol*, **411**(3):487–494.

Kiss, J. Z., Voorhuis, T. A., van Eekelen, J. A., de Kloet, E. R., de Wied, D. (1987) Organization of vasotocin-immunoreactive cells and fibers in the canary brain. *J Comp Neurol*, **263**:347–364.

Kitt, C. A., Brauth, S. E. (1982) A paleostriatal-thalamic-telencephalic path in pigeons. *Neuroscience*, **7**:2735–2751.

Kittelberger, J. M., Mooney, R. (1999) Lesions of an avian forebrain nucleus that disrupt song development alter synaptic connectivity and transmission in the vocal premotor pathway. *J Neurosci*, **19**:9385–9398.

Kittelberger, J. M., Mooney, R. (2005) Acute injections of brain-derived neurotrophic factor in a vocal premotor nucleus reversibly disrupt adult birdsong stability and trigger syllable deletion. *J Neurobiol*, **62**:406–424.

Klatt, D. H., Stefanski, R. A. (1974) How does a mynah bird imitate human speech? *J Acoust Soc Am*, **55**:822–832.

Kleinfeld, D., Sachdev, R. N., Merchant, L. M., Jarvis, M. R., Ebner, F. F. (2002) Adaptive filtering of vibrissa input in motor cortex of rat. *Neuron*, **34**:1021–1034.

Knapska, E., Kaczmarek, L. (2004) A gene for neuronal plasticity in the mammalian brain: Zif 268/Egr-1/NGFI-A/Krox-24/TIS8/ZENK? *Prog Neurobiol*, **74**:183–211.

Knipschild, M., Dorrscheidt, G. J., Rubsamen, R. (1992) Setting complex tasks to single units in the avian auditory forebrain. I. Processing of complex artificial stimuli. *Hear Res*, **57**:216–230.

Kobayashi, K., Uno, H., Okanoya, K. (2001) Partial lesions in the anterior forebrain pathway affect song production in adult Bengalese finches. *Neuroreport*, **12**:353–358.

Kohara, K., Kitamura, A., Morishima, M., Tsumoto, T. (2001) Activity-dependent transfer of brain-derived neurotrophic factor to postsynaptic neurons. *Science*, **291**:2419–2423.

Komai, S., Licznerski, P., Cetin, A. et al. (2006) Postsynaptic excitability is necessary for strengthening of cortical sensory responses during experience-dependent development. *Nat Neurosci*, **9**:1125–1133. Epub 2006 Aug 20.

Konishi, M. (1963) The role of auditory feedback in the vocal behavior of the domestic fowl. *Z Tierpsychol*, **20**:349–367.

Konishi, M. (1964a) Effects of deafening on song development in two species of juncos. *Condor*, **66**:85–102.

Konishi, M. (1964b) Song variation in a population of Oregon juncos. *Condor*, **66**:423–426.

Konishi, M. (1965a) Effects of deafening on song development in American robins and black-headed grosbeaks. *Z Tierpsychol*, **22**:584–599.

Konishi, M. (1965b) The role of auditory feedback in the control of vocalization in the white-crowned sparrow. *Z Tierpsychol*, **22**:770–783.

Konishi, M. (1969) Hearing, single-unit analysis, and vocalizations in songbirds. *Science*, **166**:1178–1181.

Konishi, M. (1970) Comparative neurophysiological studies of hearing and vocalization in songbirds. *Z Vergl Physiol*, **66**:257–272.

Konishi, M. (1973) Development of auditory neuronal responses in avian embryos. *Proc Natl Acad Sci USA*, **70**:1795–1798.

Konishi, M. (1978) Auditory environment and vocal development in birds. In *Perception and Experience* (Walk, R. D., Pick, H. L. J., eds.), pp. 105–118. New York: Plenum Press.

Konishi, M. (1985) Birdsong: from behavior to neuron. *Annu Rev Neurosci*, **8**:125–170.

Konishi, M. (1994) Pattern generation in birdsong. *Curr Opin Neurobiol*, **4**:827–831.

Konishi, M. (2004) The role of auditory feedback in birdsong. *Ann NY Acad Sci*, **1016**:463–475.

Konishi, M., Akutagawa, E. (1985) Neuronal growth, atrophy and death in a sexually dimorphic song nucleus in the zebra finch brain. *Nature*, **315**:145–147.

Konishi, M., Nottebohm, F. (1969) Experimental studies in the ontogeny of avian vocalizations. In *Bird Vocalizations: Their Relations to Current Problems in Biology and Psychology* (Hinde, R. A., ed.), pp. 29–48. Cambridge: Cambridge University Press.

Kopp, M. L. (1996) Ontogenetische Veränderung in der Zeitstruktur des Gesangs der Nachtigall (*Luscinia megarhynchos*). Ph.D. dissertation. Berlin: Freie Universität.

Korsia, S., Bottjer, S. W. (1991) Chronic testosterone treatment impairs vocal learning in male zebra finches during a restricted period of development. *J Neurosci*, **11**:2362–2371.

Kozhevnikov, A. A., Fee, M. S. (2007) Singing-related activity of identified HVC neurons in the zebra finch. *J Neurophysiol*, **97**:4271–4283.

Kraus, N., McGee, T. J., Carrell, T. D., Zecker, S. G., Nicol, T. G., Koch, D. B. (1996) Auditory neurophysiologic responses and discrimination deficits in children with learning problems. *Science*, **273**:971–973.

Krauzlis, R. J. (2004) Recasting the smooth pursuit eye movement system. *J Neurophysiol*, **91**:591–603.

Krichevsky, A. M., Kosik, K. S. (2002) RNAi functions in cultured mammalian neurons. *Proc Natl Acad Sci USA*, **99**:11926–11929.

Kröner, S., Güntürkün, O. (1999) Afferent and efferent connections of the caudolateral neostriatum in the pigeon (*Columba livia*): a retro- and anterograde pathway tracing study. *J Comp Neurol*, **407**:228–260.

Kroodsma, D. E. (1977a) A re-evaluation of song development in the song sparrow. *Anim Behav*, **25**:390–399.

Kroodsma, D. E. (1977b) Vocal virtuosity in the brown thrasher. *Auk*, **94**:783–784.

Kroodsma, D. E. (1979) Vocal dueling among male marsh wrens: evidence for ritualized expression of dominance/subordinance. *Auk*, **96**:506–515.

Kroodsma, D. E. (1982) Song repertoires: problems in their definition and use. In *Acoustic Communication in Birds*, Vol. 2. *Song Learning and its Consequences* (Kroodsma, D. E., Miller, E. H., eds.), pp. 125–146. New York: Academic Press.

Kroodsma, D. E. (1984) Songs of the alder flycatcher (*Empidonax alnorum*) and willow flycatcher (*Empidonax traillii*) are innate. *Auk*, **101**:13–24.

Kroodsma, D. E. (1985) Development and use of two song forms by the eastern phoebe. *Wilson Bull*, **97**:21–29.

Kroodsma, D. E. (1988) Song types and their use: developmental flexibility of the male blue-winged warbler. *Ethology*, **79**:235–247.

Kroodsma, D. (2004) The diversity and plasticity of birdsong. In *Nature's Music: The Science of Birdsong* (Marler, P., Slabberkoorn, H., eds.), pp. 108–131. New York: Elsevier Academic Press.

Kroodsma, D., ed. (2005) *The Singing Life of Birds. The Art and Science of Listening to Birdsong*. Boston: Houghton Mifflin Company.

Kroodsma, D. E., Baylis, J. R. (1982) A world survey of evidence for vocal learning. In *Acoustic Communication in Birds*, Vol. 2, (Kroodsma, D. E., Miller, E. H., eds.), pp. 311–337. New York: Academic Press.

Kroodsma, D. E., Canady, R. A. (1985) Differences in repertoire size, singing behavior, and associated neuroanatomy among marsh wren populations have a genetic basis. *Auk*, **102**:439–446.

Kroodsma, D. E., Konishi, M. (1991) A suboscine bird (eastern phoebe, *Sayornis phoebe*) develops normal song without auditory feedback. *Anim Behav*, **42**:477–487.

Kroodsma, D. E., Miller, E. H. (1996) *Ecology and Evolution of Acoustic Communication in Birds*. Ithaca, NY: Comstock Publishing Associates/Cornell University Press.

Kroodsma, D. E., Pickert, R. (1980) Environmentally dependent sensitive periods for avian vocal learning. *Nature*, **288**:477–479.

Kroodsma, D. E., Pickert, R. (1984) Sensitive phases for song learning: effects of social interaction and individual variation. *Anim Behav*, **32**:389–394.

Kroodsma, D., Verner, J. (1978) Complex singing behaviors among *Cistothorus* wrens. *Auk*, **95**:703–716.

Kroodsma, D. E., Vielliard, J. M. E., Stiles, F. G. (1996) Study of bird sounds in the neotropics: urgency and opportunity. In *Ecology and Evolution of Acoustic Communication in Birds* (Kroodsma, D. E., Miller, E. H., eds.), pp. 269–281. Ithaca, NY: Comstock Publishing (Cornell University Press).

Kruse, A. A., Stripling, R., Clayton, D. F. (2004) Context-specific habituation of the zenk gene response to song in adult zebra finches. *Neurobiol Learn Mem*, **82**:99–108.

Krützfeldt, N., Logerot, P., Kubke, M. F., Wild, J. M. (2007) Projections of second- and third-order auditory brainstem nuclei in a songbird. Paper at International Congress of Neuroethology, Vancouver, Canada. International Society for Neuroethology, Lawrence, KS 66044, USA.

Kubke, M. F., Yazaki-Sugiyama, Y., Mooney, R., Wild, J. M. (2005) Physiology of neuronal subtypes in the respiratory-vocal integration nucleus retroambigualis of the male zebra finch. *J Neurophysiol*, **94**:2379–2390.

Kubota, M., Saito, N. (1991) NMDA receptors participate differentially in two different synaptic inputs in neurons of the zebra finch robust nucleus of the archistriatum in vitro. *Neurosci Lett*, **125**:107–109.

Kuehl-Kovarik, M. C., Magnusson, K. R., Premkumar, L. S., Partin, K. M. (2000) Electrophysiological analysis of NMDA receptor subunit changes in the aging mouse cortex. *Mech Ageing Dev*, **115**:39–59.

Kuhl, P. K. (1986) Theoretical contributions of tests on animals to the special-mechanisms debate in speech. *Exp Biol*, **45**:233–265.

Kuhl, P. K. (1987) Perception of speech and sound in early infancy. In *Handbook of Infant Perception* (Salapatek, P., Cohen, L. B., eds.), pp. 275–382. New York: Academic Press.

Kuhl, P. K. (1989) On babies, birds, modules, and mechanisms: a comparative approach to the acquisition of vocal communication. In *The Comparative Psychology of Audition: Perceiving Complex Sounds* (Dooling, R. J., Hulse, S. H., eds.), pp. 379–419. Hillsdale, NJ: Erlbaum.

Kuhl, P. K. (1991) Human adults and human infants show a "perceptual magnet effect" for the prototypes of speech categories, monkeys do not. *Percept Psychophys*, **50**:93–107.

Kuhl, P. K. (1994) Learning and representation in speech and language. *Curr Opin Neurobiol*, **4**:812–822.

Kuhl, P. K. (1998) The development of speech and language. In *Mechanistic Relationships Between Development and Learning* (Carew, T. J., Menzel, R., Shatz, C. J., eds.), pp. 53–73. New York: Wiley.

Kuhl, P. K. (2000) A new view of language acquisition. *Proc Natl Acad Sci USA*, **97**:11850–11857.

Kuhl, P. K. (2003) Human speech and birdsong: communication and the social brain. *Proc Natl Acad Sci USA*, **100**:9645–9646.

Kuhl, P. K. (2004) Early language acquisition: cracking the speech code. *Nat Rev Neurosci*, **5**:831–843.

Kuhl, P. K., Meltzoff, A. N. (1982) The bimodal perception of speech in infancy. *Science*, **218**:1138–1141.

Kuhl, P. K., Meltzoff, A. N. (1996) Infant vocalizations in response to speech: vocal imitation and developmental change. *J Acoust Soc Am*, **100**:2425–2438.

Kuhl, P. K., Meltzoff, A. N. (1997) Evolution, nativism, and learning in the development of language and speech. In *The Inheritance and Innateness of Grammars* (Gopnik, M., ed.), pp. 7–44. New York: Oxford University Press.

Kuhl, P. K., Miller, J. D. (1975) Speech perception by the chinchilla: voiced-voiceless distinction in alveolar plosive consonants. *Science*, **190**:69–72.

Kuhl, P. K., Padden, D. M. (1983) Enhanced discriminability at the phonetic boundaries for the place feature in macaques. *J Acoust Soc Am*, **73**:1003–1010.

Kuhl, P. K., Williams, K. A., Lacerda, F., Stevens, K. N., Lindblom, B. (1992) Linguistic experience alters phonetic perception in infants by 6 months of age. *Science*, **255**:606–608.

Kuhl, P. K., Tsuzaki, M., Tohkura, Y., Meltzoff, A. N. (1994) Human processing of auditory-visual information in speech perception: potential for multimodal human-machine interfaces. In *Proceedings of the International Conference on Spoken Language Processing*, pp. 539–542. Tokyo: Acoustical Society of Japan.

Kuhl, P. K., Andruski, J. E., Chistovich, I. A., et al. (1997a) Cross-language analysis of phonetic units in language addressed to infants. *Science*, **277**:684–686.

Kuhl, P. K., Kirtani, S., Deguchi, T., Hayashi, A., Stevens, E. B., Dugger, C. D., Iverson, P. (1997b) Effects of language experience on speech perception: American and Japanese infants' perception of /ra/ and /la/. *J Acoust Soc Am*, **102**:3135–3136.

Kuhlenbeck, H. (1938) The ontogenetic development and phylogenetic significance of the cortex telencephali in the chick. *J Comp Neurol*, **69**:273–301.

Kunkel, P. (1974) Mating systems of tropical birds: the effects of weakness or absence of external reproduction-timing factors, with special reference to prolonged pair bonds. *Z Tierpsychol*, **34**:265–307.

Lade, B. L., Thorpe, W. H. (1964) Dove songs as innately coded patterns of specific behaviour. *Nature*, **212**:366–368.

Ladefoged, P. (1994) The IPA problem. *J Acoust Soc Am*, **96**:3327.

Lahiri, S., Roy, A., Baby, S. M., Hoshi, T., Semenza, G. L., Prabhakar, N. R. (2006) Oxygen sensing in the body. *Progr Biophys Mol Biol*, **91**:249–286.

Lai, C. S., Fisher, S. E., Hurst, J. A., Vargha-Khadem, F., Monaco, A. P. (2001) A forkhead-domain gene is mutated in a severe speech and language disorder. *Nature*, **413**:519–523.

Lai, C. S., Gerrelli, D., Monaco, A. P., Fisher, S. E., Copp, A. J. (2003) FOXP2 expression during brain development coincides with adult sites of pathology in a severe speech and language disorder. *Brain*, **126**:2455–2462.

Laje, R., Gardner, T. J., Mindlin, G. B. (2002) Neuromuscular control of vocalizations in birdsong: a model. *Phys Rev E Stat Nonlin Soft Matter Phys*, **65**:051921.

Laje, R., Mindlin, G. B. (2005) Modeling source-source and source-filter acoustic interaction in birdsong. *Phys Rev E Stat Nonlin Soft Matter Phys*, **72**:036218–036211.

Lane, H. (1976) *The Wild Boy of Aveyron*. Cambridge, MA: Harvard University Press.

Langmore, N. E. (1998) Functions of duet and solo songs of female birds. *Trends Ecol Evol*, **13**:136–140.

Langmore, N. E., Davies, N. B., Hatchwell, B. J., Harley, I. R. (1996) Female song attracts males in the alpine accentor, *Prunella collaris*. *Proc R Soc B*, **263**:141–146.

Larsen, O. L., Dabelsteen, T. (1990) Directionality of blackbird vocalization, implications for vocal communication and its further study. *Ornis Scandinavica*, **21**:37–45.

Larsen, O. N., Goller, F. (1999) Role of syringeal vibrations in bird vocalizations. *Proc R Soc Lond B*, **266**:1609–1615.

Larsen, O. N., Goller, F. (2002) Direct observation of syringeal muscle function in songbirds and a parrot. *J Exp Biol*, **205**:25–35.

Lashley, K. S. (1950) In search of the engram. In *Symposia of the Society for Experimental Biology, 4 – Physiological Mechanisms in Animal Behaviour* (Danielli, J. F., Brown, R., eds.). Cambridge: Cambridge University Press.

Lasky, R. E., Syrdal-Lasky, A., Klein, R. E. (1975) VOT discrimination by four to six and a half month old infants from Spanish environments. *J Exp Child Psychol*, **20**:215–225.

Lass, N. J. (1996) *Principles of Experimental Phonetics*. St. Louis: Mosby-Year Book, Inc.

Lauder, G. V. (1986) Homology, analogy, and the evolution of behavior. In *Evolution of Animal Behavior* (Nitecki, M. H., Kitchell, J. A., eds.), pp. 9–40. New York: Oxford University Press.

Leadbeater, E., Goller, F., Riebel, K. (2005) Unusual phonation, covarying song characteristics and song preferences in female zebra finches. *Anim Behav*, **70**:909–919.

Lee, B. S. (1950) Effects of delayed speech feedback. *J Acoust Soc Am*, **22**:824–826.

Lee, D. W., Smith, G. T., Tramontin, A. D., Soma, K. K., Brenowitz, E. A., Clayton, N. S. (2001) Hippocampal volume does not change seasonally in a non food-storing songbird. *Neuroreport*, **12**:1925–1928.

Lee, J. L., Everitt, B. J., Thomas, K. L. (2004a) Independent cellular processes for hippocampal memory consolidation and reconsolidation. *Science*, **304**:839–843.

Lee, C. C., Schreiner, C. E., Imaizumi, K., Winer, J. A. (2004b) Tonotopic and heterotopic projection systems in physiologically defined auditory cortex. *Neuroscience*, **128**:871–887.

Lehericy, S., Benali, H., Van de Moortele, P. F., Pelegrini-Issac, M., Waechter, T., Ugurbil, K., Doyon, J. (2005) Distinct basal ganglia territories are engaged in early and advanced motor sequence learning. *Proc Natl Acad Sci USA*, **102**:12566–12571.

Lehrman, D. S. (1953) A critique of Konrad Lorenz's theory of instinctive behavior. *Q Rev Biol*, **28**:337–363.

Leitner, S., Catchpole, C. K. (2002) Female canaries that respond and discriminate more between male songs of different quality have a larger song control nucleus (HVC) in the brain. *J Neurobiol*, **52**:294–301.

Leitner, S., Voigt, C., Gahr, M. (2001a) Seasonal changes in the song pattern of the non-domesticated island canary (*Serinus canaria*), a field study. *Behaviour*, **138**:885–904.

Leitner, S., Voigt, C., Garcia-Segura, L. M., Van't Hof, T., Gahr, M. (2001b) Seasonal activation and inactivation of song motor memories in wild canaries is not reflected in neuroanatomical changes of forebrain song areas. *Horm Behav*, **40**:160–168.

Leitner, S., Voigt, C., Metzdorf, R., Catchpole, C. K. (2005) Immediate early gene (ZENK, Arc) expression in the auditory forebrain of female canaries varies in response to male song quality. *J Neurobiol*, **64**:275–284.

Lemaire, P., Revelant, O., Bravo, R., Charnay, P. (1988) Two mouse genes encoding potential transcription factors with identical DNA-binding domains are activated by growth factors in cultured cells. *Proc Natl Acad Sci USA*, **85**:4691–4695.

Lenneberg, E. H. (1967) *Biological Foundations of Language*. New York: Wiley.

Leonard, M. L., Horn, A. G. (2005) Ambient noise and the design of begging signals. *Proc Biol Sci*, **272**:651–656.

Leonard, S. L. (1939) Induction of singing in female canaries by injections of male hormone. *Proc Soc Exp Biol*, **41**:229–230.

Leonardo, A., Konishi, M. (1999) Decrystallization of adult birdsong by perturbation of auditory feedback. *Nature*, **399**:466–470.

Leonardo, A. (2004) Experimental test of the birdsong error-correction model. *Proc Natl Acad Sci USA*, **101**:16935–16940.

Leonardo, A., Fee, M. S. (2005) Ensemble coding of vocal control in birdsong. *J Neurosci*, **25**:652–661.

Leppelsack, H. J. (1978) Unit responses to species-specific sounds in the auditory forebrain center of birds. *Fed Proc*, **37**:2236–2241.

Leppelsack, H. J. (1983) Analysis of song in the auditory pathway of song-birds. In *Advances in Vertebrate Neuroethology* (Ewert, J. P., Capranica, B. R., Ingle, D. J., eds.), pp. 783–800. New York: Plenum Press.

Leppelsack, H. J., Schwartzkopff, J. (1972) Properties of acoustic neurons in the caudal neostriatum of birds. *J Comp Physiol*, **80**:137–140.

Leppelsack, H. J., Vogt, M. (1976) Responses of auditory neurons in the forebrain of a songbird to stimulation with species-specific sounds. *J Comp Neurol*, **107**:263–274.

Levin, R. N. (1996a) Song behavior and reproductive strategies in a duetting wren, Thryothorus nigricapillus: I. Removal studies. *Animal Behviour*, **52**:1093–1106.

Levin, R. N. (1996b) Song behavior and reproductive strategies in a duetting wren, Thryothorus nigricapillus; II. Playback studies. *Animal Behaviour*, **52**:1107–1117.

Lewicki, M. S. (1996) Intracellular characterization of song-specific neurons in the zebra finch auditory forebrain. *J Neurosci*, **16**:5855–5863.

Lewicki, M. S., Arthur, B. J. (1996) Hierarchical organization of auditory temporal context sensitivity. *J Neurosci*, **16**:6987–6998.

Lewicki, M. S., Konishi, M. (1995) Mechanisms underlying the sensitivity of songbird forebrain neurons to temporal order. *Proc Natl Acad Sci USA*, **92**:5582–5586.

Lewis, J. W., Ryan, S. M., Arnold, A. P., Butcher, L. L. (1981) Evidence for a catecholaminergic projection to area X in the zebra finch. *J Comp Neurol*, **196**:347–354.

Li, X. C., Jarvis, E. D., Alvarez-Borda, B., Lim, D. A., Nottebohm, F. (2000) A relationship between behavior, neurotrophin expression, and new neuron survival. *Proc Natl Acad Sci USA*, **97**:8584–8589.

Liberman, A. M., Cooper, F. S., Shankweiler, D. P., Studdert-Kennedy, M. (1967) Perception of the speech code. *Psychol Rev*, **74**:431–461.

Liberman, A. M., Mattingly, I. G. (1985) The motor theory of speech perception revised. *Cognition*, **21**:1–36.

Lichtman, J. W., Colman, H. (2000) Synapse elimination and indelible memory. *Neuron*, **25**:269–278.

Lieberman, P. (2001) Human language and our reptilian brain. The subcortical bases of speech, syntax, and thought. *Perspect Biol Med*, **44**:32–51.

Liegeois, F., Baldeweg, T., Connelly, A., Gadian, D. G., Mishkin, M., Vargha-Khadem, F. (2003) Language fMRI abnormalities associated with FOXP2 gene mutation. *Nat Neurosci*, **6**:1230–1237.

Lill, A. (1976) Lek behavior in the golden-headed manakin, *Pipra erythrocephala* in Trinidad (West Indies). *J Comp Ethology*, **18**:1–83.

Lillie, F. R. (1916) The theory of the freemartin. *Science*, **43**:611–613.

Lind, H., Dabelsteen, T., McGregor, P. K. (1997) Female great tits can identify mates by song. *Anim Behav*, **52**:667–671.

Ling, C., Zuo, M., Alvarez-Buylla, A., Cheng, M. (1997) Neurogenesis in juvenile and adult ring doves. *J Comp Neurol*, **379**:300–312.

Link, W., Konietzko, U., Kauselmann, G., Krug, M., Schwanke, B., Frey, U., Kuhl, D. (1995) Somatodendritic expression of an immediate early gene is regulated by synaptic activity. *Proc Natl Acad Sci USA*, **92**:5734–5738.

Lipkind, D., Nottebohm, F., Rado, R., Barnea, A. (2002) Social change affects the survival of new neurons in the forebrain of adult songbirds. *Behav Brain Res*, **133**:31–43.

Lisman, J. E., McIntyre, C. C. (2001) Synaptic plasticity: a molecular memory switch. *Curr Biol*, **11**:R788–791.

Lisman, J., Schulman, H., Cline, H. (2002) The molecular basis of CaMKII function in synaptic and behavioural memory. *Nat Rev Neurosci*, **3**:175–190.

Lisman, J. (2003) Long-term potentiation: outstanding questions and attempted synthesis. *Philos Trans R Soc Lond B Biol Sci*, **358**:829–842.

Liss, B., Roeper, J. (2004) Correlating function and gene expression of individual basal ganglia neurons. *Trends Neurosci*, **27**:475–481.

Liu, W. C., Nottebohm, F. (2005) Variable rate of singing and variable song duration are associated with high immediate early gene expression in two anterior forebrain song nuclei. *Proc Natl Acad Sci USA*, **102**:10724–10729.

Liu, W. C., Gardner, T. J., Nottebohm, F. (2004) Juvenile zebra finches can use multiple strategies to learn the same song. *Proc Natl Acad Sci USA*, **101**:18177–18182.

Lively, S. E., Logan, J. S., Pisoni, D. B. (1993) Training Japanese listeners to identify English /r/ and /l/. II. The role of phonetic environment and talker variability in learning new perceptual categories. *J Acoust Soc Am*, **94**:1242–1255.

Livingston, F. S., Mooney, R. (1997) Development of intrinsic and synaptic properties in a forebrain nucleus essential to avian song learning. *J Neurosci*, **17**:8997–9009.

Livingston, F. S., Mooney, R. (2001) Androgens and isolation from adult tutors differentially affect the development of songbird neurons critical to vocal plasticity. *J Neurophysiol*, **85**:34–42.

Livingston, F. S., White, S. A., Mooney, R. (2000) Slow NMDA-EPSCs at synapses critical for song development are not required for song learning in zebra finches. *Nat Neurosci*, **3**:482–488.

Lledo, P.-E., Alonso, M., Grubb, M. S. (2006) Adult neurogenesis and functional plasticity in neuronal circuits. *Nat Rev Neurosci*, **7**:179–193.

Lobo, M. K., Karsten, S. L., Gray, M., Geschwind, D. H., Yang, X. W. (2006) FACS-array profiling of striatal projection neuron subtypes in juvenile and adult mouse brains. *Nat Neurosci*, **9**:443–452.

Locke, J. L., Pearson, D. M. (1990) Linguistic significance of babbling: evidence from a tracheostomized infant. *J Child Lang*, **17**:1–16.

Locke, J. L., Snow, C. (1997) Social influences on vocal learning in human and nonhuman primates. In *Social Influences on Vocal Development* (Snowdon, C. T., Hausberger, M., eds.), pp. 274–292. New York: Cambridge University Press.

Lois, C., Hong, E. J., Pease, S., Brown, E. J., Baltimore, D. (2002) Germline transmission and tissue-specific expression of transgenes delivered by lentiviral vectors. *Science*, **295**:868–872.

Lombardino, A. J., Nottebohm, F. (2000) Age at deafening affects the stability of learned song in adult male zebra finches. *J Neurosci*, **20**:5054–5064.

Lombardino, A. J., Li, X. C., Hertel, M., Nottebohm, F. (2005) Replaceable neurons and neurodegenerative disease share depressed UCHL1 levels. *Proc Natl Acad Sci USA*, **102**:8036–8041.

Lombardino, A. J., Hertel, M., Li, X. C., et al. (2006) Expression profiling of intermingled long-range projection neurons harvested by laser capture microdissection. *J Neurosci Methods*, **157**:195–207.

Long, K. D., Kennedy, G., Salbaum, J. M., Balaban, E. (2002) Auditory stimulus-induced changes in immediate-early gene expression related to an inborn perceptual predisposition. *J Comp Physiol A Neuroethol Sens Neural Behav Physiol*, **188**:25–38.

Lorenz, K. (1965) *Evolution and Modification of Behavior*. Chicago and London: University of Chicago Press.

Lorenz, K. (1970) *Studies in Animal and Human Behaviour*. London: Methuen.

Lowther, J. (1961) Polymorphism in the white-throated sparrow, *Zonotrichia albicollis* (Gmelin). *Canadian J Zool*, **39**:281–292.

Luine, V. N., Nottebohm, F., Harding, C., McEwen, B. S. (1980) Androgen affects cholinergic enzymes in syringeal motor neurons and muscle. *Brain Res*, **192**:89–107.

Luine, V. N., Harding, C. F., Bleisch, W. V. (1983) Specificity of gonadal hormone modulation of cholinergic enzymes in the avian syrinx. *Brain Res*, **279**:339–342.

Luo, M., Perkel, D. J. (1999a) A GABAergic, strongly inhibitory projection to a thalamic nucleus in the zebra finch song system. *J Neurosci*, **19**:6700–6711.

Luo, M., Perkel, D. J. (1999b) Long-range GABAergic projection in a circuit essential for vocal learning. *J Comp Neurol*, **403**:68–84.

Luo, M., Ding, L., Perkel, D. J. (2001) An avian basal ganglia pathway essential for vocal learning forms a closed topographic loop. *J Neurosci*, **21**:6836–6845.

Lyford, G. L., Yamagata, K., Kaufmann, W. E., et al. (1995) Arc, a growth factor and activity-regulated gene, encodes a novel cytoskeleton-associated protein that is enriched in neuronal dendrites. *Neuron*, **14**:433–445.

MacDermot, K. D., Bonora, E., Sykes, N., et al. (2005) Identification of FOXP2 truncation as a novel cause of developmental speech and language deficits. *Am J Hum Genet*, **76**:1074–1080.

MacDougall-Shackleton, S. A. (1997) Sexual selection and the evolution of song repertoires. In *Current Ornithology* (Nolan, V., Jr., Ketterson, E. D., Thompson, C. F., eds.), pp. 81–124. New York: Plenum press.

MacDougall-Shackleton, S. A., Ball, G. F. (1999) Comparative studies of sex differences in the song-control system of songbirds. *Trends Neurosci*, **22**:432–436.

MacDougall-Shackleton, S. A., Hulse, S. H., Ball, G. F. (1998a) Neural bases of song preferences in female zebra finches (*Taeniopygia guttata*). *NeuroReport*, **9**:3047–3052.

MacDougall-Shackleton, S. A., Hulse, S. H., Ball, G. F. (1998b) Neural correlates of singing behavior in male zebra finches (*Taeniopygia guttata*). *J Neurobiol*, **36**:421–430.

MacDougall-Shackleton, S. A., Hernandez, A. M., Valyear, K. F., Clark, A. P. (2003) Photostimulation induces rapid growth of song-control brain regions in male and female chickadees (*Poecile atricapilla*). *Neurosci Lett*, **340**:165–168.

MacDougall-Shackleton, S. A., Ball, G. F., Edmonds, E., Sul, R., Hahn, T. P. (2005) Age- and sex-related variation in song-control regions in Cassin's finches, *Carpodacus cassinii*. *Brain Behav Evol*, **65**:262–267.

Macklis, J. D., Madison, R. D. (1985) Unfocused laser illumination kills dye-targeted mouse neurons by selective photothermolysis. *Brain Res*, **359**:158–165.

Macphail, E. M., Bolhuis, J. J. (2001) The evolution of intelligence: adaptive specializations versus general process. *Biol Rev Camb Philos Soc*, **76**:341–364.

Madison, R., Macklis, J. D., Thies, C. (1990) Latex nanosphere delivery system (LNDS): novel nanometer-sized carriers of fluorescent dyes and active agents selectively target neuronal subpopulations via uptake and retrograde transport. *Brain Res*, **522**:90–98.

Maina, J. N. (2000) What it takes to fly: the structural and functional refinements in birds and bats. *J Exp Biol*, **203**:3045–3064.

Maney, D. L., Ball, G. F. (2003) Fos-like immunoreactivity in catecholaminergic brain nuclei after territorial behavior in free-living song sparrows. *J Neurobiol*, **56**:163–170.

Maney, D. L., Wingfield, J. C. (1998) Neuroendocrine suppression of female courtship in a wild passerine: corticotropin-releasing factor and endogenous opioids. *J Neuroendocrinol*, **10**:593–599.

Maney, D. L., Goode, C. T., Wingfield, J. C. (1997) Intraventricular infusion of arginine vasotocin induces singing in a female songbird. *J Neuroendocrinol*, **9**:487–491.

Maney, D. L., Schoech, S. J., Sharp, P. J., Wingfield, J. C. (1999) Effects of vasoactive intestinal peptide on plasma prolactin in passerines. *Gen Comp Endocrinol*, **113**:323–330.

Maney, D. L., MacDougall-Shackleton, E. A., MacDougall-Shackleton, S. A., Ball, G. F., Hahn, T. P. (2003) Immediate early gene response to hearing song correlates with receptive behavior and depends on dialect in a female songbird. *J Comp Physiol A Neuroethol Sens Neural Behav Physiol*, **189**:667–674.

Mann, N. I., Marshall-Ball, L., Slater, P. J. B. (2003) The complex song duet of the plain wren. *Condor*, **105**:672–682.

Manogue, K. R., Paton, J. A. (1982) Respiratory gating of activity in the avian vocal control system. *Brain Res*, **247**:383–387.

Mansson, H. (2000) Childhood stuttering: incidence and development. *J Fluency Disord*, **25**:47–57.

Marder, E., Calabrese, R. L. (1996) Principles of rhythmic motor pattern generation. *Physiol Rev*, **76**:687–717.

Marder, E., Bucher, D. (2001) Central pattern generators and the control of rhythmic movements. *Curr Biol*, **11**:R986–R996.

Marean, G. C., Burt, J. M., Beecher, M. D., Rubel, E. W. (1993) Hair cell regeneration in the European starling (*Sturnus vulgaris*): recovery of pure-tone detection thresholds. *Hear Res*, **71**:125–136.

Marean, G. C., Burt, J. M., Beecher, M. D., Rubel, E. W. (1998) Auditory perception following hair cell regeneration in European starling (*Sturnus vulgaris*): frequency and temporal resolution. *J Acoust Soc Am*, **103**:3567–3580.

Margoliash, D. (1983) Acoustic parameters underlying the responses of song-specific neurons in the white-crowned sparrow. *J Neurosci*, **3**:1039–1057.

Margoliash, D. (1986) Preference for autogenous song by auditory neurons in a song system nucleus of the white-crowned sparrow. *J Neurosci*, **6**:1643–1661.

Margoliash, D. (1997) Functional organization of forebrain pathways for song production and perception. *J Neurobiol*, **33**:671–693.

Margoliash, D. (2002) Evaluating theories of bird song learning: implications for future directions. *J Comp Physiol A Neuroethol Sens Neural Behav Physiol*, **188**:851–866.

Margoliash, D., Fortune, E. S. (1992) Temporal and harmonic combination-sensitive neurons in the zebra finch's HVc. *J Neurosci*, **12**:4309–4326.

Margoliash, D., Konishi, M. (1985) Auditory representation of autogenous song in the song system of white-crowned sparrows. *Proc Natl Acad Sci USA*, **82**:5997–6000.

Margoliash, D., Staicer, C., Inoue, S. A. (1994) The process of syllable acquisition in adult indigo buntings. *Behaviour*, **131**:39–64.

Marien, P., Engelborghs, S., Fabbro, F., De Deyn, P. P. (2001) The lateralized linguistic cerebellum: a review and a new hypothesis. *Brain Lang*, **79**:580–600.

Marler, P., Boatman, D. J. (1951) Observations on the birds of Pico, Azores. *Ibis*, **93**:90–99.

Marler, P. (1952) Variation in the song of the chaffinch, *Fringilla coelebs*. *Ibis*, **94**:458–472.

Marler, P. (1955a) Characteristics of some animal calls. *Nature*, **176**:6–8.

Marler, P. (1955b) Studies of fighting in chaffinches. (2) The effect on dominance relations of disguising females as males. *Brit J Anim Behav*, **3**:137–146.

Marler, P. (1956a) The voice of the chaffinch and its function as a language. *Ibis*, **98**:231–261.

Marler, P. (1956b) Uber die Eigenschaften einiger tierlicher Rufe. *J Ornithologie*, **97**:220–227.

Marler, P. (1957) Specific distinctiveness in the communication signals of birds. *Behaviour*, **11**:13–39.

Marler, P. (1961) The filtering of external stimuli during instinctive behavior. In *Current Problems in Animal Behaviour* (Thorpe, W. H., Zangwill, O. L., eds.), pp. 150–166. Cambridge: Cambridge University Press.

Marler, P. (1963) Inheritance and learning in the development of animal vocalizations. In *Acoustic Behavior of Animals* (Busnel, R. G., ed.), pp. 228–243. Amsterdam: Elsevier.

Marler, P. (1965) Communication in monkeys and apes. In *Monkeys and Apes: Field Studies of Ecology and Behavior* (DeVore, I., ed.), pp. 544–584. New York: Holt, Rinehart, and Winston.

Marler, P. (1967) Animal communication signals. *Science*, **157**:769–774.

Marler, P. (1969a) Vocalizations of wild chimpanzees. *Recent Adv Primatol*, **1**:94–100.

Marler, P. (1969b) Of foxes and hedgehogs: the interface between organismal and population biology – II. *Am Zool*, **9**: 261–267.

Marler, P. (1970a) Birdsong and speech development: could there be parallels? *Am Sci*, **58**:669–673.

Marler, P. (1970b) A comparative approach to vocal learning: song development in white-crowned sparrows. *J Comp Physiol Psychol*, **71**:Suppl. 1–25.

Marler, P. (1973) A comparison of vocalizations of red-tailed monkeys and blue monkeys, *Cercopithecus ascanius* and *C. mitis*, in Uganda. *Z Tierpsychol*, **33**(3–4): 223–247.

Marler, P. (1975) On the origin of speech from animal sounds. In *The Role of Speech in Language* (Kavanaugh, J. F., Cutting, J. E., eds.). Cambridge, MA: MIT Press.

Marler, P. (1976a) Sensory templates in species-specific behavior. In *Simpler Networks and Behavior* (Fentress, J. C., ed.), pp. 314–329. Sunderland, MA: Sinauer.

Marler, P. (1976b) Social organization, communication and graded signals: vocal behavior of the chimpanzee and the gorilla. In *Growing Points in Ethology* (Bateson, P., Hinde, R. A., eds.). Cambridge: Cambridge University Press.

Marler, P. (1981) Birdsong: the acquisition of a learned motor skill. *Trends Neurosci*, **4**:88–94.

Marler, P. (1987) Sensitive periods and the roles of specific and general sensory stimulation in birdsong learning. In *Imprinting and Cortical Plasticity* (Rauschecker, J. P., Marler, P., eds.), pp. 99–135. New York: John Wiley.

Marler, P. (1989) Learning by instinct: birdsong. *Am Speech-Lang-Hear Assoc*, **31**:75–79.

Marler, P. (1990) Song learning: the interface between behaviour and neuroethology. *Phil Trans R Soc Lond B*, **329**:109–114.

Marler, P. (1991a) Differences in behavioural development in closely related species: birdsong. In *The Development and Integration of Behaviour: Essays in Honour of Robert Hinde* (Bateson, P., ed.), pp. 41–70. Cambridge: Cambridge University Press.

Marler, P. (1991b) The instinct to learn. In *The Epigenesis of Mind: Essays on Biology and Cognition* (Gelman, S. C. R., ed.). Hillsdale, N.J.: Erlbaum.

Marler, P. (1997) Three models of song learning: evidence from behavior. *J Neurobiol*, **33**:501–516.

Marler, P. (1998) Are sparrow songs "learned" or "innate?" In *Neural Mechanisms of Communication* (Konishi, M., Hauser, M., eds.). Cambridge, MA: MIT Press.

Marler, P. (2004) Bird calls: their potential for behavioral neurobiology. In *Behavioral Neurobiology of Birdsong* (Ziegler, H. P, Marler, P., eds.), pp. 31–44. New York: Ann NY Acad Sci.

Marler, P. (2005) Ethology and the origins of behavioral endocrinology. *Horm Behav*, **47**:493–502.

Marler, P., Doupe, A. J. (2000) Singing in the brain. *Proc Natl Acad Sci USA*, **97**:2965–2967.

Marler, P., Hamilton, W. J. (1966) *Mechanisms of Animal Behavior*. New York: Wiley.

Marler, P., Nelson, D. (1992) Neuroselection and song learning in birds: species universals in a culturally transmitted behavior. *Semin Neurosci*, **4**:415–423.

Marler, P., Nelson, D. A. (1993) Action-based learning: a new form of developmental plasticity in birdsong. *Netherlands J Zool*, **43**:91–103.

Marler, P., Peters, S. (1977) Selective vocal learning in a sparrow. *Science*, **198**:519–521.

Marler, P., Peters, S. (1980) Birdsong and speech: evidence for special processing. In *Perspectives on the Study of Speech* (Eimas, P., Miller, J., eds.), pp. 75–112. Hillsdale, NJ: Erlbaum.

Marler, P., Peters, S. (1981) Sparrows learn adult song and more from memory. *Science*, **213**:780–782.

Marler, P., Peters, S. (1982a) Developmental overproduction and selective attrition: new processes in the epigenesis of birdsong. *Dev Psychobiol*, **15**:369–378.

Marler, P., Peters, S. (1982b) Subsong and plastic song: their role in the vocal learning process. In *Acoustic Communication in Birds* (Kroodsma, D. E., Miller, E. H., eds.), pp. 25–50. New York: Academic Press.

Marler, P., Peters, S. (1982c) Structural changes in song ontogeny in the swamp sparrow *Melospiza georgiana*. *Auk*, **99**:446–458.

Marler, P., Peters, S. (1982d) Long-term storage of bird songs prior to production. *Anim Behav*, **30**:479–482.

Marler, P., Peters, S. (1987) A sensitive period for song acquisition in the song sparrow, *Melospiza melodia*: a case of age-limited learning. *Ethology*, **76**:89–100.

Marler, P., Peters, S. (1988a) The role of song phonology and syntax in vocal learning preferences in the song sparrow, *Melospiza melodia*. *Ethology*, **77**:125–149.

Marler, P., Peters, S. (1988b) Sensitive periods for song acquisition from tape recordings and live tutors in the swamp sparrow. *Ethology*, **77**:76–84.

Marler, P., Peters, S. (1989) Species differences in auditory responsiveness in early vocal learning. In *The Comparative Psychology of Audition: Perceiving Complex Sounds* (Dooling, R. J., Hulse, S. H., eds.), pp. 243–273. Hillsdale, NJ: Erlbaum.

Marler, P., Pickert, R. (1984) Species: universal microstructure in the learned song of the swamp sparrow (*Melospiza georgiana*). *Anim Behav*, **32**:673–689.

Marler, P., Sherman, V. (1983) Song structure without auditory feedback: emendations of the auditory template hypothesis. *J Neurosci*, **3**:517–531.

Marler, P., Sherman, V. (1985) Innate differences in singing behaviour of sparrows reared in isolation from adult conspecific song. *Anim Behav*, **33**:57–71.

Marler, P., Slabbekoorn, H., eds. (2004) *Nature's Music: The Science of Birdsong*. San Diego, CA: Elsevier Academic Press.

Marler, P., Tamura, M. (1962) Song 'dialects' in three populations of white-crowned sparrows. *Condor*, **64**: 368–377.

Marler, P., Tamura, M. (1964) Culturally transmitted patterns of vocal behavior in sparrows. *Science*, **146**:1483–1486.

Marler, P., van Lawick-Goodall, J. (1970) *Vocalizations of Wild Chimpanzees*. Leaflet covering 40 minute color film with sound track. New York: The Rockfeller University Press.

Marler, P., Waser, M. S. (1977) Role of auditory feedback in canary song development. *J Comp Physiol Psychol*, **91**:8–16.

Marler, P., Mundinger, P., Waser, M. S., Lutjen, A. (1972) Effects of acoustical stimulation and deprivation on song development in red-winged blackbirds (*Agelaius phoeniceus*). *Anim Behav*, **20**:586–606.

Marler, P., Konishi, M., Lutjen, A., Waser, M. S. (1973) Effects of continuous noise on avian hearing and vocal development. *Proc Natl Acad Sci USA*, **70**:1393–1396.

Marler, P., Peters, S., Ball, G. F., Dufty, A. M., Jr., Wingfield, J. C. (1988a) The role of sex steroids in the acquisition and production of birdsong. *Nature*, **336**: 770–772.

Marler, P., Evans, C. S., Hauser, M. D. (1992) Animal signals: motivational, referential, or both? In *Nonverbal Vocal Communication: Comparative and Developmental Approaches* (Papousek, H., Jurgens, U., Papousek, M., eds.), pp. 66–86. New York, NY: Cambridge University Press.

Marr, D. (1982) *Vision: A Computational Investigation Into the Human Representation and Processing of Visual Information*. San Francisco: W. H. Freeman.

Marshall, R. C., Leisler, B., Catchpole, C. K., Schwabl, H. (2005) Male song quality affects circulating but not yolk steroid concentrations in female canaries (*Serinus canaria*). *J Exp Biol*, **208**:4593–4598.

Martin, T. E. (1996) Life history evolution in tropical and south temperate birds: what do we really know? *J Avian Biol*, **27**:263–272.

Martins, E. P., ed. (1996) *Phylogenies and the Comparative Method in Animal Behavior*. Oxford, UK: Oxford University Press.

Massaro, D. W. (1987) *Speech Perception by Ear and Eye: A Paradigm for Psychological Inquiry*. Hillsdale, NJ: Erlbaum.

McCardle, P., Wilson, B. E. (1990) Hormonal influence on language development in physically advanced children. *Brain Lang*, **38**:410–423.

McCasland, J. S., Konishi, M. (1981) Interaction between auditory and motor activities in an avian song control nucleus. *Proc Natl Acad Sci USA*, **78**:7815–7819.

McCasland, J. S. (1987) Neuronal control of bird song production. *J Neurosci*, **7**:23–39.

McCowan, B., Reiss, D. (1997) Vocal learning in captive bottlenose dolphins: a comparison with humans and nonhuman animals. In *Social Influences on Vocal Development* (Snowdon, C. T., Hausberger, M., eds.), pp. 178–207. New York: Cambridge University Press.

McEwen, B. S., Davis, P. G., Parsons, B., Pfaff, D. W. (1979) The brain as a target for steroid hormone action. *Annu Rev Neurosci*, **2**:65–112.

McGrew, M. J., Sherman, A., Ellard, F. M., *et al.* (2004) Efficient production of germline transgenic chickens using lentiviral vectors. *EMBO Rep*, **5**:728–733.

McGuire, P. K., Silbersweig, D. A., Frith, C. D. (1996) Functional neuroanatomy of verbal self-monitoring. *Brain*, **119**:907–917.

McGurk, H., MacDonald, J. (1976) Hearing lips and seeing voices. *Nature*, **264**:746–748.

McKenzie, T. L., Hernandez, A. M., MacDougall-Shackleton, S. A. (2006) Experience with songs in adulthood reduces song-induced gene expression in songbird auditory forebrain. *Neurobiol Learn Mem*, **86**:330–335.

Medina, L., Reiner, A. (2000) Do birds possess homologues of mammalian primary visual, somatosensory and motor cortices? *Trends Neurosci*, **23**:1–12.

Mehler, J., Jusczyk, P., Lambertz, G., Halsted, N., Bertoncini, J., Amiel-Tison, C. (1988) A precursor of language acquisition in young infants. *Cognition*, **29**: 143–178.

Meitzen, J., Perkel, D. J., Brenowitz, E. A. (2007) Seasonal changes in intrinsic electrophysiological activity of song control neurons in wild song sparrows. *J Comp Physiol A*, **193**:677–683.

Mello, C. V. (1993) Analysis of immediate early gene expression in the songbird brain following song presentation. Ph.D. dissertation. New York: Rockefeller University.

Mello, C. V. (2002a) Mapping vocal communication pathways in birds with inducible gene expression. *J Comp Physiol A*, **188**:943–959.

Mello, C. V. (2002b) Immediate early gene (IEG) expression mapping of vocal communication areas in the avian brain. In *Immediate Early Genes and Inducible Transcription Factors in Mapping of the Central Nervous System Function and Dysfunction* (Kaczmarek, L., Robertson, H. A., eds.), pp. 59–101. Amsterdam: Elsevier Science.

Mello, C. V., Clayton, D. F. (1994) Song-induced ZENK gene expression in auditory pathways of songbird brain and its relation to the song control system. *J Neurosci*, **14**: 6652–6666.

Mello, C. V., Clayton, D. F. (1995) Differential induction of the ZENK gene in the avian forebrain and song control circuit after metrazole-induced depolarization. *J Neurobiol*, **26**:145–161.

Mello, C. V., Ribeiro, S. (1998) ZENK protein regulation by song in the brain of songbirds. *J Comp Neurol*, **393**: 426–438.

Mello, C. V., Vicario, D. S., Clayton, D. F. (1992) Song presentation induces gene expression in the songbird forebrain. *Proc Natl Acad Sci USA*, **89**:6818–6822.

Mello, C. V., Nottebohm, F., Clayton, D. (1995) Repeated exposure to one song leads to a rapid and persistent decline in an immediate early gene's response to that song in zebra finch telencephalon. *J Neurosci*, **15**: 6919–6925.

Mello, C. V., Vates, G. E., Okuhata, S., Nottebohm, F. (1998a) Descending auditory pathways in the adult male zebra finch (*Taeniopygia guttata*). *J Comp Neurol*, **395**:137–160.

Mello, C. V., Pinaud, R., Ribeiro, S. (1998b) Noradrenergic system of the zebra finch brain: immunocytochemical study of dopamine-beta-hydroxylase. *J Comp Neurol*, **400**:207–228.

Mello, C. V., Velho, T. A., Pinaud, R. (2004) Song-induced gene expression: a window on song auditory processing and perception. *Ann N Y Acad Sci*, **1016**:263–281.

Meltzoff, A. N., Moore, M. K. (1977) Imitation of facial and manual gestures by human neonates. *Science*, **198**:74–78.

Meltzoff, A. N., Moore, M. K. (1997) Explaining facial imitation: a theoretical model. *Early Dev Parent*, **6**:179–192.

Mennill, D. J., Ratcliffe, L. M. (2004) Overlapping and matching in the song contests of black-capped chickadees. *Anim Behav*, **67**:441–450.

Merzenich, M. M., Jenkins, W. M., Johnston, P., Schreiner, C., Miller, S. L., Tallal, P. (1996) Temporal processing deficits of language-learning impaired children ameliorated by training. *Science*, **271**:77–81.

Messmer, E., Messmer, I. (1956) Die Entwicklung der Lautäusserungen und einiger Verhaltensweisen der Amsel (*Turdus merula merula* L) unter natuerlichen

Bedingungen und nach Einzelaufzucht in schalldichten Räumen. *Z Tierpsychol*, **13**:341–441.

Metzger, M., Jiang, S., Braun, K. (1998) Organization of the dorsocaudal neostriatal complex: a retrograde and anterograde tracing study in the domestic chick with special emphasis on pathways relevant to imprinting. *J Comp Neurol*, **395**:380–404.

Micheau, J., Riedel, G., (1999) Protein kinases: which one is the memory molecule? *Cell Mol Life Sci*, **55**:534–548.

Milbrandt, J. (1987) A nerve growth factor-induced gene encodes a possible transcriptional regulatory factor. *Science*, **238**:797–799.

Miller, D. B. (1979) Long-term recognition of father's song by female zebra finches. *Nature*, **280**:389–391.

Mindlin, G. B., Gardner, T. J., Goller, F., Suthers, R. H. (2003) Experimental support for a model of birdsong production. *Phys Rev E Stat Nonlin Soft Matter Phys*, **68**:041908.

Mindlin, G. B., Laje, R. (2005) *The Physics of Birdsong*. Berlin: Springer.

Miska, E., Alvarez-Saavedra, E., Townsend, M., et al. (2004) Microarray analysis of microRNA expression in the developing mammalian brain. *Genome Biol*, **5**:R68.

Mitani, J. C., Hasegawa, T., Gros-Louis, J., Marler, P. (1992) Dialects in wild chimpanzees? *Am J Primatol*, **27**:233–243.

Miyasato, L. E., Baker, M. C. (1999) Black-capped chickadee call dialects along a continuous habitat corridor. *Anim Behav*, **57**:1311–1318.

Miyashita, Y. (2004) Cognitive memory: cellular and network machineries and their top-down control. *Science*, **306**:435–440.

Miyashita, Y., Hayashi, T. (2000) Neural representation of visual objects: encoding and top-down activation. *Curr Opin Neurobiol*, **10**:187–194.

Miyawaki, K., Strange, W., Verbrugge, R., Liberman, A. M., Jenkins, J. J., Fujimura, O. (1975) An effect of linguistic experience: the discrimination of (r) and (l) by native speakers of Japanese and English. *Percept Psychophys*, **18**:331–340.

Moldin, S. O., Rubenstein, J. L., Hyman, S. E. (2006) Can autism speak to neuroscience? *J Neurosci*, **26**:6893–6896.

Molony, V. (1974) Classification of vagal afferents firing in phase with breathing in *Gallus domesticus*. *Resp Physiol*, **22**:57–76.

Monod, J. (1971) *Chance and Necessity*. New York: Knopf.

Monyer, H., Sprengel, R., Schoepfer, R., et al. (1992) Heteromeric NMDA receptors: molecular and functional distinction of subtypes. *Science*, **256**:1217–1221.

Monyer, H., Burnashev, N., Laurie, D. J., Sakmann, B., Seeburg, P. H. (1994) Developmental and regional expression in the rat brain and functional properties of four NMDA receptors. *Neuron*, **12**:529–540.

Moon, C., Cooper, R. P., Fifer, W. P. (1993) Two-day-olds prefer their native language. *Infant Behav Dev*, **16**:495–500.

Mooney, R., Konishi, M. (1991) Two distinct inputs to an avian song nucleus activate different glutamate receptor subtypes on individual neurons. *Proc Natl Acad Sci USA*, **88**:4075–4079.

Mooney, R. (1992) Synaptic basis for developmental plasticity in a birdsong nucleus. *J Neurosci*, **12**:2464–2477.

Mooney, R. (2000) Different subthreshold mechanisms underlie song selectivity in identified HVc neurons of the zebra finch. *J Neurosci*, **20**:5420–5436.

Mooney, R. (2004) Synaptic mechanisms for auditory-vocal integration and the correction of vocal errors. *Ann N Y Acad Sci*, **1016**:476–494.

Mooney, R., Prather, J. F. (2005) The HVC microcircuit: the synaptic basis for interactions between song motor and vocal plasticity pathways. *J Neurosci*, **25**:1952–1964.

Mooney, R., Hoese, W., Nowicki, S. (2001) Auditory representation of the vocal repertoire in a songbird with multiple song types. *Proc Natl Acad Sci USA*, **98**:12778–12783.

Mooney, R., Rosen, M. J., Sturdy, C. B. (2002) A bird's eye view: top down intracellular analyses of auditory selectivity for learned vocalizations. *J Comp Physiol A Neuroethol Sens Neural Behav Physiol*, **188**:879–895.

Moore, I. T., Walker, B. G., Wingfield, J. C. (2004a) The effects of combined aromatase inhibitor and anti-androgen on male territorial aggression in a tropical population of rufous-collared sparrows, *Zonotrichia capensis*. *Gen Comp Endocrinol*, **135**:223–229.

Moore, I. T., Wada, H., Perfito, N., Busch, D. S., Wingfield, J. C. (2004b) Territoriality and testosterone in an equatorial population of rufous-collared sparrows, *Zonotrichia capensis*. *Anim Behav*, **67**:411–420.

Moore, I. T., Wingfield, J. C., Brenowitz, E. A. (2004c) Plasticity of the avian song control system in response to localized environmental cues in an equatorial songbird. *J Neurosci*, **24**:10182–10185.

Moore, I. T., Bentley, G. E., Wotus, C., Wingfield, J. C. (2006) Photoperiod-independent changes in immunoreactive brain gonadotropin-releasing hormone

(GnRH) in a free-living, tropical bird. *Brain Behav Evol*, **68**:37–44.

Morgan, J. I., Cohen, D. R., Hempstead, J. L., Curran, T. (1987) Mapping patterns of c-fos expression in the central nervous system after seizure. *Science*, **237**:192–197.

Morgan, J. I., Curran, T. (1989) Stimulus-transcription coupling in neurons: role of cellular immediate-early genes. *Trends Neurosci*, **12**:459–462.

Morgan, J. I., Curran, T. (1991) Stimulus-transcription coupling in the nervous system: involvement of the inducible proto-oncogenes fos and jun. *Annu Rev Neurosci*, **14**:421–451.

Moriyama, K., Okanoya, K. (1996) Effect of beak movement in singing Bengalese finches. *Abstracts Acoustical Society of America and Acoustical Society of Japan*, Third Joint Meeting Honolulu, 2–6 Dec 1996: 129–130.

Morris, D. (1954) The reproductive behaviour of the zebra finch (*Poephila guttata*) with special reference to pseudofemale behaviour and displacement activities. *Behaviour*, **7**:1–31.

Morris, D. (1980) *Animal Days*. New York: William Morrow.

Morris, R. G., Anderson, E., Lynch, G. S., Baudry, M. (1986) Selective impairment of learning and blockade of long-term potentiation by an N-methyl-D-aspartate receptor antagonist, AP5. *Nature*, **319**:774–776.

Morris, R. G. (1989) Synaptic plasticity and learning: selective impairment of learning rats and blockade of long-term potentiation in vivo by the N-methyl-D-aspartate receptor antagonist AP5. *J Neurosci*, **9**:3040–3057.

Morrison, R. G., Nottebohm, F. (1993) Role of a telencephalic nucleus in the delayed song learning of socially isolated zebra finches. *J Neurobiol*, **24**:1045–1064.

Morton, E. S. (1996) A comparison of vocal behavior among tropical and temperate passerine birds. In *Ecology and Evolution of Acoustic Communication in Birds* (Kroodsma, D. E., Miller, E. H., eds.), pp. 258–268. Ithaca, NY: Comstock Publishing Associates.

Mountjoy, D. J., Lemon, R. E. (1995) Extended song learning in wild European starlings. *Anim Behav*, **49**:357–366.

Muller, C. M., Leppelsack, H. J. (1985) Feature extraction and tonotopic organization in the avian auditory forebrain. *Exp Brain Res*, **59**:587–599.

Müller-Bröse, M., Todt, D. (1991) Lokomotorische Aktivität von Nachtigallen (*Luscinia megarhynchos*) während auditorischer Stimulation in ihrer lernsensiblen Altersphase. *Verh Dtsch Zool Gese*, **84**:476–477.

Müller-Preuss, P., Ploog, D. (1981) Inhibition of auditory cortical neurons during phonation. *Brain Res*, **215**:61–76.

Mulligan, J. A. (1966) Singing behavior and its development in the song sparrow, *Melospiza melodia*. *Univ Calif Publ Zool*, **81**:1–73.

Mundinger, P. C. (1995) Behaviour-genetic analysis of canary song: inter-strain differences in sensory learning, and epigenetic rules. *Anim Behav*, **50**:1491–1511.

Mundinger, P. C. (1998) Genetics of canary song learning: innate mechanisms and some neurobiological considerations. In *Neural Mechanisms of Communication* (Konishi, M., Hauser, M., eds.). Cambridge, MA: MIT Press.

Nagel, K. I., Doupe, A. J. (2006) Temporal processing and adaptation in the songbird auditory forebrain. *Neuron*, **51**:845–859.

Naguib, M., Todt, D. (1997) Effects of dyadic vocal interactions on other conspecific receivers in nightingales. *Anim Behav*, **54**:1535–1543.

Naguib, M. (1999) Effects of song overlapping and alternating on nocturnally singing nightingales. *Anim Behav*, **58**:1061–1067.

Nakamura, K., Sakai, K., Hikosaka, O. (1999) Effects of local inactivation of monkey medial frontal cortex in learning of sequential procedures. *J Neurophys*, **82**:1063–1068.

Nakanishi, H., Kita, H., Kitai, S. T. (1990) Intracelllar study of rat entopeduncular nucleus neurons in an in vitro slice preparation: electrical membrane properties. *Brain Res*, **527**:81–88.

Nakazawa, K., Shiba, K., Satoh, I., Yoshida, K., Nakajima, Y., Konno, A. (1997) Role of pulmonary afferent inputs in vocal on-switch in the cat. *Neurosci Res*, **29**:49–54.

Narayan, R., Ergun, A., Sen, K. (2005) Delayed inhibition in cortical receptive fields and the discrimination of complex stimuli. *J Neurophysiol*, **94**:2970–2975. Epub, 2005 May 2925.

Nastiuk, K. L., Mello, C. V., George, J. M., Clayton, D. F. (1994) Immediate-early gene responses in the avian song control system: cloning and expression analysis of the canary c-jun cDNA. *Brain Res Mol Brain Res*, **27**:299–309.

Naya, Y., Yoshida, M., Miyashita, Y. (2003) Forward processing of long-term associative memory in monkey inferotemporal cortex. *J Neurosci*, **23**:2861–2871.

Nealen, P. M., Perkel, D. J. (2000) Sexual dimorphism in the song system of the Carolina wren *Thryothorus ludovicianus*. *J Comp Neurol*, **418**:346–360.

Nealen, P. M., Schmidt, M. F. (2002) Comparative approaches to avian song system function: insights into auditory and motor processing. *J Comp Physiol A Neuroethol Sens Neural Behav Physiol*, **188**:929–941.

Nealen, P. M., Schmidt, M. F. (2006) Distributed and selective auditory representation of song repertoires in the avian song system. *J Neurophysiol*, **96**(6):3433–3447.

Nelson, B. S., Beckers, G. J. L., Suthers, R. A. (2005) Vocal tract filtering and sound radiation in a songbird. *J Exp Biol*, **208**:297–308.

Nelson, D. A. (1992) Song overproduction and selective attrition lead to song sharing in the field sparrow (*Spizella pusilla*). *Behav Ecol Sociobiol*, **30**:415–424.

Nelson, D. A. (1997) Social interaction and sensitive phases for song learning: a critical review. In *Social Influences on Vocal Development* (Snowdon, C. T., Hausberger, M., eds.), pp. 7–22. New York: Cambridge University Press.

Nelson, D. A. (1998) External validity and experimental design: the sensitive phase for song learning. *Anim Behav*, **56**:487–491.

Nelson, D. A. (2000) Preference for own-subspecies' song guides vocal learning in a song bird. *Proc Natl Acad Sci USA*, **97**:13348–13353.

Nelson, D. A., Marler, P. (1993) Innate recognition of song in white-crowned sparrows: a role in selective vocal learning? *Anim Behav*, **46**:806–808.

Nelson, D. A., Marler, P. (1994) Selection-based learning in bird song development. *Proc Natl Acad Sci USA*, **91**:10498–10501.

Nelson, D. A., Marler, P., Palleroni, A. (1995) A comparative approach to vocal learning: intraspecific variation in the learning process. *Anim Behav*, **50**:83–97.

Nelson, D. A., Marler, P., Morton, M. L. (1996) Overproduction in song development: an evolutionary correlate with migration. *Anim Behav*, **51**:1127–1140.

Nelson, D. A., Marler, P., Soha, J. A., Fullerton, A. L. (1997) The timing of song memorization differs in males and females: a new assay for avian vocal learning. *Anim Behav*, **54**:587–597.

Neubauer, R. L. (1999) Super-normal length song preferences of female zebra finches (*Taeniopygia guttata*) and a theory of the evolution of bird song. *Evol Ecol*, **13**:365–380.

Newman, J., Wollberg, Z. (1978) Multiple coding of species-specific vocalizations in the auditory cortex of squirrel monkeys. *Brain Res*, **54**:287–304.

Newport, E. L. (1991) Contrasting concepts of the critical period for language. In *The Epigenesis of Mind: Essays on Biology and Cognition* (Carey, S., Gelman, R., eds.), pp. 111–130. Hillsdale, NJ: Erlbaum.

Newport, E. L., Hauser, M. D., Spaepen, G., Aslin, R. N. (2004) Learning at a distance. II. Statistical learning of non-adjacent dependencies in a non-human primate. *Cognit Psychol*, **49**:85–117.

Nick, T. A., Konishi, M. (2001) Dynamic control of auditory activity during sleep: correlation between song response and EEG. *Proc Natl Acad Sci USA*, **98**:14012–14016.

Nick, T. A., Konishi, M. (2005a) Neural song preference during vocal learning in the zebra finch depends on age and state. *J Neurobiol*, **62**:231–242.

Nick, T. A., Konishi, M. (2005b) Neural auditory selectivity develops in parallel with song. *J Neurobiol*, **62**:469–481.

Nicolai, J. (1959) Familientradition in der Gesangsentwicklung des Gimpels (*Pyrrhula pyrrhula*). *J Ornithologie*, **100**:39–46.

Nicoll, R. A., Malenka, R. C. (1999) Expression mechanisms underlying NMDA receptor-dependent long-term potentiation. *Ann N Y Acad Sci*, **868**:515–525.

Nixdorf, B. E., Davis, S. S., DeVoogd, T. J. (1989) Morphology of Golgi-impregnated neurons in hyperstriatum ventralis, pars caudalis in adult male and female canaries. *J Comp Neurol*, **284**:337–349.

Nixdorf-Bergweiler, B. E. (2001) Lateral magnocellular nucleus of the anterior neostriatum (LMAN) in the zebra finch: neuronal connectivity and the emergence of sex differences in cell morphology. *Microsc Res Tech*, **54**:335–353.

Nixdorf-Bergweiler, B. E., Lips, M. B., Heinemann, U. (1995a) Electrophysiological and morphological evidence for a new projection of LMAN-neurones towards area X. *Neuroreport*, **6**:1729–1732.

Nixdorf-Bergweiler, B. E., Wallhausser-Franke, E., DeVoogd, T. J. (1995b) Regressive development in neuronal structure during song learning in birds. *J Neurobiol*, **27**:204–215.

Nock, B. L., Leshner, A. I. (1976) Hormonal mediation of the effects of defeat on agonistic responding in mice. *Physiol Behav*, **17**:111–119.

Noctor, S. C., Flint, A. C., Weissman, T. A., Dammerman, R. S., Kriegstein, A. R. (2001) Neurons derived from radial glial cells establish radial units in neocortex. *Nature*, **409**:714–720.

Nordby, J. C., Campbell, E. S., Burt, J. M., Beecher, M. D. (2000) Social influences during song development in the song sparrow: a laboratory experiment simulating field conditions. *Anim Behav*, **59**:1187–1197.

Nordby, J., Campbell, S., Beecher, M. D. (2001) Late song learning in song sparrows. *Anim Behav*, **61**:835–846.

Nordby, J., Campbell, S., Beecher, M. D. (2002) Adult song sparrows do not alter their song repertoires. *Ethology*, **108**:39–50.

Nordeen, E. J., Nordeen, K. W., Arnold, A. P. (1987) Sexual differentiation of androgen accumulation within the zebra finch brain through selective cell loss and addition. *J Comp Neurol*, **259**:393–399.

Nordeen, K. W., Nordeen, E. J. (1988) Projection neurons within a vocal motor pathway are born during song learning in zebra finches. *Nature*, **334**:149–151.

Nordeen, K. W., Nordeen, E. J. (1992) Auditory feedback is necessary for the maintenance of stereotyped song in adult zebra finches. *Behav Neural Biol*, **57**:58–66.

Nordeen, K. W., Nordeen, E. J. (1993) Long-term maintenance of song in adult zebra finches is not affected by lesions of a forebrain region involved in song learning. *Behav Neural Biol*, **59**:79–82.

Nordeen, K. W., Nordeen, E. J. (2004) Synaptic and molecular mechanisms regulating plasticity during early learning. *Ann NY Acad Sci*, **1016**:416–437.

Nordeen, K. W., Nordeen, E. J., Arnold, A. P. (1986) Estrogen establishes sex differences in androgen accumulation in zebra finch brain. *J Neurosci*, **6**:734–738.

Northcutt, R. G. (2001) Changing views of brain evolution. *Brain Res Bull*, **55**:663–674.

Nottebohm, F. (1968) Auditory experience and song development in the chaffinch (*Fringilla coelebs*). *Ibis*, **110**:549–568.

Nottebohm, F. (1969) The "critical period" for song learning. *Ibis*, **111**:386–387.

Nottebohm, F. (1970) Ontogeny of bird song. *Science*, **167**:950–956.

Nottebohm, F. (1971a) Neural lateralization of vocal control in a passerine bird. I. Song. *J Exp Zool*, **177**:229–261.

Nottebohm, F., (1971b) Neural lateralization of vocal control in a passerine bird. II. Subsong, calls and a theory of vocal learning. *J Exp Zool*, **179**: 35–50.

Nottebohm, F. (1972) The origins of vocal learning. *Am Nat*, **106**:116–140.

Nottebohm, F. (1975) Vocal behavior in birds. In *Avian Biology*. Vol. 5 (Farner, D. S., King, J. R., eds.), pp. 287–332. New York: Academic Press.

Nottebohm, F. (1976) Phonation in the orange-winged Amazon parrot, *Amazona amazonica*. *J Comp Physiol*, **108**:157–170.

Nottebohm, F. (1977) Asymmetries in neural control of vocalization in the canary. In *Lateralization in the Nervous System* (Harnad, S., *et al.*, eds.), pp 23–44. New York: Academic Press.

Nottebohm, F. (1980a) Brain pathways for vocal learning in birds: a review of the first 10 years. In *Progress in Psychobiology and Physiological Psychology* (Sprague, J. M., Epstein, A. N., eds.), pp. 85–214. New York: Academic Press.

Nottebohm, F. (1980b) Testosterone triggers growth of brain vocal control nuclei in adult female canaries. *Brain Res*, **189**:429–436.

Nottebohm, F. (1981) A brain for all seasons: cyclical anatomical changes in song control nuclei of the canary brain. *Science*, **214**:1368–1370.

Nottebohm, F. (1984) Birdsong as a model in which to study brain processes related to learning. *Condor*, **86**:227–236.

Nottebohm, F. (1985a) *Hope for a New Neurology*. *Ann NY Acad Sci*, **457**:1–238.

Nottebohm, F. (1985b) Neuronal replacement in adulthood. In *Hope for a New Neurology* (Nottebohm, F., ed.). *Ann NY Acad Sci*, **457**:143–161.

Nottebohm, F. (1987) Plasticity in adult avian central nervous system: possible relation between hormones, learning, and brain repair. In *Handbook of Physiology, Section 1: The Nervous System*. Vol. 5. *Higher Functions of the Brain* (Plum, F., ed.), pp. 85–108. Baltimore, MD: Williams & Wilkins.

Nottebohm, F. (1989) From bird song to neurogenesis. *Sci Am*, **260**:74–79.

Nottebohm, F. (1996) The King Solomon lectures in neuroethology: a white canary on Mount Acropolis. *J Comp Physiol A*, **179**:149–156.

Nottebohm, F. (1999) The anatomy and timing of vocal learning in birds. In *The Design of Animal Communication* (Hauser, M. D., Konishi, M., eds.), pp. 63–110. Cambridge: MIT Press.

Nottebohm, F. (2002a) Why are some neurons replaced in adult brain? *J Neurosci*, **22**:624–628.

Nottebohm, F. (2002b) Neuronal replacement in adult brain. *Brain Res Bull*, **57**:737–749.

Nottebohm, F. (2004) The road we travelled: discovery, choreography, and significance of brain replaceable neurons. *Ann N Y Acad Sci*, **1016**:628–658.

Nottebohm, F., Alvarez-Buylla, A. (1993) Neurogenesis and neuronal replacement in adult birds. In *Neuronal Cell Death and Repair* (Cuello, A. C., ed.), pp. 227–236. Amsterdam: Elsevier Science Publishers.

Nottebohm, F., Arnold, A. P. (1976) Sexual dimorphism in vocal control areas of the songbird brain. *Science*, **194**:211–213.

Nottebohm, F., Nottebohm, M. E. (1971) Vocalizations and breeding behaviour of surgically deafened ring doves (*Streptopelia risoria*). *Anim Behav*, **19**:313–327.

Nottebohm, F., Nottebohm, M. E. (1976) Left hypoglossal dominance in the control of canary and white-crowned sparrow song. *J Comp Physiol A*, **108**:171–192.

Nottebohm, F., Nottebohm, M. E. (1978) Relationship between song repertoire and age in the canary, *Serinus canarius*. *Z Tierpsychol*, **46**:298–305.

Nottebohm, F., Stokes, T. M., Leonard, C. M. (1976) Central control of song in the canary, *Serinus canarius*. *J Comp Neurol*, **165**:457–486.

Nottebohm, F., Manning, F., Nottebohm, M. E. (1979) Reversal of hypoglossal dominance in canaries following syringeal denervation. *J Comp Physiol A*, **134**:227–240.

Nottebohm, F., Kasparian, S., Pandazis, C. (1981) Brain space for a learned task. *Brain Res*, **213**:99–109.

Nottebohm, F., Kelley, D. B., Paton, J. A. (1982) Connections of vocal control nuclei in the canary telencephalon. *J Comp Neurol*, **207**:344–357.

Nottebohm, F., Nottebohm, M. E., Crane, L. (1986) Developmental and seasonal changes in canary song and their relation to changes in the anatomy of song-control nuclei. *Behav Neural Biol*, **46**:445–471.

Nottebohm, F., Nottebohm, M. E., Crane, L. A., Wingfield, J. C. (1987) Seasonal changes in gonadal hormone levels of adult male canaries and their relation to song. *Behav Neural Biol*, **47**:197–211.

Nottebohm, F., Alvarez-Buylla, A., Cynx, J., et al. (1990) Song learning in birds: the relation between perception and production. *Philos Trans R Soc Lond B Biol Sci*, **329**:115–124.

Nottebohm, F., O'Loughlin, B., Gould, K., Yohay, K., Alvarez-Buylla, A. (1994) The life span of new neurons in a song control nucleus of the adult canary brain depends on time of year when these cells are born. *Proc Natl Acad Sci USA*, **91**:7849–7853.

Nowak, M. A., Komarova, N. L., Niyogi, P. (2001) Evolution of universal grammar. *Science*, **291**:114–118.

Nowicki, S. (1987) Vocal tract resonances in oscine bird sound production: evidence from birdsongs in a helium atmosphere. *Nature*, **325**:53–55.

Nowicki, S., Ball, G. F. (1989) Testosterone induction of song in photosensitive and photorefractory male sparrows. *Horm Behav*, **23**:514–525.

Nowicki, S., Capranica, R. R. (1986a) Bilateral syringeal interaction in vocal production of an oscine bird sound. *Science*, **231**:1297–1299.

Nowicki, S., Capranica, R. R. (1986b) Bilateral syringeal coupling during phonation of a songbird. *J Neurosci*, **6**:3595–3610.

Nowicki, S., Marler, P. (1988) How do birds sing? *Music Perception*, **5**:391–426.

Nowicki, S., Searcy, W. A. (2004) Song function and the evolution of female preferences: why birds sing, why brains matter. *Ann N Y Acad Sci*, **1016**:704–723.

Nowicki, S., Mitani, J. C., Nelson, D. A., Marler, P. (1989) The communicative significance of tonality in birdsong: responses to songs produced in helium. *Bioacoustics*, **2**:35–46.

Nowicki, S., Marler, P., Maynard, A., Peters, S. (1992) Is the tonal quality of birdsong learned – evidence from song sparrows. *Ethology*, **90**:225–235.

Nowicki, S., Peters, S., Podos, J. (1998) Song learning, early nutrition and sexual selection in songbirds. *Am Zool*, **38**:179–190.

Nowicki, S., Searcy, W. A., Peters, S. (2002) Brain development, song learning and mate choice in birds: a review and experimental test of the "nutritional stress hypothesis". *J Comp Physiol A*, **188**:1003–1014.

O'Loghlen, A. L., Beecher, M. D. (1997) Sexual preferences for mate song types in female song sparrows. *Anim Behav*, **53**: 835–841.

O'Loghlen, A. L., Rothstein, S. I. (1995) Delayed access to local songs prolongs vocal development in dialect populations of brown-headed cowbirds. *Condor*, **97**:402–414.

O'Loghlen, A. L., Rothstein, S. I. (2002a) Vocal development is correlated with an indicator of hatching date in brown-headed cowbirds. *Condor*, **104**:761–777.

O'Loghlen, A., Rothstein, S. I. (2002b) Ecological effects on song learning: Delayed development is widespread in wild populations of brown-headed cowbirds. *Anim Behav*, **63**:475–486.

Oberweger, K., Goller, F. (2001) The metabolic cost of birdsong production. *J Exp Biol*, **204**:3379–3388.

Ojemann, G., Mateer, C. (1979) Human language cortex: localization of memory, syntax, and sequential motor-phoneme identification systems. *Science*, **205**:1401–1403.

Ojemann, G. A. (1991) Cortical organization of language. *J Neurosci*, **11**:2281–2287.

Okada, A., Lansford, R., Weimann, J. M., Fraser, S. E., McConnell, S. K. (1999) Imaging cells in the developing nervous system with retrovirus expressing modified green fluorescent protein. *Exp Neurol*, **156**:394–406.

Okanoya, K. (1997) Voco-auditory behaviour in the Bengalese finch: a comparison with the zebra finch. *Biomed Res*, **18**:53–70.

Okanoya, K. (2004) The Bengalese finch: a window on the behavioral neurobiology of birdsong syntax. *Ann N Y Acad Sci*, **1016**:724–735.

Okanoya, K., Yamaguchi, A. (1997) Adult Bengalese finches (*Lonchura striata* var. *domestica*) require real-time auditory feedback to produce normal song syntax. *J Neurobiol*, **33**:343–356.

Okuhata, S., Saito, N. (1987) Synaptic connections of thalamo-cerebral vocal nuclei of the canary. *Brain Res Bull*, **18**:35–44.

Oller, D. K., Eilers, R. E. (1988) The role of audition in infant babbling. *Child Dev*, **59**:441–449.

Ölveczky, B. P., Andalman, A. S., Fee, M. S. (2005) Vocal experimentation in the juvenile songbird requires a basal ganglia circuit. *PLoS Biol*, **3**:e153.

Otmakhov, N., Tao-Cheng, J. H., Carpenter, S., et al. (2004) Persistent accumulation of calcium/calmodulin-dependent protein kinase II in dendritic spines after induction of NMDA receptor-dependent chemical long-term potentiation. *J Neurosci*, **24**:9324–9331.

Owens, E., Kessler, D. K., eds. (1989) *Cochlear Implants in Young Deaf Children*. Boston: College-Hill Press.

Oyama, S. (1976) A sensitive period for the acquisition of a nonnative phonological system. *J Psycholing Res*, **5**:261–283.

Oyama, S. (1978) The sensitive period and comprehension of speech. *Work Pap Biling*, **16**:1–17.

Packard, M. G., Knowlton, B. J. (2002) Learning and memory functions of the basal ganglia. *Annu Rev Neurosci*, **25**:563–593.

Palmer, T. D., Willhoite, A. R., Gage, F. H. (2000) Vascular niche for adult hippocampal neurogenesis. *J Comp Neurol*, **425**:479–494.

Panzica, G. C., Viglietti-Panzica, C., Sanchez, F., Sante, P., Balthazart, J. (1991) Effects of testosterone on a selected neuronal population within the preoptic sexually dimorphic nucleus of the Japanese quail. *J Comp Neurol*, **303**:443–456.

Panzica, G. C., Viglietti-Panzica, C., Balthazart, J. (1996) The sexually dimorphic medial preoptic nucleus of quail: a key brain area mediating steroid action on male sexual behavior. *Front Neuroendocrinol*, **17**:51–125.

Parent, A. (1997) The brain in evolution and involution. *Biochem Cell Biol*, **75**:651–667.

Park, K. H., Clayton, D. F. (2002) Influence of restraint and acute isolation on the selectivity of the adult zebra finch *zenk* gene response to acoustic stimuli. *Behav Brain Res*, **136**:185–191.

Park, K., Meitzen, J., Moore, I. T., Brenowitz, E. A., Perkel, D. J. (2005) Seasonal-like plasticity of spontaneous firing rate in a songbird pre-motor nucleus. *J Neurobiol*, **64**:181–191.

Partan, S., Marler, P. (1999) Communication goes multimodal. *Science*, **283**:1272–1273.

Paton, J. A., Nottebohm, F. N. (1984) Neurons generated in the adult brain are recruited into functional circuits. *Science*, **225**:1046–1048.

Paton, J. A., Manogue, K. R., Nottebohm, F. (1981) Bilateral organization of the vocal control pathway in the budgerigar, *Melopsittacus undulatus*. *J Neurosci*, **1**:1279–1288.

Paton, J. A., O'Loughlin, B. E., Nottebohm, F. (1985) Cells born in adult canary forebrain are local interneurons. *J Neurosci*, **5**:3088–3093.

Patterson, D. K., Pepperberg, I. M. (1994) A comparative study of human and parrot phonation: acoustic articulatory correlates of vowels. *J Acoust Soc Am*, **96**:634–648.

Patterson, D. K., Pepperberg, I. M. (1998) Acoustic and articulatory correlates of stop consonants in a parrot and a human subject. *J Acoust Soc Am*, **103**:2197–2215.

Patterson, T. A., Rose, S. P. (1992) Memory in the chick: multiple cues, distinct brain locations. *Behav Neurosci*, **106**:465–470.

Paulsen, K. (1967) *Das Prinzip der Stimmbildung in der Wirbeltierreihe und beim Menschen*. Frankfurt, am Main: Akademische Verlagsgesellschaft.

Payne, R. (1981) Song learning and social interaction in indigo buntings. *Anim Behav*, **29**:688–697.

Payne, R. B. (1983) The social context of song mimicry: song-matching dialects in indigo buntings. *Anim Behav*, **31**:788–805.

Payne, R. B., Payne, L. L. (1993) Song copying and cultural transmission in indigo buntings. *Anim Behav*, **46**:1045–1065.

Payne, R. S., McVay, S. (1971) Songs of humpback whales. *Science*, **173**:585–597.

Pearson, H. (2006) Genetics: what is a gene? *Nature*, **441**:398–401.

Pearson, K. G. (1993) Common principles of motor control in vertebrates and invertebrates. *Annu Rev Neurosci*, **16**:265–297.

Peek, F. W. (1972) An experimental study of the territorial function of vocal and visual display in the male red-winged blackbird (*Agelaius phoeniceus*). *Anim Behav*, **20**:112–118.

Pekarik, V., Bourikas, D., Miglino, N., Joset, P., Preiswerk, S., Stoeckli, E. T. (2003) Screening for gene function in chicken embryo using RNAi and electroporation. *Nat Biotechnol*, **21**:93–96.

Peltonen, L., McKusick, V. A. (2001) Genomics and medicine: dissecting human disease in the postgenomic era. *Science*, **291**:1224–1229.

Pepperberg, I. M. (1993) A review of the effects of social interaction on vocal learning in African grey parrots (*Psittacus erithacus*). *Netherlands J Zool*, **43**:104–124.

Pepperberg, I. M. (2002) In search of King Solomon's ring: cognitive and communicative studies of grey parrots (*Psittacus erithacus*). *Brain Behav Evol*, **59**:54–67.

Percival, D. B., Walden, A. T. (1993) *Spectral Analysis for Physical Applications: Multitaper and Conventional Univariate Techniques.* Cambridge: Cambridge University Press.

Perkel, D. J. (2004) Origin of the anterior forebrain pathway. *Ann N Y Acad Sci*, **1016**:736–748.

Perkel, D. J., Farries, M. A. (2000) Complementary 'bottom-up' and 'top-down' approaches to basal ganglia function. *Curr Opin Neurobiol*, **10**:725–731.

Perkell, J., Matthies, M., Lane, H., *et al.* (1997) Speech motor control: acoustic goals, saturation effects, auditory feedback and internal models. *Speech Comm*, **22**:227–250.

Perlman, W. R., Ramachandran, B., Arnold, A. P. (2003) Expression of androgen receptor mRNA in the late embryonic and early posthatch zebra finch brain. *J Comp Neurol*, **455**:513–530.

Perrins, C. (1970) The timing of birds breeding seasons. *Ibis*, **112**:242–255.

Person, A. L., Perkel, D. J. (2005) Unitary IPSPs drive precise thalamic spiking in a circuit required for learning. *Neuron*, **46**:129–140.

Pesch, A., Güttinger, H. R. (1985) Der Gesang des weiblichen Kanarienvogels. *J Ornithologie*, **126**:108–110.

Peters, S., Nowicki, S. (1996) Development of tonal quality in birdsong: further evidence from song sparrows. *Ethology*, **102**:323–335.

Peters, S., Marler, P., Nowicki, S. (1992) Song sparrows learn from limited exposure to song models. *Condor*, **94**:1016–1019.

Petersen, M. R., Beecher, M. D., Zoloth, S. R., *et al.* (1984) Neural lateralization of vocalizations by Japanese macaques: communicative significance is more important than acoustic structure. *Behav Neurosci*, **98**:779–790.

Petersen, S. E., Fox, P. T., Posner, M. I., Martin, M., Raichle, M. E. (1989) Positron emission tomographic studies of the processing of single words. *J Cogn Neurosci*, **1**:153–170.

Peterson, G. E., Barney, H. L. (1952) Control methods used in a study of vowels. *J Acoust Soc Am*, **24**:175–184.

Petitto, L. A. (1993) On the ontogenetic requirements for early language acquisition. In *Developmental Neurocognition: Speech and Face Processing in the First Year of Life* (de Boysson-Bardies, B., de Schonen, S., Jusczyk, P. W., McNeilage, P., Morton, J., eds.), pp. 365–383. New York: Kluwer Academic/Plenum Publishers.

Petitto, L. A., Marentette, P. F. (1991) Babbling in the manual mode: evidence for the ontogeny of language. *Science*, **251**:1493–1496.

Petrinovich, L., Baptista, L. F. (1987) Song development in the white-crowned sparrow: modification of learned song. *Anim Behav*, **35**:961–974.

Phan, M. L., Pytte, C. L., Vicario, D. S. (2006) Early auditory experience generates long-lasting memories that may subserve vocal learning in songbirds. *Proc Natl Acad Sci USA*, **103**:1088–1093.

Phillips, R., Peek, F. (1975) Brain organization and neuromuscular control of vocalization in birds. In *Hormones and Behaviour in Higher Vertebrates* (Wright, P., Caryl, P., Vowles, D., eds.), pp. 243–274. Amsterdam: Elsevier.

Phillmore, L. S., Bloomfield, L. L., Weisman, R. G. (2003) Effects of songs and calls on ZENK expression in the auditory telencephalon of field- and isolate-reared black capped chickadees. *Behav Brain Res*, **147**:125–134.

Phillmore, L. S., Hoshooley, J. S., Hahn, T. P., MacDougall-Shackleton, S. A. (2005) A test of absolute photorefractoriness and photo-induced neural plasticity of song-control regions in black-capped chickadees (Poecile atricapillus). *Canadian Journal of Zoology*, **83**:747–753.

Phillmore, L. S., Hoshooley, J. S., Sherry, D. F., Macdougall-Shackleton, S. A. (2006) Annual cycle of the black-capped chickadee: seasonality of singing rates and vocal-control brain regions. *J Neurobiol*, **66**:1002–1010.

Phoenix, C. H., Goy, R. W., Gerall, A. A., Young, W. C. (1959) Organizing action of prenatally administered testosterone propionate on the tissues mediating mating behavior in the female guinea pig. *Endocrinology*, **65**:369–382.

Piersma, T., Lindstrom, A. (1997) Rapid reversible changes in organ size as a component of adaptive behavior. *Trends Ecol Evol*, **12**:134–138.

Pilz, K. M., Quiroga, M., Schwabl, H., Adkins-Regan, E. (2004) European starling chicks benefit from high yolk testosterone levels during a drought year. *Horm Behav*, **46**:179–192.

Pinaud, R., Tremere, L. (2006) *Immediate Early Genes in Sensory Processing, Cognitive Performance and Neurological Disorders*. New York: Springer.

Pinaud, R., Tremere, L. A., Penner, M. R. (2000) Light-induced zif 268 expression is dependent on noradrenergic input in rat visual cortex. *Brain Res*, **882**:251–255.

Ping Tang, Y., Wade, J. (2006) Sexually dimorphic expression of the genes encoding ribosomal proteins L17 and L37 in the song control nuclei of juvenile zebra finches. *Brain Res*, **1126**:102–108.

Pinker, S. (1994) *The Language Instinct*. London: Allen Lane.

Pinker, S., Jackendoff, R. (2005) The faculty of language: what's special about it? *Cognition*, **95**:201–236.

Pinxten, R., De Ridder, E., Balthazart, J., Eens, M. (2002) Context-dependent effects of castration and testosterone treatment on song in male European starlings. *Horm Behav*, **42**:307–318.

Pittenger, C., Huang, Y. Y., Paletzki, R. F., *et al.* (2002) Reversible inhibition of CREB/ATF transcription factors in region CA1 of the dorsal hippocampus disrupts hippocampus-dependent spatial memory. *Neuron*, **34**:447–462.

Plant, G., Hammarberg, B. (1983) Acoustic and perceptual analysis of the speech of the deafened. *Speech Trans Lab Q Prog Stat Rep*, **2/3**: 85–107.

Platenik, J., Kuramoto, N., Yoneda, Y. (2000) Molecular mechanisms associated with long-term consolidation of the NMDA signals. *Life Sci*, **67**:335–364.

Podos, J. (1996) Motor constraints on vocal development in a songbird. *Anim Behav*, **51**:1061–1070.

Podos, J. (1997) A performance constraint on the evolution of trilled vocalizations in a songbird family (Passeriformes: Emberizidae). *Evolution*, **51**:537–551.

Podos, J., Nowicki, S. (2004a) Beaks, adaptation, and vocal evolution in Darwin's finches. *BioScience*, **454**:501–510.

Podos, J., Nowicki S. (2004b) Performance limits on birdsong. In: *Nature's Music. The Science of Birdsong* (Marler, P., Slabbekoorn, H., eds), pp 318–342. New York: Elsevier Academic Press.

Podos, J., Sherer, J. K., Peters, S., Nowicki, S. (1995) Ontogeny of vocal-tract movements during song production in song sparrows. *Anim Behav*, **50**:1287–1296.

Podos, J., Nowicki, S., Peters, S. (1999) Permissiveness in the learning and development of song syntax in swamp sparrows. *Anim Behav*, **58**:93–103.

Podos, J., Peters, S., Nowicki, S. (2004a) Calibration of song learning targets during vocal ontogeny in swamp sparrows, *Melospiza georgiana*. *Anim Behav*, **68**:929–940.

Podos, J., Southall, J. A., Rossi-Santos, M. R. (2004b) Vocal mechanics in Darwin's finches: correlation of beak gape and song frequency. *J Exp Biol*, **207**:607–619.

Poiani, A., Pagel, M. (1997) Evolution of avian cooperative breeding: comparative tests of the nest predation hypothesis. *Evol*, **51**:226–240.

Polley, D. B., Kvasnak, E., Frostig, R. D. (2004) Naturalistic experience transforms sensory maps in the adult cortex of caged animals. *Nature*, **429**:67–71.

Poole, J. H., Tyack, P. L., Stoeger-Horwath, A. S., Watwood, S. (2005) Animal behaviour: elephants are capable of vocal learning. *Nature*, **434**:455–456.

Poopatanapong, A., Teramitsu, I., Byun, J. S., Vician, L. J., Herschman, H. R., White, S. A. (2006) Singing, but not seizure, induces synaptotagmin IV in zebra finch song circuit nuclei. *J Neurobiol*, **66**:1613–1629.

Poremba, A., Malloy, M., Saunders, R. C., Carson, R. E., Herscovitch, P., Mishkin, M. (2004) Species-specific calls evoke asymmetric activity in the monkey's temporal poles. *Nature*, **427**:448–451.

Poulet, J. F., Hedwig, B. (2006) The cellular basis of a corollary discharge. *Science*, **311**:518–522.

Powell, F. L., Scheid P. (1989) Physiology of gas exchange in the avian respiratory system. In *Form and Function in Birds* (King, A. S., McLelland, J., eds.), pp. 393–437. London: Academic Press.

Prather, J. F., Peters, S., Nowicki, S., Mooney, R. (2008) Precise auditory–vocal mirroring in neurons for learned vocal communication. *Nature*, **451**:305–310.

Price, C. J., Wise, R. J., Warburton, E. A., *et al.* (1996) Hearing and saying: the functional neuro-anatomy of auditory word processing. *Brain*, **119** (3):919–931.

Price, P. H. (1977) Determinants of zebra finch song: studies of species song uniformity, bases of physiological determination, and developmental plasticity. Ph.D. dissertation. Philadelphia: University of Pennsylvania.

Price, P. H. (1979) Developmental determinants of structure in zebra finch song. *J Comp Physiol Psychol*, **93**:260–277.

Proulx, S. R., Phillips, P. C. (2006) Allelic divergence precedes and promotes gene duplication. *Evolution*, **60**:881–892.

Puelles, L., Kuwana, E., Puelles, E., Rubenstein, J. L. (1999) Comparison of the mammalian and avian telencephalon from the perspective of gene expression data. *Eur J Morphol*, **37**:139–150.

Puelles, L., Kuwana, E., Puelles, E., *et al.* (2000) Pallial and subpallial derivatives in the embryonic chick and mouse telencephalon, traced by the expression of the genes Dlx-2, Emx-1, Nkx- 2.1, Pax-6, and Tbr-1. *J Comp Neurol*, **424**:409–438.

Pytte, C., Suthers, R. A. (1996) Evidence for a sensitive period for sensorimotor integration during song development in the zebra finch. *Soc Neurosci Abstr*, **21**:693.

Pytte, C. L., Suthers, R. A. (2000) Sensitive period for sensorimotor integration during vocal motor learning. *J Neurobiol*, **42**:172–189.

Pytte, C., Gerson, M., Miller, J., Kirn, J. R. (2007) Increasing stereotypy in adult zebra finch song correlates with a declining rate of adult neurogenesis. *Dev Neurobiol*. **67**:1699–1720.

Raisman, G., Field, P. M. (1971) Sexual dimorphism in the preoptic area of the rat. *Science*, **173**:731–733.

Ralls, K. P., Fiorelli, P., Gish, S. (1985) Vocalizations and vocal mimicry in captive harbor seals, *Phoca vitulina*. *Can J Zool*, **63**:1050–1056.

Ramirez, J. M., Tryba, A. K., Pena, F. (2004) Pacemaker neurons and neuronal networks: an integrative view. *Curr Opin Neurobiol*, **14**:665–674.

Ramus, F., Hauser, M. D., Miller, C., Morris, D., Mehler, J. (2000) Language discrimination by human newborns and by cotton-top tamarin monkeys. *Science*, **288**:349–351.

Rasika, S., Nottebohm, F., Alvarez-Buylla, A. (1994) Testosterone increases the recruitment and/or survival of new high vocal center neurons in adult female canaries. *Proc Natl Acad Sci USA*, **91**:7854–7858.

Rasika, S., Alvarez-Buylla, A., Nottebohm, F. (1999) BDNF mediates the effects of testosterone on the survival of new neurons in an adult brain. *Neuron*, **22**:53–62.

Rasika, S. (1998) A steroid-neurotrophin pathway for the seasonal regulation of neuronal replacement in the adult canary brain. In: *Neuroscience*. New York, NY: The Rockefeller University.

Ratcliffe, L., Otter K. (1996) Sex differences in song recognition. In *Ecology and Evolution of Acoustic Communication in Birds* (Kroodsma, D. E., Miller, E. H., eds.), pp. 339–355. Ithaca, NY: Cornell University Press.

Rauschecker, J. P., Tian, B., Hauser, M. D. (1995) Processing of complex sounds in the macaque nonprimary auditory cortex. *Science*, **268**:111–114.

Rauske, P. L., Shea, S. D., Margoliash, D. (2003) State and neuronal class-dependent reconfiguration in the avian song system. *J Neurophysiol*, **89**:1688–1701.

Reiner, A., Medina, L., Veenman, C. L. (1998) Structural and functional evolution of the basal ganglia in vertebrates. *Brain Research Brain Research Reviews*, **28**:235–285.

Reiner, A., Perkel, D. J., Bruce, L. L., *et al.* (2004a) Revised nomenclature for avian telencephalon and some related brainstem nuclei. *J Comp Neurol*, **473**:377–414.

Reiner, A., Perkel, D. J., Mello, C. V., Jarvis, E. D. (2004b) Songbirds and the revised avian brain nomenclature. *Ann NY Acad Sci*, **1016**:77–108.

Reiner, A., Yamamoto, K., Karten, H. J. (2005) Organization and evolution of the avian forebrain. *Anat Rec A Discov Mol Cell Evol Biol*, **287**:1080–1102.

Reinke, H., Wild, J. M. (1997) Distribution and connections of inspiratory premotor neurons in the brainstem of the pigeon (*Columba livia*). *J Comp Neurol*, **379**:347–362.

Reinke, H., Wild, J. M. (1998) Identification and connections of inspiratory premotor neurons in songbirds and budgerigar. *J Comp Neurol*, **391**:147–163.

Reiter, L. T., Potocki, L., Chien, S., Gribskov, M., Bier, E. (2001) A Systematic analysis of human disease-associated gene sequences in *Drosophila melanogaster*. *Genome Res*, **11**:1114–1125.

Remez, R. E., Fellowes, J. M., Rubin, P. E. (1997) Talker identification based on phonetic information. *J Exp Psychol Human*, **23**:651–666.

Ressler, K., J., Paschall, G., Zhou, X. L., Davis, M. (2002) Regulation of synaptic plasticity genes during consolidation of fear conditioning. *J Neurosci*, **22**:7892–7902.

Rhodes, K. J., Trimmer, J. S. (2006) Antibodies as valuable neuroscience research tools versus reagents of mass distraction. *J Neurosci*, **26**:8017–8020.

Ribeiro, S., Mello, C. V. (2000) Gene expression and synaptic plasticity in the auditory forebrain of songbirds. *Learn Mem*, **7**:235–243.

Ribeiro, S., Cecchi, G. A., Magnasco, M. O., Mello, C. V. (1998) Toward a song code: evidence for a syllabic representation in the canary brain. *Neuron*, **21**:359–371.

Riebel, K. (2000) Early exposure leads to repeatable preferences for male song in female zebra finches. *Proc Biol Sci*, **267**:2553–2558.

Riebel, K. (2003) Developmental influences on auditory perception in female zebra finches – is there a sensitive phase for song preference learning? *Anim Biol*, **53**:73–87.

Riebel, K. (2004) The 'mute' sex revisited: vocal production and perception learning in female songbirds. *Adv Study Behav*, **33**:49–86.

Riebel, K., Smallegange, I. M., Terpstra, N. J., Bolhuis, J. J. (2002) Sexual equality in zebra finch song preference: evidence for a dissociation between song recognition and production learning. *Proc Biol Sci*, **269**:729–733.

Riede, T., Beckers, G. J., Blevins, W., Suthers, R. A. (2004) Inflation of the esophagus and vocal tract filtering in ring doves. *J Exp Biol*, **207**:4025–4036.

Riede, T., Suthers, R. A., Fletcher, N. H., Blevins, W. E. (2006) Songbirds tune their vocal tract to the fundamental frequency of their song. *Proc Natl Acad Sci USA*, **103**:5543–5548.

Riters, L. V., Alger, S. J. (2004) Neuroanatomical evidence for indirect connections between the medial preoptic nucleus and the song control system: possible neural substrates for sexually motivated song. *Cell Tissue Res*, **316**:35–44.

Riters, L. V., Ball, G. F. (1999) Lesions to the medial preoptic area affect singing in the male European starling (*Sturnus vulgaris*). *Horm Behav*, **36**:276–286.

Riters, L. V., Teague, D. P. (2003) The volume of song control nuclei, HVC and lMAN, relate to differential behavioral responses of female European starlings to male songs produced within and outside the breeding season. *Brain Res*, **978**:91–98.

Riters, L. V., Absil, P., Balthazart, J. (1998) Effects of brain testosterone implants on appetitive and consummatory components of male sexual behavior in Japanese quail. *Brain Res Bull*, **47**:69–79.

Riters, L. V., Baillien, M., Eens, M., Pinxten, R., Foidart, A., Ball, G. F., Balthazart, J. (2000a) Seasonal Variation in Androgen-metabolizing Enzymes in the Diencephalon and Telencephalon of the Male European Starling (*Sturnus vulgaris*). *J Neuroendocrinol*, **13**:985–997.

Riters, L. V., Eens, M., Pinxten, R., Duffy, D. L., Balthazart, J., Ball, G. F. (2000b) Seasonal changes in courtship song and the medial preoptic area in male European starlings (*Sturnus vulgaris*). *Hormones and Behavior*, **38**:250–261.

Riters, L. V., Teague, D. P., Schroeder, M. B., Cummings, S. E. (2004) Vocal production in different social contexts relates to variation in immediate early gene immunoreactivity within and outside of the song control system. *Behav Brain Res*, **155**:307–318.

Riters, L. V., Schroeder, M. B., Auger, C. J., Eens, M., Pinxten, R., Ball, G. F. (2005) Evidence for opioid involvement in the regulation of song production in male European starlings (*Sturnus vulgaris*). *Behav Neurosci*, **119**:245–255.

Roberts, E. B., Meredith, M. A., Ramoa, A. S. (1998) Suppression of NMDA receptor function using antisense DNA block ocular dominance plasticity while preserving visual responses. *J Neurophysiol*, **80**:1021–1032.

Roberts, J., Long, S., Malkin, C., et al. (2005) A comparison of phonological skills of boys with fragile X syndrome and Down syndrome. *J Speech Lang Hear Res*, **48**:980–995.

Roberts, T. F., Klein, M. E., Wild, J. M., Mooney, R. (2006) Lentivirus-based anterograde pathway tracing in the song-motor circuit. Program No. 44.12. 2006 Neuroscience Meeting. Atlanta, GA: Society for Neuroscience.

Roberts, T., Wild, J. M., Kubke, M. F., Mooney, R. (2007) Homogeneity of intrinsic properties of sexually dimorphic vocal motoneurons in male and female zebra finches. *J Comp Neurol*, **502**:157–169.

Robinson, G., Grozinger, C., Whitfield, C. (2005) Sociogenomics: social life in molecular terms. *Nat Rev Genet*, **6**:257–270.

Rodriguez, P. (2001) Simple recurrent networks learn context-free and context-sensitive languages by counting. *Neural Comput*, **13**:2093–2118.

Rogers, L. J., Kaplan, G. (2000) *Songs, Roars and Rituals: Communication in Birds, Mammals and Other Animals*. Cambridge, MA: Harvard University Press.

Rollenhagen, A., Bischof, H. J. (1998) Spine density changes in forebrain areas of the zebra finch by TEA-induced potentiation. *NeuroReport*, **9**:2325–2329.

Romanski, L. M., Tian, B., Fritz, J., Mishkin, M., Goldman-Rakic, P. S., Rauschecker, J. P. (1999) Dual streams of auditory afferents target multiple domains in the primate prefrontal cortex. *Nat Neurosci*, **2**:1131–1136.

Roper, A., Zann, R. (2006) The onset of song learning and song tutor selection in fledgling zebra finches. *Ethology*, **112**:458–470.

Rose, G. J., Goller, F., Gritton, H. J., Plamondon, S. L., Baugh, A. T., Cooper, B. G. (2004) Species-typical songs in white-crowned sparrows tutored with only phrase pairs. *Nature*, **432**:753–758.

Rose, M. (1914) Uber die Cytoarchitektonische Gliederung des Vorderhirns der Vogel. *J Psychol Neurol*, **21**:278–352.

Rose, S. (2003) *The Making of Memories: From Molecules to Mind*. Revised Edition. London: Vintage.

Rosen, M. J., Mooney, R. (2000) Intrinsic and extrinsic contributions to auditory selectivity in a song nucleus critical for vocal plasticity. *J Neurosci*, **20**:5437–5448.

Rosen, M. J., Mooney, R. (2003) Inhibitory and excitatory mechanisms underlying auditory responses to learned vocalizations in the songbird nucleus HVC. *Neuron*, **39**:177–194.

Rosen, M. J., Mooney, R. (2006) Synaptic interactions underlying song-selectivity in the avian nucleus HVC revealed by dual intracellular recordings. *J Neurophysiol*, **95**:1158–1175.

Rothstein, S. I., Fleisher, R. C. (1987) Vocal dialects and their possible relation to status signaling in the brown-headed cowbird. *Condor*, **89**:1–23.

Roy, A., Mooney, R. (2007) Auditory plasticity in a basal ganglia-forebrain pathway during decrystallization of adult birdsong. *J Neurosci*, **27**:6374–6387.

Rozin, P., Kalat, J. (1971) Specific hungers and poison avoidance as adaptive specializations of learning. *Psychol Rev*, **78**:459–486.

Ruan, J., Suthers, R. H. (1996) Myotopic representation of syringeal muscles in the hypoglossal nucleus of the cowbird. *Soc Neurosci Abs*, **22**:1402.

Rubinson, D. A., Dillon, C. P., Kwiatkowski, A. V., et al. (2003) A lentivirus-based system to functionally silence genes in primary mammalian cells, stem cells and transgenic mice by RNA interference. *Nat Genet*, **33**:401–406.

Rübsamen, R., Dörrscheidt, G. (1986) Tonotopic organization of the auditory forebrain in a songbird, the European starling. *J Comp Physiol*, **158**:639–646.

Rusak, B., Robertson, H. A., Wisden, W., Hunt, S. P. (1990) Light pulses that shift rhythms induce gene expression in the suprachiasmatic nucleus. *Science*, **248**:1237–1240.

Ryals, B. M., Rubel, E. W. (1982) Patterns of hair cell loss in chick basilar papilla after intense auditory stimulation. Frequency organization. *Acta Otolaryngol*, **93**:205–210.

Ryals, B. M., Rubel, E. W. (1988) Hair cell regeneration after acoustic trauma in adult *Coturnix* quail. *Science*, **240**:1774–1776.

Ryan, S. M., Arnold, A. P. (1981) Evidence for cholinergic participation in the control of bird song: acetylcholinesterase distribution and muscarinic receptor autoradiography in the zebra finch brain. *J Comp Neurol*, **202**:211–219.

Rybak, F., Gahr, M. (2004) Modulation by steroid hormones of a "sexy" acoustic signal in an oscine species, the common canary *Serinus canaria*. *An Acad Bras Cienc*, **76**:365–367.

Sachs, M. B., Woolf, N. G., Sinnott J. M. (1980) Response properties of neurons in the avian auditory system: comparisons with mammalian homologues and consideration of the neural encoding of complex stimuli. In *Comparative Studies of Hearing in Vertebrates* (Popper, A. N., Fay, R. R., eds.), pp. 323–353. Berlin: Springer.

Saffran, J. R., Aslin, R. N., Newport, E. L. (1996) Statistical learning by 8-month-old infants. *Science*, **274**:1926–1928.

Sakaguchi, H., Yamaguchi, A. (1997) Early song-deprivation affects the expression of protein kinase C in the song control nuclei of the zebra finch during a sensitive period of song learning. *NeuroReport*, **8**:2645–2650.

Sakaguchi, H., Taniguchi, I. (2000) Sex differences in the ventral paleostriatum of the zebra finch: origin of the cholinergic innervation of the song control nuclei. *NeuroReport*, **11**:2727–2731.

Sakaguchi, H. (2004) Effect of social factors on the development of PKC expression in the songbird brain. *NeuroReport*, **2004**:2819–2823.

Sakata, J. T., Brainard, M. S. (2006) Real-time contributions of auditory feedback to avian vocal motor control. *J Neurosci*, **26**:9619–9628.

Saldanha, C. J., Tuerk, M. J., Kim, Y. H., Fernandes, A. O., Arnold, A. P., Schlinger, B. A. (2000) Distribution and regulation of telencephalic aromatase expression in the zebra finch revealed with a specific antibody. *J Comp Neurol*, **423**:619–630.

Sanderson, K. J., Crouch, H. (1993) Vocal repertoire of the Australian magpie *Gymnorhina tibicen* in South Australia. *Austral Bird Watcher*, **15**:162–164.

Sandler, W., Meir, I., Padden, C., Aronoff, M. (2005) The emergence of grammar: systematic structure in a new language. *Proc Natl Acad Sci USA*, **102**:2661–2665.

Sartor, J. J., Charlier, T. D., Pytte, C. L., Ball, G. F. (2002) Converging evidence that song performance modulates seasonal changes in the avian song control system. *Soc Neurosci Abst*, **28**.

Sasaki, A., Sotnikova, T. D., Gainetdinov, R. R., Jarvis, E. D. (2006) Social context-dependent singing-regulated dopamine. *J Neurosci*, **26**:9010–9014.

Saucier, D., Cain, D. P. (1995) Spatial learning without NMDA receptor-dependent long-term potentiation. *Nature*, **378**:186–189.

Sawtell, N. B., Williams, A., Bell, C. C. (2005) From sparks to spikes: information processing in the electrosensory systems of fish. *Curr Opin Neurobiol*, **15**:437–443.

Schafe, G. E., Atkins, C. M., Swank, M. W., Bauer, E. P., Sweatt, J. D., LeDoux, J. E. (2000) Activation of ERK/MAP kinase in the amygdala is required for memory consolidation of pavlovian fear conditioning. *J Neurosci*, **20**:8177–8887.

Schafer, M., Rubsamen, R., Dorrscheidt, G. J., Knipschild, M. (1992) Setting complex tasks to single units in the avian auditory forebrain. II. Do we really need natural stimuli to describe neuronal response characteristics? *Hear Res*, **57**:231–244.

Scharff, C., Haesler, S. (2005) An evolutionary perspective on FoxP2: strictly for the birds? *Curr Opin Neurobiol*, **15**:694–703.

Scharff, C., Nottebohm, F. (1991) A comparative study of the behavioral deficits following lesions of various parts of the zebra finch song system: implications for vocal learning. *J Neurosci*, **11**:2896–2913.

Scharff, C., White, S. A. (2004) Genetic components of vocal learning. *Ann NY Acad Sci*, **1016**: 325–347.

Scharff, C., Nottebohm, F., Cynx, J. (1998) Conspecific and heterospecific song discrimination in male zebra finches with lesions in the anterior forebrain pathway. *J Neurobiol*, **36**:81–90.

Scharff, C., Kirn, J. R., Grossman, M., Macklis, J. D., Nottebohm, F. (2000) Targeted neuronal death affects neuronal replacement and vocal behavior in adult songbirds. *Neuron*, **25**:481–492.

Scheid, P., Piiper J. (1989) Respiratory mechanics and air flow in birds. In *Form and Function in Birds* (King, A. S., McLelland, J., eds.), pp. 369–392. London: Academic Press.

Schelderup-Ebbe, T. (1922) Beiträge zur Socialpyschologie des Haushuhns. *Z Psychol*, **88**:225–252.

Schleidt, W. M. (1964) Über die Spontaneität von Erbkoordinationen. *Z Tierpsychol*, **21**:235–256.

Schlinger, B. A., Arnold, A. P. (1992) Circulating estrogens in a male songbird originate in the brain. *Proc Natl Acad Sci USA*, **89**:7650–7653.

Schlinger, B. A., Brenowitz, E. A. (2002) Neural and hormonal control of birdsong. In *Hormones, Brain and Behavior* (Pfaff, D. W., Arnold, A. P., Etgen, A. M., Fahrbach, S. E., Rubin, R. T., eds.), pp. 799–839. San Diego, CA: Academic Press.

Schlinger, B. A., Lane, N. I., Grisham, W., Thompson, L. (1999) Androgen synthesis in a songbird: a study of cyp17 (17alpha-hydroxylase/C17,20-lyase) activity in the zebra finch. *Gen Comp Endocrinol*, **113**:46–58.

Schmidt, L. G., Bradshaw, S. D., Follett, B. K. (1991) Plasma levels of luteinizing hormone and androgens in relation to age and breeding status among cooperatively breeding Australian magpies (*Gymnorhina tibicen* Latham). *Gen Comp Endocrinol*, **83**:48–55.

Schmitt, M. (1995) The homology concept – still alive. In *The Nervous System of Invertebrates: An Evolutionary and Comparative Approach* (Briedbach, O., Kutsch, W., eds.), pp. 425–438. Basel, Switzerland: Birkhauser Verlag.

Schmidt, M. F. (2003) Pattern of interhemispheric synchronization in HVc during singing correlates with key transitions in the song pattern. *J Neurophysiol*, **90**:3931–3949.

Schmidt, M. F., Konishi, M. (1998) Gating of auditory responses in the vocal control system of awake songbirds. *Nat Neurosci*, **1**:513–518.

Schmidt, M. F., Perkel, D. J. (1998) Slow synaptic inhibition in nucleus HVc of the adult zebra finch. *J Neurosci*, **18**:895–904.

Schmidt, M. F., Ashmore, R. C., Vu, E. T. (2004) Bilateral control and interhemispheric coordination in the avian song motor system. *Ann N Y Acad Sci*, **1016**:171–186.

Schodde, R., Mason, J. J. (1999) *The Directory of Australian Birds*. Melbourne: CSIRO.

Schoups, A., Vogels, R., Qian, N., Orban, G. (2001) Practising orientation identification improves orientation coding in V1 neurons. *Nature*, **412**:549–553.

Schregardus, D. S., Pieneman, A. W., Ter Maat, A., Jansen, R. F., Brouwer, T. J., Gahr, M. L. (2006) A lightweight telemetry system for recording neuronal activity in freely behaving small animals. *J Neurosci Methods*, **155**:62–71.

Schultz, W. (2002) Getting formal with dopamine and reward. *Neuron*, **36**:241–263.

Schultz, W. (2004) Neural coding of basic reward terms of animal learning theory, game theory, microeconomics and behavioural ecology. *Curr Opin Neurobiol*, **14**:139–147.

Schultz, W., Dayan, P., Montague, P. R. (1997) A neural substrate of prediction and reward. *Science*, **275**:1593–1599.

Schwabl, H. (1996) Maternal testosterone in the avian egg enhances postnatal growth. *Comp Biochem Physiol A Physiol*, **114**:271–276.

Schwartzkopff, J. (1949) Über Sitz und Leistung von Gehör und Vibrationssinn bei Vögeln. *Z Vergl Physiol*, **31**:527–608.

Scott, B. B., Lois, C. (2005) Generation of tissue-specific transgenic birds with lentiviral vectors. *Proc Natl Acad Sci USA*, **102**:16443–16447.

Scott, B. B., Lois, C. (2007) Developmental origin and identity of song system neurons born during vocal learning in song birds. *J Comp Neurol*, **502**:202–214.

Scott, G. R., Milsom, W. K. (2006) Flying high: A theoretical analysis of the factors limiting exercise performance in birds at altitude. *Resp Physiol Neurobiol*, **154**:284–301.

Scott, L. L., Nordeen, E. J., Nordeen, K. W. (2000) The relationship between rates of HVc neuron addition and vocal plasticity in adult songbirds. *J Neurobiol*, **43**:79–88.

Scotto-Lomassese, S., Rochefort, C., Nshdejan, A., Scharff, C. (2007) HVC interneurons are not renewed in adult male zebra finches. *Eur J Neurosci*, **25**:1663–1668.

Scoville, W. B., Milner, B. (1957) Loss of recent memory after bilateral hippocampal lesions. *J Neurol Neurosurg PS*, **20**:11–21.

Searcy, W. A., Brenowitz, E. A. (1988) Sexual differences in species recognition of avian song. *Nature*, **332**:152–154.

Searcy, W. A., Marler, P. (1981) A test for responsiveness to song structure and programming in female sparrows. *Science*, **213**:926–928.

Searcy, W. A., Marler, P. (1987) Response of sparrows to songs of deaf and isolation-reared males: Further evidence for innate auditory templates. *Dev Psychobiol*, **20**:509–519.

Searcy, W., Nowicki S. (1998) Functions of song variation in song sparrows. In *Neural Mechanisms of Communication* (Konishi, M., Hauser, M., eds.). Cambridge, MA: MIT Press.

Searcy, W. A., Nowicki, S. (2005) *The Evolution of Animal Communication: Reliability and Deception in Signaling Systems*. Princeton, NJ: Princeton University Press.

Searcy, W. A., Yasukawa, K. (1995) *Polygyny and Sexual Selection in Red-winged Blackbirds*. Princeton, NJ: Princeton University Press.

Sen, K., Theunissen, F. E., Doupe, A. J. (2001) Feature analysis of natural sounds in the songbird auditory forebrain. *J Neurophysiol*, **86**:1445–1458.

Senghas, A., Kita, S., Ozyurek, A. (2004) Children creating core properties of language: evidence from an emerging sign language in Nicaragua. *Science*, **305**:1779–1782.

Setterwall, C. G. (1901) Studies öfver syrinx hos polymyoda passeres. Ph.D. dissertation, University of Lund.

Seyfarth, R. M., Cheney, D. L. (1997) Some general features of vocal development in nonhuman primates. In *Social Influences on Vocal Development* (Snowdon, C. T., Hausberger, M., eds.), pp. 249–273. New York: Cambridge University Press.

Seyfarth, R. M., Cheney, D. L., Marler, P. (1980) Vervet monkey alarm calls: Semantic communication in a free-ranging primate. *Anim Behav*, **28**:1070–1094.

Shaevitz, S. S., Theunissen, F. E. (2007) Functional connectivity between auditory areas field L and CLM and song system nucleus HVC in anesthetized zebra finches. *J Neurophysiol*, **98**:2747–2764.

Shannon, R. V., Zeng, F. G., Kamath, V., Wygonski, J., Ekelid, M. (1995) Speech recognition with primarily temporal cues. *Science*, **270**:303–304.

Sharma, S. K., Bagnall, M. W., Sutton, M. A., Carew, T. J. (2003) Inhibition of calcineurin facilitates the induction of memory for sensitization in Aplysia: requirement of mitogen-activated protein kinase. *Proc Nat Acad Sci USA*, **100**:4861–4866.

Shaywitz, B. A., Shaywitz, S. E., Pugh, K. R., et al. (1995) Sex differences in the functional organization of the brain for language. *Nature*, **373**:607–609.

Shea, S. D., Margoliash, D. (2003) Basal forebrain cholinergic modulation of auditory activity in the zebra finch song system. *Neuron*, **40**:1213–1226.

Sheen, V. L., Dreyer, E. B., Macklis, J. D. (1992) Calcium-mediated neuronal degeneration following singlet oxygen production. *NeuroReport*, **3**:705–708.

Shen, P., Schlinger, B. A., Campagnoni, A. T., Arnold, A. P. (1995) An atlas of aromatase mRNA expression in the zebra finch brain. *J Comp Neurol*, **360**:172–184.

Sheng, M., Greenberg, M. E. (1990) The regulation and function of c-fos and other immediate early genes in the nervous system. *Neuron*, **4**:477–485.

Sherry, D. F. (2006) Neuroecology. *Annu Rev Psychol*, **57**:167–197.

Sherry, D. F., Vaccarino, A. L. (1989) Hippocampus and memory for food caches in black-capped chickadees. *Behav Neurosci*, **103**:308–318.

Shoemaker, H. H. (1939) Effect of testosterone propionate on the behavior of the female canary. *Proc Soc Exp Biol Med*, **41**:299–302.

Shors, T. J., Miesegaes, G., Beylin, A., Zhao, M., Rydel, T., Gould, E. (2001) Neurogenesis in the adult is involved in the formation of trace memories. *Nature*, **410**:372–376.

Shu, W., Cho, J. Y., Jiang, Y., et al. (2005) Altered ultrasonic vocalization in mice with a disruption in the Foxp2 gene. *Proc Natl Acad Sci USA*, **102**:9643–9648.

Sibley, C. G., Ahlquist, J. E. (1990) *Phylogeny and Classification of Birds: A Study in Molecular Evolution*. New Haven, CT: Yale University Press.

Sibley, C. G., Monroe, B. L. (1990) *Distribution and Taxonomy of Birds of the World*. New Haven, CT: Yale University Press.

Silverin, B., Baillien, M., Foidart, A., Balthazart, J. (2000) Distribution of aromatase activity in the brain and peripheral tissues of passerine and nonpasserine avian species. *Gen Comp Endocrinol*, **117**:34–53.

Simon, H. A. (1974) How big is a chunk? *Science*, **183**:482–488.

Simonyan, K., Jürgens, U. (2005) Afferent subcortical connection into the motor cortical larynx area in the rhesus monkey. *Neuroscience*, **130**:119–131.

Simpson, H. B., Vicario, D. S. (1990) Brain pathways for learned and unlearned vocalizations differ in zebra finches. *J Neurosci*, **10**:1541–1556.

Sinclair, A., Berta, P., Palmer, M., et al. (1990) A gene from the human sex-determining region encodes a protein with homology to a conserved DNA-binding motif. *Nature*, **346**:240–244.

Singh, N. C., Theunissen, F. E. (2003) Modulation spectra of natural sounds and ethological theories of auditory processing. *J Acoust Soc Am*, **114**:3394–3411.

Singh, T. D., Basham, M. E., Nordeen, E. J., Nordeen, K. W. (2000) Early sensory and hormonal experience modulate age-related changes in NR2B mRNA within a forebrain region controlling avian vocal learning. *J Neurobiol*, **44**:82–94.

Singh, T. D., Heinrich, J. E., Wissman, A. M., Brenowitz, E. A., Nordeen, E. J., Nordeen, K. W. (2003) Seasonal regulation of NMDA receptor NR2B mRNA in the adult canary song system. *J Neurobiol*, **54**:593–603.

Singh, T. D., Nordeen, E. J., Nordeen, K. W. (2005) Song tutoring triggers CaMKII phosphorylation within a specialized portion of the avian basal ganglia. *J Neurobiol*, **65**:171–191.

Skinner, B. F. (1938) *The Behavior of Organisms: An Experimental Analysis*. New York: Appleton-Century.

Slater, P. J. (1983) Bird song learning: theme and variations. In *Perspectives in Ornithology* (Brush, G. H., Clark, Jr. G. A., eds.), pp. 475–511. Cambridge: Cambridge University Press.

Slater, P. J. B., Mann, N. I. (2004) Why do the females of many bird species sing in the tropics? *J Avian Biol*, **35**:289–294.

Slater, P. J. B., Eales, L. A., Clayton, N. S. (1988) Song learning in zebra finches (*Taeniopygia guttata*): progress and prospects. *Adv Study Behav*, **18**:1–34.

Slater, P. J. B., Richards, C., Mann, N. I. (1991) Song learning in zebra finches exposed to a series of tutors during the sensitive phase. *Ethology*, **88**:163–171.

Slater, P. J. B., Jones, A., ten Cate, C. (1993) Can lack of experience delay the end of the sensitive phase for song learning? *Netherlands J Zool*, **43**:80–90.

Slater, P. J. B., Gil, D., Barlow, C. R., Graves, J. A. (2002) Male led duets in the moho (*Hypergerus atriceps*) and yellow-crowned gonolek (*Lanarius barbarus*): mate guarding by females. *Ostrich*, **73**:49–51.

Smith, G. T., Brenowitz, E. A., Beecher, M. D., Wingfield, J. C. (1997a) Seasonal changes in testosterone, neural attributes of song control nuclei, and song structure in wild songbirds. *J Neurosci*, **17**:6001–6010.

Smith, G. T., Brenowitz, E. A., Wingfield, J. C. (1997b) Roles of photoperiod and testosterone in seasonal plasticity of the avian song control system. *J Neurobiol*, **32**:426–442.

Smith, M. A., Brandt, J., Shadmehr, R. (2000) Motor disorder in Huntington's disease begins as a dysfunction in error feedback control. *Nature*, **403**:544–549.

Smotherman, M., Kobayasi, K., Ma, J., Zhang, S. Y., Metzner, W. (2006) A mechanism for vocal-respiratory coupling in the mammalian parabrachial nucleus. *J Neurosci*, **26**:4860–4869.

Smulders, T. V. (2002) Natural breeding conditions and artificial increases in testosterone have opposite effects on the brains of adult male songbirds: a meta-analysis. *Horm Behav*, **41**:156–169.

Smulders, T., Lisi, M. D., Tricomi, E., Otter, K. A., Chruszcz, B., Ratcliffe, L. M., DeVoogd, T. (2006) Failure to detect seasonal changes in the song system nuclei of the black-capped chickadee. *J Neurobiol*, **66**: 991–1001.

Snow, C. E., Hoefnagel-Hohle, M. (1978) The critical period for language acquisition: evidence from second language learning. *Child Dev*, **49**:1114–1128.

Snow, C. E. (1987) Relevance of the notion of a critical period to language acquisition. In *Sensitive Periods in Development: Interdisciplinary Perspectives* (Bornstein, M. H., ed.), pp. 183–209. Hillsdale, NJ: Erlbaum.

Sockman, K. W., Gentner, T. Q., Ball, G. F. (2002) Recent experience modulates forebrain gene-expression in response to mate-choice cues in European starlings. *Proc Biol Sci*, **269**:2479–2485.

Sockman, K. W., Gentner, T. Q., Ball, G. F. (2005) Complementary neural systems for the experience-dependent integration of mate-choice cues in European starlings. *J Neurobiol*, **62**:72–81.

Soha, J. A. (1995) Cues for selective learning of conspecific song by young white-crowned sparrows. M.Sc. thesis, University of California, Davis.

Soha, J. A., Shimizu, T., Doupe, A. J. (1996) Development of the catecholaminergic innervation of the song system of the male zebra finch. *J Neurobiol*, **29**:473–489.

Soha, J. A., Marler, P. (2000) A species-specific acoustic cue for selective song learning in the white-crowned sparrow. *Anim Behav*, **60**:297–306.

Soha, J. A., Marler, P. (2001a) Cues for early discrimination of conspecific song in the white-crowned sparrow (*Zonotrichia leucophrys*). *Ethology*, **107**:813–826.

Soha, J. A., Marler, P. (2001b) Vocal syntax development in the white-crowned sparrow. *J Comp Psychol*, **115**:172–180.

Sohrabji, F., Nordeen, K. W., Nordeen, E. J. (1989) Projections of androgen-accumulating neurons in a nucleus controlling avian song. *Brain Res*, **488**:253–259.

Sohrabji, F., Nordeen, E. J., Nordeen, K. W. (1990) Selective impairment of song learning following lesions of a forebrain nucleus in the juvenile zebra finch. *Behav Neural Biol*, **53**:51–63.

Sohrabji, F., Nordeen, E. J., Nordeen, K. W. (1993) Characterization of neurons born and incorporated into a vocal control nucleus during avian song learning. *Brain Res*, **620**:335–338.

Solis, M. M., Doupe, A. J. (1997) Anterior forebrain neurons develop selectivity by an intermediate stage of birdsong learning. *J Neurosci*, **17**:6447–6462.

Solis, M. M., Doupe, A. J. (1999) Contributions of tutor and bird's own song experience to neural selectivity in the songbird anterior forebrain. *J Neurosci*, **19**:4559–4584.

Solis, M. M., Doupe, A. J. (2000) Compromised neural selectivity for song in birds with impaired sensorimotor learning. *Neuron*, **25**:109–121.

Solis, M. M., Perkel, D. J. (2005) Rhythmic activity in a forebrain vocal control nucleus in vitro. *J Neurosci*, **25**:2811–2822.

Solis, M. M., Brainard, M. S., Hessler, N. A., Doupe, A. J. (2000) Song selectivity and sensorimotor signals in vocal learning and production. *Proc Natl Acad Sci USA*, **97**:11836–11842.

Soma, K. K., Wingfield, J. C. (2001) Dehydroepiandrosterone in songbird plasma: seasonal regulation and relationship to territorial aggression. *Gen Comp Endocrinol*, **123**:144–155.

Soma, K. K., Hartman, V. N., Wingfield, J. C., Brenowitz, E. A. (1998) Seasonal changes in androgen receptor immunoreactivity in the song nucleus HVc of a wild bird. *J Comp Neurol*, **409**:224–236.

Soma, K. K., Tramontin, A. D., Wingfield, J. C. (2000) Oestrogen regulates male aggression in the non-breeding season. *Proc Biol Sci*, **267**:1089–1096.

Soma, K. K., Alday, N. A., Schlinger, B. A. (2002) 3 Beta-HSD and aromatase in songbird brain: DHEA metabolism, aggression, and song. *Soc Neurosci Abstr*, **28**:189.1.

Soma, K. K., Wissman, A. M., Brenowitz, E. A., Wingfield, J. C. (2002) Dehydroepiandrosterone (DHEA) increases territorial song and the size of an associated brain region in a male songbird. *Horm Behav*, **41**:203–212.

Soma, K. K., Tramontin, A. D., Featherstone, J., Brenowitz, E. A. (2004) Estrogen contributes to seasonal plasticity of the adult avian song control system. *J Neurobiol*, **58**:413–422.

Sommer, M. A., Wurtz, R. H. (2004a) What the brain stem tells the frontal cortex. I. Oculomotor signals sent from superior colliculus to frontal eye field via mediodorsal thalamus. *J Neurophysiol*, **91**:1381–1402.

Sommer, M. A., Wurtz, R. H. (2004b) What the brain stem tells the frontal cortex. II. Role of the SC-MD-FEF

Sonnenschein, E., Reyer, H.-U. (1983) Mate-guarding and other functions of antiphonal duets in the slate-coloured boubou (*Laniarius funebris*). *Z Tierpsychol*, **63**:112–140.

Sossinka, R., Böhner, J. (1980) Song types in the zebra finch (*Poephila guttata castanotis*). *Z Tierpsychol*, **53**:123–132.

Spector, D. A., McKim, L. K., Kroodsma, D. E. (1989) Yellow warblers are able to learn songs and situations in which to use them. *Anim Behav*, **38**:723–725.

Spiro, J. E., Dalva, M. B., Mooney, R. (1999) Long-range inhibition within the zebra finch song nucleus RA can coordinate the firing of multiple projection neurons. *J Neurophysiol*, **81**:3007–3020.

Squire, L. R., Kandel, E. R. (1999) *Memory: From Mind to Molecules*. New York: Scientific American Library.

Stacey, P. D., Koenig, W. D. (1990) *Cooperative Breeding in Birds: Long-Term Studies of Ecology and Behavior*. Cambridge: Cambridge University Press.

Stamps, J. A., Krishnan, V. V. (1999) A learning-based model of territory establishment. *Quart Review Biol*, **74**:291–307.

Stark, L. L., Perkel, D. J. (1999) Two-stage, input-specific synaptic maturation in a nucleus essential for vocal production in the zebra finch. *J Neurosci*, **19**:9107–9116.

Stefansson, H., Helgason, A., Thorleifsson, G., et al. (2005) A common inversion under selection in Europeans. *Nat Genet*, **37**:129–137.

Stevens, K. N. (1994) Scientific substrates of speech production. In *Introduction to Communication Sciences and Disorders* (Minifie, F. D., ed.), pp. 399–437. San Diego, CA: Singular.

Steward, O., Wallace, C. S., Lyford, G. L., Worley, P. F. (1998) Synaptic activation causes the mRNA for the IEG Arc to localize selectively near activated postsynaptic sites on dendrites. *Neuron*, **21**:741–751.

Stoddard, P. K. (1996) Vocal recognition of neighbors by territorial passerines. In *Ecology and Evolution of Acoustic Communication in Birds* (Kroodsma, D. E., Miller, E. H., eds.), pp. 356–374. Ithaca, NY: Cornell University Press.

Stoddard, P. K., Beecher, M. D., Campbell, S. E. (1992) Song-type matching in the song sparrow. *Can J Zool*, **70**:1440–1444.

Stoel-Gammon, C., Otomo, K. (1986) Babbling development of hearing-impaired and normally hearing subjects. *J Speech Hear Disord*, **51**:33–41.

Stoel-Gammon, C. (1992) Prelinguistic vocal development: measurement and predictions. In *Phonological Development: Models, Research, Implications* (Ferguson, C. A., Menn, L., Stoel-Gammon, C., eds.), pp. 439–456. Timonium, MD: York.

Stokes, A. W. (1968) *Animal Behavior in Laboratory and Field*. San Francisco, CA: W. H. Freeman.

Stokes, T. M., Leonard, C. M., Nottebohm, F. (1974) The telencephalon, diencephalon, and mesencephalon of the canary, *Serinus canaria*, in stereotaxic coordinates. *J Comp Neurol*, **156**:337–374.

Strack, S., Colbran, R. J. (1998) Autophosphorylation-dependent targeting of calcium/calmodulin-dependent protein kinase II by the NR2B subunit of the N-methyl-D-aspartate receptor. *J Biol Chem*, **273**:20689–20692.

Streeter, L. A. (1976) Language perception of 2-month-old infants shows effects of both innate mechanisms and experience. *Nature*, **259**:39–41.

Striedter, G. F. (1994) The vocal control pathways in budgerigars differ from those in songbirds. *J Comp Neurol*, **343**:35–56.

Striedter, G. F. (1997) The telencephalon of tetrapods in evolution. *Brain Behav Evol*, **49**:179–213.

Striedter, G. F., Beydler, S. (1997) Distribution of radial glia in the developing telencephalon of chicks. *J Comp Neurol*, **387**:399–420.

Striedter, G. F., Northcutt, R. G. (1991) Biological hierarchies and the concept of homology. *Brain Behav Evol*, **38**:177–189.

Striedter, G. F., Vu, E. T. (1998) Bilateral feedback projections to the forebrain in the premotor network for singing in zebra finches. *J Neurobiol*, **34**:27–40.

Stripling, R., Volman, S. F., Clayton, D. F. (1997) Response modulation in the zebra finch neostriatum: relationship to nuclear gene regulation. *J Neurosci*, **17**:3883–3893.

Stripling, R., Kruse, A. A., Clayton, D. F. (2001) Development of song responses in the zebra finch caudomedial neostriatum: role of genomic and electrophysiological activities. *J Neurobiol*, **48**:163–180.

Strote, J., Nowicki, S. (1996) Responses to songs with altered tonal quality by adult song sparrows (*Melospiza melodia*). *Behaviour*, **133**:161–172.

Struhsaker, T. T. (1967) Auditory communication among vervet monkeys (*Cernopithecus aethiops*). In *Social Communication Among Primates* (Altman, S. A., ed.), pp. 281–324. Chicago, IL: Chicago University Press.

Struhsaker, T. T. (1997) *Ecology of an African Rain Forest: Logging in Kibale and the Conflict Between Conservation and Exploitation*. Gainesville, FL: University Press of Florida.

Sturdy, C. B., Wild, J. M., Mooney, R. (2003) Respiratory and telencephalic modulation of vocal motor neurons in the zebra finch. *J Neurosci*, **23**:1072–1086.

Stutchbury, B. J., Morton, E. S. (2001) *Behavioral Ecology of Tropical Birds*. San Diego, CA: Academic Press.

Sugiura, H. (1998) Matching of acoustic features during the vocal exchange of coo calls by Japanese macaques. *Anim Behav*, **55**:673–687.

Sukhatme, V. P., Cao, X. M., Chang, L. C., *et al.* (1988) A zinc finger-encoding gene coregulated with c-fos during growth and differentiation, and after cellular depolarization. *Cell*, **53**:37–43.

Sundberg, J. (1975) Formant technique in a professional female singer. *Acustica*, **32**:89–96.

Suthers, R. A. (1990) Contributions to birdsong from the left and right sides of the intact syrinx. *Nature*, **347**:473–477.

Suthers, R. A. (1992) Lateralization of sound production and motor action on the left and right sides of the syrinx during bird song. In *Proceedings, 14th International Congress on Acoustics*, Beijing, pp. I 1–5. IUPAP.

Suthers, R. A. (1994) Variable asymmetry and resonance in the avian vocal tract: a structural basis for individually distinct vocalizations. *J Comp Physiol A*, **175**:457–466.

Suthers, R. A. (1997) Peripheral control and lateralization of birdsong. *J Neurobiol*, **33**:632–652.

Suthers, R. A. (1999) The motor basis of vocal performance in songbirds. In *The Design of Animal Communication* (Hauser, M. D., Konishi, M., eds.), pp. 37–62. Cambridge: MIT Press.

Suthers, R. A., Goller, F. (1997) Motor correlates of vocal diversity in songbirds. In *Current Ornithology* (Nolan, V., Jr., Ketterson, E., Thompson, C. F., eds.), pp. 235–288. New York: Plenum Press.

Suthers, R. A. (2004) How birds sing and why it matters. In *Nature's Music: The Science of Birdsong* (Marler, P., Slabbekoorn, H., eds.), pp. 272–295. New York: Elsevier Academic Press.

Suthers, R. A., Margoliash, D. (2002) Motor control of birdsong. *Curr Opin Neurobiol*, **12**:684–690.

Suthers, R. A., Wild, J. M. (2000) Real-time modulation of the syringeal motor program in response to externally imposed respiratory perturbations in adult songbirds. *Soc Neurosci Abs*, **26**:723.

Suthers, R. A., Zollinger, S. A. (2004) Producing song: the vocal apparatus. *Ann NY Acad Sci*, **1016**:109–129.

Suthers, R. A., Goller, F., Hartley, R. S. (1994) Motor dynamics of song production by mimic thrushes. *J Neurobiol*, **25**:917–936.

Suthers, R. A., Goller, F., Bermejo, R., Wild, J. M., Zeigler, H. P. (1996a) Relationship of beak gape to the lateralization, acoustics and motor dynamics of song in cardinals. In *Abstracts of the Nineteenth Midwinter Meeting for Research in Otolaryngology*. (Popelka, G. ed.), p. 158. Des Moines: Association.

Suthers, R. A., Goller, F., Hartley, R. S. (1996b) Motor stereotypy and diversity in songs of mimic thrushes. *J Neurobiol*, **30**:231–245.

Suthers, R. A., Goller, F., Pytte, C. (1999) The neuromuscular control of birdsong. *Philos Trans R Soc Lond B Biol Sci*, **354**:927–939.

Suthers, R. A., Goller, F., Wild, J. M. (2002) Somatosensory feedback modulates the respiratory motor program of crystallized birdsong. *Proc Natl Acad Sci USA*, **99**:5680–5685.

Suthers, R. A., Vallet, E. M., Tanvez, A., Kreutzer, M. (2004) Bilateral song production in domestic canaries. *J Neurobiol*, **60**:381–393.

Suthers, R. A., Beckers, G. J. L., Nelson, B. S. (2006) Vocal mechanisms for avian communication. In *Behavior and Neurodynamics for Auditory Communication* (Kanwal, J., Ehret, G., eds.), pp. 3–35. Cambridge: Cambridge University Press.

Sutter, M. L., Margoliash, D. (1994) Global synchronous response to autogenous song in zebra finch HVc. *J Neurophysiol*, **72**:2105–2123.

Sutton, R. S., Barto, A. G. (1998) *Reinforcement Learning: An Introduction*. Cambridge, MA: MIT Press.

Suzuki, R., Buck, J. R., Tyack, P. L. (2006) Information entropy of humpback whale songs. *J Acoust Soc Am*, **119**:1849–1866.

Svenningsson, P., Nishi, A., Fisone, G., Girault, J. A., Nairn, A. C., Greengard, P. (2004) DARPP-32: an integrator of neurotransmission. *Annu Rev Pharmacol Toxicol*, **44**:269–296.

Swanson, L. W. (2000) Cerebral hemisphere regulation of motivated behavior. *Brain Res*, **886**:113–164.

Székely, A. D., Boxer, M. I., Stewart, M. G., Csillag, A. (1994) Connectivity of the lobus parolfactorius of the domestic chicken (*Gallus domesticus*): an anterograde and retrograde pathway tracing study. *J Comp Neurol*, **348**:374–393.

Székely, T., Catchpole, C. K., DeVoogd, A., Marchl, Z., DeVoogd, T. J. (1996) Evolutionary changes in a song control area of the brain (HVC) are associated with evolutionary changes in song repertoire among European warblers (Sylviidae). *Proc R Soc Lond B*, **263**:607–610.

Tahta, S., Wood, M., Loewenthal, K. (1981) Age changes in the ability to replicate foreign pronunciation and intonation. *Lang Speech*, **24**:363–372.

Takahashi, K., Liu, F. C., Hirokawa, K., Takahashi, H. (2003) Expression of Foxp2, a gene involved in speech and language, in the developing and adult striatum. *J Neurosci Res*, **73**:61–72.

Tallal, P., Miller, S. L., Bedi, G., et al. (1996) Language comprehension in language-learning impaired children improved with acoustically modified speech. *Science*, **271**:81–84.

Tamura, S., Morikawa, Y., Iwanishi, H., Hisaoka, T., Senba, E. (2004) FoxP1 gene expression in projection neurons of the mouse striatum. *Neuroscience*, **124**:261–267.

Tanji, J. (2001) Sequential organization of multiple movements: involvement of cortical motor areas. *Annu Rev Neurosci*, **24**:631–651.

Taubenfeld, S. M., Milekic, M. H., Monti, B., Alberini, C.M., (2001) The consolidation of new but not reactivated memory requires hippocampal C/EBPbeta. *Nat Neurosci*, **4**:813–818.

Taylor, E. W., Jordan, D., Coote, J. H. (1999) Central control of the cardiovascular and respiratory systems and their interactions in vertebrates. *Physiol Rev*, **79**:855–916.

Tchernichovski, O., Schwabl, H., Nottebohm, F. (1998) Context determines the sex appeal of male zebra finch song. *Anim Behav*, **55**:1003–1010.

Tchernichovski, O., Lints, T., Mitra, P. P., Nottebohm, F. (1999) Vocal imitation in zebra finches is inversely related to model abundance. *Proc Natl Acad Sci USA*, **96**:12901–12904.

Tchernichovski, O., Nottebohm, F., Ho, C. E., Pesaran, B., Mitra, P. P. (2000) A procedure for an automated measurement of song similarity. *Anim Behav*, **59**:1167–1176.

Tchernichovski, O., Mitra, P. P., Lints, T., Nottebohm, F. (2001) Dynamics of the vocal imitation process: how a zebra finch learns its song. *Science*, **291**:2564–2569.

Tchernichovski, O., Lints, T. J., Deregnaucourt, S., Cimenser, A., Mitra, P. P. (2004) Studying the song development process: rationale and methods. *Ann NY Acad Sci*, **1016**:348–363.

Teichmann, M., Dupoux, E., Kouider, S., Bachoud-Levi, A. C. (2006) The role of the striatum in processing language rules: evidence from word perception in Huntington's disease. *J Cogn Neurosci*, **18**:1555–1569.

Tekumalla, P. K., Tontonoz, M., Hesla, M. A., Kirn, J. R. (2002) Effects of excess thyroid hormone on cell death, cell proliferation, and new neuron incorporation in the adult zebra finch telencephalon. *J Neurobiol*, **51**:323–341.

ten Cate, C. (1989) Behavioural development: toward understanding processes. In *Perspectives in Ethology*. Vol. 8. *Whither Ethology?* (Bateson, P. P. G., Klopfer, P. H., eds.), pp. 243–269. New York: Plenum Press.

ten Cate, C. (1991) Behavior-contingent exposure to taped song and zebra finch song learning. *Anim Behav*, **42**:857–859.

ten Cate, C. (1994) Perceptual mechanisms in imprinting and song learning. In *Causal Mechanisms of Behavioural Development* (Hogan, J. A., Bolhuis, J. J., eds.), pp. 116–146. Cambridge: Cambridge University Press.

ten Cate, C., Vos D. R., Mann, N. (1993) Sexual imprinting and song learning: two of one kind. *Netherlands J Zool*, **43**:34–45.

Teramitsu, I., White, S. A. (2006) FoxP2 regulation during undirected singing in adult songbirds. *J Neurosci*, **26**:7390–7394.

Teramitsu, I., Kudo, L. C., London, S. E., Geschwind, D. H., White, S. A. (2004) Parallel FoxP1 and FoxP2 expression in songbird and human brain predicts functional interaction. *J Neurosci*, **24**:3152–3163.

Terleph, T. A., Mello, C. V., Vicario, D. S. (2006) Auditory topography and temporal response dynamics of canary caudal telencephalon. *J Neurobiol*, **66**:281–292.

Terpstra, N. J., Bolhuis, J. J., den Boer-Visser, A. M. (2004) An analysis of the neural representation of birdsong memory. *J Neurosci*, **24**:4971–4977.

Terpstra, N. J., Bolhuis, J. J., den Boer-Visser, A. M., ten Cate, C. (2005) Neuronal activation related to auditory perception in the brain of a non-songbird, the ring dove. *J Comp Neurol*, **488**:342–351.

Terpstra, N. J., Bolhuis, J. J., Riebel, K., van der Burg, J. M., den Boer-Visser A. M. (2006) Localized brain activation specific to auditory memory in a female songbird. *J Comp Neurol*, **494**:784–791.

Tettamanti, M., Moro, A., Messa, C., et al. (2005) Basal ganglia and language: phonology modulates dopaminergic release. *NeuroReport*, **16**:397–401.

Theunissen, F. E., Doupe, A. J. (1998) Temporal and spectral sensitivity of complex auditory neurons in the nucleus HVc of male zebra finches. *J Neurosci*, **18**:3786–3802.

Theunissen, F. E., Shaevitz, S. S. (2006) Auditory processing of vocal sounds in birds. *Curr Opin Neurobiol*, **16**:400–407.

Theunissen, F. E., Sen, K., Doupe, A. J. (2000) Spectral-temporal receptive fields of nonlinear auditory neurons obtained using natural sounds. *J Neurosci*, **20**:2315–2331.

Theunissen, F. E., David, S. V., Singh, N. C., Hsu, A., Vinje, W., Gallant, J. L. (2001) Estimating spatio-temporal receptive fields of auditory and visual neurons from their responses to natural stimuli. *Network: Comp Neural Syst*, **12**:1–28.

Theunissen, F. E., Amin, N., Shaevitz, S. S., Woolley, S. M., Fremouw, T., Hauber, M. E. (2004a) Song selectivity in the song system and in the auditory forebrain. *Ann N Y Acad Sci*, **1016**:222–245.

Theunissen, F. E., Woolley, S. M., Hsu, A., Fremouw, T. (2004b) Methods for the analysis of auditory processing in the brain. *Ann NY Acad Sci*, **1016**:187–207.

Thielcke, G. (1970) Lernen von Gesang als möglicher Schrittmacher der Evolution. *Z Zool Syst Evol*, **8**:309–320.

Thode, C., Bock, J., Braun, K., Darlison, M. G. (2005) The chicken immediate-early gene ZENK is expressed in the medio-rostral neostriatum/hyperstriatum ventrale, a brain region involved in acoustic imprinting, and is up-regulated after exposure to an auditory stimulus. *Neuroscience*, **130**:611–617.

Thompson, C. K., Brenowitz, E. A. (2005) Seasonal change in neuron size and spacing but not neuronal recruitment in a basal ganglia nucleus in the avian song control system. *J Comp Neurol*, **481**:276–283.

Thompson, C. K., Bentley, G. E., and Brenowitz, E. A. (2007) Rapid regression of the avian song control system following a transition from breeding to non-breeding physiology. *Proc Natl Acad Sci USA*, **104**:15520–15525.

Thompson, R. F. (2005) In search of memory traces. *Ann Rev Psychol*, **56**:1–23.

Thompson, R. R., Adkins-Regan, E. (1994) Photoperiod affects the morphology of a sexually dimorphic nucleus within the preoptic area of male Japanese quail. *Brain Res*, **667**:201–208.

Thorneycroft, H. (1966) Chromosomal polymorphism in the white-throated sparrow, *Zonotrichia albicollis*. *Science*, **154**:1571–1572.

Thorneycroft, H. (1975) A cytogenetic study of the white-throated sparrow, *Zonotrichia albicollis* (Gmelin). *Evolution*, **29**:611–621.

Thorpe, W. H. (1951) The learning abilities of birds. *Ibis*, **93**:1–52, 252–296.

Thorpe, W. H. (1954) The process of song-learning in the chaffinch as studied by means of the sound spectrograph. *Nature*, **173**:465.

Thorpe, W. H. (1956) *Learning and Instinct in Animals*. London: Methuen.

Thorpe, W. H. (1958) The learning of song patterns by birds, with especial reference to the song of the chaffinch *Fringilla coelebs*. *Ibis*, **100**:535–570.

Thorpe, W. H. (1961) *Bird Song: The Biology of Vocal Communication and Expression in Birds*. Cambridge: Cambridge University Press.

Thorpe, W. H. (1972) Duetting and antiphonal song in birds, its extent and significance. *Behaviour*, **18**:1197.

Thorpe, W. H. (1974) *Gifford Lectures: Animal Nature and Human Nature 1969–1971*. New York: Harvard University Press.

Timmons, B. A. (1982) Physiological factors related to delayed auditory feedback and stuttering: a review. *Percept Mot Skills*, **55**:1179–1189.

Tinbergen, N. (1951) *The Study of Instinct*. Oxford, UK: Oxford University Press.

Tischmeyer, W., Grimm, R. (1999) Activation of immediate early genes and memory formation. *Cell Mol Life Sci*, **55**:564–574.

Titze, I. R. (1994) *Principles of Voice Production*. Englewood Cliffs, NJ: Prentice Hall.

Tobari, Y., Nakamura, K. Z., Okanoya, K. (2005) Sex differences in the telencephalic song control circuitry in Bengalese finches (*Lonchura striata* var. *domestica*). *Zoolog Sci*, **22**:1089–1094.

Todt, D. (1968) Zur Steuerung unregelmässiger Verhaltensabläufe. In *Kybernetik* (Mittelstaedt F., ed.), pp. 465–485. Munich: Oldenbourg.

Todt, D. (1970) [Singing and vocal correspondence of the blackbird]. *Naturwissenschaften*, **57**:61–66.

Todt, D. (1971) Äquivalente und konvalente gesangliche Reaktion einer extrem regelmäig singenden Nachtigall (*Luscinia megarhynchos* L.). *Z vergl Physiol*, **71**:262–285.

Todt, D. (1974) [On the relevance of the "right" syntax of auditory patterns to vocal responses of blackbirds (author's transl.)]. *Z Naturforsch C*, **29**:157–160.

Todt, D., Hultsch, H., Heike, D. (1979) Conditions affecting song acquisition in nightingales (*Luscinia megarhynchos* L.). *Z Tierpsychol*, **51**:23–35.

Todt, D. (1981) On functions of vocal matching: effect of counter-replies on song post choice and singing. *Z Tierpsychol*, **57**:73–93.

Todt, D. (2004) From birdsong to speech: a plea for comparative approaches. *An Acad Bras Cienc*, **76**:201–208.

Todt, D., Böhner, J. (1994) Former experience can modify social selectivity during song learning in the nightingale (*Luscinia megarhynchos*). *Ethology*, **97**:169–176.

Todt, D., Geberzahn, N. (2003) Age dependent effects of song exposure: song crystallization sets a boundary between fast and delayed vocal imitation. *Anim Behav*, **65**:971–979.

Todt, D., Hultsch H. (1980) Functional aspects of sequence and hierarchy in song structure. In *Acta XVII Congressus Internationalis Ornithologici* (Nöhring, R., ed.), pp. 663–669. Berlin: Verlag der deutschen Ornithologen-Gesellschaft.

Todt, D., Hultsch, H. (1996) Acquisition and performance of song repertoires: ways of coping with diversity and versatility. In *Ecology and Evolution of Acoustic Communication in Birds* (Kroodsma, D. E., Miller, E. H., eds.), pp. 79–96. Ithaca, NY: Cornell University Press.

Todt, D., Hultsch, H. (1998a) Hierarchical learning, development and representation of song. In *Animal Cognition in Nature: The Convergence of Psychology and Biology in Laboratory and Field* (Balda, R. P., Pepperberg, I. M., Kamil, A. C., eds.), pp. 275–303. San Diego, CA: Academic Press.

Todt, D., Hultsch, H. (1998b) How songbirds deal with large amounts of serial information: retrieval rules suggest a hierarchical song memory. *Biol Cybernet*, **79**:487–500.

Todt, D., Naguib, M. (2000) Vocal interactions in birds: the use of song as a model in communication. *Adv Study Behav*, **29**:247–296.

Todt, D., Cirillo, N., Geberzahn, N., Schleuss, F. (2001) The role of hierarchy levels in vocal imitations of birds. *Cybernet Syst*, **32**:257–283.

Tononi, G., Sporns, O., Edelman, G. M. (1996) A complexity measure for selective matching of signals by the brain. *Proc Natl Acad Sci USA*, **93**:3422–3427.

Tovar, K. R., Westbrook, G. L. (1999) The incorporation of NMDA receptors with a distinct subunit composition at nascent hippocampal synapses in vitro. *J Neurosci*, **19**:4180–4188.

Tramontin, A. D., Brenowitz, E. A. (1999) A field study of seasonal neuronal incorporation into the song control system of a songbird that lacks adult song learning. *J Neurobiol*, **40**:316–326.

Tramontin, A. D., Brenowitz, E. A. (2000) Seasonal plasticity in the adult brain. *Trends Neurosci*, **23**:251–258.

Tramontin, A. D., Brenowitz, E. A., Wingfield, J. C. (1997) Contributions of social influences and photoperiod to seasonal plasticity in avian song nuclei. *Soc Neurosci Abstr*, **22**:1328.

Tramontin, A. D., Smith, G. T., Breuner, C. W., Brenowitz, E. A. (1998) Seasonal plasticity and sexual dimorphism in the avian song control system: stereological measurement of neuron density and number. *J Comp Neurol*, **396**:186–192.

Tramontin, A. D., Wingfield, J. C., Brenowitz, E. A. (1999) Contributions of social cues and photoperiod to seasonal plasticity in the adult avian song control system. *J Neurosci*, **19**:476–483.

Tramontin, A. D., Hartman, V. N., Brenowitz, E. A. (2000). Breeding conditions induce rapid and sequential growth in adult avian song control circuits: a model of seasonal plasticity in the brain. *J Neurosci*, **20**:854–861.

Tramontin, A. D., Perfito, N., Wingfield, J. C., Brenowitz, E. A. (2001) Seasonal growth of song control nuclei precedes seasonal reproductive development in wild adult song sparrows. *Gen Comp Endocrinol*, **122**:1–9.

Tramontin, A. D., Wingfield, J. C., Brenowitz, E. A. (2003) Androgens and estrogens induce seasonal-like growth of song nuclei in the adult songbird brain. *J Neurobiol*, **57**:130–140.

Treisman, R. (1996) Regulation of transcription by MAP kinase cascades. *Curr Opin Cell Biol*, **8**:205–215.

Trevisan, M. A., Mendez, J. M., Mindlin, G. B. (2006a) Respiratory patterns in oscine birds during normal respiration and song production. *Phys Rev E*, **73**.

Trevisan, M. A., Mindlin, G. B., Goller, F. (2006b) Nonlinear model predicts diverse respiratory patterns of birdsong. *Phys Rev Lett*, **96**.

Troyer, T. W., Bottjer, S. W. (2001) Birdsong: models and mechanisms. *Curr Opin Neurobiol*, **11**:721–726.

Troyer, T. W., Doupe, A. J. (2000a) An associational model of birdsong sensorimotor learning. I. Efference copy and the learning of song syllables. *J Neurophysiol*, **84**:1204–1223.

Troyer, T., Doupe, A. J. (2000b) An associational model of birdsong sensorimotor learning. II. Temporal hierarchies and the learning of song sequence. *J Neurophysiol*, **84**:1224–1239.

Tsien, J. Z., Huerta, P. T., Tonegawa, S. (1996) The essential role of hippocampal CA1 NMDA receptor-dependent synaptic plasticity in spatial memory. *Cell*, **87**:1327–1338.

Tucci, D. L., Rubel, E. W. (1990) Physiologic status of regenerated hair cells in the avian inner ear following aminoglycoside ototoxicity. *Otolaryngol Head Neck Surg*, **103**:443–450.

Tucker, V. A. (1968) Respiratory physiology of house sparrows in relation to high-altitude flight. *J Exp Biol*, **48**:55–66.

Tuttle, E. M. (2003) Alternative reproductive strategies in the white-throated sparrow: behavioral and genetic evidence *Behavioral Ecology*, **14**:425–432.

Tyler, R. S. (1993) Speech perception by children. In *Cochlear Implants: Audiological Foundations* (Tyler, R. S., ed.), pp. 191–256. San Diego, CA: Singular.

Ullman, M. T. (2004) Contributions of memory circuits to language: the declarative/procedural model. *Cognition*, **92**:231–270.

Uno, H., Maekawa, M., Kaneko, H. (1997) Strategies for harmonic structure discrimination by zebra finches. *Behav Brain Res*, **89**:225–228.

Vallet, E., Beme, I. I., Kreutzer, M. (1998) Two-note syllables in canary songs elicit high levels of sexual display. *Anim Behav*, **55**:291–297.

van de Lavoir, M. C., Diamond, J. H., Leighton, P. A., *et al.* (2006) Germline transmission of genetically modified primordial germ cells. *Nature*, **441**:766–769.

Van Meir, V., Boumans, T., De Groof, G., *et al.* (2005) Spatiotemporal properties of the BOLD response in the songbirds' auditory circuit during a variety of listening tasks. *NeuroImage*, **25**:1242–1255.

van Praag, H., Schinder, A. F., Christie, B. R., Toni, N., Palmer, T. D., Gage, F. H. (2002) Functional neurogenesis in the adult hippocampus. *Nature*, **415**:1030–1034.

van Tuinen, M., Hedges, S. (2001) Calibration of avian molecular clocks. *Mol Biol Evol*, **18**:206–213.

Vanderhorst, V. G., Terasawa, E., Ralston, H. J., 3rd, Holstege, G. (2000) Monosynaptic projections from the lateral periaqueductal gray to the nucleus retroambiguus in the rhesus monkey: implications for vocalization and reproductive behavior. *J Comp Neurol*, **424**:251–268.

Vanson, A., Arnold, A. P., Schlinger, B. A. (1996) 3 beta-hydroxysteroid dehydrogenase/isomerase and aromatase activity in primary cultures of developing zebra finch telencephalon: dehydroepiandrosterone as substrate for synthesis of androstenedione and estrogens. *Gen Comp Endocrinol*, **102**:342–350.

Vargha-Khadem, F., Carr, L. J., Isaacs, E., Brett, E., Adams, C., Mishkin, M. (1997) Onset of speech after left hemispherectomy in a nine-year-old boy. *Brain*, **120**(1):159–182.

Vargha-Khadem, F., Watkins, K. E., Price, C. J., *et al.* (1998) Neural basis of an inherited speech and language disorder. *Proc Natl Acad Sci USA*, **95**:12695–12700.

Vargha-Khadem, F., Gadian, D. G., Copp, A., Mishkin, M. (2005) FOXP2 and the neuroanatomy of speech and language. *Nat Rev Neurosci*, **6**:131–138.

Vates, G. E., Nottebohm, F. (1995) Feedback circuitry within a song-learning pathway. *Proc Natl Acad Sci USA*, **92**:5139–5143.

Vates, G. E., Broome, B. M., Mello, C. V., Nottebohm, F. (1996) Auditory pathways of caudal telencephalon and their relation to the song system of adult male zebra finches. *J Comp Neurol*, **366**:613–642.

Vates, G. E., Vicario, D. S., Nottebohm, F. (1997) Reafferent thalamo- "cortical" loops in the song system of oscine songbirds. *J Comp Neurol*, **380**:275–290.

Veenman, C. L., Wild, J. M., Reiner, A. (1995) Organization of the avian "corticostriatal" projection system: a retrograde and anterograde pathway tracing study in pigeons. *J Comp Neurol*, **354**:87–126.

Velho, T. A., Pinaud, R., Rodrigues, P. V., Mello, C. V. (2005) Co-induction of activity-dependent genes in songbirds. *Eur J Neurosci*, **22**:1667–1678.

Velho, T. A., Lovell, P., Mello, C. V. (2006) Enriched expression and developmental regulation of the middle-weight neurofilament (NF-m) gene in song control nuclei of the zebra finch. *J Comp Neurol*, **500**:477–497.

Veltman, C. J. (1984) The social system and reproduction in a New Zealand magpie population, and a test of the cooperative breeding hypothesis. Ph.D. dissertation. Massey University, Palmerston North, NZ.

Veltman, C. J. (1989) Flock, pair and group-living lifestyles without cooperative breeding by Australian magpies, *Gymnorhina tibicen*. *Ibis*, **131**:6016–6018.

Veltman, C. J., Carrick, R. (1990) Male-biased dispersal in Australian magpies. *Anim Behav*, **40**:190–192.

Verner, J. (1976) Complex song repertoire of male long-billed marsh wrens in eastern Washington. *Living Bird*, **14**:263–300.

Vicario, D. S. (1991a) Organization of the zebra finch song control system. II. Functional organization of outputs from nucleus robustus archistriatalis. *J Comp Neurol*, **309**:486–494.

Vicario, D. S. (1991b) Neural mechanisms of vocal production in songbirds. *Curr Opin Neurobiol*, **1**:595–600.

Vicario, D. S. (1993) A new brain stem pathway for vocal control in the zebra finch song system. *NeuroReport*, **4**:983–986.

Vicario, D. S. (2004) Using learned calls to study sensory-motor integration in songbirds. *Ann NY Acad Sci*, **1016**:246–262.

Vicario, D. S., Nottebohm, F. (1988) Organization of the zebra finch song control system. I. Representation of syringeal muscles in the hypoglossal nucleus. *J Comp Neurol*, **271**:346–354.

Vicario, D. S., Simpson, H. B. (1995) Electrical stimulation in forebrain nuclei elicits learned vocal patterns in songbirds. *J Neurophysiol*, **73**:2602–2607.

Vicario, D. S., Yohay, K. H. (1993) Song-selective auditory input to a forebrain vocal control nucleus in the zebra finch. *J Neurobiol*, **24**:488–505.

Vicario, D. S., Naqvi, N. H., Raksin, J. N. (2001) Behavioral discrimination of sexually dimorphic calls by male zebra finches requires an intact vocal motor pathway. *J Neurobiol*, **47**:109–120.

Vicario, D. S., Raksin, J. N., Naqvi, N. H., Thande, N., Simpson, H. B. (2002) The relationship between perception and production in songbird vocal imitation: what learned calls can teach us. *J Comp Physiol A*, **188**:897–908.

Vignal, C., Attia, J., Mathevon, N., Beauchaud, M. (2004a) Background noise does not modify song-induced genic activation in the bird brain. *Behav Brain Res*, **153**:241–248.

Vignal, C., Mathevon, N., Mottin, S. (2004b) Audience drives male songbird response to partner's voice. *Nature*, **430**:448–451.

Vignal, C., Andru, J., Mathevon, N. (2005) Social context modulates behavioural and brain immediate early gene responses to sound in male songbird. *Eur J Neurosci*, **22**:949–955.

Volman, S. F. (1993) Development of neural selectivity for birdsong during vocal learning. *J Neurosci*, **13**:4737–4747.

Volman, S. F. (1996) Quantitative assessment of song-selectivity in the zebra finch "high vocal center". *J Comp Physiol A*, **178**:849–862.

Volman, S. F., Khanna, H. (1995) Convergence of untutored song in group-reared zebra finches (*Taeniopygia guttata*). *J Comp Psychol*, **109**:211–221.

von Bartheld, C. S, Byers, M. R., Williams, R., Bothwell, M. (1996) Anterograde transport of neurotrophins and axodendritic transfer in the developing visual system. *Nature*, **379**:830–833.

Voorhuis, T. A., De Kloet E. R. (1992) Immunoreactive vasotocin in the zebra finch brain (*Taeniopygia guttata*). *Brain Res Dev Brain Res*, **69**:1–10.

Voorhuis, T. A., De Kloet, E. R., De Wied, D. (1991) Effect of a vasotocin analog on singing behavior in the canary. *Horm Behav*, **25**:549–559.

Vu, E. T., Coleman, M. J. (2001) Song recovery by adult zebra finches following unilateral Uva lesion requires nucleus mMAN. *Soc Neurosci Abstract*, **538**.13.

Vu, E. T., Mazurek, M. E., Kuo, Y.-C. (1994) Identification of a forebrain motor programming network for the learned song of zebra finches. *J Neurosci*, **14**:6924–6934.

Vu, E. T., Schmidt, M. F., Mazurek, M. E. (1998) Interhemispheric coordination of premotor neural activity during singing in adult zebra finches. *J Neurosci*, **18**:9088–9098.

Wada, K., Sakaguchi, H., Jarvis, E. D., Hagiwara, M. (2004) Differential expression of glutamate receptors in avian neural pathways for learned vocalization. *J Comp Neurol*, **476**:44–64.

Wada, K., Howard, J. T., McConnell, P., et al. (2006) A molecular neuroethological approach for identifying and characterizing a cascade of behaviorally regulated genes. *Proc Natl Acad Sci USA*, **103**:15212–15217.

Wade, J., Arnold, A. P. (1996) Functional testicular tissue does not masculinize development of the zebra finch song system. *Proc Natl Acad Sci USA*, **93**:5264–5268.

Wade, J., Arnold, A. P. (2004) Sexual differentiation of the zebra finch song system. *Ann NY Acad Sci*, **1016**:540–559.

Wade, J., Buhlman, L. (2000) Lateralization and effects of adult androgen in a sexually dimorphic neuromuscular system controlling song in zebra finches. *J Comp Neurol*, **426**:154–164.

Wade, J., Buhlman, L., Swender, D. (2002) Post-hatching hormonal modulation of a sexually dimorphic neuromuscular system controlling song in zebra finches. *Brain Res*, **929**:191–201.

Wade, J., Peabody, C., Coussens, P., et al. (2004) A cDNA microarray from the telencephalon of juvenile male and female zebra finches. *J Neurosci Methods*, **138**:199–206.

Waelti, P., Dickinson, A., Schultz, W. (2001) Dopamine responses comply with basic assumptions of formal learning theory. *Nature*, **412**:43–48.

Waldron, I. (1964) Courtship sound production in two sympatric sibling *Drosophila* species. *Science*, **144**:191–193.

Waldstein, R. S. (1990) Effects of postlingual deafness on speech production: implications for the role of auditory feedback. *J Acoust Soc Am*, **88**:2099–2114.

Wallhausser-Franke, E., Nixdorf-Bergweiler, B. E., DeVoogd, T. J. (1995) Song isolation is associated with maintaining high spine frequencies on zebra finch lMAN neurons. *Neurobiol Learn Mem*, **64**:25–35.

Walsh, P. J., Mommsen, T. P., Bass, A. H. (1995) Biochemical and molecular aspects of singing in batrachoidid fishes. In *Biochemistry and Molecular Biology of Fishes* (Hochachka, P. W., Mommsen, T. P., eds.), pp. 279–289. Amsterdam: Elsevier Science.

Walters, M. J., Harding, C. F. (1988) The effects of an aromatization inhibitor on the reproductive behavior of male zebra finches. *Horm Behav*, **22**:207–218.

Walters, M. J., Collado, D., Harding, C. F. (1991) Oestrogenic modulation of singing in male zebra finches: differential effects on directed and undirected songs. *Anim Behav*, **42**:445–452.

Wan, H., Warburton, E. C., Kusmierek, P., Aggleton, J. P., Kowalska, D. M., Brown, M. W. (2001) Fos imaging reveals differential neuronal activation of areas of rat temporal cortex by novel and familiar sounds. *Eur J Neurosci*, **14**:118–124.

Wang, J., Sakaguchi, H., Sokabe, M. (1999a) Sex differences in the vocal motor pathway of the zebra finch revealed by real-time optical imaging technique. *NeuroReport*, **10**:2487–2491.

Wang, N., Aviram, R., Kirn, J. R. (1999b) Deafening alters neuron turnover within the telencephalic motor pathway for song control in adult zebra finches. *J Neurosci*, **19**:10554–10561.

Wang, N., Hurley, P., Pytte, C., Kirn, J. R. (2002) Vocal control neuron incorporation decreases with age in the adult zebra finch. *J Neurosci*, **22**:10864–10870.

Wang, X., Merzenich, M. M., Beitel, R., Schreiner, C. E. (1995) Representation of a species-specific vocalization in the primary auditory cortex of the common marmoset: temporal and spectral characteristics. *J Neurophysiol*, **74**:2685–2706.

Wang, X., Grus, W. E., Zhang, J. (2006) Gene losses during human origins. *PLoS Biol*, **4**:e52.

Ward, B. C., Nordeen, E. J., Nordeen, K. W. (1998) Individual variation in neuron number predicts differences in the propensity for avian vocal imitation. *Proc Natl Acad Sci USA*, **95**:1277–1282.

Warren, D. K., Patterson, D. K., Pepperberg, I. M. (1996) Mechanisms of American English vowel production in a grey parrot (*Psittacus erithacus*). *Auk*, **113**:41–58.

Waser, M. S., Marler, P. (1977) Song learning in canaries. *J Comp Physiol Psychol*, **91**:1–7.

Watkins, K. E., Dronkers, N. F., Vargha-Khadem, F. (2002a) Behavioural analysis of an inherited speech and language disorder: comparison with acquired aphasia. *Brain*, **125**:452–464.

Watkins, K. E., Vargha-Khadem, F., Ashburner, J., et al. (2002b) MRI analysis of an inherited speech and language disorder: structural brain abnormalities. *Brain*, **125**:465–478.

Watkins, K. E., Strafella, A. P., Paus, T. (2003) Seeing and hearing speech excites the motor system involved in speech production. *Neuropsychologia*, **41**:989–994.

Watson, M. (1969) Significance of antiphonal song in the eastern whipbird *Psophodes olivaceus*. *Behaviour*, **35**:157–178.

Wehr, M., Zador, A. M. (2003) Balanced inhibition underlies tuning and sharpens spike timing in auditory cortex. *Nature*, **426**:442–446.

Wei, F., Qiu, C. S., Liauw, J., Robinson, D. A., Ho, N., Chatila, T., Zhuo, M. (2002) Calcium calmodulin-dependent protein kinase IV is required for fear memory. *Nat Neurosci*, **5**:573–579.

Weichel, K., Schwager, G., Heid, P., Guttinger, H. R., Pesch, A. (1986) Sex differences in plasma steroid concentrations and singing behaviour during ontogeny in canaries (*Serinus canaria*). *Ethology*, **73**:281–294.

Weir, A. A., Chappell, J., Kacelnik, A. (2002) Shaping of hooks in New Caledonian crows. *Science*, **297**:981.

Weisman, R., Ratcliffe, L. (2004) Relative pitch and the song of black-capped chickadees. *Am Sci*, **92**:532–539.

Wennstrom, K. L., Reeves, B. J., Brenowitz, E. A. (2001) Testosterone treatment increases the metabolic capacity of adult avian song control nuclei. *J Neurobiol*, **48**:256–264.

Werker, J. F., Polka L. (1993) The ontogeny and developmental significance of language-specific phonetic perception. In *Developmental Neurocognition: Speech and Face Processing in the First Year of Life* (de Boysson-Bardies, B., de Schonen, S., Jusczyk, P. W., McNeilage, P., Morton, J., eds.), pp. 275–288. New York: Kluwer Academic/Plenum Publishers.

Werker, J. F., Tees, R. C. (1984) Cross-language speech perception: evidence for perceptual reorganization during the first year of life. *Infant Behav Dev*, **7**:49–63.

Werker, J. F., Tees, R. C. (1992) The organization and reorganization of human speech perception. *Annu Rev Neurosci*, **15**:377–402.

West, M. J., King, A. P. (1988) Female visual displays affect the development of male song in the cowbird. *Nature*, **334**:244–246.

Westneat, M. W., Long, J. H., Jr., Hoese, W., Nowicki, S. (1993) Kinematics of birdsong: functional correlation of cranial movements and acoustic features in sparrows. *J Exp Biol*, **182**:147–171.

Wever, E. G., Bray, C. W. (1930) Action currents in the auditory nerve in response to acoustical stimulation. *Proc Natl Acad Sci USA*, **16**:344–350.

Whaling, C. S., Nelson, D. A., Marler, P. (1995) Testosterone-induced shortening of the storage phase of song development in birds interferes with vocal learning. *Dev Psychobiol*, **28**:367–376.

Whaling, C. S., Solis, M. M., Doupe, A. J., Soha, J. A., Marler, P. (1997) Acoustic and neural bases for innate recognition of song. *Proc Natl Acad Sci USA*, **94**:12694–12698.

Whaling, C. S., Soha, J. A., Nelson, D. A., Lasley, B., Marler, P. (1998) Photoperiod and tutor access affect the process of vocal learning. *Anim Behav*, **56**:1075–1082.

White, G. F. (1789) *The Natural History and Antiquities of Selbourne*. London: Benjamin White & Son.

White, S. A., Livingston, F. S., Mooney, R. (1999) Androgens modulate NMDA receptor-mediated EPSCs in the zebra finch song system. *J Neurophysiol*, **82**:2221–2234.

Whitfield-Rucker, M. G., Cassone, V. M. (1996) Melatonin binding in the house sparrow song control system: sexual dimorphism and the effect of photoperiod. *Horm Behav*, **30**:528–537.

Whitney, O., Johnson, F. (2005) Motor-induced transcription but sensory-regulated translation of ZENK in socially interactive songbirds. *J Neurobiol*, **65**:251–259.

Wilbrecht, L., Kirn, J. R. (2004) Neuron addition and loss in the song system: regulation and function. *Ann NY Acad Sci*, **1016**:659–683.

Wilbrecht, L., Nottebohm, F. (2003) Vocal learning in birds and humans. *Ment Retard Dev D R*, **9**:135–148.

Wilbrecht, L., Petersen, T., Nottebohm, F. (2002a) Bilateral LMAN lesions cancel differences in HVC neuronal recruitment induced by unilateral syringeal denervation. Lateral magnocellular nucleus of the anterior neostriatum. *J Comp Physiol A Neuroethol Sens Neural Behav Physiol*, **188**:909–915.

Wilbrecht, L., Crionas, A., Nottebohm, F. (2002b) Experience affects recruitment of new neurons but not adult neuron number. *J Neurosci*, **22**:825–831.

Wilbrecht, L., Williams, H., Gangadhar, N., Nottebohm, F. (2006) High levels of new neuron addition persist when the sensitive period for song learning is experimentally prolonged. *J Neurosci*, **26**:9135–9141.

Wild, J. M. (1981) Identification and localization of the motor nuclei and sensory projections of the glossopharyngeal, vagus, and hypoglossal nerves of the cockatoo (*Cacatua roseicapilla*), Cacatuidae. *J Comp Neurol*, **203**:351–377.

Wild, J. M. (1987) Thalamic projections to the paleostriatum and neostriatum in the pigeon (*Columba livia*). *Neuroscience*, **20**:305–327.

Wild, J. M. (1993a) Descending projections of the songbird nucleus robustus archistriatalis. *J Comp Neurol*, **338**:225–241.

Wild, J. M. (1993b) The avian nucleus retroambigualis: a nucleus for breathing, singing and calling. *Brain Res*, **606**:319–324.

Wild, J. M. (1994a) Visual and somatosensory inputs to the avian song system via nucleus uvaeformis (Uva) and a comparison with the projections of a similar thalamic nucleus in a nonsongbird, *Columba livia*. *J Comp Neurol*, **349**:512–535.

Wild, J. M. (1994b) The auditory-vocal-respiratory axis in birds. *Brain Behav Evol*, **44**:192–209.

Wild, J. M. (1997a) Neural pathways for the control of birdsong production. *J Neurobiol*, **33**:653–670.

Wild, J. M. (1997b) The avian somatosensory system: the pathway from wing to Wulst in a passerine (*Chloris chloris*). *Brain Res*, **759**:122–134.

Wild, J. M. (2004a) Functional neuroanatomy of sensorimotor control of singing. *Ann NY Acad Sci*, **1016**:438–462.

Wild, J. M. (2004b) Pulmonary and tracheosyringeal afferent inputs to the avian song system. In *Seventh Congress of the International Society for Neuroethology*. Nyborg, Denmark.

Wild, J. M., Arends, J. J. (1987) A respiratory-vocal pathway in the brainstem of the pigeon. *Brain Res*, **407**:191–194.

Wild, J. M., Farabaugh, S. M. (1996) Organization of afferent and efferent projections of the nucleus basalis prosencephali in a passerine, *Taeniopygia guttata*. *J Comp Neurol*, **365**:306–328.

Wild, J. M., Williams, M. N. (1999) Rostral wulst in passerine birds. II. Intratelencephalic projections to nuclei associated with the auditory and song systems. *J Comp Neurol*, **413**:520–534.

Wild, J. M., Williams, M. N. (2000) Rostral wulst in passerine birds. I. Origin, course, and terminations of an avian pyramidal tract. *J Comp Neurol*, **416**:429–450.

Wild, J. M., Zeigler, H. P. (1980) Central representation and somatotopic organization of the jaw muscles within the facial and trigeminal nuclei of the pigeon (*Columba livia*). *J Comp Neurol*, **192**:175–201.

Wild, J. M., Arends, J. J. A., Zeigler, H. P. (1985) Telencephalic connections of the trigeminal system in the pigeon (*Columba livia*): a trigeminal sensorimotor circuit. *J Comp Neurol*, **234**:441–464.

Wild, J. M., Arends, J. J., Zeigler, H. P. (1990) Projections of the parabrachial nucleus in the pigeon (*Columba livia*). *J Comp Neurol*, **293**:499–523.

Wild, J. M., Karten, H. J., Frost, B. J. (1993) Connections of the auditory forebrain in the pigeon (*Columba livia*). *J Comp Neurol*, **337**:32–62.

Wild, J. M., Li, D., Eagleton, C. (1997) Projections of the dorsomedial nucleus of the intercollicular complex (DM) in relation to respiratory-vocal nuclei in the brainstem of pigeon (*Columba livia*) and zebra finch (*Taeniopygia guttata*). *J Comp Neurol*, **377**:392–413.

Wild, J. M., Goller, F., Suthers, R. A. (1998) Inspiratory muscle activity during bird song. *J Neurobiol*, **36**:441–453.

Wild, J. M., Williams, M. N., Suthers, R. A. (2000) Neural pathways for bilateral vocal control in songbirds. *J Comp Neurol*, **423**:413–426.

Wild, J. M., Williams, M. N., Suthers, R. A. (2001) Parvalbumin-positive projection neurons characterise the vocal premotor pathway in male, but not female, zebra finches. *Brain Res*, **917**:235–252.

Wild, J. M., Williams, M. N., Howie, G. J., Mooney, R. (2005) Calcium-binding proteins define interneurons in HVC of the zebra finch (*Taeniopygia guttata*). *J Comp Neurol*, **483**:76–90.

Wiley, R. H. (1973) Territoriality and non-random mating in sage grouse, *Centrocercus urophasianus*. *Anim Behav Mono*, **6**:87–169.

Wiley, R. H., Hatchwell, B. J., Davies, N. B. (1991) Recognition of individual males' songs by female dunnocks: a mechanism increasing the number of copulatory partners and reproductive success. *Ethology*, **88**:145–153.

Williams, H. (1985) Interhemispheric coordination of bird song. *Soc Neurosci Abstr*, **11**:871.

Williams, H. (1989) Multiple representations and auditory-motor interactions in the avian song system. *Ann NY Acad Sci*, **563**:148–164.

Williams, H. (1990) Models for song learning in the zebra finch: fathers or others? *Anim Behav*, **39**:745–757.

Williams, H. (2001) Choreography of song, dance and beak movements in the zebra finch (*Taeniopygia guttata*). *J Exp Biol*, **204**:3497–3506.

Williams, H. (2004) Birdsong and singing behavior. *Ann NY Acad Sci*, **1016**:1–30.

Williams, H., McKibben, J. R. (1992) Changes in stereotyped central motor patterns controlling vocalization are induced by peripheral nerve injury. *Behav Neural Biol*, **57**:67–78.

Williams, H., Mehta, N. (1999) Changes in adult zebra finch song require a forebrain nucleus that is not necessary for song production. *J Neurobiol*, **39**:14–28.

Williams, H., Nottebohm, F. (1985) Auditory responses in avian vocal motor neurons: a motor theory for song perception in birds. *Science*, **229**:279–282.

Williams, H., Staples, K. (1992) Syllable chunking in zebra finch (*Taeniopygia guttata*) song. *J Comp Psychol*, **106**:278–286.

Williams, H., Vicario, D. S. (1993) Temporal patterning of song production: participation of nucleus uvaeformis of the thalamus. *J Neurobiol*, **24**:903–912.

Williams, H., Cynx, J., Nottebohm, F. (1989) Timbre control in zebra finch (*Taeniopygia guttata*) song syllables. *J Comp Psychol*, **103**:366–380.

Williams, H., Crane, L. A., Hale, T. K., Esposito, M. A., Nottebohm, F. (1992) Right-side dominance for song control in the zebra finch. *J Neurobiol*, **23**:1006–1020.

Williams, H., Kilander, K., Sotanski, M. L. (1993) Untutored song, reproductive success and song learning. *Anim Behav*, **45**:695–705.

Williams, H., Connor, D. M., Hill J. W. (2003) Testosterone decreases the potential for song plasticity in adult male zebra finches. *Horm Behav*, **44**:402–412.

Wilson, D. M. (1961) The central nervous control of flight in a locust. *J Exp Biol*, **38**:471–490.

Winer, J. A. (2005) Decoding the auditory corticofugal systems. *Hear Res*, **207**:1–9.

Wingfield, J. C. (1984) Environmental and endocrine control of reproduction in the song sparrow, *Melospiza melodia*. I. Temporal organization of the breeding cycle. *Gen Comp Endocrinol*, **56**:406–416.

Wingfield, J. C. (2005) Historical contributions of research on birds to behavioral neuroendocrinology. *Horm Behav*, 48:395–402.

Wingfield, J. C., Farner, D. S. (1993) Endocrinology of reproduction in wild species. In *Avian Biology* (Farner, D. S., King, J., Parkes, K. C., eds.), pp. 163–327. New York: Academic Press.

Wingfield, J. C., Marler P. (1988) Endocrine basis of communication in reproduction and aggression. In *The Physiology of Reproduction*, Vol. 2 (Knobil E., Neill, J. D., Ewing, L. L., Greenwald, G. S., Markert, C. L., Pfaff, D. W., eds.), pp. 1647–1677. New York: Raven Press.

Wingfield J. C., Moore, M. C. (1987) Hormonal, social and environmental factors in the reproductive biology of free-living male birds. In *Psychobiology of Reproductive behavior: An Evolutionary Perspective* (Crews, D., ed.), pp. 149–175. Englewood Cliffs, NJ: Prentice-Hall.

Wingfield, J. C., Jacobs, J. D., Soma, K. K., *et al.* (1999) Testosterone, aggression, and communication: ecological bases of endocrine phenomenon. In *The Design of Animal Communication* (Hauser, M., Konishi, M., eds.), pp. 255–283. Cambridge, MA: MIT Press.

Wise, R. A. (2004) Dopamine, learning and motivation. *Nat Rev Neurosci*, 5:483–494.

Wise, R. J. (2003) Language systems in normal and aphasic human subjects: functional imaging studies and inferences from animal studies. *Br Med Bull*, 65:95–119.

Wissman, A. M., Brenowitz, E. A. (2003) Regulation of BDNF expression in the song system in response to seasonal cues. *Soc Neurosci Abst*, 29.

Wissman, A. M., Brenowitz, E. A. (2004) Intracerebral administration of BDNF promotes seasonal-like growth of an avian song control nucleus. *Soc Neurosci Abst*.

Wolffgramm, J., Todt D. (1982) Pattern and time specificity in vocal responses of blackbirds *Turdus merula* L. *Behaviour*, 65:264–287.

Wood, R. I. (1998) Integration of chemosensory and hormonal input in the male Syrian hamster brain. *Ann NY Acad Sci*, 855:362–372.

Woods, B. T. (1983) Is the left hemisphere specialized for language at birth? *Trends Neurosci*, 6:115–117.

Woolley, S. M., Casseday, J. H. (2004) Response properties of single neurons in the zebra finch auditory midbrain: response patterns, frequency coding, intensity coding, and spike latencies. *J Neurophysiol*, 91, 136–151.

Woolley, S. M., Casseday, J. H. (2005) Processing of modulated sounds in the zebra finch auditory midbrain: responses to noise, frequency sweeps, and sinusoidal amplitude modulations. *J Neurophysiol*, 94, 1143–1157.

Woolley, S. M., Rubel, E. W. (1997) Bengalese finches *Lonchura striata domestica* depend upon auditory feedback for the maintenance of adult song. *J Neurosci*, 17:6380–6390.

Woolley, S. M., Rubel, E. W. (1999) High-frequency auditory feedback is not required for adult song maintenance in Bengalese finches. *J Neurosci*, 19:358–371.

Woolley, S. M., Rubel, E. W. (2002) Vocal memory and learning in adult Bengalese finches with regenerated hair cells. *J Neurosci*, 22:7774–7787.

Woolley, S. M., Wissman, A. M., Rubel, E. W. (2001) Hair cell regeneration and recovery of auditory thresholds following aminoglycoside ototoxicity in Bengalese finches. *Hear Res*, 153:181–195.

Woolley, S. M., Fremouw, T. E., Hsu, A., Theunissen, F. E. (2005) Tuning for spectro-temporal modulations as a mechanism for auditory discrimination of natural sounds. *Nat Neurosci*, 8:1371–1379.

Woolley, S. M., Gill, P. R., Theunissen, F. E. (2006) Stimulus-dependent auditory tuning results in synchronous population coding of vocalizations in the songbird midbrain. *J Neurosci*, 26:2499–2512.

Worley, P. F., Christy, B. A., Nakabeppu, Y., Bhat, R. V., Cole, A. J., Baraban, J. M. (1991) Constitutive expression of zif268 in neocortex is regulated by synaptic activity. *Proc Natl Acad Sci USA*, 88:5106–5110.

Wurtz, R. H., Sommer, M. A., Cavanaugh J. (2005) Drivers from the deep: the contribution of collicular input to thalamocortical processing. *Progr Brain Res*, 149:207–225.

Yamada, R. A., Tohkura, Y. (1992) The effects of experimental variables on the perception of American English /r/ and /l/ by Japanese listeners. *Percept Psychophys*, 52:376–392.

Yamaguchi, A. (1998) A sexually dimorphic learned birdsong in the Northern cardinal. *Condor*, 100:504–511.

Yazaki-Sugiyama, Y., Mooney, R. (2004) Sequential learning from multiple tutors and serial retuning of auditory neurons in a brain area important to song learning. *J Neurophysiol*, 92:2771–2788.

Yin, J., Wallach, J., Del Vecchio M., *et al.* (1994) Induction of a dominant negative CREB transgene specifically blocks long-term memory in *Drosophila*. *Cell*, 79:49–58.

Yoshida, M., Naya, Y., Miyashita, Y. (2003) Anatomical organization of forward fiber projections from area TE to perirhinal neurons representing visual long-term memory in monkeys. *Proc Natl Acad Sci USA*, **100**:4257–4262.

Yu, A. C., Margoliash, D. (1996) Temporal hierarchical control of singing in birds. *Science*, **273**:1871–1875.

Zaccarelli, R., Elemans, C. P. H., Fitch, W. T., Herzel, H. (2006) Modelling bird songs: voice onset, overtones and registers. *Acta Acustica*, **92**:741–748.

Zajchowski, L. D., Etches, R. J. (2000) Transgenic chicken: past, present and future. *Avian Poult Biol Rev*, **11**:63–80.

Zann, R. (1984) Structural variation in the zebra finch distance call. *Z Tierpsychol*, **66**:328–345.

Zann, R. (1985) Ontogeny of the zebra finch distance call. I. Effects of cross-fostering to bengalese finches. *Z Tierpsychol*, **68**:1–23.

Zann, R. A. (1996) *The Zebra Finch: A Synthesis of Field and Laboratory Studies*. Oxford, UK: Oxford University Press.

Zann R. (1997) Vocal learning in wild and domesticated zebra finches: signature cues for kin recognition or epiphenomena? In *Social Influences on Vocal Development* (Snowden, C. T., Hausberger M., eds.), pp. 85–97. Cambridge: Cambridge University Press.

Zaretsky, M. D., Konishi, M. (1976) Tonotopic organization in the avian telencephalon. *Brain Res*, **111**:167–171.

Zeier, H., Karten, H. J. (1971) The archistriatum of the pigeon: organization of afferent and efferent connections. *Brain Res*, **31**:313–326.

Zeigler, H. P., Marler, P., eds. (2004) Behavioral neurobiology of birdsong. *Ann N Y Acad Sci*, **1016**:xiii–xvii, 1–788.

Zevin, J. D., Seidenberg, M. S., Bottjer, S. W. (2004) Limits on reacquisition of song in adult zebra finches exposed to white noise. *J Neurosci*, **24**:5849–5862.

Zigmond, R. E., Nottebohm, F., Pfaff, D. W. (1973) Androgen-concentrating cells in the midbrain of a songbird. *Science*, **179**:1005–1007.

Zollinger, S. A., Suthers, R. A. (2004) Motor mechanisms of a vocal mimic: implications for birdsong production. *Proc R Soc Lond B*, **271**:483–491.

Zupanc, G. K. H., Horschke, I. (1995) Proliferation zones in the brain of adult gymnotiform fish: a quantitative mapping study. *J Comp Neurol*, **353**:213–233.

Zweers, G. (1982) The feeding system of the pigeon (*Columba livia* L.). *Adv Anat Embryol Cell Biol*, **73**:1–108.

Author index

Aamodt, S. M. 30, 247, 262, 263, 272, 273, 299
Abraham, W. C. 267
Abramson, A. S. 16
Abrous, D. N. 350
Abs, M. 141
Absil, P. 330
Adams, J. P. 376
Adkins-Regan, E. 46, 315, 320, 324, 328, 332, 342
Adret, P. 21, 34, 55, 161, 202, 203, 282
Adret-Hausberger, M. 188, 189
Agate, R. J. 313, 378
Ahlquist, J. E. 326, 395, 396
Ahumada, J. A. 56
Airey, D. C. 37, 328
Aitkin, L. 293
Aitman, T. J. 373
Akutagawa, E. 245, 324, 339, 357, 422, 423
Alcock, K. J. 398, 401
Alger, S. J. 124, 133, 134, 330, 331
Allan, S. E. 84, 89, 101, 104, 105
Alm, P. A. 406
Altar, C. A. 394
Altman, J. 434
Alvarez-Borda, B. 343, 354, 358, 359, 364, 394, 437, 441, 442, 443, 446
Alvarez-Buylla, A. 119, 312, 350, 351, 353, 354, 356, 357, 359, 360, 361, 439, 440, 441, 445
Ames, A. 345
Ames, N. 447
Amin, N. 157, 166, 167, 168, 176, 181
Andalman, A. 403
Andersson, M. 320
Andrews, G. 380
Ang, C. W.-Y. 385
Appeltants, D. 157, 315, 330
Archer, A. 11
Arends, J. J. 143
Ariëns-Kappers, C. U. 58, 59, 61
Armstrong, E. A. 320, 411, 414

Arnold, A. P. 23, 33, 46, 50, 52, 57, 66, 86, 118, 138, 143, 148, 164, 228, 246, 249, 305, 310, 311, 313, 315, 323, 324, 325, 327, 330, 333, 342, 353, 369, 371, 377, 378, 379, 431, 459
Arthur, B. J. 166, 168, 176, 177, 371
Ascenzi, M. 46
Aschoff, J. 420
Asher, J. J. 25
Ashmore, R. C. 3, 66, 76, 77, 115, 118, 119, 120, 122, 126, 128, 130, 131, 149, 151
Ashraf, S. I. 400
Aston-Jones, G. 134
Athos, J. 376
Atkins, C. M. 376
Aubin, T. 157
Avey, M. T. 387, 388

Backstrom, N. 373
Bailey, D. J. 259, 273, 276, 278, 388
Baker, A. M. 53, 56
Baker, M. C. 47, 427, 450
Bakin, J. S. 192
Baldwin, F. M. 431
Ball, G. F. 3, 46, 50, 131, 133, 134, 164, 275, 303, 313, 320, 321, 323, 326, 328, 329, 330, 331, 333, 334, 335, 337, 342, 393
Ballam, G. O. 148
Ballentine, B. 94
Ballintijn, M. R. 79, 93
Balthazart, J. 303, 320, 323, 324, 328, 329, 330
Bankes, S. C. 358
Bannerman, D. M. 273
Banta-Lavenex, P. 350
Baptista, L. F. 17, 20, 23, 29, 34, 61, 205, 206, 207, 326, 346
Barami, K. 356
Barclay, S. R. 313, 315
Barkan, S. 445
Barker, F. K. 379
Barlow, H. 452
Barnea, A. 445
Barney, H. L. 91
Barria, A. 266

Barto, A. G. 247
Basham, M. E. 247, 250, 260, 262, 263, 266, 267, 272, 273, 298, 299, 395
Bass, A. H. 311, 341
Bates, E. 15, 25, 187
Bateson, P. 413
Bäumer, E. 419
Bayer, K. U. 266
Baylis, J. R. 398, 461
Beach, F. 31, 454
Beach, F. A. 320, 417
Bear, M. F. 267
Beckers, G. J. L. 87, 91, 92, 97, 112
Beecher, M. D. 188, 204, 332, 345, 346, 347
Belton, E. 401
Benali, H. 406
Bendix, J. 56
Benedetta, L. 350
Benoit, T. C. 21
Bentley, G. E. 317, 320, 334, 337
Benton, S. 177
Bergman, A. 157
Berman, D. E. 376
Bermejo, M. 97
Bernard, D. J. 46, 50, 323, 324, 325, 329, 334, 335, 343
Berridge, C. W. 134
Beydler, S. 66
Bickerton, D. 15
Biedermann, W. 417
Bigalke-Kunz, B. 166
Birkhead, T. R. 373
Bischof, H. J. 44, 204, 267
Bleisch, W. 315
Bliss, T. V. 262
Bloodstein, O. 380
Blum, S. 376
Boatman, D. J. 450
Bock, J. 267, 395
Boettiger, C. A. 182, 247, 262, 263, 267, 298
Bohland, J. W. 406
Böhner, J. 36, 40, 42, 43, 115, 122, 131, 205, 206, 207, 253, 263, 317, 332
Boland, C. R. J. 56

Boles, W. E. 55
Bolhuis, J. J. xvi, 33, 34, 167, 181, 202, 250, 259, 271, 272, 273, 274, 275, 276, 277, 278, 279, 280, 282, 294, 299, 327, 354, 376, 382, 384, 386, 388
Bonke, B. A. 388
Bonke, D. 172, 190
Bonner, J. 420
Borror, D. J. 86
Boseret, G. 46, 317
Bottjer, S. W. 29, 46, 50, 52, 66, 68, 69, 70, 118, 124, 127, 141, 148, 157, 164, 177, 180, 182, 211, 217, 228, 240, 242, 245, 247, 253, 261, 262, 263, 267, 269, 270, 294, 298, 332, 333, 342, 354, 376, 389, 391, 397, 431
Boughman, J. W. 9
Boumans, T. 280
Bourtchuladze, R. 376
Bout, R. G. 141, 145
Bower, G. 212, 215
Bozon, B. 376
Braaten, R. F. 286
Bradbury, J. W. 96, 457
Bragg, W. 413
Brainard, M. S. 29, 66, 118, 122, 131, 150, 161, 201, 202, 210, 217, 228, 229, 232, 236, 239, 240, 242, 244, 247, 249, 253, 257, 258, 273, 332, 352, 353, 354, 362, 366
Brann, D. W. 306
Brannon, E. M. 198
Braun, K. 143, 267
Brauth, S. E. 70, 384, 388, 395, 397
Bray, C. W. 421
Brennan, P. A. 262
Brenner, S. 420
Brenowitz, E. A. 3, 57, 165, 174, 250, 275, 303, 315, 320, 323, 324, 325, 326, 327, 328, 330, 332, 333, 334, 335, 338, 339, 340, 342, 343, 344, 345, 347, 354, 358, 359, 361, 365, 446
Bricker, P. D. 189
Broadbent, D. 452
Broca, P. 81
Brooks, R. J. 188
Brown, A. M. 264
Brown, C. H. 452
Brown, E. D. 51
Brown, J. L. 56, 323, 374
Brown, S. D. 91
Brown, W. R. 373
Brumm, H. 40, 208, 209, 211
Brummett, R. E. 231
Bucher, D. 134
Buckley, M. J. 273
Buhlman, L. 138, 341
Bull, J. J. 313

Buller, A. L. 266
Burchuladze, R. 262
Burd, G. D. 433
Burek, M. J. 353, 354, 356
Burgeon, J. K. 215
Burgoyne, P. 378, 379
Burmeister, S. S. 397
Burnham, D. K. 14
Burt, J. M. 165, 327
Buskirk, D. 439
Butler, A. B. 63, 66, 158
Butler, P. J. 106

Cain, D. P. 273
Calabrese, R. L. 134
Calder, W. A. 82, 104, 345
Callan, D. E. 270
Calvin, W. H. 341
Campbell, C. B. 59, 60
Canady, R. A. 16, 46, 323, 326, 433
Capranica, R. R. 87
Capsius, B. 166, 190
Carcraft, J. 60
Cardin, J. A. 115, 118, 119, 131, 159, 162, 163, 164, 166, 177, 183, 260, 293, 390
Carleton, A. 365
Carlsson, P. 401
Carmignoto, G. 266
Caro, S. P. 337
Carr, C. E. 190
Carrick, R. 51, 52, 54, 56
Carruth, L. 379
Casey, R. M. 79
Casseday, J. H. 232
Cassone, V. M. 56
Castelino, C. B. 131, 393
Castellani, G. C. 262
Catchpole, C. K. 8, 10, 17, 53, 56, 131, 285, 286, 306, 307, 308, 315, 317, 318, 320, 327, 328
Cecchi, G. A. 385
Cepko, C. L. 365
Chaiken, M. 26, 35, 189, 205, 206, 207, 346
Chappell, J. 61
Chaudhuri, A. 381
Chen, X. 378
Cheney, D. L. 9, 10, 457
Cheng, H. Y. 376, 383
Cheng, M.-F. 46, 316
Chew, S. J. 167, 181, 190, 250, 258, 274, 376, 385, 388, 389
Chi, Z. 122, 131
Chomsky, N. A. 15, 194, 197, 398
Christy, B. A. 383
Cirelli, C. 387

Clark, C. W. 208, 416
Clark, J. T. 318, 411
Clayton, D. F. 32, 165, 190, 250, 258, 274, 313, 376, 377, 380, 381, 383, 385, 387, 389, 395
Clayton, N. S. 206, 207, 228, 236
Clower, R. P. 315
Clucas, B. A. 198
Cockburn, A. 56
Cohen, D. 134, 145, 299
Colbran, R. J. 264, 266
Coleman, M. J. 68, 118, 124, 127, 128, 131, 142, 150, 159, 162, 164, 165, 167, 176, 177, 178, 180, 229, 293
Collingridge, G. L. 262
Collins, P. W. 231
Collins, S. 320
Collins, T. 447
Colman, H. 267
Colombo, M. 279
Contanche, D. A. 231
Cooper, B. G. 79, 101, 104, 108, 109, 112, 113, 116, 120, 122, 131, 132, 133, 141, 145, 148
Corwin, J. T. 231
Cotanche, D. A. 231
Cotman, C. W. 262
Cousillas, H. 170, 189
Cowan, N. 212
Cowie, R. 10, 11
Craigie, E. H. 58
Crair, M. C. 266
Creutzfeld, O. 421
Crouch, H. 51
Crow, T. 376
Crowder, R. G. 212
Cruz, R. M. 231
Cull-Candy, S. 266
Cunningham, M. A. 450
Curran, T. 381
Cynx, J. 8, 9, 11, 40, 46, 91, 228, 229, 232, 353, 366

Dabelsteen, T. 91, 97
Daley, M. 81, 95, 110, 116
Dantzker, M. S. 96
Das, G. D. 434
Das, R. M. 373, 394
Dave, A. S. 163, 164, 251, 260, 271, 390
Davidson, W. R. 54
Davies, D. C. 69
Davis, H. P. 376
Davis, S. 262
Daw, N. W. 263, 266
Dawson, D. A. 316, 317, 320, 373
de Boysson-Bardies, B. 21
de Kloet, E. R. 316

DeCasper, A. J. 14, 16
Dehaene-Lambertz, G. 280
Del Negro, C. 132, 134, 162, 163, 164, 174, 275, 276, 318, 327
den Boer, P. J. 141
Deng, C. 52, 55, 57
Denisenko-Nehrbass, N. I. 399
Dennis, M. 25
Deregnaucourt, S. 201, 217, 218, 219, 220, 224, 226
Derrickson, K. C. 346
Deviche, P. 325, 335, 343
DeVoogd, T. J. 37, 46, 50, 119, 137, 138, 164, 273, 312, 313, 326, 327, 333, 336, 354, 357, 433
De Vries, G. 379
Dewing, P. 379
DeWulf, V. 354
Dietrich, K. 228
Dilger, W. C. 375
Ding, L. 68, 253, 262, 264, 266
Dittrich, F. 339
Dobzhansky, T. 373, 453
Doetsch, F. 354, 440
Dominey, P. F. 270
Donham, R. S. 335
Dooling, R. J. 14, 91, 205, 286, 399, 421, 427, 461
Dörrscheidt, G. 172, 190
Douglas-Cowie, E. 10, 11
Doupe, A. J. 20, 66, 68, 99, 118, 119, 150, 151, 161, 162, 163, 164, 165, 166, 167, 176, 182, 183, 187, 190, 192, 210, 211, 216, 217, 218, 236, 239, 242, 244, 247, 249, 250, 251, 252, 253, 257, 258, 261, 262, 263, 267, 270, 271, 273, 276, 282, 294, 295, 296, 298, 299, 332, 352, 353, 354, 362, 366, 380, 385, 389, 390, 391, 393, 398, 404, 423
Doya, K. 247, 249, 265, 270
Doyon, J. 406
Dubbeldam, J. L. 69, 70, 141
Duchowny, M. 25
Duffy, D. L. 276, 278
Dugatkin, L. A. 56
Dumas, T. C. 395
Dunham, M. L. 53
Dunn, P. O. 56
Durand, S. E. 46, 62, 312, 397
Durant, W. 425
Dutar, P. 118, 179
Dykxhoorn, D. M. 400

Eakin, R. 453
Eales, L. A. 21, 29, 206, 207, 236, 258, 264, 266, 267, 332, 352, 360, 445
Ecklund, H. 448

Eda-Fujiwara, H. 273, 274, 276, 281, 395
Edelman, G. M. 267
Edinger, L. 58, 61
Edwards, C. J. 198
Eens, M. 133, 188, 189, 328, 329, 330
Ehrlich, P. 184
Eibl-Eibesfeldt, I. 453
Eichenbaum, H. 299
Eilers, R. E. 10
Eimas, P. D. 13, 14, 17
Elbrecht, A. 313
Elemans, C. P. H. 79, 81
Eley, M. 455
Elman, J. L. 15
Emlen, J. 420, 455
Enard, W. 406
English, J. D. 376
Epling, C. 373
Esser, K. H. 398
Evans, C. S. 461
Evans, M. R. 373

Fadiga, L. 279, 401
Falls, J. B. 42, 188, 326
Fant, G. 91, 92
Farabaugh, S. M. 33, 51, 56, 145, 150
Faraci, F. M. 106
Farner, D. S. 320
Farries, M. A. 4, 32, 63, 68, 69, 134, 332, 397
Fedde, M. R. 81, 106, 109, 138, 148
Fee, M. S. 7, 79, 87, 112, 120, 122, 126, 131, 183, 184, 226, 239, 403, 424, 429
Feldman, J. L. 132, 134
Ferland, R. J. 398, 402
Fernald, A. 19, 397
Feuk, L. 398, 401
Ficken, M. S. 198
Field, P. M. 311
Fields, R. D. 262
Fiete, I. R. 118, 120, 122, 131, 217
Fifer, W. P. 16
Fitch, W. T. 91, 98, 109, 195, 197, 398
Fitzpatrick, M. J. 399
Flege, J. E. 25
Fleischer, R. C. 47
Fletcher, N. H. 79, 91, 111, 112
Flint, A. C. 266
Floody, O. R. 86, 118
Fodor, J. A. 15
Foidart, A. 328
Follett, B. K. 306
Ford, H. A. 56
Fortin, G. 148
Fortune, E. S. 20, 52, 69, 158, 159, 161, 176, 177, 249, 294, 389

Foster, E. F. 68, 127, 142, 176, 177, 180, 261, 389
Fox, K. 266, 401
Fraley, D. S. 335
Franz Huber 420
Franz, M. 84, 106, 116, 148, 149
Freeberg, T. M. 49
Freking, F. T. 343
Friederici, A. D. 270
Friedländer, B. 417
Fritz, J. 279
Fromkin, V. 10, 25
Fry, C. H. 56
Funabiki, Y. 229, 232, 239, 263, 272, 294
Fusani, L. 54, 313, 318

Gaffan, D. 273
Gage, F. H. 350
Gahr, M. 46, 54, 62, 164, 250, 259, 271, 272, 273, 274, 276, 277, 279, 280, 282, 318, 323, 324, 325, 326, 327, 333, 335, 350, 354, 356, 376, 382, 384, 397, 396
Galaburda, A. M. 27
Gale, S. D. 68
Galeotti, P. 318
Gallagher, T. M. 380
Gallen, J. G. 452
Galoch, Z. 44
Garcia, R. 25
Gardiner, S. 411
Gardner, T. J. 34, 90, 151, 287
Gaunt, A. S. 7, 79, 92, 93, 344
Gaunt, S. L. L. 81
Geberzahn, N. 205, 206, 209, 210, 216, 278, 287, 292, 346
Gehr, D. D. 172
Gehring, W. 380
Gell, C. 40
Gentner, T. Q. 133, 155, 157, 165, 166, 167, 168, 174, 180, 181, 187, 188, 189, 190, 191, 192, 193, 195, 196, 197, 198, 250, 276, 278, 330, 354, 358, 366, 387, 388
George, J. M. 380
Gerrits, P. O. 143
Geschwind, N. 27, 134
Ghazanfar, A. A. 169
Gil, D. 318
Gilbert, C. D. 192
Gill, P. 170
Glaze, C. M. 40, 122, 131, 132
Gleeson, M. 109, 148
Gobes, S. M. H. 280
Godard, R. 188
Goelet, P. 381
Goldin-Meadow, S. 15, 399

Goldman, S. A. 246, 312, 353, 365, 433, 434, 438, 441, 459
Goldschmidt, R. B. 453
Goller, F. 3, 7, 35, 78, 79, 80, 81, 82, 83, 84, 86, 92, 93, 94, 95, 97, 98, 99, 100, 101, 102, 103, 104, 105, 106, 107, 108, 109, 110, 111, 112, 113, 116, 118, 120, 122, 131, 132, 133, 136, 138, 141, 145, 148, 149, 150, 342, 429, 458
Goodall, J. 456
Goodson, J. L. 311, 316, 317, 319, 331
Gorski, R. A. 323
Gottlieb, G. 422
Gould, E. 350
Gould, J. L. 374
Gould, S. J. 198
Gouzoules, H. 457
Gouzoules, S. 457
Goy, R. W. 323
Grabatin, O. 141
Grace, J. A. 169, 176, 181, 190
Gray, J. 412, 417
Gray, R. D. 61
Gray, T. 449
Graybiel, A. M. 254, 270, 299
Green, S. 456
Greenberg, M. E. 381
Greenewalt, C. H. 79, 86, 92, 109, 427
Gregory, R. 452
Grieser, D. L. 19
Griffin, D. 457
Grimm, D. 381, 400
Grisham, W. 66, 143
Gross, C. G. 176, 350
Grossman, M. 406
Grubb, B. 106
Guenther, F. H. 406
Gulledge, C. C. 325, 335, 343
Güntürkün, O. 69
Guo, Z. 403
Gurney, M. E. 153, 174, 310, 313, 324, 358, 377, 389, 423, 432, 434
Güttinger, H. R. 33, 47, 325, 326, 353, 385
Gutzke, W. H. 313
Guzowski, J. F. 376, 381, 394
Gwinner, H. 304, 329
Gyger, M. 461

Hackett, T. A. 279
Haesler, S. xvi, 216, 380, 394, 398, 399, 402, 403, 404
Hahnloser, R. H. 99, 118, 119, 120, 122, 124, 131, 177, 183, 184, 341, 356, 358, 390, 431, 443
Hailman, J. P. 47, 198, 374
Haldane, J. B. S. 449, 453

Hall, M. L. 54, 57
Halle, F. 174, 275
Hamilton, K. S. 327, 374, 455
Hamilton, W. D. 455
Hamilton, W. J. 455
Hammarberg, B. 11
Hand, C. 417
Hannon, G. 372
Hansson, B. 373
Harding, C. F. 303, 304, 306, 307, 313, 314, 315, 316, 317, 320, 329
Haripal, S. 448
Harris, J. O. 431
Harris, M. 380
Hartley, R. S. 81, 82, 83, 86, 101, 104, 106, 144, 429
Hartman, V. N. 380
Harvey, P. H. 325
Hassenstein, B. 453
Hau, M. 56
Hauber, M. E. 157, 286, 325
Hausberger, M. 141, 189
Hauser, M. D. 157, 187, 195, 197, 457
Haüsler, U. 190
Hayashi, T. 298
He, L. 372
Hedges, S. 379
Hedwig, B. 183
Hegner, R. E. 307
Heil, P. 172, 388
Heiligenberg, W. 420
Heimovics, S. A. 56, 330, 331, 393
Hein, A. M. 264, 265
Heinrich, J. E. 250, 262, 266, 267, 268, 269, 272, 273, 395
Heinz, R. D. 91
Helekar, S. A. 406
Hermann, K. 246
Hernandez, A. M. 388
Herrera, D. G. 381
Herrick, C. J. 58, 60, 61
Herrick, E. H. 431
Herrmann, K. 246, 249, 315
Hersch, G. L. 92
Hessler, N. A. 119, 163, 183, 251, 252, 253, 257, 391, 393
Hestrin, S. 266
Hewer, S. J. 333
Hidalgo, A. 336
Hikosaka, O. 261, 270
Hile, A. G. 61
Hinde, R. A. 53, 374, 409, 411, 450, 451, 455
Hinds, J. W. 434
Hirano, S. 24, 252
Ho, C. E. 131
Hochstadt, J. 406

Hodos, W. 59, 60, 63, 158, 430
Hoefnagal-Hohle, M. 26
Hoese, W. J. 43, 97, 98, 111, 141, 145, 150, 429
Hofer, M. 266
Hoffmann, H. 266
Hoke, K. L. 397
Holland, J. 198
Holloway, C. C. 313, 377
Holstege, G. 143
Holy, T. E. 403
Homberger, D. G. 95, 97
Hopcroft, J. 194, 198
Hopkins, C. D. 460
Horn, G. 157, 271, 279, 280, 299
Horschke, I. 356
Houde, J. F. 11
Hough, G. E. 46, 287, 292, 340, 346, 347, 353, 366
Houk, J. C. 254
Houx, B. B. 258
Hovatta, I. 400
Howell, P. 11
Hsu, A. 170, 190
Huang, H. 380
Hudson, W. H. 425
Huesmann, G. 376
Hughes, M. 208
Hull, E. M. 323
Hulse, S. H. 133, 188, 189, 330
Hultsch, H. xvi, 20, 37, 50, 201, 204, 205, 206, 207, 208, 209, 210, 211, 212, 213, 214, 215, 216, 278, 282, 346
Hunt, G. R. 61
Hunt, S. P. 381
Hutt, S. J. 14

i Cancho, R. F. 398
Ikeo, K. 380
Immelmann, K. 9, 17, 20, 23, 29, 34, 204, 228, 258, 263, 345, 352, 420, 442, 445
Imms, A. D. 411
Ince, S. A. 214
Ingvar, D. H. 24
Itasaki, I. 374
Itoh, Y. 378
Iyengar, S. 246, 247, 262, 267, 269, 294, 297

Jackendoff, R. 398
Jackson, A. L. 400
Jackson, D. 448
Jakobson, R. 8
Janata, P. 142, 159, 162, 164, 166, 168, 176, 177, 249
Janik, V. M. 398

Jarvis, E. D. 42, 58, 62, 64, 131, 133, 158, 187, 190, 202, 253, 254, 257, 258, 259, 261, 273, 274, 279, 294, 381, 382, 383, 384, 387, 389, 390, 391, 393, 395, 396, 397, 405, 406, 441
Jenkins, P. F. 188, 189, 346
Jin, H. 250, 258, 380, 387, 395
Johnson, F. 50, 240, 245, 246, 247, 253, 254, 262, 267, 294, 307, 309, 339, 358
Johnson, J. S. 25
Johnston, J. B. 58
Joliveau, E. 92
Jones, M. W. 29, 30, 190, 376, 383
Jordan, M. I. 11, 251, 253
Jost, A. 377
Jürgens, U. 134, 143
Jurisevic, M. A. 51
Jusczyk, P. W. 14, 16, 211

Kaas, J. H. 279
Kacsoh, B. 305
Kaczmarek, L. 381, 383
Kaestner, K. H. 398
Kalat, J. 374
Källén, B. 58
Kamen, R. S. 11
Kandel, E. R. 283, 299
Kao, M. H. 66, 122, 131, 247, 249, 253, 354, 393
Kaplan, G. xvi, 50, 51, 52, 53, 56, 57
Kaplan, M. S. 434
Karten, H. J. 59, 61, 66, 69, 143, 145, 148, 384, 385, 389, 430
Kasparian, S. 433, 448
Katz, D. M. 143, 148
Katz, L. C. 155, 174, 358, 389, 422, 423, 434
Kawaguchi, Y. 68, 178
Kawata, M. 311
Kay, L. M. 192
Kelley, D. B. 69, 91, 177, 323, 384, 389
Kennedy, D. 417
Kent, R. D. 114
Kessler, D. K. 28
Khanna, H. 30, 40
Kiang, N. Y. 421
Kilgard, M. P. 21, 192
Kim, K. H. 26
Kim, M. 262
Kim, Y. H. 313
Kimpo, R. R. 165, 390, 391
Kimura, M. 406
King, A. P. 23, 40, 42, 54,
King, A. S. 78, 81
King, J. R. 345
Kipper, S. 214
Kirn, J. R. 119, 304, 312, 350, 353, 354, 355, 356, 357, 359, 361, 437, 438, 441, 443, 445

Kiss, J. Z. 316
Kitt, C. A. 70
Kittelberger, J. M. 182, 229, 245, 246, 247, 249
Klatt, D. H. 91
Kleinfeld, D. 180
Knipschild, M. 172
Knowlton, B. J. 270
Knudsen, E. I. 29
Kobayashi, K. 255, 393
Koehler, O. 420, 453
Koenig, W. D. 56
Kohara, K. 339
Komai, S. 400
Konishi, M. xvi, 8, 10, 11, 14, 15, 19, 24, 46, 84, 119, 142, 147, 150, 157, 158, 161, 163, 164, 167, 172, 174, 182, 183, 184, 201, 202, 205, 217, 228, 229, 232, 233, 236, 238, 239, 242, 245, 249, 251, 256, 257, 258, 260, 261, 262, 263, 271, 273, 280, 282, 283, 284, 285, 286, 287, 294, 295, 296, 298, 299, 310, 313, 324, 339, 347, 353, 357, 377, 385, 389, 390, 397, 404, 409, 413, 415, 417, 419, 420, 421, 422, 423, 424, 427, 428, 432, 447, 452, 454, 458
Kopachena, J. G. 326
Kopp, M. L. 205, 206, 209
Korsia, S. 29, 46, 211
Kosik, K. S. 400
Kotz, S. A. 270
Kozhevnikov, A. A. 183
Kraus, N. 28
Krauzlis, R. J. 134
Krichevsky, A. M. 400
Krishnan, V. V. 53
Kroeber, A. L. 454
Kröner, S. 69
Kroodsma, D. E. 10, 16, 17, 24, 29, 34, 35, 50, 119, 142, 157, 187, 206, 213, 216, 266, 284, 285, 345, 346, 395, 397, 398, 427, 428
Kruse, A. A. 376, 387
Kubke, M. F. 118, 126, 142, 144, 148, 151
Kubota, M. 246, 298
Kuehl-Kovarik, M. C. 266
Kuhl, P. K. 3, 5, 13, 14, 16, 17, 18, 19, 21, 22, 23, 26, 28, 99, 151, 187, 211, 215, 216, 271, 273, 279, 280, 282, 353, 380, 398, 399
Kuhlenbeck, H. 58
Kunkel, P. 56
Kuo, Z. Y. 422

Lack, D. 412, 450
Lade, B. L. 375
Ladefoged, P. 18

Lahiri, S. 132
Lai, C. S. 394, 398, 401, 402, 403
Laje, R. 79, 87, 90, 92, 112, 151
Lane, H. 10, 25
Langmore, N. E. 54, 326, 327
Larsen, O. N. 7, 78, 80, 93, 97, 100, 104, 105, 118, 136, 429
Lashley, K. S. 291, 446
Lasky, R. E. 13
Lass, N. J. 114
Lauder, G. V. 60
Laurent, G. 192
Leadbeater, E. 81
Lee, B. S. 11
Lee, C. C. 158, 376
Lee, D. W. 344
Lefroy, M. 411
Lehericy, S. 261, 270
Lehrman, D. S. 375, 422
Leitner, S. 276, 327, 334, 335, 340, 352, 353, 361, 388
Lemaire, P. 383
Lemon, R. E. 26, 284, 330
Lenneberg, E. H. 25, 27
Lent, K. 315, 339, 343
Leonard, C. 430
Leonard, C. M. 61
Leonard, M. L. 157
Leonard, S. L. 431
Leonardo, A. 11, 46, 84, 122, 131, 161, 183, 184, 228, 229, 232, 233, 236, 239, 242, 252, 347, 353, 404, 419, 424
Leppelsack, H. J. 166, 168, 172, 190, 388
Lesher, A. I. 306
Levin, R. N. 323
Lewicki, M. S. 166, 167, 168, 176, 178, 179, 423, 424
Lewis, J. W. 68, 261, 265
Lewontin, R. C. 198
Li, S. 403
Li, X. C. 246, 247, 339, 342, 359, 362, 393, 394, 441, 443, 447
Liberman, A. M. 13, 17, 21, 401, 417
Lichtman, J. W. 267
Licznerski, P. 406
Lieberman, P. 406
Lill, A. 460
Lillie, F. R. 377
Lim, W. A. 350
Lind, H. 188
Lindstrom, A. 344
Ling, W. 350
Link, W. 394
Linsley, P. S. 400
Lipkind, D. 354, 358, 364, 445
Lisker, L. 16

Lisman, J. E. 262, 264, 266
Liss, B. 399
Lissmann, H. 412
Liu, W. C. 42, 218, 393
Lively, S. E. 29
Livingston, F. S. 176, 182, 247, 250, 262, 263, 267, 268, 272, 273
Lledo, P. -E. 350
Lobo, M. K. 399
Locke, J. L. 11, 21
Lois, C. 400, 437, 441
Lombardino, A. J. 229, 233, 249, 347, 352, 353, 361, 365, 366, 393, 394, 420, 445, 446
Long, K. D. 395
Lorenz, K. 374, 375, 412, 413, 420, 450, 453, 455
Lowther, J. 379
Lucas, J. R. 49
Luine, V. N. 138, 315
Luo, M. 66, 68, 247, 254, 261, 391, 397
Lyford, G. L. 394

MacDermot, K. D. 398, 401, 403
MacDonald, J. 22
MacDougall-Shackleton, S. A. 52, 166, 168, 174, 190, 275, 276, 278, 303, 320, 325, 326, 327, 334, 388
Macklis, J. D. 357, 443
Macphail, E. M. 271, 273
Madison, R. 357
Magrath, R. D. 54, 57
Mahlapuu, M. 401
Maina, J. N. 106
Makatsch, W. 416
Malenka, R. C. 266
Malinow, R. 266
Maney, D. L. 276, 278, 316, 317, 329, 330, 331, 387, 393
Manley, G. 421
Mann, N. I. 50, 54, 57
Manning, A. 450
Manogue, K. R. 61, 144
Mansson, H. 380
Marder, E. 134
Marean, G. C. 231
Marentette, P. F. 399
Margoliash, D. 20, 24, 52, 69, 90, 99, 118, 119, 120, 122, 126, 131, 134, 142, 158, 159, 161, 162, 163, 164, 165, 166, 168, 174, 176, 177, 180, 181, 182, 183, 187, 191, 192, 197, 198, 249, 250, 253, 257, 258, 271, 273, 276, 294, 296, 341, 346, 353, 358, 385, 388, 389, 390, 423, 424, 431
Marien, P. 406

Marler, P. xv, xvi, 3, 5, 9, 10, 14, 15, 17, 19, 20, 23, 26, 32, 33, 34, 46, 47, 50, 51, 57, 61, 62, 63, 75, 92, 93, 97, 110, 111, 115, 136, 155, 157, 174, 184, 187, 201, 204, 205, 206, 207, 208, 210, 212, 216, 217, 218, 228, 229, 238, 240, 250, 253, 256, 263, 271, 273, 276, 278, 281, 282, 283, 284, 285, 286, 287, 289, 291, 292, 294, 299, 303, 320, 332, 333, 343, 345, 346, 347, 349, 350, 351, 353, 366, 369, 371, 374, 380, 398, 399, 409, 410, 411, 412, 413, 415, 416, 417, 418, 419, 420, 421, 424, 425, 426, 427, 428, 429, 445, 447, 449, 450, 451, 452, 454, 455, 456, 457, 460, 461, 462
Marr, D. 176
Marshall, R. C. 318
Martin, T. E. 55
Martins, E. P. 325, 326
Massaro, D. W. 22
Mateer, C. 24
Matthews, G. V. T. 412
Mattingly, I. G. 17, 401
Mayr, E. 412, 457
McCardle, P. 27
McCasland, J. S. 24, 119, 158, 161, 174, 183, 184, 249, 251, 257, 295, 389, 390, 391, 423, 431
McClearn, G. 454
McCowan, B. 9
McEwen, B. S. 46, 323
McGaugh, J. L. 262, 376
McGrew, M. J. 373, 400
McGuire, P. K. 24, 252
McGurk, H. 22
McHale, C. M. 384, 388
McIntyre, C. C. 266
McKenzie, T. L. 388
McKibben, J. R. 46, 229, 242
McKusick, V. A. 380
McVay, S. 398
Medina, L. 65
Mehler, J. 22
Mehta, N. 41, 66, 229, 242, 245, 354
Mello, C. V. 58, 62, 158, 160, 164, 165, 166, 167, 190, 202, 250, 258, 259, 273, 274, 293, 294, 354, 369, 376, 381, 382, 383, 384, 385, 386, 387, 389, 390, 393, 395
Meltzof, A. N. 21, 22, 23, 28
Mennill, D. J. 42
Merzenich, M. M. 21, 28
Messmer, E. 418
Metizen, J. 333
Metzdorf, R. 335
Metzger, M. 69, 384
Micheau, J. 268

Milbrandt, J. 383
Miller, A. 453
Miller, D. B. 10, 13, 17, 187, 274, 416, 417, 418
Milner, B. 273
Milsom, W. K. 106
Mindlin, G. B. 79, 87, 90, 92, 112, 151
Miska, E. 372
Mitani, J. C. 456
Miyasato, L. E. 47, 293, 298
Miyawaki, K. 13, 16, 26
Möhres, F. 453
Moldin, S. O. 399
Molnar 66
Molony, V. 81, 109, 148
Monroe, B. L. 379
Monyer, H. 266
Moon, C. 16
Mooney, R. 118, 142, 144, 150, 155, 158, 159, 162, 163, 164, 165, 167, 174, 176, 177, 178, 179, 182, 183, 184, 229, 240, 245, 246, 247, 249, 250, 251, 262, 263, 264, 267, 273, 280, 293, 294, 296, 297, 298, 358, 362, 389, 422
Moore, I. T. 22, 54, 56, 333, 335
Morgan, J. I. 381
Moriyama, K. 97
Morris, D. 43, 451
Morris, R. G. 262, 273
Morrison, R. G. 29, 242, 267, 360
Morton, E. S. 20, 50, 56
Mountjoy, D. J. 26, 330
Mueller-Preuss, P. 24
Muller, C. M. 168, 172, 388, 190
Müller-Bröse, M. 211
Mulligan, J. A. 284, 285, 421, 454
Mundinger, P. C. 16, 399
Mylander, C. 15, 399

Nagel, K. I. 192
Naguib, M. 42, 214, 215
Nakamura, K. 254
Nakanishi, H. 68
Nakazawa, K. 149
Narayan, R. 170
Nastiuk, K. L. 258, 383, 389
Naya, Y. 298
Nealen, P. M. 184, 325, 326, 327
Nelson, B. S. 95, 97, 111
Nelson, D. A. 14, 16, 17, 19, 20, 23, 26, 34, 51, 205, 206, 210, 218, 238, 253, 278, 286, 346, 399, 462
Neubauer, R. L. 133
Neville, N. H. 98
Newman, J. 168
Newport, E. L. 25, 29, 195

Nick, T. A. 163, 182, 183, 184, 260, 261, 271, 273, 280, 294, 296, 299, 390
Nicolai, J. 47, 206, 453
Nicolelis, M. A. 169
Nicoll, R. A. 266
Nixdorf, B. E. 118
Nixdorf-Bergweiler, B. E. 260, 262, 267, 431
Nock, B. L. 306
Noctor, S. C. 440
Nordby, J. C. 340, 345, 346
Nordeen, E. J. 11, 46, 84, 201, 202, 228, 233, 236, 241, 242, 244, 245, 246, 256, 261, 272, 315, 333, 347, 353, 356, 358, 359, 419
Nordeen, K. W. 11, 46, 84, 201, 202, 228, 233, 236, 241, 242, 244, 245, 246, 256, 261, 272, 324, 347, 353, 356, 358, 359, 419
Northcutt, R. G. 58, 60
Nottebohm, F. xv, xvi, 10, 11, 26, 27, 29, 33, 34, 35, 36, 37, 38, 42, 46, 52, 61, 66, 68, 69, 76, 86, 91, 115, 118, 119, 122, 124, 136, 137, 138, 141, 142, 143, 144, 150, 157, 164, 165, 177, 180, 205, 206, 217, 218, 223, 229, 233, 240, 242, 246, 247, 251, 257, 261, 267, 271, 273, 274, 282, 283, 284, 286, 287, 294, 304, 310, 311, 312, 313, 320, 322, 323, 324, 325, 327, 328, 330, 332, 333, 334, 336, 340, 343, 344, 345, 347, 350, 351, 352, 353, 354, 356, 357, 358, 359, 360, 361, 366, 376, 377, 380, 381, 382, 384, 389, 390, 391, 393, 394, 395, 397, 402, 404, 409, 410, 413, 418, 420, 422, 423, 425, 428, 429, 430, 431, 432, 433, 434, 436, 437, 438, 439, 440, 441, 442, 443, 445, 447, 458, 459, 461
Nottebohm, M. E. 35, 86, 332, 333, 347, 427, 428, 445, 448, 458
Nowak, M. A. 398
Nowicki, S. 7, 8, 46, 87, 91, 92, 93, 94, 97, 98, 99, 104, 109, 110, 111, 320, 321, 328, 366, 427, 454, 461

Oberweger, K. 35, 106
Ojemann, G. A. 24
Okada, A. 365
Okanoya, K. 11, 84, 97, 228, 233, 325, 353, 419
Oller, D. K. 10
O'Loghlen, A. L. 188, 346
O'Loughlin, B. 448
Olveczky, B. P. 66, 131, 218, 240, 242, 247
Orians, G. 453
Osten, P. 406

Otmakhov, N. 264
Otomo, K. 10
Otter, K. 327
Owens, E. 28
Oyama, S. 25

Packard, M. G. 270
Padden, D. M. 13
Pagel, M. 56
Pagel, M. D. 325
Palmer, T. D. 436
Pandazis, G. 448
Pantin, C. 413
Panzica, G. C. 323, 328
Parent, A. 60
Park, K. 333, 331
Park, K. H. 387
Park, V. 293
Partan, S. 457
Paton, J. A. 62, 144, 357, 358, 434, 436
Patterson, D. K. 97, 112
Patterson, T. A. 279
Paulsen, K. 92
Pawel, P. 406
Payne, L. L. 346
Payne, R. B. 42, 206, 346, 457
Payne, R. S. 398
Pearsall, W. H. 450
Pearson, D. M. 11
Pearson, H. 372
Pearson, K. G. 251
Peek, F. W. 143, 188
Pekarik, V. 400
Peltonen, L. 380
Penny, R. 457
Pepperberg, I. M. 61, 97, 112, 216
Percival, D. B. 219
Perkel, D. J. 4, 58, 65, 66, 68, 118, 134, 179, 240, 246, 247, 253, 254, 262, 264, 325, 326, 327, 332, 333, 339, 391
Perkell, J. 22
Perlman, W. R. 333
Perrins, C. 320
Person, A. L. 391
Pesch, A. 33
Peters, S. 14, 19, 20, 23, 34, 91, 205, 206, 207, 208, 210, 253, 278, 287, 289, 345, 346, 399, 428, 445, 461
Petersen, M. R. 457
Petersen, S. E. 24
Peterson, G. E. 91
Petherbridge, F. R. 411
Petitto, L. A. 10, 15, 399
Petrinovich, L. 17, 20, 23, 29, 34, 205, 206, 207, 346
Pfaff, D. W. 323

Phan, M. L. 167, 202, 259, 274, 275, 278, 294, 299, 354, 376, 388
Phillips, R. 143, 373
Phillmore, L. S. 325, 334, 335, 387, 388
Phipps, H. 447
Phoenix, C. H. 311, 377
Pickert, R. 29, 184, 206, 266, 345, 346, 417
Piersma, T. 344
Piiper, J. 81
Pilz, K. M. 318
Pinaud, R. 381, 387
Ping Tang, Y. 399
Pinker, S. 398, 399
Pinxten, R. 307
Pitelka, F. 453
Pittenger, C. 376
Plant, G. 11
Platenik, J. 395
Ploog, D. 24
Podos, J. 13, 94, 97, 98, 99, 104, 110, 112, 141, 150, 208, 287, 292
Poiani, A. 56
Polka, L. 26
Polley, D. B. 180
Poopatanapong, A. 394
Poremba, A. 279
Poulet, J. F. 183
Powell, F. L. 106
Prather, J. F. 156, 178, 180, 184, 297, 362
Price, C. J. 24
Price, P. H. 34, 40, 217, 228, 284, 353
Proulx, S. R. 373
Pruzansky, S. 189
Puelles, L. 65, 66
Pytte, C. L. 27, 46, 304, 350, 353, 362, 365, 366

Rafiqpoor, M. D. 56
Raisman, G. 311
Ralls, K. P. 398
Ramirez, J. M. 134
Ramus, F. 195
Rasika, S. 46, 246, 247, 339, 342, 364, 394, 441, 443, 444
Ratcliffe, L. 42, 327
Rauschecker, J. P. 168
Rauske, P. L. 162, 163, 164, 177, 183, 184, 260, 390
Reese, C. R. 86
Reid, I. C. 273
Reiner, A. 4, 32, 58, 62, 64, 65, 66, 68, 160, 240, 254, 279, 351, 383, 390, 404
Reinke, H. 115, 126, 138, 142, 143, 148, 150
Reiss, D. 9
Reiter, L. T. 380
Remez, R. E. 189

Ressler, K. J. 376
Reyer, H. -U. 57
Reynolds, K. 286
Rhodes, K. J. 309
Ribeiro, S. 169, 190, 250, 276, 278, 385, 386, 387, 389
Riebel, K. 54, 274, 279
Riede, T. 43, 87, 91, 93, 95, 96, 111, 141, 145
Riedel, G. 268
Riters, L. V. 56, 124, 133, 134, 303, 313, 320, 327, 328, 329, 330, 331, 393
Roberts, E. B. 262
Roberts, J. 380
Roberts, T. 137, 138, 142
Robertson, H. A. 381, 383
Rodriguez, P. 198
Roeper, J. 399
Rogers, L. J. 56
Rollenhagen, A. 267
Romanski, L. M. 279
Roper, A. 296
Rose, G. J. 34, 291
Rose, M. 58, 430
Rose, S. 262, 279, 299
Rosen, M. J. 165, 176, 178, 179, 182
Rosenblatt, R. 299
Rosenzweig, M. 454
Rothstein, S. I. 47, 346
Rowe, S. A. 316
Rozin, P. 374
Ruan, J. 137
Rubel, E. W. 11, 228, 231, 232, 233, 236, 347, 353, 419
Rubinson, D. A. 400
Rübsamen, R. 172, 190
Rusak, B. 381
Ryals, B. M. 231
Ryan, S. M. 164
Rybak, F. 318

Saar, S. 201, 210, 217, 242
Sachs, M. B. 158
Saffran, J. R. 16, 193
Saine, T. 215
Saito, N. 246, 298
Sakagami, S. 416
Sakaguchi, H. 269, 298, 324
Sakata, J. T. 228, 229, 232
Saldanha, C. J. 329
Saltiel, A. 323, 324
Sanderson, K. J. 51
Sandler, W. 399
Sartor, J. J. 342
Sasaki, A. 254, 393
Saucier, D. 273
Sawtell, N. B. 183

Schafe, G. E. 376
Schafer, M. 169, 172
Scharff, C. xvi, 38, 66, 119, 165, 184, 216, 217, 218, 240, 242, 244, 246, 250, 261, 327, 353, 354, 355, 356, 358, 376, 380, 391, 398, 399, 402, 404, 442, 443, 446
Scheich, H. 172, 388
Scheid, P. 81, 106
Schelderup-Ebbe, T. 419
Schlinger, B. A. 305, 315, 320, 328, 333, 339, 343, 377
Schmidt, M. F. 51, 76, 77, 99, 118, 119, 120, 121, 125, 126, 131, 138, 145, 159, 162, 163, 164, 166, 177, 179, 183, 184, 251, 260, 271, 390
Schmitt, M. 60
Schneider, D. 420
Schoups, A. 192
Schregardus, D. S. 280
Schuchmann, K. L. 61
Schultz, W. 134, 270, 362
Schwabl, H. 318
Schwartz, M. S. 24
Schwartzkopff, J. 190, 418, 420, 428
Scott, B. B. 400, 437, 441, 445
Scott, G. R. 106
Scott, L. L. 229, 233, 445
Scoville, W. B. 273
Searcy, M. 8
Searcy, W. A. 14, 104, 205, 283, 286, 327, 366, 399, 427, 461
Sejnowski, T. 247, 249, 265
Sen, K. 170, 181, 190, 192, 389
Senghas, A. 399
Sepe, S. 448
Setterwall, C. G. 79
Seyfarth, R. M. 9, 10, 457
Shaevitz, S. S. 157, 187
Shannon, R. V. 232
Sharma, S. K. 376
Shaywitz, B. A. 27
Shea, S. D. 164, 177
Shen, P. 329
Sheng, M. 381
Sherman, V. 10, 34, 256, 283, 284, 285, 428, 461
Sherry, D. F. 273
Shimizu, T. 385
Shoemaker, H. H. 377, 431
Shors, T. J. 365
Shu, W. 402, 403, 406
Sibley, C. G. 326, 379, 395, 396
Silverin, B. 313
Simon, H. A. 212, 215
Simonyan, K. 134
Simpson, H. B. 49, 66, 118, 119, 120, 143
Sinclair, A. 378

Singer, H. 447
Singh, N. C. 168, 170
Singh, T. D. 262, 264, 265, 267, 298, 376, 403
Skinner, B. F. 374
Slabbekoorn, H. 282, 461
Slate, M. 373
Slater, P. J. B. 10, 17, 21, 29, 50, 54, 56, 57, 131, 204, 205, 206, 207, 214, 217, 285, 286, 296, 306, 307, 308, 315, 317, 318, 320, 345
Smith, G. T. 46, 55, 330, 333, 334, 335, 337, 338, 340, 343, 446
Smith, M. A. 254
Smith, R. G. 313
Smotherman, M. 134
Smulders, T. V. 310, 334, 335, 340
Snow, C. E. 21, 25, 26
Sockman, K. W. 133, 274, 387
Soha, J. A. 15, 68, 164, 206, 261, 289, 291, 362, 462
Sohrabji, F. 66, 240, 242, 244, 247, 261, 333, 354, 355, 391
Solis, M. M. 20, 118, 161, 162, 163, 182, 183, 249, 258, 261, 271, 273, 274, 294, 295, 296, 299, 358
Soma, K. K. 316, 335, 336, 337, 338, 343
Sommer, M. A. 134
Sonnenschein, E. 57
Sossinka, R. 36, 40, 42, 115, 122, 131, 253, 317
Spector, D. A. 216
Spence, M. J. 16
Spiro, J. E. 142, 143
Squire, L. R. 283, 299, 376
Stacey, P. D. 56
Stamps, J. A. 53
Staples, K. 36, 37
Stark, L. L. 240, 246, 249
Stebbins, W. 456
Stefanski, R. A. 91
Stefansson, H. 373
Stern, C. 453
Stevens, K. N. 6
Steward, O. 394
Stoddard, P. K. 42, 187
Stoel-Gammon, C. 10, 21
Stokes, A. W. 454
Stokes, T. M. 61, 124, 430, 439
Strack, S. 266
Streeter, L. A. 13
Striedter, G. F. 60, 62, 66, 68, 115, 124, 126, 137, 143, 148, 150
Stripling, R. 167, 190, 250, 258, 259, 376, 385, 386, 388
Strote, J. 91
Struhsaker, T. T. 457, 458

Sturdy, C. B. 118, 126, 142, 144, 151
Stutchbury, B. J. 56
Sugiura, H. 9
Sukhatme, V. P. 383
Sundberg, J. 92
Suomi, S. J. 204
Suthers, R. A. 7, 27, 46, 51, 76, 78, 79, 80, 81, 82, 83, 84, 85, 86, 87, 89, 90, 97, 98, 100, 101, 102, 103, 104, 106, 108, 109, 110, 111, 116, 118, 136, 137, 138, 141, 144, 145, 147, 148, 208, 284, 458, 342, 429
Sutter, M. L. 161
Sutton, R. S. 247
Suzuki, R. 194
Svenningsson, P. 264
Swanson, L. W. 60
Sweatt, J. D. 376
Székely, A. D. 70
Székely, T. 210, 328, 433

Tahta, S. 25
Takahashi, K. 402
Tallal, P. 28
Tamashige, M. 417
Tamura, M. 10
Tamura, S. 374, 403, 428, 445, 454
Taniguchi, I. 324
Tanji, J. 134
Tarnopolsky, A. 97, 111
Taubenfeld, S. M. 376
Taylor, E. W. 106
Tchernichovski, O. xv, 34, 37, 40, 131, 201, 208, 210, 217, 218, 219, 220, 222, 224, 235, 236, 287, 318, 351
Teague, D. P. 327
Tecumseh, W. 98
Tees, R. C. 16
Teichmann, M. 406
Tekumalla, P. K. 316, 317, 337, 354
Tellerday, P. 448
ten Cate, C. 79, 87, 204, 258, 283
Teramitsu, I. 380, 393, 394, 402, 405
Terleph, T. A. 385, 389
Terpstra, N. J. 167, 168, 181, 259, 273, 274, 276, 278, 279, 281, 294, 299, 376, 388, 395
Terrace, H. S. 198
Theunissen, F. E. 157, 161, 162, 164, 165, 166, 167, 168, 170, 176, 181, 187, 190
Thielcke, G. 206
Thode, C. 395
Thompson, C. K. 338, 365
Thompson, R. F. 291
Thompson, R. R. 328
Thorneycroft, H. 379

Thorpe, W. H. xv, xvi, 9, 10, 19, 26, 27, 29, 34, 56, 61, 203, 205, 282, 375, 409, 411, 412, 419, 420, 425, 427, 428, 445, 447, 451, 452, 453, 458, 462
Timmons, B. A. 233
Tinbergen, N. 374, 412, 414, 416, 420, 450, 451, 455
Tischmeyer, W. 381
Titze, I. R. 91, 92, 95
Todt, D. xvi, 20, 37, 40, 42, 50, 201, 204, 205, 206, 207, 209, 211, 212, 213, 214, 215, 216, 282
Tohkura, Y. 18
Tononi, G. 267
Tovar, K. R. 266
Tramontin, A. D. 30, 46, 324, 330, 332, 333, 334, 336, 337, 338, 339, 340, 341, 343, 344, 354, 359, 361, 365, 446
Treisman, R. 381
Tremere, L. A. 381
Trevisan, M. A. 84
Trimmer, J. S. 309
Troyer, T. W. 40, 122, 131, 132, 163, 242, 253, 270, 297
Tsien, J. Z. 298
Tucci, D. L. 231
Tucker, V. A. 106
Tuttle, E. M. 379
Tyler, R. S. 11

Ullman, M. T. 194, 198, 270
Uno, H. 327

Vaccarino, A. L. 273
Vallet, E. 318, 385
van de Lavoir, M. C. 373
van Lawick-Goodall, J. 456
Van Meir, V. 167, 280, 294
van Praag, H. 365, 436
van Tuinen, M. 379
Vanderhorst, V. G. 143
Vanson, A. 343
Vargha-Khadem, F. 25, 398, 401, 406
Vates, G. E. 69, 115, 118, 122, 124, 126, 142, 150, 158, 159, 177, 180, 181, 189, 260, 293, 384, 385, 388, 389, 397
Veenman, C. L. 70
Vehrencamp, S. L. 457
Velho, T. A. 382, 383, 385, 389, 393, 394, 399
Veltman, C. J. 54, 56
Verner, J. 34, 214
Vicario, D. S. 47, 49, 66, 82, 116, 118, 119, 120, 122, 128, 137, 142, 143, 160, 167, 180, 202, 258, 294, 389
Vicini, S. 266
Vignal, C. 49, 157, 387

Vince, M. 413
Vogt, M. 168, 190
Volman, S. F. 20, 30, 40, 46, 161, 176, 182, 183, 258, 260, 271, 294, 296
von Bartheld, C. S. 339
von Frisch, K. 412, 426
von Holst, E. 412, 417
von Rad, U. 46, 228, 229, 232, 353
von Stresemann, E. 452, 453
Voorhuis, T. A. 316
Vowles, D. 453
Vu, E. T. 66, 68, 84, 115, 119, 124, 126, 127, 128, 131, 143, 148, 150, 229

Wada, K. 390, 393, 394, 395, 397, 399, 400, 406
Waddington, C. H. 460
Wade, J. 138, 259, 273, 274, 313, 315, 341, 377, 378, 388, 399
Waelti, P. 362
Walden, A. T. 219
Waldron, I. 455
Waldstein, R. S. 10, 11, 353
Wallhausser-Franke, E. 30, 246, 250, 267
Walsh, P. J. 341
Walters, M. J. 313, 317, 329
Wan, H. 279, 397
Wang, J. 324, 353, 356, 364
Wang, N. 354, 356, 358, 364, 442, 445
Wang, X. 168, 373
Ward, B. C. 37, 359
Warren, D. K. 91, 97
Waser, M. S. 205, 206, 229, 284, 351, 399, 457, 460
Waser, P. 457
Washburn, S. 454, 455
Waterhouse, B. D. 134
Watkin, K. L. 380
Watkins, K. E. 398, 401, 403, 406
Watson, M. 11, 54
Wehr, M. 179
Wei, F. 376
Weichel, K. 351, 352
Weinberger, N. M. 192
Weir, A. A. 61
Weiskrantz, L. 452
Weisman, R. 42
Wennstrom, K. L. 336, 345
Werker, J. F. 16, 26
West, M. J. 23, 40, 42, 54, 253, 287
Westbrook, G. L. 266
Westneat, M. W. 7, 43, 97, 110, 141, 150
Wever, E. G. 421
Whaling, C. S. 14, 15, 27, 29, 211, 289, 462
Whitaker, H. A. 25

White, G. F. 320
White, S. A. 182, 262, 267, 380, 393, 394, 399, 402, 405
Whitfield-Rucker, M. G. 55
Whitney, O. 307, 309
Wilbrecht, L. 246, 282, 294, 304, 342, 350, 352, 354, 358, 359, 360, 361, 363, 364, 445
Wild, J. M. 44, 69, 70, 82, 97, 100, 101, 104, 114, 115, 116, 118, 122, 124, 126, 127, 133, 136, 137, 138, 140, 141, 142, 143, 144, 145, 147, 148, 150, 151, 159, 165, 167, 178, 180, 293, 384, 388
Wiley, R. H. 188, 460
Williams, H. 32, 36, 37, 40, 41, 44, 46, 50, 66, 86, 91, 97, 110, 118, 119, 126, 128, 141, 144, 145, 150, 160, 164, 180, 206, 207, 229, 242, 245, 293, 354, 389, 423, 431
Wilson, D. M. 27, 417, 420, 454, 455
Winer, J. A. 158

Wingfield, J. C. 306, 307, 308, 316, 317, 320, 335, 338, 460
Wise, R. A. 270
Wise, R. J. 279
Wissman, A. M. 339, 348
Wolffgramm, J. 214
Wollberg, Z. 168
Wood, R. I. 330
Woods, B. T. 25
Woolley, S. M. 11, 157, 170, 172, 190, 201, 228, 231, 232, 233, 236, 238, 347, 353, 419
Worley, P. F. 381
Wurtz, R. H. 115, 134

Yamada, R. A. 18
Yamaguchi, A. 11, 84, 228, 233, 269, 353, 419, 462
Yasukawa, K. 461
Yazaki-Sugiyama, Y. 183, 261, 264, 294, 296
Yin, J. 376
Yohay, K. H. 258, 389

Yoshida, M. 298
Young, J. Z. 450
Yu, A. C. 99, 118, 119, 120, 126, 134, 341, 390, 431

Zaccarelli, R. 79
Zador, A. M. 179
Zajchowski, L. D. 400
Zangwill, O. 413, 452
Zann, R. A. 35, 42, 47, 49, 116, 184, 205, 296, 358, 391
Zaretsky, M. D. 158, 172
Zeier, H. 69
Zeigler, H. P. xvi, 62, 97, 141, 187, 299, 409
Zevin, J. D. 228, 229, 236, 237
Zigmond, R. E. 431
Zollinger, S. A. 76, 78, 79, 83, 84, 86, 87, 90, 104, 116, 118, 136, 138, 145
Zupanc, G. K. H. 356
Zweers, G. 141
Zychlinsky, A. 406

Subject index

"grind and bind" assays 308
"pidgin" languages 399
acetylcholine 164, 315, 387
acquired auditory template 217, 256, 269, 283
action-based learning 33, 210, 238, 287
activation by hormones 312, 377
activity-dependent genes 381, 382, 397
activity-dependent regulatory cascades 381
adult plasticity 46
adult song degradation 233
adult song variability 35
AFP (see "anterior forebrain pathway")
age-independent learners 345, 352
Ai (see "intermediate acropalliulm")
air sac pressure 101, 102
air sacs 81, 102, 138, 148
alarm calls 461
Alzheimer's disease 394, 446
American cardinal 78, 80, 87, 95, 104, 111, 462
American Sign Language 7, 25
AMPA glutamate receptors 394
amygdala 58, 65, 66
androgen 26, 323
androgen receptors (AR) 303, 323
androgen receptors in HVC and RA 310
androstenedione 329
anesthesia and behavioral state 260
anesthesia effects 180, 251
animal communication 461
animal semantics 461
anterior forebrain pathway (AFP) 66, 68, 115, 182, 218, 240, 256, 260, 293, 332, 423
anterior hypothalamus 331
anterior thalamus 406
antibody specificity 309
apraxia 394
AR (see "androgen receptors")
Arc gene 394
archicortex 58
archistriatum 58, 65, 69, 293
Area X 68, 163, 164, 165, 177, 202, 402, 403

Area X lesions 243, 247
Area X of the medial striatum 66, 430
areas lacking song-induced gene expression 389
aromatase 313, 328, 377
aromatase blockade 328
aromatase inhibition effects 336
arousal states 163
artificially accelerated song 211
auditory connectivity similar in learners and non-learners 395
auditory feedback xv, 24, 31, 46, 76, 84, 99, 113, 147, 155, 156, 157, 161, 172, 201, 217, 228, 242, 351, 418, 419, 459
auditory feedback and song maintenance 238
auditory feedback of self 24
auditory forebrain 376
auditory grandmother neuron 176
auditory inputs to HVC 180
auditory template 202
auditory template theory 283
Australian magpie xvi, 50, 57
autism 31, 380, 399
autoradiography 303, 308, 323
avian and mammalian brain homologies 60, 64
avian brain nomenclature revised 58
avian hyperpallium 65
avian pallium 65, 66
avian respiratory system 109
avian telencephalon 58, 60
awake birds 390

babbling 9, 10, 21, 23, 399
barbary dove 54
barn swallow 318
Bas (see "nucleus basorostralis")
basal ganglia 46, 58, 68, 270, 398, 406
basilar papilla 230
bat echolocation 457
bats 32, 398
bay wren 33, 57, 323
BDNF (see "brain derived neurotrophic factor")
beak xv, 75, 99, 110

beak gape 7, 43, 97, 112, 141, 145, 150
beak-dance coordination 44
begging call measures 15
behavioral endocrinology 455
behavioral primatology 455
bell crickets 415
Bengalese finch 29, 33, 97, 228, 233, 255, 347, 393, 415, 445
beta-actin 394
bilateral motor skills 86
biological control 411
biological races 412
biomechanical coupling 92
bird calls 46, 91, 94
bird's own song (see "BOS")
birdsong as a model system 63
black grouse 42
black-capped chickadee 33, 48, 87, 325, 334, 340, 445
blot tests 308
blueprint 282
blue-tailed monkey 456
BOS (bird's own song) 20, 155, 158, 160, 174, 202, 217, 256, 257, 271, 385, 389, 423
BOS-selectivity 161, 174, 260, 296
BOS-selectivity and conspecific song 164, 165
BOS-selectivity development 161
BOS-selectivity in the auditory forebrain 166
BOS-selectivity modulation 163
BOS-selectivity 172
BOS-sensitive neurons 156
brain areas and song-induced IEG expression 383
brain derived neurotrophic factor (BDNF) 246, 339, 362, 394, 441
brain structure homology 59, 69
brainstem instructive mechanisms 134
brainstem vocal respiratory network 116
breathing 136
breathing and singing 115, 126
Broca's aphasia 24
Broca's area 279, 401
broken syntax 287

Subject index

brown thrasher 78, 80, 81, 84, 86, 87, 346
brown-headed cowbird 33, 40, 42, 82, 84, 87, 104, 105, 107, 286, 346
budgerigar 273, 276, 312, 384, 395, 397
buff-breasted wren 323
bullfinch xv, 206, 418

calcium binding proteins 142
calcium/calmodulin-dependent protein kinase II 263
canary 26, 27, 33, 35, 81, 82, 83, 86, 87, 97, 106, 111, 142, 148, 161, 163, 169, 174, 205, 206, 229, 267, 274, 276, 286, 304, 312, 316, 317, 322, 324, 327, 332, 333, 335, 340, 343, 344, 347, 350, 351, 353, 354, 359, 361, 385, 399, 402, 404, 421, 423, 429, 430, 431, 432, 442, 444, 445, 446, 459, 460
canary brain 430, 442
canary song 37, 150
candidate gene approach 399
canonical song 35, 42
carboxy-terminal hydrolase L1 (UCHL1) 394
Carolina wren 326
carolling 51
castration effects 26, 46, 307, 347
catbird 86
catecholamines 393
categorical perception 13, 17
cats 434
caudal lateral mesopallium (CLM) 158, 166, 168, 190, 388
caudal lateral mesopallium lesions 388
caudal medial nidopallium 155, 166, 167, 190, 191, 202, 258, 274, 388
caudal medial nidopallium and female song memory 276
caudal medial nidopallium and song memory 274, 278
caudal medial nidopallium and tutor song memory 280
caudal medial nidopallium tuning 173
caudal mesopallium (CM) 180, 185, 293, 388
caudal mesopallium song perception 385
caudate nucleus 58, 401
caudolateral belt of nidopallium (NCL) 69, 70
cell death 354
central auditory pathways are conserved 384
central motor programs 86
central motor programs and song patterning 283
central pattern generator (CPG) 11, 285
cerebellum 401, 406
cerebral lateralization 456
cetaceans 32

c-fos gene 165
chaffinch 26, 29, 34, 86, 203, 218, 233, 282, 283, 412, 418, 419, 429, 442, 445, 450, 451, 452, 458
chaffinch call types 451
chaffinch song dialects 450
Charles Darwin 398, 426
chickadee 42, 47, 198, 335
chicken 10, 66, 69, 109, 283, 314, 372, 373, 384, 395, 399, 415, 419, 422, 428
chicken vocalizations 419
chicks 69
chimpanzee 456
chinchilla 13
chipping sparrow 42, 393
cholecystokinin 317
cholinergic innervation 164
choreography of cell replacement 443
chromatography 308
chunking 212
chunking in human serial learning 215
cHV (see "HVC and cHV")
classical conditioning 376
claustrum 65, 66
CLM (see "caudal lateral mesopallium")
closed beak vocalization 145
closed-ended learners 26, 345
CM (see "caudal mesopallium")
cochlea 418
cochlear implants 11, 28
cochlear nucleus 158, 180
cochlear nucleus angularis 150
cockatiels 78
combinatorial calls 49
communication disorders 31
communication in monkeys and apes 455
communication signals 187
comparator 298
comparator circuit 280
comparator system 296
computer-assisted analysis xv
conditioned head turn 13
conditioned taste aversion 376
congenitally deaf infants 10
connections between the preoptic area and the song system 330
conspecific song recognition 160, 164
conspecific song tuning 169
constraints on vocal performance 78
context effect 214
context-free grammar 155, 194
cooperative breeding 56
coot 413
copying speech sounds 91
cordon bleu 326
corollary discharge 186, 251
corticotropin releasing hormone 317

countersinging 42
cow 426
cowbird 23, 87, 108, 109, 287, 327
CPG (see "central pattern generator")
critical (sensitive) periods 5, 27, 30, 398
critical period closure 27, 29, 246
critical period closure and experience 28
critical period extension 30
critical-period learners 34, 38
crocodiles 402
crow 61, 78
crystallized song 23, 201
cuckoo 416

DAF (see "delayed auditory feedback")
Darwin's finches 97
daylength 303, 306, 348
deaf children 11, 15
deafening abolishes zenk induction 384
deafening before puberty 11
deafening effects 217, 228, 231, 242, 253, 353, 358, 391, 445, 454, 460
deafening effects on the anterior forebrain pathway 252
deafness 11
dehydroepiandrosterone (DHEA) 316
dehydroepiandrosterone effects on HVC 316
delayed auditory feedback (DAF) 11, 201, 229, 347, 424
dendritic spines 247
descending motor pathway 117, 118
descending motor pathway lesion effects 118
developmental plasticity 461
developmental trajectories 208
developmental verbal dyspraxia (DVD) 398
devocalized birds 342
DHEA (see "dehydroepiandrosterone")
dialects 9, 24, 47, 387, 417, 427
diencephalon 63
directed song 40, 42, 83, 89, 122, 253, 391, 393, 394
distorted auditory feedback 183
DLM (see "dorsolateral anterior thalamic nucleus")
DM (see "dorsomedial nucleus of the intercollicular complex")
DMP (see "dorsomedial posterior nucleus of the thalamus")
DNA/RNA hybridization 372
dogs 415
donkey 415
dopamine 362, 393
dopaminergic and tutor-driven auditory inputs 270

dopaminergic input 68, 270
dopaminergic neurons 253
dorsal lateral nucleus of the mesencephalon (MLd) 158
dorsolateral anterior thalamic nucleus (DLM) 66, 68, 390
dorsomedial nucleus of the intercollicular complex 143
dorsomedial nucleus of the intercollicular complex (DM) 126
dorsomedial nucleus of the interollicular complex 69
dorsomedial posterior nucleus of the thalamus (DMP) 68, 70, 126, 131
doves 81, 96
dragonflies 460
Drosophila 380, 400
Drosophila songs 455
ducks 69, 109, 141
duetting 33, 56, 324, 326, 332, 413
DVD (see "developmental verbal dyspraxia")
dynamic feedback controls 304
dynamic vocal development articulation problems 406
dynamic vocal development maps 220
dyslexia 27, 28, 31

early deafening effects 283
early effectors 381
early isolation effects 267
early perceptual learning 30
early phonetic learning 17
early tuning of IMAN neurons 264
early vocal motor learning 22
eastern phoebe 119, 428
eastern towhee 95, 97
eastern whip bird 54
effects of a hearing female 40
efference copy 120, 251
EIA (see "enzymeimmunoassays")
electric fish 460
electromyograms 80, 101
electrophoresis 308
elephants 398
ELISA (see "enzyme-linked immunosorbant assays")
elongated trachea 91
encoding template 264
endothelia 436
enhancers 372
ensemble and sparse coding 122
entropy 220
enzymeimmunoassays (EIA) 308
enzyme-linked immunosorbant assays (ELISA) 308
error signals 217

estradiol and neural circuit masculinization 377
estrogen receptor subtypes 309
estrogenic metabolites 313
estrogens and the brain 305
ethology 412
European blackbird 97
European nightingale 282
European robin 316
European starling 33, 104, 106, 155, 187, 188, 195, 328, 346
eventual variety in singing 35
exogenous hormone studies 310
exogenous testosterone effects 320
experience effects on BOS-selectivity 182
experience-dependent plasticity 192
exposure to sound in utero 16
exposure-induced song associations 215
eye saccade 134

facial nuclei 141
fear conditioning 376
feedback pathways 112, 136, 151
feed-forward pathways 158
FEF (see "frontal eye fields")
female brains xv
female canaries 431
female preferences 188
female song 54, 326, 462
female song discrimination 165
female song systems 275
female zebra finches 142, 168
female-directed song 405
field L 69, 155, 158, 166, 168, 173, 176, 190, 258, 293, 388, 430
field L and primary auditory cortex 279
field L subdivisions 158
field sparrow 316, 317, 346
finite-state grammars 194
finite-state patterns 195
fireflies 415
fitness 106
fixed action patterns 374
flycatchers 24
food calls 461
foot shock 387
forebrain 63
forgetting 366, 445
formants 91, 97, 109
FoxP2 gene 370, 380, 394, 398, 399
FoxP2 gene and avian vocal communication 380
FoxP2 gene and developmental verbal dyspraxia patients 406
FoxP2 gene and singing activity 402
FoxP2 gene and speech 401

FoxP2 gene functions in songbirds 403
FoxP2 gene in Area X 404
FoxP2 gene knockout mouse 402
FoxP2 gene KO mouse vocalizations 403
FoxP2 gene-dependent developmental verbal dyspraxia 401
fragile X syndrome 380
frequency modulation 220
frontal and temporoparietal lobes 23
frontal eye fields (FEF) 134
frontal lobes 25
functional homology 60
functional lateralization 86
functional reference 457
functions of adult neurogenesis 364

GABA in HVC 142
GABA$_A$ receptor 306
GABAergic projections 66
galliforms 461
gape and sound frequency 110
gape and vocal tract resonance 97
gape size 110
gender differences 138
gene activation markers 294
gene expression xv, 372
gene microarrays 365
gene responses to song stimulation 376
genes and sexual brain differentiation 377
genes and visual system development 380
genetic and phenotypic differences 375
genetic constraints 369
genetic linkage maps 373
genetic manipulations 371
genetic structure of songbird populations 373
genetically modified songbirds 373
genome sequencing 372
genomic action potential 381
genomic revolution xvi, 369
gestural imitation 22
glia 436
globus pallidus (GP) 58, 64, 68
globus pallidus external segment (Gpe) 69
globus pallidus internal segment (Gpi) 69
glossopharyngeal nerve 136
glutamate receptor subtypes 399
glutamate receptors 202, 394
glutamate receptors and singing-regulated gene expression 395
glyceraldehyde-3-phosphate dehydrogenase 394
'go/no-go' procedure 188
goats 415
goldfish 415
gonadal hormones 164
goose syrinx 92

Subject index

GP (see "globus pallidus")
Gpe (see "globus pallidus external segment")
Gpi (see "globus pallidus internal segment")
graded vocal signals 456
grammatical rules 157
grasshoppe warbler 201
graylag goose 374
great reed warbler 416
great tit 409
greater sage grouse 96
grey catbird 81
ground squirrel 15
ground squirrel alarm call 289
gynandromorphism 378

habituation 388
habituation rates in nidopallium caudal medial 259
hair cell loss and song degradation 233
hair cell loss and tonotopy 231
hair cell regeneration 231
hairpins 400
harmonic structures 109
heart rate changes 206
heart rate measures 14, 15
Hebbian networks 261
heliox 92
hemispherectomy 25
hemispheric dominance reversal 430
herring gull 375
hierarchical model of song retrieval 213
hierarchical processing of auditory information 164, 173
high repetition trills 83
high-amplitude sucking paradigm 13
higher vocal center (see "HVC")
hill myna 78
hindbrain 63
hippocampus 58, 65, 273, 279, 350
hormonal modulation of singing 305
hormone action xvi
hormone action at multiple sites 315
hormone production sites 314
hormone receptor blockers 310
hormones and motivation 44
hormones and neurotransmitter receptors 306
hormones and seasonal growth 339
hotspots 297
house sparrow 307, 330
human brain 155
human disease 380
human infants 5
human language xvi, xvii, 9, 52, 57, 187, 270, 282, 370, 398

human medicine 371
human professional sopranos 92
human speech xv, 5, 33, 81, 91, 114, 176, 198, 279, 353, 369, 380, 401
human speech learning 157
human speech perception 187
human talker recognition 189
human vocal behavior 99
humans 17, 151, 318, 350, 402
hummingbirds 9, 32, 61, 312, 369, 382, 395, 459
Huntington's disease 254, 394, 406, 446
HVC (higher vocal center) 66, 69, 76, 115, 141, 156, 158, 159, 177, 178, 180, 185, 422, 423, 430
HVc (see "song premotor nucleus")
HVC activity synchronization 128
HVC and cHV 250
HVC and song repertoires 432
HVC as a clock 131
HVC filtering of auditory input 177
HVC has three neuron classes 118
HVC lesion effects 322
HVC motor output 120
HVC neuron replacement 356
HVC premotor activity 126
HVC-RA neuron ablation and song 357
HVC_{RA} neurons 177
HVC_{X} neurons 177
hyoid apparatus 111, 141, 145
hyperpallium 65
hyperpallium homologous to mammalian isocortex 65
hyperstriatum 58, 58
hypoglossal dominance 27, 429
hypoglossal motoneurons 137, 138
hypoglossal nerve 78, 80, 136, 458
hypoglossal nucleus (nXIIts) 115, 423, 430
hypothalamus 46, 63, 65

ICC (see "immunocytochemistry")
IEG (see "immediate early genes")
imitation 34, 351, 413
imitation accuracy 224
imitative learning 49
IMM (see "intermediate and medial mesopallium")
immediate early gene expression 167, 258, 332, 389
immediate early gene expression in mammals 387
immediate early gene zenk 42, 165, 168, 383
immediate early gene zenk in area X and MAN 391
immediate early gene zenk in canary nidopallium caudal medial 385

immediate early gene zenk in ring dove caudal medial nidopallium 259
immediate early genes (IEG) 190, 273, 381, 402
immediate variety in singing 35
immunocytochemistry (ICC) 308, 323, 382
imprinting 271, 280, 283, 395, 413
imprinting and intermediate and medial mesopallium 279
improvisation 34
improvisations 23
in utero learning 17
inborn sensory recognition 14
indigo bunting 42, 206, 346
individual recognition 413
individual variation 37
inducible early genes 369
infant-directed speech 19, 29
inflatable vocal sacs 96
infra-olivarus superior (IOS) 143
innate auditory template 202, 284
innate biases for vocal learning 16, 460
innate learning preferences 15
innate mental concepts 212
innate population differences 462
innate predispositions 5, 13, 20, 30, 34
innate song 202, 375
innate song recognition 15, 462
innate template 256
insect parasites 411
insight learning 413
inspiration 124
inspiratory syllables 81
instinct to learn 428
instinctive behaviour 413
instinctive motor behavior 60
instructive and permissive models 242
instructive and permissive models are not mutually exclusive 246
instructive experience 17
instructive models 202, 242
instructive signal 217
interclavicular air sac 95
intermediate and medial mesopallium (IMM) 279
intermediate arcopallium (Ai) 69
interneurons 355
intracellular hormone receptors 305
intra-specific variation in HVC 323
inventions 23
IOS (see "infra-olivarus superior")
island canaries 340
isocortex 64, 69
isolate songs 10, 284
isolating mechanisms 412

Japanese monkeys 456
Japanese quail 328
Japanese speech 16
jaw motoneurons 141
jaw muscle reflexes 145

kin recognition 188
knockouts 400

language skills and mental retardation 398
larynx 99, 136, 141
laser ablation techniques 357
laser capture microdissection 446
lateral geniculate nucleus 65
lateral lemniscus (LL) 150, 158, 180
lateral magnocellular nucleus of the anterior nidopallium (LMAN) 66, 68, 70, 163, 164, 182, 202, 218, 298, 422
lateral magnocellular nucleus of the anterior nidopallium lesion effects 240
lateral magnocellular nucleus of the anterior nidopallium lesions in juveniles 240, 247
lateral septum 331
lateralization of birdsong 81
learned and non-learned vocalizations 119
learned calls 47
learned vocalizations 70
learning in the womb 20
learning long word strings 201
learning-related gene expression 447
lek 42
lentiviral vectors 370, 373, 400
lexicon 8
limbic system 46
lingual motoneurons 137
linguistic space 19
lion-tailed macaque 456
lip 109
listener effects 41
LL (see "lateral lemniscus")
LMAN (see "lateral magnocellular nucleus of the anterior nidopallium")
LoC (see "locus coeruleus")
locus coeruleus (LoC) 134, 393
locust flight rhythms 454
long songs 106
long-term depression (LTD) 258, 261
long-term potentiation (LTP) 246, 258, 261, 298, 376
loudness and song stereotypy 209
low frequency auditory feedback and normal song 232
lPs (see "nucleus parasolitarius lateralis")
LTD (see "long-term depression")
LTP (see "long-term potentiation")

lung-air sac system 104
lyrebirds 346

M. syringealis ventralis 137
magpie lark 54
magpie tanager 87
mallard 422
mamallian inferior colliculus 383
mammalian and avian forebrains compared 63
mammalian auditory cortex 168, 172, 385
mammalian basal ganglia xv
mammalian cortico-basal ganglia and the anterior forebrain pathway 254
mammalian larynx 136
mammalian neocortex 61, 279
mammalian sex determination 378
mammalian visual cortex 170
mammals xvi, 70, 90, 136, 140, 143, 158, 273, 330, 350, 371, 378, 381, 383, 384, 394, 397, 398, 440
manakin 42, 460
mangabey 457
marsh wren 213, 323, 326, 346
matched countersinging 23, 318
mate-guarding 57
McGurk effect 22
MD (see "mediodorsal nucleus of the thalamus")
meadowlark 42
mechanoreceptive feedback 86
medaka fish 415
medial bed nucleus of the stria terminalis 331
medial geniculate body (MGB) 158
medial magnocellular nucleus of the anterior nidopallium (mMAN) 68, 70, 126, 127, 180
medial tympaniform membrane (MTM) 78, 95
mediodorsal nucleus of the thalamus (MD) 134
melatonin 317, 337
mesopallium 65
Metrazole 389
MGB (see "medial geniculate body")
mice 375, 400, 446
microarray 446
microarrays 399
midbrain 63
mimicry 51
minibreath 75, 82, 104, 144
MLd (see "dorsal lateral nucleus of the mesencephalon")
mockingbird 8, 33, 87, 316
models (instructive and selective) 19
models and mimics 89

models of learning and memory 376
modulation transfer function (MTF) 170
MO-like, oval nucleus of the mesopallium 390
monk parakeet 97, 112
monkey calls 279
monkeys 13
moorhen 413
motif 228
motif repertoires 189
motifs 8, 35, 187, 188
motivation to sing 321
motivational drive and song control 133
motor command theory 21
motor constraints 13
motor control of complex behaviors 115
motor driven immediate early gene expression 391
motor memory 273
motor output variability 247
motor phase of song development 204
motor programs for beak movements 113
motor programs for respiration and syrinx 113
motor sequence generation 115
motor stereotypy 84
motor variability 249
MTF (see "modulation transfer function")
MTM (see "medial tympaniform membrane")
multimodal information 22
multimodal signaling 457
multiple representations of song-related auditory information 269
multiple song types 165
multi-process memory system 299
multitaper spectral analysis 219
muscle spindles 114, 148
muting effects 46, 391
mynah 33, 91
myotopic organization 137

NADH dehydrogenase 394
nature and nurture 204, 369, 374, 454
NCL (see "caudolateral belt of nidopallium")
neocortex 58
neostriatum 58, 58, 60
neural basis of birdsong 70
neural circuitry and complex behaviors 459
neural migration 438
neuroendocrine control of song 320, 323

neuroendocrine control of the motivation to sing 331
neuroendocrine mechanisms and the singing probability 328
neurogenesis xvi, 246, 298, 303, 304, 312, 332, 334, 350, 380, 394, 427
neurogenesis and deafening 364
neurogenesis and projection neurons 436
neurogenesis and song learning 365
neurogenesis in adult brains 459
neurogenesis; adult significance 356
neurogenomics of song learning 376
neuron addition and perceptual memory 358
neuron addition and song stereotypy 362
neuron addition in adults 353
neuron lifespan 356
neuron types formed after hatching 355
neuronal activation and gene expression 381
neuronal life cycle 353
neuronal recruitment in HVC 354
neuronal replacement 350, 447
neuronal replacement and perception 358
neuronal stem 439, 441
new neurons xv
new neurons and singing 357
nidopallium 65, 69
nidopallium caudal medial (NCM) 155, 158, 169, 181, 190, 202, 250, 258, 274, 274, 293, 299, 385
nidopallium caudal medial and song memory 274
nidopallium caudal medial and song perception 385
nidopallium caudal medial and song processing 385
nidopallium caudal medial and the acquired song template 259
nidopallium caudal medial and tutor song memory 278, 280
nidopallium caudal medial tuning 173
nidopallium caudal medial/caudal medial nidopallium 269
Nif (see "nucleus interfacialis")
nightingale 36, 37, 40, 41, 206, 207, 209, 211, 213, 346
NMDA (see "N-methyl-D-aspartate receptors")
NMDAR (see "N-methyl-d-aspartate glutamate receptor")
N-methyl-d-aspartate glutamate receptor (NMDAR) 262
N-methyl-d-aspartate glutamate receptor developmental changes 266
N-methyl-d-aspartate glutamate receptors and sensory acquisition 263

N-methyl-d-aspartate glutamate receptors and synaptic plasticity 262
N-methyl-D-aspartate receptors (NMDA) 246, 272, 298, 395, 422
noise masking 239
nonauditory feedback 99, 147
nonhuman primates 10, 30, 399
nonlinear behavior 112
nonlinear dynamics 87
nonpasserines 142
nonspeech gurgles and cries 21
nonvocal "motor pathways" 70
noradrenaline 387, 393
noradrenergic innervation 164
noradrenergic song system innervation 313
norepinephrine 164
northern cardinal 78, 82, 87, 95, 97, 102
northern mockingbird 83, 87, 346
note 8
notes 35, 188, 224
nTS (see "nucleus tractus solitarius")
nucleus ambiguous 142
nucleus basorostralis (Bas) 145, 150
nucleus interfacialis (Nif) 66, 69, 150, 158, 159, 164, 165, 177, 180, 180, 185
nucleus interfacialis and HVC 293
nucleus laminaris 150
nucleus ovoidalis (Ov) 158, 384, 391
nucleus parambigualis (PAm) 115, 140, 142, 148
nucleus parambigualis as integrator of motor commands and respiration 132
nucleus parasolitarius lateralis (lPs) 148
nucleus retroambigualis 77, 138, 140, 142, 143
nucleus tractus solitarius (nTS) 143, 148
nucleus uvaeformis (Uva) 68, 69, 77, 126, 127, 143, 148, 150, 158, 159, 167, 180, 185
nucleus uvaformis-lesioned birds 127
numbers of bird genes 371
nXIIts (see "hypoglossal nucleus")
NXIIts (see "tracheosyringeal nerve")

OEC (see "oropharyngeal-esophageal cavity")
olfactory bulb 58, 350, 440
olfactory conditioning 412
open-ended learners 11, 26, 34, 46, 84, 286, 332, 345, 352, 445
operant conditioning 188
Oregon junco 417
organizational actions of steroids 377
organizational hormone effects 310
orofacial dyspraxia 401

oropharyngeal cavity 75, 95
oropharyngeal-esophageal cavity (OEC) 95
oscine songbirds 32, 201, 312, 369, 374
osmotic minipumps 309
ototoxic drugs 230
Ov (see "nucleus ovoidalis")
overlapping developmental phases 298
overproduction in plastic song 461
own song memory storage 238
own species' song 14

PA (see "paleostriatum augmentatum")
package formation 213
PAG (see "periaqueductal grey in mammals")
pair bonding 57
paleocortex 58
paleostriatum 58
paleostriatum augmentatum (PA) 58
paleostriatum primitivum 58
pallial-basal-thalamic-pallial loop 397
pallium 58, 61, 64, 155
PAm (see "nucleus parambigualis")
para-HVC 141
parallel memory stores 279
parentese 19, 29
parietal lobes 25
Parkinson's disease 254, 380, 394, 406, 446
parrots 9, 32, 61, 66, 91, 96, 97, 112, 274, 312, 369, 382, 390, 395, 458, 459
pars reticulata (SNr) 68
parvalbumin (PV) 178
pattern memorization 216
pattern recognition 187
pattern rule learning 194
PBvI (see "ventrolateral parabrachial nucleus")
pCaMKII template 264
perceptual learning 16, 358
perceptual magnet effect (PME) 18
performance-associated hypertrophy 344
periaqueductal grey in mammals (PAG) 143
perirhinal cortex 279
permissive models 202, 245
phoebes 24
phonation 75, 80, 99
phonemes 8
phonetic units in languages 13
phonology 8, 287
photoperiod 55, 333, 346
phrases 8, 35
pidgin languages 15
pigeon 66, 69, 78, 141, 145, 148, 259, 314, 384, 430
piriform cortex 65
PKC (see "protein kinase C")

plainfin midshipman 341
plasma membrane receptors 305
plastic song 23, 33, 201
PME (see "perceptual magnet effect")
PMMAI (see "production memory maintenance")
PMTEM (see "production memory template")
POM (see "preoptic medial nucleus")
postlingually deaf adults 11
postnatal neurogenesis 434
post-sleep deterioration 224
post-sleep deterioration effect 226
predator recognition 462
preference for conspecific song 173
preference for conspecific tutor 21
pre-imaginal conditioning 411
premature song crystallization 247
premotor activity in HVC 119
preoptic area of the hypothalamus 313, 331
preoptic medial nucleus (POM) 133, 303, 328
prespecified motor programs 15
primary auditory areas 157
primate behavior 455
primates 279, 399
production memory 202, 291, 303
production memory maintenance (PM$_{MAI}$) 292
production memory template 296
production memory template (PM$_{TEM}$) 292
promoters 372
proprioception and vocal control 458
proprioceptive feedback 86, 147, 181, 428
prosody 8
protein kinase C (PKC) 268
prototypes 18
pruning of AFP connections 246
PrV (see "trigeminal nucleus")
Pt (see "putamen")
pure-tonality 98
putamen (Pt) 58
PV (see "parvalbumin")

quail 314, 329, 330, 395, 400
quality of singing 321
quiet breathing 144

RA (robust nucleus of the archistriatum) 66, 69, 76, 77, 115, 118, 137, 138, 142, 144, 158, 177, 247, 422, 430
RA functional compartments 122
RA lesioning effects 49
RA projection neurons 142
rabbit 415

radial glia 354, 439
radioimmunoassays 303, 308
RAm (see "retroambigualis medulla")
randomly ordered song bouts 193
raphe nucleus 134
rapid song matching 214
rapid trill rates 104
rats 434
rattlesnake 81
recognition memory 202, 291, 303, 304
recovery of original song following hearing restoration 233
recurrent song circuit 126
recursive embedding 194
recursive syntactic patterning 155, 197
red-cheeked cordon blue 326
red-tailed monkey 456
redwinged blackbird 35, 307, 313, 461
reentrant feedback 293
reinforcement signals 173
replaceable neurons and replaceable memories 437
resonance 109
resonance filters 96
respiration 75, 99
respiratoration and singing 81
respiratory circuits 44
respiratory motor nuclei 138
respiratory rhythms 138
respiratory system 76, 106
respiratory-syringeal coordination 99, 126, 143
responses to song in auditory areas 166
retroambigualis medulla (RAm) 115, 136
retroambigualis medulla as an integrator of respiratory-vocal activity 145
retroambigualis medulla projection neurons 144
retroviruses 365
reversible inactivation of lateral magnocellular nucleus of the anterior nidopallium 250
reversible paradigm for induced plasticity 239
revised nomenclature 62
reward learning 404
right hemisphere 25
ring dove 81, 92, 95, 273, 395, 428, 458
RNA interference (RNAi) 370, 400
RNAi (see "RNA interference")
robust nucleus of the archistriatum (see "RA")
robust nucleus of the Arcopallium 430
rodents 380, 436
roller and border canaries 16
rostral hyperpallium 65

rufous-collared sparrow 56, 333
RVL (see "ventrolateral nucleus of the rostral medulla")

saddleback 346
sage grouse 96, 460
salary men 416
Savannah sparrow 34
SC (see "superior colliculus")
sea mollusks 381
seasonal and hormonal plasticity xv
seasonal brain plasticity 332, 333, 348, 459
seasonal breeders 34
seasonal changes in the preoptic medial nucleus 329
seasonal changes in the song system 332
seasonal growth and regression rates 337
seasonal modulation 164
seasonal plasticity and gonadal steroids 335
seasonal plasticity benefits 344
seasonal singing changes 303
seasonal song production 56
second language 25, 26, 31, 280
secondary auditory areas 164
sedge warbler 340
sedge wren 34
selective attrition 218
selective auditory activity of HVC$_X$ cells 184
selective experience 17
selective responsiveness 191, 294
selectivity to familiar songs in nidopallium caudal medial and caudal mesopallium 167
semantics 8
sensitive period closure 250, 267
sensitive period independence 238
sensitive period learners 445
sensitive periods 24, 25, 266, 351, 387
sensitive periods in humans 25
sensitive phase 204, 216, 286
sensitive phase closure 206
sensitive phase timing 204
sensorimotor integration 99, 136, 163
sensorimotor learning 11, 19, 23, 207, 242, 253
sensorimotor learning and dopamine 270
sensory acquisition mechanisms 256, 265
sensory exposures necessary for learning 20
sensory feedback 75, 84, 94
sensory learning 11
sensory learning period 17
sensory period 17
sensory phase 11
sensory phases of song learning 23

sensory response gating 251
sensory template 17
sensory-to-motor conversion 287
sex chromosomes 377
sex difference coevolution in brain and singing 324
sex differences in mammals 377
sex differences in phenotypes 379
sex differences in song behavior 320, 331
sex differences in the song system 323, 324
sex steroid receptors 46
sex steroids 28, 46
sexual differences in song 305
sexual differences in the brain anatomy 318
sexual differences in the brain structure 311
sexual differentiation 369
sexual differentiation in mammals 379
sexual differentiation of the brain 376
sexual dimorphism in brain 431, 459
sexual imprinting 204
sexually attractive songs 276
sexy syllables 318
sign language 10, 399
signal categories 456
silastic capsules 46, 309
silent song 322
singing behavior xvi, xvi, 23
singing frogs 415
singing rate 46, 348
singing-induced gene expression 383
singing-induced IEG expression 276, 395
singing-induced ZENK expression 258
singing-regulated gene expression 369, 394
singing-regulated genes 393
singing-regulated IEG expression 390
single nucleotide polymorphisms (SNPs) 373, 375
single-unit activity 156
slate colored junco 420
sleep 163, 260, 296
social behavior 319, 369
social effects 23, 29, 30, 31, 131, 206, 216, 253, 303, 330, 387, 391, 393, 398
somatosensorimotor feedback 147
somatosensory feedback 75, 86, 150
sonagraph 421
song acquisition xvi
song and FoxP2 gene knockdowns 404
song and mate choice 366
song and visual display 106
song auditory memories 388
song crystallization 46, 209
song dialects 9, 35, 450, 454
song displays 109
song feature developmental maps 226

song instability 361
song lateralization 27
song learning xv, 32, 33, 187, 204
song learning and speech 211
song learning feature maps 219
song maintenance 353, 358, 365
song matching 41, 346
song memorization 202, 204, 250
song memory xvi, 369
song perception 327, 385
song perception and production 391
song perception in females 327
song premotor nucleus (HVc) 24
song production xvi
song production is lateralized 458
song recognition xvi, 174, 369
song recovery after cell death 443
song repertoires 8
song representations in HVC 183
song selectivity 160
song selectivity in the AFP 249
song sequence stereotypy 404
song sparrow 14, 33, 35, 42, 97, 205, 206, 209, 283, 285, 289, 316, 331, 333, 337, 338, 340, 343, 344, 346, 347, 354, 359, 361, 393, 445, 446, 454, 461
song stereotypy 46, 330, 343, 353, 358, 361, 366, 460
song stereotypy and repertoires 35
song syntax 8, 102, 106
song system xv, xv, 50, 70, 76, 157, 256, 271, 293, 402, 427, 430, 458
song system absence 327
song system and mammalian system commonalities 68
song system nuclei 157
song system nuclei and song repertoires 459
song system plasticity 332
song system regression 338, 348
song template 217, 224, 283, 428
song tempo 40, 83
song type matching 213
song variability 340
songbirds 61, 66, 382, 395
songs and sentences 215
song-selective auditory responses 66
song-selective neurons 174
song-specific habituation 385
song-specific neurons 176
sound amplitude 112
sound localization 451
sound spectrograms xv, 219
sound spectrograph 413, 451
source-filter theory 92
spandrels 198
sparrows 26, 150, 365, 415, 458

sparse code models 122
spatial memory 376
species universals 19
species-typical phonology 289
spectral properties 91
spectro-temporal receptive fields (STRFs) 155, 169
speech 79
speech perception 24
speech processing 252
speech production 6, 24
speech representation 192
sperm competition 373
spinal cord 63
spindles 141
spotted towhee 340
squirrel monkey 24
starling 26, 35, 133, 190, 194, 206, 276, 278, 306, 307, 327, 329, 330, 332, 335, 387, 393
state-dependent gating 150
state-dependent mechanisms 296
sternotrachealis muscles 80
steroid hormones 303
steroid receptors in the preoptic medial nucleus 323
steroid sex hormone receptors 333
steroid-sensitive neurons 323
stimulus-specific sensory responses 174, 180
stress and intonation 16
striatopallidothalamic pathways 261
striatum 64
stuttering 27, 31, 233, 380, 393
suboscine 119
suboscines 24, 30, 142
subpallium 64, 64
subsong 9, 23, 33, 112, 201, 207
substance P 317
substantia nigra 64, 68
superfast muscles 81
superior colliculus (SC) 134
superior olive (SO) 158
superior temporal gyri 24
surgical deafening 232
swallow 415
swamp sparrow 14, 15, 33, 97, 111, 156, 174, 184, 205, 206, 209, 283, 289, 343, 347, 417, 461
syllable 8, 35, 115, 131, 229
syllable deterioration 232
syllable transitions 38
syllable type development 201
syllable type emergence 220
sympathetic preganglionic neurons 140
synapse-related molecules 393
synaptic mechanisms underlying BOS-selectivity 178, 180

synaptic selection 298
synaptotagmin IV 394
synelfin and song learning 380
syntax 8, 82, 195, 287
synthesized songs 461
synthetic sounds 168
synthetic tutor songs 14
syringeal labia 100
syringeal muscles 78, 102
syringeal nerve section 46
syrinx xv, 6, 32, 75, 76, 78, 90, 99, 115, 118, 136, 138, 315, 341, 426, 428
syrinx and respiration 118
syrinx duplex 80

tabula rasa 374
telencephalon 58, 63, 66
telencephalon and premotor song control 141
telencephalon in birds 60
telencephalon in mammals 60
temperate zone species 306
template 11, 19, 31, 157, 161, 163, 201, 202, 238, 282, 350, 413, 419, 420, 428
template concept 282
template learning 212
template matching 226
temporal organization 76
temporal song characteristics 105
territoriality 188
testosterone 46, 320, 323
testosterone and dendritic length 433
testosterone and new neuron survival 441
testosterone effects on females 431
thalamus 63
the auditory system 158
the avian auditory telencephalon 181
the genome and social behaviors 379
the genomics of birdsong 371
the KE-family 401
the motor theory of speech perception 401
the mouse 399
theoretical models 90
thrush 201
thyroid hormones 316, 337
timbre 170, 173, 189
toadfish 81
tongue 109, 112, 137, 141
tongue movement 96, 141
tonotopic organization 190
tonotopy 172
trachea 75, 110
tracheal resonance 95
tracheobronchial syrinx 78
tracheobronchialis muscles 80
tracheobronchialis ventralis muscle 137
tracheostomies 11
tracheostomized children 26

tracheosyringeal motoneurons 137
tracheosyringeal nerve (NXIIts) 148, 428
tracheosyringeal nerve section 242
transcription factors 372, 381
transgenic chicken 400
transgenic knock-out methods 378
transgenic mice 373
transgenic songbirds 373
tree-creepers 206
trial and error 226
trigeminal nucleus (PrV) 141, 145, 148, 150
trill 289
trophic inputs to the motor pathway 245
tropical birds 56
turtles 415
tutor song memory 271, 273
tutor template 173
two separate clocks for timing 131
two-alternative choice tasks 188
two-voice syllables 86
tyrosine hydroxylase (TH) 362

ubiquitin 394
undirected song 40, 42, 122, 253, 314, 317, 391, 393, 394, 405
universal grammar 398
unlearned calls 47
untutored song 40
Uva (see "nucleus uvaformis")

vagus nerve 136, 148
variability in song structure 393
variability of plasma testosterone in breeding canaries 335
vasoactive intestinal peptide (VIP) 317
vasotocin 303, 316
ventral tegmental area (VTA) 134, 393
ventricular zone (VZ) 354
ventrolateral nucleus of the rostral medulla (RVL) 143
ventrolateral parabrachial nucleus (PBvl) 143, 148
ventromedial nucleus of the hypothalamus (VMH) 331
vertebrate embryonic pallium 65
viscerosensory feedback 75
visual displays 42
visual-motor learning 22
vocal change trajectories 223
vocal control pathway 376
vocal motor pathway (VMP) 256, 260
vocal non-learners 395, 397
vocal pathways in parrots and hummingbirds 395
vocal respiratory network (VRN) 76, 115, 134

vocal respiratory network group (VRG) 140
vocal respiratory network, VRN/Uva pathway 128
vocal tract 75, 76, 78, 90, 95, 109
vocal tract filter tuning 93
vocal tract filtering 109
vocal tract resonance 91
voice-onset time (VOT) 8, 13

warblers 8, 216, 317
waterslager canary 83, 429, 458
Wernicke's aphasias 24
Wernicke's area 24, 279
whisker representations in rodents 180
white noise 236, 460
white-browed robinchat 323
white-crowned sparrow 8, 15, 17, 26, 27, 29, 33, 86, 161, 166, 206, 209, 260, 271, 276, 283, 285, 286, 289, 295, 326, 332, 333, 335, 336, 338, 339, 340, 343, 344, 346, 347, 374, 375, 387, 419, 421, 427, 428, 428, 429, 442, 445, 454, 458, 462
white-crowned sparrow deafening effects 343
white-crowned sparrow innate song 454
white-crowned sparrow isolate songs 15
white-crowned sparrow song 15
white-crowned sparrow subspecies 17
white-throated sparrow 97, 111, 326
white-throated sparrow plumage morphs 379
wren 198, 325, 326
Wulst 58

x-ray cinematography 95

zebra finch: auditory feedback 186, 228, 233, 347, 406, 419, 445
zebra finch: auditory physiology 166, 174, 184, 390
zebra finch: bird's own song (BOS) 156, 160, 163, 166, 168, 169, 176, 183
zebra finch: directed and undirected song 83, 97, 112, 122, 308, 393
zebra finch: female responses to song 33, 167, 274
zebra finch: genetic studies 371, 373, 399, 400, 402, 404
zebra finch: hormonal effects 29, 30, 46, 133, 316, 318, 319, 328
zebra finch: immediate early gene studies 190, 273, 274, 276, 387, 388
zebra finch: neurogenesis 354, 359, 361, 365, 437, 442, 445, 446

zebra finch: sensitive periods 29, 246, 332, 351, 445
zebra finch: sexual dimorphism 324, 326, 330, 377, 378, 431, 459
zebra finch: social effects on learning 20, 37, 206
zebra finch: song learning 17, 150, 217, 219, 250, 267, 271, 273, 286, 296, 298, 350, 420
zebra finch: song structure and variability 8, 33, 35, 37, 40, 49, 110, 116, 224, 442
zebra finch: song system physiology 119, 142, 145, 165, 240, 388, 437, 442
zebra finch: syrinx, phonation and respiration 78, 81, 81, 92, 95, 97, 104, 106, 110, 111, 112, 113, 115, 144, 148, 150